# THE INFINITE-DIMENSIONAL
# TOPOLOGY OF FUNCTION SPACES

# North-Holland Mathematical Library

*Board of Honorary Editors:*

M. Artin, H. Bass, J. Eells, W. Feit, P.J. Freyd, F.W. Gehring, H. Halberstam, L.V. Hörmander, J.H.B. Kemperman, W.A.J. Luxemburg, F.P. Peterson, I.M. Singer and A.C. Zaanen

*Board of Advisory Editors:*

A. Björner, R.H. Dijkgraaf, A. Dimca, A.S. Dow, J.J. Duistermaat, E. Looijenga, J.P. May, I. Moerdijk, S.M. Mori, J.P. Palis, A. Schrijver, J. Sjöstrand, J.H.M. Steenbrink, F. Takens and J. van Mill

VOLUME 64

ELSEVIER
Amsterdam – London – New York – Oxford – Paris – Shannon – Tokyo

# The Infinite-Dimensional Topology of Function Spaces

Jan van Mill
*Faculteit der Exacte Wetenschappen, Amsterdam, The Netherlands*

2001
ELSEVIER
Amsterdam – London – New York – Oxford – Paris – Shannon – Tokyo

ELSEVIER SCIENCE B.V.
Sara Burgerhartstraat 25
P.O. Box 211, 1000 AE Amsterdam, The Netherlands

© 2001 Elsevier Science B.V. All rights reserved.

This work is protected under copyright by Elsevier Science, and the following terms and conditions apply to its use:

Photocopying
Single photocopies of single chapters may be made for personal use as allowed by national copyright laws. Permission of the Publisher and payment of a fee is required for all other photocopying, including multiple or systematic copying, copying for advertising or promotional purposes, resale, and all forms of document delivery. Special rates are available for educational institutions that wish to make photocopies for non-profit educational classroom use.

Permissions may be sought directly from Elsevier Science Global Rights Department, PO Box 800, Oxford OX5 1DX, UK; phone: (+44) 1865 843830, fax: (+44) 1865 853333, e-mail: permissions@elsevier.co.uk. You may also contact Global Rights directly through Elsevier's home page (http://www.elsevier.nl), by selecting 'Obtaining Permissions'.

In the USA, users may clear permissions and make payments through the Copyright Clearance Center, Inc., 222 Rosewood Drive, Danvers, MA 01923, USA; phone: (+1) (978) 7508400, fax: (+1) (978) 7504744, and in the UK through the Copyright Licensing Agency Rapid Clearance Service (CLARCS), 90 Tottenham Court Road, London W1P 0LP, UK; phone: (+44) 207 631 5555; fax: (+44) 207 631 5500. Other countries may have a local reprographic rights agency for payments.

Derivative Works
Tables of contents may be reproduced for internal circulation, but permission of Elsevier Science is required for external resale or distribution of such material.
Permission of the Publisher is required for all other derivative works, including compilations and translations.

Electronic Storage or Usage
Permission of the Publisher is required to store or use electronically any material contained in this work, including any chapter or part of a chapter.

Except as outlined above, no part of this work may be reproduced, stored in a retrieval system or transmitted in any form or by any means, electronic, mechanical, photocopying, recording or otherwise, without prior written permission of the Publisher.
Address permissions requests to: Elsevier Science Global Rights Department, at the mail, fax and e-mail addresses noted above.

Notice
No responsibility is assumed by the Publisher for any injury and/or damage to persons or property as a matter of products liability, negligence or otherwise, or from any use or operation of any methods, products, instructions or ideas contained in the material herein. Because of rapid advances in the medical sciences, in particular, independent verification of diagnoses and drug dosages should be made.

First edition 2001

Library of Congress Cataloging in Publication Data
A catalog record from the Library of Congress has been applied for.

ISBN:   0 444 50557 1
ISSN:   0924-6509

∞ The paper used in this publication meets the requirements of ANSI/NISO Z39.48-1992 (Permanence of Paper).
Printed in The Netherlands.

*Aan de nagedachtenis van Maarten Maurice.*

# Contents

Introduction .................................................................. xi

Chapter 1. Basic topology ................................................. 1
   1.1. Linear spaces ...................................................... 1
   1.2. Extending continuous functions ................................. 21
   1.3. Function spaces ................................................... 29
   1.4. The Borsuk homotopy extension theorem ....................... 37
   1.5. Topological characterization of some familiar spaces .............. 41
   1.6. The inductive convergence criterion and applications .............. 58
   1.7. Bing's shrinking criterion ......................................... 66
   1.8. Isotopies ........................................................... 70
   1.9. Homogeneous zero-dimensional spaces ........................... 73
   1.10. Inverse limits .................................................... 80
   1.11. Hyperspaces ..................................................... 95

Chapter 2. Basic combinatorial topology ................................. 111
   2.1. Affine notions ..................................................... 111
   2.2. Barycenters and subdivisions ..................................... 125
   2.3. The nerve of an open covering ................................... 132
   2.4. Simplices in $\mathbb{R}^n$ .................................................. 138
   2.5. The Lusternik-Schnirelman-Borsuk theorem ...................... 148

Chapter 3. Basic dimension theory ....................................... 151
   3.1. The covering dimension ........................................... 151
   3.2. Translation into open covers ..................................... 157
   3.3. The imbedding theorem .......................................... 168
   3.4. The inductive dimension functions ind and Ind .................. 176
   3.5. Dimensional properties of compactifications ..................... 183
   3.6. Mappings into spheres ............................................ 193
   3.7. Dimension of subsets of $\mathbb{R}^n$ and certain generalizations ........... 204
   3.8. Higher-dimensional hereditarily indecomposable continua ........ 210
   3.9. Totally disconnected spaces ...................................... 216
   3.10. The origins of dimension theory ................................. 221
   3.11. The dimensional kernel of a space ............................... 227
   3.12. Colorings of maps ............................................... 237

3.13. Various kinds of infinite-dimensionality ........................... 251
3.14. The Brouwer fixed-point theorem revisited ....................... 257

Chapter 4. Basic ANR theory ............................................. 263
4.1. Some properties of ANR's ......................................... 263
4.2. A characterization of ANR's and AR's ............................. 277
4.3. Open subspaces of ANR's .......................................... 301

Chapter 5. Basic infinite-dimensional topology .......................... 307
5.1. $Z$-sets ........................................................... 307
5.2. Extending homeomorphisms in $s$ .................................. 311
5.3. The estimated homeomorphism extension theorem ............... 320
5.4. The compact absorption property ................................. 329
5.5. Absorbing systems ................................................ 343

Chapter 6. Function spaces .............................................. 367
6.1. Notation ......................................................... 368
6.2. The spaces $C_p(X)$: Introductory remarks ........................ 369
6.3. The Borel complexity of function spaces ......................... 372
6.4. The Baire property in function spaces ........................... 377
6.5. Filters and the Baire property in $C_p(\mathbb{N}_\mathcal{F})$ ........................ 387
6.6. Extenders ........................................................ 393
6.7. The topological dual of $C_p(X)$ .................................. 399
6.8. The support function ............................................. 404
6.9. Nonexistence of linear homeomorphisms ......................... 411
6.10. Bounded functions ............................................... 416
6.11. Nonexistence of homeomorphisms ................................ 426
6.12. Topological equivalence of certain function spaces ............. 434
6.13. Examples ........................................................ 445

Appendix A. Preliminaries ............................................... 457
A.1. Prerequisites and notation ....................................... 457
A.2. Separable metrizable topological spaces ......................... 465
A.3. Limits of continuous functions .................................. 468
A.4. Normality type properties ....................................... 469
A.5. Compactness type properties ..................................... 473
A.6. Completeness type properties .................................... 479
A.7. A covering type property ........................................ 485
A.8. Extension type properties ....................................... 490
A.9. Wallman compactifications ....................................... 494
A.10. Connectivity .................................................... 500
A.11. The quotient topology .......................................... 505
A.12. Homotopies ..................................................... 510
A.13. Borel and similar sets ......................................... 517

| | |
|---|---|
| Appendix B. Answers to selected exercises | 527 |
| Appendix C. Notes and comments | 579 |
| Bibliography | 597 |
| Special Symbols | 613 |
| Author Index | 615 |
| Subject Index | 619 |

# Introduction

In this book we study function spaces of low Borel complexity. This is a particularly interesting class of spaces; to investigate it one needs a mix of methods and techniques from areas as diverse as general topology, infinite-dimensional topology, functional analysis and descriptive set theory. A striking result is the theorem of Dobrowolski, Marciszewski and Mogilski, which states that all function spaces of low Borel complexity are topologically homeomorphic. A major feature of this book is a complete and self-contained proof of this remarkable fact.

In order to understand these details, a solid background in infinite-dimensional topology is needed. And for that one needs to know a fair amount of dimension theory as well as ANR theory. The necessary material was partially covered in my previous book *'Infinite-dimensional topology, prerequisites and introduction'*. A selection of what was done in that volume can also be found here, but completely revised and in many places expanded with recent results. I chose to take a 'scenic' route towards the Dobrowolski-Marciszewski-Mogilski Theorem, that is linking the results needed for the theorem's proof to interesting recent research developments in dimension theory and infinite-dimensional topology.

The first five chapters of this book are intended as a text for graduate courses in topology. For a course in dimension theory, one should cover Chapters 2 and 3 and part of Chapter 1. For a course in infinite-dimensional topology, one should cover Chapters 1, 4 and 5. In Chapter 6, which deals with function spaces, I discuss recent research results. It can also be used for a graduate course in topology but its focus is more suited to that of a research monograph than of a textbook; it would, therefore, be more appropriate to use it as a text for a research seminar. This book, consequently, has the character of a textbook as well as a research monograph.

In Chapters 1 through 5, unless stated otherwise, all spaces under discussion are separable and metrizable. In Chapter 6 results for more general classes of spaces are presented.

In Appendix A we collected for easy reference and for sake of completeness some basic facts that are important in the book. The reader will see that it is not intended as the basis for a course in topology; its purpose is to collect what one should know about general topology, nothing more nothing less.

The exercises in the book serve three purposes: to test the reader's understanding of the material, to supply proofs of statements that are used in the text and are not proven therein, and to provide additional information not covered by the text. We included the solutions to selected exercises in

Appendix B. These exercises are important or difficult; they are marked in the text by the symbol ▶.

If the reader wants to find the meaning of some unfamiliar term in this book, it is best to check Appendix A first, since many basic concepts are defined there. To simplify the search process, in the index all page numbers for terms from Appendix A are italicized. For example, if the reader would like to know the definition of the term *'topologically complete'*, she or he should look at Page 480. For in the index, the first italicized page number under 'topologically complete' is 480.

Finally, I express my indebtedness to Jan Baars, Stoyu Barov, Jan Dijkstra, Tadek Dobrowolski, Klaas Pieter Hart, Michael van Hartskamp, Henryk Michalewski, Witek Marciszewski, Roman Pol, Ruud Salomon and Jan de Vries for their critical reading of parts of the manuscript and their many valuable suggestions for improvements. None of these distinguished colleagues is responsible for the remaining errors, which are mine.

*Jan van Mill*
Bussum, March 29, 2001

CHAPTER 1

# Basic topology

In this chapter we present some basic facts on the topology of separable metrizable spaces. We discuss linear spaces, inverse limits, hyperspaces, Bing's Shrinking Criterion, etc. Questions about the possibility of extending continuous functions, or creating new continuous functions from old ones, are central in this chapter. Applications include a proof of the topological homogeneity of the Hilbert cube and proofs of topological characterizations of various interesting spaces such as the Cantor set, the unit interval and the spaces of rational and irrational numbers, respectively. Many of the results presented in this chapter are geometrically motivated, although this is not always clear at first sight.

For background information, see Appendix A. Our conventions with respect to notation can be found in §A.1.

*All topological spaces under discussion are separable and metrizable.*

## 1.1. Linear spaces

A *linear space* is a real vector space $L$ carrying a (separable metrizable) topology with the property that the algebraic operations of addition and scalar multiplication are continuous (warning: a vector space is an algebraic structure which may or may not carry a topology while a linear space is automatically a topological space). Observe that the continuity of the algebraic operations on a linear space show that it is a topological group.

A subset $A$ of a linear space $L$ is called *convex* if for all $x, y \in A$ and $\alpha \in \mathbb{I}$ we have $\alpha x + (1 - \alpha)y \in A$.

Let $L$ be a linear space. If $A \subseteq L$ then $\operatorname{conv}(A)$ denotes the smallest convex subset of $L$ containing $A$; this set is called the *convex hull* of $A$. A *convex combination* of elements of $A$ is a vector of the form $\sum_{i=1}^{n} \lambda_i a_i$ with $a_1, \ldots, a_n \in A$, $\lambda_1, \ldots, \lambda_n \in \mathbb{I}$ and $\sum_{i=1}^{n} \lambda_i = 1$.

For each $n \in \mathbb{N}$, let $\operatorname{conv}_n(A)$ denote the set of vectors $x \in L$ that can be written as a convex combination of at most $n$ vectors from $A$. Also, put

$$\operatorname{conv}_\infty(A) = \bigcup_{n=1}^{\infty} \operatorname{conv}_n(A)$$

and observe that $\mathrm{conv}_\infty(A)$ is the set of all convex combinations of elements of $A$.

It is left as an exercise to the reader to present a proof of the following basic lemma (see Exercise 1.1.5).

**Lemma 1.1.1.** *Let $L$ be a linear space with subset $A$. Then*

(1) $\mathrm{conv}(A) = \mathrm{conv}_\infty(A)$,
(2) *if $A$ is finite then $\mathrm{conv}(A)$ is compact.*

A linear space $L$ is called *locally convex* if the origin of $L$ (which we shall always denote by $\underline{0}$) has arbitrarily small convex neighborhoods. Obviously, the spaces $\mathbb{R}^n$ for $n \in \mathbb{N} \cup \{\infty\}$ with their usual product topologies are locally convex linear spaces under coordinatewise defined addition and scalar multiplication.

Let $L$ be a vector space. A *norm* on $L$ is a function $\|\cdot\| : L \to [0, \infty)$ having the following properties:

(1) $\quad \|x + y\| \leq \|x\| + \|y\| \quad$ for all $x, y \in L$,
(2) $\quad \|tx\| = |t| \cdot \|x\| \quad$ for all $t \in \mathbb{R}$ and $x \in L$,
(3) $\quad \|x\| = 0 \quad$ if and only if $x = 0$.

If $\|\cdot\|$ is a norm on $L$ then the function

$$\varrho(x, y) = \|x - y\|$$

defines a metric on $L$; it is called the *metric derived from the norm* $\|\cdot\|$. We call a linear space $L$ *normable* provided that there exists a norm on it such that the metric derived from this norm is admissible; for obvious reasons such a norm is called *admissible*. Observe that each normable linear space is locally convex (See Exercise 1.1.6). A *normed linear space* is a pair $(L, \|\cdot\|)$, where $L$ is a vector space and $\|\cdot\|$ is a norm on $L$. We shall always endow the underlying vector space of a normed linear space with the topology derived from its norm.

So we make a formal distinction between *normed linear space* and *normable linear space*: a normable linear space may possess many different norms that generate its topology, see Exercise 1.1.8, whereas in a normed linear space the norm is fixed. In topology we make a similar distinction between *metric* and *metrizable* spaces. A metrizable space may posses many admissible metrics generating the same topology, whereas in a metric space the metric under consideration is fixed.

If $(V, \|\cdot\|)$ is a normed linear space then

$$B = \{x \in V : \|x\| \leq 1\}$$

and

$$S = \{x \in B : \|x\| = 1\}$$

denote its *unit ball* and *unit sphere*, respectively.

A subset $A$ of a normed linear space $L$ is called *bounded* if there exists an $\varepsilon \in [0, \infty)$ such that $\|a\| \leq \varepsilon$ for every $a \in A$.

A *Banach space* is a normable linear space for which there exists an admissible norm such that the metric derived from it is complete.

So a Banach space is topologically complete by definition, and hence is a Baire space by Theorem A.6.6.

**Examples of linear spaces.** We will now discuss various important examples of linear spaces.

**Example 1.1.2.** *The Euclidean spaces $\mathbb{R}^n$.*
The standard norm on $\mathbb{R}^n$ is the Euclidean norm which is defined by

$$\|x\| = \sqrt{\sum_{i=1}^{n} x_i^2}.$$

As noticed on Page 461, $B^n$ and $\mathbb{S}^{n-1}$ abbreviate the unit ball and unit sphere in $\mathbb{R}^n$. It is clear that $\mathbb{R}^n$ is a Banach space since the metric derived from its Euclidean norm is the well-known Euclidean metric which is complete.

**Example 1.1.3.** *The space s.*
The space $s$ is the vector space $\mathbb{R}^\infty$ endowed with the Tychonoff product topology. It is a classical object both in topology and functional analysis and will play a central role in the remaining part of this book.

Observe that $s$ is topologically complete by Lemma A.6.2. Its standard complete metric is the following one:

$$\varrho(x, y) = \sum_{n=1}^{\infty} 2^{-n} \frac{|x_n - y_n|}{1 + |x_n - y_n|}.$$

(See Exercise 1.1.1 for the verification.)

Since each $\mathbb{R}^n$ is normable and since $s$ is in many respects their 'limit', the question naturally arises whether $s$ is normable. We will show below that it is not. Define

$$\sigma = \{x \in s : x_n = 0 \text{ for all but finitely many } n \in \mathbb{N}\}.$$

It is clear that $\sigma$ is a linear subspace of $s$.

**Lemma 1.1.4.** *If $L$ is a linear subspace of $s$ with $\sigma \subseteq L$ then $L$, endowed with the subspace topology it inherits from $s$, is not normable.*

**Proof.** Assume, to the contrary, that $\|\cdot\|$ is an admissible norm on $L$. Then

$$U = \{x \in L : \|x\| < 1\}$$

is an open neighborhood of the origin of $L$. By definition of the product topology on $s$ there are an open neighborhood $V$ of $0$ in $\mathbb{R}$ and an $n \in \mathbb{N}$ such that

$$(*) \qquad \left(\prod_{i=1}^{n} V_i \times \prod_{i=n+1}^{\infty} \mathbb{R}_i\right) \cap L \subseteq U,$$

where $V_i = V$ for $i \leq n$ and $\mathbb{R}_i = \mathbb{R}$ for $i > n$. Let $y \in s$ be defined by $y_i = 0$ if $i \neq n+1$ and $y_{n+1} = 1$. Since $y \in \sigma \subseteq L$ and $y \neq \underline{0}$ it follows that $\varepsilon = \|y\| > 0$. By $(*)$, $ty \in U$ for every $t \in \mathbb{R}$. In particular, $\|y/\varepsilon\| < 1$ but also $\|y/\varepsilon\| = \varepsilon/\varepsilon = 1$, which is a contradiction. $\square$

From the proof of Lemma 1.1.4 it is clear that the interplay between the topology and the linear structure on $s$ prevents it from being normable. (The question naturally arises whether every vector space can be endowed with a norm which is compatible with its linear structure. The answer to this question is in the affirmative, see Exercise 1.1.8.) Consequently, although $s$ seems a natural 'limit' of the spaces $\mathbb{R}^n$, it is notably different from any of its finite dimensional analogs $\mathbb{R}^n$.

**Example 1.1.5.** *The spaces $C(X)$ and $C^*(X)$.*
For a nonempty compact space $X$ we let $C(X)$ denote the set of all continuous real-valued functions on $X$. Obviously, $C(X)$ is a vector space; addition of functions and scalar multiplication are defined pointwise. If $f \in C(X)$ then define its norm, $\|f\|$, by

$$\|f\| = \sup\{|f(x)| : x \in X\}.$$

(Observe that by compactness of $X$ this supremum is attained. That is: there is an element $x \in X$ such that $\|f\| = |f(x)|$.) It is easily seen that $\|\cdot\| : C(X) \to [0, \infty)$ is indeed a norm; it is called the *sup-norm* on $C(X)$. Consequently, the function

$$(*) \qquad \hat{\varrho}(f, g) = \|f - g\|$$

defines a metric on $C(X)$ and therefore generates a topology. From now on we shall endow $C(X)$ with this topology.

There are other useful and interesting topologies on $C(X)$. In Chapter 6 we shall endow $C(X)$ with the so-called *topology of pointwise convergence*. It will be clear from our notation which topology on $C(X)$ we are using. For example, $C(X)$ denotes the set $C(X)$ endowed with the above topology, and $C_p(X)$ denotes the set $C(X)$ endowed with the topology of pointwise convergence, etc.

We claim that $C(X)$ is a Banach space. Let $(f_n)_n$ be a $\hat{\varrho}$-Cauchy sequence. Then $(f_n)_n$ clearly converges pointwise, so $f = \lim_{n \to \infty} f_n$ exists and belongs to $C(X)$ by Lemma A.3.1. It therefore suffices to prove that $f_n \to f$ in $C(X)$. But this follows easily from the proof of Lemma A.3.1.

The spaces $C(X)$ have the following interesting property that will be used quite frequently in our Chapter 4 on ANR-theory.

**Lemma 1.1.6.** *For every compact space $X$ there are a compact space $A$ and an imbedding $i\colon X \hookrightarrow C(A)$ such that $i[X]$ is linearly independent.*

**Remark 1.1.7.** The linear independence of the set $i[X]$ in the above result is quite interesting, and has proved to be useful in several research papers in infinite-dimensional and related topology.

**Proof.** Let $Y$ be the topological sum of $X$ and a point $x_0 \notin X$, and let $\varrho$ be an admissible metric for $Y$. Let $A$ be the subspace of $C(Y)$ consisting of all Lipschitz functions $f\colon Y \to \mathbb{R}$ such that $f(x_0) = 0$. Observe that if $f \in A$ then
$$f[Y] \subseteq [-\operatorname{diam}(Y), \operatorname{diam}(Y)].$$
This is clear since if $y \in Y$ then
$$|f(y)| = |f(y) - 0| = |f(y) - f(x_0)| \le \varrho(y, x_0) \le \operatorname{diam}(Y).$$
It is easy to show that $A$ is a closed subspace of $C(Y)$. But even more is true.

**Claim 1.** *$A$ is compact.*

*Proof.* Let $(f_m)_m$ be any sequence in $A$. It suffices to prove that it has a convergent subsequence in $A$ (Theorem A.5.1). For every $n$ let $\mathcal{U}_n$ be a finite open cover of $[-\operatorname{diam}(Y), \operatorname{diam}(Y)]$ consisting of open sets with diameter at most $2^{-n}$. This cover has a Lebesgue number, say $\lambda_n > 0$ (Lemma A.5.3). By compactness of $X$ there is a finite open cover $\mathcal{V}_n$ of $X$ such that
$$\operatorname{mesh}(\mathcal{V}_n) < {}^1\!/_2 \lambda_n.$$
Now fix any $f \in A$ and $V \in \mathcal{V}_n$. Since $f$ is Lipschitz, the diameter of $f[V]$ is at most $\lambda_n$. Hence $f[V]$ is contained in an element $U \in \mathcal{U}_n$. So for each $f \in A$ there is a function
$$\xi_n(f)\colon \mathcal{V}_n \to \mathcal{U}_n$$
such that for every $V \in \mathcal{V}_n$ we have
$$f[V] \subseteq \xi_n(f)(V) \in \mathcal{U}_n.$$
Let $n = 1$. Observe that there are finitely many functions $\mathcal{V}_1 \to \mathcal{U}_1$ only. Hence there is an infinite subset $N_1 \subseteq \mathbb{N}$ such that $\xi_1(f_m) = \xi_1(f_k)$ for all integers $m, k \in N_1$. Let $m$ and $k$ be two arbitrary elements of $N_1$, and pick an arbitrary $x \in X$. Pick an element $V \in \mathcal{V}_1$ such that $x \in V$. Then both $f_m(x)$ and $f_k(x)$ belong to the same element of $\mathcal{U}_1$, i.e., $|f_m(x) - f_k(x)| < 2^{-1}$. Since $x$ was arbitrary, this shows that $\|f_m - f_k\| \le 2^{-1}$. So now it is clear how to proceed. Let $n_1 = \min N_1$ and consider the infinite set $N_1 \setminus \{n_1\}$. There is an infinite subset $N_2 \subseteq N_1 \setminus \{n_1\}$ such that for all $m$ and $k$ in $N_2$ the functions $\xi_2(f_m)$ and $\xi_2(f_k)$ agree. Then by a similar argument as the

one above, $\|f_m - f_k\| \leq 2^{-2}$ for all $m, k \in N_2$. Let $n_2 = \min N_2$. Continuing in this way resursively, we can construct an infinite sequence
$$n_1 < n_2 < \cdots < n_m < \cdots$$
of natural numbers and a Cauchy sequence $(f_{n_m})_m$ in $C(Y)$. Since $C(Y)$ is a Banach space, this sequence has a limit and since $A$ is closed, this limit belongs to $A$. So we conclude that $(f_{n_m})_m$ is the desired convergent subsequence of $(f_n)_n$. $\diamond$

Define $i \colon X \to C(A)$ by the following formula:
$$i(x)(f) = f(x).$$
Observe that $i$ is well-defined. For $i(x)$ should be an element of $C(A)$, hence should be a function $i(x) \colon A \to \mathbb{R}$. But an element of $A$ is a function $f$ from $Y$ to $\mathbb{R}$. So the formula tells us that $i(x)$ sends the function $f$ onto its evaluation in the point $x \in X \subseteq Y$.

**Claim 2.** $i \colon X \to i[X]$ is an isometry.

*Proof.* Let $x_1, x_2 \in X$. Then

(1)
$$\begin{aligned}\|i(x_1) - i(x_2)\| &= \sup\{|i(x_1)(f) - i(x_2)(f)| : f \in A\} \\ &= \sup\{|f(x_1) - f(x_2)| : f \in A\} \\ &\leq \varrho(x_1, x_2).\end{aligned}$$

Observe that the last inequality follows from the fact that $f$ is Lipschitz. Define the function $g \colon Y \to \mathbb{R}$ by
$$g(y) = \varrho(y, x_2) - \varrho(x_0, x_2).$$
If $y, y' \in Y$ are arbitrary then
$$\begin{aligned}|g(y) - g(y')| &= |\varrho(y, x_2) - \varrho(x_0, x_2) - \varrho(y', x_2) + \varrho(x_0, x_2)| \\ &= |\varrho(y, x_2) - \varrho(y', x_2)| \\ &\leq \varrho(y, y').\end{aligned}$$
Also, $g(x_0) = 0$. Hence $g \in A$ and
$$\begin{aligned}|i(x_1)(g) - i(x_2)(g)| &= |\varrho(x_1, x_2) - \varrho(x_0, x_2) - \varrho(x_2, x_2) + \varrho(x_0, x_2)| \\ &= \varrho(x_1, x_2).\end{aligned}$$
Hence by (1) we get $\|i(x_1) - i(x_2)\| = \varrho(x_1, x_2)$, as required. $\diamond$

It remains to prove that $i[X]$ is a linearly independent subset of $C(A)$. To this end, let $x_1, \ldots, x_{n+1}$ be distinct elements of $X$. We claim that $i(x_{n+1})$ is not a linear combination of the $i(x_1), \ldots, i(x_n)$.

Define $h \colon Y \to \mathbb{R}$ by
$$h(y) = \min_{0 \leq i \leq n} \varrho(y, x_i).$$

**Claim 3.** $h \in A$.

*Proof.* We shall first prove that $h$ is Lipschitz. To this end, pick arbitrary $a, b \in Y$. We may assume without loss of generality that $h(b) \leq h(a)$. Fix $i$ and $j$ such that $h(a) = \varrho(a, x_i)$ and $h(b) = \varrho(b, x_j)$. Since $h(a) = \varrho(a, x_i)$ we get by the definition of $h$ that $\varrho(a, x_i) \leq \varrho(a, x_j)$. From this we conclude that

$$\begin{aligned}|h(a) - h(b)| &= h(a) - h(b) \\ &= \varrho(a, x_i) - \varrho(b, x_j) \\ &\leq \varrho(a, x_j) - \varrho(b, x_j) \\ &\leq \varrho(a, b),\end{aligned}$$

as required.

Since it is trivial that $h(x_0) = 0$, this completes the proof. $\diamond$

Observe that $i(x_k)(h) = h(x_k) = 0$ for every $k \leq n$, but since the $x_0, \ldots, x_{n+1}$ are distinct elements of $Y$,

$$i(x_{n+1})(h) \neq 0.$$

So $i(x_{n+1})$ is not a linear combination of the $i(x_1), \ldots, i(x_n)$, and this is what we had to prove. $\square$

**Corollary 1.1.8.** *Every space is homeomorphic to a linearly independent bounded and closed subspace of a normed linear space.*

**Proof.** Let $X$ be a space, and let $aX$ be a compactification of $X$ (Corollary A.4.8). By Lemma 1.1.6 there are a compact space $A$ and an imbedding $i\colon aX \to C(A)$ such that $i[aX]$ is linearly independent. By linear independence, if $L$ is the linear hull of $i[X]$ then $L \cap i[aX] = i[X]$. Hence $i[X]$ is a closed linearly independent subset of the normed linear space $L$.

Since $\|\cdot\|\colon C(A) \to [0, \infty)$ is continuous, and $aX$ is compact, it is clear that $i[aX]$ is a bounded subset of $C(A)$, hence so is $i[X]$. $\square$

If $X$ is not compact then $C(X)$ contains unbounded functions (see Exercise A.5.14 below), and so the formula

$$\|f\| = \sup\{|f(x)| : x \in X\}$$

does not define a norm on $C(X)$. By considering the subset $C^*(X)$ of $C(X)$ consisting of all *bounded* functions, this problem does not occur; $C^*(X)$ endowed with the sup-norm is a Banach space for the same reasons $C(Y)$ is for compact $Y$.

The topology defined here on $C^*(X)$ is called the *topology of uniform convergence*.

**Example 1.1.9.** *The space $c_0$.*
Put $c_0 = \{x \in s : \lim_{n \to \infty} x_n = 0\}$, and endow it with the norm
$$\|x\| = \sup\{|x_n| : n \in \mathbb{N}\}.$$
It follows by straightforward calculations that this is indeed a norm compatible with the linear structure on $c_0$. There is however another way of proving this. Let $S$ be a nontrivial convergent sequence including its limit $t$, and consider
$$L = \{f \in C(S) : f(t) = 0\}.$$
Then $L$ is a closed linear subspace of $C(S)$ which clearly can be identified with $c_0$. So $c_0$ is a closed linear subspace of the function space $C(S)$.

The set $c_0$ endowed with a different vector space topology will play a prominent role in our analysis of function spaces later. See Chapter 6 for details.

**Example 1.1.10.** *The Hilbert space $\ell^2$.*
We saw that the topology on $s$ is very different from the topology on any of its finite dimensional analogs $\mathbb{R}^n$. We shall now construct another natural 'limit' of the spaces $\mathbb{R}^n$ which behaves better (in this respect).

Consider the usual inner product on $\mathbb{R}^n$ given by
$$\langle x, y \rangle = \sum_{i=1}^{n} x_i y_i.$$
If we try to generalize this inner product for $\mathbb{R}^\infty$ then we have to deal with infinite series and it is therefore quite natural to restrict ourselves to the following subset of $\mathbb{R}^\infty$:
$$\ell^2 = \{x \in s : \sum_{i=1}^{\infty} x_i^2 < \infty\}.$$
We shall first prove that $\ell^2$ is a linear subspace of $\mathbb{R}^\infty$. For every $x \in \ell^2$ we write $p(x) = \sqrt{\sum_{i=1}^{\infty} x_i^2}$. If $x, y \in \ell^2$, then Schwarz's inequality applied to $\mathbb{R}^n$ shows that $|\sum_{i=1}^{n} x_i y_i| \leq p(x) \cdot p(y)$. From this it follows that
$$\left|\sum_{i=1}^{\infty} x_i y_i\right| \leq p(x) \cdot p(y) < \infty,$$
and so
$$\sum_{i=1}^{\infty} (x_i + y_i)^2 = \sum_{i=1}^{\infty} x_i^2 + \sum_{i=1}^{\infty} y_i^2 + 2\sum_{i=1}^{\infty} x_i y_i < \infty$$
since all infinite series considered are convergent. We conclude that for every $x, y \in \ell^2$ we have $x + y \in \ell^2$. If $x \in \ell^2$ and $t \in \mathbb{R}$ then trivially $tx \in \ell^2$. Consequently, $\ell^2$ is indeed a linear subspace of $\mathbb{R}^\infty$.

Since $\sum_{i=1}^{\infty} x_i y_i < \infty$ for all $x, y \in \ell^2$ we have a well-defined function $\langle \cdot, \cdot \rangle \colon \ell^2 \times \ell^2 \to \mathbb{R}$,

$$\langle x, y \rangle = \sum_{i=1}^{\infty} x_i y_i,$$

which is easily seen to be an inner product. Consequently, $\|x\| = p(x)$ defines a norm on $\ell^2$ and the metric derived from this norm is:

$$\varrho(x, y) = \sqrt{\sum_{i=1}^{\infty}(x_i - y_i)^2}.$$

We endow $\ell^2$ with the topology generated by this metric and refer to $\ell^2$ with this topology as *Hilbert space*.

It is clear that $\sigma$ is a subset of $\ell^2$ and so Lemma 1.1.4 shows that the topology that $\ell^2$ inherits from $s$ is different from the topology on $\ell^2$ which we just defined. We will comment on the precise relation between these topologies later.

**Lemma 1.1.11.** *The metric $\varrho$ on $\ell^2$ defined above is complete.*

So $\ell^2$ is also a Banach space. But there is a difference with the spaces $C(X)$ that we discussed before. The norm on $\ell^2$ is derived from an inner product, while no inner product on $C(\mathbb{I})$ yields its standard norm (Exercise 1.1.9).

The topology on $s$ is the topology of 'coordinatewise convergence', see Exercise A.2.2. Topologists usually find such product topologies easier to handle than topologies derived from a norm. However, convergence in $\ell^2$ can be handled with the same ease, as is stated in the next result.

**Lemma 1.1.12** (Exercise 1.1.26). *Suppose that $(x(n))_n$ is a sequence in $\ell^2$, and $x \in \ell^2$. The following statements are equivalent:*

(1) $\lim_{n \to \infty} x(n) = x$ *(in $\ell^2$)*,
(2) $\lim_{n \to \infty} \|x(n)\| = \|x\|$ *and for every $i \in \mathbb{N}$, $\lim_{n \to \infty} x(n)_i = x_i$.*

From Lemma 1.1.12 it immediately follows that the topology on $\ell^2$ is finer than the topology that $\ell^2$ inherits from $s$. However, more can be concluded. For example, consider the unit sphere $S = \{x \in \ell^2 : \|x\| = 1\}$. Since all points in $S$ have the same norm, the topology that $S$ inherits from $\ell^2$ is precisely the same as the topology that $S$ inherits from $s$, i.e., the topology of 'coordinatewise convergence'. This remark plays an important role in the proof of the Anderson Theorem from ANDERSON [15] that $\ell^2$ and $s$ are topologically homeomorphic (see also VAN MILL [298, Chapter 6] for a complete proof of this result).

10                                1. BASIC TOPOLOGY

**Classical theorems.** We now present some classical theorems on Banach spaces that will be important later.

**Open Mapping Theorem 1.1.13.** *Let $T$ be a continuous linear mapping of a Banach space $E$ onto a Banach space $F$. Then $T$ is open.*

**Proof.** The proof is in three steps.

**Claim 1.** There exists $\alpha > 0$ such that such that
$$\{y \in F : \|y\| \leq 1\} \subseteq \overline{T[\{x \in E : \|x\| \leq \alpha\}]}.$$

*Proof.* For each $\alpha > 0$ put $B_\alpha = \{x \in E : \|x\| \leq \alpha\}$. Since
$$F = \bigcup_{n=1}^{\infty} \overline{T[B_n]},$$
and $F$ is a Baire space (see Page 3), there exists $m \in \mathbb{N}$ such that $\overline{T[B_m]}$ has nonempty interior. Since $T$ is linear, it follows easily that $T[B_m]$ is convex, and by the continuity of the algebraic operations on $F$, so is $\overline{T[B_m]}$. In addition, $B_m$ is symmetric, i.e., $-B_m = B_m$. Again since $T$ is linear, it follows that $T[B_m]$ is symmetric, from which it follows easily that $\overline{T[B_m]}$ is symmetric as well. Now choose $y \in F$ and $\beta > 0$ such that $D(y, \beta) \subseteq \overline{T[B_m]}$. Let $z \in F$ be such that $\|z\| \leq \beta$. Then $\|(z+y)-y\| \leq \beta$, hence $z+y \in \overline{T[B_m]}$. Similarly, $\|(y-z)-y\| \leq \beta$ from which it follows that $y - z \in \overline{T[B_m]}$ and so since $\overline{T[B_m]}$ is symmetric that $z - y \in \overline{T[B_m]}$. Now observe that since $\underline{0} \in \overline{T[B_m]}$ by convexity we get
$$z = \tfrac{1}{2}(z+y) + \tfrac{1}{2}(z-y) \in \overline{T[B_m]}.$$
This proves that $D(\underline{0}, \beta) \subseteq \overline{T[B_m]}$ from which it follows easily that
$$D(\underline{0}, 1) \subseteq \overline{T[B_\alpha]},$$
where $\alpha = m/\beta$. $\Diamond$

In the remaining part of the proof we adopt the notation introduced in Claim 1.

**Claim 2.** $\{y \in F : \|y\| \leq 1\} \subseteq T[B_{2\alpha}]$.

*Proof.* Let $y \in F$ with $\|y\| \leq 1$. We shall define recursively a sequence $(y_n)_n$ in $T[B_\alpha]$ such that for all $n$,

(∗)
$$\left\|y - \sum_{k=1}^{n} 2^{-(k-1)} y_k\right\| \leq 2^{-n}.$$

By Claim 1, there exists $y_1 \in T[B_\alpha]$ such that $\|y - y_1\| \leq 1/2$. Suppose that $y_1, \ldots, y_n$ are chosen properly. Then
$$\left\| 2^n \left( y - \sum_{k=1}^n 2^{-(k-1)} y_k \right) \right\| \leq 1$$
and therefore, again by Claim 1, there exists $y_{n+1} \in T[B_\alpha]$ such that
$$\left\| 2^n \left( y - \sum_{k=1}^n 2^{-(k-1)} y_k \right) - y_{n+1} \right\| < 1/2.$$
It is clear that $y_{n+1}$ is as required.

From $(*)$ it easily follows that
$$y = \lim_{n \to \infty} \sum_{k=1}^n 2^{-(k-1)} y_k.$$
For every $n$ choose a point $x_n \in B_\alpha$ with $T(x_n) = y_n$. Since $\|x_n\| \leq \alpha < \infty$ for every $n$, it is easily seen that the sequence
$$\left( \sum_{k=1}^n 2^{-(k-1)} x_k \right)_n$$
is Cauchy, and hence that
$$x = \lim_{n \to \infty} \sum_{k=1}^n 2^{-(k-1)} x_k$$
exists. Since $\|\cdot\| \colon E \to \mathbb{R}$ is continuous,
$$\|x\| \leq \sum_{k=1}^\infty 2^{-(k-1)} \|x_k\| \leq \sum_{k=1}^\infty 2^{-(k-1)} \cdot \alpha = 2\alpha.$$
By observing that $T$ is continuous and linear we obtain
$$T(x) = \lim_{n \to \infty} \sum_{k=1}^\infty 2^{-(k-1)} T(x_k) = y,$$
from which we conclude that $y \in T[B_{2\alpha}]$.  $\diamond$

We are now in a position to prove that $T$ is open. Indeed, let $U$ be a nonempty open subset of $E$. Let $x \in U$ and choose $\varepsilon > 0$ with $D(x, \varepsilon) \subseteq U$. Consequently, $B_\varepsilon \subseteq U - x$. It follows by Claim 2 that
$$\{y \in F : \|y\| \leq \varepsilon/2\alpha\} \subseteq T[B_\varepsilon].$$
Consequently, since $T[B_\varepsilon] \subseteq T[U - x] = T[U] - T(x)$, we have
$$\{y \in F : \|y - T(x)\| \leq \varepsilon/2\alpha\} \subseteq T[U].$$
So $T[U]$ is a neighborhood of $T(x)$ and since $x$ was an arbitrarily chosen point from $U$, we are done. $\square$

**Corollary 1.1.14.** *A bijective continuous linear function between Banach spaces is a (linear) homeomorphism.*

**Proof.** This is clear since such a function is open by Theorem 1.1.13. □

The following corollary to Theorem 1.1.13 will be of particular importance later. Its converse is trivial (Exercise A.1.8).

**Closed Graph Theorem 1.1.15.** *Let $E$ and $F$ be Banach spaces and let*
$$T \colon E \to F$$
*be linear. If the graph*
$$\Gamma = \{(x, Tx) : x \in E\}$$
*of $T$ is a closed subset of $E \times F$ then $T$ is continuous.*

**Proof.** Observe that $\Gamma$ is a closed linear subspace of the Banach space $E \times F$ (Exercise A.1.8) and hence is a Banach space itself. In addition, the function $S_1 \colon \Gamma \to E$ defined by $S_1(x, Tx) = x$ is clearly continuous (since it is the restriction of a projection). As a consequence, $S_1$ is a linear homeomorphism by Corollary 1.1.14. Observe that the function $S_2 \colon \Gamma \to F$ defined by $S_2(x, T(x)) = T(x)$ is continuous as well. So the function $E \to F$ defined by $x \mapsto (x, Tx) \mapsto Tx$ is continuous, being a composition of continuous functions. So $T$ is continuous. □

**Corollary 1.1.16.** *Let $E$ and $F$ be Banach spaces additionally endowed with weaker vector space topologies. If $T \colon E \to F$ is a linear homeomorphism with respect to the weaker topologies then it is also a linear homeomorphism with respect to the Banach space topologies.*

**Proof.** All we need to observe is that by continuity the graph of $T$ is closed in the weaker topology on $E \times F$ (Exercise A.1.8) and hence also in $E \times F$ endowed with the product of the Banach space topologies. Simply observe that the product of the weaker topologies is weaker than the product of the Banach space topologies. □

**The Michael selection theorem.** We shall prove that certain set-valued functions admit a continuous selection (for definitions, see below).

Let $X$ and $Y$ be sets. A *set-valued function* $F$ from $X$ to $Y$ is defined to be a function from $X$ to $\mathcal{P}(Y) \setminus \{\emptyset\}$, i.e., $F \colon X \to \mathcal{P}(Y) \setminus \{\emptyset\}$. By the symbol
$$F \colon X \rightrightarrows Y$$
we shall mean that $F$ is a set-valued function from $X$ to $Y$.

Let $X$ and $Y$ be topological spaces and let $F \colon X \rightrightarrows Y$. For every $V \subseteq Y$ we put
$$F^{\Leftarrow}[V] = \{x \in X : F(x) \cap V \neq \emptyset\}.$$

We say that $F$ is *lower semi-continuous* (abbreviated LSC) provided that for every open subset $U$ of $Y$, $F^{\Leftarrow}[U]$ is open in $X$. Observe that if $\mathcal{U}$ is a covering of $Y$ then
$$F^{\Leftarrow}[\mathcal{U}] = \{F^{\Leftarrow}[U] : U \in \mathcal{U}\}$$
covers $X$ since for every $x \in X$, $F(x) \neq \emptyset$.

A basic example of an LSC set-valued function is the following one. If $X$ and $Y$ are spaces and $f\colon X \to Y$ is an open surjection then we define
$$F\colon Y \Rightarrow X$$
by $F(y) = f^{-1}(y)$. The function $F$ is LSC since for every open $U \subseteq Y$ we have
$$F^{\Leftarrow}[U] = \{y \in Y : F(y) \cap U \neq \emptyset\} = \{y \in Y : f^{-1}(y) \cap U \neq \emptyset\} = f[U].$$
Other examples of LSC set-valued mappings will be presented later.

Let $X$ and $Y$ be sets and let $F\colon X \Rightarrow Y$. A function $f\colon X \to Y$ is called a *selection* for $F$ if for all $x \in X$, $f(x) \in F(x)$. Since for all $x \in X$ the set $F(x)$ is nonempty, by the Axiom of Choice such a selection exists. The question naturally arises whether it is possible to find a *continuous* selection if $X$ and $Y$ are topological spaces. This question is natural but rather naive. Simple examples show that the answer in general is in the negative.

**Example 1.1.17.** Define $F\colon \mathbb{I} \Rightarrow \mathbb{I}$ by
$$F(x) = \begin{cases} \{0\} & (0 \leq x \leq 1/3), \\ \{0,1\} & (1/3 < x < 2/3), \\ \{1\} & (2/3 \leq x \leq 1). \end{cases}$$

There does not exist a continuous function $f\colon \mathbb{I} \to \mathbb{I}$ the graph of which is contained in the 'graph' of $F$, which is left as an exercise to the reader as well as to prove that $F$ is LSC.

The values of $F$ in the above example are too 'small' for $F$ to admit a *continuous* selection. If we enlarge these values by for example to require that $F(x) = \mathbb{I}$ for all $1/3 < x < 2/3$ then $F$ *does* admit a continuous selection.

The following result shows that although not every LSC map has a continuous selection any map with sufficiently many continuous selections must be LSC.

**Proposition 1.1.18.** *Let $X$ and $Y$ be spaces and let $F\colon X \Rightarrow Y$ be a set-valued function with the following property:*

> *for all $x \in X$ and $y \in F(x)$ there exists a continuous selection $f$ for $F$ such that $f(x) = y$.*

*Then $F$ is LSC.*

**Proof.** Let $U \subseteq Y$ be open and take $x \in F^{\Leftarrow}[U]$. Pick $y \in F(x) \cap U$. By assumption there exists a continuous selection $f \colon X \to Y$ for $F$ such that $f(x) = y$. Put $V = f^{-1}[U]$. By continuity of $f$, $V$ is a neighborhood of $x$. In addition, $V \subseteq F^{\Leftarrow}[U]$ since if $x' \in V$ then
$$f(x') \in F(x') \cap U.$$
We conclude that $F^{\Leftarrow}[U]$ is open. $\square$

One of the main results here is the Michael Selection Theorem which states that in normed linear spaces, convex valued LSC set-valued functions admit continuous selections. The following three technical lemmas are needed in the proof of this result.

**Lemma 1.1.19.** *Let $X$ and $Y$ be spaces. If $F \colon X \rightrightarrows Y$ is LSC then*
  (1) *the function $F_c \colon X \rightrightarrows Y$ defined by $F_c(x) = \overline{F(x)}$ is LSC,*
  (2) *if $f \colon X \to Y$ is continuous and $\varrho$ is an admissible metric for $Y$ and the number $r > 0$ is such that*
$$\forall x \in X : \varrho\big(f(x), F(x)\big) < r,$$
*then the function $G \colon X \rightrightarrows Y$ defined by*
$$G(x) = \overline{F(x) \cap B(f(x), r)}$$
*is LSC.*

**Proof.** For (1), simply observe that for all $x \in X$ and open $U \subseteq Y$ we have that $F(x) \cap U \neq \emptyset$ if and only if $\overline{F(x)} \cap U \neq \emptyset$. Consequently, for every open subset $U \subseteq Y$ the equality $F^{\Leftarrow}[U] = F_c^{\Leftarrow}[U]$ holds.

For (2) we use (1) to conclude that it suffices to prove that the set-valued mapping $\vec{G} \colon X \rightrightarrows Y$ defined by
$$\vec{G}(x) = F(x) \cap B(f(x), r)$$
is LSC. Let $V \subseteq Y$ be open and take a point $x \in \vec{G}^{\Leftarrow}[V]$. There exists $y \in Y$ such that
$$y \in \big(F(x) \cap B(f(x), r)\big) \cap V.$$
Let $\varepsilon = r - \varrho\big(y, f(x)\big)$ and choose $0 < \delta < \varepsilon$ such that
$$B(y, \delta) \subseteq B(f(x), r) \cap V.$$
Since $F(x) \cap B(y, \delta/2) \neq \emptyset$ and $F$ is LSC, $U_0 = F^{\Leftarrow}[B(y, \delta/2)]$ is a neighborhood of $x$. In addition, since $f$ is continuous, $U_1 = f^{-1}[B(f(x), \delta/2)]$ is also a neighborhood of $x$. Put $U = U_0 \cap U_1$. We claim that $U \subseteq \vec{G}^{\Leftarrow}[V]$.

To this end, take an arbitrary $x' \in U$. Since $x' \in U_0$, there exists a point $y' \in F(x') \cap B(y, \delta/2)$. In addition, $f(x') \in B(f(x), \delta/2)$ since $x' \in U_1$.

Consequently,

$$\varrho(y', f(x')) \le \varrho(y', y) + \varrho(y, f(x)) + \varrho(f(x), f(x'))$$
$$< \delta/2 + r - \varepsilon + \delta/2$$
$$< \delta/2 + r - \delta + \delta/2$$
$$= r$$

and therefore $y' \in F(x') \cap B(f(x'), r) \cap V = \vec{G}(x') \cap V$. We conclude that the point $x'$ belongs to $\vec{G}^{\Leftarrow}[V]$. □

**Lemma 1.1.20.** *Let $L$ be a normed linear space, let $X$ be a space and let $F\colon X \Rightarrow L$ be LSC such that $F(x)$ is convex for every $x \in X$. Then for every $r > 0$ there exists a continuous function $f\colon X \to L$ such that for every $x \in X$ we have $\varrho(f(x), F(x)) < r$.*

**Proof.** Put

$$\mathcal{B} = \{B(y, r) : y \in L\}.$$

By Theorem A.7.5 there exists a partition of unity $\mathcal{P}$ on $X$ which is subordinated to $F^{\Leftarrow}[\mathcal{B}]$. Consequently, for each $p \in \mathcal{P}$ there exists $b_p \in L$ such that

$$p^{-1}[(0, 1]] \subseteq F^{\Leftarrow}[B(b_p, r)].$$

Define $f\colon X \to L$ by

$$f(x) = \sum_{p \in \mathcal{P}} p(x) \cdot b_p.$$

For each $x \in X$ there are a neighborhood $U_x$ and a finite subset $G(x)$ of $\mathcal{P}$ such that $U_x \cap p^{-1}[(0, 1]] \ne \emptyset$ if and only if $p \in G(x)$.

Now fix an arbitrary $x \in X$. Observe that the restriction of $f$ to $U_x$ is given by

$$f(y) = \sum_{p \in G(x)} p(y) \cdot b_p,$$

which is a continuous expression in $y$ since $G(x)$ is finite. From this we conclude that $f$ is well-defined and continuous at $x$.

Put $\mathcal{G}(x) = \{p \in G(x) : x \in p^{-1}[(0, 1]]\}$. For each $p \in \mathcal{G}(x)$ we have

$$x \in p^{-1}[(0, 1]] \subseteq F^{\Leftarrow}[B(b_p, r)]$$

and so there exists $y_p \in F(x) \cap B(b_p, r)$. Observe that

$$f(x) = \sum_{p \in \mathcal{G}(x)} p(x) \cdot b_p, \quad \sum_{p \in \mathcal{G}(x)} p(x) = 1.$$

This implies that
$$\left\|f(x) - \sum_{p \in \mathcal{G}(x)} p(x) \cdot y_p\right\| \le \sum_{p \in \mathcal{G}(x)} p(x) \cdot \|b_p - y_p\|$$
$$< \sum_{p \in \mathcal{G}(x)} p(x) \cdot r$$
$$= r.$$

Since $F(x)$ is convex and $\sum_{p \in \mathcal{G}(x)} p(x) = 1$ we have
$$\sum_{p \in \mathcal{G}(x)} p(x) \cdot y_p \in F(x)$$
(Lemma 1.1.1), and consequently, $\varrho\bigl(f(x), F(x)\bigr) < r$. □

We need one more technical lemma.

**Lemma 1.1.21.** *Let $X$ and $Y$ be spaces and let $F\colon X \Rightarrow Y$ be LSC. Suppose that $A \subseteq X$ is closed and that $f\colon A \to Y$ is a continuous selection for the function $F \restriction A\colon A \Rightarrow Y$. Define $G\colon X \Rightarrow Y$ by*
$$G(x) = \begin{cases} \{f(x)\} & (x \in A), \\ F(x) & (x \in X \setminus A). \end{cases}$$
*Then $G$ is LSC.*

**Proof.** Let $U \subseteq Y$ be open. We first claim that $f^{-1}[U] \subseteq F^{\Leftarrow}[U]$. Indeed, take an arbitrary $x \in f^{-1}[U]$. Then $f(x) \in F(x) \cap U$, hence $x \in F^{\Leftarrow}[U]$. Observe that by continuity of $f$, the set $f^{-1}[U]$ is open in $A$. So there is an open $V \subseteq X$ such that $V \cap A = f^{-1}[U]$. Since $f^{-1}[U] \subseteq F^{\Leftarrow}[U]$ and $F^{\Leftarrow}[U]$ is open, without loss of generality, $V \subseteq F^{\Leftarrow}[U]$. Consequently,
$$G^{\Leftarrow}[U] = V \cup (F^{\Leftarrow}[U] \setminus A)$$
is open. □

Let $(L, \|\cdot\|)$ be a normed linear space. A subset $A$ of $L$ is called *complete with respect to* $\|\cdot\|$ if the restriction of the metric $\varrho(x, y) = \|x - y\|$ to $A$ is complete. Observe that such an $A$ is automatically closed in $L$ and also that every compact subset of $L$ is complete with respect to $\|\cdot\|$.

We are now in a position to prove the announced classical result.

**The Michael Selection Theorem 1.1.22.** *let $X$ be a space and let*
$$F\colon X \Rightarrow L$$
*be LSC, where $(L, \|\cdot\|)$ is a normed linear space. Assume that each $F(x)$ is convex in $L$ and complete with respect to $\|\cdot\|$. Then for every closed subset $A$ of $X$ and every continuous selection $f\colon A \to L$ for the function*
$$F \restriction A\colon A \Rightarrow L,$$

there exists a continuous selection $g\colon X \to L$ for $F$ which extends $f$.

**Proof.** We shall first prove the theorem in the special case $A = \emptyset$.

By induction on $n$ we shall construct a sequence $(f_n)_n$ in $C(X, L)$ such that

(1) $\hat{\varrho}(f_n, f_{n+1}) < 2^{-(n-1)}$,
(2) $\hat{\varrho}(f_n(x), F(x)) < 2^{-n}$ for every $x \in X$.

Apply Lemma 1.1.20 with $r = 1/2$ to find $f_1\colon X \to L$ such that for every $x \in X$,
$$\varrho(f_1(x), F(x)) < 1/2.$$
Suppose that $f_n$ has been defined. Define $F_n\colon X \rightrightarrows L$ by
$$F_n(x) = \overline{F(x) \cap B(f_n(x), 2^{-n})}.$$
Then $F_n$ is LSC by Lemma 1.1.19. By another appeal to Lemma 1.1.20 we find $h\colon X \to L$ such that $\varrho(h(x), F_n(x)) < 2^{-n}$ for every $x \in X$. It is easy to see that $f_{n+1} = h$ is as required.

**Claim 1.** For every $x \in X$, $g(x) = \lim_{n\to\infty} f_n(x)$ exists and belongs to $F(x)$.

*Proof.* Take an arbitrary $x \in X$. By (2) there exists for every $n \in \mathbb{N}$ a point $a_n \in F(x)$ such that $\varrho(f_n(x), a_n) < 2^{-n}$. Consequently, by (1) we have
$$\varrho(a_n, a_{n+1}) \leq \varrho(a_n, f_n(x)) + \varrho(f_n(x), f_{n+1}(x)) + \varrho(f_{n+1}(x), a_{n+1})$$
$$< 2^{-n} + 2^{-(n-1)} + 2^{-(n+1)}$$
$$< 2^{-(n-2)}.$$

We conclude that the sequence $(a_n)_n$ is Cauchy and so by completeness of $F(x)$, $a = \lim_{n\to\infty} a_n$ exists. Since $\varrho(f_n(x), a_n) < 2^{-n}$ for every $n$, this implies that $\lim_{n\to\infty} f_n(x) = g(x)$ also exists and is equal to $a \in F(x)$. ◇

By Lemma A.3.1 we consequently conclude that the function $g$ is the required continuous selection.

Now if $A \neq \emptyset$, the above special case and Lemma 1.1.21 yield the desired result. □

Theorem 1.1.13 and the Michael Selection Theorem imply the following

**Corollary 1.1.23.** *Let $E$ and $F$ be Banach spaces and let $T\colon E \to F$ be a surjective continuous and linear function. In addition, let $\operatorname{Ker} T$ denote the kernel of $T$. Then there exists a continuous function $f\colon F \to E$ such that $T \circ f = 1_F$. The function $h\colon E \to \operatorname{Ker} T \times F$ defined by*
$$h(x) = (x - f(Tx), Tx)$$
*is a homeomorphism.*

**Proof.** Define $H\colon F \Rightarrow E$ by $H(y) = T^{-1}(y)$. Then for each open $U \subseteq E$ we have $H^{\Leftarrow}[U] = T[U]$, so by Theorem 1.1.13, $F$ is LSC. Since the fibers of $T$ are convex and closed by linearity and continuity of $T$, respectively, the existence of $f$ follows directly from Theorem 1.1.22.

The easy proof that $h$ is a homeomorphism is left as an exercise to the reader. □

**Remarks.** We conclude this section by some remarks. We introduced linear spaces, locally convex linear spaces, normable linear spaces and implicitly also inner product spaces. Obviously, the 'underlying' linear space of an inner product space is normable and each normable linear space is locally convex. Since $s$ is locally convex but not normable (Lemma 1.1.4), we see that the class of normable linear spaces is a proper subclass of the class of all locally convex linear spaces. The sup-norm on $C(\mathbb{I})$ violates the parallelogram law and therefore cannot be derived from an inner product (Exercise 1.1.9). As a consequence, the class of normed spaces is strictly larger than the class of inner product spaces.

In Exercise 1.1.13 we will present examples of linear spaces that are not locally convex.

It is also possible to introduce *topological* versions of the above concepts. One may ask for example whether every linear space is homeomorphic to a locally convex linear space, whether every locally convex linear space is homeomorphic to a normed linear space and whether every normed linear space is homeomorphic to an inner product space. There is a lot of research in infinite-dimensional topology on these questions, especially when the linear spaces under consideration are (absolutely) Borel. For general (separable metrizable) linear spaces, the answers to these questions are in the negative, see MARCISZEWSKI [269].

**Exercises for §1.1.** Let $L$ be a linear space. A *maximal* linearly independent subset $B$ of $L$ (i.e., a subset of $L$ which is maximal with respect to the property of being linearly independent) is called a *Hamel basis* for $L$. The Kuratowski-Zorn Lemma easily implies that every linear space has a Hamel basis. It is well-known, and easy to prove, that if $B$ is a Hamel basis for $L$ then each $x \in L \setminus \{\underline{0}\}$ can be written uniquely in the form

$$x = \alpha_1 x_1 + \alpha_2 x_2 + \cdots + \alpha_n x_n,$$

with $x_i \in B$ and $\alpha_i \in \mathbb{R} \setminus \{0\}$ for every $i \leq n$ (for details, consult any (good) textbook on Linear Algebra).

A linear space $L$ is called *finite dimensional* if it has a finite Hamel basis. Otherwise it is called *infinite-dimensional*.

## 1.1. LINEAR SPACES

If $L$ is a linear space and $x, y \in L$ then $I(x,y)$ denotes the *straight-line segment from $x$ to $y$*, i.e.,
$$I(x,y) = \{\alpha x + (1-\alpha)y : \alpha \in \mathbb{I}\}.$$

Let $X$ be a space. For a compact subset $K$ in $X$ and an open $U$ in $\mathbb{R}$ define
$$[K, U] = \{f \in C(X) : f[K] \subseteq U\}.$$
Topologize $C(X)$ by taking the collection
$$\{[K, U] : K \subseteq X \text{ is compact and } U \subseteq \mathbb{R} \text{ is open}\}$$
as an open subbase. This topology is called the *compact-open topology* on $C(X)$.

1. Prove that the formula
$$\varrho(x,y) = \sum_{n=1}^{\infty} 2^{-n} \frac{|x_n - y_n|}{1 + |x_n - y_n|}$$
defines an admissible metric on $s$.

2. Let $L$ be a linear space. Prove that for every $x \in L$ and every convex $C \subseteq L$ the set $x + C$ is convex as well.

3. Let $L$ be a linear space. Prove that every translation of $L$ is a homeomorphism.

4. Let $L$ be a linear space and let $x \in L$. In addition, let $\mathcal{W}$ be a local base at the origin of $L$. Show that $x$ has arbitrarily small neighborhoods of the form $x + W$, where $W \in \mathcal{W}$.

5. Prove Lemma 1.1.1.

6. Prove that every normable linear space is locally convex.

7. Define a function $|||\cdot|||\colon \mathbb{R}^n \to \mathbb{R}$ by $|||x||| = \max\{|x_i| : 1 \leq i \leq n\}$. Prove that $|||\cdot|||$ is a norm on $\mathbb{R}^n$ and that it generates the Euclidean topology.

8. Let $L$ be a linear space and let $B$ be a Hamel basis for $L$. If
$$x = \alpha_1 x_1 + \alpha_2 x_2 + \cdots + \alpha_n x_n,$$
with $x_i \in B$ and $\alpha_i \in \mathbb{R}$ for every $i \leq n$, then put
$$\|x\| = |\alpha_1| + |\alpha_2| + \cdots + |\alpha_n|.$$
Prove that $\|\cdot\|$ defines a norm on $L$.

9. Prove that the sup-norm on $C(\mathbb{I})$ cannot be derived from an inner product by showing that it violates the parallelogram law.

▶10. Prove that for every compact space $(X, \varrho)$ there exists an isometry
$$i\colon X \hookrightarrow C(X)$$
such that
   (1) for every subset $Y \subseteq X$ the image set $i[Y]$ is closed in $\text{conv}(i[Y])$,
   (2) $i[X] \subseteq \{f \in C(X) : \|f\| \leq \text{diam } X\}$.

11. Prove that there does not exist a homeomorphism $h\colon \ell^2 \to s$ which is *linear*, i.e., has the property that
$$h(\lambda x + \mu y) = \lambda h(x) + \mu h(y)$$
for all $x, y \in \ell^2$ and $\lambda, \mu \in \mathbb{R}$.

12. Let $A$ be a bounded subset of a normed linear space $L$. Prove that if $(t_n)_n$ is a sequence in $\mathbb{R}$ such that $t_n \to 0$ and $(a_n)_n$ is any sequence of elements in $A$ then $t_n a_n \to \underline{0}$.

▶13. Let $0 < p < 1$ and put $\ell^p = \{x \in s : \sum_{n=1}^{\infty} |x_n|^p < \infty\}$. Prove that $\ell^p$ is a linear subspace of $s$ (in the algebraic sense). Define the following metric on $\ell^p$:
$$\varrho(x, y) = \sum_{n=1}^{\infty} |x_n - y_n|^p.$$
Prove that $\ell^p$ with the topology derived from this metric is a linear space which is not locally convex.

14. Let $X$ be a compact space. Prove that the compact-open topology on $C(X)$ coincides with the topology derived from the sup-norm $\|\cdot\|$.

15. Prove that $\ell^2$ and $s$ are separable.

16. Prove that $\ell^2$ and $\ell^2 \times \mathbb{R}$ are linearly homeomorphic.

17. Let $X = \{x \in \ell^2 : (\forall n \in \mathbb{N})(|x_n| \leq 1/n)\}$. Prove that $X$ and $Q$ are homeomorphic.

▶18. Let $\|\cdot\|$ be a norm on $\mathbb{R}^n$. Prove that the topology derived from this norm is the Euclidean topology. In particular, for every $x \in \mathbb{R}^n$ and $\varepsilon > 0$ we have that the set $\{y \in \mathbb{R}^n : \|x - y\| \leq \varepsilon\}$ is compact.

19. Let $V$ be a normed linear space and let $W$ be a finite dimensional linear subspace. Prove that $W$ is closed in $V$.

▶20. Let $\|\cdot\|$ be a norm on $\mathbb{R}^n$ and let $f\colon \mathbb{R}^n \to \mathbb{R}$ be linear but not identically 0. Prove that there exists $t > 0$ such that
   (1) $f^{-1}(t) \cap B(\underline{0}, 1) = \emptyset$,
   (2) $f^{-1}(t) \cap D(\underline{0}, 1) \neq \emptyset$.

▶21. Let $V$ be an infinite-dimensional normed linear space. Prove that there is a sequence $(e_n)_n$ in $V$ such that
   (1) $\{e_n : n \in \mathbb{N}\}$ is linearly independent,
   (2) $\|e_n\| = 1$ $(n \in \mathbb{N})$,
   (3) $\|e_m - e_n\| \geq 1$ $(m, n \in \mathbb{N}, m \neq n)$.

22. Let $V$ be an infinite-dimensional normed linear space. Prove that the unit sphere $S = \{x \in V : \|x\| = 1\}$ is not compact.

23. Let $x \in \mathbb{R}^n$ and let
$$K = \bigcup \{I(x, y) : y \in \mathbb{S}^{n-1}\}.$$
Prove that for every $\alpha \in [0, 1)$ there exists $\varepsilon > 0$ such that $B(\alpha x, \varepsilon) \subseteq K$.

▶24. Prove that every compact convex subset $A \subseteq \mathbb{R}^n$ with nonempty interior is homeomorphic to $B^n$ and its boundary is homeomorphic to $\mathbb{S}^{n-1}$.

25. Prove that the metric $\varrho$ on $\ell^2$ defined on Page 9 is complete.

▶26. Prove Lemma 1.1.12.

27. Let $X$ be a space and let $f, g\colon X \to \mathbb{R}$ with $f$ **lsc** and $g$ **usc**. Prove that if $g \le f$, i.e., $g(x) \le f(x)$ for every $x \in X$, then the function $F\colon X \rightrightarrows \mathbb{R}$ defined by $F(x) = [g(x), f(x)]$ is LSC. Use this to conclude that there exists a continuous $h\colon X \to \mathbb{R}$ such that $g \le h \le f$. (See Corollary A.7.6 for a stronger result.)

28. Prove that the function in Corollary 1.1.23 is a homeomorphism.

▶29. Let $X$ be a closed and bounded subspace of a normed linear space $L$. Prove that $\triangle(X)$ is homeomorphic to the subspace
$$\{((1-t)x, t) : t \in \mathbb{I}, x \in X\}$$
of $L \times \mathbb{I}$.

30. Let $E$ and $F$ be Banach spaces and let $\varphi\colon E \to F$ be linear and continuous. Prove that there exists $k \in \mathbb{N}$ such that
$$\|\varphi(x)\| \le k\|x\|.$$
for every $x \in E$.

31. Let $E$ and $F$ be Banach spaces and let $\varphi\colon E \to F$ be a linear homeomorphism. Prove that there exists $k \in \mathbb{N}$ such that for every $x \in E$ we have
$$\|x\|/k \le \|\varphi(x)\| \le k\|x\|.$$

## 1.2. Extending continuous functions

Suppose that $X$, $Y$ and $Z$ are topological spaces with $Y$ a subspace of $X$ and let $f\colon Y \to Z$ be continuous. In topology it is often of interest to know whether $f$ is the restriction to $Y$ of a continuous function $\bar{f}\colon X \to Z$. Easy examples show that in general this need not be the case. If $f$ is the restriction to $Y$ of a continuous function $\bar{f}\colon X \to Z$ then we say that $f$ is *continuously extendable* over $X$ or that $\bar{f}$ is a *continuous extension* of $f$. In this section we shall present examples of spaces $Z$ having the property that if $Y$ is closed in an arbitrary space $X$, then every continuous function $f\colon Y \to Z$ is continuously extendable over $X$ (respectively, over some neighborhood of $Y$).

Observe that Urysohn's Lemma can be looked at as a result on extending continuous functions. For let $X$ be a space and let $A$ and $B$ be disjoint closed subsets of $X$. If $E = A \cup F$ then by Lemma A.4.1 it follows that the continuous function $f\colon E \to \mathbb{I}$ defined by $f \upharpoonright A \equiv 0$ and $f \upharpoonright B \equiv 1$ can be extended to a continuous function $\bar{f}\colon X \to \mathbb{I}$. The function $f\colon E \to \mathbb{I}$ is not very interesting and the question naturally arises whether it is also possible to extend more interesting functions. In order to study this question with success, let us first formulate and prove the following technical result.

**Lemma 1.2.1.** Let $X$ be a space, $A$ a closed subset of $X$, and let $A' \subseteq A$ be dense in $A$. Then there exist a locally finite open cover $\mathcal{U}$ of $X \setminus A$ and a sequence of points $\{a_U : U \in \mathcal{U}\}$ in $A'$ such that

(1) for all $U \in \mathcal{U}$ and $x \in U$, $\varrho(x, a_U) \leq 2\varrho(x, A)$,
(2) if $U_n \in \mathcal{U}$ for every $n$ and
$$\lim_{n \to \infty} \varrho(U_n, A) = 0$$
then
$$\lim_{n \to \infty} \operatorname{diam}(U_n) = 0.$$

**Proof.** Let
$$\mathcal{V} = \{B(x, \tfrac{1}{4}\varrho(x, A)) : x \in X \setminus A\}.$$
Since $A$ is closed, $\mathcal{V}$ is an open cover of $X \setminus A$. By paracompactness of $X \setminus A$ (Corollary A.7.3) there exists a locally finite open cover $\mathcal{U}$ of $X \setminus A$ that refines $\mathcal{V}$. Since $\mathcal{U} < \mathcal{V}$, for each $U \in \mathcal{U}$ there exists $x_U \in X \setminus A$ with
$$U \subseteq B(x_U, \tfrac{1}{4}\varrho(x_U, A)).$$
In addition, since $A'$ is dense in $A$, for each $U \in \mathcal{U}$ there exists $a_U \in A'$ with
$$\varrho(x_U, a_U) \leq \tfrac{5}{4}\varrho(x_U, A).$$
We claim that the $U$'s and the $a_U$'s are as required.

**Claim 1.** For every $U \in \mathcal{U}$ and $x \in U$ the following inequalities hold:
$$\varrho(x, a_U) \leq \tfrac{3}{2}\varrho(x_U, A) \leq 2\varrho(x, A).$$

*Proof.* The first inequality is easy since
$$\varrho(x, a_U) \leq \varrho(x, x_U) + \varrho(x_U, a_U)$$
$$\leq \tfrac{1}{4}\varrho(x_U, A) + \tfrac{5}{4}\varrho(x_U, A)$$
$$= \tfrac{3}{2}\varrho(x_U, A).$$
Also,
$$\varrho(x_U, A) \leq \varrho(x_U, x) + \varrho(x, A) \leq \tfrac{1}{4}\varrho(x_U, A) + \varrho(x, A),$$
from which it follows that
$$\tfrac{3}{4}\varrho(x_U, A) \leq \varrho(x, A),$$
as required. ◇

So it remains to verify (2). To this end, assume that $U_n \in \mathcal{U}$ for all $n$ and that
$$\lim_{n \to \infty} \varrho(U_n, A) = 0.$$
For each $n$ pick $p_n \in U_n$ such that $\lim_{n \to \infty} \varrho(p_n, A) = 0$. By the claim we obtain
$$\lim_{n \to \infty} \varrho(x_{U_n}, A) \leq \lim_{n \to \infty} \tfrac{4}{3}\varrho(p_n, A) = 0.$$

Since $U_n \subseteq B(x_{U_n}, 1/4 \varrho(x_{U_n}, A))$ for all $n$, we get $\lim_{n\to\infty} \operatorname{diam}(U_n) = 0$. So we are done. □

An open cover $\mathcal{U}$ and a sequence of points $\{a_U : U \in \mathcal{U}\}$ such as in this lemma is called a *Dugundji system* for $X$ and $A$.

We now come to the main result in this section.

**The Dugundji Theorem (Part 1) 1.2.2.** Let $L$ be a *locally convex linear space* and let $C \subseteq L$ be convex. Then for every space $X$ with closed subspace $A$, every continuous function $f: A \to C$ can be extended to a continuous function $\bar{f}: X \to C$.

**Remark 1.2.3.** For Part 2 of The Dugundji Theorem, see Page 394.

**Proof.** Let the open cover $\mathcal{U}$ of $X \setminus A$ and the sequence of points $(a_U)_{U \in \mathcal{U}}$ in $A(= A')$ be a Dugundji system for $X$ and $A$. In addition, let $\kappa_U : X \setminus A \to \mathbb{R}$ for $U \in \mathcal{U}$ be the $\kappa$-functions with respect to $\mathcal{U}$.

Define $\bar{f} : X \to L$ by

$$(*) \qquad \bar{f}(x) = \begin{cases} f(x) & (x \in A), \\ \sum_{U \in \mathcal{U}} \kappa_U(x) \cdot f(a_U) & (x \in X \setminus A). \end{cases}$$

We will first prove that $\bar{f}$ is well-defined and continuous at all points of $X \setminus A$. To this end, fix an arbitrary $x \in X \setminus A$. Since $\mathcal{U}$ is locally finite, there is a neighborhood $W$ of $x$ in $X \setminus A$ meeting finitely many elements of $\mathcal{U}$ only, say $U_1, \ldots, U_n$. For $U \in \mathcal{U}$ missing $W$ we clearly have

$$\kappa_U \restriction W \equiv 0.$$

As a consequence, for every $y \in W$ we have

$$\bar{f}(y) = \sum_{n=1}^{n} \kappa_{U_n}(y) \cdot f(a_{U_n}).$$

So $\bar{f}(y)$ is a convex combination of points in $C$ and therefore belongs to $C$. By continuity of the $\kappa$-functions, it also follows that $\bar{f} \restriction W$ is continuous.

It suffices to prove continuity of $\bar{f}$ at the points of $A$. Pick an arbitrary element $a \in A$. An arbitrary neighborhood of $f(a)$ is without loss of generality of the form $f(a) + W$, where $W$ is a convex neighborhood of $\underline{0}$ (Exercise 1.1.4). So let $f(a) + W$ be such a neighborhood. The continuity of $f$ at $a$ implies that there exists $\delta > 0$ such that $B(a, \delta) \cap A \subseteq f^{-1}[f(a) + W]$.

**Claim 1.** $\bar{f}[B(a, \delta/3)] \subseteq f(a) + W$.

*Proof.* Pick an arbitrary $x \in B(a, \delta/3)$. If $x \in A$ then there is nothing to prove. So assume without loss of generality that $x \notin A$. Then

$$\varrho(x, A) \leq \varrho(x, a) < \delta/3;$$

as a consequence, if $x \in U \in \mathcal{U}$ then
$$\varrho(a, a_U) \leq \varrho(a, x) + \varrho(x, a_U)$$
$$\leq \varrho(a, x) + 2\varrho(x, A)$$
$$< \delta$$
by (1) of Lemma 1.2.1. From this we conclude that if $x \in U \in \mathcal{U}$ then we have $a_U \in B(a, \delta) \cap A$ and so $f(a_U) \in f(a) + W$. Consequently,
$$\bar{f}(x) - \bar{f}(a) = \Big( \sum_{x \in U \in \mathcal{U}} \kappa_U(x) \cdot f(a_U) \Big) - f(a)$$
$$= \sum_{x \in U \in \mathcal{U}} \kappa_U(x) \cdot \big(f(a_U) - f(a)\big).$$
Since $f(a_U) - f(a) \in W$ for every $U \in \mathcal{U}$ we see that $\bar{f}(x) - \bar{f}(a)$ is a convex combination of elements of $W$. Hence by convexity of $W$ we obviously get that $\bar{f}(x) - \bar{f}(a) \in W$, as required. ◇

We conclude that $\bar{f}$ is continuous at $a$. □

**Remark 1.2.4.** It is a natural problem whether the local convexity assumption in Theorem 1.2.2 can be dropped. This was a fundamental open problem ever since DUGUNDJI's paper [140] appeared in 1951. It was finally solved in 1994 by CAUTY [87] in the negative. His construction used in an essential way DRANIŠNIKOV's result in [139] about the existence of an infinite-dimensional compactum with finite cohomological dimension.

As a corollary to the Dugundji Theorem we get (cf. Theorem A.4.6):

**The Tietze Theorem 1.2.5.** *For every space $X$ with closed subspace $A$, every continuous function from $A$ to $\mathbb{R}$ or $\mathbb{I}$ can be extended over $X$.*

A space $X$ is called an *Absolute Retract* (abbreviated AR) provided it is a retract of every space $Y$ containing it as a closed subspace. If $X$ is an AR and $f: X \to Y$ is a homeomorphism then $Y$ is an AR as well. Consequently, $X$ is an AR if and only if for every space $Y$ containing a closed subspace $Z$ which is homeomorphic to $X$, there exists a retraction $r: Y \to Z$. Theorem 1.2.7 below implies that a retract of an AR is an AR.

A space $X$ is called an *Absolute Neighborhood Retract* (abbreviated ANR) provided it is a neighborhood retract of every space $Y$ containing it as a closed subspace. The space $X = \{0, 1\}$ is easily seen to be an ANR but is not an AR; simply observe that by continuity retractions preserve connectivity, and so there does not exist a retraction $r: \mathbb{I} \to X$. Notice that every AR is an ANR.

As above, Theorem 1.2.7 below easily implies that $X$ is an ANR if and only if for every space $Y$ containing a closed subspace $Z$ which is homeomorphic to $X$, $Z$ is a neighborhood retract of $Y$. Also, every neighborhood retract of an ANR is again an ANR (Proposition 1.2.10).

We call a space $X$ an *Absolute (Neighborhood) Extensor* (abbreviated A(N)E) provided that for every space $Y$ and for every closed subspace $A$ of $Y$, every continuous function $f: A \to X$ can be extended over $Y$ (over a neighborhood (depending on $f$) of $A$ in $Y$). We shall prove in Theorem 1.2.7 below that $X$ is an A(N)R if and only if $X$ is an A(N)E. This is of fundamental importance. In the sequel we shall not always conscientiously refer to Theorem 1.2.7 when dealing with A(N)R's. The reader should keep this in mind.

We first prove an important fact.

**Lemma 1.2.6.** *Every space can be imbedded as a closed subspace of some* AE.

**Proof.** This is easy. First observe that every space is homeomorphic to a closed subspace of some normed linear space (Corollary 1.1.8). Now apply Theorem 1.2.2. □

We shall now present the announced characterization of A(N)R's.

**Theorem 1.2.7.** *Let $X$ be a space. The following statements are equivalent:*

(1) *$X$ is an A(N)R,*
(2) *$X$ is an A(N)E.*

**Proof.** The implication (2) ⇒ (1) is trivial.

For (1) ⇒ (2), let us assume that $X$ is an ANR. The proof for AR's is entirely similar, and shall therefore be omitted.

Let $Y$ be a space, $A \subseteq Y$ be closed, and $f: A \to X$ be continuous. By Lemma 1.2.6, we may assume that $X$ is a closed subspace of some AE, say $Z$. Let $G: Y \to Z$ be a continuous extension of $f$. Since $X$ is a closed subspace of $Z$, it follows that $X$ is a neighborhood retract of $Z$. So let $U$ be a neighborhood of $X$ in $Z$ for which there exists a retraction $r: U \to X$. Put $V = G^{-1}[U]$. Since $G[A] \subseteq X$, $V$ is clearly a neighborhood of $A$ in $Y$. Let $h$ denote the restriction of $r \circ G$ to $V$ (observe that $h$ is well-defined since $G[V] \subseteq U$). Then $h$ is continuous and extends $f$ since for every $y \in A$ we have $h(y) = r(G(y)) = f(y)$. □

**Corollary 1.2.8.** *Every space is homeomorphic to a closed subspace of an* AR.

**Proof.** This follows directly from Lemma 1.2.6 and Theorem 1.2.7. □

**Corollary 1.2.9.** *Let $C$ be a convex subset of a locally convex linear space. Then $C$ is an* AR.

**Proof.** This follows from Theorems 1.2.2 and 1.2.7. □

Notice that Cauty's linear space mentioned in Remark 1.2.4 shows that the local convexity assumption in this corollary cannot be dropped.

**Proposition 1.2.10.** *A neighborhood retract of an ANR is an ANR. As a consequence, an open subspace of an ANR is an ANR.*

**Proof.** Let $X$ be an ANR, $A \subseteq X$ closed, $U$ an open neighborhood of $A$, and $r \colon U \to A$ a retraction. In addition, let $B$ be a closed subspace of a space $Y$ and let $f \colon B \to A$ be continuous. Since $X$ is an ANR there are a neighborhood $V$ of $B$ in $Y$ and a continuous extension $g \colon V \to X$ of $f$. Then clearly $W = g^{-1}[U]$ is a neighborhood of $A$. The function $\bar{f} = r \circ g \colon W \to A$ is continuous and extends $f$ since for every $y \in B$ we have

$$\bar{f}(y) = r\bigl(g(y)\bigr) = r\bigl(f(y)\bigr) = f(y).$$

Since an open subspace of a space is clearly a neighborhood retract, the second statement is indeed a consequence of the first. □

Proving that a given space is an AR or an ANR is usually a difficult task. We will come back to this in Chapter 4. For the moment we shall prove a few elementary results about AR's and ANR's only.

**Proposition 1.2.11.** *A product of countably many nonempty spaces is an AR iff all factors are.*

**Corollary 1.2.12.** $\mathbb{R}^n$, $\mathbb{I}^n$, $Q$ *and* $s$ *are AR's.*

**Proof.** Apply Proposition 1.2.11 and Corollary 1.2.9. □

**Corollary 1.2.13.** *For each* $n \geq 0$, $\mathbb{S}^n$ *is an ANR.*

**Proof.** Let $U = \mathbb{R}^{n+1} \setminus \{(0, 0, \ldots, 0)\}$. Then $U$ is an ANR by Corollary 1.2.12 and Proposition 1.2.10. The function $r \colon U \to \mathbb{S}^n$ defined by

$$r(x) = \frac{x}{\|x\|}$$

is clearly a retraction. Consequently, $\mathbb{S}^n$ is a neighborhood retract of an ANR and is therefore an ANR itself (Proposition 1.2.10). □

In Chapter 2 we shall prove that no $\mathbb{S}^n$ is an AR.

**Proposition 1.2.14.** *The product of finitely many ANR's is an ANR.*

**Proof.** The simple proof is left as an exercise to the reader. □

A product of a countably infinite number of ANR's need not be an ANR (in contrast to Proposition 1.2.11 on AR's). The following result gives more information on this.

## 1.2. EXTENDING CONTINUOUS FUNCTIONS

**Theorem 1.2.15.** *Let $X_n$ be a nonempty space for every $n \in \mathbb{N}$. For the product $X = \prod_{n=1}^{\infty} X_n$, the following statements are equivalent:*

(1) *$X$ is an ANR,*
(2) *each $X_n$ is an ANR and there is an $n \in \mathbb{N}$ such that $X_m$ is an AR for every $m \geq n$.*

**Proof.** Since for arbitrary $n$, the product $X$ can be factorized as
$$\prod_{i=1}^{n} X_i \times \prod_{i=n+1}^{\infty} X_i,$$
the implication (2) $\Rightarrow$ (1) follows from Propositions 1.2.11 and 1.2.14.

For (1) $\Rightarrow$ (2), first observe that each $X_n$ can be viewed as a retract of $X$ (cf. the proof of Proposition 1.2.11), hence $X_n$ is an ANR for every $n$. Now take points $x_n \in X_n$, $n \in \mathbb{N}$, and let $C_n$ be an AR which contains the space $X_n$ as a closed subset (Corollary 1.2.8). Since $X$ is an ANR and is closed in $C = \prod_{n=1}^{\infty} C_n$ (Exercise A.1.13), there is a neighborhood $U$ of $X$ in $C$ for which there exists a retraction $r\colon U \to X$. Put $x = (x_1, x_2, \ldots)$. Since $U$ is a neighborhood of $x$, there are an $n \in \mathbb{N}$ and neighborhoods $V_i$ of $x_i$ in $C_i$ for every $i \leq n-1$ such that
$$x \in W = V_1 \times \cdots \times V_{n-1} \times C_n \times C_{n+1} \times \cdots \subseteq U.$$
Observe that by Proposition 1.2.11, the product $\prod_{m=n}^{\infty} C_m$ is an AR. Now define a function
$$s\colon \prod_{i=n}^{\infty} C_n \longrightarrow \prod_{i=n}^{\infty} X_n$$
by
$$s(y) = \bigl(r(x_1, \ldots, x_{n-1}, y)_n, r(x_1, \ldots, x_{n-1}, y)_{n+1}, \ldots\bigr).$$
An easy check shows that $s$ is a retraction, from which it follows that the product $\prod_{m=n}^{\infty} X_m$ is an AR. Now apply Proposition 1.2.11. $\square$

The following result enables us to create many new A(N)R's from old ones in yet another way.

**Theorem 1.2.16.** *Let $X = X_1 \cup X_2$, where $X_1$ and $X_2$ are closed in $X$, and let $X_0 = X_1 \cap X_2$. Then*

(1) *If $X_0$, $X_1$ and $X_2$ are A(N)R's then $X$ is an A(N)R.*
(2) *If $X$ and $X_0$ are A(N)R's then $X_1$ and $X_2$ are A(N)R's.*

**Proof.** For (1), assume that $X_0$, $X_1$ and $X_2$ are ANR's. We shall prove that $X$ is an ANR. The proof of (1) for AR's is similar.

Assume that $X$ is a closed subset of a space $Z$. Our task is to construct a neighborhood $U$ of $X$ in $Z$ and a retraction $r: U \to X$. To this end, define

$$Z_0 = \{z \in Z : \varrho(z, X_1) = \varrho(z, X_2)\},$$
$$Z_1 = \{z \in Z : \varrho(z, X_1) \leq \varrho(z, X_2)\},$$
$$Z_2 = \{z \in Z : \varrho(z, X_1) \geq \varrho(z, X_2)\},$$

respectively (cf. the proof of Lemma A.8.1). It is clear that $Z_i \cap X = X_i$ for $i \in \{0, 1, 2\}$, that $Z_1 \cap Z_2 = Z_0$ and finally that $Z_1 \cup Z_2 = Z$. Also observe that the $Z_i$ are closed in $Z$.

Since $X_0$ is closed in $Z_0$ there are a closed neighborhood $W_0$ of $X_0$ in $Z_0$ and a retraction $r_0: W_0 \to X_0$. Now for $i \in \{1, 2\}$ define $r_i: W_0 \cup X_i \to X_i$ by

$$r_i(z) = \begin{cases} r_0(z) & (z \in W_0), \\ z & (x \in X_i). \end{cases}$$

Observe that $r_i$ is a retraction. By Theorem 1.2.7 there exists a closed neighborhood $V_i$ of $W_0 \cup X_i$ in $Z_i$ such that $r_i$ can be extended to a continuous function

$$\bar{r}_i: V_i \to X_i \qquad (i \in \{1, 2\}).$$

It is clear that there exists for $i \in \{1, 2\}$ a closed neighborhood $U_i$ of $X_i$ in $Z_i$ such that $U_i \subseteq V_i$ and $U_i \cap Z_0 \subseteq W_0$. Then

$$U_1 \cap U_2 \subseteq W_0$$

from which it follows that the function $\bar{r}: U_1 \cup U_2 \to X$ defined by

$$\bar{r}(z) = \begin{cases} \bar{r}_1(z) & (z \in U_1), \\ \bar{r}_2(z) & (z \in U_2), \end{cases}$$

is a well-defined retraction. Since $U_1 \cup U_2$ is a neighborhood of $X$ in the space $Z$, we are done.

For (2), let $X_0$ and $X$ be ANR's. Again, the proof for AR's is similar. Since $X_0$ is an ANR, there are a neighborhood $U_0$ of $X_0$ in $X$ and a retraction $r: U_0 \to X_0$. Define $\bar{r}: X_1 \cup U_0 \to X_1$ by

$$\bar{r}(x) = \begin{cases} x & (x \in X_1), \\ \bar{r}(x) & (x \in U_0 \cap X_2). \end{cases}$$

An easy check shows that $\bar{r}$ is a retraction. Since $X_1 \cup U_0$ is a neighborhood of $X_1$ and $X$ is an ANR, it now follows that $X_1$ is an ANR as well. The proof for $X_2$ is the same. $\square$

**Exercises for §1.2.**

1. Prove that every ANR is locally contractible.
2. Prove that if $\Delta(X)$ is an ANR then so is $X$.

3. Prove Proposition 1.2.11.

4. Let $X = \{0\} \cup \{1/n : n \in \mathbb{N}\}$. Prove that $\triangle(X)$ is an example of a contractible space which is not an ANR.

5. Let $A \subseteq \mathbb{S}^{n-1}$. Observe that $B^n \setminus A$ is a convex subset of $\mathbb{R}^n$ and hence is an AR. Present a direct proof of this.

6. Prove that each AR is path-connected. Prove that every ANR is locally path-connected. Observe that, in particular, every ANR is locally connected.

7. Give an example of a subspace $X$ of $\mathbb{R}$, such that $X$ is an ANR, but $X \cup \{p\}$ is not an ANR for some $p \in \mathbb{R}$.

8. Let $X$ be the $\sin(1/x)$-continuum in the plane. Prove that $X$ is not an ANR.

▶9. Let $A_1$ and $A_2$ be closed subsets of the locally convex linear spaces $L_1$ and $L_2$, respectively. Prove that for each homeomorphism $h\colon A_1 \to A_2$ there exists a homeomorphism $H\colon L_1 \times L_2 \to L_1 \times L_2$ such that
$$H(a, 0) = \bigl(0, h(a)\bigr)$$
for every $a \in A$.

10. Let $L$ be a linear space. In addition, let $X$ be a space, $A \subseteq X$ be closed and $f\colon A \to L$ be continuous. Finally, let $\mathcal{U}$ and $\{a_U : U \in \mathcal{U}\}$ be a Dugundji system for $X$ and $A$. For each $x \in X \setminus A$ let
$$\mathcal{E}(x) = \{U \in \mathcal{U} : x \in U\}.$$
Define a function $F \Rightarrow L$ by
$$F(x) = \begin{cases} \{f(x)\} & (x \in A), \\ \mathrm{conv}(\{f(a_U) : U \in \mathcal{E}(x)\}) & (x \in X \setminus A). \end{cases}$$
Prove that $F$ is LSC.

11. Use Exercise 1.2.10 to prove that a normed linear space satisfies the conclusion of the Dugundji Theorem 1.2.2.

12. Let $L$ be a linear space. Assume that for every space $X$, every LSC function $F\colon X \Rightarrow L$ such that for every $x \in X$ the set $F(x)$ is compact and convex, has a continuous selection. Prove that every linear subspace of $L$ is an AR.

## 1.3. Function spaces

The space $C^*(X)$ from Example 1.1.5 is a special case of a more general construction. Let $X$ and $Y$ be spaces and let $\varrho$ be an admissible metric on $Y$. Define
$$C_\varrho(X, Y) = \{f \in C(X, Y) : \mathrm{diam}_\varrho(f[X]) < \infty\}.$$
It is sometimes convenient to refer to the elements of $C_\varrho(X, Y)$ as *bounded* functions. We shall endow $C_\varrho(X, Y)$ with a useful topology.

In §A.2 we observed that $\hat{\varrho}$ need not be a metric on $C(X,Y)$. Fortunately, on $C_\varrho(X,Y)$ it is a metric.

**Lemma 1.3.1.** *Let $X$ and $Y$ be spaces and let $\varrho$ be an admissible metric on $Y$. Then*

(1) *for all $f,g \in C_\varrho(X,Y)$ we have $\hat{\varrho}(f,g) < \infty$,*
(2) *the function $\hat{\varrho}\colon C_\varrho(X,Y) \times C_\varrho(X,Y) \to [0,\infty)$ is a metric.*

**Proof.** For (1), take an arbitrary point $z \in X$ and observe that for all functions $f,g \in C_\varrho(X,Y)$ the following holds:
$$\hat{\varrho}(f,g) \leq \operatorname{diam}_\varrho(f[X]) + \varrho\bigl(f(z),g(z)\bigr) + \operatorname{diam}_\varrho\bigl(g(X)\bigr) < \infty.$$
The proof of (2) is routine and is left as an exercise to the reader. □

From now on we shall endow $C_\varrho(X,Y)$ with the topology induced by $\hat{\varrho}$. Let us emphasize that the set $C_\varrho(X,Y)$ as well as its topology depend on the choice of the metric $\varrho$. It is a natural question to ask when the choice of $\varrho$ is irrelevant.

**Lemma 1.3.2.** *Let $X$ and $Y$ be spaces with $X$ compact. In addition, let $\varrho_1$ and $\varrho_2$ be admissible metrics for $Y$. Then*

(1) $C_{\varrho_1}(X,Y) = C_{\varrho_2}(X,Y) = C(X,Y)$,
(2) *the topologies on $C(X,Y)$ induced by $\hat{\varrho}_1$ and $\hat{\varrho}_2$ are the same.*

**Proof.** (1) is trivial.

For each $\varepsilon > 0, y \in Y$ and $i \in \{1,2\}$ we put
$$B_i(y,\varepsilon) = \{z \in Y : \varrho_i(y,z) < \varepsilon\}.$$

For (2), take $f \in C_{\varrho_1}(X,Y)$ and $\varepsilon > 0$, arbitrarily. Our aim is to prove that there exists $\delta > 0$ such that
$$\{g \in C_{\varrho_2}(X,Y) : \hat{\varrho}_2(f,g) < \delta\} \subseteq \{g \in C_{\varrho_1}(X,Y) : \hat{\varrho}_1(f,g) < \varepsilon\},$$
i.e., that the $\hat{\varrho}_2$-ball about $f$ with radius $\delta$ is contained in the $\hat{\varrho}_1$-ball about $f$ with radius $\varepsilon$.

To this end, observe that since $f[X]$ is compact, the open cover
$$\mathcal{U} = \{B_1(f(x), {}^1\!/_2\varepsilon) : x \in X\}$$
has a $\varrho_2$-Lebesgue number, say $\delta$ (Lemma A.5.3). We claim that this $\delta$ is as required. To see that this is indeed the case, take an arbitrary $g \in C_{\varrho_2}(X,Y)$ such that $\hat{\varrho}_2(f,g) < \delta$. For each $x \in X$ we have $\hat{\varrho}_2\bigl(f(x),g(x)\bigr) < \delta$, so there exists $p_x \in X$ such that $\{f(x),g(x)\} \subseteq B_1(f(p_x), {}^\varepsilon\!/_2)$. Consequently, for each element $x \in X$ we have
$$\varrho_1\bigl(f(x),g(x)\bigr) < \varepsilon,$$
from which it follows by Exercise A.5.4 that $\hat{\varrho}_1(f,g) < \varepsilon$.

So we conclude that the topology on $C(X,Y)$ induced by $\hat{\varrho}_2$ is finer than the topology on $C(X,Y)$ induced by $\hat{\varrho}_1$. By interchanging the roles of $\varrho_1$ and $\varrho_2$ in the above argument we find that the induced topologies are indeed the same. □

Let $X$ and $Y$ be spaces with $X$ compact. From the above lemma we conclude that all the topologies we defined on the set $C(X,Y)$ coincide. So for compact $X$ and any $(Y,\varrho)$ we shall denote the space $C_\varrho(X,Y)$ simply by $C(X,Y)$. The topology on $C(X,Y)$ is called *the topology of uniform convergence*.

Observe that the norm topology on $C(X)$ for compact $X$ defined in Example 1.1.5 coincides with the just defined topology on $C(X,\mathbb{R})$.

On Page 19 we defined the so-called compact-open topology on $C(X)$. This is again a special case of a more general construction. Indeed, let $X$ and $Y$ be spaces and for an arbitrary compact subset $K$ in $X$ and an arbitrary open subset $U$ in $Y$ define

$$[K,U] = \{f \in C(X,Y) : f[K] \subseteq U\}.$$

Topologize $C(X,Y)$ by taking the collection

$$\{[K,U] : K \subseteq X \text{ compact and } U \subseteq Y \text{ open}\}$$

as an open subbase. This topology is called the *compact-open topology* on the set $C(X,Y)$.

We will now show that for compact $X$ and arbitrary $Y$ the compact-open topology on $C(X,Y)$ coincides with the topology of uniform convergence on $C(X,Y)$. This allows us to prove quite easily that $C(X,Y)$ is separable.

**Proposition 1.3.3.** *Let $X$ and $Y$ be spaces with $X$ compact. The topology of uniform convergence on $C(X,Y)$ coincides with the compact-open topology on $C(X,Y)$. As a consequence, $C(X,Y)$ is separable.*

**Proof.** Let $\varrho$ be an admissible metric on $Y$. In addition, let $K \subseteq X$ be compact and $U \subseteq Y$ open. If $f \in [K,U]$ then $f[K]$ is a compact subset of $U$. By Corollary A.5.4 there exists $\delta > 0$ such that

$$B(f[K], \delta) \subseteq U.$$

This clearly implies that if $g \in C(X,Y)$ is such that $\hat{\varrho}(f,g) < \delta$ then $g[K]$ is contained in $U$, i.e.,

$$\{g \in C(X,Y) : \hat{\varrho}(f,g) < \delta\} \subseteq [K,U].$$

From this we conclude that $[K,U]$ is open in the topology of uniform convergence on $C(X,Y)$ and hence that the topology of uniform convergence is finer than the compact-open topology.

For the converse, let $\mathcal{B}$ and $\mathcal{E}$ be countable open bases for $X$ and $Y$, respectively, which are both closed under finite unions. For $B \in \mathcal{B}$ and $E \in \mathcal{E}$ put
$$A(B, E) = [\overline{B}, E].$$
By the above, each $A(B, E)$ is open in the topology of uniform convergence on $C(X, Y)$. Let $\mathcal{A}$ be the (countable) collection of all $A(B, E)$'s. We claim that the family $\mathcal{A}^*$ of all finite intersections of elements of $\mathcal{A}$ is an open base for $C(X, Y)$ endowed with the topology of uniform convergence. This proves on the one hand that compact-open topology is finer than the topology of uniform convergence and on the other hand that that $C(X, Y)$ has a countable base.

Let $f \in C(X, Y)$ and $\varepsilon > 0$. We shall prove that there exists an element $F \in \mathcal{A}^*$ such that $f \in F \subseteq \{g \in C(X, Y) : \hat{\varrho}(f, g) < \varepsilon\}$. By compactness of the set $f[X]$, there are finitely many elements of $\mathcal{E}$, say $E_1, E_2, \ldots, E_n$, such that

(1) $f[X] \subseteq \bigcup_{i=1}^n E_i$,
(2) for every $i \leq n$, $\operatorname{diam}(E_i) < \varepsilon$.

Let $\mathcal{U} = \{f^{-1}[E_i] : i \leq n\}$. Since $\mathcal{B}$ is a base, there clearly is a a cover $\mathcal{V}$ of $X$ consisting of elements of $\mathcal{B}$ such that $\overline{\mathcal{V}} < \mathcal{U}$. By compactness of $X$, we may assume that $\mathcal{V}$ is finite. For each $i \leq n$ let $W_i$ be the union of the elements of $\mathcal{V}$ the closures of which are contained in $f^{-1}[E_i]$. Since $\mathcal{B}$ is closed under finite unions, $\mathcal{W} = \{W_i : i \leq n\}$ is a subcollection of $\mathcal{B}$, $\mathcal{W}$ covers $X$, and $\mathcal{W}$ has the property that the closure of each $W_i$ is contained in $f^{-1}[E_i]$. (This is so since $W_i$ is a *finite* union of sets the closures of which are contained in $f^{-1}[E_i]$.) Now put
$$F = \bigcap_{i=1}^n A(W_i, E_i).$$
It is clear that $f \in F$. In addition, $F$ is open in the topology of uniform convergence. We claim that $F \subseteq \{g \in C(X, Y) : \hat{\varrho}(f, g) < \varepsilon\}$. To this end, take $g \in F$ and $x \in X$. There exists $i \leq n$ with $x \in W_i$. Since $f, g \in F$, it follows that $f[W_i] \cup g[W_i] \subseteq E_i$. Consequently, both $f(x)$ and $g(x)$ belong to $E_i$ from which we get by (2) that $\varrho(f(x), g(x)) < \varepsilon$. So we are done. $\square$

This result can be used to estimate the number of continuous functions.

**Corollary 1.3.4.** *Let $X$ and $Y$ be spaces with $X$ compact. Then $C(X, Y)$ has cardinality at most $\mathfrak{c}$.*

**Proof.** This follows from Proposition 1.3.3 and Exercise A.2.16. $\square$

We now turn to completeness properties of $C(X, Y)$ and $C_\varrho(X, Y)$.

**Proposition 1.3.5.** Let $X$ and $(Y, \varrho)$ be spaces. Let $(f_n)_n$ be a $\hat{\varrho}$-Cauchy sequence in $C(X,Y)$ such that for every $x \in X$, $\lim_{n\to\infty} f_n(x)$ exists. Then the function $f\colon X \to Y$ defined by $f(x) = \lim_{n\to\infty} f_n(x)$ is continuous. In addition, if $f_n \in C_\varrho(X,Y)$ for every $n$ then $f \in C_\varrho(X,Y)$ and $f = \lim_{n\to\infty} f_n$ (in $C_\varrho(X,Y)$).

**Proof.** For the first part of the proposition, see Lemma A.3.1.

For the remaining part, suppose that $f_n \in C_\varrho(X,Y)$ for every $n$. We shall prove that $\operatorname{diam}(f[X]) < \infty$. By Claim 1 in the proof of Lemma A.3.1 there exists an $M \in \mathbb{N}$ such that for every $x \in X$, $\varrho(f(x), f_M(x)) < 1$. Take arbitrary $x, z \in X$. Then

$$\varrho(f(x), f(z)) \leq \varrho(f(x), f_M(x)) + \varrho(f_M(x), f_M(z)) + \varrho(f_M(z), f(z))$$
$$< 2 + \operatorname{diam}(f_M[X]),$$

so $\operatorname{diam}(f[X]) < \infty$.

It remains to prove that $f = \lim_{n\to\infty} f_n$ (in $C_\varrho(X,Y)$). But this follows again easily from Claim 1 in the proof of Lemma A.3.1. □

**Corollary 1.3.6.** Let $X$ and $(Y, \varrho)$ be spaces. Then $\varrho$ is complete if and only if $\hat{\varrho}$ is complete.

**Proof.** Suppose that $(Y, \varrho)$ is complete and let $(f_n)_n$ be a $\hat{\varrho}$-Cauchy sequence in $C_\varrho(X,Y)$. Fix $z \in X$ arbitrarily and let $\varepsilon > 0$. There exists $N \in \mathbb{N}$ such that $\hat{\varrho}(f_n, f_m) < \varepsilon$ for all $n, m \geq N$. Since for all $n$ and $m$,

$$\varrho(f_n(z), f_m(z)) \leq \hat{\varrho}(f_n, f_m),$$

we conclude that $(f_n(z))_n$ is Cauchy in $(Y, \varrho)$. The completeness of $(Y, \varrho)$ and Proposition 1.3.5 now yield that the sequence $(f_n)_n$ converges.

Now assume that $\hat{\varrho}$ is complete. Let $(y_n)_n$ be a $\varrho$-Cauchy sequence in $Y$. For each $n$ let $f_n\colon X \to Y$ be the constant function with value $y_n$. It is easy to see that $(f_n)_n$ is a $\hat{\varrho}$-Cauchy sequence in $C_\varrho(X,Y)$. So $f = \lim_{n\to\infty} f_n$ exists and belongs to $C_\varrho(X,Y)$. It is trivial to prove that $f$ is constant and that the sequence $(y_n)_n$ converges to the unique point in the range of $f$. □

**Corollary 1.3.7.** Let $X$ be compact and $Y$ topologically complete. Then the space $C(X,Y)$ is topologically complete.

**Groups of homeomorphisms.** We will now restrict our attention to spaces of homeomorphisms and will derive some basic properties of them.

Let $X$ and $Y$ be spaces and define

$$\mathcal{S}(X,Y) = \{f \in C(X,Y) : f \text{ is surjective}\}.$$

There are spaces $X$ and $Y$ for which $\mathcal{S}(X,Y)$ is empty, for example this is the case if $X$ has smaller cardinality than $Y$.

**Proposition 1.3.8.** *Let $X$ and $Y$ be spaces with $X$ compact. Then $\mathcal{S}(X,Y)$ is closed in $C(X,Y)$.*

**Proof.** Take an arbitrary $f \notin \mathcal{S}(X,Y)$. There exists a point $y \in Y \setminus f[X]$. So
$$f \in [X, Y \setminus \{y\}] \subseteq C(X,Y) \setminus \mathcal{S}(X,Y).$$
Since $[X, Y \setminus \{y\}]$ is open in $C(X,Y)$ (Proposition 1.3.3), this shows that
$$C(X,Y) \setminus \mathcal{S}(X,Y)$$
is a neighborhood of $f$. Hence $C(X,Y) \setminus \mathcal{S}(X,Y)$ is open in $C(X,Y)$. □

Let $X$ and $Y$ be spaces with $X$ compact, and let $\varepsilon > 0$. A continuous function $f: X \to Y$ is called an $\varepsilon$-map if for every $y \in Y$,
$$\operatorname{diam}\left(f^{-1}(y)\right) < \varepsilon.$$
Let $\varepsilon > 0$ and put
$$C_\varepsilon(X,Y) = \{f \in C(X,Y) : f \text{ is an } \varepsilon\text{-map}\}$$
and
$$\mathcal{S}_\varepsilon(X,Y) = C_\varepsilon(X,Y) \cap \mathcal{S}(X,Y),$$
respectively. In addition, put
$$\mathcal{G}_\varepsilon(X,Y) = \mathcal{S}(X,Y) \setminus C_\varepsilon(X,Y).$$

**Lemma 1.3.9.** *Let $X$ and $Y$ be spaces with $X$ compact. Then $C_\varepsilon(X,Y)$ is open in $C(X,Y)$ for every $\varepsilon > 0$. Consequently, $\mathcal{G}_\varepsilon(X,Y)$ is a closed subset of $C(X,Y)$.*

**Proof.** Take $f \in C_\varepsilon(X,Y)$. Since $X$ is compact, $f: X \to f[X]$ is a closed map (Exercise A.5.5). We claim that for every $y \in f[X]$ there exists an open neighborhood $U_y$ (in $f[X]$) such that
$$\operatorname{diam}\left(f^{-1}[U_y]\right) < \varepsilon.$$
This will be achieved in two steps. Take an arbitrary $y \in f[X]$. Then
$$\operatorname{diam}\left(f^{-1}(y)\right) < \varepsilon$$
since $f$ is an $\varepsilon$-map. There clearly exists an open neighborhood $U$ of $f^{-1}(y)$ such that $\operatorname{diam}(U) < \varepsilon$. Now by using that $f$ is a closed map, Exercise A.1.15 gives us the required neighborhood $U_y$.

Let $\delta > 0$ be a Lebesgue number for the open covering
$$\{U_y : y \in f[X]\}$$
of $f[X]$ (Lemma A.5.3). Let $g \in C(X,Y)$ be such that $\hat{\varrho}(g,f) < \delta/2$. We claim that $g \in C_\varepsilon(X,Y)$. This clearly suffices. To this end, take an arbitrary $y \in Y$. Since we have $\hat{\varrho}(f,g) < \delta/2$ it follows easily that $\operatorname{diam}\left(fg^{-1}(y)\right) < \delta$. There

consequently exists a point $z \in f[X]$ such that $fg^{-1}(y) \subseteq U_z$. This implies that
$$\operatorname{diam}\left(f^{-1}fg^{-1}(y)\right) < \varepsilon.$$
Since $g^{-1}(y) \subseteq f^{-1}fg^{-1}(y)$, we conclude that $\operatorname{diam}\left(g^{-1}(y)\right) < \varepsilon$, i.e., $g$ is an $\varepsilon$-map. That $\mathcal{G}_\varepsilon(X, Y)$ is closed now follows by Proposition 1.3.8. □

Let $X$ and $Y$ be spaces. We introduce a few more interesting subsets of $C(X, Y)$. Let $\mathcal{I}(X, Y)$ denote the subset of $C(X, Y)$ consisting of all imbeddings of $X$ into $Y$, and let $\mathcal{H}(X, Y)$ denote the set of all homeomorphisms from $X$ onto $Y$. If $X = Y$ then for $\mathcal{H}(X, X)$ we shall simply write $\mathcal{H}(X)$. As usual, $\mathcal{H}(X)$ is called the *autohomeomorphism group* of $X$.

**Lemma 1.3.10.** *Let $X$ and $Y$ be spaces with $X$ compact. Then*
$$\mathcal{I}(X, Y) = \bigcap_{n=1}^{\infty} C_{1/n}(X, Y).$$
*As a consequence, $\mathcal{I}(X, Y)$ is a $G_\delta$-subset of $C(X, Y)$, and $\mathcal{H}(X, Y)$ is a $G_\delta$-subset of both $C(X, Y)$ and $\mathcal{S}(X, Y)$.*

**Proof.** That $\mathcal{I}(X, Y) \subseteq \bigcap_{n=1}^{\infty} C_{1/n}(X, Y)$ is a triviality. Pick an arbitrary
$$f \in \bigcap_{n=1}^{\infty} C_{1/n}(X, Y).$$
Then $f$ is a $1/n$-map for every $n$, hence $f$ is one-to-one. So the compactness of $X$ implies that $f$ is an imbedding (Exercise A.5.9). The remaining statements are obvious. □

**Corollary 1.3.11.** *Let $X$ and $Y$ be spaces with $X$ compact and $Y$ topologically complete. Then both $\mathcal{I}(X, Y)$ and $\mathcal{H}(X, Y)$ are topologically complete.*

**Proof.** Since $C(X, Y)$ is completely metrizable by Corollary 1.3.6, this follows immediately from Lemma 1.3.10 and Theorem A.6.3. □

So $\mathcal{H}(X)$ is complete if $X$ is compact. It will be convenient to explicitly describe a complete metric for $\mathcal{H}(X)$ that generates its topology.

**Proposition 1.3.12.** *Let $X$ be a compact space. For $f, g \in \mathcal{H}(X)$ define*
$$\sigma(f, g) = \hat{\varrho}(f, g) + \hat{\varrho}(f^{-1}, g^{-1}).$$
*Then $\sigma$ is a complete metric on $\mathcal{H}(X)$ that generates its topology.*

**Proof.** That $\sigma$ is a metric is left as an exercise to the reader. We shall first prove that $\hat{\varrho}$ and $\sigma$ generate the same topology on $\mathcal{H}(X)$. Since for all $f, g \in \mathcal{H}(X)$ we have $\hat{\varrho}(f,g) \leq \sigma(f,g)$, the only thing to verify is that for every $\varepsilon > 0$ and every $f \in \mathcal{H}(X)$ there exists $\delta > 0$ such that

$$\text{if } g \in \mathcal{H}(X) \text{ and } \hat{\varrho}(f,g) < \delta \text{ then } \sigma(f,g) < \varepsilon.$$

Choose arbitrary $\varepsilon > 0$ and $f \in \mathcal{H}(X)$. By compactness, $f^{-1}$ is uniformly continuous (Exercise A.5.18) and consequently there exists $\gamma > 0$ such that for all $x, y \in X$ with $\varrho(x,y) < \gamma$ we have $\varrho(f^{-1}(x), f^{-1}(y)) < \varepsilon/2$. Let

$$\delta = \min\{\gamma, \varepsilon/2\}.$$

Take $g \in \mathcal{H}(X)$ such that $\hat{\varrho}(f,g) < \delta$. Pick an arbitrary $x \in X$ and put

$$z = g^{-1}(x).$$

Since $\hat{\varrho}(f,g) < \delta$, it follows that

$$\varrho(f(z), g(z)) = \varrho(fg^{-1}(x), x) < \delta \leq \gamma.$$

As a consequence,

$$\varrho(g^{-1}(x), f^{-1}(x)) = \varrho(f^{-1}fg^{-1}(x), f^{-1}(x)) < \varepsilon/2.$$

We conclude that $\hat{\varrho}(g^{-1}, f^{-1}) < \varepsilon/2$ by Exercise A.5.4. Therefore,

$$\sigma(f,g) = \hat{\varrho}(f,g) + \hat{\varrho}(f^{-1}, g^{-1}) < \varepsilon/2 + \varepsilon/2 = \varepsilon.$$

Now let $(f_n)_n$ be a $\sigma$-Cauchy sequence in $\mathcal{H}(X)$. Then $(f_n)_n$ is a $\hat{\varrho}$-Cauchy sequence in $C(X, X)$ and therefore the limit $f = \lim_{n \to \infty} f_n$ exists and belongs to $C(X, X)$ (Corollary 1.3.6). Similarly, by the definition of $\sigma$, the limit $g = \lim_{n \to \infty} f_n^{-1}$ exists and belongs to $C(X, X)$. It is easily seen that $f \circ g = 1_X = g \circ f$ from which it follows that $f \in \mathcal{H}(X)$. □

Lemma 1.3.10 and Proposition 1.3.12 both imply by Theorem A.6.6 the following

**Corollary 1.3.13.** *If $X$ is compact then $\mathcal{H}(X)$ is a Baire space.*

**Exercises for §1.3.**

1. Let $\mathbb{N}$ denote the discrete space of natural numbers. Prove that $C(\mathbb{N}, \mathbb{R}; |\cdot|)$ is not separable.

2. Let $X$, $Y$ and $Z$ be compact spaces. For $f \in C(Z, X)$ and $g, h \in C(X, Y)$ prove that
$$\hat{\varrho}(g \circ f, h \circ f) \leq \hat{\varrho}(g, h).$$
In addition, show that if $f$ is surjective then
$$\hat{\varrho}(g \circ f, h \circ f) = \hat{\varrho}(g, h).$$

3. Let $X$ be compact, let $f, g \in C(X, X)$ such that $g$ is a homeomorphism. Prove that
$$\hat{\varrho}(1_X, f \circ g^{-1}) = \hat{\varrho}(f, g).$$

4. Let $X$ be a compact space. Prove that the function
$$\xi \colon \mathcal{H}(X) \times \mathcal{H}(X) \to \mathcal{H}(X)$$
defined by
$$\xi(f,g) = f \circ g^{-1}$$
is continuous (i.e., $\mathcal{H}(X)$ is a topological group).

5. Prove that the function $f \colon \mathbb{I} \to \mathbb{I}$ defined by
$$f(x) = \begin{cases} 2x & (0 \le x \le 1/4), \\ 1/2 & (1/4 \le x \le 3/4), \\ 2x - 1 & (3/4 \le x \le 1). \end{cases}$$
belongs to the closure of $\mathcal{H}(\mathbb{I})$ in $C(\mathbb{I},\mathbb{I})$. (Hence $\mathcal{H}(X)$ is even for compact $X$ not necessarily a closed subspace of $C(X,X)$.)

6. Prove that $\mathcal{H}(\mathbb{I})$ has exactly two components.

7. Give an example of two nontrivial continua $X$ and $Y$ such that $\mathcal{S}(X,Y)$ is empty.

## 1.4. The Borsuk homotopy extension theorem

The aim of the present section is to prove that a continuous function $f$ from a closed subspace $A$ of a space $X$ into an ANR $Z$ is extendable over $X$ if and only if $f$ is homotopic to an extendable function $g \colon A \to Z$.

We shall need the following simple lemma:

**Lemma 1.4.1.** *Let $A$ be a closed subset of a space $X$. Then for every neighborhood $V$ of $B = (X \times \{0\}) \cup (A \times \mathbb{I})$ in $X \times \mathbb{I}$ there is a continuous function $\alpha \colon X \times \mathbb{I} \to V$ which is the identity on $B$.*

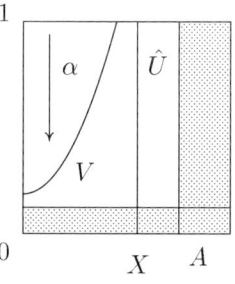

Figure 1.

**Proof.** Let $\pi \colon X \times \mathbb{I} \to X$ be the projection. Then by compactness of $\mathbb{I}$, $\pi$ is closed by Exercise A.5.8. We may assume without loss of generality that $V$ is open. So $F = (X \times \mathbb{I}) \setminus U$ is a closed set which misses $A \times \mathbb{I}$. Hence
$$U = X \setminus \pi[F]$$

is an open neighborhood of $A$ in $X$ such that $\hat{U} = U \times \mathbb{I} \subseteq V$. Now let
$$\beta \colon X \to \mathbb{I}$$
be a Urysohn function (Lemma A.4.1) such that
$$\beta \restriction A \equiv 1, \quad \beta \restriction X \setminus U \equiv 0.$$
Define $\alpha \colon X \times \mathbb{I} \to X \times \mathbb{I}$ by
$$\alpha(x, t) = (x, \beta(x) \cdot t).$$
Then $\alpha$ is clearly continuous. It is easily seen that it restricts to the identity on $B$. We shall prove that $\alpha[X \times \mathbb{I}] \subseteq V$, thereby showing that $\alpha$ is as required. To this end, take $(x, t) \in X \times \mathbb{I}$ arbitrarily. If $x \notin U$ then $\beta(x) = 0$ and consequently, $\alpha(x, t) = (x, 0) \in B \subseteq V$. On the other hand, if $x \in U$ then $\alpha(x, t) = (x, \beta(x) \cdot t) \in U \times \mathbb{I} \subseteq V$. $\square$

We now come to the main result in this section.

**The Borsuk Homotopy Extension Theorem 1.4.2.** *Let $A$ be a closed subspace of a space $X$, let $Z$ be an ANR and let $H \colon A \times \mathbb{I} \to Z$ be a homotopy such that $H_0$ is extendable to a continuous function $f \colon X \to Z$. Then there is a homotopy $F \colon X \times \mathbb{I} \to Z$ such that*

(1) $F_0 = f$,
(2) *for every* $t \in \mathbb{I}$, $F_t \restriction A = H_t$.

**Proof.** Using the notation as in Lemma 1.4.1, define a function $\xi \colon B \to Z$ by
$$\xi(x, t) = \begin{cases} f(x) & (x \in X, t = 0), \\ H(x, t) & (x \in A, t \in \mathbb{I}). \end{cases}$$
Since $X \times \{0\}$ and $A \times \mathbb{I}$ are closed in $B$, it follows easily that $\xi$ is continuous. As $Z$ is an ANR and as $B$ is closed in $X \times \mathbb{I}$, we can find a neighborhood $V$ of $B$ in $X \times \mathbb{I}$ such that $\xi$ can be extended to a continuous function $\xi' \colon V \to Z$. Let $\alpha \colon X \times \mathbb{I} \to V$ be as in Lemma 1.4.1.

Define $F \colon X \times \mathbb{I} \to Z$ by the formula
$$F(x, t) = \xi'\big(\alpha(x, t)\big).$$
It is easily seen that $F$ is as required. $\square$

Theorem 1.4.2 has some interesting corollaries.

**Corollary 1.4.3.** *Let $A$ be a closed subset of a space $X$. Let $Z$ be an ANR and let $f \colon A \to Z$ be continuous. The following statements are equivalent:*

(1) *$f$ can be extended over $X$,*
(2) *$f$ is homotopic to an extendable function $g \colon A \to Z$.*

**Proof.** That (1) ⇒ (2) is clear since $f \simeq f$ by Lemma A.12.1(1).

For (2) ⇒ (1), let $H\colon A \times \mathbb{I} \to Z$ be a homotopy with $H_0 = g$ and $H_1 = f$. Since $g$ is extendable, $f$ is extendable as well (Theorem 1.4.2). □

Our next goal is to investigate contractibility within the class of ANR's.

**Theorem 1.4.4.** *Let $X$ be an AR. Then $X$ is contractible.*

**Proof.** By Corollary 1.1.8, $X$ can be thought of as a closed subset of a normed linear space $L$. Fix an arbitrary point $p \in L$. Since the function $F\colon L \times \mathbb{I} \to L$ defined by the formula
$$(x,t) \mapsto t \cdot p + (1-t) \cdot x$$
is a contraction, it follows that $X$ is contractible, being a retract of $L$ (Theorem A.12.4). □

These results yield the following interesting characterization of AR's.

**Corollary 1.4.5.** *Let $X$ be a space. The following statements are equivalent:*

(1) *$X$ is an AR,*
(2) *$X$ is a contractible ANR.*

**Proof.** The implication (1) ⇒ (2) is trivial by Theorem 1.4.4.

For (2) ⇒ (1), let
$$H\colon X \times \mathbb{I} \to X$$
be a homotopy such that $H_0 = 1$ and $H_1$ is constant, say with constant value $c$. In addition, let $Y$ be a space, $A \subseteq Y$ be closed and $f\colon A \to X$ be continuous. Define $F\colon A \times \mathbb{I} \to X$ by
$$F(a,t) = H(f(a),t).$$
Then $F$ is clearly a homotopy with $F_0 = f$ and $F_1$ a constant function. Trivially, $F_1$ can be extended to a continuous function from $Y$ to $X$. As $X$ is an ANR, by Corollary 1.4.3 it follows that $F_0 = f$ can be extended over $Y$. We conclude that $X$ is an AR. □

For a nontrivial generalization of Corollary 1.4.5, see Theorem 4.2.20.

**Cones.** Let $X$ be a space and consider its cone $\triangle(X)$. We observed in Exercise 1.2.2 that if $\triangle(X)$ is an ANR, then so is $X$. It is an interesting question whether the converse holds.

**Theorem 1.4.6.** *For a space $X$ the following statements are equivalent.*

(1) *$X$ is an ANR,*
(2) *$\triangle(X)$ is an ANR,*

(3) $\Delta(X)$ is an AR.

**Proof.** Observe that (3) $\Rightarrow$ (2) is trivial, and (2) $\Rightarrow$ (1) follows from the above observation. So we only need to prove the implication (1) $\Rightarrow$ (3). To this end, first observe that we may think of $X$ as a closed and bounded subspace of a normed linear space $L$ (Corollary 1.1.8). The cone $\Delta(X)$ is by Exercise 1.1.29 homeomorphic to the subspace

$$K = \{(t, tx) : t \in \mathbb{I}, x \in X\}$$

of $\mathbb{I} \times L$ (for notational simplicity, we switched the coordinates of $L$ and $\mathbb{I}$, and applied the homeomorphism $t \mapsto 1 - t$ of $\mathbb{I}$). Since $X$ is an ANR, there are an open neighborhood $U$ of $X$ in $L$ and a retraction $r \colon U \to X$. Let $V$ be an open neighborhood of $L \setminus U$ such that $\overline{V} \cap X = \emptyset$ (Corollary A.4.3). By Lemma A.4.1 there is a continuous function $\alpha \colon L \to \mathbb{I}$ such that $\alpha \restriction X \equiv 1$ and $\alpha \restriction \overline{V} \equiv 0$. Define a function $\beta \colon \mathbb{I} \times L \to K$ by

$$\beta(t, x) = \begin{cases} (0, \underline{0}) & (t = 0 \text{ or } {}^x\!/_t \notin U), \\ \bigl(t\alpha({}^x\!/_t), t\alpha({}^x\!/_t)r({}^x\!/_t)\bigr) & \text{(otherwise)}. \end{cases}$$

We claim that $\beta$ is a retraction from $\mathbb{I} \times L$ onto $K$. This suffices for the proof, since clearly $\mathbb{I} \times L$ is an AR (Corollary 1.2.9 and Proposition 1.2.11) and a retract of an AR is an AR.

Notice that $\beta(0, \underline{0}) = (0, \underline{0})$. In addition, if $x \in X$ and $t \in \mathbb{I} \setminus \{0\}$ are arbitrary then ${}^{tx}\!/_t = x \in U$ and so

$$\beta(t, tx) = \bigl(t\alpha(x), t\alpha(x)r(x)\bigr) = (t, tx).$$

This proves that $\beta \restriction K$ is the identity. So all there remains to prove is that $\beta$ is continuous, and this is left to the reader in Exercise 1.4.2. $\square$

Notice that the implication (2) $\Rightarrow$ (3) also follows from Corollary 1.4.5. Simply observe that $\Delta(X)$ is contractible (Exercise A.12.5).

**Exercises for §1.4.**

1. Let $A$ be a closed subspace of a space $X$ and let $T$ be the subspace

   $$(X \times \{0\}) \cup (A \times \mathbb{I})$$

   of $X \times \mathbb{I}$. Prove that for every space $Y$, if a continuous function $f \colon T \to Y$ can be extended over $(X \times \{0\}) \cup U$, where $U$ is a neighborhood of $A \times \mathbb{I}$ in $X \times \mathbb{I}$, then $f$ can be extended over $X \times \mathbb{I}$.

▶2. Prove that the function $\beta$ in the proof of Theorem 1.4.6 is continuous.

## 1.5. Topological characterization of some familiar spaces

The aim of this section is to present topological characterizations of two familiar spaces, namely, the Cantor middle-third set $\mathbb{C}$ and the closed unit interval $\mathbb{I}$. (See §1.9 for topological characterizations of $\mathbb{Q}$ and $\mathbb{P}$.)

**Topological characterization of the Cantor set.** A space $X$ is called *zero-dimensional* if it is nonempty and has a base consisting of clopen sets, i.e., if for every point $x \in X$ and for every neighborhood $U$ of $x$ there exists a clopen subset $C \subseteq X$ such that $x \in C \subseteq U$. It is clear that a nonempty subspace of a zero-dimensional space is again zero-dimensional and that products of zero-dimensional spaces are zero-dimensional. Observe that by Exercise A.2.12 it follows that a space is zero-dimensional if and only if it has a *countable* base consisting of clopen sets.

It is clear that no nontrivial connected space is zero-dimensional. As a consequence, if a space contains a nontrivial connected subspace then it is not zero-dimensional.

**Proposition 1.5.1.** *Let $X$ be zero-dimensional. Then for every open cover $\mathcal{U}$ of $X$ there exists a refinement $\mathcal{V}$ of $\mathcal{U}$ consisting of pairwise disjoint clopen sets.*

**Proof.** Let $\mathcal{U}$ be an open cover of $X$. Since $X$ is zero-dimensional, it can be refined by a clopen cover $\mathcal{E}$. We may assume without loss of generality that $\mathcal{E}$ is countable, say $\mathcal{E} = \{E_n : n \in \mathbb{N}\}$ (Corollary A.2.3). Now put $V_1 = E_1$ and put
$$V_n = E_n \setminus \bigcup_{i<n} E_i$$
for every $n \geq 2$. Then $\mathcal{V} = \{V_n : n \in \mathbb{N}\}$ is as required. □

**Corollary 1.5.2.** *Let $X$ be a space. Then $X$ is zero-dimensional if and only if for all disjoint closed sets $A$ and $B$ there exists a clopen subset $U \subseteq X$ such that $A \subseteq U$ and $B \cap U = \emptyset$.*

**Proof.** First assume that $X$ is zero-dimensional. For every $x \in X$ let $U_x$ be a neighborhood of $x$ such that $U_x \cap A = \emptyset$ or $U_x \cap B = \emptyset$ (here we use that $A$ and $B$ are disjoint closed sets). By Proposition 1.5.1 there is a refinement $\mathcal{V}$ of $\{U_x : x \in X\}$ consisting of pairwise disjoint clopen sets. Now put
$$U = \bigcup\{V \in \mathcal{V} : V \cap A \neq \emptyset\}.$$
It is easy to see that $U$ is as required.

If for all disjoint closed subsets $A$ and $B$ of $X$ there exists a clopen set $C$ with $A \subseteq C \subseteq X \setminus B$, then the clopen subsets of $X$ clearly form a base. So then $X$ is zero-dimensional. □

We shall now characterize the zero-dimensional subspaces of the real line $\mathbb{R}$.

**Proposition 1.5.3.** *A nonempty subspace $X$ of $\mathbb{R}$ is zero-dimensional if and only if it does not contain any (nondegenerate) interval.*

**Proof.** Assume that $X \subseteq \mathbb{R}$ is zero-dimensional. If $X$ contains a nontrivial interval $E$ then $X$ contains a nontrivial connected subspace and is therefore not zero-dimensional.

Now assume that $X \subseteq \mathbb{R}$ is such that it contains no nontrivial intervals. Then the set $D = \mathbb{R} \setminus X$ is dense in $\mathbb{R}$. The collection
$$\{(d_1, d_2) \cap X : d_1 < d_2 \text{ and } d_1, d_2 \in D\}$$
is easily seen to be a base for $X$ consisting of clopen subsets of $X$. Consequently, $X$ is zero-dimensional. □

This proposition gives us a rich supply of zero-dimensional spaces. For example, the rational numbers $\mathbb{Q}$, the irrational numbers $\mathbb{P}$, the product $\mathbb{Q} \times \mathbb{P}$, etc., are all zero-dimensional. The following example is of particular interest:

**Example 1.5.4.** The Cantor middle-third set C.

From $\mathbb{I} = [0, 1]$ remove the interval $(1/3, 2/3)$, i.e., the 'middle-third' interval. From the remaining two intervals, again remove their 'middle-thirds', and continue in this way. At stage $i$ of the construction we have a disjoint family $\mathcal{F}_i$ of $2^i$ closed subintervals of $\mathbb{I}$, each of length $3^{-i}$. The union of $\mathcal{F}_i$ is denoted by $H_i$. Observe that $H_i$ is a closed subset of $\mathbb{I}$ and hence is compact. The intersection of the $H_i$'s is called the *Cantor middle-third set*, C.

Clearly, C is closed in $\mathbb{I}$ and hence compact. Also, C is zero-dimensional by Proposition 1.5.3. For if $(a, b)$ is a nondegenerate subinterval of $\mathbb{R}$ then there exists $i \in \mathbb{N}$ such that $3^{-i} < b - a$ which implies that $(a, b)$ cannot be contained in C since C can be covered by a disjoint family of closed intervals each of length $3^{-i}$.

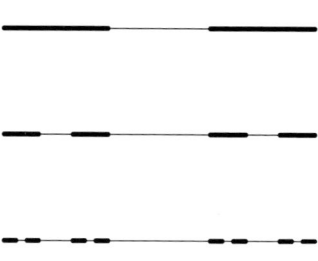

Figure 2.

## 1.5. TOPOLOGICAL CHARACTERIZATION OF SOME FAMILIAR SPACES

Observe that if $F \in \mathcal{F}_i$ then $F \cap C \neq \emptyset$. For at stage $i+1$ of the construction there exists an interval $A_1 \in \mathcal{F}_{i+1}$ such that $A_1 \subseteq F$. In addition, at stage $i+2$ of the construction, there exists an interval $A_2 \in \mathcal{F}_{i+2}$ such that $A_2 \subseteq A_1$. Proceeding in this way inductively, it is easy to construct a decreasing sequence $(A_n)_n$ of elements of $\mathcal{F}_{i+n}$ such that $A_n \subseteq F$ for every $n$. By compactness of $\mathbb{I}$, $\emptyset \neq \bigcap_{n=1}^{\infty} A_n \subseteq C$, as desired.

It is easy to see that C is the subspace of $\mathbb{I}$ consisting of all points that have a triadic expansion in which the digit 1 does not occur, i.e., the set

$$\left\{x = \sum_{i=1}^{\infty} \frac{x_i}{3^i} : x_i \in \{0,2\} \text{ for every } i\right\}.$$

We claim that the set C has no isolated points. If $x$ belongs to C and $a < x < b$ then choose $i$ so large that $3^{-i} < \min\{x-a, b-x\}$. Take $F \in \mathcal{F}_i$ with $x \in F$. Then $F \subseteq (a,b)$. There are precisely two disjoint elements of $\mathcal{F}_{i+1}$ that are contained in $F$. By the above, both these elements contain at least one point of C. We conclude that $(a,b) \cap C$ contains at least two points.

The Cantor set C is a very interesting space. There exists a simple topological characterization of it which we shall now present.

**Theorem 1.5.5.** C *is (topologically) the unique nonempty zero-dimensional compact space without isolated points.*

**Proof.** Since in Example 1.5.4 we showed that C is a zero-dimensional compact space without isolated points, it suffices to prove that if $X$ and $Y$ are such spaces then they are homeomorphic. To this end, let $X$ and $Y$ be compact zero-dimensional spaces without isolated points. By induction on $n$, we shall construct elements $m(n) \in \mathbb{N}$ and finite disjoint clopen covers

$$\mathcal{U}_n = \{U_{n,i} : i \leq m(n)\}, \quad \mathcal{V} = \{V_{n,i} : i \leq m(n)\}$$

of $X$ and $Y$, respectively, consisting of nonempty sets, such that

$$m(1) = 1, \quad \mathcal{U}_1 = \{X\}, \quad \mathcal{V}_1 = \{Y\},$$

and for $n \geq 2$,

(1) $\operatorname{mesh}(\mathcal{U}_n) < 1/n$, $\operatorname{mesh}(\mathcal{V}_n) < 1/n$,
(2) $\mathcal{U}_n < \mathcal{U}_{n-1}$, $\mathcal{V}_n < \mathcal{V}_{n-1}$,
(3) for $i \leq m(n-1)$ and $j \leq m(n)$: $U_{n,j} \subseteq U_{n-1,i}$ iff $V_{n,j} \subseteq V_{n-1,i}$.

Suppose that we constructed the covers $\mathcal{U}_m$ and $\mathcal{V}_m$ for all $m \leq n$, $n \geq 1$. Pick an arbitrary $i \leq m(n)$. By Proposition 1.5.1 there exist covers $\mathcal{E}$ and $\mathcal{F}$ of $U_{n,i}$ and $V_{n,i}$, respectively, consisting of pairwise disjoint clopen sets, such that

(4) $\mathcal{E} < \{B(x, \frac{1}{2(n+1)}) : x \in U_{n,i}\}$,
(5) $\mathcal{F} < \{B(y, \frac{1}{2(n+1)}) : y \in V_{n,i}\}$.

By compactness of $U_{n,i}$ and $V_{n,i}$, the covers $\mathcal{E}$ and $\mathcal{F}$ are finite. We assume that each element of $\mathcal{E}$ as well as $\mathcal{F}$ is nonempty. We claim that without loss of generality we may assume that $\mathcal{E}$ and $\mathcal{F}$ have the same cardinality. To see this, pick an arbitrary $E \in \mathcal{E}$. Since $X$ has no isolated points, there exist distinct $x, y \in E$. In addition, since $X$ is zero-dimensional, its clopen sets form a base and so there is a clopen subset $C \subseteq E$ such that $x \in C$ and $y \notin C$. So $E$ can be split into two nonempty clopen sets, namely, $C$ and $E \setminus C$. So we can replace $\mathcal{E}$ by a cover having $|\mathcal{E}| + 1$ elements. By repeating the same procedure if necessary, we can therefore ensure that $\mathcal{E}$ and $\mathcal{F}$ have the same cardinality.

We conclude that for $i \leq m(n)$ there are covers $\mathcal{E}_i$ and $\mathcal{F}_i$ of $U_{n,i}$ and $V_{n,i}$, respectively, such that

(6) $|\mathcal{E}_i| = |\mathcal{F}_i|$,
(7) $\mathcal{E}_i$ and $\mathcal{F}_i$ are pairwise disjoint and consist of nonempty clopen sets, each of diameter less than $1/n$.

Put $\mathcal{U}_{n+1} = \bigcup_{i=1}^{m(n)} \mathcal{E}_i$ and $\mathcal{V}_{n+1} = \bigcup_{i=1}^{m(n)} \mathcal{F}_i$. By construction, we can enumerate $\mathcal{U}_{n+1}$ and $\mathcal{V}_{n+1}$ in such a way that (3) is satisfied for $n+1$. This completes the inductive construction. Observe that $\bigcup_{n=1}^{\infty} \mathcal{U}_n$ is a base for $X$ and, similarly, that $\bigcup_{n=1}^{\infty} \mathcal{V}_n$ is a base for $Y$ (Exercise A.2.15).

We will now use these covers for the construction of the required homeomorphism. If $x \in X$ then for each $n \in \mathbb{N}$ there is a unique $i(n,x) \leq m(n)$ such that $x \in U_{n,i(n,x)}$. By (3) it follows that the collection $\{V_{n,i(n,x)}\}_n$ is decreasing and since $Y$ is compact, we get $\bigcap_{n=1}^{\infty} V_{n,i(n,x)} \neq \emptyset$. So by (1), the intersection $\bigcap_{n=1}^{\infty} V_{n,i(n,x)}$ consists of precisely one point. By interchanging the roles of $X$ and $Y$, these remarks imply that the function $f \colon X \to Y$ defined by

$$f(x) = y \iff \text{for every } n \in \mathbb{N} \text{ and } i \leq m(n) \quad (x \in U_{n,i} \text{ iff } y \in V_{n,i}),$$

is a bijection. Also $f$ is continuous since for $n \in \mathbb{N}$ and $i \leq m(n)$, $f^{-1}[V_{n,i}]$ is equal to $U_{n,i}$. Similarly, $f^{-1}$ is continuous. So we conclude that $f$ is a homeomorphism. $\square$

A space homeomorphic to C is called a *Cantor set* from now on. The characterization of C allows us to recognize many spaces as Cantor sets.

**Corollary 1.5.6.** *Let $X$ be any compact nonempty zero-dimensional space. Then $X \times$ C is a Cantor set.*

**Proof.** Since both $X$ and C are zero-dimensional it easily follows that $X \times$ C is compact and zero-dimensional. Since C has no isolated points it follows that $X \times$ C has no isolated points. An appeal to Theorem 1.5.5 now yields the desired result. $\square$

## 1.5. TOPOLOGICAL CHARACTERIZATION OF SOME FAMILIAR SPACES

Since $\{0,1\}^\infty$ is compact and zero-dimensional, and has clearly no isolated points, we conclude that, in particular, $\mathsf{C} \approx \{0,1\}^\infty$. This consequence can also be verified directly: the function $f\colon \{0,1\}^\infty \to \mathsf{C}$ defined by

$$f(x) = \sum_{i=1}^{\infty} \frac{2x_i}{3^i}$$

is a homeomorphism (see Page 43).

**Corollary 1.5.7.** *Every zero-dimensional space can be imbedded in* $\mathsf{C}$.

**Proof.** Let $X$ be zero-dimensional and let $\mathcal{U}$ be a countable base for $X$ consisting entirely of clopen sets. Let $\{U_n : n \in \mathbb{N}\}$ enumerate $\mathcal{U}$ and for every $n$ define $f_n \colon X \to \{0,1\}$ by

$$f_n(x) = 0 \iff x \in U_n.$$

Now define an imbedding $e \colon X \to \{0,1\}^\infty$ by putting

$$e(x)_n = f_n(x) \qquad (n \in \mathbb{N})$$

(this is in fact the same imbedding as in the proof of Corollary A.4.4). Since by the above $\mathsf{C} \approx \{0,1\}^\infty$, this completes the proof. □

So every zero-dimensional space can be imbedded in $\mathbb{R}$. This fact admits a nontrivial generalization that will be proved in §3.3.

By Cantor's standard diagonal argument, it follows that $\{0,1\}^\infty$ is uncountable. So $\mathsf{C}$ is uncountable as well. (To be precise, its cardinality is $\mathfrak{c}$.) Since by Corollary 1.5.6, $\mathsf{C}$ is homeomorphic to $\mathsf{C} \times \mathsf{C}$, this remark implies that there is a family $\mathcal{A}$ consisting of uncountably many pairwise disjoint Cantor sets in $\mathbb{R}$. (Let $\varphi \colon \mathsf{C} \times \mathsf{C} \to \mathsf{C}$ be a homeomorphism and put $\mathcal{A} = \{\varphi[\{c\} \times \mathsf{C}] : c \in \mathsf{C}\}$.) Since there are only countably many rational numbers in $\mathbb{R}$, one of the elements of $\mathcal{A}$ misses $\mathbb{Q}$, i.e., there exists a Cantor set $K$ in $\mathbb{R}$ consisting entirely of irrational numbers.

Let $\mathbb{P}$ denote the set of irrational numbers in $\mathbb{R}$. Since as was observed in Corollary 1.5.7, every zero-dimensional space can be imbedded in $\mathsf{C}$, we consequently get:

**Corollary 1.5.8.** *Every zero-dimensional space can be imbedded in* $\mathbb{P}$.

We now aim at proving that for every compact space $X$ there exists a continuous surjection from $\mathsf{C}$ onto $X$.

**Lemma 1.5.9.** *The function* $f \colon \mathsf{C} \times \mathsf{C} \to \mathbb{J}$ *defined by* $f(x,y) = x - y$ *is a continuous surjection.*

**Proof.** Since $\mathsf{C} \subseteq \mathbb{I}$, $f$ is well-defined and clearly continuous. Consider the set $F_1 = [0, \frac{1}{3}] \cup [\frac{2}{3}, 1]$, i.e., the first approximation to $\mathsf{C}$. It is a triviality to verify that $F_1 - F_1$, i.e., the set $\{x - y : x, y \in F_1\}$, is equal to $\mathbb{J}$. By induction one can prove that all approximations $F_n$, $n \in \mathbb{N}$, to C that were used to define C, have the same property (the induction step can be performed by noting that $F_n = (1/3 \cdot F_{n-1}) \cup (1/3 \cdot F_{n-1} + 2/3)$, a union of two shrunken copies of $F_{n-1}$.). Since the $F_n$ decrease and $C = \bigcap_{n=1}^{\infty} F_n$, an easy compactness-type argument shows that

$$C - C = \mathbb{J},$$

as required. □

We now come to the announced

**Theorem 1.5.10.** *Every compact space is a continuous image of* C.

**Proof.** By Lemma 1.5.9 and Exercise A.1.13 it follows that there exists a continuous surjection $f: C^\infty \to Q$. Now let $X$ be any compact space. According to Corollary A.4.4, we may assume that $X \subseteq Q$. Put $Y = f^{-1}[X]$. The restriction

$$g = f \restriction Y : Y \to X$$

is a continuous surjection. Observe that $Y$ is zero-dimensional. By Corollary 1.5.6, $C \times Y$ and C are homeomorphic. The required continuous surjection $\alpha : C \to X$ can therefore be obtained as the composition

$$\alpha = \beta \circ \pi \circ g,$$

where $\pi : C \times Y \to Y$ is the projection and $\beta : C \to C \times Y$ is an arbitrary homeomorphism. □

**Remark 1.5.11.** Theorem 1.5.10 can be used to construct so-called '*space filling curves*', i.e., continuous surjections from $\mathbb{I}$ onto $\mathbb{I}^2$. For let $g: C \to \mathbb{I}^2$ be a continuous surjection (Theorem 1.5.10). Then, by Corollary 1.2.12, $g$ can be extended to a continuous surjection $\bar{g}: \mathbb{I} \to \mathbb{I}^2$.

The Cantor set can be imbedded in 'many' spaces. Since it is uncountable, it cannot be a subspace of any countable space. There are also examples of uncountable spaces of which it is not a subspace, but it is not so easy to construct such a space. That such spaces exist follows however from Corollary 1.5.14 below.

Before we present the next theorem, we first fix some notation. By $\Sigma$ we mean the set consisting of all finite sequences consisting of 0's and 1's. (This includes the empty sequence.) If $s \in \Sigma$ then $|s|$ denotes its *length*, i.e., if $s = \emptyset$ then $|s| = 0$ and if $s = i_1 i_2 \ldots i_n$, where $i_j \in \{0,1\}$ for each $j \leq n$, then $|s| = n$. If $s = i_1 i_2 \ldots i_n \in \Sigma$ and $k \in \{0,1\}$ then $s^\frown k \in \Sigma$ denotes the sequence $i_1 i_2 \cdots i_n k$. If $s \in \Sigma \cup \{0,1\}^\infty$ then $t \in \Sigma$ is an *initial segment* of $s$ if there is an $n \in \mathbb{N}$ such that $t$ is equal to the sequence consisting of the first $n$ elements of $s$. Finally, if $s \in \Sigma \cup \{0,1\}^\infty$ then $s$ *extends* $t \in \Sigma$ if $t$ is an initial segment of $s$.

## 1.5. TOPOLOGICAL CHARACTERIZATION OF SOME FAMILIAR SPACES

**Theorem 1.5.12.** *Let $X$ be topologically complete and let $f\colon X \to Y$ be a continuous surjection. If $Y$ is uncountable then $X$ contains a Cantor set $K$ such that $f \upharpoonright K$ is one-to-one (and hence an imbedding).*

**Proof.** Let $\varrho$ be an admissible complete metric on $X$ which is bounded by $1/2$ (Exercise A.1.6). Since $Y$ is uncountable, $X$ contains an uncountable subset on which $f$ is one-to-one. This set contains an uncountable dense-in-itself subspace, say $E$ (Exercise A.2.13).

For $s \in \Sigma$ we shall define a nonempty open subset $U_s$ of $X$ having the following properties:

(1) $U_s \cap E \neq \emptyset$,
(2) if $t \in \Sigma$ and $t$ extends $s$ then $\overline{U}_t \subseteq U_s$,
(3) $f[\overline{U}_{s\frown 0}] \cap f[\overline{U}_{s\frown 1}] = \emptyset$,
(4) $\operatorname{diam} U_s < 2^{|s|}$.

We put $U_\emptyset = X$. Now suppose that we constructed $U_s$ for all $s \in \Sigma$ such that $|s| \leq n$. Fix an arbitrary $s \in \Sigma$ such that $|s| = n$. By (1) and the fact that $E$ is dense-in-itself, we may pick two distinct points $x, y \in U_s \cap E$. There exists $\varepsilon > 0$ such that $D(x,\varepsilon) \cup D(y,\varepsilon) \subseteq U_s$. Since $f(x) \neq f(y)$ and $f$ is continuous, we may additionally assume that $f[D(x,\varepsilon)] \cap f[D(y,\varepsilon)] = \emptyset$ and that $\varepsilon < 2^{|s|}$. Define $U_{s\frown 0} = B(x,\varepsilon)$ and $U_{s\frown 1} = B(y,\varepsilon)$, respectively. It is easily seen that our choices are as required.

Now for $s \in \{0,1\}^\infty$ put

$$A_s = \bigcap \{U_t : s \text{ extends } t\}.$$

By (1), (2) and (4) and the fact that $\varrho$ is complete it follows easily that $A_s$ consists of a single point of $X$, say $x_s$. Define a function $\xi\colon \{0,1\}^\infty \to X$ by

$$\xi(s) = x_s.$$

We claim that $\xi$ is an imbedding. That $\xi$ is one-to-one is clear since (3) implies that $f \circ \xi$ is one-to-one. By Exercise A.5.9 it therefore suffices to prove that $\xi$ is continuous. Since by Exercise A.2.15 the collection $\{U_s : s \in \Sigma\}$ is a local base at every point of the range of $\xi$, it suffices to prove that $\xi^{-1}[U_t]$ is open for every $t \in \Sigma$. But this is clear since

$$\xi^{-1}[U_t] = \{s \in \{0,1\}^\infty : s \text{ extends } t\}$$

is a basic clopen subset of $\{0,1\}^\infty$. □

Since the cardinality of the Cantor set is $\mathfrak{c}$, we immediately obtain from this that 'for topologically complete spaces the Continuum Hypothesis holds'. By this we mean that a topologically complete space cannot be of cardinality strictly between $\omega$ and $\mathfrak{c}$. (Observe that by Exercise A.2.16 it also cannot be of cardinality larger than $\mathfrak{c}$.)

**Corollary 1.5.13.** *Let $X$ be an uncountable analytic space. Then $X$ contains a Cantor set. As a consequence, if $X$ is analytic then either $|X| \leq \omega$ or $|X| = \mathfrak{c}$.*

**Proof.** Let $Y$ be topologically complete, and let $f\colon Y \to X$ be a continuous surjection. If $|X| \leq \omega$ then there is nothing to prove. So assume that $X$ is uncountable. Then $Y$ is uncountable, and hence by Theorem 1.5.12 there exists a Cantor set $K \subseteq Y$ such that $f\restriction K$ is one-to-one, and hence a topological imbedding (Exercise A.5.9). From this we get $|X| \geq \mathfrak{c}$. Finally, $|X| \leq \mathfrak{c}$ by Exercise A.2.16. □

This result enables us to prove the following interesting result.

**Corollary 1.5.14.** *Let $X$ be a space. Then we can decompose $X$ into two sets neither of which contains an uncountable topologically complete subspace (in particular, neither contains an uncountable compact subspace).*

**Proof.** If $|X| < \mathfrak{c}$ then by Corollary 1.5.13 there is nothing to prove. So assume that $|X| = \mathfrak{c}$.

First observe that the number of closed subsets of $X$ is at most $\mathfrak{c}$ (Exercise A.2.16). So there are also at most $\mathfrak{c}$ closed subsets of $X$ of cardinality $\mathfrak{c}$ and we can list them as $\{A_\alpha : \alpha < \lambda\}$ (no repetitions permitted) for some cardinal $\lambda \leq \mathfrak{c}$. By transfinite induction on $\alpha < \lambda$ we will pick distinct points $x_\alpha, y_\alpha \in A_\alpha$ such that

$$\{x_\alpha, y_\alpha\} \cap \{x_\beta, y_\beta : \beta < \alpha\} = \emptyset.$$

This is easily done since each $|A_\alpha| = \mathfrak{c}$ and the number of points that has to be avoided at stage $\alpha$ is smaller than $\mathfrak{c}$.

Now put $A = \{x_\alpha : \alpha < \lambda\}$ and $B = X \setminus A$, respectively. We claim that $A$ and $B$ are as required. To this end, let $Y \subseteq A$ be a topologically complete subspace. If $Y$ were uncountable, then it would contain a Cantor set by Theorem 1.5.12, and hence it would contain some $A_\alpha$. But this would contradict $y_\alpha \in A_\alpha \setminus A$. The proof for $B$ is similar. □

We now present an interesting application of Corollary 1.3.11, which is a partial generalization of Theorem 1.5.12.

**Corollary 1.5.15.** *Let $X$ be a topologically complete space, let $K$ be compact, and assume that $X$ contains an uncountable family of pairwise disjoint homeomorphs of $K$. Then $X$ contains a copy of the product $\mathsf{C} \times K$.*

**Remark 1.5.16.** Observe that the uncountably many pairwise disjoint homeomorphs of $K$ may be irregularly imbedded. This result shows that they may be replaced by a Cantor set of 'regularly' imbedded copies of $K$.

**Proof.** By Corollary 1.3.11 it follows that $\mathfrak{I}(K, X)$ is topologically complete, so we can pick an arbitrary complete admissible metric $\rho$ for it and use the same technique as in the proof of Theorem 1.5.12 to create a Cantor set subspace of it. By assumption, $\mathfrak{I}(K, X)$ contains an uncountable subspace $\mathcal{A}$ consisting of functions having pairwise disjoint images. We perform our construction with that subspace.

Since $\mathcal{A}$ is uncountable it contains an uncountable dense-in-itself subspace, say $\mathcal{E}$ (Exercise A.2.13). We proceed as in the proof of Theorem 1.5.12, with one slight adaptation only. Recall that in each step of the construction we picked in some nonempty open subset $\mathcal{U} \subseteq \mathfrak{I}(K, X)$ meeting $\mathcal{E}$ two distinct elements $f_0, f_1 \in \mathcal{E}$, together with small enough neighborhoods $\mathcal{V}_0, \mathcal{V}_1 \subseteq \mathcal{U}$ of $f_0$ and $f_1$, respectively. Here we make the neighborhoods even smaller (see below). An inspection of the proof of Theorem 1.5.12 shows that this does not create problems. We do not only want to get a Cantor set $\mathcal{C} \subseteq \mathfrak{I}(K, X)$ at the end of the process, but we additionally want the collection

(∗) $$\{h[K] : h \in \mathcal{C}\}$$

to be pairwise disjoint. But this can easily be achieved. For recall that

$$f_0[K] \cap f_1[K] = \emptyset.$$

So by compactness of $K$ there exists $\varepsilon > 0$ such that

$$B_\varepsilon(f_0[K]) \cap B_\varepsilon(f_1[K]) = \emptyset$$

(Corollary A.5.4). Put

$$\mathcal{F}_i = \{h \in \mathfrak{I}(X, Y) : \hat{\varrho}(f_i, h) < \varepsilon\} \qquad (i = 0, 1).$$

If $h_i \in \mathcal{F}_i$ for $i = 0, 1$ then $h_0[K] \cap h_1[K] = \emptyset$. So now we replace $\mathcal{V}_i$ by the set $\mathcal{V}_i \cap \mathcal{F}_i$ for $i = 0, 1$. These neighborhoods are clearly as required.

So at the end of our process we indeed get a Cantor set $\mathcal{C} \subseteq \mathfrak{I}(K, X)$ having the property that the collection in (∗) is pairwise disjoint. Now define the function $\xi \colon \mathcal{C} \times K \to X$ by

$$\xi(f, x) = f(x) \qquad (f \in \mathcal{C}, x \in K).$$

Then $\xi$ is clearly continuous. It is moreover one-to-one by the fact that the collection in (∗) is pairwise disjoint. So it is an imbedding by an application of Exercise A.5.9. □

**Remark 1.5.17.** An uncountable analytic space contains a Cantor set by Corollary 1.5.13. So if the compact space $K$ contains one point only, then for $K$ Corollary 1.5.15 not only holds for topologically complete spaces but also for the broader class of analytic spaces. The question therefore naturally arises whether Corollary 1.5.15 is also true for analytic spaces. This question was considered in BECKER, VAN ENGELEN and VAN MILL [45]; it was shown there that it is undecidable.

We say that a space $X$ is *totally disconnected* if for all distinct points $x, y$ in $X$ there exists a clopen set $C$ in $X$ such that $x \in C$ but $y \notin C$. It is clear that every zero-dimensional space is totally disconnected. Even more is true. Every space $X$ that admits a one-to-one continuous function into a zero-dimensional space is totally disconnected. The question naturally arises whether every totally disconnected space is zero-dimensional. If this were true then checking whether a given space is zero-dimensional would be simpler. However, the answer to this question is in the negative, as the following example shows.

**Example 1.5.18.** *Erdős' space.*
Put $E = \{x \in \ell^2 : x_i \in \mathbb{Q} \text{ for every } i\}$. We claim that $E$ is totally disconnected but not zero-dimensional. On Page 9 it was observed that the topology on $\ell^2$ is finer than the topology that $\ell^2$ inherits from $\mathbb{R}^\infty$. We conclude that $E$ admits a one-to-one continuous function into the product $\mathbb{Q}^\infty$, which is zero-dimensional (see Exercise 1.5.2). This implies that $E$ is totally disconnected.

We shall now prove that $E$ is not zero-dimensional by showing that if $U$ is open in $\ell^2$ and $\underline{0} \in U \subseteq \{x \in \ell^2 : \|x\| < 1\}$ then $U \cap E$ is not clopen in $E$. To this end, let $U$ be such a neighborhood of $\underline{0}$. We shall inductively define a sequence $q_i \in \mathbb{Q}$, $i \in \mathbb{N}$, such that for every $i$,

$$(*) \qquad x_i = (q_1, \ldots, q_i, 0, 0, \ldots) \in U, \quad \varrho(x_i, \ell^2 \setminus U) \leq 1/i.$$

Let $q_1 = 0$ and assume that the $q_j$ are defined for $j \leq i-1$ for some $i \geq 2$. For $0 \leq m \leq i$ put

$$x(i, m) = (q_1, \ldots, q_{i-1}, m/i, 0, 0, \ldots).$$

Observe that $x(i, 0) = x_{i-1} \in U$ and that $x(i, i) = (q_1, \ldots, q_{i-1}, 1, 0, 0, \ldots)$ has norm at least one and therefore does not belong to $U$. So there exists an element $0 \leq m_0 < i$ such that $x(i, m_0) \in U$ but $x(i, m_0 + 1) \notin U$. Now define $q_i = m_0/i$. Then $q_i$ is as required since

$$x_i = (q_1, \ldots, q_i, 0, 0, \ldots) = x(i, m_0) \in U,$$

and

$$\begin{aligned}\varrho(x_i, \ell^2 \setminus U) &\leq \varrho\bigl(x(i, m_0), x(i, m_0 + 1)\bigr) \\ &= \bigl\|(0, 0, \ldots, 0, -1/i, 0, 0, \ldots)\bigr\| \\ &= 1/i.\end{aligned}$$

This completes the inductive construction.

By $(*)$ we have

$$\sum_{j=1}^{i} q_j^2 \leq 1 \qquad (\forall \, i \in \mathbb{N}),$$

## 1.5. TOPOLOGICAL CHARACTERIZATION OF SOME FAMILIAR SPACES

and hence
$$\sum_{i=1}^{\infty} q_i^2 \leq 1.$$
We conclude that the point $q = (q_1, q_2, \dots)$ belongs to $\ell^2$, and hence to $E$. Since clearly,
$$\lim_{i \to \infty} x_i = q \qquad (\text{in } \ell^2)$$
(Lemma 1.1.12), it follows by $(*)$ that on the one hand

$q$ belongs to the closure of $U \cap E$ in $E$,

while on the other hand
$$\varrho(q, \ell^2 \setminus U) = \lim_{i \to \infty} \varrho(x_i, \ell^2 \setminus U) = 0,$$
so $q \notin U$. We conclude that $U \cap E$ is not closed in $E$, hence not clopen.

**Topological characterization of the closed unit interval.** We shall now present a topological characterization of the closed unit interval $\mathbb{I}$. The characterization below in Theorem 1.5.19 is not very appealing and its proof is rather involved. But it turns out to be very useful as will be demonstrated in the proof of the Mazurkiewicz Theorem 1.5.22 about the existence of topological copies of $\mathbb{I}$ in certain spaces.

**Theorem 1.5.19.** *Let $Y$ be a continuum containing two distinct points $x$ and $y$. Assume that for every $z \in Y \setminus \{x, y\}$ there exist closed sets $L_z$ and $U_z$ of $Y$ such that*

(1) $L_z \cup U_z = Y$,
(2) $x \in L_z$, $y \in U_z$,
(3) $L_z \cap U_z = \{z\}$.

*Then $Y$ is homeomorphic to $\mathbb{I}$.*

**Remark 1.5.20.** In Exercise A.10.6 we proved that every continuum has at least two noncut points. From the conditions in Theorem 1.5.19 it obviously follows that $Y \setminus \{x, y\}$ consists entirely of cut points. So the points $x$ and $y$ are the only noncut points of $Y$.

**Proof.** We first establish the following

**Claim 1.** *For every $z \in Y \setminus \{x, y\}$, $L_z$ and $U_z$ are continua.*

*Proof.* Suppose that for certain $z \in Y \setminus \{x, y\}$, $L_z$ is not a continuum. Then $L_z$ can be written as the union of two disjoint nonempty relatively open sets, say $E$ and $F$. Without loss of generality, $z \subset E$. Then $F$ is contained in $A = Y \setminus U_z$ and since $A$ is open in $Y$, this easily implies that $F$ is open in $Y$. However, since $F$ is closed in $L_z$ and $L_z$ is closed in $Y$, it also follows that $F$ is closed in $Y$. Consequently, $F$ is a nonempty clopen subset of $Y$, which contradicts the fact that $Y$ is connected. (See also Exercise A.10.5.) The proof for $U_z$ is similar. ◇

Define an order $\preccurlyeq$ on $Y$ by putting:
$$\begin{cases} x \preccurlyeq d & (d \in Y), \\ d \preccurlyeq e \text{ iff } L_d \subseteq L_e & (d, e \in Y \setminus \{x, y\}), \\ d \preccurlyeq y & (d \in Y). \end{cases}$$

**Claim 2.** Let $d, e \in Y \setminus \{x, y\}$. The following statements are equivalent:

(4) $d \preccurlyeq e$,
(5) $d \in L_e$,
(6) $L_d \subseteq L_e$,
(7) $e \in U_d$,
(8) $U_e \subseteq U_d$.

*Proof.* The implication (4) $\Rightarrow$ (5) is trivial. For (5) $\Rightarrow$ (6) and (5) $\Rightarrow$ (8), assume that $d \in L_e$. If $d = e$ then there is nothing to prove, so assume that $d \neq e$. This implies $d \notin U_e$, for otherwise
$$d \in L_e \cap U_e = \{e\}.$$
Observe that $Y \setminus \{d\}$ is the disjoint union of the open sets
$$L_d \setminus \{d\}, \quad U_d \setminus \{d\}.$$
Since $y \in U_e \cap (U_d \setminus \{d\})$, it therefore follows by connectivity of $U_e$ (Claim 1) that $U_e \subseteq (U_d \setminus \{d\})$ (hence (8) follows) which also implies that
$$L_d \subseteq X \setminus U_e \subseteq L_e$$
(hence (6) follows). Observe that (8) $\Rightarrow$ (7) is trivial. The implication (7) $\Rightarrow$ (6) can be proved in precisely the same way as (5) $\Rightarrow$ (8). Since (6) $\Rightarrow$ (4) follows by the definition of $\preccurlyeq$, we are done. ◇

**Claim 3.** $\preccurlyeq$ is a linear order on $Y$.

*Proof.* This follows easily from Claims 1 and 2. ◇

Let $D$ be the set of all rational numbers between 0 and 1, i.e.,
$$D = \mathbb{Q} \cap (0, 1).$$

**Claim 4.** There is a subset $E \subseteq D$ and a one-to-one function
$$f \colon E \to Y \setminus \{x, y\}$$
such that

(9) for all $d, e \in E$, $f(d) \preccurlyeq f(e)$ iff $d \leq e$,
(10) $E$ is dense in $\mathbb{I}$,
(11) $f[E]$ is dense in $Y$.

## 1.5. TOPOLOGICAL CHARACTERIZATION OF SOME FAMILIAR SPACES   53

*Proof.* Let $\mathcal{B} = \{B_n : n \in \mathbb{N}, n \text{ even}\}$ be a base for $Y$ consisting of nonempty open sets. In addition, let $\mathcal{F} = \{F_n : n \in \mathbb{N}, n \text{ odd}\}$ be a base for $D$, also consisting of nonempty open sets. Observe that by connectivity, every $B_n$ is infinite. In addition, since $\mathbb{Q}$ has no isolated points, the same holds for every $F_n$.

By induction on $n$ we shall pick a point $e_n \in D$ and a point $f(e_n) \in Y$ such that the following conditions are satisfied:

(12) if $n$ is odd then $e_n \in F_n$,
(13) if $n$ is even then $f(e_n) \in B_n$,
(14) $f \upharpoonright \{e_1, \ldots, e_n\}$ is one-to-one and order preserving (in the sense of (9)).

For $n = 1$ pick an arbitrary point $e_1$ in $F_1$ and an arbitrary point $f(e_1)$ in $Y \setminus \{x, y\}$. Observe that by connectivity of $Y$ such a choice is possible. Now suppose that we completed the construction of the points

$$e_1, \ldots, e_n, \quad f(e_1), \ldots, f(e_n).$$

There are two cases to consider of course. Suppose first that $n + 1$ is odd. Pick an arbitrary point $e_{n+1} \in F_{n+1} \setminus \{e_1, \ldots, e_n\}$ and define

$$G = \{m \leq n : e_m \preccurlyeq e_{n+1}\}, \quad H = \{m \leq n : e_m \succcurlyeq e_{n+1}\},$$

respectively. Let $a = \max\{f(e_m) : m \in G\}$ and $b = \min\{f(e_m) : m \in H\}$. Since $f$ is one-to-one, $a \neq b$ and so $L_a \cap U_b = \emptyset$. In addition, $L_a$ and $U_b$ are closed so that by connectivity of $Y$ there exists a point $f(e_{n+1})$ in $Y$ strictly between $a$ and $b$ (with respect to the order $\preccurlyeq$). It is easy to see that $e_{n+1}$ and $f(e_{n+1})$ are as required. For $n + 1$ even we can proceed by a similar argument. ◇

Now fix $t \in \mathbb{I}$ for a moment and put

$$\mathcal{A}_t = \{L_{f(d)} : d \in E \text{ and } t < d\} \cup \{U_{f(d)} : d \in E \text{ and } d < t\}.$$

**Claim 5.** $\mathcal{A}_t$ has the finite intersection property and $\bigcap \mathcal{A}_t$ consists of precisely one point.

*Proof.* That $\mathcal{A}_t$ has the finite intersection property follows easily from the definition of $\preccurlyeq$. Consequently, the compactness of $Y$ implies that $\bigcap \mathcal{A}_t \neq \emptyset$. Now suppose that there exist two distinct points $a$ and $b \in \bigcap \mathcal{A}_t$. Without loss of generality, $a \preccurlyeq b$. Since both $L_a$ and $U_b$ are closed in $Y$ and disjoint, by connectivity and by the fact that $f[F]$ is dense, there exists $d \in E \setminus \{t\}$ such that $f(d)$ lies strictly between $a$ and $b$. Now if $t < d$ then $L_{f(d)} \subseteq \mathcal{A}_t$ which implies that $b \in \bigcap \mathcal{A}_t \subseteq L_{f(d)}$, which is a contradiction. However, if $d < t$ then $U_{f(d)} \in \mathcal{A}_t$ which implies that $a \in \bigcap \mathcal{A}_t \subseteq U_{f(d)}$, which is also a contradiction. ◇

Define a function $g\colon \mathbb{I} \to Y$ by
$$\{g(t)\} = \bigcap A_t.$$

**Claim 6.** $g$ is a homeomorphism.

*Proof.* First observe that $g$ is order-preserving. Therefore, since $g \upharpoonright E = f$, $f$ is one-to-one and $E \subseteq \mathbb{I}$ is dense, it follows that $g$ is injective. Also observe that the range of $g$ is dense in $Y$ since it contains $f[E]$. By compactness of $\mathbb{I}$ it therefore suffices to prove that $g$ is continuous (Exercise A.5.9). To this end, let $U \subseteq Y$ be open, and take an arbitrary point $t \in g^{-1}[U]$. By compactness of $Y$, Claim 5 implies that there exist $d_0, d_1 \in E$ such that $d_0 < t < d_1$ and
$$L_{f(d_1)} \cap U_{f(d_0)} \subseteq U.$$
A straightforward verification shows that $(d_0, d_1) \subseteq g^{-1}[U]$, i.e., $g^{-1}[U]$ is a neighborhood of $t$. ◇

So we are done with the proof of the theorem. □

As an application we shall now prove the Mazurkiewicz Theorem about the existence of topological copies of $\mathbb{I}$ in certain spaces.

Let $X$ be a space and let $a, b \in X$. A *simple chain connecting* $a$ and $b$ is a collection $U_1, \ldots, U_n$ of open subsets of $X$ such that

(1) $a \in U_1 \setminus \bigcup_{i=2}^{n} U_i$,
(2) $b \in U_n \setminus \bigcup_{i=1}^{n-1} U_i$,
(3) $U_i \cap U_j \neq \emptyset$ iff $|i - j| \leq 1$.

**Lemma 1.5.21.** *Let $X$ be a connected space and let $\mathcal{U}$ be an open cover of $X$. For any two points $x$ and $y$ in $X$ there exists a simple chain connecting $x$ and $y$ consisting of elements of $\mathcal{U}$.*

**Proof.** Let $V$ be the set of all points in $X$ which are connected to $x$ by a simple chain of elements of $\mathcal{U}$. Then $V$ is clearly open and $x \in V$. We shall prove that $V$ is closed as well. The connectivity of $X$ will then imply that $V = X$.

Take an arbitrary $v \in \overline{V}$. There exists $U \in \mathcal{U}$ containing $v$; pick a point $t \in U \cap V$. Then $x$ and $t$ can be connected by a simple chain $U_1, \ldots, U_n$ of elements of $\mathcal{U}$. Observe that $U \cap U_n \neq \emptyset$ and define
$$k = \min\{i \leq n : U \cap U_i \neq \emptyset\}.$$
If $v$ belongs to $U_k$ then $U_1, \ldots, U_k$ is a simple chain from $x$ to $v$. If $v$ does not belong to $U_k$ then $U_1, \ldots, U_k, U$ does the job for us. □

We now come to the announced application.

## 1.5. TOPOLOGICAL CHARACTERIZATION OF SOME FAMILIAR SPACES

**The Mazurkiewicz Theorem 1.5.22.** *Let $X$ be topologically complete, connected and locally connected. If $x$ and $y$ are distinct points of $X$ then there exists an imbedding $\Phi\colon \mathbb{I} \to X$ such that $\Phi(0) = x$ and $\Phi(1) = y$.*

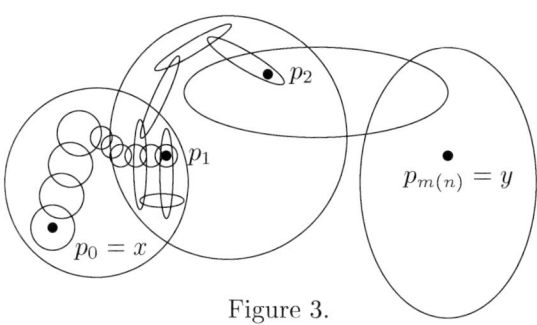

Figure 3.

**Proof.** Let $\varrho$ be an admissible complete metric on $X$. For each $n \in \mathbb{N}$ we shall construct a simple chain $U_{n,1}, \ldots, U_{n,m(n)}$ from $x$ to $y$ having the following properties

(1) for each $i \leq m(n)$, $U_{n,i}$ is connected and $\operatorname{diam}(U_{n,i}) < 1/n$,
(2) for all $i_1 \leq i_2 \leq m(n+1)$ there exist $j_1 \leq j_2 \leq m(n)$ such that $\overline{U}_{n+1,i_1} \subseteq U_{n,j_1}$ and $\overline{U}_{n+1,i_2} \subseteq U_{n,j_2}$.

The existence of $U_{1,1}, \ldots, U_{1,m(1)}$ follows from Lemma 1.5.21. Assume that the sets $U_{n,1}, \ldots, U_{n,m(n)}$ have been constructed for certain $n$. For $j$ smaller than $m(n)$ pick a point $p_j \in U_{n,j} \cap U_{n,j+1}$; put $p_0 = x$ and $p_{m(n)} = y$. By another application of Lemma 1.5.21 for each $1 \leq j \leq m(n)$ there exists a simple chain $\mathcal{V}_j$ from $p_{j-1}$ to $p_j$ in $U_{n,j}$ consisting of connected sets such that

(3) if $V \in \mathcal{V}_j$ then $\overline{V} \subseteq U_{n,j}$,
(4) for every $V \in \mathcal{V}_j$, $\operatorname{diam}(V) < 1/{n+1}$.

Unfortunately, we cannot simply join these chains together, because of doubling back (see Figure 3).

We can obtain the desired simple chain by the following procedure. For every $1 \leq j \leq m(n)$ let $\mathcal{V}_j = \{V_{j,1}, \ldots, V_{j,n(j)}\}$. Put

$$\pi = \min\{\kappa \leq n(1) : (\exists \lambda \leq n(2))(V_{1,\kappa} \cap V_{2,\lambda} \neq \emptyset)\}.$$

Let $\lambda = \max\{\mu \leq n(2) : V_{1,\pi} \cap V_{2,\mu} \neq \emptyset\}$. Now replace the collection $\mathcal{V}_1$ by $\{V_{1,1}, \ldots, V_{1,\pi}\}$, and, similarly, $\mathcal{V}_2$ by $\{V_{2,\lambda}, \ldots, V_{2,n(2)}\}$. Repeat this with the 'new' $\mathcal{V}_1 \cup \mathcal{V}_2$ and the 'old' $\mathcal{V}_3$, etc. At the end of the process, the union of the 'new' $\mathcal{V}_j$'s is clearly the required simple chain. This completes the inductive construction.

For every $n \in \mathbb{N}$ put
$$C_n = \bigcup_{i=1}^{m(n)} \overline{U}_{n,i},$$
and let
$$Y = \bigcap_{n=1}^{\infty} C_n.$$
We claim that $Y$ is homeomorphic to the closed unit interval $\mathbb{I}$. Observe that $Y$ is a closed subset of $X$ and that $x, y \in Y$.

**Claim 1.** $Y$ is a continuum.

*Proof.* That $Y$ is compact follows from the construction and Corollary A.5.2. Now apply Proposition A.10.7. ◇

We claim that $Y$ satisfies the conditions mentioned in Theorem 1.5.19. Let $z \in Y \setminus \{x, y\}$. For each $n$, at least one and at most two of the $U_{n,i}$'s contain $z$. Let $A_n$ be the union of all the $U_{n,i}$'s preceding these and $B_n$ be the union of all the $U_{n,i}$'s following these. Put
$$A = Y \cap \bigcup_{n=1}^{\infty} A_n, \quad B = Y \cap \bigcup_{n=1}^{\infty} B_n.$$
Observe that both $A$ and $B$ are open subsets of $Y$. Now if $d \in Y \setminus \{z\}$ then (1) easily implies that for certain $n$, $d \in A_n \cup B_n$. So $A \cup B = Y \setminus \{z\}$. We claim that $A \cap B = \emptyset$. To the contrary, assume that for some $n, p \in \mathbb{N}$ there exists $d \in A_n \cap B_p$. Without loss of generality, $n \leq p$. Let $i_1 \leq m(p)$ be such that $z \in U_{p,i_1}$. Since $d \in B_p$ there exists $i$ such that $i_1 < i \leq m(p)$ and $d \in U_{p,i}$. By (2) there exist $j_1 \leq j \leq m(n)$ such that
$$U_{p,i_1} \subseteq U_{n,j_1}, \quad U_{p,i} \subseteq U_{n,j}.$$
Suppose that $z \in U_{n,j}$. Then by construction, $U_{n,j} \cap A_n = \emptyset$, which is impossible since $d \in A_n \cap U_{n,j}$. Therefore $z \notin U_{n,j}$. Now since
$$z \in U_{p,i_1} \subseteq U_{n,j_1}, \quad j_1 \leq j,$$
we obtain $U_{n,j} \subseteq B_n$. But this is clearly a contradiction since $d \in U_{n,j} \cap A_n$ and $A_n \cap B_n = \emptyset$.

Put $L_z = A \cup \{z\}$ and $U_z = B \cup \{z\}$, respectively.

By Theorem 1.5.19 we conclude that $Y \approx \mathbb{I}$. □

**Corollary 1.5.23.** *Let $X$ be topologically complete and locally connected. If $U \subseteq X$ is open and connected and $x$ and $y$ are distinct points of $U$ then there exists an imbedding $\Phi \colon \mathbb{I} \to U$ such that $\Phi(0) = x$ and $\Phi(1) = y$.*

*Proof.* By Theorem A.6.3, $U$ is topologically complete. So it satisfies the conditions in Theorem 1.5.22. □

## 1.5. TOPOLOGICAL CHARACTERIZATION OF SOME FAMILIAR SPACES

**Corollary 1.5.24.** *Let $X$ be topologically complete, connected and locally connected. Then $X$ is path-connected and locally path-connected.*

**Remark 1.5.25.** A disconnected space cannot be path-connected. So connectivity is essential in the Mazurkiewicz Theorem. The same holds for both completeness and local connectivity. Examples:

- *A connected and locally connected space which is not path-connected.*
  For every $i \in \mathbb{N}$ let
  $$W_i = \prod_{n=1}^{i-1}[-1 + 1/i, 1 - 1/i]_n \times \{1\}_i \times \prod_{n=i+1}^{\infty}[-1 + 1/i, 1 - 1/i]_n.$$
  It will be shown in Exercise 1.5.17 that the subspace
  $$W = \bigcup_{i=1}^{\infty} W_i$$
  of $Q$ is as required.

- *A connected, topologically complete space which is not locally connected.*
  The $\sin(1/x)$-continuum in the plane is clearly as required.

**Exercises for §1.5.** A spaxe $X$ is called *nowhere locally compact* if no point has a neighborhood with compact closure.

A metric $\varrho$ on a set $X$ is called *non-Archimedean* if
$$\varrho(x, y) \leq \max\{\varrho(x, z), \varrho(z, y)\}$$
for all $x, y, z \in X$.

1. Prove that every space of cardinality less than the cardinality of $\mathbb{R}$ is zero-dimensional.
2. Let $X_n$ be zero-dimensional for every $n \in \mathbb{N}$. Prove that $\prod_{n=1}^{\infty} X_n$ is zero-dimensional.
3. Let $X$ be a space. Suppose that for every $n \in \mathbb{N}$ there is a discrete collection $\mathcal{F}_n$ of closed subsets of $X$ such that
   (1) $\bigcup \mathcal{F}_n = X$,
   (2) $\operatorname{mesh}(\mathcal{F}_n) < 1/n$.
   Prove that $X$ is zero-dimensional.
4. Let $X$ be a compact zero-dimensional space. Prove that the collection of clopen subsets of $X$ is countable. (Observe that $\mathbb{N}$ has uncountably many clopen subsets, so the compactness assumption on $X$ is essential.)
5. Let $X$ be a zero-dimensional space and let $f \colon X \to Y$ be an open and closed continuous surjection. Prove that $Y$ is zero-dimensional.
6. Let $X$ be zero-dimensional. Prove that $X$ admits a zero-dimensional compactification.

7. Let $X$ be a zero-dimensional $\sigma$-compact space. Prove that there is a null-sequence $(A_n)_n$ of compact subsets of $X$ such that $A_n \cap A_m = \emptyset$ if $n \neq m$ and $\bigcup_{n=1}^{\infty} A_n = X$.

8. Let $X$ be a locally connected and locally compact space. In addition, let $U \subseteq X$ be connected and open. Prove that if $K \subseteq U$ is compact then there is a continuum $C \subseteq U$ which contains $K$.

9. Let $X$ be a Peano continuum and let $U$ and $V$ be connected open subsets of $X$ such that $U \cup V = X$. Prove that for all compact subsets $A$ and $B$ of $X$ with $A \subseteq U$ and $B \subseteq V$ there are subcontinua $K$ and $L$ in $X$ such that $A \subseteq K \subseteq U$ and $B \subseteq L \subseteq V$ while moreover $K \cup L = X$.

10. Prove that for every space $X$ there exist a zero-dimensional space $Y$ and a continuous surjection $f \colon Y \to X$.

11. Let $X$ be a space. Prove that the following statements are equivalent:
    (1) $X$ is zero-dimensional,
    (2) $X$ has an admissible metric which is non-Archimedean.

12. Let $X$ be locally compact and totally disconnected. Prove that $X$ is zero-dimensional.

13. Let $X$ be totally disconnected. Prove that if $x \in X$ and $K \subseteq X \setminus \{x\}$ is compact then there is a clopen $C \subseteq X$ such that $x \in C \subseteq X \setminus K$.

14. Let $E$ be Erdős' space. Prove that $E$ is homeomorphic to $E \times E$.

15. Prove that both $\mathbb{Q}$ are $\mathbb{P}$ are nowhere locally compact.

16. Prove that if infinitely many factors in a product $\prod_{n=1}^{\infty} X_n$ are not compact then it is nowhere locally compact.

17. Prove that the space $W$ described on Page 57 is connected and locally connected, but not path-connected.

## 1.6. The inductive convergence criterion and applications

In this section we present a method enabling us to construct new homeomorphisms from old ones, namely, the Inductive Convergence Criterion. The method is illustrated with applications: we deduce the topological homogeneity of $Q$ and we present results on countable dense homogeneity.

**Lemma 1.6.1.** *Let $(X, \varrho)$ be a complete metric space and let $(A_n)_n$ be a sequence of subsets of $X$. Suppose that $(x_n)_n$ is a Cauchy sequence in $X$ such that for every $n$,*

$$(*) \qquad \varrho(x_{n+1}, x_n) < 3^{-n} \cdot \min\{\varrho(x_i, A_i) : 1 \leq i \leq n\}.$$

*Then $\lim_{n \to \infty} x_n$ does not belong to $\bigcup_{n=1}^{\infty} A_n$.*

**Proof.** Take an arbitrary $n \in \mathbb{N}$ and let $x = \lim_{n \to \infty} x_n$. We shall prove that $x \notin A_n$. To this end, observe that condition $(*)$ implies that for arbitrary $m$ we have
$$\varrho(x_{m+n}, x_{(m-1)+n}) < 3^{-((m-1)+n)} \cdot \varrho(x_n, A_n) \leq 3^{-m} \cdot \varrho(x_n, A_n).$$
From this it follows that
$$\varrho(x_{m+n}, x_n) < \sum_{i=1}^{m} 3^{-i} \cdot \varrho(x_n, A_n),$$
and as
$$\sum_{i=1}^{\infty} 3^{-i} = 1/2,$$
we obtain
$$\varrho(x, x_n) = \lim_{m \to \infty} \varrho(x_{m+n}, x_n) \leq \sum_{i=1}^{\infty} 3^{-i} \cdot \varrho(x_n, A_n) = 1/2 \varrho(x_n, A_n),$$
so $x \notin A_n$. $\square$

Observe that in the above lemma the distance between $x_{n+1}$ and $x_n$ 'depends' only on the points $x_1, \ldots, x_n$. Hence if one wishes to choose the points $(x_n)_n$ inductively so that $(*)$ is satisfied, at stage $n+1$ the choice of $x_{n+1}$ is subject to $n$ – hence finitely many – conditions.

Let $X$ be a compact space and let $(h_n)_n$ be a sequence in $\mathcal{H}(X)$. It is clear that for each $n \in \mathbb{N}$ the function
$$f_n = h_n \circ \cdots \circ h_1$$
belongs to $\mathcal{H}(X)$. If $f = \lim_{n \to \infty} f_n$ exists then it will be denoted by
$$\lim_{n \to \infty} h_n \circ \cdots \circ h_1$$
and is called the *infinite left product* of the sequence $(h_n)_n$. We want to find conditions on the sequence $(h_n)_n$ which ensure that $\lim_{n \to \infty} h_n \circ \cdots \circ h_1$ exists and belongs to $\mathcal{H}(X)$.

**The Inductive Convergence Criterion 1.6.2.** Let $(h_n)_n$ be a sequence in $\mathcal{H}(X)$ such that $X$ is compact. In addition, assume that for all $n \in \mathbb{N}$,

(1) $\hat{\varrho}(h_{n+1}, 1_X) < 2^{-n}$,
(2) $\hat{\varrho}(h_{n+1}, 1_X) < 3^{-n} \cdot \min\{\varrho(h_i \circ \cdots \circ h_1, \mathcal{G}_{1/i}(X, X)) : 1 \leq i \leq n\}$.

Then $h = \lim_{n \to \infty} h_n \circ \cdots \circ h_1$ exists and belongs to $\mathcal{H}(X)$.

**Proof.** For every $n$, put $f_n = h_n \circ \cdots \circ h_1$. Condition (1) and Exercise 1.3.2 imply that for all $n$,

(3) $\quad \hat{\varrho}(f_{n+1}, f_n) = \hat{\varrho}(h_{n+1} \circ \cdots \circ h_1, h_n \circ \cdots \circ h_1) = \hat{\varrho}(h_{n+1}, 1_X) < 2^{-n}.$

We conclude that the sequence $(f_n)_n$ is $\hat{\varrho}$-Cauchy and consequently,
$$f = \lim_{n\to\infty} f_n$$
exists (Corollary 1.3.6). Since $f_n \in \mathcal{S}(X,X)$ for every $n$, Proposition 1.3.8 implies that $f \in \mathcal{S}(X,X)$. Now by (2), (3), Lemmas 1.6.1 and 1.3.10 we obtain
$$f \in \bigcap_{n=1}^{\infty} \mathcal{S}_{1/n}(X,X) = \mathcal{H}(X),$$
as required. $\square$

The above result tells us that if the sequence $\left(\hat{\varrho}(h_n,1)\right)_n$ converges rapidly to 0, then the infinite left product $\lim_{n\to\infty} h_n \circ \cdots \circ h_1$ is a homeomorphism. It turns out that we are never interested in the precise speed at which $\left(\hat{\varrho}(h_n,1)\right)_n$ converges. We are always in the pleasant situation that while inductively defining the sequence $(h_n)_n$, we are able to choose the next homeomorphism 'sufficiently close' to the identity. This simplifies life considerably.

**Application 1: Topological homogeneity of the Hilbert cube.**
Every nontrivial topological group has fixed-point free homeomorphisms. Simply observe that if $(G, \circ)$ is a topological group then every translation of the form
$$x \mapsto g \circ x,$$
where $g$ is not the identity element of $G$, is a fixed-point free homeomorphism.

If $Q$ were a topological group then the proof of its homogeneity would be trivial by the above remarks. But $Q$ is not a topological group since it has the fixed-point property, as we will see later in Corollary 2.4.6.

It is geometrically obvious that no $\mathbb{I}^n$, $n \in \mathbb{N}$, is homogeneous. For $n = 1$ this is a triviality. Simply observe that $\mathbb{I}\setminus\{0\}$ is connected while $\mathbb{I}\setminus\{1/2\}$ is not. For larger $n$ the proof is more complicated. See the proof of Theorem 2.4.14 for more details.

As we announced, we will prove that $Q$ is homogeneous, thereby demonstrating a striking difference between the 'finite dimensional' and the 'infinite-dimensional' situation.

We put
$$s = \prod_{n=1}^{\infty} (-1,1)_n, \quad B(Q) = Q \setminus s,$$
respectively. We call $s$ the *pseudo-interior* and $B(Q)$ the *pseudo-boundary* of $Q$.

## 1.6. THE INDUCTIVE CONVERGENCE CRITERION AND APPLICATIONS

This notation is slightly confusing since $s$ was already our abbreviation for $\mathbb{R}^\infty$. But $\mathbb{R}^\infty$ and $\prod_{n=1}^{\infty}(-1,1)_n$ are homeomorphic and so this abuse of notation is not as bad as it seems at first glance.

**Lemma 1.6.3.** *Suppose that $x, y \in s \subseteq Q$. Then there is a homeomorphism $h \in \mathcal{H}(Q)$ with $h(x) = y$.*

**Proof.** Since for each $i$, $x_i, y_i \in (-1,1)_i$, there is an element $h_i \in \mathcal{H}(\mathbb{J}_i)$ such that $h_i(x_i) = y_i$, we may take $h = h_1 \times h_2 \times \cdots$ (Exercise A.1.13). $\square$

We see that all points in the pseudo-interior are topologically equivalent. Consequently, if we show that every point in $Q$ can be 'homeomorphed' into $s$, then we have shown that $Q$ is homogeneous. We need a preliminary lemma.

**Lemma 1.6.4.** *Suppose that $x \in Q$, that $m \in \mathbb{N}$, and that $\varepsilon > 0$. Then there is an element $h \in \mathcal{H}(Q)$ such that*

(1) $\hat{\varrho}(h, 1_Q) < \varepsilon$,
(2) $h(x)_m \in (-1,1)_m$,
(3) $h$ *does not affect the first $m-1$ coordinates of any point, i.e.,*

$$h(y)_i = y_i$$

*for all $i \leq m - 1$ and $y \in Q$.*

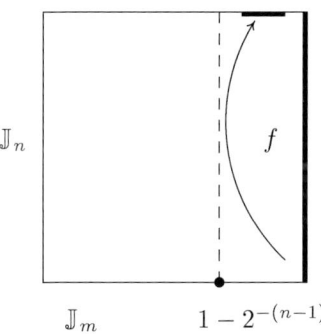

Figure 4.

**Proof.** If $|x_m| \neq 1$, let $h = 1_Q$. Therefore, without loss of generality assume that $x_m = 1$. Let $n > m$ be such that $2^{-(n-2)} < \varepsilon$. It is geometrically obvious that there is a homeomorphism $f \colon \mathbb{J}_m \times \mathbb{J}_n \to \mathbb{J}_m \times \mathbb{J}_n$ such that

(4) $f(p,q) = (p,q)$ if $p \leq 1 - 2^{-(n-1)}$,
(5) $f[\{1\} \times \mathbb{J}_n]$ is contained in $\mathbb{J}_m \times \{1\}$ and its projection on $\mathbb{J}_m$ is contained in the open interval $(1 - 2^{-(n-1)}, 1)$.

For a moment we think of $Q$ as $(\mathbb{J}_m \times \mathbb{J}_n) \times R$, where $R$ denotes the product of all remaining factors. We define $h = f \times 1_R$. So $h$ only affects the $n$-th and $m$-th coordinate of any point. To verify that $h$ is as required, first observe that (2) and (3) are clearly satisfied. For (1), notice that for every $x \in Q$,

$$\varrho(h(x), x) = 2^{-m} \cdot |h(x)_m - x_m| + 2^{-n} \cdot |h(x)_n - x_n|$$
$$\leq 1 \cdot 2^{-(n-1)} + 2^{-n} \cdot 2$$
$$= 2^{-(n-2)}$$
$$< \varepsilon.$$

From this it follows that $\hat{\varrho}(h, 1_Q) < \varepsilon$ (Exercise A.5.4). □

**Lemma 1.6.5.** Let $x \in Q$. There is an $h \in \mathcal{H}(Q)$ with $h(x) \in s$.

**Proof.** The homeomorphism $h$ shall be of the form

$$\lim_{n \to \infty} h_n \circ \cdots \circ h_1,$$

where $(h_n)_n$ is an inductively constructed sequence of homeomorphisms of $Q$. To ensure convergence in the limit, at each stage of the construction the next homeomorphism is constructed in accordance with the smallness conditions in the Inductive Convergence Criterion 1.6.2. Specifically, the sequence $(h_n)_n$ shall satisfy the following conditions:

$(1)_n$ $h_n$ does not affect the first $n - 1$ coordinates of any point,
$(2)_n$ $\bigl(h_n \circ \cdots \circ h_1(x)\bigr)_i \in (-1, 1)$ for every $i \leq n$,
$(3)_n$ $\hat{\varrho}(h_n, 1_Q)$ is so small that the conditions of the Inductive Convergence Criterion 1.6.2 are satisfied.

Apply Lemma 1.6.4 to find a homeomorphism $f \in \mathcal{H}(Q)$ such that

$$f(x)_1 \in (-1, 1)$$

and define $h_1 = f$. Then $h_1$ is as required since $(1)_1$ and $(3)_1$ are empty conditions. Now suppose that $h_i$ has been defined for every $i \leq n$. The Inductive Convergence Criterion 1.6.2 gives us a magic $\varepsilon > 0$ and tells us that we must choose the next homeomorphism $\varepsilon$-close to the identity. We are not interested at all in the precise value of $\varepsilon$. The only thing we need is that such $\varepsilon > 0$ exists. By Lemma 1.6.4 there is an element $h_{n+1} \in \mathcal{H}(Q)$ such that $\hat{\varrho}(h_{n+1}, 1_Q) < \varepsilon$, $h_{n+1}$ does not affect the first $n$ coordinates of any point, and

$$\bigl(h_{n+1}(h_n \circ \cdots \circ h_1(x))\bigr)_{n+1} \in (-1, 1).$$

It is evident that $h_{n+1}$ is as required.

Now put $h = \lim_{n\to\infty} h_n \circ \cdots \circ h_1$. By construction, $h \in \mathcal{H}(Q)$. By continuity of the projections $\pi_m \colon Q \to \mathbb{J}_m$ we find that for every $m$,

$$\begin{aligned} h(x)_m &= \Big( \lim_{n\to\infty} h_n \circ \cdots \circ h_1(x) \Big)_m \\ &= \lim_{n\geq m} \big( h_n \circ \cdots \circ h_1(x) \big)_m \\ &= \big( h_m \circ \cdots \circ h_1(x) \big)_m \\ &\in (-1,1), \end{aligned}$$

i.e., $h(x) \in s$. □

We now come to the announced result:

**Theorem 1.6.6.** *The Hilbert cube $Q$ is homogeneous.*

**Proof.** Take $x, y \in Q$. By Lemma 1.6.5 we find $h_1, h_2 \in \mathcal{H}(Q)$ such that both $h_1(x)$ and $h_2(y)$ belong to $s$. In addition, by Lemma 1.6.3, there is an element $f \in \mathcal{H}(Q)$ with $f(h_1(x)) = h_2(y)$. Then

$$g = h_2^{-1} \circ f \circ h_1 \in \mathcal{H}(Q)$$

takes $x$ onto $y$. □

**Application 2: Countable dense homogeneity.** A space $X$ is called *countable dense homogeneous* provided that for all countable and dense subsets $D, E \subseteq X$ there exists an element $f \in \mathcal{H}(X)$ with $f[D] = E$.

The topological sum of $\mathbb{S}^1$ and $\mathbb{S}^2$ is an example of a countable dense homogeneous space which is not homogeneous. But for connected spaces it will turn out that countable dense homogeneity implies homogeneity.

Let $X$ be a space and fix an arbitrary point $x \in X$. Let $\tau(x)$ denote the *type* of $x$, i.e.,

$$\{ y \in X : (\exists h \in \mathcal{H}(X))(h(x) = y) \}.$$

Observe that $\tau(x)$ is *invariant* under $\mathcal{H}(X)$, that is, for all $y \in \tau(x)$ and all $h \in \mathcal{H}(X)$ the image $h(y)$ also belong to $\tau(x)$.

**Theorem 1.6.7.** *If $X$ is countable dense homogeneous and $x \in X$ then $\tau(x)$ is clopen in $X$.*

**Proof.** We will show that $\tau(x)$ is clopen.

**Claim 1.** $\tau(x)$ *is closed.*

*Proof.* To the contrary, assume that there exists $p \in \overline{\tau(x)} \setminus \tau(x)$. Let $E$ be a countable dense subset of $\tau(x)$, and let $F$ be a countable dense subset of $U = X \setminus \overline{\tau(x)}$. Then both

$$A = E \cup F, \quad B = A \cup \{p\}$$

are countable dense subsets of $X$ and hence there exists by assumption an element $h \in \mathcal{H}(X)$ with $h[A] = B$. Since $p \notin h[E]$ because $h[\tau(x)] \subseteq \tau(x)$, there exists $b \in F$ with $h(b) = p$. But then $h[U]$ is a neighborhood of $p$ and hence intersects $\tau(x)$. But this implies that for some $q \in U$ we have that $h(q) \in \tau(x)$, i.e., $q \in \tau(x)$. This is a contradiction. ◇

**Claim 2.** $\tau(x)$ is open.

*Proof.* It is easy to see that $\tau(x)$ is open if and only if Int $\tau(x) \neq \emptyset$. Striving for a contradiction, suppose therefore that Int $\tau(x) = \emptyset$. By Claim 1 it follows that $\tau(x)$ is a closed subset of $X$ with empty interior, hence is nowhere dense. Let $E \subseteq U = X \setminus \tau(x)$ be countable and dense in $X$. There is an element $h \in \mathcal{H}(X)$ with $h[E] = E \cup \{x\}$. Pick $e \in E$ such that $h(e) = x$. But then $e \in \tau(x)$, which is a contradiction. ◇

So we are done. □

**Corollary 1.6.8.** *A connected countable dense homogeneous space is homogeneous.*

So for connected spaces, countable dense homogeneity is a strong form of homogeneity. There is an example of a connected homogeneous space which is not countable dense homogeneous, see VAN MILL [294] for details.

A space $X$ is called *strongly locally homogeneous* provided that it has an open base $\mathcal{U}$ such that for all $U \in \mathcal{U}$ and points $x, y \in U$ there exists an element $h \in \mathcal{H}(X)$ such that $h(x) = y$ and $h(z) = z$ for $z \notin U$. Such a homeomorphism is said to be *supported* on $U$. It is geometrically obvious that the Euclidean spaces $\mathbb{R}^n$, $n \in \mathbb{N}$, are strongly locally homogeneous.

Before we come to the main result in this Application, we make a few remarks. Let $X$ be a locally compact space. Observe that if $f \in \mathcal{H}(X)$ then the function $\alpha f \colon \alpha X \to \alpha X$ defined by

$$\alpha f(x) = \begin{cases} f(x) & (x \in X), \\ \infty & (x = \infty), \end{cases}$$

belongs to $\mathcal{H}(\alpha X)$.

It will sometimes be convenient to think of a locally compact space as subspace of its Alexandroff compactification.

**Theorem 1.6.9.** *Let $X$ be locally compact and strongly locally homogeneous. Then $X$ is countable dense homogeneous.*

**Proof.** Let $A = \{a_1, a_2, \dots\}$ and $B = \{b_1, b_2, \dots\}$ be faithfully indexed dense subsets of $X$. The hypothesis of strong local homogeneity implies that for each neighborhood $U$ of a point $x \in X$, and for any dense $G \subseteq X$, there exists a homeomorphism of $\alpha X$ which is supported on $U$ and takes $x$ into $G$

## 1.6. THE INDUCTIVE CONVERGENCE CRITERION AND APPLICATIONS 65

(use that $G \cap U \neq \emptyset$). Using the Inductive Convergence Criterion 1.6.2, we construct a sequence $(h_n)_n$ of homeomorphisms of $\alpha X$ such that its infinite left product $h$ is a homeomorphism and such that the following conditions (which ensure $h[A] = B$ and $h \restriction X \in \mathcal{H}(X)$) are satisfied:

(1) $h_n \circ \cdots \circ h_1(a_i) = h_{2i} \circ \cdots \circ h_1(a_i) \in B$ for each $i$ and $n \geq 2i$,
(2) $(h_n \circ \cdots \circ h_1)^{-1}(b_i) = (h_{2i+1} \circ \cdots \circ h_1)^{-1}(b_i) \in A$ for each $i$ and each $n \geq 2i + 1$,
(3) for all $n$, $h_n(\infty) = \infty$.

Assume $h_1, \ldots, h_{2i-1}$ have been defined for certain $i$.

If $h_{2i-1} \circ \cdots \circ h_1(a_i) \in B$, take $h_{2i} = 1_{\alpha X}$. Otherwise, choose a small neighborhood $U_{2i}$ of $h_{2i-1} \circ \cdots \circ h_1(a_i)$ having compact closure in $X$ which moreover is disjoint from the finite set
$$\{b_1, \ldots, b_{i-1}\} \cup h_{2i-1} \circ \cdots \circ h_1[\{a_1, \ldots, a_{i-1}\}].$$
Take $f_{2i}$ to be a homeomorphism of $X$ supported on $U_{2i}$ and such that
$$f_{2i} \circ h_{2i-1} \circ \cdots \circ h_1(a_1) \in B.$$
Finally, let $h_{2i} = \alpha f_{2i}$.

If $\left(h_{2i} \circ \cdots \circ h_1(b_i)\right)^{-1} \in A$, take $h_{2i+1} = 1_{\alpha X}$. Otherwise, choose a small neighborhood $U_{2i+1}$ of $b_i$ having compact closure in $X$ which moreover is disjoint from the finite set
$$\{b_1, \ldots, b_{i-1}\} \cup h_{2i} \circ \cdots \circ h_1[\{a_1, \ldots, a_{i-1}\}].$$
Take $f_{2i+1}$ to be a homeomorphism of $X$ supported on $U_{2i+1}$ and such that
$$f_{2i+1}^{-1}(b_i) \in (h_{2i} \circ \cdots \circ h_1)[A].$$
Finally, let $h_{2i+1} = \alpha f_{2i+1}$.

If the neighborhoods $U_{2i}$ and $U_{2i+1}$ are chosen small enough, the conditions of the Inductive Convergence Criterion are satisfied. In addition, (3) easily implies that $h(\infty) = \infty$, i.e., $h \restriction X \in \mathcal{H}(X)$. $\square$

Since the Euclidean spaces $\mathbb{R}^n$, $n \in \mathbb{N}$, are strongly locally homogeneous, we get:

**Corollary 1.6.10.** *The Euclidean spaces $\mathbb{R}^n$, $n \in \mathbb{N}$, are countable dense homogeneous.*

**Exercises for §1.6.** Let $X$ be a space with cover $\mathcal{U}$. We say that a function $f \colon X \to X$ is *limited* by $\mathcal{U}$ provided that for every $x \in X$ there exists $U \in \mathcal{U}$ such that $\{x, h(x)\} \subseteq U$.

▶1. Let $(X, \varrho)$ be a complete space, and let $(\mathcal{U}_n)_n$ be a sequence of open covers and $(h_n)_n$ a sequence of homeomorphisms of $X$ satisfying the following conditions:

(1) $\mathcal{U}_n$ is a star-refinement of $\mathcal{U}_{n-1}$,
   (2) $\mathrm{mesh}(\mathcal{U}_n) < 2^{-n}$,
   (3) $(h_{n-1} \circ \cdots \circ h_1)[\mathcal{U}_n]$ has mesh less that $2^{-n}$,
   (4) $h_n$ is limited by $\mathcal{U}_n$.
   Prove that $h = \lim_{n\to\infty} h_n \circ \cdots \circ h_1$ is a homeomorphism of $X$.

2. Use Exercise 1.6.1 to show that a topologically complete space which is strongly locally homogeneous is countable dense homogeneous.

3. Prove that both $s$ and $B(Q)$ are dense in $Q$. Prove that $B(Q)$ is $\sigma$-compact. Are $s$ and $B(Q)$ homeomorphic?

4. Prove that an open subspace of a strongly locally homogeneous space is strongly locally homogeneous.

5. Let $X$ be a compact space with closed subspace $A$. Prove that there is a continuous function $f\colon X \to Q$ such that
   (1) $f^{-1}(\mathbf{0}) = A$,
   (2) $f \upharpoonright X \setminus A$ is one-to-one.

6. Let $X$ be a homogeneous space. Prove that the following statements are equivalent:
   (1) $X$ is a Baire space.
   (2) $X$ is not meager in itself.

7. Let $X$ be a space and $x \in X$. Prove that for every $h \in \mathcal{H}(X)$ we have $h[\tau(x)] = \tau(x)$.

## 1.7. Bing's shrinking criterion

In this section we provide a second method for constructing new homeomorphisms from old ones, namely Bing's Shrinking Criterion, which is one of the most powerful tools in geometric topology. As an application we will prove that $Q$ is homeomorphic to its own cone.

Let $X$ and $Y$ be compact spaces. A continuous surjection $f\colon X \to Y$ is called a *near homeomorphism* provided that for every $\varepsilon > 0$ there is a homeomorphism $h\colon X \to Y$ such that $\hat{\varrho}(h, f) < \varepsilon$. Clearly, $f \in C(X, Y)$ is a near homeomorphism if and only if $f$ belongs to the closure of $\mathcal{H}(X, Y)$ in $C(X, Y)$. Simple examples show that a near homeomorphism need not be a homeomorphism (Exercise 1.3.5). Near homeomorphisms are very important in geometric topology.

Let $X$ and $Y$ be compact spaces. A continuous surjection $f\colon X \to Y$ is called *shrinkable* provided that for every $\varepsilon > 0$ there is a homeomorphism $h\colon X \to X$ satisfying

   (1) $\mathrm{diam}\bigl(hf^{-1}(y)\bigr) < \varepsilon$ for every $y \in Y$ (i.e., $f \circ h^{-1}$ is an $\varepsilon$-map),
   (2) $\hat{\varrho}(f, f \circ h) < \varepsilon$.

Shrinkability means that the point-inverses of $f$ can be uniformly shrunk to small sets by a homeomorphism of $X$ that, from the standpoint of the space $Y$, does not change $f$ very much. The reader should pause to get used to this definition.

Bing's Shrinking Criterion says that shrinkablity is the same as being a near homeomorphism. If the domain of the function under consideration has a rich supply of homeomorphisms, e.g., if it is a manifold, then verifying shrinkability is usually easier than verifying directly that the function in question is a near homeomorphism.

**Lemma 1.7.1.** *Let $X$ and $Y$ be compact spaces and let $f\colon X \to Y$ be shrinkable. Then for each $\varepsilon > 0$ and for each $\varphi \in \mathcal{H}(X)$ there exists an element $\psi \in \mathcal{H}(X)$ such that*

(1) $f \circ \psi$ *is an $\varepsilon$-map,*
(2) $\hat{\varrho}(f \circ \varphi, f \circ \psi) < \varepsilon$.

**Proof.** Let $\gamma > 0$ be such that for $A \subseteq X$, if $\operatorname{diam}(A) < \gamma$ then

$$(*) \qquad \operatorname{diam}(\varphi^{-1}[A]) < \varepsilon$$

(use the fact that $\varphi^{-1}$ is uniformly continuous (Exercise A.5.18)). Since $f$ is shrinkable, there is a homeomorphism $h \in \mathcal{H}(X)$ such that

(3) $\hat{\varrho}(f, f \circ h) < \varepsilon$,
(4) $\forall y \in Y : \operatorname{diam}\left(hf^{-1}(y)\right) < \gamma$.

Since $\hat{\varrho}(f, f \circ h) < \varepsilon$, it follows that $\hat{\varrho}(fh^{-1}\varphi, fhh^{-1}\varphi) < \varepsilon$ (Exercise 1.3.2), so that

$$\hat{\varrho}(f\varphi, fh^{-1}\varphi) < \varepsilon.$$

We claim that $fh^{-1}\varphi$ is an $\varepsilon$-map. If so, then $\psi = h^{-1}\varphi$ is as required. To this end, take an arbitrary $y \in Y$ and consider $(fh^{-1}\varphi)^{-1}(y) = \varphi^{-1}hf^{-1}(y)$. Since $\operatorname{diam}\left(hf^{-1}(y)\right) < \gamma$ we see by $(*)$ that $\operatorname{diam}\left(\varphi^{-1}hf^{-1}(y)\right) < \varepsilon$, and this is what we had to prove. $\square$

**Bing's Shrinking Criterion 1.7.2.** *Let $X$ and $Y$ be compact spaces and let $f\colon X \to Y$ be a continuous surjection. Then $f$ is a near homeomorphism if and only if $f$ is shrinkable.*

**Proof.** First assume that $f$ is a near homeomorphism. Let $\varepsilon > 0$. There is a homeomorphism $g\colon X \to Y$ such that $\hat{\varrho}(g, f) < \varepsilon/2$. Find $\gamma > 0$ such that if $A \subseteq Y$ has diameter less than $\gamma$ then $g^{-1}[A]$ has diameter less than $\varepsilon$ (use the fact that $g^{-1}$ is uniformly continuous (Exercise A.5.18)). Let $p\colon X \to Y$ be a homeomorphism such that $\hat{\varrho}(p, f) < \min\{\gamma/2, \varepsilon/2\}$ and put $h = g^{-1} \circ p$. We claim that $h$ is a shrinking homeomorphism. Since $\hat{\varrho}(f, p) < \varepsilon/2$ and

$$\begin{aligned}\hat{\varrho}(p, fh) &= \hat{\varrho}(p, fg^{-1}p) = \varrho(1_X, fy^{-1}) \\ &= \hat{\varrho}(gg^{-1}, fg^{-1}) = \hat{\varrho}(g, f) \\ &< \varepsilon/2\end{aligned}$$

(we used Exercise 1.3.2 twice), it follows that
$$\hat{\varrho}(f,fh) \leq \hat{\varrho}(f,p) + \varrho(p,fh) < \varepsilon/2 + \varepsilon/2 = \varepsilon.$$
Now take an arbitrary $y \in Y$. If $z \in p(f^{-1}(y))$ then since $\hat{\varrho}(p,f) < \gamma/2$ we get $\varrho(y,z) < \gamma/2$. So, $\operatorname{diam}\bigl(p(f^{-1}(y))\bigr) < \gamma$ from which it follows that $\operatorname{diam}\bigl(g^{-1}p(f^{-1}(y))\bigr) < \varepsilon$. We conclude that $\operatorname{diam}\bigl(h(f^{-1}(y))\bigr) < \varepsilon$.

Now assume that $f$ is shrinkable. Let $\varepsilon > 0$ and let $h_0$ be the identity homeomorphism on $X$. By applying Lemma 1.7.1 inductively, we find that there exists a sequence $(h_n)_{n \geq 1}$ in $\mathcal{H}(X)$ such that if we put
$$p_n = f \circ h_n$$
for each $n \geq 1$, then the following conditions are satisfied:

(1) $p_n$ is a $1/n$-map,
(2) $\hat{\varrho}(p_{n+1}, p_n) < 3^{-n} \cdot \varepsilon$ and $\hat{\varrho}(p_1, f) < \varepsilon/2$,
(3) $\hat{\varrho}(p_{n+1}, p_n) < 3^{-n} \cdot \min\{\varrho(p_i, \mathcal{G}_{1/i}(X,Y)) : 1 \leq i \leq n\}$.

By (2), the sequence $(p_n)_n$ is Cauchy. Since $\varrho$ is complete, the function
$$p = \lim_{n \to \infty} p_n$$
is well-defined and is continuous by Lemma A.3.1. Observe that by Proposition 1.3.8, $p \in \mathcal{S}(X,Y)$. Again by (2), it follows that
$$\hat{\varrho}(f,p) < \varepsilon/2 + \sum_{i=1}^{\infty} 3^{-i} \cdot \varepsilon = \varepsilon/2 + \varepsilon/2 = \varepsilon.$$
Now apply Lemma 1.6.1 to the sequence $(p_n)_n$ to obtain
$$p \notin \bigcup_{n=1}^{\infty} \mathcal{G}_{1/n}(X,Y),$$
i.e., $p \in \mathcal{H}(X,Y)$ (Lemma 1.3.10). We conclude that $f$ is a near homeomorphism. □

**Corollary 1.7.3.** *Let $X$ and $Y$ be compact spaces and let $f \colon X \to Y$ be shrinkable. Then $X$ and $Y$ are homeomorphic.*

**Application: The cone over the Hilbert cube is the Hilbert cube.** Recall that for a compact space the cone over $X$ is the space we get from $X \times \mathbb{I}$ by identifying $X \times \{1\}$ to a single point $\infty$ (see Page 515). It is geometrically clear that for each $n$ the cone over $\mathbb{I}^n$ is homeomorphic to $\mathbb{I}^{n+1}$ and so naively one would expect, by taking the 'limit' as $n$ goes to infinity, that $\triangle(Q) \approx Q$. That this is indeed true will be demonstrated here as an application of Bing's Shrinking Criterion.

**Lemma 1.7.4.** *For every $n \in \mathbb{N}$, $t \in [-1,1)$ and $\varepsilon > 0$ there exists a homeomorphism $h \colon \mathbb{J}^n \times \mathbb{J} \to \mathbb{J}^n \times \mathbb{J}$ such that*

(1) $h \upharpoonright \mathbb{J}^n \times [-1, t]$ is the identity function,
(2) the diameter of $h[\mathbb{J}^n \times \{1\}]$ is less than $\varepsilon$.

**Proof.** The lemma is geometrically obvious. Alternatively, observe that the Lemma is clear for $n = 1$ since then $h$ can be found such as in the proof of Lemma 1.6.4, and proceed inductively. □

Lemma 1.7.4 allows us to conclude that:

**Theorem 1.7.5.** $\triangle(Q)$ is homeomorphic to $Q$.

**Proof.** Let $p \colon Q \times \mathbb{I} \to \triangle(Q)$ be the natural quotient map. We will show that $p$ is shrinkable. From Corollary 1.7.3 it will then follow that
$$Q \approx Q \times \mathbb{I} \approx \triangle(Q).$$
Let $\varepsilon > 0$ and let $U$ be a neighborhood of $\infty \in \triangle(Q)$ having diameter less than $\varepsilon$. There exists $t \in (0, 1)$ such that $(Q \times [t, 1)) \cup \{\infty\} \subseteq U$. Find $n \in \mathbb{N}$ such that
$$\sum_{m=n}^{\infty} 2^{-m} < \varepsilon/2$$
and applying Lemma 1.7.4 find a homeomorphism $h \colon \mathbb{J}^n \times \mathbb{I} \to \mathbb{J}^n \times \mathbb{I}$ such that

(1) $h \upharpoonright \mathbb{J}^n \times [0, t] = 1$,
(2) $\operatorname{diam}(h[\mathbb{J}^n \times \{1\}]) < \varepsilon/2$.

Define $H \in \mathcal{H}(Q \times \mathbb{I})$ by
$$H((x_1, \ldots, x_n, x_{n+1}, \ldots), t) = (h(x_1, \ldots, x_n, t), x_{n+1}, x_{n+2}, \cdots)$$
(we make a few obvious identifications here). By the special choice of $n$ it follows easily that $\operatorname{diam}(H[Q \times \{1\}]) < \varepsilon$. We claim that $\hat{\varrho}(p, p \circ H) < \varepsilon$. Take a point $(x, s) \in Q \times \mathbb{I}$. If $s \leq t$ then $H(x, s) = (x, s)$, so
$$\varrho(p(x, s), pH(x, s)) = 0.$$
If $s > t$ then $\{p(x, s), pH(x, s)\} \subseteq U$. Therefore, since $\operatorname{diam}(U) < \varepsilon$, we find that $\varrho(p(x, s), pH(x, s)) < \varepsilon$. We conclude that $p$ is shrinkable. □

Consider the point $\infty \in \triangle(Q)$. Basic closed neighborhoods of this point have the form
$$U_t = (Q \times [t, 1)) \cup \{\infty\} \qquad (t \in (0, 1)).$$
Observe that the boundary of $U_t$ is equal to $Q \times \{t\}$. Consequently, Theorem 1.7.5 implies that $\infty$ has arbitrarily small closed neighborhoods $U$ such that

(1) $U$ is homeomorphic to $Q$,
(2) $\operatorname{Fr}(U)$ is homeomorphic to $Q$.

By the homogeneity of $Q$ (Theorem 1.6.6), *each* point of $Q$ has arbitrarily small closed neighborhoods $U$ satisfying (1) and (2). In particular, *each* point of $Q$ has arbitrarily small closed neighborhoods with contractible boundaries.

This demonstrates a striking difference with the finite dimensional situation.

**Exercises for §1.7.**

1. Let $X$ be a compact space. Let $Z$ be the space we obtain from $\triangle(X) \times \mathbb{I}$ by identifying $\triangle(X) \times \{1\}$ to a single point, and let $\pi \colon \triangle(X) \times \mathbb{I} \to Z$ be the decomposition map. Prove that $\pi$ is shrinkable and conclude from this that $\triangle\bigl(\triangle(X)\bigr) \approx \triangle(X) \times \mathbb{I}$.

2. Let $X$ be a compact space and let $Y = \triangle(X) \times Q$. Prove that $Y$ is homeomorphic to its own cone $\triangle(Y)$.

3. Let $X, Y$ and $Z$ be compact spaces and let $f \colon X \to Y$ and $g \colon Y \to Z$ be near homeomorphisms. Prove that $g \circ f \colon X \to Z$ is a near homeomorphism.

4. Let $p \colon Q \to \prod_{i=2}^{\infty} \mathbb{J}_i$ be the projection. Prove that $p$ is a near homeomorphism.

## 1.8. Isotopies

We will now present the third and last method for constructing new homeomorphisms from old ones by means of so-called isotopies. In comparison to the Inductive Convergence Criterion and Bing's Shrinking Criterion, at first glance this method does not seem to be as important as the other two since direct appealing applications are hard to find. However, it turns out that this method is as fundamental as the preceding two.

Let $X$ and $Y$ be spaces. Recall that a homotopy from $X$ to $Y$ is a continuous function $H \colon X \times \mathbb{I} \to Y$. We call a continuous function

$$H \colon X \times K \to Y,$$

where $K$ is any compact space, a *K-homotopy*. If $t \in K$ then the function $H_t \colon X \to Y$ defined by $H_t(x) = H(x,t)$ is called the *t-th level* of $H$. A *K-isotopy* from $X$ to $Y$ is a $K$-homotopy $H \colon X \times K \to Y$ each level of which is a homeomorphism from $X$ onto $Y$. If $K \subseteq \mathbb{R}$ is a compact interval then a $K$-homotopy (resp. $K$-isotopy) is simply called a homotopy (resp. isotopy).

The proof of the following proposition is a triviality and is left as an exercise to the reader.

**Proposition 1.8.1.** *For each $n \in \mathbb{N}$ let $H_n \colon X_n \times K_n \to Y_n$ be a $K_n$-isotopy. Then the function*

$$H \colon \prod_{n=1}^{\infty} X_n \times \prod_{n=1}^{\infty} K_n \to \prod_{n=1}^{\infty} Y_n,$$

*defined by*

$$H(x,t)_n = H_n(x_n, t_n) \qquad (n \in \mathbb{N}),$$

*is a $\prod_{n=1}^{\infty} K_n$-isotopy.*

We shall now present our last method for constructing new homeomorphisms from old ones.

**Theorem 1.8.2.** *Let $K$ be a compact space. If $H \colon X \times K \to X$ is a $K$-homotopy and $\alpha \colon Y \to K$ is continuous then so is the function*

$$f \colon X \times Y \to X \times Y$$

*defined by*

(1) $$f(x,y) = \bigl(H(x, \alpha(y)), y\bigr).$$

*Moreover, if $X$ and $Y$ are compact and $H$ is a $K$-isotopy then $f$ is a homeomorphism.*

**Remark 1.8.3.** Observe that the function $f$ defined in (1) is 'level preserving', i.e., $f$ does not change the second coordinate of any point.

**Proof.** That $f$ is continuous is trivial. Assume next that $X$ and $Y$ are compact and that $H$ is a $K$-isotopy. Since $f$ is 'level preserving' and since each 'level' of $f$ is a homeomorphism from $X$ onto $X$, it follows that $f$ is surjective. By compactness we therefore only need to show that $f$ is one-to-one (Exercise A.5.9). To this end, take distinct

$$(a,b), (a',b') \in X \times Y.$$

If $b \neq b'$ then clearly $f(a,b) \neq f(a',b')$. If $b = b'$ then $\alpha(b) = \alpha(b')$ and $a \neq a'$. Consequently, $H\bigl(a, \alpha(b)\bigr) \neq H\bigl(a', \alpha(b')\bigr)$ since $H_{\alpha(b)}$ is one-to-one. $\square$

**Application: Extending certain homeomorphisms.** Isotopies are very useful in situations where one wants to extend a given homeomorphism. This will be demonstrated in a simple situation here. The same ideas will be used to derive much more complicated results in §5.2.

Let $E, F \subseteq \mathbb{R}$ be compact subsets and let $f \colon E \to F$ be a homeomorphism. It is easy to see that it is not always possible to extend $f$ to a homeomorphism $\bar{f} \colon \mathbb{R} \to \mathbb{R}$. For example, let $E = F = \{0,1,2\}$ and define $f$ by $f(0) = 1, f(1) = 0$ and $f(2) = 2$.

Identify $\mathbb{R}$ and the $x$-axis in $\mathbb{R}^2$. Although in general a map $f$ between compact subsets of $\mathbb{R}$ cannot be extended over $\mathbb{R}$, an easy application of

Exercise 1.2.9 yields that if $f$ is a homeomorphism then it can be extended to a homeomorphism $\bar{f} \colon \mathbb{R}^2 \to \mathbb{R}^2$. We shall now present a different proof of this which much more than the solution of Exercise 1.2.9 illustrates the technique of extending homeomorphisms that we are going to use in §5.2.

**Theorem 1.8.4.** *Let $E, F \subseteq \mathbb{R}$ be compact and let $f \colon E \to F$ be a homeomorphism. Then $f$ can be extended to a homeomorphism $\bar{f} \colon \mathbb{R}^2 \to \mathbb{R}^2$.*

**Proof.** We shall prove the theorem for the interval $(-1,1)$ instead of $\mathbb{R}$. Since $(-1,1)$ and $\mathbb{R}$ are homeomorphic, we are allowed to do this.

So let $E, F \subseteq (-1,1)$ be compact and let $f \colon E \to F$ be a homeomorphism. Let $K \subseteq (-1,1)$ be a compact interval containing both $E$ and $F$. For each $t \in K$ let $H_t \colon \mathbb{J} \to \mathbb{J}$ be the unique homeomorphism taking $[-1,t]$ linearly onto $[-1,0]$ and $[t,1]$ linearly onto $[0,1]$. It is easily seen that the function $H \colon \mathbb{J} \times K \to \mathbb{J}$ defined by $H(x,t) = H_t(x)$ is an isotopy (Exercise 1.8.1).

Let $\Gamma \subseteq (-1,1)^2$ be the graph of the function $f$, i.e.,
$$\Gamma = \{(x, f(x)) : x \in E\}.$$

Our aim is to find a homeomorphism $h \colon \mathbb{J}^2 \to \mathbb{J}^2$ that takes $\Gamma$ onto $\{0\} \times F$. We achieve this by applying Theorem 1.8.2. Define the function $\varphi \colon F \to K$ as follows: $\varphi(x) = f^{-1}(x)$. By the Tietze Extension Theorem 1.2.5, we can extend $\varphi$ to a continuous function $\alpha \colon \mathbb{J} \to K$. Now define $h \colon \mathbb{J}^2 \to \mathbb{J}^2$ by
$$h(x,y) = \bigl(H(x, \alpha(y)), y\bigr).$$

By Theorem 1.8.2 it follows that $h$ is a homeomorphism. If $(x, f(x)) \in \Gamma$ then

(1) $\qquad h\bigl(x, f(x)\bigr) = \bigl(H(x, \alpha(f(x))), f(x)\bigr) = \bigl(H(x,x), f(x)\bigr) = \bigl(0, f(x)\bigr).$

We conclude that $h$ is as required. By precisely the same argumentation we can find a homeomorphism $g \colon \mathbb{J}^2 \to \mathbb{J}^2$ such that for every $(x, f(x)) \in \Gamma$,

(2) $\qquad\qquad\qquad g\bigl(x, f(x)\bigr) = (x, 0).$

Now define $\xi \colon \mathbb{J}^2 \to \mathbb{J}^2$ by $\xi(x,y) = (y,x)$ and put $\bar{f} = \xi \circ h \circ g^{-1}$. Then $\bar{f}$ is a homeomorphism of $\mathbb{J}^2$ and it is easily seen that $\bar{f}[(-1,1)^2] = (-1,1)^2$. We claim that $\bar{f}$ extends $f$ and is therefore as required. Take an arbitrary $x \in E$. Then
$$\bar{f}(x,0) = \xi h g^{-1}(x,0) = \xi h\bigl(x, f(x)\bigr) = \xi\bigl(0, f(x)\bigr) = (f(x), 0)$$
(apply (1) and (2)). $\square$

**Exercises for §1.8.**

1. Let $-1 < a < b < 1$ and for each $t \in K = [a, b]$ let $H_t \colon \mathbb{J} \to \mathbb{J}$ be the unique homeomorphism taking $[-1, t]$ linearly onto $[-1, 0]$ and $[t, 1]$ linearly onto $[0, 1]$. Prove that the function
$$H \colon \mathbb{J} \times K \to \mathbb{J}$$
defined by $H(x, t) = H_t(x)$ is an isotopy.

▶2. Prove that $Q$ is strongly locally homogeneous.

3. Let $D \subseteq Q$ be countable and dense. Prove that $Q \setminus D$ is homogeneous.

## 1.9. Homogeneous zero-dimensional spaces

Homeomorphism theory in the class of zero-dimensional spaces is particularly elegant. It turns out that various classical zero-dimensional spaces such as $\mathbb{Q}$ and $\mathbb{P}$ can be topologically characterized as nicely as the Cantor set.

**Homogeneous zero-dimensional spaces.** Our first aim is to characterize the class of all homogeneous zero-dimensional spaces.

**Lemma 1.9.1.** *Let $X$ be a zero-dimensional space and let $x, y \in X$. If for each $\varepsilon > 0$ there are clopen neighborhoods $U$ and $V$ of $x$ and $y$, respectively, such that*

(1) $\operatorname{diam}(U) < \varepsilon$, $\operatorname{diam}(V) < \varepsilon$,
(2) *$U$ is homeomorphic to $V$,*

*then there is a homeomorphism $h \colon X \to X$ with $h(x) = y$.*

**Proof.** For each $n \geq 0$ we will construct a clopen neighborhood $U_n$ of $y$ and a clopen neighborhood $V_n$ of $x$ and for each $n \geq 1$ an element $h_n \in \mathcal{H}(X)$ such that

(3) $\operatorname{diam}(U_n) < 2^{-n}$, $\operatorname{diam}(V_n) < 2^{-n}$,
(4) $U_n \cup h_{n-1} \circ \cdots \circ h_1[V_n] \subseteq U_{n-1}$ and $h_1^{-1} \circ \cdots \circ h_{n-1}^{-1}[U_n] \cup V_n \subseteq V_{n-1}$,
(5) $h_n \circ \cdots \circ h_1[V_n] = U_n$,
(6) $h_n$ is supported on $U_{n-1}$.

Put $U_0 = V_0 = X$ and suppose that for some $n \geq 0$, $V_i$ and $U_i$ have been constructed for all $0 \leq i \leq n-1$ and $h_i$ for all $1 \leq i \leq n-1$. If
$$h_{n-1} \circ \cdots \circ h_1(x) = y$$
then let $V$ be any clopen neighborhood of $x$ of diameter less than $2^{-n}$ such that both $V$ and $U = h_{n-1} \circ \cdots \circ h_1[V]$ have diameter less than $2^{-n}$. Put
$$V_n = V, \quad U_n = U, \quad h_n = 1_X.$$

Then our choices are obviously as required. Suppose therefore that
$$h_{n-1} \circ \cdots \circ h_1(x) \neq y.$$
Let $V \subseteq V_{n-1}$ be a clopen neighborhood of $x$ and $U \subseteq U_{n-1}$ be a clopen neighborhood of $y$ such that

(7) $\operatorname{diam}(V) < 2^{-n}$, $\operatorname{diam}(U) < 2^{-n}$,
(8) $V$ is homeomorphic to $U$,
(9) $h_{n-1} \circ \cdots \circ h_1[V] \subseteq U_{n-1}$,
(10) $h_{n-1} \circ \cdots \circ h_1[V] \cap U = \emptyset$.

Let $f \colon h_{n-1} \circ \cdots \circ h_1[V] \to U$ be a homeomorphism. Define $U_n = U$, $V_n = V$ and $h_n \colon X \to X$ by

$$h_n(x) = \begin{cases} x & (x \notin U_n \cup h_{n-1} \circ \cdots \circ h_1[V_n]), \\ f(x) & (x \in h_{n-1} \circ \cdots \circ h_1[[V_n]]), \\ f^{-1}(x) & (x \in U_n). \end{cases}$$

It is clear that $U_n$, $V_n$ and $h_n$ satisfy our inductive requirements.

If $p \in X$ and $\varrho(p, x) > 2^{-n}$ then $p \notin V_n$ and so
$$h_n \circ \cdots \circ h_1(p) \notin h_n \circ \cdots \circ h_1[V_n] = U_n.$$
By (6) this implies that
(11) $$h_k \circ \cdots \circ h_1(p) = h_n \circ \cdots \circ h_1(p)$$
for all $k \geq n$. Consequently, if we define $h \colon X \to X$ by
$$h = \lim_{n \to \infty} h_n \circ \cdots \circ h_1$$
then $h$ is well-defined. Observe that by (4) and (5), $h(x) = y$. We claim that $h$ is a homeomorphism. If $\varrho(p, x) > 2^{-n}$ then, as remarked above,
$$h_{n-1} \circ \cdots \circ h_1(p) \notin U_n$$
which implies by (11) that $h(p) \notin U_n$. Since $h(x) = y$, this implies that $h$ is one-to-one. In addition, if $\varrho(p, y) > 2^{-n}$ then $p \notin U_n$ and therefore by (4) if
$$q = h_1^{-1} \circ \cdots \circ h_n^{-1}(p)$$
then $h(q) = p$. We conclude that $h$ is surjective. By (3) and (6), $h$ is continuous (Lemma A.3.1). It is clear that $h$ is open at all points but $x$. We therefore only check openness of $h$ at $x$. This will give us that $h$ is a homeomorphism. To this end, let $V$ be any neighborhood of $x$. Let $n \geq 0$ be so large that $V_n \subseteq V$. Then clearly $y = h(x) \in U_n = h[V_n] \subseteq h[V]$. We therefore conclude that $h[V]$ is a neighborhood of $y$, which implies that $h[V]$ is open. □

Lemma 1.9.1 will be used among other things in Exercises 1.9.1 and 1.9.2 for the construction of a partition of $\mathbb{R}$ into two homeomorphic homogeneous parts.

## 1.9. HOMOGENEOUS ZERO-DIMENSIONAL SPACES

**Corollary 1.9.2.** *Let $X$ be zero-dimensional. The following statements are equivalent:*

(1) *$X$ is homogeneous.*
(2) *if $x, y \in X$ then $x$ and $y$ have arbitrarily small homeomorphic clopen neighborhoods.*

**Strongly homogeneous zero-dimensional spaces.** Let $X$ be zero-dimensional. We say that $X$ is *strongly homogeneous* provided that all nonempty clopen subsets of $X$ are homeomorphic (and hence are homeomorphic to $X$). By Corollary 1.9.2 it follows that a strongly homogeneous (zero-dimensional) space is homogeneous.

**Corollary 1.9.3.** *A strongly homogeneous (zero-dimensional) space is homogeneous.*

A homogeneous zero-dimensional space need not be strongly homogeneous (VAN DOUWEN [128]), but at least the following weaker statement holds.

**Lemma 1.9.4.** *A homogeneous zero-dimensional space is strongly locally homogeneous.*

**Proof.** Let $x \in X$ and let $U$ be an arbitrary clopen neighborhood of $x$. In addition, pick an arbitrary $y \in U \setminus \{x\}$. We claim that there is a homeomorphism $h \colon X \to X$ sending $x$ onto $y$ and which moreover restricts to the identity on the complement of $U$. Since $X$ is homogeneous, $x$ and $y$ have arbitrary small clopen neighborhoods which are homeomorphic by homeomorphisms sending $x$ onto $y$. So we can find disjoint clopen neighborhoods $E$ and $F$ of $x$ and $y$, respectively, and a homeomorphism $f \colon E \to F$ such that

(1) $E \cup F \subseteq U$,
(2) $f(x) = y$.

Now define $h \colon X \to X$ by

$$h(p) = \begin{cases} p & (p \notin E \cup F), \\ f(p) & (p \in E), \\ f^{-1}(p) & (p \in F). \end{cases}$$

Then $h$ is clearly as required. □

**Lemma 1.9.5.** *The Cantor set $\mathsf{C}$ is strongly homogeneous and strongly locally homogeneous.*

**Proof.** This is clear from the topological characterization of $\mathsf{C}$ (see Theorem 1.5.5) since every nonempty clopen subspace of $\mathsf{C}$ is compact, zero-dimensional and has no isolated points. □

**Topological characterization of $\mathbb{Q}$ and $\mathbb{P}$.** We now apply the results obtained so far to get interesting topological characterizations of various classical zero-dimensional spaces.

**Theorem 1.9.6.** *The space of all rational numbers $\mathbb{Q}$ is topologically the unique nonempty countable space without isolated points.*

**Proof.** It is clear that $\mathbb{Q}$ is countable, nonempty, and has no isolated points. It therefore suffices to prove that all nonempty countable spaces without isolated points are homeomorphic. To this end, let $X$ and $Y$ be such spaces. By Exercise 1.5.1, both $X$ and $Y$ are zero-dimensional. They consequently have zero-dimensional compactifications $aX$ and $aY$ by Exercise 1.5.6. Both $aX$ and $aY$ clearly have no isolated points because $X$ and $Y$ both have no isolated points. So $aX \approx aY \approx \mathsf{C}$ by Theorem 1.5.5. Since $\mathsf{C}$ is strongly locally homogeneous (Lemma 1.9.5), it is countable dense homogeneous by Theorem 1.6.9. So there is a homeomorphism $f \colon aX \to aY$ with $f[X] = Y$. In particular, $X \approx Y$. □

**Remark 1.9.7.** Our next result is a topological characterization of the space of irrational numbers $\mathbb{P}$. There is a proof of this characterization along the same lines as the proof of Theorem 1.9.6. But it is simpler by using the completeness of $\mathbb{P}$ to follow the method in the proof of Theorem 1.5.5.

**Theorem 1.9.8.** *The space of all irrational numbers $\mathbb{P}$ is topologically the unique nonempty, topologically complete, nowhere locally compact and zero-dimensional space.*

**Proof.** That $\mathbb{P}$ is topologically complete, zero-dimensional and nowhere locally compact, follows easily from Remark A.6.5, Proposition 1.5.3 and Exercise 1.5.15.

**Claim 1.** Every nonempty open subset of a nowhere locally compact zero-dimensional space can be decomposed into infinitely many pairwise disjoint nonempty clopen sets.

*Proof.* Let $X$ be nowhere locally compact and zero-dimensional. Let $U \subseteq X$ be nonempty and open. The proof of Proposition 1.5.1 shows that we can decompose $U$ into nonempty clopen subsets of $X$ (possibly, a finite number). So without loss of generality we may assume that $U$ is clopen. Since $X$ is nowhere locally compact (Exercise 1.5.15), $U$ is not compact. It consequently has an open cover $\mathcal{U}$ without finite subcover. The proof of Proposition 1.5.1 shows that this cover can be refined by a clopen partition (here clopen means clopen in $X$). This partition is necessarily infinite since if it were finite then $\mathcal{U}$ would have a finite subcover. ◇

Observe that in the proof of Theorem 1.5.5 we proved that if $X$ and $Y$ are zero-dimensional compact spaces without isolated points then they have isomorphic trees of clopen sets which form bases of $X$ and $Y$, respectively. A path $P$ through these trees has a unique point as its interesting because of compactness and the fact that the elements of $P$ have smaller and smaller diameters. The trees are finitely branching, i.e., each node has finitely many successors only.

With topologically complete, zero-dimensional nowhere locally compact spaces similar trees of clopen sets can be constructed (their construction is easier than in the case of Cantor sets). By applying Claim 1 we can ensure that all levels of the trees are infinite, that they have countable height and that they are infinitely branching, i.e., each node has infinitely many (hence $\omega$ many) successors. So the trees are obviously isomorphic. By using complete metrics, one can again ensure that for each path $P$ the intersection of its elements is a single point. So the proof can then be completed such as in the proof of Theorem 1.5.5. □

**Corollary 1.9.9.** $\mathbb{P}$ and $\mathbb{N}^\infty$ are homeomorphic.

**Proof.** Use Exercise 1.5.16 to conclude that $\mathbb{N}^\infty$ is nowhere locally compact. Zero-dimensionality and completeness follow from Proposition 1.5.3, Exercise 1.5.2, Remark A.6.5 and Lemma A.6.2. So we are done by Theorem 1.9.8. □

**Corollary 1.9.10.** If $X$ is topologically complete then it is a continuous image of $\mathbb{P}$.

**Proof.** By Theorem A.6.3, $X$ is homeomorphic to a closed subspace of $\mathbb{R}^\infty$.

**Claim 1.** $\mathbb{R}^\infty$ is a continuous image of $\mathbb{P}$.

*Proof.* Since $\mathbb{P}^\infty \approx (\mathbb{N}^\infty)^\infty \approx \mathbb{N}^\infty \approx \mathbb{P}$ (Corollary 1.9.9), it suffices to prove that $\mathbb{R}$ is a continuous image of $\mathbb{P}$. Consider the zero-dimensional product space $Y = \mathbb{N} \times \mathsf{C}$. Since $\{n\} \times \mathsf{C}$ maps onto $[-n, n]$ by Theorem 1.5.10, $Y$ maps onto $\mathbb{R}$. But $\mathbb{P} \times Y$ satisfies the conditions mentioned in Theorem 1.9.8 (cf. the proofs of Theorem 1.5.10 and Corollary 1.9.9) and hence is homeomorphic to $\mathbb{P}$. So we conclude that $\mathbb{P}$ maps onto $Y$ and so in turn maps onto $\mathbb{R}$. ◇

So we are done. □

This yields the following characterization of analytic spaces.

**Corollary 1.9.11.** Let $X$ be a space. The following statements are equivalent:

(1) $X$ is analytic,
(2) $X$ is a continuous image of $\mathbb{P}$.

**A Hurewicz-type theorem.** The space $\mathbb{Q}$ is clearly not a Baire space. It is in fact a 'test space' for the Baire property.

**Theorem 1.9.12.** *Let $X$ be a space. If $X$ is not a Baire space then $X$ contains a closed subspace homeomorphic to $\mathbb{Q}$.*

**Proof.** Let $\mathcal{G}$ be a countable collection of dense open subsets of $X$ such that $\bigcap \mathcal{G}$ is not dense. There exists a nonempty open subset $V \subseteq X$ such that
$$\overline{V} \cap \bigcap \mathcal{G} = \emptyset.$$
Then $\overline{V}$ contains no isolated points. For if $x$ is an isolated point of $\overline{V}$ then clearly $x \in V$ and so, $V$ being open in $X$, $x$ is an isolated point of $X$. But then $x$ belongs to every dense subset of $X$, i.e.,
$$x \in \overline{V} \cap \bigcap \mathcal{G} = \emptyset.$$

So by Theorem 1.9.6 it suffices to prove the following:

**Claim 1.** *Let $Y$ be a space having no isolated points. If $\mathcal{G}$ is a countable collection of dense open sets in $Y$ then $Y$ has a countable subspace $K$ having no isolated points such that $\overline{K} \setminus K \subseteq \bigcap \mathcal{G}$.*

*Proof.* For $y \in Y$, if $\mathcal{A}$ is a collection of subsets of $Y$ we say that $\mathcal{A}$ *converges to $y$* if every neighborhood of $y$ contains all but finitely many $A \in \mathcal{A}$. It is easy to see that if $y \in Y$ and $U$ is a neighborhood of $y$ then there is a pairwise disjoint infinite collection $\mathcal{V}$ of nonempty open subsets of $U \setminus \{y\}$ such that $\mathcal{V}$ converges to $y$ and $\overline{\bigcup \mathcal{V}} \subseteq U$ (Exercise 1.9.4).

Enumerate $\mathcal{G}$ as $\{G_n : n \in \mathbb{N}\}$. We will construct a sequence
$$\{\mathcal{U}_n : n \in \mathbb{N}\}$$
of pairwise disjoint collections of nonempty open sets in $Y$ and a sequence
$$\{K_n : n \in \mathbb{N}\}$$
of countable subsets of $Y$, as follows.

Let $\mathcal{U}_1 = \{Y\}$. At stage $n+1$, where $n \geq 0$, for each $U \in \mathcal{U}_n$, pick an arbitrary element $k_n(U) \in U \cap G_n$ and choose an infinite pairwise disjoint collection $\mathcal{V}_n(U)$ of nonempty open subsets of $Y \setminus \{k_n(U)\}$ that converges to $k_n(U)$, and satisfies

(1) $\overline{\bigcup \mathcal{V}_n(U)} \subseteq U \cap G_n.$

Let
$$\mathcal{U}_{n+1} = \bigcup \{\mathcal{V}_n(U) : U \in \mathcal{U}_n\}$$
and
$$K_n = \{k_n(U) : U \in \mathcal{U}_n\}.$$
Note that

(2) $K_n \subseteq \bigcup \mathcal{U}_n$, $\bigcup \mathcal{U}_{n+1} \subseteq \bigcup \mathcal{U}_n \setminus K_n$ and $\bigcup \mathcal{U}_{n+1} \subseteq G_n$.

This completes the construction of the desired sequences.

Put $K = \bigcup_{n=1}^{\infty} K_n$. Then $K$ is clearly countable. It remains to show that $K$ has no isolated points and has the property that $\overline{K} \setminus K \subseteq \bigcap \mathcal{G}$.

For each $n$ and $U \in \mathcal{U}_n$ the subset $\{k_{n+1}(V) : V \in \mathcal{V}_{n+1}(U)\}$ of $K_{n+1}$ converges to $k_n(U)$ since $\mathcal{V}_n(U)$ converges to $k_n(U)$, and it does not contain $k_n(U)$ since $k_n(U) \notin K_{n+1}$. So $k_n(U)$ is not an isolated point of $K$.

We will next show that $\overline{K} \setminus K \subseteq \bigcap \mathcal{G}$. It follows from (2) that

(3) For all $j \in \mathbb{N}$ and $U \in \mathcal{U}_j$:
$$U \cap K \subseteq \{k_j(U)\} \cup \bigcup \mathcal{V}_j(U).$$

To see this, consider any $j, s \in \mathbb{N}$ and $U \in \mathcal{U}_j$. If $s > j$ then
$$K_s \subseteq \bigcup \mathcal{U}_s \subseteq \bigcup \mathcal{U}_{j+1}$$
and so $K_s \cap U \subseteq \bigcup\{V \in \mathcal{U}_{j+1} : V \subseteq U\} = \bigcup \mathcal{V}_j(U)$. Moreover, if $s = j$ then
$$k_j(U) \in K_s \cap U = \{k_j(V) : V \cap U \neq \emptyset\} = \{k_j(U)\}.$$
Finally, if $s < j$ then $K_s \cap U \subseteq K_s \cap \bigcup \mathcal{U}_{j+1} \subseteq K_s \cap \bigcup \mathcal{U}_{s+1} = \emptyset$.

The reason why the proof works is that by (1) and (3) we get:

(4) For every $j \in \mathbb{N}$ and $U \in \mathcal{U}_j$ we have $\overline{K \cap U} \subseteq U$.

Now consider any $x \in \overline{K} \setminus K$. We claim that for every $j$ there is an element $U_j \in \mathcal{U}_j$ which contains $x$. Since $\bigcup \mathcal{U}_j \subseteq G_j$ this implies that
$$x \in \bigcap_{j=1}^{\infty} G_j,$$
which is as required. Let $U_1 = Y$. Next, consider any $j \in \mathbb{N}$ and assume that we already found $U_j$. Since $x \in \overline{K}$ but $x \neq k_j(U_j)$ and $\mathcal{V}_j(U_j)$ converges to $k_j(U_j)$ we conclude by (3) that there is an element $U_{j+1} \in \mathcal{V}_j(U_j)$ with
$$x \in \overline{U_{j+1} \cap K}.$$
Then $x \in U_{j+1}$ because of (4). This completes the construction of the sequence $\{U_n : n \in \mathbb{N}\}$. ◇

So we are done. ☐

**Corollary 1.9.13.** *Let $X$ be a space. Then every closed subspace of $X$ is a Baire space if and only if $X$ has no closed subspace homeomorphic to $\mathbb{Q}$.*

**Proof.** Necessity is clear. To prove the sufficiency, it is enough to prove that if $X$ is not a Baire space then it contains a closed homeomorph of $\mathbb{Q}$. But this is the statement of Theorem 1.9.12. ☐

**Exercises for §1.9.**

1. Let $A$ be a subset of $\mathbb{R}$ such that $A+\mathbb{Q} = A$. Prove that $A$ is homogeneous.

2. Let
$$\mathcal{A} = \Big\{ A \subseteq \mathbb{R} : \mathbb{Q} \subseteq A, A + \mathbb{Q} = A, \ A \cap \bigcup_{n \in \mathbb{Z}\setminus\{0\}} A + n\pi = \emptyset \Big\}.$$

   (1) Prove that $\mathcal{A} \neq \emptyset$.
   (2) Let $\mathcal{K}$ be a chain (under inclusion) of elements of $\mathcal{A}$. Prove that $\bigcup \mathcal{K} \in \mathcal{A}$.
   (3) Apply the Kuratowski-Zorn Lemma to conclude that $\mathcal{A}$ contains a maximal element, say $A_0$.
   (4) Prove that if $x \in \mathbb{R} \setminus A_0$ then there exist $n \in \mathbb{Z} \setminus \{0\}$ and $a \in A_0$ such that $x = a + n\pi$.
   (5) For each $n \in \mathbb{Z} \setminus \{0\}$ let
   $$A_n = \{x \in \mathbb{R} \setminus A_0 : (\exists a \in A_0)(x = a + n\pi)\}.$$
   Prove that the collection $\{A_n : n \in \mathbb{Z}\}$ partitions $\mathbb{R}$.
   (6) Prove that for every $n \in \mathbb{Z}$ we have $A_n = A_n + \mathbb{Q}$.
   (7) Prove that $\mathbb{R}$ can be decomposed into two homeomorphic homogeneous sets.

3. (1) Prove that there is a Borel subset $E \subseteq \mathbb{R}$ such that $E \approx \mathbb{R} \setminus E$.
   (2) Prove that if $E$ is a homogeneous subset of $\mathbb{R}$ such that $E \approx \mathbb{R} \setminus E$ then $E$ is not Borel.

4. Let $Y$ be a space without isolated points. Prove that if $y \in Y$ and $U$ is a neighborhood of $y$ then there is a pairwise disjoint infinite collection $\mathcal{V}$ of nonempty open subsets of $U \setminus \{y\}$ such that $\mathcal{V}$ converges to $y$ and $\overline{\bigcup \mathcal{V}} \subseteq U$.

## 1.10. Inverse limits

Sometimes it is possible to 'approximate' a complicated space 'arbitrarily closely' by less complicated objects. In this section we shall formally define what we mean by this and we shall derive an interesting 'approximation theorem'. On Page 137 we will discuss Freudenthal's Approximation Theorem about approximations of spaces by polyhedra.

An *inverse sequence* is a sequence of pairs $(X_n, f_n)_n$ of spaces $X_n$ and continuous functions $f_n \colon X_{n+1} \to X_n$. The spaces $X_n$ are called *coordinate spaces* and the mappings $f_n$ are called *bonding maps*. The *inverse limit* of the inverse sequence $(X_n, f_n)_n$, denoted by
$$\varprojlim (X_n, f_n)_n,$$
is defined to be the following subspace of the product of the $X_n$:
$$\Big\{ x \in \prod_{n=1}^{\infty} X_n : (\forall n \in \mathbb{N})(f_n(x_{n+1}) = x_n) \Big\}.$$

It will sometimes be convenient to let $X_\infty$ denote the inverse limit of the sequence $(X_n, f_n)_n$.

For every $n \in \mathbb{N}$, the restriction to $\varprojlim(X_n, f_n)_n$ of the projection onto the $n$-th factor of the product $\prod_{n=1}^\infty X_n$ shall be denoted by $f_n^\infty$. By definition it follows that for every $n$, $f_n \circ f_{n+1}^\infty = f_n^\infty$. The functions $f_n^\infty$ are called the *projections* of the inverse sequence. In addition, for $m > n$ the composition

$$f_m \circ \cdots \circ f_n \colon X_{m+1} \to X_n$$

shall be denoted by $f_n^{m+1}$. Observe that $f_n^{n+1} = f_n$. Finally, $f_m^m$ is defined to be the identity function from $X_m$ to $X_m$.

**Lemma 1.10.1.** *Let $(X_n, f_n)$ be an inverse sequence. Then the collection*

$$\{(f_n^\infty)^{-1}[U] : U \subseteq X_n \text{ open}, n \in \mathbb{N}\}$$

*is a base for the topology of $\varprojlim(X_n, f_n)_n$. In addition, this base is closed under finite unions and finite intersections.*

**Proof.** Let $U$ be a relatively open subset of $X_\infty$. In addition, let $x \in U$. By the definition of the product topology, there exist $N \in \mathbb{N}$ and for every $i \leq N$ an open subset $V_i \subseteq X_i$ such that

$$x \in \Big(\prod_{i=1}^N V_i \times \prod_{i=N+1}^\infty X_i\Big) \cap X_\infty \subseteq U.$$

Since $x_1 \in V_1$, $x_2 \in V_2$ and $f_1(x_2) = x_1$, it follows that

$$x_2 \in V_2 \cap f_1^{-1}[V_1].$$

Continuing in this way, we obtain that

$$x_N \in W = \bigcap_{i=1}^N (f_i^N)^{-1}[V_i].$$

An easy verification shows that

$$x \in (f_N^\infty)^{-1}[W] \subseteq U,$$

which is as required.

The second part of the lemma follows easily from the fact that

$$f_n^m \circ f_m^\infty = f_n^\infty$$

for all $m \geq n$. Simply observe that a basic open subset of $X_\infty$ that 'depends' on the $n$-th element of the inverse sequence, i.e., is of the form

$$(f_n^\infty)^{-1}[U]$$

for certain open $U \subseteq X$, also 'depends' on the $m$-th element of the sequence for every $m \geq n$. This is true since for every $m \geq n$ and $U \subseteq X_n$ we have

$$(f_n^\infty)^{-1}[U] = (f_m^\infty)^{-1}\big[(f_n^m)^{-1}[U]\big].$$

So for a finite collection of basic open sets $\mathcal{U}$ in $X_\infty$ it is possible to identify an index $N \in \mathbb{N}$ such that every $U \in \mathcal{U}$ 'depends' on $X_N$. □

We will now present a few illuminating examples of inverse sequences and their limits.

**Example 1.10.2.** Consider the inverse sequence $(X_n, f_n)_n$, where $X_n = X$ and $f_n$ is the identity on $X$ for every $n$. This is a trivial inverse sequence. A moments reflection shows that its inverse limit is (homeomorphic to) $X$.

The next example shows that inverse limits are generalizations of products (see the exercises for details).

**Example 1.10.3.** For every $n$ let $X_n$ be a space. The sequence

$$\left( \prod_{m=1}^{n} X_m, \tau_n \right)_n,$$

where $\tau_n \colon \prod_{m=1}^{n+1} X_m \to \prod_{m=1}^{n} X_m$ is the projection, is an inverse sequence with the infinite product

$$\prod_{n=1}^{\infty} X_n$$

as its inverse limit.

**Example 1.10.4.** Let $X$ be a space and for every $n$ let $X_n$ be a subspace of $X$ such that $X_{n+1} \subseteq X_n$. Form an inverse sequence $(X_n, i_n)_n$ by letting $i_n$ be the inclusion $X_{n+1} \hookrightarrow X_n$. Then $\varprojlim(X_n, i_n)_n$ is homeomorphic to $\bigcap_{n=1}^{\infty} X_n$.

From this example we see that an inverse limit of nonempty spaces can be empty. Since the intersection of a decreasing sequence of nonempty compact spaces is nonempty, for compact spaces one cannot get such an example (see Lemma 1.10.10(2) below).

**Example 1.10.5.** Let $X$ be a compact space. By Corollary A.4.4 we may assume that $X$ is a subspace of $Q$. For every $n$ let

$$X_n = \tau_n[X],$$

where $\tau_n \colon Q \to \mathbb{J}^n$ is the projection. In addition, let $f_n \colon X_{n+1} \to X_n$ be defined by

$$f_n(x_1, \ldots, x_{n+1}) = (x_1, \ldots, x_n).$$

A trivial verification shows that $\varprojlim(X_n, f_n)_n = X$.

There are several simple but useful observations that we have to make about inverse limits. First, let $(X_n, f_n)_n$ be an inverse system, and let $A \subseteq \mathbb{N}$ be infinite. Write $A$ as $\{m_1, \ldots, m_n, \ldots\}$ in such a way that

$$m_1 < m_2 < \cdots < m_n < \cdots.$$

The sequence $(X_{m_i}, f_{m_i}^{m_{i+1}})_i$ is also an inverse sequence. By abuse of notation it will be denoted by $(X_n, f_n)_{n \in A}$. It is called an *inverse subsequence* of $(X_n, f_n)_n$.

**Lemma 1.10.6.** *If $(X_n, f_n)_{n \in A}$ is an inverse subsequence of $(X_n, f_n)_n$ then the limits $\varprojlim(X_n, f_n)_{n \in A}$ and $\varprojlim(X_n, f_n)_n$ are homeomorphic.*

**Proof.** The function $\varphi \colon \varprojlim(X_n, f_n)_n \to \varprojlim(X_n, f_n)_{n \in A}$ defined by

$$\varphi\bigl((x_n)_n\bigr) = (x_n)_{n \in A}$$

is easily seen to be a homeomorphism. □

So passing to a subsequence makes no difference in the limit (as with ordinary limits).

**Lemma 1.10.7.** *Let $(X_n, f_n)_n$ and $(Y_n, g_n)_n$ be inverse sequences. Suppose that for every $n \in \mathbb{N}$ there is a continuous function $\tau_n \colon X_n \to Y_n$ such that the diagram*

$$\begin{array}{ccc} X_n & \xleftarrow{f_n} & X_{n+1} \\ {\scriptstyle \tau_n}\downarrow & & \downarrow{\scriptstyle \tau_{n+1}} \\ Y_n & \xleftarrow{g_n} & Y_{n+1} \end{array}$$

*commutes. Then the function $\varphi \colon \varprojlim(X_n, f_n)_n \to \varprojlim(Y_n, g_n)_n$ defined by*

$$\varphi\bigl((x_n)_n\bigr) = \bigl(\tau_n(x_n)\bigr)_n$$

*is continuous.*

**Proof.** This is clear since $\varphi$ is the restriction to $\varprojlim(X_n, f_n)_n$ of a product map from $\prod_{n=1}^\infty X_n \to \prod_{n=1}^\infty Y_n$ (Exercise A.1.13). □

These simple results will sometimes be used in the following situation. Suppose that we are inductively defining an inverse sequence $(X_n, f_n)_n$ and that we would like to construct it in such a way that $\varprojlim(X_n, f_n)_n$ admits a selfmap $\varphi$ with certain special properties. This is sometimes achieved by constructing for every $n$ a function $\varphi_n \colon X_{n+1} \to X_n$ such that the diagram

$$\begin{array}{ccc} X_{n+1} & \xleftarrow{f_{n+1}} & X_{n+2} \\ {\scriptstyle \varphi_n}\downarrow & & \downarrow{\scriptstyle \varphi_{n+1}} \\ X_n & \xleftarrow{f_n} & X_{n+1} \end{array}$$

commutes. By Lemma 1.10.6 it follows that the inverse sequences $(X_n, f_n)_n$ and $(X_n, f_n)_{n \geq 2}$ have the same limit, and Lemma 1.10.7 shows that the functions $\varphi_n$ induce a canonical selfmap of that limit.

**Lemma 1.10.8.** *Let $(X_n, f_n)_n$ be an inverse sequence with surjective bonding maps. Then the projections of $(X_n, f_n)_n$ are also surjective.*

**Proof.** Fix $n \in \mathbb{N}$ and take an arbitrary $x \in X_n$. Since $f_n$ is surjective, there exists $x_{n+1} \in X_{n+1}$ with $f_n(x_{n+1}) = x$. Similarly, there exists $x_{n+2} \in X_{n+2}$ with $f_{n+1}(x_{n+2}) = x_{n+1}$, etc. Define
$$x_n = x, \quad x_{n-1} = f_{n-1}(x_n), \quad x_{n-2} = f_{n-2}(x_{n-1}),$$
etc. The sequence $(x_n)_n \in \prod_{n=1}^{\infty} X_n$ belongs to $\varprojlim(X_n, f_n)_n$ and since
$$f_n^{\infty}((x_n)_n) = x_n = x,$$
we are done. □

**Remark 1.10.9.** Let $(X_n, f_n)_n$ be an inverse sequence. For every $i$ define the subspace $X(i)$ of $\prod_{n=1}^{\infty} X_n$ by
$$X(i) = \left\{ x \in \prod_{n=1}^{\infty} X_n : (\forall n < i)(f_n(x_{n+1}) = x_n) \right\}.$$
It is clear that $X(i)$ is closed in $\prod_{n=1}^{\infty} X_n$ and nonempty and that
$$X(i+1) \subseteq X(i)$$
for every $i$, and, finally, that
$$\bigcap_{i=1}^{\infty} X(i) = \varprojlim(X_n, f_n)_n.$$
So $X_\infty$ is a closed subspace of $\prod_{n=1}^{\infty} X_n$.

This shows that if $\mathcal{P}$ is a topological property which is 'closed hereditary and (countably) productive' then $X_\infty$ has $\mathcal{P}$ if all $X_n$ have $\mathcal{P}$.

This simple observation allows us to prove many properties of inverse limits.

**Compactness and completeness properties of inverse limits.** It is natural to ask when an inverse limit is compact or topologically complete.

**Lemma 1.10.10.**

(1) *The inverse limit of an inverse sequence of topologically complete spaces is topologically complete.*
(2) *The inverse limit of an inverse sequence of nonempty compact spaces is nonempty and compact.*

**Proof.** Observe that compactness and completeness are closed hereditary and productive (Lemma A.6.2). So we are done by Remark 1.10.9. □

## 1.10. INVERSE LIMITS

**Connectivity properties of inverse limits.** Detecting connectivity in topological spaces is important. For products this is easy: a product is connected if and only if all factors are. But inverse limits do not share this property with products, as the following example shows.

**Example 1.10.11.** *A disconnected inverse limit of connected spaces.*
Let $\{q_n : n \in \mathbb{N}\}$ be an enumeration of the rational numbers in $\mathbb{I}$. For every $n \in \mathbb{N}$ define
$$X_n = (\mathbb{I} \times \{0,1\}) \cup \bigcup_{m=n}^{\infty} (\{q_m\} \times \mathbb{I}).$$
Then $X_1 \supseteq X_2 \supseteq \cdots \supseteq X_n \supseteq \cdots$, every $X_n$ is connected and
$$\bigcap_{n=1}^{\infty} X_n = \mathbb{I} \times \{0,1\}$$
is disconnected. So we are done by Example 1.10.4.

Observe that the inverse limit in this example is compact. So compactness of the inverse limit is not enough to guarantee connectivity. If all spaces involved are continua, then there are no problems.

**Lemma 1.10.12.** *The inverse limit of an inverse sequence of continua is a continuum.*

**Proof.** Let $(X_n, f_n)_n$ be an inverse sequence of continua. Let the spaces $X(i)$ for every $i$ be defined as in Remark 1.10.9. We claim that every $X(i)$ is a continuum. Indeed, fix $i$ and define a function $\varphi \colon \prod_{j=i}^{\infty} X_j \to X(i)$ by
$$\varphi\big((x_i, x_{i+1}, \ldots)\big) = \big(f_1^i(x_i), f_2^i(x_i), \ldots, f_i^i(x_i), x_{i+1}, \ldots\big).$$
Then $\varphi$ is evidently a continuous surjection. It therefore follows that $X(i)$ is a continuum since $\prod_{j=i}^{\infty} X_j$ is.

It now suffices to apply Proposition A.10.7. □

**The dyadic solenoid.** As an example of the power of inverse limits, we construct a very peculiar space here.

Consider the following inverse system
$$\mathbb{S}^1 \xleftarrow{f} \mathbb{S}^1 \xleftarrow{f} \cdots \xleftarrow{f} \mathbb{S}^1 \xleftarrow{f} \cdots,$$
where $f \colon \mathbb{S}^1 \to \mathbb{S}^1$ is the function $f(z) = z^2$ (complex multiplication). The inverse limit of this system is denoted by $\Sigma_2$ and is called the *dyadic solenoid*. We list a few properties of $\Sigma_2$ that we can derive easily from previous results. First, it is a continuum by Lemma 1.10.12. In addition, since $\mathbb{S}^1$ is a topological group under complex multiplication, and $f$ is a continuous homomorphism, it follows easily that $\Sigma_2$ is a closed subgroup of the product group $(\mathbb{S}^1)^{\infty}$. So $\Sigma_2$ is a topological group and hence a homogeneous space.

A continuum $X$ is called *indecomposable* if it is not the union of two proper subcontinua. So $\mathbb{I}$ is not indecomposable (= *decomposable*) since it is the union of $[0, 1/2]$ and $[1/2, 1]$.

**Lemma 1.10.13.** *Let $X$ be a continuum. The following statements are equivalent:*

(1) *$X$ is indecomposable,*
(2) *every proper subcontinuum of $X$ is nowhere dense.*

**Proof.** That (2) $\Rightarrow$ (1) is clear since no space is the union of two nowhere dense subsets.

For (1) $\Rightarrow$ (2), suppose that $X$ contains a proper subcontinuum $C$ with nonempty interior $U$. Let $\mathcal{E}$ be the collection of components of $Y = X \setminus U$. Then every $E \in \mathcal{E}$ meets $\overline{U} \setminus U$ by Corollary A.10.5. So every $E \in \mathcal{E}$ meets $C$ but misses $U$. If $|\mathcal{E}| = 1$ then clearly $X$ is the union of two proper subcontinua. Suppose therefore that $\mathcal{E}$ has more than one element. Then $Y$ contains a proper nonempty relatively clopen subspace $W$ by Proposition A.10.3(1). Every member from $\mathcal{E}$ either is contained in $W$ or is disjoint from it. So $W$ is the union of a family of subcontinua of $Y$ which all meet $C$. So $C \cup W$ is a proper subcontinuum of $X$. Similarly, $C \cup (Y \setminus W)$ is a proper subcontinuum of $X$. Since they cover $X$, we arrive at a contradiction. $\square$

**Corollary 1.10.14.** *A nondegenerate locally connected continuum is decomposable.*

**Proof.** Let $x$ and $y$ be distinct points in $X$ and let $U$ be a connected neighborhood of $x$ whose closure misses $y$. Then $\overline{U}$ is a proper subcontinuum of $X$ which is not nowhere dense. So $X$ is decomposable by Lemma 1.10.13. $\square$

**Theorem 1.10.15.** *The dyadic solenoid is indecomposable.*

**Proof.** Let $C \subseteq \Sigma_2$ be a proper subcontinuum. For every $n$ put $C_n = f_n^\infty[C]$. Since $C$ is proper there exists $n \in \mathbb{N}$ such that $C_n$ is a proper subset of $\mathbb{S}^1$. Observe that $C_n$ is a subinterval of $\mathbb{S}^1$. Now $f^{-1}[C_n]$ is the union of two pairwise disjoint nonempty subintervals of $\mathbb{S}^1$ of arc-length exactly one half the arc-length of $C_n$. In addition, it contains $C_{n+1}$. By connectivity, $C_{n+1}$ is contained in one of them. We conclude that the arc-length of $C_{n+1}$ is less than or equal to half of the arc-length of $C_n$. So for large $N$ the arc-length of $C_N$ is arbitrarily small.

Now assume that $C$ has nonempty interior. Then there are $i \in \mathbb{N}$ and a nonempty open interval $U \subseteq \mathbb{S}^1$ such that $(f_i^\infty)^{-1}[U] \subseteq C$. But the arc-length of $f^{-1}[U]$ is the same as the arc-length of $U$, etc. So the arc-length of $C_N$ does not tend to 0 for large $N$. This is a contradiction.

So we conclude that $C$ has empty interior. Since $C$ was arbitrary, we conclude that $\Sigma_2$ is indecomposable by Lemma 1.10.13. □

**Remark 1.10.16.** It is easy to show that the dyadic solenoid contains an arc. For every $n$ put
$$\mathbb{I}_n = \{e^{i\pi t} : 0 \leq t \leq 2^{-n}\}.$$
Then $g_n = f \restriction \mathbb{I}_{n+1} \to \mathbb{I}_n$ is a homeomorphism. The inverse limit of the inverse sequence $(\mathbb{I}_n, g_n)_n$ is therefore an arc (Exercise 1.10.6). In addition, this inverse limit is a subspace of $\Sigma_2$ by Exercise 1.10.7. So $\Sigma_2$ indeed contains an arc. This shows that although $\Sigma_2$ is indecomposable, it contains a decomposable proper subcontinuum. This motivates the following definition. A continuum $X$ is called *hereditarily indecomposable* if every subcontinuum containing more than one point is indecomposable. It is not clear at all that such continua exist. We will come back to this in §3.8, where it will be shown that there are in fact 'many' hereditarily indecomposable continua.

Let $X$ be a continuum, and let $x \in X$. The *composant* of $x$ in $X$ is the union of all proper subcontinua of $X$ containing $x$.

**Lemma 1.10.17.** *Let $X$ be an indecomposable continuum. Then every composant of $X$ is a meager $F_\sigma$-subset of $X$.*

**Proof.** Let $x \in X$ and let $\mathcal{B}$ be a countable base for the space $X \setminus \{x\}$. We may assume without loss of generality that every $B \in \mathcal{B}$ is nonempty.

For every $B \in \mathcal{B}$ let $C(B)$ be the component of $x$ in $X \setminus B$. Then $C(B)$ is clearly a proper subcontinuum of $X$ which contains $x$. We claim that
$$C = \bigcup_{B \in \mathcal{B}} C(B)$$
is equal to the composant $E$ of $x$. It is clear that $C$ is a subset of $E$. Now if $y \in E$ then there is a proper subcontinuum $K$ of $X$ such that both $x$ and $y$ belong to $K$. Since $K$ is proper, there is an element $B \in \mathcal{B}$ which misses $K$. But then $K \subseteq C(B)$ and hence $y \in C(B) \subseteq C$. So $E$ is the union of countably many proper subcontinua of $X$. An application of Lemma 1.10.13 therefore yields the desired result. □

**A continuum $Z$ with $Z^2 \approx Z$ but $Z^\infty \not\approx Z$.** As a second example of the power of inverse limits, we construct another very peculiar space here.

Most of the known examples of compact spaces $X$ such that $X^2 \approx X$ (which implies that $X^n \approx X$ for every $n$) have the property that $X^\infty \approx X$. Think of the Cantor set and the Hilbert cube as examples of such spaces. We will use inverse limits to demonstrate that this is not always the case.

**Theorem 1.10.18.** *Let $Y$ be a locally compact space homeomorphic to its own square. Then $Y$ has a compactification homeomorphic to its own square.*

Moreover, if the one-point compactification of $Y$ is a continuum then this compactification can be chosen to be a continuum as well.

**Proof.** The compactification $X$ of $Y$ is the inverse limit of a suitable inverse sequence $(a_n Y, f_n)_n$ of compactifications $a_n Y$ of $Y$, where for each $n$, $f_n$ is a continuous function from $a_{n+1} Y \to a_n Y$ which restricts to the identity on $Y$. This implies that $f[a_{n+1} Y \setminus Y] = a_n Y \setminus Y$ by Exercise A.4.2. In order to ensure that $X^2 \approx X$ we will construct the inverse sequence in such a way that for each $n \in \mathbb{N}$ there is a homeomorphism $h_n \colon a_{n+1} Y \to a_n Y \times a_n Y$ so that the diagram

$$a_1 Y \xleftarrow{f_1} a_2 Y \xleftarrow{f_2} a_3 Y \xleftarrow{\phantom{f_3}} \cdots$$
$$\phantom{a_1 Y}\downarrow h_1 \phantom{a_2 Y \xleftarrow{f_2} a_3 Y} \downarrow h_2$$
$$a_1 Y \times a_1 Y \xleftarrow{f_1 \times f_1} a_2 Y \times a_2 Y \xleftarrow{\phantom{f_2}} \cdots$$

commutes. Let $a_1 Y$ be the one-point compactification of $Y$ and let $h$ be any homeomorphism from $Y$ onto $Y^2$. There is a compactification $a_2 Y$ of $Y$ such that $h$ extends to a homeomorphism $h_1 \colon a_2 Y \to a_1 Y \times a_1 Y$. Since $a_1 Y$ is the smallest compactification of $Y$, the identity mapping on $Y$ extends to a mapping $f_1 \colon a_2 Y \to a_1 Y$. Similarly, there is a compactification $a_3 Y$ of $Y$ such that $h$ extends to a homeomorphism $h_2 \colon a_3 Y \to a_2 Y \times a_2 Y$. Define

$$f_2 \colon a_3 Y \to a_2 Y$$

by

$$f_2 = h_1^{-1} \circ (f_1 \times f_1) \circ h_2.$$

It is easily seen that $f_2 \upharpoonright Y$ is the identity on $Y$. In the same way define the compactification $a_4 Y$, and the functions $h_3$ and $f_3$, etc.

Since the $h_n$'s are homeomorphisms, it follows from Lemma 1.10.7 applied twice and Lemma 1.10.10(2) that $X = \varprojlim (a_n Y, f_n)_n$ is a compactification of $Y$ homeomorphic to its own square (cf. Exercise 1.10.9).

Observe that if $a_1 Y$ is a continuuum then so are all the other $a_n Y$'s. As a consequence, $X$ is a continuum as well by Lemma 1.10.12. $\square$

Now consider the Cantor set $\mathsf{C}$, and let $B$ be the topological sum of $\mathbb{N} \times \mathsf{C}$ and a one-point space. Clearly,

$$\begin{aligned} B^2 &= \big((\mathbb{N} \times \mathsf{C}) \oplus \{\mathrm{pt}\}\big) \times \big((\mathbb{N} \times \mathsf{C}) \oplus \{\mathrm{pt}\}\big) \\ &\approx (\mathbb{N}^2 \times \mathsf{C}^2) \oplus (\mathbb{N} \times \mathsf{C}) \oplus (\mathbb{N} \times \mathsf{C}) \oplus \{\mathrm{pt}\} \\ &\approx (\mathbb{N} \times \mathsf{C}) \oplus \{\mathrm{pt}\} \\ &\approx B. \end{aligned}$$

Put $Y = B \times [0,1) \times Q$.

**Lemma 1.10.19.** $Y^2 \approx Y$.

**Proof.** Simply observe that
$$Y^2 = (B \times [0,1) \times Q) \times (B \times [0,1) \times Q)$$
$$\approx (B \times B) \times ([0,1) \times [0,1)) \times (Q \times Q)$$
$$\approx B \times ([0,1) \times \mathbb{I}) \times Q$$
$$\approx B \times [0,1) \times Q$$
$$= Y.$$
So we are done. □

Let $Z$ be the compactification of $Y$ we get from Lemma 1.10.19. Observe that the one-point compactification of $Y$ is connected since it is the cone over $B \times Q$. From this it follows that $Z$ is a continuum as well.

**Lemma 1.10.20.** $Z \not\approx Z^\infty$.

**Proof.** Striving for a contradiction, suppose that $h$ is a homeomorphism from $Z$ onto $Z^\infty$. Since $Y$, being locally compact, is open in $Z$, $h[Y]$ is open in $Z^\infty$. In addition, since $B$ has an isolated point, $Y$ contains an open copy of $[0,1) \times Q$. Therefore, $Z^\infty$ contains an open copy of $[0,1) \times Q$, say $F$. Since $F$ is open and nonempty, it contains a basic open subset. So there are an integer $n$ and for each $i \leq n$ a nonempty open set $U_i \subseteq Z$ so that
$$U_1 \times U_2 \times \cdots \times U_n \times Z \times Z \times \cdots \subseteq F.$$
Hence $Z$ is (homeomorphic to) a retract of some open subset of $F$, or, equivalently, $Z$ is homeomorphic to a retract of some open subset of $[0,1) \times Q$. Since $[0,1) \times Q$ is locally connected, this implies that $Z$ is locally connected (Exercise A.12.4). Its open subspace $Y$ is therefore locally connected as well. But this implies that $C$ is locally connected, which is a contradiction. □

**Remark 1.10.21.** In view of the existence of a compact space homeomorphic to its own square but not to its countable infinite product, the question naturally arises whether there is a compact space $X$ which can be mapped onto $X^2$ but not onto $X^\infty$. Interestingly, the answer to this question is in the negative (see Exercise 1.10.8).

**Brown's Approximation Theorem.** Before we are in a position to formulate and prove an interesting 'approximation theorem' for inverse sequences, we have to fix some notation. Let $(X_n, f_n)_n$ be an inverse sequence with inverse limit $X_\infty$. It is sometimes useful to have an explicit admissible metric for $X_\infty$. For every $n \in \mathbb{N}$ let $d_n$ be an admissible metric for $X_n$ which is bounded by 1 (Exercise A.1.6). The formula
$$\varrho(x,y) = \sum_{n=1}^\infty 2^{-n} d_n(x_n, y_n)$$

defines an admissible metric for the product of the $X_n$ (Exercise A.1.10), and hence also for $X_\infty$. Observe that $\varrho$ is also bounded by 1.

**Lemma 1.10.22.** *Let $(X_n, f_n)_n$ be an inverse sequence. In addition, for every $n$ let $d_n$ be an admissible metric for $X_n$ which is bounded by 1. Then with respect to the metric $\varrho$ for*

$$X_\infty = \varprojlim (X_n, f_n)_n$$

*defined above, we have that every $f_n^\infty \colon X_\infty \to X_n$ is a $2^{-(n-1)}$-mapping.*

**Proof.** This is easy. Fix $n \in \mathbb{N}$ and let $p, q \in X_\infty$ be such that

$$f_n^\infty(p) = f_n^\infty(q).$$

Observe that $p_i = q_i$ for every $i \leq n$. Consequently,

$$\varrho(p, q) = \sum_{i=1}^\infty 2^{-i} d_i(p_i, q_i) = \sum_{i=n+1}^\infty 2^{-i} d_i(p_i, q_i) \leq \sum_{i=n+1}^\infty 2^{-i} = 2^{-n}.$$

We conclude that $f_n^\infty$ is a $2^{-(n-1)}$-mapping. $\square$

**Brown's Approximation Theorem 1.10.23.** *Let $(X_n, f_n)_n$ be an inverse sequence of compact spaces with inverse limit $X_\infty$. If each $f_n$ is a near homeomorphism, then so is each $f_n^\infty$. In particular, if each $f_n$ is a near homeomorphism then $X_\infty$ is homeomorphic to $X_1$ (and hence to every $X_n$).*

**Proof.** We shall prove that $f_1^\infty \colon X_\infty \to X_1$ is a near homeomorphism. For each $n$, let $d_n$ be an admissible metric for $X_n$ which is bounded by 1 and let $\varrho$ be the metric for $X_\infty$ introduced above. Let $\varepsilon > 0$. Inductively, we shall construct for every $n \in \mathbb{N}$ a homeomorphism $h_n \colon X_{n+1} \to X_n$ such that the functions $g_n \colon X_\infty \to X_1$ defined by $g_n = h_1 \circ \cdots \circ h_n \circ f_{n+1}^\infty$ have the following properties:

(1) $\hat{d}_1(g_1, f_1^\infty) < \varepsilon/2$,
(2) $\hat{d}_1(g_n, g_{n+1}) < 3^{-n} \cdot \varepsilon$,
(3) $\hat{d}_1(g_n, g_{n+1}) < 3^{-n} \cdot \min\{\hat{d}_1(g_i, \mathcal{G}_{1/i}(X_\infty, X_1)) : 1 \leq i \leq n\}$.

Since $f_1$ is a near homeomorphism and $f_1^\infty = f_1 \circ f_2^\infty$, it is clear that $h_1$ can be constructed so that condition (1) is met. Suppose that the homeomorphisms $h_i$ are defined for $1 \leq i \leq n$. Observe that for every $1 \leq i \leq n$, the function $g_i \in \mathcal{S}(X_\infty, X_1) \setminus \mathcal{G}_{1/i}(X_\infty, X_1)$ (Lemmas 1.10.8 and 1.10.22). Let

$$\delta = 3^{-n} \cdot \min\left\{\varepsilon, \min\{\hat{d}_1(g_i, \mathcal{G}_{1/i}(X_\infty, X_1)) : 1 \leq i \leq n\}\right\}$$

and notice that $\delta > 0$. Since

$$g_n = h_1 \circ \cdots \circ h_n \circ f_{n+1}^\infty = h_1 \circ \cdots \circ h_n \circ f_{n+1} \circ f_{n+2}^\infty$$

and $f_{n+1}$ is a near homeomorphism, we can find a homeomorphism
$$h_{n+1}\colon X_{n+2}\to X_{n+1}$$
such that
$$\hat{\varrho}(h_1\circ\cdots\circ h_n\circ f_{n+1}\circ f_{n+2}^\infty, h_1\circ\cdots\circ h_n\circ h_{n+1}\circ f_{n+2}^\infty)<\delta$$
(use the fact that $h_1\circ\cdots\circ h_n$ is uniformly continuous (Exercise A.5.18)). It is clear that $h_{n+1}$ is as required, which completes the inductive construction of the $h_n$.

By (2) it follows that the sequence $(g_n)_n$ is $\hat{d}_1$-Cauchy and consequently,
$$g=\lim_{n\to\infty}g_n$$
exists. Since $g_n\in\mathcal{S}(X_\infty,X_1)$ for every $n$, Proposition 1.3.8 implies that
$$g\in\mathcal{S}(X_\infty,X_1).$$
Observe that (1) and (2) imply that
$$\hat{d}_1(f_1^\infty,g)<\varepsilon.$$
Also, (3), Lemmas 1.6.1 and 1.3.10 show that
$$g\in\bigcap_{n=1}^\infty\mathcal{S}_{1/n}(X_\infty,X_1)=\mathcal{H}(X_\infty,X_1),$$
which is as required. $\square$

**Dynamical systems.** A *dynamical system* is a pair $(X,f)$ where $X$ is a space and $f\colon X\to X$ is continuous. In topological dynamics one studies the behaviour of the iterates of $f$. Inverse limits play a role in certain considerations there. We will not go into this, but we remark that the following considerations that will play a role at one point in our chapter about dimension theory were motivated by well-known ideas in topological dynamics.

Let $(X,f)$ be a dynamical system, and consider the following inverse sequence
$$X\xleftarrow{f}X\xleftarrow{f}\cdots\xleftarrow{f}X\xleftarrow{f}\cdots$$
with inverse limit $X_\infty$. So the same map is repeated infinitely often in the inverse system under consideration. Define the so-called *shift* $\sigma\colon X^\infty\to X^\infty$ as follows:
$$\sigma(x_1,x_2,\dots)=(f(x_1),x_1,x_2,\dots).$$
It is clear that $\sigma$ is well-defined. The compositions
$$\pi_n\circ\sigma=\begin{cases}f\circ\pi_1 & (n=1),\\ \pi_{n-1} & (n>1),\end{cases}$$
are all continuous from which it follows that $\sigma$ is continuous. It is in general not true that $\sigma$ is a homeomorphism.

Define another shift function $\tau\colon X^\infty \to X^\infty$ by
$$\tau(x_1, x_2, \ldots) = (x_2, x_3, \ldots).$$
It is clear that $\tau$ is continuous as well.

A moments reflection shows that
$$\sigma[X_\infty] \subseteq X_\infty, \quad \tau[X_\infty] \subseteq X_\infty.$$
Moreover, if $x \in X_\infty$ then
$$\sigma\bigl(\tau(x)\bigr) = \sigma(x_2, x_3, \ldots) = (f(x_2), x_2, x_3, \ldots) = (x_1, x_2, \ldots) = x.$$
Similarly,
$$\tau\bigl(\sigma(x)\bigr) = \tau(f(x_1), x_1, x_2, \ldots) = (x_1, x_2, \ldots) = x.$$
We conclude that $\sigma \restriction X_\infty$ is a homeomorphism, with inverse $\tau \restriction X_\infty$.

By abuse of notation, we will denote $\sigma \restriction X_\infty$ also by $\sigma$.

**Lemma 1.10.24.** *Let $(X, f)$ be a dynamical system, and let $\sigma\colon X_\infty \to X_\infty$ denote the shift homeomorphism. Then for each $n$ the diagram*

$$\begin{array}{ccc} X & \xleftarrow{f_n^\infty} & X_\infty \\ {\scriptstyle f}\downarrow & & \downarrow{\scriptstyle \sigma} \\ X & \xleftarrow{f_n^\infty} & X_\infty \end{array}$$

*commutes. As a consequence, if $f$ is fixed-point free then so is $\sigma$.*

**Proof.** This is clear since if $x \in X_\infty$ then $f \circ f_n^\infty(x) = f(x_n)$ and
$$f_n^\infty \circ \sigma(x) = f_n^\infty(f(x_1), x_1, x_2, \ldots),$$
which is equal to $f(x_1)$ if $n = 1$ and $x_{n-1} = f(x_n)$ if $n > 1$.

This implies that if $x$ is a fixed point of $\sigma$ then $f_n^\infty(x)$ is a fixed point of $f$, which establishes the second statement. $\square$

So in some vague sense, every continuous function $f\colon X \to X$ can be 'approximated' by a homeomorphism (cf. Lemma 1.10.22).

**Exercises for §1.10.** A space $X$ is called *isotopically homogeneous* provided that for every $x, y \in X$ there exists an isotopy $H\colon X \times \mathbb{I} \to X$ such that $H_0 = 1_X$ and $H_1(x) = y$.

1. Let $X$ be the inverse limit of an inverse sequence of copies of $\mathbb{I}$. Prove that $X$ has the fixed-point property.

2. Prove that the circle $\mathbb{S}^1$ is not homeomorphic to the inverse limit of an inverse sequence of copies of $\mathbb{I}$.

3. Prove that the statement in Example 1.10.3 is correct.

4. Prove that the statement in Example 1.10.4 is correct.
5. Let $(X_n, f_n)$ be an inverse system with open surjective (continuous) functions. Prove that for every $n \in \mathbb{N}$ the function
$$f_n^\infty \colon \varprojlim(X_n, f_n) \to X_n$$
is open.
6. Let $(X_n, f_n)_n$ be an inverse sequence such that every $f_n$ is a homeomorphism. Prove that $\varprojlim(X_n, f_n)_n$ is homeomorphic to $X_1$ (and hence to every $X_n$).
7. Let $(X_n, f_n)_n$ be an inverse sequence. In addition, for every $n$ let $A_n \subseteq X_n$ be such that
$$f_n[A_{n+1}] \subseteq A_n;$$
put $g_n = f_n \restriction A_{n+1} \colon A_{n+1} \to A_n$. Prove that the inverse limit
$$\varprojlim(A_n, g_n)_n$$
and the subspace
$$\{x \in \varprojlim(X_n, f_n)_n : (\forall n \in \mathbb{N})(x_n \in A_n)\}$$
of $\varprojlim(X_n, f_n)_n$ are homeomorphic.
▶8. Assume that $X$ is a compact space. Prove that $X$ maps onto $X^2$ if and only if it maps onto $X^\infty$.
9. Let $(a_n X, f_n)_n$ be an inverse sequence consisting of compactifications of the space $X$. Assume moreover that for every $n$ the function $f_n$ restricts to the identity on $X$. Hence
$$a_1 X \le a_2 X \le \cdots \le a_n X \le \cdots.$$
Prove that the inverse limit $a_\infty X$ of this sequence is a compactification of $X$, and that every projection $f_n^\infty \colon a_\infty X \to a_n X$ restricts to the identity on $X$. (Hence $a_\infty X \ge a_n X$ for every $n$.)
10. Let $(X_n, f_n)$ be an inverse sequence and let $Y$ be a space such that for each $n$ there is a continuous function $g_n \colon Y \to X_n$ such that $f_n \circ g_{n+1} = g_n$. Prove that the function $g \colon Y \to \varprojlim(X_n, f_n)_n$ defined by
$$g(y) = \bigl(g_n(y)\bigr)_n$$
is well-defined and continuous.
11. Let $X$ be a space and let
$$\mathcal{F}_1 \subseteq \mathcal{F}_2 \subseteq \cdots \subseteq \mathcal{F}_n \subseteq \cdots$$
be an increasing sequence of Wallman bases of $X$.
   (1) Prove that for every $n$ there is a continuous function
$$f_n \colon \omega(X, \mathcal{F}_{n+1}) \to \omega(X, \mathcal{F}_n)$$
   which extends the identity on $X$.
   (2) Prove that $\mathcal{F} = \bigcup_{n=1}^\infty \mathcal{F}_n$ is a Wallman base of $X$.
   (3) Prove that the inverse limit of the sequence $(\omega(X, \mathcal{F}_n), f_n)_n$ and $\omega(X, \mathcal{F})$ are equivalent compactifications of $X$.

12. Let $(X_n, f_n)_n$ be an inverse sequence with compact spaces. Prove that if $A$ and $B$ are disjoint closed subsets of $X_\infty$ then there exists an $N \in \mathbb{N}$ such that for each $n \geq N$ the sets $f_n^\infty[A]$ and $f_n^\infty[B]$ are disjoint.

13. Let $X$ and $Y$ be compact spaces and let $f\colon X \to Y$ and $g\colon Y \to X$ be continuous functions such that both $g \circ f\colon X \to X$ and $f \circ g\colon Y \to Y$ are near homeomorphisms. Prove that $X$ and $Y$ are homeomorphic.

14. Prove that the statement in Example 1.10.5 is correct.

15. Prove that in the statement in Remark 1.10.5 the compactness assumption is essential.

16. Prove that an indecomposable continuum is not path-connected.

17. Give an example of a homogeneous continuum which is not isotopically homogeneous.

18. Let $X$ be a compact space and let $f\colon X \to X$ be continuous. Prove that
$$K = \bigcap_{i=1}^{\infty} f^i[X]$$
is nonempty and also that $f[K] = K$.

19. Let $X$ be an indecomposable continuum and let $E$ be one of its composants. Prove that if $E$ is not dense then $E$ is closed.

20. Let $X$ be an indecomposable continuum and let $E$ and $F$ be distinct composants of $X$. Prove that $E$ and $F$ are disjoint.

21. Prove that for a continuum $X$ the following statements are equivalent:
    (1) $X$ is hereditarily indecomposable.
    (2) If $C, D$ are arbitrary subcontinua of $X$ such that $C \cap D \neq \emptyset$ then $C \subseteq D$ or $D \subseteq C$.

22. Let $X$ be a hereditarily indecomposable continuum, and let $f\colon X \to Y$ be monotone. Prove that $Y$ is a hereditarily indecomposable continuum.

23. Let $X$ be an indecomposable continuum. Prove that the family of composants of $X$ is uncountable.

It can be shown that the dyadic solenoid $\Sigma_2$ cannot be imbedded in the plane. It is an interesting question whether there is a planar indecomposable continuum. This question will be considered in the remaining part of this section.

Let $a, b$ and $c$ be three distinct points in the plane. Inductively, construct finite simple chains $\mathcal{C}_1, \mathcal{C}_2, \ldots$ of open balls having among other things the following properties:

(1) $\mathrm{mesh}(\mathcal{C}_n) < 2^{-n}$,
(2) $\overline{\mathcal{C}}_n < \mathcal{C}_{n-1}$.

But these are not all conditions. We require the $\mathcal{C}_n$'s to follow a certain pattern. The collection $\mathcal{C}_1$ is a simple chain from $a$ to $c$ through $b$, $\mathcal{C}_2$ is a simple chain from $a$ to $b$ through $c$, $\mathcal{C}_3$ is a simple chain from $b$ to $c$ through $a$ (see Figure 5). Then repeat this process at infinitum. For every $n$ let $C_n = \bigcup \mathcal{C}_n$ and let $P = \bigcap_{n=1}^{\infty} C_n$.

24. Prove that $P$ is a continuum.

25. Prove that $P$ is indecomposable.

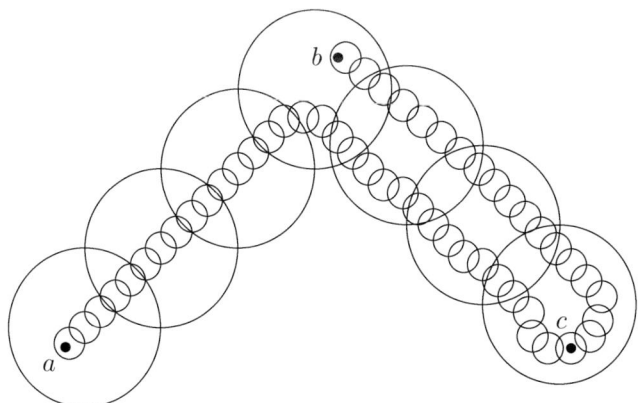

Figure 5.

## 1.11. Hyperspaces

In this section we shall introduce the *hyperspace* of a topological space. Hyperspaces are used later to construct various interesting examples.

Let $X$ be a space. The collection of all non-empty *compact* subsets of $X$ shall be denoted by $2^X$. We will endow $2^X$ with a suitable topology. For $E, F \in 2^X$ define the *Hausdorff distance* $\varrho_H(E, F)$ between $E$ and $F$ by

$$\varrho_H(E, F) = \inf\{\varepsilon > 0 : E \subseteq B(F, \varepsilon) \text{ and } F \subseteq B(E, \varepsilon)\}.$$

The compactness of $E$ and $F$ implies that $\varrho_H(E, F) \in [0, \infty)$ (Exercise 1.11.1).

**Lemma 1.11.1.** $\varrho_H : 2^X \times 2^X \to [0, \infty)$ *is a metric.*

**Proof.** We shall only present a proof of the triangle inequality, the other verifications being trivial.

Take $E, F, G \in 2^X$, let $\varepsilon > 0$ and put $\delta = \varepsilon/2$. By definition we have

(∗) $\qquad E \subseteq B\bigl(F, \delta + \varrho_H(E, F)\bigr)$ and $F \subseteq B\bigl(G, \delta + \varrho_H(F, G)\bigr)$.

Take an arbitrary point $x \in E$. By (∗) there exists $y \in F$ such that

$$\varrho(x, y) < \delta + \varrho_H(E, F).$$

Analogously there exists $z \in G$ such that

$$\varrho(y, z) < \delta + \varrho_H(F, G).$$

We conclude that

$$\varrho(x, z) < \varepsilon + \varrho_H(E, F) + \varrho_H(F, G).$$

Therefore, since $x$ was an arbitrary point of $E$,
$$E \subseteq B\big(G, \varepsilon + \varrho_H(E,F) + \varrho_H(F,G)\big).$$
Similarly,
$$G \subseteq B\big(E, \varepsilon + \varrho_H(E,F) + \varrho_H(F,G)\big).$$
Since $\varepsilon$ was an arbitrary positive number, we obtain
$$\varrho_H(E,G) \leq \varrho_H(E,F) + \varrho_H(F,G),$$
as required. □

The metric $\varrho_H$ is called the *Hausdorff metric* on $2^X$. We endow $2^X$ with the topology derived from this metric and call it the *hyperspace* of $X$.

Define $e\colon X \to 2^X$ by $e(x) = \{x\}$. It is easily seen that $e$ is an isometry, hence an imbedding, and that $e[X]$ is a closed subset of $2^X$ (Exercise 1.11.3). It will sometimes be convenient to identify $X$ and $e[X]$.

**Compactness and completeness properties of hyperspaces.** We will show that compactness and completeness are 'preserved' by the hyperspace operator. We need a simple lemma first.

**Lemma 1.11.2.** *Let $(X, \varrho)$ be a space and $(A_n)_n$ a Cauchy sequence in $2^X$. Let $\tilde{A} = \overline{\bigcup_{n=1}^\infty A_n}$. Then*

(1) *The metric $\rho = \varrho \restriction (\tilde{A} \times \tilde{A})$ is totally bounded.*
(2) *If $A_\infty = \lim_{n\to\infty} A_n$ exists in $2^X$ then it is equal to*

(†) $\qquad \{x \in X : (\forall n \in \mathbb{N})(\exists x_n \in A_n)(\lim_{n\to\infty} x_n = x)\},$

*and also to*

(‡) $\qquad \bigcap_{n=1}^\infty \overline{\bigcup_{m \geq n} A_m}.$

(3) *If $A_\infty = \lim_{n\to\infty} A_n$ exists in $2^X$ then $\tilde{A}$ is compact and*
$$\tilde{A} = A_\infty \cup \bigcup_{n=1}^\infty A_n.$$

(4) *If $\varrho$ is complete then $\lim_{n\to\infty} A_n$ exists in $2^X$ (and hence has to be equal to the sets (†) and (‡) in (2)).*

**Proof.** For (1), let $\varepsilon > 0$. There exists $N \in \mathbb{N}$ such that $\varrho_H(A_n, A_m) < \tfrac{1}{4}\varepsilon$ for all $n, m \geq N$. Consider $A_N$ and let $F \subseteq A_N$ be a finite $\tfrac{1}{4}\varepsilon$-net (by Theorem A.6.1, $\varrho\restriction(A_N \times A_N)$ is totally bounded, whence the desired net exists by Exercise A.6.2). Fix $n \geq N$. Since $\varrho_H(A_N, A_n) < \tfrac{1}{4}\varepsilon$, $A_n \subseteq B(A_N, \tfrac{1}{4}\varepsilon)$. So for every $x \in A_n$ there exists $y \in A_N$ such that $\varrho(x, y) < \tfrac{1}{4}\varepsilon$. For such a $y$ there exists $z \in F$ such that $\varrho(z, y) < \tfrac{1}{4}\varepsilon$. We conclude that $F$ is a $\tfrac{1}{2}\varepsilon$-net for $A_n \cup A_N$. Since $n$ was arbitrary, $F$ is even a $\tfrac{1}{2}\varepsilon$-net for $\bigcup_{n \geq N} A_n$, and hence an $\varepsilon$-net for $\overline{\bigcup_{n \geq N} A_n}$. Since $\bigcup_{n=1}^{N-1} A_n$ is compact, it has a finite $\varepsilon$-net by the same reason $A_N$ has one. The union of these two finite $\varepsilon$-nets is a finite $\varepsilon$-net for $\tilde{A}$. So $\rho$ is totally bounded by Exercise A.6.2.

For (2), let $B = \overline{\{x \in X : (\forall n \in \mathbb{N})(\exists x_n \in A_n)(\lim_{n \to \infty} x_n = x)\}}$ and $C = \bigcap_{n=1}^{\infty} \overline{\bigcup_{m \geq n} A_m}$, respectively. We will prove that
$$A_\infty \subseteq B \subseteq C \subseteq A_\infty.$$
Pick an arbitrary $x \in A_\infty$. For each $n \in \mathbb{N}$ there exists a point $x_n \in A_n$ such that $\varrho(x, x_n) \leq 2\varrho(x, A_n)$. We claim that the sequence $(x_n)_n$ converges to $x$. To this end, let $\varepsilon > 0$. Since $(A_n)_n$ converges to $A_\infty$, there exists $N \in \mathbb{N}$ such that for each $m \geq N$, $A_\infty \subseteq B(A_m, {}^1\!/_2\varepsilon)$. From this it follows that for $m \geq N$ we have $\varrho(x, A_m) < {}^1\!/_2\varepsilon$, i.e., $\varrho(x, x_m) < \varepsilon$. Hence $A_\infty \subseteq B$.

That $B \subseteq C$ is trivial.

Striving for a contradiction, assume that there exists a point $x \in C \setminus A_\infty$. Let $\varepsilon = \varrho(x, A_\infty)$ and observe that $\varepsilon > 0$. Since $A_n \to A_\infty$ in $2^X$, there exists $N \in \mathbb{N}$ such that for every $n \geq N$ we have $A_n \subseteq B(A_\infty, {}^1\!/_2\varepsilon)$. As a consequence,
$$x \in \overline{\bigcup_{n \geq N} A_n} \subseteq \overline{B(A_\infty, {}^1\!/_2\varepsilon)} \subseteq B(A_\infty, \varepsilon) \subseteq X \setminus \{x\},$$
which is a contradiction.

For (3), let $\mathcal{U}$ be a cover of $A^* = A_\infty \cup \bigcup_{n=1}^{\infty} A_n$ by open subsets of $X$. Since $A_\infty$ is compact, there is a finite $\mathcal{F} \subseteq \mathcal{U}$ such that $A_\infty \subseteq \bigcup \mathcal{F}$. By Corollary A.5.4 there exists $\varepsilon > 0$ such that $B(A_\infty, \varepsilon) \subseteq \bigcup \mathcal{F}$. Since $(A_n)_n$ converges to $A_\infty$, there exists $N \in \mathbb{N}$ such that $A_n \subseteq B(A_\infty, \varepsilon) \subseteq \bigcup \mathcal{F}$ for $n \geq N$, and so
$$\bigcup_{n \geq N} A_n \subseteq B(A_\infty, \varepsilon) \subseteq \bigcup \mathcal{F}.$$
Since $\mathcal{F}$ is finite and $\bigcup_{n=1}^{N-1} A_n$ is compact, we conclude that $A^*$ can be covered by finitely many elements of $\mathcal{U}$. So $A^*$ is compact, and since $A_\infty \subseteq \tilde{A}$ by (2), it follows that $A^* \subseteq \tilde{A}$. But since $\bigcup_{n=1}^{\infty} A_n$ is dense in $\tilde{A}$ and is contained in $A^*$, it also follows that $\tilde{A} \subseteq A^*$.

For (4), we first observe that by (1) the metric $\rho = \varrho \restriction (\tilde{A} \times \tilde{A})$ is totally bounded. Since $\tilde{A}$ is closed, $\rho$ is also complete. So $\tilde{A}$ is compact by Theorem A.6.1. Put $K = \bigcap_{n=1}^{\infty} \overline{\bigcup_{m \geq n} A_m}$. Then $K$ is the intersection of a decreasing sequence of nonempty closed subsets of the compact space $\tilde{A}$, and hence is nonempty and compact itself. Hence $K \in 2^X$.

To prove that $A_n \to K$ in $2^X$, let $\varepsilon > 0$. By the compactness of $K$, there exists $m_1 \in \mathbb{N}$ such that

(5) $$\overline{\bigcup_{n \geq m_1} A_n} \subseteq B(K, \varepsilon).$$

Since $(A_n)_n$ is Cauchy, there exists $m_2 \in \mathbb{N}$ such that $\varrho_H(A_n, A_m) < \varepsilon/2$ for $n, m \geq m_2$. As a consequence, for $n \geq m_2$,

(6) $$K \subseteq \overline{\bigcup_{m \geq m_2} A_m} \subseteq \overline{B(A_n, \varepsilon/2)} \subseteq B(A_n, \varepsilon).$$

From (5) and (6) it now follows that $\varrho_H(K, A_n) < \varepsilon$ for $n \geq \max(m_1, m_2)$, which is as desired. □

**Theorem 1.11.3.** *Let $X$ be a space with admissible metric $\varrho$. Then*

(1) *If $\varrho$ is complete then so is $\varrho_H$.*
(2) *If $\varrho$ is totally bounded then so is $\varrho_H$.*

**Proof.** Observe that (1) follows from Lemma 1.11.2(4). For (2), assume that $\varrho$ is totally bounded. Let $\varepsilon > 0$ and let $\mathcal{F}$ be a finite open cover of $X$ with nonempty sets of diameter less than $\varepsilon$. For every $F \in \mathcal{F}$ pick $x_F \in F$. For $A \in 2^X$, put
$$B_A = \{x_F : F \cap A \neq \emptyset\}.$$
Since $\mathcal{F}$ covers $X$, $A \subseteq \bigcup\{F \in \mathcal{F} : F \cap A \neq \emptyset\}$, from which it easily follows that $\varrho_H(A, B_A) < \varepsilon$. Since $\mathcal{F}$ is finite, this shows that $2^X$ can be covered by finitely many $\varepsilon$-balls. □

**Corollary 1.11.4.** *If $X$ is topologically complete then so is $2^X$. If $X$ is compact then so is $2^X$.*

**Proof.** Both statements follow from Theorems 1.11.3 and A.6.1. □

**Hyperspace maps.** Let $X$ and $Y$ be spaces and let $f \colon X \to Y$ be continuous. By Exercise 1.11.7 we may regard $X$ and $Y$ to be (canonical) subspaces of $2^X$ and $2^Y$, respectively. There exists a canonical continuous extension $2^f$ of $f$ over $2^X$, which is called the *hyperspace map* of $f$. Indeed, define $2^f \colon 2^X \to 2^Y$ in the obvious way by
$$2^f(A) = f[A].$$
We shall prove below that hyperspace maps are continuous. They are of crucial importance in the process of analyzing hyperspaces.

**Lemma 1.11.5.** *Let $X$ and $Y$ be spaces and let $f \colon X \to Y$ be continuous. Then the hyperspace map $2^f \colon 2^X \to 2^Y$ is continuous.*

**Proof.** Suppose that $(A_n)_n$ is a sequence in $2^X$ converging to $A \in 2^X$. Then obviously $(A_n)_n$ is Cauchy in $2^X$, and hence $\tilde{A} = \overline{\bigcup_{n=1}^{\infty} A_n}$ is compact by Lemma 1.11.2(3). As a consequence $f \restriction \tilde{A}$ is uniformly continuous (Exercise A.5.18). Now let $\varepsilon > 0$ and pick $\delta > 0$ such that if $x, y \in \tilde{A}$ are arbitrary points with $\varrho(x, y) < \delta$ then $\varrho(f(x), f(y)) < \varepsilon$. If $D, E \subseteq \tilde{A}$ are nonempty and compact and $\varrho_H(D, E) < \delta$ then $\varrho_H(f[D], f[E]) < \varepsilon$. So, $f[A_n] \to f[A]$ in $2^Y$. □

**The union operator.** We shall now consider a more complicated 'hyperspace map'. Let $X$ be a space. Consider the space $2^{2^X}$, i.e., the hyperspace of the hyperspace of $X$. Each point $\mathcal{A} \in 2^{2^X}$ is a compact subset of $2^X$ and therefore can be regarded as a 'compact' family of compact subsets of $X$. We shall prove that the union of such a family is compact. From this it follows that the 'union-operator'

$$\bigcup : 2^{2^X} \to 2^X,$$

defined by

$$\bigcup(\mathcal{A}) = \bigcup \mathcal{A},$$

is well-defined. We shall also prove that this operator is continuous.

**Lemma 1.11.6.** *Let $X$ be a space and let $\mathcal{A}$ belong to $2^{2^X}$. Then $\bigcup \mathcal{A}$ is a compact subset of $X$.*

**Proof.** Take a sequence $(x_n)_n$ in $\bigcup \mathcal{A}$. We shall prove that it has a convergent subsequence in $\bigcup \mathcal{A}$. For each $n$ there exists $A_n \in \mathcal{A}$ such that $x_n \in A_n$. By compactness of the subspace $\mathcal{A}$ of $2^X$, we may assume without loss of generality that the sequence $(A_n)_n$ converges to a point $A \in \mathcal{A}$. From Lemma 1.11.2(3) it follows that $B = A \cup \bigcup_{n=1}^{\infty} A_n$ is compact. Hence $(x_n)_n$ has a convergent subsequence in $B$ (Theorem A.5.1). Since $B \subseteq \bigcup \mathcal{A}$, it follows that this convergent subsequence is convergent in $\bigcup \mathcal{A}$. Now apply Theorem A.5.1. □

So from Lemma 1.11.6 we conclude that for every $X$, the union-operator

$$\bigcup : 2^{2^X} \to 2^X$$

is well-defined. We shall now prove that it is continuous.

**Proposition 1.11.7.** *Let $X$ be a space. Then the union-operator*

$$\bigcup : 2^{2^X} \to 2^X$$

*is Lipschitz and hence is continuous.*

**Proof.** Let $\varrho_{HH}$ denote the Hausdorff metric on $2^{2^X}$. Take arbitrary points $\mathcal{E}$ and $\mathcal{F}$ in $2^{2^X}$ and let $\varrho_{HH}(\mathcal{E}, \mathcal{F}) = \delta$. Fix $\varepsilon > 0$ and pick an arbitrary element $x \in \bigcup \mathcal{E}$. There exists $E \in \mathcal{E}$ such that $x \in E$. Since $E \in \mathcal{E}$ there exists $F \in \mathcal{F}$ such that $\varrho_H(E, F) < \delta + \varepsilon$. This implies that $E \subseteq B(F, \delta + \varepsilon)$ and so there exists $y \in F \subseteq \bigcup \mathcal{F}$ such that $\varrho(x, y) < \delta + \varepsilon$. We conclude that

$$\bigcup \mathcal{E} \subseteq B(\bigcup \mathcal{F}, \delta + \varepsilon).$$

By interchanging the roles of $\mathcal{E}$ and $\mathcal{F}$ we therefore conclude that

$$\varrho_H(\bigcup \mathcal{E}, \bigcup \mathcal{F}) < \delta + \varepsilon.$$

Since $\varepsilon$ was chosen arbitrarily, we get $\varrho_H(\bigcup \mathcal{E}, \bigcup \mathcal{F}) \leq \delta$, as required. □

**Corollary 1.11.8.** *Let $X$ and $Y$ be spaces and let $g \colon X \to 2^Y$ be continuous. Then the function $\bar{g} \colon 2^X \to 2^Y$ defined by*

$$\bar{g}(A) = \bigcup_{x \in A} g(x)$$

*is continuous.*

**Proof.** The hyperspace map $2^g \colon 2^X \to 2^{2^Y}$ is continuous by Lemma 1.11.5. In addition, by Proposition 1.11.7, the union-operator $\bigcup \colon 2^{2^Y} \to 2^Y$ is also continuous. Since $\bar{g} = \bigcup \circ 2^g$, we are done. □

Let $X$ be a space and for each $n \in \mathbb{N}$ define

$$\mathcal{F}_n(X) = \{A \in 2^X : |A| \leq n\}.$$

Observe that $\mathcal{F}_1(X) \subseteq \mathcal{F}_2(X) \subseteq \cdots \subseteq \mathcal{F}_n(X) \subseteq \cdots$.

The following simple consequence of Exercise 1.11.6, is basic for hyperspaces.

**Theorem 1.11.9.** *Let $X$ be a space containing more than one point and let $n \in \mathbb{N}$. Then the function $f_n \colon X^n \to \mathcal{F}_n(X)$ defined by*

$$f_n(x_1, \ldots, x_n) = \{x_1, \ldots, x_n\}$$

*is continuous.*

**Corollary 1.11.10.** *Let $X$ be a space and let $\mathcal{A}_1, \ldots, \mathcal{A}_n$ be nonempty subsets of $2^X$. Then the function $f \colon \prod_{i=1}^n \mathcal{A}_i \to 2^X$ defined by*

$$f(A_1, \ldots, A_n) = A_1 \cup \cdots \cup A_n$$

*is continuous.*

**Proof.** Again, let $\bigcup \colon 2^{2^X} \to 2^X$ denote the union-operator. By Theorem 1.11.9 we get that the function $g \colon \prod_{i=1}^n \mathcal{A}_i \to 2^{2^X}$ defined by

$$g(A_1, \ldots, A_n) = \{A_1, \ldots, A_n\}$$

is continuous. Since $f = \bigcup \circ g$, we are done. □

**A canonical base for $2^X$.** We used the Hausdorff distance to endow $2^X$ with a natural topology. There is a different way to do this in terms of a canonical base for $2^X$ yielding the so-called *Vietoris topology* on $2^X$. We will show that both topologies on $2^X$ are the same. The alternative description of the topology on $2^X$ in terms of a canonical base is sometimes very convenient.

Let $X$ be a space. For every *finite* collection $\mathcal{A}$ of subsets of $X$ put

$$\langle \mathcal{A} \rangle = \{B \in 2^X : B \subseteq \bigcup \mathcal{A} \text{ and for every } A \in \mathcal{A}, A \cap B \neq \emptyset\}.$$

**Lemma 1.11.11.** *Let $X$ be a space. Then*

(1) *if $\mathcal{A}$ is a finite family of open subsets of $X$ then $\langle \mathcal{A} \rangle$ is open in $2^X$,*
(2) *the collection*

$$\mathcal{B}(X) = \{\langle \mathcal{F} \rangle : \mathcal{F} \text{ is a finite family of open subsets of } X\}$$

*is an open base for $2^X$ which is closed under finite intersections.*

**Proof.** For (1), let $\mathcal{A}$ be a finite family of open subsets of $X$ and let $D \in \langle \mathcal{A} \rangle$. For each $A \in \mathcal{A}$ pick an arbitrary point $x_A \in A \cap D$. Since $\mathcal{A}$ is finite and consists of open sets, and $D$ is compact, there exists by Corollary A.5.4 an $\varepsilon > 0$ such that

(3) for every $A \in \mathcal{A}$, $B(x_A, \varepsilon) \subseteq A$,
(4) $B(D, \varepsilon) \subseteq \bigcup \mathcal{A}$.

Now take $E \in 2^X$ such that $\varrho_H(D, E) < \varepsilon$. By (4), $E \subseteq \bigcup \mathcal{A}$. Fix an arbitrary element $A \in \mathcal{A}$. Since $D \subseteq B(E, \varepsilon)$, there exists an $x \in E$ such that $\varrho(x_A, x) < \varepsilon$. By (3), $x$ belongs to $A$, i.e., $E$ intersects $A$. This proves that the ball (in $2^X$) about $D$ of radius $\varepsilon$ is contained in $\langle \mathcal{A} \rangle$.

For the first part of (2), take arbitrary $D \in 2^X$ and $\varepsilon > 0$. By compactness of $D$, there is a finite family $\mathcal{A}$ of open subsets of $X$ such that

(5) $D \subseteq \bigcup \mathcal{A} \subseteq B(D, \varepsilon)$,
(6) for every $A \in \mathcal{A}$, $A \cap D \neq \emptyset$,
(7) for every $A \in \mathcal{A}$, $\operatorname{diam}(A) < \varepsilon$.

So $D \in \langle \mathcal{A} \rangle$. Observe that $\langle \mathcal{A} \rangle$ is open by (1). We claim that $\langle \mathcal{A} \rangle$ is contained in the $\varepsilon$-ball in $2^X$ about $D$. To this end, take an arbitrary $E \in \langle \mathcal{A} \rangle$. By (5), we need only show that $D \subseteq B(E, \varepsilon)$. But this is trivial. For take an arbitrary element $x \in D$. By (5) there exists $A \in \mathcal{A}$ containing $x$. Since $E \in \langle \mathcal{A} \rangle$ we have $E \cap A \neq \emptyset$, which implies by (7) that there exists $x' \in E$ with $\varrho(x, x') < \varepsilon$. We conclude that $x \in B(E, \varepsilon)$.

Finally, let $\mathcal{F}_0 = \{U_1, \ldots, U_n\}$ and $\mathcal{F}_1 = \{V_1, \ldots, V_m\}$ be arbitrary collections open subsets of $X$. Put $U = \bigcup \mathcal{F}_0$ and $V = \bigcup \mathcal{F}_1$, respectively. An easy verification shows that

$$\langle \mathcal{F}_0 \rangle \cap \langle \mathcal{F}_1 \rangle = \langle \{U \cap V_1, \ldots, U \cap V_m, U_1 \cap V, \ldots, U_n \cap V\} \rangle,$$

as required. □

**Corollary 1.11.12.** *Let $X$ be a space. Then*

$$\mathcal{F}_\infty(X) = \bigcup_{n=1}^{\infty} \mathcal{F}_n(X)$$

*is dense in $2^X$.*

**Connectivity in hyperspaces.** We now turn to connectivity properties of $2^X$.

**Proposition 1.11.13.** *Let $X$ be a space. Then*

(1) *$X$ is connected iff $2^X$ is connected,*
(2) *$X$ is locally connected iff $2^X$ is locally connected.*

**Proof.** *We prove* (1).

First suppose that $X$ is connected. Since $\mathcal{F}_\infty(X)$ is dense in $2^X$ (Corollary 1.11.12), it suffices to prove that it is connected, and for this it suffices to prove that every $\mathcal{F}_n(X)$ is connected. However, this follows directly by Theorem 1.11.9 since for each $n$, $\mathcal{F}_n(X)$ is the image of $X^n$ under the map $f_n \colon X^n \to 2^X$ defined by

$$f_n(x_1, \ldots, x_n) = \{x_1, \ldots, x_n\}.$$

Conversely, assume that $2^X$ is connected. We shall prove that $X$ is connected. Striving for a contradiction, assume that $X$ is not connected. So it is possible to write $X$ as $U \cup V$, where $U$ and $V$ are disjoint nonempty open sets. Clearly,

$$2^X = \langle\{U\}\rangle \cup \langle\{V\}\rangle \cup \langle\{U,V\}\rangle.$$

In addition, $\langle\{U\}\rangle, \langle\{V\}\rangle$, and $\langle\{U,V\}\rangle$ are pairwise disjoint, nonempty and open by Lemma 1.11.11(1). This contradicts the connectivity of $2^X$.

*We prove* (2).

Now assume that $X$ is locally connected. Take $A \in 2^X$ and let $\mathcal{U}$ be a neighborhood of $A$ in $2^X$. By Lemma 1.11.11(2) there exists a finite family $\mathcal{V}$ of open subsets of $X$ such that $A \in \langle \mathcal{V} \rangle \subseteq \mathcal{U}$. Since $X$ is locally connected, every component of every $V \in \mathcal{V}$ is open (Exercise A.2.8). By compactness of $A$ we may therefore assume without loss of generality that $\mathcal{V}$ consists of connected open sets. We claim that $\langle \mathcal{V} \rangle$ is connected. First observe that for every $V \in \mathcal{V}$ we may identify $2^V$ and the subspace $\{A \in 2^X : A \subseteq V\}$ of $2^X$ (Exercise 1.11.7). Let $\mathcal{V} = \{V_1, \ldots, V_n\}$. Now define the function

$$g \colon \prod_{i=1}^{n} 2^{V_i} \to 2^X$$

by
$$g(A_1,\ldots,A_n) = \bigcup_{i=1}^{n} A_i.$$
Then $g$ is continuous by Corollary 1.11.10. By the first part of the Theorem, $2^{V_i}$ is connected for every $i \leq n$. So $\langle \mathcal{V} \rangle$ is connected, being the range of $g$.

Conversely, assume that $2^X$ is locally connected. Let $x \in X$ and let $U$ be an open neighborhood of $x$ in $X$. Since $\langle \{U\} \rangle$ is an open neighborhood of $\{x\}$ in $2^X$ (Lemma 1.11.11(1)), by local connectivity of $2^X$ there is a connected neighborhood $\mathcal{C}$ of $\{x\}$ in $2^X$ such that $\{x\} \in \mathcal{C} \subseteq \langle \{U\} \rangle$. Then $V = \mathcal{C} \cap X$ is a neighborhood of $x$ in $X$ and clearly
$$V \subseteq C = \bigcup \mathcal{C} \subseteq U.$$
We claim that $C$ is connected. Suppose that this is not true. Let $E$ and $F$ be disjoint nonempty relatively open nonempty subsets of $C$ that cover $C$. There are disjoint open subsets $E', F' \subseteq X$ such that $E' \cap C = E$ and $F' \cap C = F$ (Lemma A.8.1). Now observe that $\langle \{E'\} \rangle, \langle \{F'\} \rangle$ and $\langle \{E', F'\} \rangle$ are disjoint nonempty open subsets of $2^X$ (Lemma 1.11.11(1)), that $\mathcal{C} \cap \langle \{E'\} \rangle$ and $\mathcal{C} \cap \langle \{F'\} \rangle$ are nonempty and that $\mathcal{C} \subseteq \langle \{E'\} \rangle \cup \langle \{F'\} \rangle \cup \langle \{E', F'\} \rangle$. This contradicts the connectivity of $\mathcal{C}$. □

**Hyperspaces of subcontinua.** Let $X$ be a space and put
$$\mathcal{C}(X) = \{A \in 2^X : A \text{ is connected}\}.$$
It is called the *hyperspace of subcontinua of $X$* and is of particular interest in the case that $X$ is a continuum.

**Lemma 1.11.14.** *Let $X$ be a space. Then $\mathcal{C}(X)$ is a closed subspace of $2^X$. So if $X$ is compact then so is $\mathcal{C}(X)$.*

**Proof.** Let $A$ be an element of $2^X$ which does not belong to $\mathcal{C}(X)$. Write $A$ as the union of two disjoint relatively clopen nonempty subsets $E$ and $F$. By Corollary A.4.3 there are disjoint open subsets $E'$ and $F'$ in $X$ with $E \subseteq E'$ and $F \subseteq F'$. Then $\langle \{E', F'\} \rangle$ is an open neighborhood of $A$ which clearly misses $\mathcal{C}(X)$.

The second part of the lemma follows from the first part and Corollary 1.11.4. □

**Whitney maps.** A fundamental tool in hyperspace theory is the concept of a Whitney map which we will now discuss. Let $X$ be a compact space. A *Whitney map* for $2^X$ is a continuous function $g\colon 2^X \to [0, \infty)$ having the following properties:

(1) if $A, B \in 2^X$ and $A$ is a proper subset of $B$ then $g(A) < g(B)$,

(2) $g(\{x\}) = 0$ for every $x \in X$.

Observe that (1) implies that if $A \in 2^X$ and $|A| > 1$ then $g(A) > 0$.

There are several ways to construct such functions. We will discuss only one of them. Consider the function space $C(X, \mathbb{I})$. It is separable by Proposition 1.3.3. So let us fix a sequence $(f_i)_i$ of functions in $C(X, \mathbb{I})$ which form a dense subset of $C(X, \mathbb{I})$. Define $\omega \colon 2^X \to \mathbb{I}$ by

$$(*) \qquad \omega(A) = \sum_{i=1}^{\infty} 2^{-i} \operatorname{diam} f_i[A].$$

It is clear that $\omega$ is well-defined.

**Proposition 1.11.15.** *Let $X$ be a compact space. The function $\omega \colon 2^X \to \mathbb{I}$ defined in $(*)$ is a Whitney map for $2^X$.*

**Proof.** Let us first prove that for a fixed $i$ the function $2^X \to \mathbb{I}$ defined by

$$A \mapsto \operatorname{diam} f_i[A]$$

is continuous. This is clear since it is the composition of the functions

$$2^X \xrightarrow{2^{f_i}} 2^{\mathbb{I}} \xrightarrow{\operatorname{diam}} \mathbb{I};$$

here $2^{f_i}$ is the hyperspace map of $f_i$ (which is continuous by Lemma 1.11.5) and $\operatorname{diam} \colon 2^{\mathbb{I}} \to \mathbb{I}$ is the function that sends every element $A \in 2^{\mathbb{I}}$ onto its diameter (which is continuous by Exercise 1.11.12).

So $\omega$ is continuous by Exercise A.3.1.

Let $A, B \in 2^X$ be such that $A$ is a proper subset of $B$. Fix an arbitrary element $x \in B \setminus A$. There is by Lemma A.3.1 a continuous function $f \colon X \to \mathbb{I}$ such that $f[A] = \{0\}$ and $f(x) = 1$. Let $n \geq 2$ be such that

$$\|f - f_n\| < 1/4.$$

Observe that $0, 1 \in f[B]$ since $A \neq \emptyset$ which implies that

$$\operatorname{diam} f_n[B] \geq 1/2.$$

In addition, $f[A] = \{0\}$ so that

$$\operatorname{diam} f_n[A] \leq 1/4.$$

From this we conclude that $\operatorname{diam} f_n[A] < \operatorname{diam} f_n[B]$. Also,

$$\operatorname{diam} f_i[A] \leq \operatorname{diam} f_i[B]$$

for every $i$ since $A$ is a subset of $B$.

This shows that

$$\omega(A) = \sum_{i=1}^{\infty} 2^{-i} \operatorname{diam} f_i[A]$$
$$\leq \sum_{i=1}^{n-1} 2^{-i} \operatorname{diam} f_i[B] + 2^{-n} \operatorname{diam} f_n[A] + \sum_{i=n+1}^{\infty} \operatorname{diam} f_i[B]$$
$$< \sum_{i=1}^{n-1} 2^{-i} \operatorname{diam} f_i[B] + 2^{-n} \operatorname{diam} f_n[B] + \sum_{i=n+1}^{\infty} \operatorname{diam} f_i[B]$$
$$= \omega(B),$$

which is as required.

Since clearly $\omega(\{x\}) = 0$ for every $x \in X$, this completes the proof. □

Let $\omega\colon 2^X \to [0, \infty)$ be a Whitney function. For each $t \in [0, \omega(X)]$ the collections
$$\{A \in 2^X : \omega(A) = t\}$$
and
$$\{A \in \mathcal{C}(X) : \omega(A) = t\}$$
are called *Whitney levels*.

**Lemma 1.11.16.** *Let $X$ be a continuum, and $\omega\colon 2^X \to [0, \infty)$ a Whitney function. Then for every $t \in [0, \omega(X)]$ the Whitney level*
$$\mathcal{E} = \{A \in \mathcal{C}(X) : \omega(A) = t\}$$
*is closed in $\mathcal{C}(X)$ and moreover covers $X$.*

**Proof.** The first statement is trivial since $\mathcal{E}$ is equal to
$$\omega^{-1}(t) \cap \mathcal{C}(X)$$
which is an intersection of two closed subsets of $2^X$ (use that $\omega$ is continuous and Lemma 1.11.14).

To check that $\mathcal{E}$ covers $X$, let us consider any $x \in X$, and let

(∗) $\qquad \mathcal{K} = \{K \in \mathcal{C}(X) : x \in K, \omega(K) \leq t\}.$

Observe that $\mathcal{K} \neq \emptyset$ since $\omega(\{x\}) = 0$ and hence $\{x\} \in \mathcal{K}$. Since by continuity of $\omega$ and Exercise 1.11.2 the collection $\mathcal{K}$ is closed in $\mathcal{C}(X)$ (and hence compact), there is an $E \in \mathcal{K}$ with
$$\omega(E) = \max\{\omega(K) : K \in \mathcal{K}\}.$$
To see that $\omega(E) = t$, let us fix any $\varepsilon > 0$, and let $U$ be an open neighborhood of $E$ with $\omega(C) \leq \omega(E) + \varepsilon$ for any $E \subseteq C \subseteq \overline{U}$ (Exercise 1.11.22). The

component $H$ of $\overline{U}$ which contains $x$ intersects $\overline{U} \setminus U$ by Corollary A.10.5. So $E$ is a proper subcontinuum of the continuum $C = E \cup H$. Hence

$$\omega(E) < \omega(C) \leq \omega(E) + \varepsilon.$$

By $(*)$, $C \notin \mathcal{K}$ for otherwise $\omega(C) \leq \omega(E)$. Hence $\omega(C) > t$ and so

$$\omega(E) > t - \varepsilon.$$

Since $\varepsilon > 0$ was arbitrary, this shows that, indeed, $x \in E \in \mathcal{E}$. $\square$

**Corollary 1.11.17.** *Let $X$ be a hereditarily indecomposable continuum, and $\omega \colon 2^X \to [0, \infty)$ a Whitney function. Then for every $t \in [0, \omega(X)]$ the Whitney level*

$$\mathcal{E} = \{A \in \mathcal{C}(X) : \omega(A) = t\}$$

*is an upper semi-continuous decomposition $X$.*

**Proof.** Suppose $C$ and $D$ are intersecting continua in $X$ with $\omega(C) = \omega(D)$. Since $X$ is hereditarily indecomposable, $C \subseteq D$ or $D \subseteq C$ (Exercise 1.10.21). If e.g., $C \subseteq D$ then $C = D$ for otherwise $\omega(C) < \omega(D)$. This shows by Lemma 1.11.16 that $\mathcal{E}$ is a partition of $X$.

To prove that $\mathcal{E}$ is upper semi-continuous, let $A \subseteq X$ be an arbitrary closed set. Then

$$\langle\{A, X\}\rangle \cap \mathcal{E} = \{E \in \mathcal{E} : E \cap A \neq \emptyset\}$$

is closed in $2^X$ since both $\langle\{A, X\}\rangle$ and $\mathcal{E}$ are (use Exercise 1.11.2 and Lemma 1.11.16). As a consequence,

$$\bigcup\{E \in \mathcal{E} : E \cap A \neq \emptyset\}$$

is closed in $X$ by Lemma 1.11.6. $\square$

**Exercises for §1.11.**

1. Let $X$ be a space and let $E$ and $F$ be compact nonempty subsets of $X$. Prove that $\varrho_H(E, F) \in [0, \infty)$.

2. Let $X$ be a space and let $F \subseteq X$ be closed. Prove that the sets
   $$\{A \in 2^X : A \cap F \neq \emptyset\}, \quad \{A \in 2^X : A \subseteq F\}$$
   are both closed in $2^X$.

3. Prove that $\mathcal{F}_n(X)$ is closed in $2^X$ for every $n \in \mathbb{N}$.

4. Prove that $2^X$ is separable.

5. Let $(A_n)_n$ be a decreasing sequence of nonempty compact subsets of $X$. Prove that $A_n \to \bigcap_{n=1}^{\infty} A_n$ in $2^X$.

6. Let $X$ be a space containing more than one point and $n \in \mathbb{N}$. Prove that the function $f_n \colon X^n \to \mathcal{F}_n(X)$ defined by
$$f_n(x_1, \ldots, x_n) = \{x_1, \ldots, x_n\}$$
is continuous, and a homeomorphism if and only if $n = 1$.

7. Let $X$ be a subspace of $Y$. Prove that the natural inclusion $2^X \hookrightarrow 2^Y$ is an imbedding.

8. Let $X$ be a space. In addition, let $(A_n)_n$ and $(B_n)_n$ be sequences in $2^X$ converging to $A$ and $B$, respectively. Prove that if $A_n \subseteq B_n$ for every $n$ then $A \subseteq B$.

9. Let $X$ be a space and let $(A_n)_n$ be a sequence in $2^X$ converging to an element $A \in 2^X$. Prove that if $a_n \in A_n$ for every $n$ then there is a subsequence of $(a_n)_n$ converging to an element $a \in A$.

10. Let $A \in 2^X$, and for every $n$ let $A_n \in 2^X$ be such that $A \subseteq A_n \subseteq D(A, 1/n)$. Prove that $(A_n)_n$ converges to $A$ in $2^X$.

11. Prove that $\mathcal{C}(\mathbb{I})$ is homeomorphic to $\mathbb{I}^2$.

12. Let $X$ be a space. Prove that the function $\operatorname{diam}\colon 2^X \to [0, \infty)$ sending each element of $2^X$ onto its diameter, is continuous.

13. Let $X$ be compact and let $E, F \in 2^X$. Prove that
$$\varrho_H(E, F) = \max\{\sup_{x \in E} \varrho(x, F), \sup_{x \in F} \varrho(x, E)\}.$$

14. Prove that $\mathbb{I}^n$ can be imbedded in $2^{\mathbb{I}}$ for every $n$.

15. Let $X$ be compact and let $\mathcal{B} = \{B_n : n \in \mathbb{N}\}$ be an open base for it such that $B_1 = X$. Let $(A_i)_i$ be a sequence in $2^X$.
    (1) Prove that there exists a decreasing sequence $(X_n)_n$ of closed subsets of $X$ with the following properties:
        1. $X_1 = X$,
        2. every neighborhood of $X_n$ contains infinitely many $A_i$'s, and
        3. if $X_n \cap B_n \neq \emptyset$ then $X_n \setminus B_n$ has a neighborhood $V$ containing at most finitely many $A_i$'s.
    (2) Put $D = \bigcap_{n=1}^{\infty} X_n$. Prove that there is a subsequence of the sequence $(A_i)_i$ converging to $D$.
    (So this is an alternative proof of the second part of Corollary 1.11.4.)

16. Let $X$ and $Y$ be compact, and let $f \colon X \to Y$ be a continuous surjection. Prove that the following statements are equivalent:
    (1) $f$ is open.
    (2) The function $F \colon Y \to 2^X$ defined by
    $$F(y) = f^{-1}(\{y\})$$
    is continuous.

17. Let $(X, \varrho)$ be a compact space and let $f\colon X \to Y$ be a surjection which is both continuous and open. Prove that $Y$ is homeomorphic to the subspace
$$\{f^{-1}(y) : y \in Y\}$$
of $2^X$. Conclude that formula
$$d(y_1, y_2) = \varrho_H\left(f^{-1}(y_1), f^{-1}(y_2)\right)$$
defines an admissible metric on $Y$.

18. Prove that the functions $s, t\colon 2^{\mathbb{I}} \to \mathbb{I}$ defined by
$$s(A) = \min A \quad \text{and} \quad t(A) = \max A$$
are continuous.

▶19. Let $\varrho$ be an admissible convex metric on the compact space $X$. Prove that the function
$$e\colon 2^X \times [0, \infty) \to 2^X$$
defined by
$$e(A, t) = D_t(A)$$
is continuous.

▶20. Let $X$ be a space and let $k \in \mathbb{N}$. In addition, let $\mathcal{A} = \{A_i : i \in \mathbb{N}\}$ be a sequence of subsets of $X$, each of cardinality $k$. Prove by induction on $k$ that there are an infinite subset $E \subseteq \mathbb{N}$ and a (possibly empty) subset $A^0$ of $X$ and for every $i \in E$ a partition $A_i^0, A_i^1$ of $A_i$ such that that one of the following statements is true:
   (A) Every $x \in X$ has a neighborhood meeting finitely many terms of the sequence $(A_i)_{i \in E}$ only.
   (B) $A^0 \neq \emptyset$ and
       (1) the sequence $(A_i^0)_{i \in E}$ converges to $A^0$ in the Vietoris topology.
       (2) Every $x \in X$ has a neighborhood meeting finitely many terms of the sequence $(A_i^1)_{i \in E}$ only.

21. Let $X$ be a compact space with disjoint closed subsets $A$ and $B$. Define
$$\mathcal{C} = \{C \in 2^X : C \text{ is a continuum from } A \text{ to } B\}.$$
Prove that $\mathcal{C}$ is a closed subspace of $2^X$.

22. Let $X$ be a space, and let $\omega$ be a Whitney function for $2^X$. Let $\varepsilon > 0$ be arbitrary. Prove that for every $E \in 2^X$ there is a neighborhood $U$ of $E$ in $X$ such that for every $C \in 2^X$ with $E \subseteq C \subseteq \overline{U}$ we have $\omega(C) \leq \omega(E) + \varepsilon$.

▶23. Let $X$ be a hereditarily indecomposable continuum, and $\omega\colon 2^X \to [0, \infty)$ a Whitney function. For $t \in [0, \omega(X)]$ consider the Whitney level
$$\mathcal{E} = \{A \in \mathcal{C}(X) : \omega(A) = t\}.$$
Prove that the natural quotient map $\pi\colon X \to X/\mathcal{E}$ is open.

24. Let $X$ be a compact space and let $\omega\colon 2^X \to [0,\infty)$ be a Whitney map. Prove that if $t \in (0,\omega(X)]$ is such that its corresponding Whitney level
$$\mathcal{E} = \{A \in 2^X : \omega(A) = t\}$$
is nonempty then
$$\inf\{\operatorname{diam}(E) : E \in \mathcal{E}\} > 0.$$

25. Let $X$ be a compact space and let $\omega\colon 2^X \to [0,\infty)$ be a Whitney map. Prove that for every $\varepsilon > 0$ there is an element $t \in (0,\infty)$ such that if $A \in 2^X$ is such that $\omega(A) \leq t$ then $\operatorname{diam}(A) < \varepsilon$.

26. Let $T$ and $X$ be spaces and let $\varphi\colon T \to 2^X$ be continuous. In addition, let $\pi\colon X \times T \to T$ be the projection. Prove that if
$$E = \{(x,t) : x \in \varphi(t)\} \subseteq X \times T$$
then $\pi \restriction E\colon E \to T$ is a closed surjection.

CHAPTER 2

# Basic combinatorial topology

In this chapter we present some elementary combinatorial results and apply these to get nontrivial information about the topology of the Euclidean spaces $\mathbb{R}^n$, $n \in \mathbb{N}$. The main result is the Brouwer Fixed-Point Theorem.

## 2.1. Affine notions

In this section we will introduce simplicial complexes and present some basic results on them. So-called subdivisions of simplicial complexes are important in the proof of Brouwer's Fixed-Point Theorem as well as in ANR-theory.

For completeness sake, we begin by reviewing some elementary Linear Algebra. The proofs of these results are left to the reader.

Let $V$ denote a fixed vector space.

Let $F$ be a finite subset of $V$, say $F = \{v_1, \ldots, v_n\}$. An *affine combination* of $v_1, \ldots, v_n$ is a vector $v$ that can be written in the form $\sum_{i=1}^n t_i v_i$ with $t_1, \ldots, t_n \in \mathbb{R}$ and $\sum_{i=1}^n t_i = 1$. Such a $v$ is called *geometrically dependent* on $F$.

A *linear (affine) subspace* of $V$ is a subset of $V$ closed under the formation of linear (affine) combinations.

**Theorem 2.1.1.** *Let $A$ be a subset of $V$ and let $a \in A$. Then $A$ is an affine subspace if and only if $A - a$ is a linear subspace.*

If $S \subseteq V$, then the intersection of all affine (linear) subspaces of $V$ containing $S$ is the smallest affine (linear) subspace of $V$ containing $S$. This subset is called *affine hull (linear hull)* of $S$ and is denoted by $\mathrm{aff}(S)$ ($\mathrm{lin}(S)$).

**Theorem 2.1.2.** $\mathrm{aff}(S)$ *is the set of all affine combinations of elements of $S$. Moreover, for every $a \in S$ the following equality holds:*

$$\mathrm{aff}(S) - a = \mathrm{lin}(S - a).$$

We see that if $S \subseteq V$ then $\mathrm{aff}(S)$ is a translated linear subspace of $V$, since for every $a \in S$ the equality

$$\mathrm{aff}(S) = a + \mathrm{lin}(S - a)$$

holds. We say that $S$ *spans* the affine subspace $\mathrm{aff}(S)$.

Let $v_1, \ldots, v_n \in V$. Then

(1) the vectors $v_1, \ldots, v_n$ are said to be *geometrically independent* if for all elements $t_1, \ldots, t_n \in \mathbb{R}$ with $\sum_{i=1}^n t_i = 0$ and $\sum_{i=1}^n t_i v_i = 0$ we have $t_1 = \cdots = t_n = 0$,
(2) a subset $S \subseteq V$ is *linearly (geometrically) independent* if and only if every finite subset is.

**Theorem 2.1.3.** *Let $S \subseteq V$. The following statements are equivalent:*

(1) *$S$ is geometrically independent,*
(2) *no $x \in S$ is geometrically dependent on a finite subset $F \subseteq S \setminus \{x\}$,*
(3) *for every $s \in S$, $\{x - s : x \in S, x \neq s\}$ is linearly independent,*
(4) *for some $s \in S$, $\{x - s : x \in S, x \neq s\}$ is linearly independent.*

**Corollary 2.1.4.** *Let $S \subseteq \mathbb{R}^n$ be geometrically independent. Then $S$ contains at most $n + 1$ points.*

Consider the affine subspace $\mathrm{aff}(S)$ and fix an arbitrary $a \in S$. We saw that $\mathrm{aff}(S) = a + \mathrm{lin}(S - a)$. So there exists a subset $T \subseteq S$ such that $\mathrm{lin}(S - a) = \mathrm{lin}(T - a)$ while moreover $T - a$ is linearly independent. By Theorem 2.1.3 it follows that $T' = \{a\} \cup T$ is geometrically independent. Then

$$\mathrm{aff}(T') = a + \mathrm{lin}(T' - a) = a + \mathrm{lin}(T - a) = a + \mathrm{lin}(S - a) = \mathrm{aff}(S).$$

So we conclude that $S$ and its geometrically independent subset $T'$ span the same affine subspace.

An affine subspace of a linear space is called *m-dimensional* if it is spanned by $m + 1$ geometrically independent vectors.

**Proposition 2.1.5.** *Let $S \subseteq V$ be geometrically independent. If $A, B \subseteq S$ then*

$$\mathrm{aff}(A) \cap \mathrm{aff}(B) = \mathrm{aff}(A \cap B).$$

The following result is a nice test for proving that subsets of $V$ are geometrically independent.

**Theorem 2.1.6.** *Let $a_0, a_1, \ldots, a_n$ be elements of $V$ such that*

$$a_{i+1} \notin \mathrm{aff}(\{a_0, \ldots, a_i\})$$

*for every $i < n$. Then $\{a_0, \ldots, a_n\}$ is geometrically independent.*

A function between affine subspaces of linear spaces is called *affine* if it preserves affine combinations. Images and preimages of affine sets under affine functions are again affine.

**Theorem 2.1.7.** *Let $V_1$ and $V_2$ be linear spaces, let $A_1 \subseteq V_1$ and $A_2 \subseteq V_2$ be affine subspaces and let $f\colon A_1 \to A_2$ be a function. Then the following statements are equivalent:*

(1) *$f$ is affine,*
(2) *the composition*

$$A_1 - a_1 \xrightarrow{+a_1} A_1 \xrightarrow{f} A_2 \xrightarrow{-a_2} A_2 - a_2 \quad (a_1 \in A_1, a_2 = f(a_1)),$$

*i.e., the function $\xi\colon A_1 - a_1 \to A_2 - a_2$ defined by*

$$\xi(x) = f(x + a_1) - a_2,$$

*is linear.*

From Theorem 2.1.7 it follows that an affine function, the domain and range of which are both contained in a finite dimensional normed linear space, is continuous. This can be seen as follows.

We first claim that for arbitrary $n, m \in \mathbb{N}$, linear functions

$$\mathbb{R}^n \longrightarrow \mathbb{R}^m$$

are continuous. First observe that each linear function $f\colon \mathbb{R} \to \mathbb{R}$ is a multiplication and hence is continuous. Assume that every linear function $f\colon \mathbb{R}^i \to \mathbb{R}$ is continuous for $i \leq n$, and let $F\colon \mathbb{R}^{n+1} \to \mathbb{R}$ be linear. If $f_1 = F \restriction \mathbb{R}^n \times \{0\}$ and

$$f_2 = F \restriction \underbrace{\{0\} \times \{0\} \times \cdots \times \{0\}}_{n-1} \times \mathbb{R}.$$

then they are continuous by assumption. But then $F$ is continuous by linearity, being the sum of the continuous functions

$$(x_1, \ldots, x_{n+1}) \mapsto f_1(x_1, \ldots, x_n, 0)$$

and

$$(x_1, \ldots, x_{n+1}) \mapsto f_2(0, \ldots, 0, x_{n+1}).$$

Consequently, each linear function $f\colon \mathbb{R}^n \to \mathbb{R}^m$ is continuous since $\mathbb{R}^m$ is endowed with the product topology. By Exercise 1.1.18 it follows that each finite dimensional normed linear space is topologically isomorphic to some $\mathbb{R}^n$. So we conclude that a linear function between finite dimensional normed linear spaces is indeed continuous.

This gives us what we want since by Theorem 2.1.7 each affine function $f$, the domain and range of which are both contained in a finite dimensional normed linear space, is of the form

$$f = \xi \circ F \circ \eta,$$

where both $\xi$ and $\eta$ are translations and hence homeomorphisms (Exercise 1.1.3), and $F$ is linear.

**Simplices.** Let $V$ be a linear space. An *n-simplex* in $V$ is a geometrically independent subset of $V$ having precisely $n+1$ points. Simplices are denoted by Greek letters. If $\sigma$ and $\tau$ are simplices and $\sigma \subseteq \tau$ then $\sigma$ is called a *face* of $\tau$; to indicate that $\sigma$ is a face of $\tau$ we shall also use the notation: $\sigma \preccurlyeq \tau$. An $n$-simplex in $V$ is sometimes also called an *n-dimensional simplex*.

**Theorem 2.1.8.** *Let $\sigma$ be a simplex in $V$ and let $A = \mathrm{aff}(\sigma)$. Then every element $x \in A$ can be written uniquely as an affine combination $\sum_{v \in \sigma} t_v \cdot v$ of $\sigma$. In addition, the functions $\alpha_v \colon A \to \mathbb{R}$ defined by $\alpha_v(x) = t_v$ are affine.*

**Corollary 2.1.9.** *Let $\sigma$ and $\tau$ be simplices in $V$ such that $\tau \preccurlyeq \sigma$ and let*
$$A = \mathrm{aff}(\sigma), \quad B = \mathrm{aff}(\tau).$$
*If $x \in B$ then $\alpha_v(x) = 0$ for every $v \in \sigma \setminus \tau$.*

The real numbers $\alpha_v(x)$ for $v \in \sigma$ are called the *affine coordinates* of $x$ with respect to $\sigma$. We call the $\alpha_v$ the *coordinate functions* of $\mathrm{aff}(\sigma)$. This notation will remain in force throughout the book.

Observe that by the above remarks and Theorem 2.1.8 it follows that the coordinate functions of $\mathrm{aff}(\sigma)$ are continuous on $\mathrm{aff}(\sigma)$.

A *geometric simplex* is the convex hull of a simplex. We use $|\sigma|$ as an abbreviation for $\mathrm{conv}(\sigma)$ and sometimes say that $|\sigma|$ is the geometric simplex spanned by $\sigma$. If $x \in |\sigma|$ then it can be written as a convex combination of the elements of $\sigma$ (Lemma 1.1.1). But a convex combination is a special case of an affine combination, and affine coordinates are unique by Theorem 2.1.8. So we conclude that the affine coordinates of $x \in |\sigma|$ are non-negative.

The elements of $\sigma$ are called the vertices of $|\sigma|$. If $\tau$ is a face of $\sigma$ then $|\tau|$ is also called a face of $|\sigma|$. The union of all proper faces of $|\sigma|$ is called the *(geometric) boundary* $\partial|\sigma|$ of $|\sigma|$ and its complement is the (geometric) interior $|\sigma|^\circ$ of $|\sigma|$.

It is clear that the geometric interior of a simplex is non-empty. Just observe that a point in $|\sigma|$ all whose affine coordinates are strictly positive belongs to the geometric interior of $|\sigma|$.

Since a geometric simplex $|\sigma|$ is the convex hull of the finite set $\sigma$ it follows that $|\sigma|$ is compact (Lemma 1.1.1(2)); this remark will be used without explicit reference in the remaining part of this book.

**Theorem 2.1.10.** *If $\sigma$ is a simplex and $\sigma_1, \sigma_2 \subseteq \sigma$ then $|\sigma_1| \cap |\sigma_2| = |\sigma_1 \cap \sigma_2|$.*

The diameter of a geometric simplex is attained at its vertices, as the next result shows.

**Theorem 2.1.11.** *Let $(V, \|\cdot\|)$ be a normed linear space, and let $\sigma \subseteq V$ be a simplex. Then $\mathrm{diam}(\sigma) = \mathrm{diam}(|\sigma|)$.*

**Proof.** We prove that $\operatorname{diam}(|\sigma|) \leq \operatorname{diam}(\sigma)$. Take arbitrary $p, q \in |\sigma|$. Then since $\sum_{v \in \sigma} \alpha_v(p) = 1$ we get

$$\|p - q\| = \left\| \sum_{v \in \sigma} \alpha_v(p) \cdot v - \sum_{v \in \sigma} \alpha_v(p) \cdot q \right\|$$
$$= \left\| \sum_{v \in \sigma} \alpha_v(p)(v - q) \right\|$$

from which it follows by using standard properties of the norm that

$$\|p - q\| \leq \sum_{v \in \sigma} \alpha_v(p) \|v - q\|$$
$$\leq \sum_{v \in \sigma} \alpha_v(p) \cdot \max_{v \in \sigma} \|v - q\|$$
$$= \max_{v \in \sigma} \|v - q\|.$$

As a consequence we obtain for every $w \in \sigma$ that $\|q - w\| \leq \max_{v \in \sigma} \|v - w\|$. So

$$\|p - q\| \leq \max_{w \in \sigma} \|w - q\| \leq \max_{v, w \in \sigma} \|v - w\|,$$

as required. □

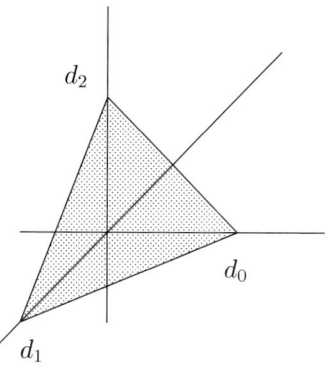

Figure 6.

For each $0 \leq i \leq k$ let $d_i \in \mathbb{R}^{k+1}$ be the vector all coordinates of which are 0, except the $(i+1)$-th coordinate which equals 1. The vectors $d_0, \ldots, d_k$ are clearly linearly independent (hence geometrically independent). For $x$ in $\operatorname{aff}(\{d_0, \ldots, d_k\})$ we have that the affine coordinates of $x = (x_0, \ldots, x_k)$ with respect to $d_0, \ldots, d_k$ are equal to $x_0, \ldots, x_k$, respectively. Put

$$\tau_k = \{d_0, \ldots, d_k\}.$$

Observe that $|\tau_k| = \{x \in \mathbb{I}^{k+1} : \sum_{i=1}^{n} x_i = 1\}$ is a closed and bounded subset of $\mathbb{R}^{k+1}$ and hence is compact (alternatively, use that $|\tau_k|$ is a geometric simplex).

**Theorem 2.1.12.** *Let* $(V, \|\cdot\|)$ *be a normed linear space, and let*
$$\sigma = \{v_0, \ldots, v_k\} \subseteq V$$
*be a $k$-simplex. Then the function $f \colon |\sigma| \to |\tau_k|$ defined by*
$$f(a) = (\alpha_{v_0}(a), \ldots, \alpha_{v_k}(a))$$
*is a homeomorphism.*

**Proof.** First observe that the functions $a_{v_i}$ are affine (Theorem 2.1.8). Consequently, $f$ is continuous by the remark following Theorem 2.1.7, and one-to-one by the definition of affine coordinates (Theorem 2.1.8). Moreover, $f$ is clearly surjective since if $y \in |\tau_k|$ then $f(x) = y$, where
$$x = \sum_{i=1}^{n} y_i \cdot v_i.$$
By compactness of $|\sigma|$ it therefore follows that $f$ is a homeomorphism (Exercise A.5.9). $\square$

**Triangulation.** A *simplicial complex* is a countable collection $\mathcal{S}$ of non-empty, finite sets such that:

(SC)$_{\text{abstract}}$  if $\sigma \in \mathcal{S}$ and $\emptyset \neq \tau \subseteq \sigma$ then $\tau \in \mathcal{S}$.

The elements of the set
$$S = \bigcup \mathcal{S}$$
are called the *vertices* of the simplicial complex $\mathcal{S}$.

Although $S$ need not be a subset of a normed linear space, without loss of generality we may assume that this is the case. In fact, we can think of $S$ as being a linearly independent subset of some normed linear space. This can be seen as follows. For every $n \in \mathbb{N}$ let $x_n \in \ell^2$ be the vector all whose coordinates are 0 except for the $n$-th coordinate which equals 1. Observe that the set
$$D = \{x_n : n \in \mathbb{N}\}$$
is linearly independent. Let $f \colon S \to D$ be an injection, and identify every simplex $\sigma \in \mathcal{S}$ with $f[\sigma] \subseteq \ell^2$. Observe that $S$, now regarded to be a subset of $\ell^2$, is linearly independent and hence geometrically independent. By Theorem 2.1.10 it therefore follows that for every $\sigma_1, \sigma_2 \in \mathcal{S}$ we have

(SC)  $\qquad\qquad |\sigma_1| \cap |\sigma_2| = |\sigma_1 \cap \sigma_2|.$

In the sequel, when dealing with a simplicial complex $\mathcal{S}$, we shall always require that $S = \bigcup \mathcal{S}$ is contained in a normed linear space $L$, and that (SC) holds. The set $|\mathcal{S}| = \bigcup\{|\sigma| : \sigma \in \mathcal{S}\}$ is called the *geometric realization* of $\mathcal{S}$

and $|S|$ is said to be *triangulated* by $S$ (or $S$ is a *triangulation* of $|S|$). Observe that the collection $S_{\|} = \{|\sigma| : \sigma \in S\}$ consists of geometric simplices in $L$ and has the following properties:

(SC)$_{\text{geometric}}$    (1)   if $|\sigma| \in S_{\|}$ and $\tau$ is a face of $\sigma$ then $|\tau| \in S_{\|}$,
                    (2)   for all $|\sigma_1|, |\sigma_2| \in S_{\|}$ with $|\sigma_1| \cap |\sigma_2| \neq \emptyset$, $|\sigma_1| \cap |\sigma_2|$ is a face of $|\sigma_1|$ as well as $|\sigma_2|$.

Let $|\sigma|$ be a geometric simplex in $L$. The collection of all faces of $|\sigma|$ is clearly a simplicial complex which triangulates $|\sigma|$. It is called the *standard triangulation* of $|\sigma|$ and is denoted by $\mathcal{F}(\sigma)$.

**Topologizing a simplicial complex.** From this moment on, we will no longer distinguish between an abstract simplicial complex and its geometric realization. So a simplicial complex in a normed linear space $L$ is a collection $S$ of geometric simplices (which by abuse of terminology we will call simplices for short in the sequel) in $L$ such that for all $\sigma_1, \sigma_2 \in S$ we have

(1) if $\sigma \in S$ and $\tau$ is a face of $\sigma$ then $\tau \in S$,
(2) $\sigma_1 \cap \sigma_2 \neq \emptyset \Rightarrow \sigma_1 \cap \sigma_2$ is a face of $\sigma_1$ as well as $\sigma_2$.

A subcollection $\mathcal{T}$ of $S$ which is also a simplicial complex is called a *subcomplex* of $S$.

Let $\mathcal{T}$ be a simplicial complex in $L$. For each $m \geq 0$ define

$$\mathcal{T}^{(m)} = \{s \in \mathcal{T} : \sigma \text{ is at most } m\text{-dimensional}\}.$$

So $\mathcal{T}^{(0)}$ is the set of vertices of all simplices in $\mathcal{T}$, $\mathcal{T}^{(1)}$ is the collection of all at most one-dimensional faces of all simplices in $\mathcal{T}$, etc. The elements of $\mathcal{T}^{(0)}$ are called the *vertices* of $\mathcal{T}$. It is a triviality that $\mathcal{T}^{(m)}$ is a subcomplex of $\mathcal{T}$; we call it the *m-skeleton* of $\mathcal{T}$.

Let $L$ be a normed linear space and let $\mathcal{T}$ be a simplicial complex in $L$. We are interested in useful topologies on the set $X = \bigcup \mathcal{T}$. There are several natural candidates for such a topology. First, $X$ is a subset of $L$ and therefore carries the subspace topology inherited from $L$. It turns out that this is in general an interesting but not a useful topology. We shall now describe a better topology. To this end, let us first agree that there cannot be ambiguity about the topology that each simplex of $\mathcal{T}$ should carry. By Theorem 2.1.12, each $k$-dimensional simplex in $L$ with its subspace topology is naturally homeomorphic to a standard $k$-dimensional simplex in $\mathbb{R}^{k+1}$; each simplex in $L$ from this moment on is endowed with its subspace topology. It would be very unnatural if the useful topology we want to define on $X$ should induce a different topology on one of the simplices of $\mathcal{T}$. So there is a natural candidate for a topology on $X$, namely, the largest topology that induces the 'right' topology on every simplex of $\mathcal{T}$, i.e.,

$$U \subseteq X \text{ is open iff for every simplex } \tau \in \mathcal{T}, U \cap \tau \text{ is open in } \tau.$$

This is called the *Whitehead topology* on $X$ and $X$ with this topology is usually denoted by $|\mathcal{T}|$.

**Lemma 2.1.13.** *Let $\mathcal{T}$ be a simplicial complex in $L$ and let $X = \bigcup \mathcal{T}$. Then*

(1) *The Whitehead topology on $X$ is a topology and is finer than the topology that $X$ inherits from $L$, i.e., the inclusion*
$$|\mathcal{T}| \hookrightarrow X \subseteq L$$
*is continuous,*

(2) *On every simplex of $\mathcal{T}$ the Whitehead topology and the topology on $L$ induce the same subspace topology.*

**Proof.** The proof of (1) is a triviality the verification of which we leave to the reader.

For (2), let $\tau$ be a simplex of $\mathcal{T}$. It suffices to prove that for every relatively open subset $U \subseteq \tau$ there exists a Whitehead-open $V \subseteq |\mathcal{T}|$ such that $V \cap \tau = U$. We claim that $V = U \cup (|\mathcal{T}| \setminus \tau)$ is as required. It is clear that $V \cap \tau = U$, and so it remains to prove that $V$ is Whitehead-open. To this end, let $\sigma \in \mathcal{T}$ and put $V' = V \cap (\tau \cup \sigma)$. Then $V'$ is open in $\tau \cup \sigma$ since its complement is equal to $\tau \setminus U$ which is compact and hence closed in $\tau \cup \sigma$. Consequently, $V' \cap \sigma = V \cap \sigma$ is open in $\sigma$, as required. □

Observe that a subset $A$ of $|\mathcal{T}|$ is closed if and only if $A \cap \tau$ is closed for every $\tau \in \mathcal{T}$. This immediately implies that $|\mathcal{T}|$ is a $T_1$-space, i.e., a space in which every singleton is closed (this follows also from Lemma 2.1.13(1)).

Due to our self-chosen (sometimes unpleasant) restriction to deal with separable metrizable spaces exclusively for the time being, there is a small problem with the Whitehead topology since it need not be metrizable. However, there is a simple combinatorial condition on $\mathcal{T}$ that ensures that $|\mathcal{T}|$ is separable and metrizable; call a simplicial complex $\mathcal{T}$ *locally finite* if every vertex of $\mathcal{T}$ is contained in at most finitely many simplices of $\mathcal{T}$. In turns out that for our considerations it always suffices to consider locally finite simplicial complexes and these are precisely the simplicial complexes that are separable metrizable when given the Whitehead topology, see Proposition 2.1.21 below. For the moment, we shall not worry about the metrizability of $|\mathcal{T}|$ but we shall prove a few easy but important lemmas from which metrizability in the locally finite case will follow rather easily later.

**Lemma 2.1.14.** *Let $\mathcal{T}$ be a simplicial complex with subcollection $\mathcal{S}$. Then*

(1) $\bigcup \mathcal{S}$ *is a closed subspace of $|\mathcal{T}|$,*

(2) *if $\mathcal{S}$ is a subcomplex then the topology that $\bigcup \mathcal{S}$ inherits from $|\mathcal{T}|$ coincides with the Whitehead topology on $\bigcup \mathcal{S}$.*

**Proof.** For (1), for convenience, put $Y = \bigcup \mathcal{S}$. Since each simplex of $\mathcal{T}$ has finitely many faces only, and $\mathcal{T}$ is a simplicial complex, it is clear that for every $\tau \in \mathcal{T}$, $Y \cap \tau$ is closed in $\tau$. By definition of the Whitehead topology, we therefore conclude that $Y$ is closed in $|\mathcal{T}|$.

For (2) observe that since $\mathcal{S} \subseteq \mathcal{T}$, the topology that $\bigcup \mathcal{S}$ inherits from $|\mathcal{T}|$ is coarser than its Whitehead topology. Now let $A \subseteq |\mathcal{S}|$ be closed. We shall prove that $A$ is a closed subset of $|\mathcal{T}|$. To this end, let $\tau$ be an arbitrary simplex in $\mathcal{T}$. Since $\tau$ has finitely many faces only, there is a finite subcollection $\mathcal{F}$ of $\mathcal{S}$ such that
$$\bigcup \mathcal{F} \cap \tau = \bigcup \mathcal{S} \cap \tau.$$
Consequently, $A \cap \tau = (A \cap \bigcup \mathcal{S}) \cap \tau = (A \cap \bigcup \mathcal{F}) \cap \tau$. Now since $A \subseteq |\mathcal{S}|$ is closed, for every $\sigma \in \mathcal{F}$, $A \cap \sigma$ is closed in $\sigma$, and consequently, $A \cap \sigma \cap \tau$ is closed in $\tau$. Since $\mathcal{F}$ is finite, it therefore follows that $(A \cap \bigcup \mathcal{F}) \cap \tau$ is closed in $\tau$. We conclude that $A \cap \tau$ is closed in $\tau$. □

From the above lemma it follows that if $\mathcal{T}$ is a simplicial complex and if $\mathcal{S}$ is a subcomplex of $\mathcal{T}$ then we can identify $|\mathcal{S}|$ and the subspace $\bigcup \mathcal{S}$ of $|\mathcal{T}|$. It will be convenient to do that.

**Corollary 2.1.15.** *Let $\mathcal{T}$ be a simplicial complex. Then $|\mathcal{T}^{(0)}|$ carries the discrete topology.*

**Proof.** Since $\mathcal{T}^{(0)}$ is a subcollection of $\mathcal{T}$, by Lemma 2.1.14(1) each subset of $\mathcal{T}^{(0)}$ is closed in $|\mathcal{T}|$. So the subspace topology of $\bigcup \mathcal{T}^0$ is the discrete topology. An application of Lemma 2.1.14(2) therefore does the job. □

This corollary easily shows that the Whithead topology on a simplicial complex is usually different from the topology it inherits from its ambient linear space.

**Example 2.1.16.** The collection $\mathcal{T} = \{\{0\}, \{1\}, \{^1\!/_2\}, \{^1\!/_3\}, \dots\}$ consists of 0-dimensional simplices of $\mathbb{R}$ and hence is a simplicial complex. The Whitehead topology on $\bigcup \mathcal{T}$ is the discrete topology by Corollary 2.1.15. But the subspace topology on $\bigcup \mathcal{T}$ is not discrete.

For each $x \in |\mathcal{T}|$ define the *star*, $\operatorname{St} x$, and the *carrier*, $\operatorname{car} x$, of $x$ by
$$\operatorname{St} x = |\mathcal{T}| \setminus \bigcup \{\tau \in \mathcal{T} : x \notin \tau\},$$
and
$$\operatorname{car} x = \bigcap \{\tau \in \mathcal{T} : x \in \tau\},$$
respectively. We shall prove that $\operatorname{car} x \in \mathcal{T}$ and it is therefore the smallest simplex of $\mathcal{T}$ that contains $x$.

**Corollary 2.1.17.** *Let $\mathcal{T}$ be a simplicial complex. Then for every $x \in |\mathcal{T}|$,*

(1) *$\operatorname{St} x$ is open in $|\mathcal{T}|$,*

(2) car $x$ belongs to $\mathcal{T}$.

In addition, the family $\{\operatorname{St} x : x \in \mathcal{T}^{(0)}\}$ covers $|\mathcal{T}|$.

**Proof.** For (1), let $\mathcal{S} = \{\tau \in \mathcal{T} : x \notin \tau\}$. From Lemma 2.1.14 we obtain that $\bigcup \mathcal{S}$ is closed in $|\mathcal{T}|$, hence $\operatorname{St} x$ is open.

For (2), first take $\tau \in \mathcal{T}$ such that $x \in \tau$. Since for every $\sigma \in \mathcal{T}$ such that $x \in \sigma$ we have $\sigma \cap \tau \preccurlyeq \tau$, and $\tau$ has only finitely many faces, car $x \in \mathcal{T}$.

Take an arbitrary $y \in |\mathcal{T}|$. If car $y$ is a zero-dimensional simplex then $y$ is a vertex of $\mathcal{T}$ and hence belongs to $\bigcup\{\operatorname{St} x : x \in \mathcal{T}^{(0)}\}$. Assume that car $y$ is not zero-dimensional and let $x$ be a vertex of car $y$. Observe that car $y$ is the smallest simplex containing $y$, hence $y$ does not belong to the boundary of car $y$. We claim that $y \in \operatorname{St} x$. If not, then there is a simplex $\tau \in \mathcal{T}$ containing $y$ but not containing $x$. Then $\tau \cap $ car $y$ is a proper face of car $y$ containing $y$, contradiction. $\square$

**Lemma 2.1.18.** *Let $\mathcal{T}$ be a simplicial complex and Let $K \subseteq |\mathcal{T}|$ be compact. Then $K$ is metrizable and is contained in the union of finitely many simplices of $\mathcal{T}$.*

**Proof.** That $K$ is metrizable follows immediately from Lemma 2.1.13(1) and Exercise A.5.9.

Suppose that $K$ is not contained in the union of finitely many simplices of $\mathcal{T}$. Then we can find an infinite collection $\mathcal{S} = \{\tau_n : n \in \mathbb{N}\}$ of simplices in $\mathcal{T}$ and for every $n \in \mathbb{N}$ a point

$$(*) \qquad x_n \in (K \cap \tau_{n+1}) \setminus \bigcup_{i=1}^{n} \tau_n.$$

Since $K$ is compact, and metrizable, we may by Theorem A.5.1 assume without loss of generality that $x = \lim_{n \to \infty} x_n$ exists (and belongs to $K$). Now let $\tau \in \mathcal{T}$ be an arbitrary simplex. Since $\tau_n \cap \tau$ is either empty or a face of $\tau$, there exists $m \in \mathbb{N}$ such that

$$\bigcup_{n=1}^{\infty} \tau_n \cap \tau = \bigcup_{n=1}^{m} \tau_n \cap \tau,$$

from which it follows that

$$\{x_n : n \in \mathbb{N}\} \cap \tau \subseteq \{x_1, \ldots, x_m\}.$$

By the definition of the Whitehead topology, this implies that $\{x_n : n \in \mathbb{N}\}$ is closed in $|\mathcal{T}|$ (recall that $|\mathcal{T}|$ is $T_1$). Since the sequence $(x_n)_n$ converges, it therefore has to be eventually constant, which contradicts $(*)$. $\square$

One of the reasons that the Whitehead topology on a simplicial complex is relatively simple to deal with is that it is easy to check that certain functions are continuous.

**Lemma 2.1.19.** *Let $\mathfrak{T}$ be a simplicial complex and let $X$ be a space. A function $f\colon |\mathfrak{T}| \to X$ is continuous if and only if the restriction of $f$ to every simplex $\tau \in \mathfrak{T}$ is continuous on $\tau$.*

**Proof.** Suppose that $f\colon |\mathfrak{T}| \to X$ is such that the restriction of $f$ to every simplex $\tau \in \mathfrak{T}$ is continuous. Let $U \subseteq X$ be open. For each $\tau \in \mathfrak{T}$,
$$f^{-1}[U] \cap \tau = (f \restriction \tau)^{-1}[U]$$
is open in $\tau$. By the definition of the Whitehead topology, it therefore follows that $f^{-1}[U]$ is open in $|\mathfrak{T}|$. □

**Corollary 2.1.20.** *Let $\mathfrak{T}$ be a simplicial complex. Then $|\mathfrak{T}|$ is a normal topological space, i.e., for every pair of disjoint closed subsets $A$ and $B$ of $|\mathfrak{T}|$ there exists a continuous function $f\colon |\mathfrak{T}| \to \mathbb{I}$ such that $f \restriction A \equiv 0$ and $f \restriction B \equiv 1$.*

**Proof.** Since $\mathfrak{T}$ is countable, we can enumerate it as $\{\tau_n : n \in \mathbb{N}\}$. By Lemma A.4.1 there exists a continuous function $f_1\colon \tau_1 \to \mathbb{I}$ such that
$$f_1 \restriction (\tau_1 \cap A) \equiv 0, \quad f \restriction (\tau_1 \cap B) \equiv 1.$$
By induction on $n$, we shall construct continuous functions $f_n\colon \bigcup_{i=1}^{n} \tau_i \to \mathbb{I}$ such that

(1) $f_n \restriction (A \cap \bigcup_{i=1}^{n} \tau_i) \equiv 0$,
(2) $f_n \restriction (B \cap \bigcup_{i=1}^{n} \tau_i) \equiv 1$,
(3) if $n > 1$ then $f_n$ extends $f_{n-1}$.

Since we already defined $f_1$, assume $f_i$ to be constructed for $1 \leq i \leq n$. We shall construct $f_{n+1}$. Let $A' = A \cap \bigcup_{i=1}^{n+1} \tau_i$ and $B' = B \cap \bigcup_{i=1}^{n+1} \tau_i$, respectively. Define $g\colon A' \cup B' \cup \bigcup_{i=1}^{n} \tau_i \to \mathbb{I}$ by
$$g(x) = \begin{cases} 0 & (x \in A'), \\ 1 & (x \in B'), \\ f_n(x) & (x \in \bigcup_{i=1}^{n} \tau_i). \end{cases}$$
Then $g$ is clearly continuous and extends $f_n$. Since $\bigcup_{i=1}^{n+1} \tau_i$ is compact, it is metrizable by Lemma 2.1.18. By Corollary 1.2.5 there consequently exists a continuous extension $f_{n+1}\colon \bigcup_{i=1}^{n+1} \tau_i \to \mathbb{I}$ of $g$. It is clear that $f_{n+1}$ is as required.

Now put $f = \bigcup_{i=1}^{\infty} f_n$. Then $f$ is a well-defined function from $|\mathfrak{T}|$ into $\mathbb{I}$ having the property that $f \restriction A \equiv 0$ and $f \restriction B \equiv 1$. Moreover, $f$ is continuous by Lemma 2.1.19. □

These simple results enable us to prove the following

**Proposition 2.1.21.** *Let $\mathfrak{T}$ be a locally finite simplicial complex in the linear space $L$. Then $|\mathfrak{T}|$ is a locally compact separable metrizable space.*

**Proof.** We shall first prove that each point of $|\mathcal{T}|$ has an open neighborhood which is separable, metrizable, and has compact closure in $|\mathcal{T}|$. To this end, take an arbitrary $x \in |\mathcal{T}|$. By Corollary 2.1.17(1), St $x$ is open. Since $\mathcal{T}$ is locally finite, St $x$ is contained in the union of finitely many simplices of $\mathcal{T}$. This can be seen as follows. First observe that St $x$ is contained in the union of all simplices that contains $x$. But, as we will show, there are only finitely many of such simplices. Striving for a contradiction, assume that $\mathcal{S}$ is an infinite subcollection of $\mathcal{T}$ such that $x \in \bigcap \mathcal{S}$. Fix $\sigma \in \mathcal{S}$. Then for $\sigma' \in \mathcal{S} \setminus \{\sigma\}$ we have that $\sigma' \cap \sigma$ is a face of $\sigma$. So there are an infinite subcollection $\mathcal{S}'$ of $\mathcal{S}$ and a face $\tau$ of $\sigma$ such that for every $\sigma' \in \mathcal{S}'$ we have $\sigma' \cap \sigma = \tau$. But then any vertex of $\tau$ is contained in infinitely many simplices of $\mathcal{T}$ which contradicts the local finiteness condition.

Consequently, St $x$ is contained in a subspace of $|\mathcal{T}|$ which is compact and therefore metrizable (Lemma 2.1.18). We conclude that $|\mathcal{T}|$ is locally compact.

Since $\mathcal{T}$ is countable, $|\mathcal{T}|$ is the union of countably many compact subspaces. So $|\mathcal{T}|$ is Lindelöf and from the above it therefore follows that $|\mathcal{T}|$ can be covered by a countable family $\mathcal{B}$ consisting of open, separable metrizable subspaces of $|\mathcal{T}|$. For each $B \in \mathcal{B}$ let $\mathcal{F}(B)$ be a countable base for $B$ consisting of open subsets of $B$. Then $\mathcal{F} = \bigcup_{B \in \mathcal{B}} \mathcal{F}(B)$ is countable, consists of open subsets of $|\mathcal{T}|$, and is easily seen to be a base for $|\mathcal{T}|$. So by Corollary 2.1.20 it follows that $|\mathcal{T}|$ is a second countable regular $T_1$-space and is therefore a separable metrizable space (see Page 465). □

In Exercise 2.1.15 we shall see that if $|\mathcal{T}|$ is metrizable then $\mathcal{T}$ is locally finite.

We shall now prove that the Whitehead topology on $|\mathcal{T}|$ is in fact 'intrinsic', i.e., that it has nothing to do with the topology on $L$. To begin with, let us say that a subset $F$ of $\mathcal{T}^{(0)}$, the 0-skeleton of $\mathcal{T}$, spans a simplex in $\mathcal{T}$ if there exists a simplex $\tau = |\{x_0, \ldots, x_k\}| \in \mathcal{T}$ such that $F = \{x_0, \ldots, x_k\}$.

**Lemma 2.1.22.** *Let $\mathcal{T}$ be a simplicial complex in $L$ and let $F$ be a subset of $\mathcal{T}^{(0)}$. Then $F$ spans a simplex in $\mathcal{T}$ if and only if $\bigcap_{x \in F} \text{St } x \neq \emptyset$.*

**Proof.** Suppose that $F$ spans a simplex $\tau$ in $\mathcal{T}$ and let $b$ be a point in $|\tau|^\circ$. We claim that $b \in \bigcap_{x \in F} \text{St } x$. If not, then there exists $x \in F$ such that $b \notin \text{St } x$, or, equivalently, there is a simplex $\sigma \in \mathcal{T}$ containing $b$ but not containing $x$. Then $\sigma \cap \tau$ is a face of $\tau$. However, $b$ is not contained in a proper face of $\tau$, which implies that $\sigma \cap \tau = \tau$ and hence that $x \in \tau \subseteq \sigma$, contradiction. Conversely, assume that $F$ is such that there exists $y \in \bigcap_{x \in F} \text{St } x$. We claim that $F$ spans a face of car $y$. If for certain $x \in F$, $x \notin \text{car } y$ then by definition, St $x \cap \text{car } y = \emptyset$, which is a contradiction. Consequently, $F \subseteq \text{car } y$. Since $\mathcal{T}$ is a simplicial complex, and $F \subseteq \mathcal{T}^{(0)}$, each $x \in F$ is a vertex of car $y$. We conclude that $F$ spans a simplex in $\mathcal{T}$. □

Now let $L$ and $E$ be normed linear spaces and let $\mathcal{T}$ and $\mathcal{S}$ be simplicial complexes in $L$ and $E$, respectively. We say that $\mathcal{T}$ and $\mathcal{S}$ are *combinatorially equivalent* if there exists a bijection $f\colon \mathcal{T}^{(0)} \to \mathcal{S}^{(0)}$ such that $F \subseteq \mathcal{T}^{(0)}$ spans a simplex in $\mathcal{T}$ if and only if $f[F]$ spans a simplex in $\mathcal{S}$. Let us also say that $|\mathcal{T}|$ and $|\mathcal{S}|$ are *simplicially homeomorphic* if there exists a homeomorphism $h\colon |\mathcal{T}| \to |\mathcal{S}|$ such that for all $\tau \in \mathcal{T}$ and $\sigma \in \mathcal{S}$, we have $h[\tau] \in \mathcal{S}$ as well as $h^{-1}[\sigma] \in \mathcal{T}$.

**Proposition 2.1.23.** *Let $L$ and $E$ be normed linear spaces and let $\mathcal{T}$ and $\mathcal{S}$ be simplicial complexes in $L$ and $E$, respectively. The following statements are equivalent:*

(1) *$\mathcal{T}$ and $\mathcal{S}$ are combinatorially equivalent,*
(2) *$|\mathcal{T}|$ and $|\mathcal{S}|$ are simplicially homeomorphic.*

**Proof.** For (1) $\Rightarrow$ (2), let $f\colon \mathcal{T}^{(0)} \to \mathcal{S}^{(0)}$ be a bijection such that $F \subseteq \mathcal{T}^{(0)}$ spans a simplex in $\mathcal{T}$ if and only if $f[F]$ spans a simplex in $\mathcal{S}$. So for every simplex $\tau = |\{x_0, \ldots, x_k\}| \in \mathcal{T}$, the set $\{f(x_0), \ldots, f(x_k)\}$ spans a simplex $\sigma$ of $\mathcal{S}$. Define a homeomorphism $f_\tau \colon \tau \to \sigma$ by

$$f_\tau\left(\sum_{i=0}^{k} \alpha_i(x) x_i\right) = \sum_{i=0}^{k} \alpha_i(x) f(x_i);$$

here $\alpha_0(x), \ldots, \alpha_k(x)$ denote the affine coordinates with respect to $\tau$ of an arbitrarily chosen point $x \in \tau$ of course. It is easily seen that $f_\tau$ is indeed a homeomorphism, cf. Theorem 2.1.12. Now define $\bar{f}\colon |\mathcal{T}| \to |\mathcal{S}|$ by

$$\bar{f} = \bigcup_{\tau \in \mathcal{T}} f_\tau.$$

By unicity of affine coordinates (Theorem 2.1.8), $\bar{f}$ is well-defined. In addition, $\bar{f}$ is continuous since the restriction of $\bar{f}$ to every simplex of $\mathcal{T}$ is continuous (Lemma 2.1.19). It is easily seen that $f$ is one-to-one and surjective and that $\bar{f}^{-1}$ is continuous for the same reason $\bar{f}$ is. We conclude that $f$ is a homeomorphism. Clearly both $\bar{f}$ and $\bar{f}^{-1}$ are 'simplex preserving'.

Conversely, let $\bar{f}\colon |\mathcal{T}| \to |\mathcal{S}|$ be a homeomorphism such as given by (2). The function

$$f \restriction \mathcal{T}^{(0)} \colon \mathcal{T}^{(0)} \to \mathcal{S}^{(0)}$$

is clearly as required. $\square$

From the above proposition we conclude that the topology of a simplicial complex depends only on the combinatorial properties of its vertex set. For that reason when discussing a simplicial complex, it is no longer necessary to mention the normed linear space it is a subset of.

Let $X$ be a space. We say that $X$ is a *polytope* if there exists a locally finite simplicial complex $\mathcal{T}$ such that $X$ and $|\mathcal{T}|$ are homeomorphic. In case $\mathcal{T}$

is finite we say that $X$ is a *polyhedron*. Observe that each polyhedron is compact. Polytopes are very important since they are the 'bridge' between abstract topological spaces and concrete ones.

If $X$ is a polytope, say $X$ is homeomorphic to $|\mathcal{T}|$, then we will find it sometimes convenient to not distinguish between $X$ and $|\mathcal{T}|$. This will never cause confusion because the triangulation under consideration will always be explicitly defined but the reader should keep in mind that there usually are many different triangulations of the same polytope.

If $|\mathcal{T}|$ is a polytope then a subset $Y \subseteq |\mathcal{T}|$ is called a *subpolytope* if there is a subcomplex $\mathcal{S} \subseteq \mathcal{T}$ such that $Y = |\mathcal{S}|$. Similarly for *subpolyhedron*.

**Theorem 2.1.24.** *Each polyhedron is an ANR.*

**Proof.** Let $\mathcal{T}$ be a finite simplicial complex. By induction on the cardinality of $\mathcal{T}$ we shall prove that $|\mathcal{T}|$ is an ANR. If $\mathcal{T}$ consists of one simplex only then $|\mathcal{T}|$ is an AR by Theorems 2.1.12 and 1.2.9. Suppose that for every simplicial complex $\mathcal{T}$ of cardinality at most $n - 1 \geq 1$, $|\mathcal{T}|$ is an ANR, and let $\mathcal{T}$ be a simplicial complex of cardinality $n$. Since $\mathcal{T}$ is finite, there exists a simplex $\tau \in \mathcal{T}$ such that for every $\sigma \in \mathcal{T}$ with $\tau \preccurlyeq \sigma$, $\sigma = \tau$ (let $\tau$ be any simplex of maximal dimension). Put $\mathcal{S} = \mathcal{T} \setminus \{\tau\}$. Then $\mathcal{S}$ is a simplicial complex and $|\mathcal{T}| = \tau \cup |\mathcal{S}|$. In addition, clearly $\tau \cap |\mathcal{S}| = |\{\tau \cap \sigma : \sigma \in S\}|$. So by our inductive assumptions and by Theorem 1.2.16(1) it follows that $|\mathcal{T}|$ is an ANR. □

In Corollary 4.3.6 we shall prove that every polytope is an ANR, which generalizes Theorem 2.1.24.

**Exercises for §2.1.**

1. Prove Theorem 2.1.1.
2. Prove Theorem 2.1.2.
3. Prove Theorem 2.1.3.
4. Prove Proposition 2.1.5.
5. Prove Theorem 2.1.6.
6. Prove Theorem 2.1.8.
7. Prove Theorem 2.1.10.
8. Let $A$ be an affine subspace of a finite dimensional linear space $L$. Prove that $A$ is closed in $L$.
▶9. Prove that if $\sigma$ is an $n$-simplex in $\mathbb{R}^n$ then $\mathrm{Int}(|\sigma|)$ is equal to $|\sigma|^\circ$.
10. Let $\sigma$ be a simplex and let $p \in \sigma$. Prove that $\sigma \setminus \{p\}$ is convex if and only if $p$ is one of the vertices.

11. Let $\sigma$ be a simplex and let $C$ be a convex set which is contained in $\partial \sigma$. Prove that $C$ is contained in a proper face of $\sigma$.

12. Let $X$ be a set triangulated by the simplicial complex $S$. Prove that the collection of all geometric interiors of elements of $|S|$ is a partition of $X$.

13. Let $\mathcal{U}$ be a countable collection of sets. Prove that the collection
$$\{\mathcal{F} : \mathcal{F} \subseteq \mathcal{U} \text{ is finite and } \bigcap \mathcal{F} \neq \emptyset\}$$
is a simplicial complex.

14. Let $\mathcal{T}$ be a simplicial complex. Prove that $|\mathcal{T}|$ is paracompact (we do not assume that $\mathcal{T}$ is locally finite).

15. Let $\mathcal{T}$ be a simplicial complex that is not locally finite. Prove that $|\mathcal{T}|$ is not metrizable (this is the converse to Proposition 2.1.21).

▶16. Let $\mathcal{T}$ be a simplicial complex and $S$ a subcomplex of $\mathcal{T}$. Prove that the topological interior and topological boundary of $|S|$ in $|\mathcal{T}|$ are given by
$$\{x \in |S| : x \in \sigma \in \mathcal{T} \Rightarrow \sigma \in S\}$$
and
$$\bigcup (\mathcal{T} \setminus S) \cap |S|,$$
respectively.

## 2.2. Barycenters and subdivisions

Througout this section, $L$ is a fixed normed linear space. All simplices under consideration are subsets of $L$. If $\sigma$ is a simplex in $L$ then $\dot{\sigma}$ denotes the set of its vertices.

**Barycenters.** Let $\sigma$ be a simplex. The *barycenter* $b_\sigma$ of $\sigma$ is the point in $\sigma$ whose affine coordinates (with respect to $\dot{\sigma}$) are all equal: if
$$\dot{\sigma} = \{v_0, \ldots, v_n\}$$
then
$$b_\sigma = \sum_{i=0}^{n} \frac{1}{n+1} \cdot v_i.$$
Observe that $b_\sigma$ belongs to the geometric interior $\sigma^\circ$ of $\sigma$.

**Lemma 2.2.1.** *Let $S$ be a simplicial complex. Then*

*if $\sigma, \sigma' \in S$ and $b_\sigma \in \sigma'$ then $\sigma$ is a face of $\sigma'$.*

**Proof.** By (SC), $\sigma \cap \sigma'$ is a face of $\sigma$. This face contains $b_\sigma$ so it must be $\sigma$ itself. □

**Lemma 2.2.2.** *Let $\tau_1 \subset \tau_2 \subset \cdots \subset \tau_k$ be a strictly increasing collection of faces of a simplex $\sigma$. Then the barycenters $b_{\tau_1}, b_{\tau_2}, \ldots, b_{\tau_k}$ are geometrically independent.*

**Proof.** By Theorem 2.1.6 it suffices to show that for $i < k$,
$$b_{\tau_{i+1}} \notin \mathrm{aff}(\{b_{\tau_1}, b_{\tau_2}, \ldots, b_{\tau_i}\}).$$
To this end, take an arbitrary vertex $v \in \dot\tau_{i+1} \setminus \dot\tau_i$. Then
$$\alpha_v(b_{\tau_{i+1}}) = \frac{1}{|\dot\tau_{i+1}|} \neq 0,$$
Since
$$\{b_{\tau_1}, b_{\tau_2}, \ldots, b_{\tau_i}\} \subseteq \mathrm{aff}(\dot\tau_i)$$
we get
$$\mathrm{aff}(\{b_{\tau_1}, b_{\tau_2}, \ldots, b_{\tau_i}\}) \subseteq \mathrm{aff}(\dot\tau_i).$$
This and Corollary 2.1.9 imply that for every $x \in \mathrm{aff}(\{b_{\tau_1}, b_{\tau_2}, \ldots, b_{\tau_i}\})$ we have $\alpha_v(x) = 0$. This is clearly as required. □

Let $\sigma$ be a simplex and let $\mathcal{K} = \{\tau_1, \tau_2, \ldots, \tau_k\}$ be a chain of faces of $\sigma$ (by a chain we mean of course a chain with respect to inclusion) with corresponding chain of vertices $\dot{\mathcal{K}} = \{\dot\tau_1, \dot\tau_2, \ldots, \dot\tau_k\}$. If $\mathcal{K}$ is a *maximal* chain of faces then clearly $\sigma \in \mathcal{K}$ and consequently $\bigcup \dot{\mathcal{K}} = \dot\sigma$. This rather trivial remark will be used without explicit reference a few times in the remaining part of this section.

For every $v \in \bigcup \dot{\mathcal{K}}$ define its *height* $ht(v)$ to be the first $i \leq k$ with $v \in \dot\tau_i$. Define a *quasi-order* '$\preccurlyeq$' on $\bigcup \dot{\mathcal{K}}$ by putting: $v \preccurlyeq w \Leftrightarrow ht(v) \leq ht(w)$.

**Lemma 2.2.3.** *Let $\sigma$ be an $n$-simplex and let $\mathcal{K}$ be a maximal chain of faces of $\sigma$ with corresponding quasi-order $\preccurlyeq$. Then*

(1) *$\mathcal{K}$ has cardinality $n+1$ and the corresponding quasi-order $\preccurlyeq$ is a linear order on $\dot\sigma = \bigcup \dot{\mathcal{K}}$.*

(2) *If $\beta$ is the geometrically independent set of all barycenters of elements of $\mathcal{K}$ then it is geometrically independent and*
$$|\beta| = \{x \in \sigma : (\forall v, w \in \dot\sigma)(v \preccurlyeq w \Rightarrow \alpha_v(x) \geq \alpha_w(x))\}.$$

(3) *If $v \preccurlyeq w$ and for certain $x \in |\beta|$ we have $\alpha_v(x) = \alpha_w(x)$ then for every $\tau \in \mathcal{K}$ with $v \in \dot\tau$ and $w \notin \dot\tau$ we have that the affine coordinate of $x$ in $|\beta|$ with respect to $b_\tau$ is equal to 0.*

**Proof.** We claim that if $\alpha$ is the direct successor of $\beta$ in $\mathcal{K}$ then $\dot\alpha \setminus \dot\beta$ is a singleton: if it contains two distinct points $v$ and $w$ then
$$\ldots, \beta, |\dot\beta \cup \{v\}|, \alpha, \ldots$$
would be a chain extending $\mathcal{K}$; likewise one sees that the first element of $\mathcal{K}$ is a singleton. Thus $\mathcal{K} = \{\tau_0, \tau_1, \ldots, \tau_n\}$ where $\dot\tau_i$ has $i+1$ points. Now write $\dot\sigma = \{v_0, v_1, \ldots, v_n\}$ such that for every $0 \leq i \leq n$, $\dot\tau_i = \{v_0, v_1, \ldots, v_i\}$. Then $v_i \preccurlyeq v_j$ if and only if $i \leq j$. This proves (1).

For (2), observe that $\beta$ is geometrically independent by Lemma 2.2.2. An arbitrary element $x \in |\beta|$ can be written as

$$x = \sum_{i=0}^{n} t_i \cdot b_{\tau_i} \text{ where } t_i \geq 0 \text{ for every } i \text{ and } \sum_{i=0}^{n} t_i = 1$$

(Lemma 1.1.1). Observe that

$$x = \sum_{i=0}^{n} t_i \cdot b_{\tau_i} = \sum_{i=0}^{n} t_i \cdot \sum_{j=0}^{i} \frac{v_j}{j+1} = \sum_{i=0}^{n} \left( \sum_{j=i}^{n} \frac{t_j}{j+1} \right) v_i.$$

For every $0 \leq i \leq n$ define

$$s_i = \sum_{j=i}^{n} \frac{t_j}{j+1}$$

and observe that

$$\sum_{i=0}^{n} s_i = \sum_{i=0}^{n} \left[ t_i \cdot \sum_{j=0}^{i} \frac{1}{i+1} \right] = \sum_{i=0}^{n} t_i = 1.$$

From this we conclude that the $s_i$ are the affine coordinates of $x$ (Theorem 2.1.8) and satisfy $s_0 \geq s_1 \geq \cdots \geq s_n$.

Conversely, let $p$ be a point in $\sigma$ whose affine coordinates $s_0, \ldots, s_n$ have the property that $s_0 \geq s_1 \geq \cdots \geq s_n$. Then the numbers

$$t_i = (i+1)(s_i - s_{i+1}) \qquad (0 \leq i < n),$$
$$t_n = (n+1)s_n$$

are non-negative and have the property that

$$\sum_{i=0}^{n} t_i \cdot b_{\tau_i} = (s_0 - s_1)v_0 +$$
$$2(s_1 - s_2)[{}^1\!/_2 v_0 + {}^1\!/_2 v_1] +$$
$$\vdots$$
$$n(s_{n-1} - s_n)[{}^1\!/_n v_0 + \cdots + {}^1\!/_n v_{n-1}] +$$
$$(n+1)s_n[{}^1\!/_{(n+1)} v_0 + \cdots + {}^1\!/_{(n+1)} v_n]$$
$$= \sum_{i=0}^{n} s_i \cdot v_i$$
$$= p.$$

In addition, $\sum_{i=0}^{n} t_i = \sum_{i=0}^{n} s_i = 1$. The $t_i$ are consequently the affine coordinates of $p$ in $|\beta|$ (Theorem 2.1.8). This proves (2).

For (3), simply observe that if for $p$ above we have that if $s_i = s_j$ for certain $i < j \leq n$ then all $t_k$ for $i \leq k < j$ are equal to 0. $\square$

**Lemma 2.2.4.** *Let $\sigma$ be a simplex and let $\mathcal{K}_1$ and $\mathcal{K}_2$ be chains of faces of $\sigma$. Let $\beta_1$ and $\beta_2$ denote the sets of barycenters of elements of $\mathcal{K}_1$ and $\mathcal{K}_2$, respectively. Then*
$$||\beta_1| \cap |\beta_2| = |\beta_1 \cap \beta_2|.$$

**Proof.** Since $|\beta_1| \cap |\beta_2|$ is convex and contains $\beta_1 \cap \beta_2$, it is clear that
$$|\beta_1 \cap \beta_2| = \text{conv}(\beta_1 \cap \beta_2) \subseteq |\beta_1| \cap |\beta_2|.$$
For the reverse inclusion, take an arbitrary $x \in |\beta_1| \cap |\beta_2|$. Since
$$(\mathcal{K}_1 \setminus \mathcal{K}_2) \cap (\mathcal{K}_2 \setminus \mathcal{K}_1) = \emptyset,$$
without loss of generality we may assume that $\sigma \notin \mathcal{K}_1 \setminus \mathcal{K}_2$. We assume without loss of generality that $\mathcal{K}_1 \setminus \mathcal{K}_2 \neq \emptyset$ for otherwise we get what we want from Lemma 2.2.2. It is clear that there exists a linear order $\preccurlyeq_1$ on the set $\dot\sigma$ such that for every element of $\tau \in \mathcal{K}_1$ we have that $\dot\tau$ is an initial $\preccurlyeq_1$-segment. This order corresponds to a maximal chain of faces of $\sigma$. Pick an arbitrary element $\tau \in \mathcal{K}_1 \setminus \mathcal{K}_2$; observe that $\tau \neq \sigma$. The collection $\mathcal{K}_2$ can easily be extended to a maximal chain $\mathcal{L}$ of simplices such that $\tau \notin \mathcal{L}$ (Exercise 2.2.1). Consequently, $\dot\tau$ is not an initial segment with respect to the corresponding linear order $\preccurlyeq_2$ (Lemma 2.2.3(1)), i.e., there exist distinct elements $v, w \in \dot\sigma$ such that $v \in \dot\tau$, $w \notin \dot\tau$ and $w \preccurlyeq_2 v$; consequently,
$$\alpha_w(x) \geq \alpha_v(x)$$
(Lemma 2.2.3(2)). Observe that $v \preccurlyeq_1 w$ from which it follows similarly that
$$\alpha_v(x) \geq \alpha_w(x).$$
By Lemma 2.2.3(3) we therefore conclude that the affine coordinate of $x$ in the simplex $|\beta_1|$ with respect to $b_\tau$ is equal to 0. $\square$

**Subdivisions.** Let $X$ be a subset of $L$ which is triangulated by $\mathcal{S}$. A triangulation $\mathcal{T}$ of $X$ is called a *subdivision* of $\mathcal{S}$ if for every simplex $\sigma \in \mathcal{S}$ we have that the collection
$$\mathcal{T}(\sigma) = \{\tau \in \mathcal{T} : \tau \subseteq \sigma\}$$
is finite and is a triangulation of $\sigma$.

**Lemma 2.2.5.** *Let $\mathcal{S}$ and $\mathcal{T}$ be triangulations of $X$. If $\mathcal{T}$ is a subdivision of $\mathcal{S}$ then for every $\tau \in \mathcal{T}$ and $\sigma \in \mathcal{S}$ such that $b_\tau \in \sigma$ we have $\tau \subseteq \sigma$.*

**Proof.** Take an element $\tau' \in \mathcal{T}(\sigma)$ such that $b_\tau \in \tau'$. Then by Lemma 2.2.1 it follows that $\tau \subseteq \tau' \subseteq \sigma$, as required. $\square$

**Theorem 2.2.6.** *Let $\mathcal{S}, \mathcal{T}$ and $\mathcal{R}$ be simplicial complexes. If $\mathcal{S}$ is a subdivision of $\mathcal{T}$ and $\mathcal{T}$ is a subdivision of $\mathcal{R}$ then $\mathcal{S}$ is a subdivision of $\mathcal{R}$.*

**Proof.** We have to prove that $S(\rho)$ is a finite triangulation of $\rho \in \mathcal{R}$. That the collection $S(\rho)$ is a simplicial complex is clear since condition (SC) holds for all simplices in $S$.

Let $\sigma \in S(\rho)$. We claim that there exists an element $\tau_\sigma \in \mathcal{T}(\rho)$ such that $\sigma \subseteq \tau_\sigma$. This is easy. Indeed, since $\bigcup \mathcal{T}(\rho) = \rho$ there exists $\tau_\sigma \in \mathcal{T}(\rho)$ such that $b_\sigma \in \tau_\sigma$. So by Lemma 2.2.5 it follows that $\tau_\sigma$ is as required.

Since $\mathcal{T}(\rho)$ is finite and for every $\tau \in \mathcal{T}(\rho)$, $S(\tau)$ is finite, we obtain that the collection $S(\rho)$ is finite. To conclude the proof, observe that $\bigcup S(\rho) = \rho$ is trivial since $\bigcup \mathcal{T}(\rho) = \rho$ and $\bigcup S(\tau) = \tau$ for every $\tau \in \mathcal{T}(\rho)$. □

Let $\sigma$ be a geometric simplex in $L$. We shall define a special triangulation of $\sigma$, the so-called *barycentric triangulation*. Let $B(\sigma)$ denote the set of all barycenters of faces of $\sigma$. We shall define a simplicial complex $\mathcal{B}(\sigma)$ consisting of subsets of $B(\sigma)$ as follows:

> A nonempty subset $\beta$ of $B(\sigma)$ belongs to $\mathcal{B}(\sigma)$ if and only if the faces of $\sigma$ the barycenters of which belong to $\beta$ form a chain.

It is clear that if $\beta \in \mathcal{B}(\sigma)$ and $\gamma \subseteq \beta$ then $\gamma \in \mathcal{B}(\sigma)$. So $\mathcal{B}(\sigma)$ is an abstract simplicial complex. By Lemma 2.2.4 it follows that the collection

$$|\mathcal{B}(\sigma)| = \{|\beta| : \beta \in \mathcal{B}(\sigma)\}$$

satisfies condition (SC) (see Page 116). So the geometric realization of $\mathcal{B}(\sigma)$ is $|\mathcal{B}(\sigma)|$ and, in particular, is contained in $\sigma$. In order to prove that $|\mathcal{B}(\sigma)|$ is a subdivision of $\sigma$, it therefore remains to prove that

$$\bigcup\{|\beta| : \beta \in \mathcal{B}(\sigma)\} = \sigma.$$

Let $x \in \sigma$. Without loss of generality we have $\dot\sigma = \{v_0, \ldots, v_n\}$ and

$$\alpha_{v_n}(x) \leq \alpha_{v_{n-1}}(x) \leq \cdots \leq \alpha_{v_0}(x).$$

This ordering corresponds to a (maximal) chain of faces in $\sigma$, which in turn corresponds to a (maximal) simplex $\beta \in \mathcal{B}(\sigma)$. By Lemma 2.2.3(2), $x \in |\beta|$.

Let $X$ be a set in $L$ which is triangulated by the simplicial complex $S$. As above, for every $\sigma \in S$ let $\mathcal{B}(\sigma)$ be the barycentric triangulation of $\sigma$. The collection

$$\mathcal{B}(S) = \{|\tau| \in |\mathcal{B}(\sigma)| : \sigma \in S\}$$

is called the *barycentric subdivision* of $S$. That $\mathcal{B}(S)$ is a collection geometric simplices with union $X$ is clear. We claim that it is a simplicial complex and is a subdivision of $S$. For that we only need to verify condition (SC) (see Page 116) because for every $\sigma \in S$,

$$\{|\tau| \in |\mathcal{B}(S)| : |\tau| \subseteq \sigma\} = |\mathcal{B}(\sigma)|$$

and $|\mathcal{B}(\sigma)|$ is a finite triangulation of $\sigma$.

Let $\beta_i \in \mathcal{B}(\sigma_i)$ for $\sigma_i \in \mathcal{S}$, $i = 1, 2$. If $\sigma_1 \cap \sigma_2 = \emptyset$ then there is nothing to prove. If $\sigma_1 \cap \sigma_2 \neq \emptyset$ then $\sigma = \sigma_1 \cap \sigma_2$ is a face of $\sigma_1$ as well as $\sigma_2$. Now put

$$\gamma_i = \beta_i \cap \sigma \quad (i = 1, 2).$$

Then $\gamma_1$ and $\gamma_2$ belong to $\mathcal{B}(\sigma)$ and we claim that

$$|\gamma_i| = |\beta_i| \cap \sigma \quad (i = 1, 2).$$

That $|\gamma_i| \subseteq |\beta_i| \cap \sigma$ is clear. Take an arbitrary $x \in |\beta_i| \cap \sigma$. There exists an element $\beta \in \mathcal{B}(\sigma)$ such that $x \in |\beta|$. Observe that

$$\beta \in \mathcal{B}(\sigma_1) \cap \mathcal{B}(\sigma_2).$$

Consequently,

$$x \in |\beta_i| \cap |\beta| = |\beta_i \cap \beta| \subseteq |\beta_i \cap \sigma| = |\gamma_i|.$$

By Lemma 2.2.4 we therefore conclude that

$$|\beta_1| \cap |\beta_2| = |\beta_1| \cap |\beta_2| \cap \sigma = |\gamma_1| \cap |\gamma_2| = |\gamma_1 \cap \gamma_2| = |\beta_1 \cap \beta_2|,$$

which is as required.

Sometimes it will be convenient to denote $\mathcal{S}$ by $\mathrm{sd}^{(0)}\mathcal{S}$ and $\mathcal{B}(\mathcal{S})$ by $\mathrm{sd}^{(1)}\mathcal{S}$. We define the second barycentric subdivision $\mathrm{sd}^{(2)}\mathcal{S}$ of $\mathcal{S}$ by $\mathcal{B}(\mathcal{B}(\mathcal{S}))$. Similarly, one defines $\mathrm{sd}^{(n)}\mathcal{S}$, the $n$-th barycentric subdivision of $\mathcal{S}$. Notice that by Theorem 2.2.6, $\mathrm{sd}^{(n)}\mathcal{S}$ is a subdivision of $\mathcal{S}$ for all $n$.

Let $\mathcal{S}$ be a triangulation of a subset of $L$. The *mesh* of $\mathcal{S}$, denoted by $\mathrm{mesh}(\mathcal{S})$, is defined to be the number

$$\sup_{\sigma \in \mathcal{S}} \mathrm{diam}(\sigma).$$

We allow $\mathrm{mesh}(\mathcal{S})$ to be equal to $\infty$.

**Theorem 2.2.7.** *Let $\sigma$ be an $n$-simplex in the normed linear space $L$ and let $\mathcal{B}(\sigma)$ be the barycentric triangulation of $\sigma$. Then*

$$\mathrm{mesh}(|\mathcal{B}(\sigma)|) \leq \frac{n}{n+1} \cdot \mathrm{diam}(\sigma).$$

**Proof.** First observe that for every $\beta \in \mathcal{B}(\sigma)$ we have $\mathrm{diam}(\beta) = \mathrm{diam}(|\beta|)$ (Theorem 2.1.11). Fix $\beta \in \mathcal{B}(\sigma)$ and let $d, c \in \beta$. We will estimate the distance between $d$ and $c$. Choose an enumeration

$$\{v_0, v_1, \ldots, v_n\}$$

of $\dot{\sigma}$ such that $c$ is the barycenter of $\{v_0, v_1, \ldots, v_k\}$ and $d$ is the barycenter of $\{v_0, v_1, \ldots, v_m\}$, respectively. We may assume without loss of generality that $k \leq m$. Then

$$\|d - c\| = \left\|\sum_{i=0}^{m} \frac{v_i}{m+1} - \sum_{j=0}^{k} \frac{v_j}{k+1}\right\|$$

$$= \frac{1}{k+1}\left\|\sum_{i=0}^{m} \frac{k+1}{m+1} v_i - \sum_{j=0}^{k} v_j\right\|$$

$$= \frac{1}{k+1}\left\|\sum_{j=0}^{k}\left(\left(\frac{1}{m+1}\sum_{i=0}^{m} v_i\right) - v_j\right)\right\|$$

$$\leq \frac{1}{k+1}\sum_{j=0}^{k}\left\|\left(\frac{1}{m+1}\sum_{i=0}^{m} v_i\right) - v_j\right\|$$

$$\leq \max_{0 \leq j \leq k}\left\|\left(\frac{1}{m+1}\sum_{i=0}^{m} v_i\right) - v_j\right\|.$$

Moreover,

$$\left\|\left(\frac{1}{m+1}\sum_{i=0}^{m} v_i\right) - v_j\right\| = \frac{1}{m+1}\left\|\sum_{i=0}^{m}(v_i - v_j)\right\|$$

$$\leq \frac{1}{m+1}\sum_{i=0}^{m}\|v_i - v_j\|$$

$$\leq \frac{m}{m+1} \cdot \operatorname{diam}(\sigma).$$

In the last inequality we used that one of the terms (if $i = j$) is equal to 0. We conclude that

$$\operatorname{diam}(\beta) \leq \frac{m}{m+1} \cdot \operatorname{diam}(\sigma).$$

Finally observe that $m \leq n$. □

**Corollary 2.2.8.** *Let $\mathcal{S}$ be a triangulation of a set $X$ in a normed linear space $L$ such that every $\sigma \in \mathcal{S}$ is at most $n$-dimensional. If $\mathcal{B}$ is the barycentric subdivision of $\mathcal{S}$ then*

$$\operatorname{mesh}(\mathcal{B}) \leq \frac{n}{n+1} \cdot \operatorname{mesh}(\mathcal{S}).$$

**Corollary 2.2.9.** *Let $|\tau|$ be a geometric simplex in a normed linear space $L$. Then for every $\varepsilon > 0$ there is an $m \in \mathbb{N}$ such that*

$$\operatorname{mesh}\left(\operatorname{sd}^{(m)} \mathcal{F}(\tau)\right) < \varepsilon.$$

**Proof.** Assume that $\tau$ is an $n$-simplex. By Corollary 2.2.8 it follows that for every $m \in \mathbb{N}$ we have

$$\mathrm{mesh}\left(\mathrm{sd}^{(m)}\mathcal{F}(\tau)\right) \leq \left(\frac{n}{n+1}\right)^m \cdot \mathrm{diam}(\tau).$$

We conclude that for a sufficiently large $m$, $\mathrm{mesh}(\mathrm{sd}^{(m)}\mathcal{F}(\tau)) < \varepsilon$. □

We shall now formulate a very important property of polyhedra.

**Theorem 2.2.10.** *Let $\mathcal{T}$ be a finite simplicial complex. Then for every open cover $\mathcal{U}$ of $|\mathcal{T}|$ there exists $m \in \mathbb{N}$ such that $\mathrm{sd}^{(m)}\mathcal{T}$ refines $\mathcal{U}$, i.e., for every simplex $\sigma \in \mathrm{sd}^{(m)}\mathcal{T}$ there exists an element $U \in \mathcal{U}$ with $\sigma \subseteq U$.*

**Proof.** Fix $\tau \in \mathcal{T}$ for a moment. By Corollary 2.2.9 and Lemma A.5.3 there exists $n(\tau) \in \mathbb{N}$ such that every $\sigma \in \mathrm{sd}^{(n(\tau))}\mathcal{F}(\tau)$ is contained in an element of $\mathcal{U}$. Let $m = \max\{n(\tau) : \tau \in \mathcal{T}\}$. Now take an arbitrary $\kappa \in \mathrm{sd}^{(m)}\mathcal{T}$. By the definition of $\mathrm{sd}^{(m)}\mathcal{T}$ it is clear that there exists $\tau \in \mathcal{T}$ with $\kappa \in \mathrm{sd}^{(m)}\mathcal{F}(\tau)$. Since every element of $\mathrm{sd}^{(m)}\mathcal{F}(\tau)$ is contained in an element of $\mathrm{sd}^{(n(\tau))}\mathcal{F}(\tau)$, the simplex $\kappa$ is contained in an element of $\mathcal{U}$. □

The above theorem can be generalized as follows: for every polytope $|\mathcal{T}|$ and for every open cover $\mathcal{U}$ of $|\mathcal{T}|$ there exists a subdivision $\mathcal{S}$ of $\mathcal{T}$ such that each simplex $\sigma \in \mathcal{S}$ is contained in an element of $\mathcal{U}$. However, for noncompact polytopes, $\mathcal{S}$ generally cannot be chosen to be an iterated barycentric subdivision. For details, see WHITEHEAD [408].

**Exercise for §2.2.**

▶1. Let $\sigma$ be a simplex and let $\mathcal{K}$ be a chain of faces of $\sigma$. Prove that if $\tau$ is a proper face of $\sigma$ such that $\tau \notin \mathcal{K}$ then there is a maximal chain of faces $\mathcal{L}$ such that $\mathcal{K} \subseteq \mathcal{L}$ but $\tau \notin \mathcal{L}$.

## 2.3. The nerve of an open covering

Let $X$ be a space and let $\mathcal{U}$ be a countable open cover of $X$. We say that a simplicial complex $\mathcal{T}$ is a *nerve* of $\mathcal{U}$ if $\mathcal{T}^{(0)}$ can be indexed as $\{x(U) : U \in \mathcal{U}\}$ such that for every $n \geq 0$,

$$(*) \qquad x(U_0), \ldots, x(U_n) \text{ spans a simplex in } \mathcal{T} \text{ iff } \bigcap_{i=0}^{n} U_i \neq \emptyset.$$

If $\mathcal{T}$ is a nerve of $\mathcal{U}$ then it is convenient to adopt the notation $\{x(U) : U \in \mathcal{U}\}$ for $\mathcal{T}^{(0)}$, where it is implicitly assumed that the 'indexing' is such that $(*)$ holds.

**Proposition 2.3.1.** *Let $X$ be a space. Each countable open cover of $X$ has a nerve.*

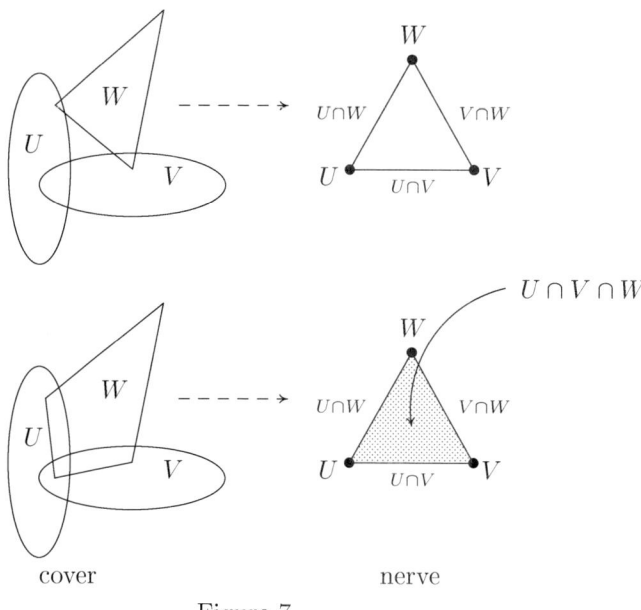

Figure 7.

**Proof.** Let $\mathcal{U} = \{U_n : n \in \mathbb{N}\}$ be a countable open cover of $X$. For each $n$ let $x_n \in \ell^2$ be the vector all coordinates of which are 0, except for the $n$-th coordinate which equals 1. The sequence $(x_n)_n$ is linearly independent, hence geometrically independent. Define

$$\mathcal{F} = \{F \subseteq \mathbb{N} : F \text{ is finite and } \bigcap_{n \in F} U_n \neq \emptyset\}.$$

For $F \in \mathcal{F}$ let $\tau(F)$ be the simplex in $\ell^2$ spanned by $\{x_n : n \in F\}$. Put

$$\mathcal{T} = \{\tau(F) : F \in \mathcal{F}\}.$$

It is easy to verify that $\mathcal{T}$ is a simplicial complex and that it moreover is a nerve for $\mathcal{U}$. □

Clearly, any two nerves of the same open cover are combinatorially equivalent (hence 'isomorphic'), so by Proposition 2.1.23 we can now speak of *the* nerve $N(\mathcal{U})$ of the cover $\mathcal{U}$. If $L$ is a normed linear space containing a simplicial complex $\mathcal{T}$ which is combinatorially equivalent to the nerve $N(\mathcal{U})$ then we say that $N(\mathcal{U})$ can be *realized* in $L$.

Let $N(\mathcal{U})$ be the nerve of the locally finite open cover $\mathcal{U}$ (recall that $\mathcal{U}$ is countable, see Page 486). We shall use the $\kappa$-functions with respect to $\mathcal{U}$ to define a canonical continuous function $\kappa\colon X \to |N(\mathcal{U})|$. This function is the 'bridge' between the 'abstract' space $X$ and the 'concrete' space $|N(\mathcal{U})|$ and is called the $\kappa$-*function* of the cover $\mathcal{U}$.

Indeed, define $\kappa\colon X \to |N(\mathcal{U})|$ by

$$\kappa(x) = \sum_{U \in \mathcal{U}} \kappa_U(x) \cdot x(U).$$

We claim that $\kappa$ is well-defined. To check this, for $x \in X$ put

$$\mathcal{F}(x) = \{U \in \mathcal{U} : x \in U\}.$$

Observe that by the local finiteness condition, $\mathcal{F}(x)$ is finite, that $\kappa_U(x) = 0$ if $U \notin \mathcal{F}(x)$, and that $\sum_{U \in \mathcal{F}(x)} \kappa_U(x) = 1$. This implies that

$$\kappa(x) = \sum_{U \in \mathcal{F}(x)} \kappa_U(x) \cdot x(U).$$

Hence the infinite sum in the definition of $\kappa$ reduces to a finite sum. In addition, the

$$\{\kappa_U(x) : U \in \mathcal{F}(x)\}$$

are the affine coordinates of $\kappa(x)$ in the simplex $\tau(x)$ spanned by the vertices

$$\{x(U) \in \mathcal{U} : U \in \mathcal{F}(x)\}$$

of $N(\mathcal{U})$. This shows that $\kappa(x)$ is well-defined, and also that $\kappa(x) \in \tau(x)$.

**Lemma 2.3.2.** *Let $X$ be a space and let $\mathcal{U}$ be a locally finite open cover of $X$. Then for each $x \in X$, $\tau(x)$ is the carrier of $\kappa(x)$ in $N(\mathcal{U})$.*

**Proof.** As remarked above, $\tau(x)$ is a simplex of $N(\mathcal{U})$ that contains $\kappa(x)$. As remarked above, the $\kappa_U(x)$'s are the affine coordinates of $\kappa(x)$ with respect to the simplex in $N(\mathcal{U})$ spanned by the vertices $\{x(U) : U \in \mathcal{F}(x)\}$. Now since for every $U \in \mathcal{F}(x)$ we have that $\kappa_U(x) \neq 0$, $\tau(x)$ is the smallest simplex containing $\kappa(x)$, i.e., $\tau(x) = \operatorname{car}\kappa(x)$. □

**Theorem 2.3.3.** *Let $X$ be a space and let $\mathcal{U}$ be a locally finite open cover of $X$. Then the $\kappa$-function $\kappa\colon X \to |N(\mathcal{U})|$ has the following properties:*

(1) *$\kappa$ is continuous;*
(2) *for every $U \in \mathcal{U}$: $\kappa^{-1}[\operatorname{St} x(U)] = U$.*

**Proof.** We shall first prove (2). Observe that for every $x \in X$ and $U \in \mathcal{U}$,

$$x \notin U \iff U \notin \mathcal{F}(x) \iff x(U) \notin \tau(x) = \operatorname{car}\kappa(x)$$

(Lemma 2.3.2). Now if $x(U) \notin \tau(x)$ then since $\kappa(x) \in \tau(x)$, $\kappa(x) \notin \operatorname{St} x(U)$. This proves that $\kappa^{-1}[\operatorname{St} x(U)] \subseteq U$. Conversely, if $\kappa(x) \notin \operatorname{St} x(U)$ then there is a simplex $\tau \in N(\mathcal{U})$ which contains $\kappa(x)$ but not $x(U)$. Since $\operatorname{car}\kappa(x) \subseteq \tau$, this gives us that $x(U) \notin \operatorname{car}\kappa(x) = \tau(x)$ and hence that $x \notin U$. This proves that $\kappa[U] \subseteq \operatorname{St} x(U)$.

For (1), let $L$ be a normed linear space which realizes $N(\mathcal{U})$ (e.g., $\ell^2$, see the proof of Proposition 2.3.1). Take an arbitrary $x \in X$. There is a neighborhood $W$ of $x$ meeting only finitely many elements of $\mathcal{U}$. Since the infinite sum in the definition of $\kappa$ for points of $W$ clearly reduces to a finite sum, and the functions $\kappa_U$ are all continuous, by the continuity of the algebraic operations on $L$ it follows that $\kappa$ regarded as a mapping from $X$ into $L$ is continuous. Now let $\mathcal{W} = \{U \in \mathcal{U} : U \cap W \neq \emptyset\}$ and let $Y$ be the union of all simplices of $N(\mathcal{U})$ the vertices of which correspond to elements of the collection $\mathcal{W}$. Observe that $Y$ is a finite union of elements of $N(\mathcal{U})$ and hence is compact. Since $x \in W$ it follows that $\mathcal{F}(x) \subseteq \mathcal{W}$ and so

$$\kappa(x) \in \tau(x) \subseteq Y.$$

Since $Y$ is compact, by Lemma 2.1.13(1) it follows that the topology that $Y$ inherits from $L$ is the same as the topology that $Y$ inherits from $|N(\mathcal{U})|$. We conclude that $\kappa \colon X \to |N(\mathcal{U})|$ is continuous at all points of $W$, and hence at the point $x$. □

Let $X$ and $Y$ be spaces and let $\mathcal{U}$ be an open cover of $X$. A continuous function $f \colon X \to Y$ is called a $\mathcal{U}$-mapping if there is an open cover $\mathcal{V}$ of $Y$ such that $f^{-1}[\mathcal{V}] < \mathcal{U}$. This concept is related to the concept of a $\varepsilon$-mapping on Page 34. For let $f \colon X \to Y$ be an $\varepsilon$-map, where $X$ is compact, and let $\mathcal{U}$ be the open cover of $X$ consisting of all open sets of diameter less than $\varepsilon$. We claim that $f$ is a $\mathcal{U}$-mapping. To this end, first observe that if

$$y \in V = Y \setminus f[X]$$

then $V$ is an open neighborhood of $y$ such that $f^{-1}[V] = \emptyset$ is contained in every $U \in \mathcal{U}$. Next, if $y \in f[X]$ then let $U$ be an open neighborhood of $f^{-1}(y)$ such that $\mathrm{diam}(U) < \varepsilon$. Since by compactness of $X$ the function $f$ is closed (Exercise A.5.5), there is a neighborhood $V$ of $y$ such that $f^{-1}[V] \subseteq U$ (Exercise A.1.15). So we conclude that $f$ is a $\mathcal{U}$-mapping. We see that the concept of an $\varepsilon$-map is indeed the 'compact' version of the concept of a $\mathcal{U}$-map.

**Corollary 2.3.4.** *Let $X$ be a space and let $\mathcal{U}$ be a locally finite open cover of $X$. Then $\kappa \colon X \to |N(\mathcal{U})|$ is a $\mathcal{U}$-mapping.*

**Proof.** This follows from Theorem 2.3.3(2) and Corollary 2.1.17. □

It is unfortunately not the case that a cover $\mathcal{U}$ of a space $X$ is locally finite if and only if the simplicial complex $N(\mathcal{U})$ is locally finite. Since, as we pointed out above, our main interest is in locally finite simplicial complexes, it is natural to wonder when $N(\mathcal{U})$ is locally finite.

First a definition. An open cover $\mathcal{U}$ of $X$ is said to be *star-finite* if for every $U \in \mathcal{U}$ the set

$$\{V \in \mathcal{U} : V \cap U \neq \emptyset\}$$

is finite. Observe that a star-finite open cover is locally finite. Consequently, every star-finite cover is countable.

This definition leads us to the following

**Theorem 2.3.5.** *Let $X$ be a space and let $\mathcal{U}$ be an open cover of $X$. Then*

(1) *there exists an open refinement $\mathcal{V}$ of $\mathcal{U}$ such that $\mathcal{V}$ is star-finite,*
(2) *$N(\mathcal{U})$ is locally finite if and only if $\mathcal{U}$ is star-finite.*

**Proof.** For (1), we may assume without loss of generality that $X$ is a subspace of the Hilbert cube $Q$ (Corollary A.4.4). For every $U \in \mathcal{U}$ let $\hat{U} \subseteq Q$ be open such that $\hat{U} \cap X = U$ and put $\hat{\mathcal{U}} = \{\hat{U} : U \in \mathcal{U}\}$. Then $V = \bigcup \hat{\mathcal{U}}$ is an open subset of $Q$ and hence is locally compact and $\sigma$-compact. There are compacta $F_n \subseteq V$ for $n \in \mathbb{N}$ such that

$$F_1 \subseteq \operatorname{Int} F_2 \subseteq F_2 \subseteq \operatorname{Int} F_3 \subseteq \cdots \subseteq F_{n-1} \subseteq \operatorname{Int} F_n \subseteq F_n \subseteq \cdots$$

while moreover $V = \bigcup_{n=1}^{\infty} F_n$. Put $F_0 = \emptyset$. For every $n$ let $\mathcal{E}_n$ be a finite subcollection of $\hat{\mathcal{U}}$ with $F_n \subseteq \bigcup \mathcal{E}_n$. Now put

$$\mathcal{W} = \{\hat{U} \cap (Q \setminus F_{n-1}) \cap \operatorname{Int} F_{n+1} : \hat{U} \in \mathcal{E}_n, n \in \mathbb{N}\}.$$

It is clear that $\mathcal{W}$ covers $V$, is star-finite and that $\mathcal{W} < \hat{\mathcal{U}}$. As a consequence, the collection $\mathcal{V} = \mathcal{W} \upharpoonright X$ is as required.

For (2), first assume that $N(\mathcal{U})$ is locally finite. If there exists $U \in \mathcal{U}$ such that the set

$$\{V \in \mathcal{U} : V \cap U \neq \emptyset\}$$

is infinite then the vertex $x(U)$ of $N(\mathcal{U})$ is contained in infinitely many one-dimensional simplices of $N(\mathcal{U})$, which contradicts the local finiteness condition on $N(\mathcal{U})$. The verification of the reverse implication is a triviality which we leave to the reader. □

We shall now prove that every space can be 'approximated arbitrarily closely' by a polytope.

**Corollary 2.3.6.** *Let $X$ be a space. For every open cover $\mathcal{U}$ of $X$ there exists a polytope $P$ and a $\mathcal{U}$-mapping $f \colon X \to P$. If $X$ is compact then $P$ can be chosen to be a polyhedron.*

**Proof.** Let $\mathcal{U}$ be an open cover of $X$. By Theorem 2.3.5(1), there exists a star-finite open cover $\mathcal{V}$ of $X$ which refines $\mathcal{U}$. Let

$$\kappa \colon X \to |N(\mathcal{V})|$$

be the $\kappa$-mapping of $\mathcal{V}$. Since $\mathcal{V}$ is star-finite, $|N(\mathcal{V})|$ is locally finite by Theorem 2.3.5(2). By Corollary 2.3.4, $\kappa$ is a $\mathcal{V}$-mapping. Since $\mathcal{V} < \mathcal{U}$, we are done.

If $X$ is compact then $\mathcal{V}$ is a finite subcover of $\mathcal{U}$ and proceed as above. □

## 2.3. THE NERVE OF AN OPEN COVERING

**Freudenthal's Approximation Theorem.** We now come to an interesting 'Approximation Theorem'. See §1.10 for results that are in the same spirit.

**Freudenthal's Approximation Theorem 2.3.7.** *Every compact space is homeomorphic to the inverse limit of an inverse sequence consisting of polyhedra.*

**Proof.** We start as in Example 1.10.5. By Corollary A.4.4 we may restrict ourselves to a compact subspace $X$ of $Q$. For every $n$ let
$$X_n = \tau_n[X],$$
where $\tau_n \colon Q \to \mathbb{J}^n$ is the projection. In addition, let $f_n \colon X_{n+1} \to X_n$ be defined by
$$f_n(x_1, \ldots, x_{n+1}) = (x_1, \ldots, x_n).$$
Then $\varprojlim(X_n, f_n)_n = X$ (Example 1.10.5).

Unfortunately, the sets $X_n$ need not be polyhedra. It will require a little extra work to take care of this problem.

**Claim 1.** *Let $X$ be a compact subspace of $\mathbb{J}^n$ for some $n$. Then for every neighborhood $U$ of $X$ there exists a polyhedron $P$ such that $X \subseteq P \subseteq U$.*

*Proof.* By Corollary A.5.4 there exists $\delta > 0$ such that if $A \subseteq \mathbb{J}^n$, diam $A < \delta$ and $A \cap X \neq \emptyset$ then $A \subseteq U$. Since $\mathbb{J}^n$ is a polyhedron, it has by Theorem 2.2.10 a triangulation $\mathcal{S}$ such that for every $\sigma \in \mathcal{S}$ we have diam $\sigma < \delta$. Let $P$ be the union of all simplices in $\mathcal{S}$ that intersect $X$. An easy check show that $P$ is as required. ◇

Now for every $n \in \mathbb{N}$ we shall define a polyhedron $P_n$ in $\mathbb{J}^n$ with $X_n \subseteq P_n$, as follows. Let $P_1 = \mathbb{J}$. By Claim 1 there exists a polyhedron $P_2$ in $\mathbb{J}^2$ such that $X_2 \subseteq P_2 \subseteq B(X_2, 1/2)$. By the Dugundji Theorem 1.2.2, we can extend the function $f_1 \colon X_2 \to X_1 \subseteq P_1$ to a continuous function $\xi_1 \colon P_2 \to P_1$. For $P_3$ we have to do a little extra work. Since $P_2$ is an ANR (Theorem 2.1.24), there is a neighborhood $U$ of $X_3$ such that the function $f_2 \colon X_3 \to X_2 \hookrightarrow P_2$ can be extended to a continuous function $\eta \colon U \to P_2$. Pick a neighborhood $V \subseteq U$ of $X_3$ such that
$$\eta[V] \subseteq B(X_2, 1/3), \quad \xi_1 \circ \eta[V] \subseteq B(X_1, 1/3).$$
By Claim 1, we may pick a polyhedron $P_3$ such that $X_3 \subseteq P_3 \subseteq V$. Let $\xi_3$ be the function $\eta \restriction P_3$.

Continuing in this way inductively, yields an inverse sequence of polyhedra $(P_n, \xi_n)$ such that for every $n$, $X_n \subseteq P_n \subseteq \mathbb{J}^n$, $\xi_n \restriction X_{n+1} = f_n$ and

(∗) $$\bigcap_{m=n}^{\infty} \xi_n \circ \cdots \circ \xi_m[P_{m+1}] = X_n.$$

We think of $\varprojlim(P_n, \xi_n)$ and $\varprojlim(X_n, f_n) = X$ as subspaces of $\prod_{n=1}^{\infty} \mathbb{J}^n$. We claim that $\varprojlim(P_n, \xi_n) = \varprojlim(X_n, f_n) = X$. It is clear that

$$\varprojlim(X_n, f_n) \subseteq \varprojlim(P_n, \xi_n).$$

Now take an arbitrary $x \in \varprojlim(P_n, \xi_n)$ and fix $N \in \mathbb{N}$. Then,

$$x_N \in \bigcap_{m=N}^{\infty} \xi_N \circ \cdots \circ \xi_m[P_{m+1}]$$

and hence $x_N \in X_N$ by $(*)$. Since $\xi_n$ extends $f_n$ for every $n$, this easily implies that $x \in \varprojlim(X_n, f_n)$. □

**Exercises for §2.3.**

1. Let $X$ be a space and let $\mathcal{U}$ and $\mathcal{V}$ be open covers of $X$ such that $\mathcal{V} < \mathcal{U}$. For every $V \in \mathcal{V}$ pick $U(V) \in \mathcal{U}$ such that $V \subseteq U(V)$. This defines a function

$$f_0 \colon N(\mathcal{V})^{(0)} \to N(\mathcal{U})^{(0)}.$$

Extend this function 'linearly' over each simplex of $N(\mathcal{V})$ and prove that the resulting function from $|N(\mathcal{V})| \to |N(\mathcal{U})|$ is continuous.

2. Prove that the following statements for a space $X$ are equivalent:
   (1) $X$ is compact.
   (2) Every locally finite open cover of $X$ is star-finite.

## 2.4. Simplices in $\mathbb{R}^n$

In this section we shall formulate and prove some fundamental properties of simplices in $\mathbb{R}^n$. Among other things, we shall present a proof of the Brouwer Fixed-Point Theorem and some of its applications.

**Lemma 2.4.1.** *Let $\sigma_1$ and $\sigma_2$ be two (geometric) $n$-simplices in $\mathbb{R}^n$ which intersect in a common $(n-1)$-face $\tau$. Then $\sigma_1 \cup \sigma_2$ is a neighborhood of the barycenter of $\tau$.*

**Proof.** Observe that the lemma is trivial if $n = 1$; we may therefore assume that $n > 1$. Let $\dot{\sigma}_1 = \{a_0, a_1, \ldots, a_{n-1}, a_n^1\}$ and $\dot{\sigma}_2 = \{a_0, a_1, \ldots, a_{n-1}, a_n^2\}$, where

$$\tau = \sigma_1 \cap \sigma_2 = |\{a_0, a_1, \ldots, a_{n-1}\}|.$$

The points $\{a_0, a_1, \ldots, a_{n-1}, a_n^1\}$ are geometrically independent, from which it follows by Theorem 2.1.3 that the points

$$a_1 - a_0, a_2 - a_0, \ldots, a_{n-1} - a_0, a_n^1 - a_0$$

are linearly independent. Now consider the standard basis $e_1, e_2, \ldots, e_n$ in $\mathbb{R}^n$ and a linear isomorphism $g^1 \colon \mathbb{R}^n \to \mathbb{R}^n$ with the properties:
$$\begin{cases} g^1(e_i) = a_i - a_0 & (i < n), \\ g^1(e_n) = a_n^1 - a_0. \end{cases}$$
This isomorphism followed by the translation $x \mapsto x + a_0$ is an affine isomorphism (Theorem 2.1.7) and hence a homeomorphism (see Page 113) which we denote by $f^1$. Observe that $f^1$ has the following properties:
$$\begin{cases} f^1(\underline{0}) = a_0, \\ f^1(e_i) = a_i & (1 \leq i \leq n-1), \\ f^1(e_n) = a_n^1. \end{cases}$$
In the standard basis we now change $e_n$ into $-e_n$. By a similar argumentation as above we obtain an affine isomorphism $f^2 \colon \mathbb{R}^n \to \mathbb{R}^n$ such that
$$\begin{cases} f^2(\underline{0}) = a_0, \\ f^2(e_i) = a_i & (1 \leq i \leq n-1), \\ f^2(-e_n) = a_n^2. \end{cases}$$
Observe that $f^1$ and $f^2$ agree on the plane $P = \{x \in \mathbb{R}^n : x_n = 0\}$. Also observe that $f^1 \restriction P = f^2 \restriction P$ is an affine isomorphism between $P$ and the hyperplane spanned by the elements of $\dot{\tau} = \{a_0, a_1, \ldots, a_{n-1}\}$. Now define a homeomorphism $f \colon \mathbb{R}^n \to \mathbb{R}^n$ as follows
$$f(x) = \begin{cases} f^1(x) & (x_n \geq 0), \\ f^2(x) & (x_n \leq 0). \end{cases}$$
Let $b$ be the barycenter of $\tau$ and let $c = (1/n, 1/n, \ldots, 1/n, 0)$. Then $f(c) = b$ and the set
$$U = \left(0, \frac{1}{n} + \frac{1}{n^2}\right)^{n-1} \times \left(\frac{-1}{n^2}, \frac{1}{n^2}\right)$$
is a neighborhood of $c$ of which of which it is geometrically obvious that it is mapped by $f$ into $\sigma_1 \cup \sigma_2$ (see Exercise 2.4.1). We conclude that $\sigma_1 \cup \sigma_2$ is a neighborhood of $b$. $\square$

The next result is the key in the proof of the Brouwer Fixed-Point Theorem.

**Theorem 2.4.2.** *Let $\tau$ be an $n$-simplex in $\mathbb{R}^n$, and let $\mathcal{S}$ be a finite triangulation of $\tau$ which subdivides its standard triangulation. Finally, let $\mu \in \mathcal{S}$ be an $(n-1)$-simplex and let $b$ be the barycenter of $\mu$.*

(1) *If $b \in \partial \tau$ then $\mu \subset \partial \tau$ and $\mu$ is a face of precisely one $n$-simplex in $\mathcal{S}$.*
(2) *If $b \notin \partial \tau$ then $\mu$ is a face of precisely two $n$-simplices in $\mathcal{S}$.*

**Proof.** Assume that $b \in \partial \tau$. We will show that $\mu \subseteq \partial \tau$. Since $\mathcal{S}$ subdivides the standard triangulation of $\tau$ there is a simplex $\sigma \in \mathcal{S}$ with $b \in \sigma$ such that $\sigma$ is contained in a proper face of $\tau$, i.e., $\sigma \subseteq \partial \tau$. By Lemma 2.2.1 we obtain that $\mu$ is a face of $\sigma$ so that $\mu \subseteq \partial \tau$.

Now we shall prove that depending on whether $b \in \partial \sigma$ or not, $\mu$ is a face of at most one or at most two $n$-simplices in $\mathcal{S}$.

To this end, first assume that $b \in \partial \tau$. If there exist two $n$-simplices in $\mathcal{S}$ having $\mu$ as a common face then the union of these two simplices is a neighborhood of $b$ (Lemma 2.1.5). But this contradicts the fact that $b \in \partial \tau$ which is equal to the *topological* boundary of $\tau$ (Exercise 2.1.9).

If $b \notin \partial \tau$ and there exist three distinct $n$-simplices, say $\sigma_1, \sigma_2$ and $\sigma_3 \in \mathcal{S}$, having $\mu$ as a common face then since $\mathcal{S}$ is a simplicial complex,

$$(\sigma_1 \cup \sigma_2) \cap (\sigma_2 \cup \sigma_3) \cap (\sigma_3 \cup \sigma_1) = (\sigma_1 \cap \sigma_2) \cup (\sigma_2 \cap \sigma_3) \cup (\sigma_3 \cap \sigma_1)$$
$$= \mu.$$

For the last equality observe e.g., that $\mu$ is an $(n-1)$-simplex contained in the simplex $\sigma_1 \cap \sigma_2$. If $\sigma_1 \cap \sigma_2$ is an $n$-simplex then $\sigma_1 = \sigma_2$ which is a contradiction. So $\sigma_1 \cap \sigma_2$ is an $(n-1)$-simplex since it contains $\mu$ as a face, hence it must be equal to $\mu$.

By Lemma 2.4.1 the above intersection is also a neighborhood of $b$ in $\mathbb{R}^n$. Consequently, $\mu$ is a neighborhood of $b$ in $\mathbb{R}^n$ which is impossible because $\mu$ is contained in an $(n-1)$-dimensional hyperplane.

For the remaining part of the proof, put

$$U = \tau \setminus \bigcup \{\sigma \in \mathcal{S} : b \notin \sigma\}.$$

Observe that $U$ is a nonempty relatively open subset of $\tau$.

We shall now prove that $\mu$ is a face of at least one $n$-simplices of $\mathcal{S}$. Striving for a contradiction, assume that there does not exist an $n$-simplex in $\mathcal{S}$ having $\mu$ as a face. According to Lemma 2.2.1 this implies that $\mu$ is the only simplex in $\mathcal{S}$ containing $b$ which implies that $U \subseteq \mu$. Now if $b \in \partial \tau$ then we arrive at the desired contradiction because then $U \subseteq \mu \subseteq \partial \tau$ and $\partial \tau$ has empty interior in $\tau$. If $b \notin \partial \tau$ then $U \cap \text{Int}(\tau)$ is a neighborhood of $b \in \mathbb{R}^n$ which is contained in $\mu$. But this contradicts the fact that $\mu$ is contained in an $(n-1)$-dimensional hyperplane of $\mathbb{R}^n$.

All there remains to prove is that if $b \notin \partial \tau$ then there are at least two $n$-simplices having $\mu$ as a face. Striving again for a contradiction, assume that this is not the case. We already know from the above that there is an $n$-simplex $\sigma \in \mathcal{S}$ having $\mu$ as a face.

Pick an arbitrary $p \in U$ and let $\gamma \in \mathcal{S}$ be such that $p \in \gamma$. Then $b \in \gamma$ by the definition of $U$ and so by the same argumentation as above, $\mu$ is a face of $\gamma$. But then either $\gamma = \mu$ or $\gamma = \sigma$. In either case, $p \in \sigma$. We conclude that $U \subseteq \sigma$. So $\text{Int}(\tau) \cap U$ is a neighborhood of $b \in \mathbb{R}^n$ which is contained in $\sigma$. Since $\mu$ is a proper face of $\sigma$ we also have $b \in \partial\sigma$. This is a contradiction. □

Let $\tau = |\{x_0, \ldots, x_k\}|$ be an arbitrary $k$-simplex in $\mathbb{R}^n$. An $m$-*Sperner map* for $\tau$ is a function from the vertices of $\mathrm{sd}^{(m)}\mathcal{F}(\tau)$ to $\{0, \ldots, k\}$ such that if $v$ is such a vertex and
$$v \in |\{x_{i_0}, \ldots, x_{i_\ell}\}|$$
then
$$h(v) \in \{i_0, \ldots, i_\ell\}.$$
We call a $k$-dimensional simplex $\sigma$ in $\mathrm{sd}^{(m)}\mathcal{F}(\tau)$ *full* if
$$h[\dot\sigma] = \{0, \ldots, k\}.$$

There are always full simplices as the following result shows. As to be expected, its proof is of a combinatorial nature.

**Sperner's Lemma 2.4.3.** *Let $\tau = |\{x_0, \ldots, x_k\}|$ be a $k$-simplex in $\mathbb{R}^k$ and let $h$ be an $m$-Sperner map for $\tau$. The number of full simplices in $\mathrm{sd}^{(m)}\mathcal{F}(\tau)$ is odd and hence non-zero.*

**Proof.** We shall prove the lemma by induction on $k$. If $k = 0$ then $\tau = |\{x_0\}|$, and there is one full simplex. Assume that the theorem is true for all $(k-1)$-simplices, $k - 1 \geq 0$. We shall prove the theorem for $\tau$. Put

$\beta = \{x_0, \ldots, x_{k-1}\}$,
$\mathcal{P}_1 = \{\sigma \in \mathrm{sd}^{(m)}\mathcal{F}(\tau) : \sigma$ is $(k-1)$-dimensional, $h[\dot\sigma] = \{0, \ldots, k-1\}\}$,
$\mathcal{P}_2 = \{\sigma \in \mathrm{sd}^{(m)}\mathcal{F}(\tau) : \sigma$ is $k$-dimensional, $\{0, \ldots, k-1\} \subseteq h[\dot\sigma]\}$,
$\mathcal{P}_1(0) = \{\sigma \in \mathcal{P}_1 : \sigma \subseteq |\beta|\}$,
$\mathcal{P}_1(1) = \mathcal{P}_1 \setminus \mathcal{P}_1(0)$,
$\mathcal{P}_2(0) = \{\sigma \in \mathcal{P}_2 : h[\dot\sigma] = \{0, \ldots, k\}\}$,
$\mathcal{P}_2(1) = \mathcal{P}_2 \setminus \mathcal{P}_2(0)$,

respectively. Observe that
$$\mathrm{sd}^{(m)}\mathcal{F}(\beta) \subseteq \mathrm{sd}^{(m)}\mathcal{F}(\tau).$$
By the definition of an $m$-Sperner map it therefore follows that the restriction of $h$ to the vertices of $\mathrm{sd}^{(m)}\mathcal{F}(\beta)$ is an $m$-Sperner map for $\beta$. As a consequence, the collection $\mathcal{P}_1(0)$ is the set of full simplices in the $m$-th barycentric subdivision of $|\beta|$, so $|\mathcal{P}_1(0)|$ is odd by our inductive assumptions. Also, $\mathcal{P}_2(0)$ is the set of full simplices in the $m$-th barycentric subdivision of $\tau$, so we have to prove that $|\mathcal{P}_2(0)|$ is an odd number.

Put
$$\mathcal{R} = \{(\kappa, \mu) \in \mathcal{P}_1 \times \mathcal{P}_2 : \kappa \text{ is a face of } \mu\}.$$

We compute the cardinality of $\mathcal{R}$ twice. For each $\kappa \in \mathcal{P}_1$ put
$$\mathcal{R}[\kappa] = \{\mu \in \mathcal{P}_2 : (\kappa, \mu) \in \mathcal{R}\}.$$

**Claim 1.** $|\mathcal{R}| = |\mathcal{P}_1(0)| + 2 \cdot |\mathcal{P}_1(1)|$.

*Proof.* Clearly,
$$|\mathcal{R}| = \sum_{\kappa \in \mathcal{P}_1} |\mathcal{R}[\kappa]|.$$
We claim that if $\kappa \in \mathcal{P}_1(0)$ then $|\mathcal{R}[\kappa]| = 1$. Simply observe that such $\kappa$ is a face of precisely one $k$-dimensional simplex in $\mathrm{sd}^{(m)} \mathcal{F}(\tau)$ by Theorem 2.4.2, and this simplex clearly belongs to $\mathcal{P}_2$. We next claim that if $\kappa \in \mathcal{P}_1(1)$ then $|\mathcal{R}[\kappa]| = 2$. For take an arbitrary $\kappa \in \mathcal{P}_1(1)$ and assume first that it is contained in a $(k-1)$-dimensional face of $\tau$, say $\kappa \subseteq |\{x_{i_0}, \ldots, x_{i_{k-1}}\}|$. Then
$$h[\dot\kappa] \subseteq \{i_0, \ldots, i_{k-1}\}$$
which implies that $\{i_0, \ldots, i_{k-1}\} = \{0, \ldots, k-1\}$ and so $\kappa \subseteq |\beta|$, contradiction. So such a $\kappa$ is not contained in a $(k-1)$-dimensional face of $\tau$, and therefore, again by Theorem 2.4.2, is a $(k-1)$-dimensional face of precisely two $k$-simplices in $\mathrm{sd}^{(m)} \mathcal{F}(\tau)$, and these simplices clearly belong to $\mathcal{P}_2$. ◇

For each $\mu \in \mathcal{P}_2$ put
$$\mathcal{R}^{-1}[\mu] = \{\kappa \in \mathcal{P}_1 : (\kappa, \mu) \in \mathcal{R}\}.$$

**Claim 2.** $|\mathcal{R}| = |\mathcal{P}_2(0)| + 2 \cdot |\mathcal{P}_2(1)|$.

*Proof.* Clearly,
$$|\mathcal{R}| = \sum_{\mu \in \mathcal{P}_2} |\mathcal{R}^{-1}[\mu]|.$$
Take $\mu \in \mathcal{P}_2(0)$. We claim that $|\mathcal{R}^{-1}[\mu]| = 1$. Indeed, $h[\dot\mu] = \{0, \ldots, k\}$, i.e., the function $h$ is one-to-one on $\dot\mu$. So $\mu$ has exactly one $(k-1)$-dimensional face $\kappa$ such that $\{0, \ldots, k-1\} = h[\dot\kappa]$ and this face is clearly the only element of $|\mathcal{R}^{-1}[\mu]|$. If $\mu \in \mathcal{P}_2(1)$ then $h[\dot\mu] = \{0, \ldots, k-1\}$. Since $\dot\mu$ has size $k+1$, there exist precisely two subsets $E$ and $F$ of $\dot\mu$ of cardinality $k$ with the property that
$$h[E] = \{0, \ldots, k-1\} = h[F].$$
Consequently, there exist precisely two $(k-1)$-dimensional faces $\kappa_1$ and $\kappa_2$ of $\mu$ with $h[\dot\kappa_1] = \{0, \ldots, k-1\} = h[\dot\kappa_2]$. These simplices are obviously the only elements of $|\mathcal{R}^{-1}[\mu]|$. ◇

We find that
$$|\mathcal{P}_2(0)| - |\mathcal{P}_1(0)| = 2(|\mathcal{P}_2(1)| - |\mathcal{P}_1(1)|)$$
is even. As was remarked at the beginning of the proof, $|\mathcal{P}_1(0)|$ is odd, hence so is $|\mathcal{P}_2(0)|$. □

## 2.4. SIMPLICES IN $\mathbb{R}^n$

This result has interesting consequences.

**Lemma 2.4.4.** *Let $\tau$ be a k-simplex in $\mathbb{R}^k$, say $\tau = |\{x_0, \ldots, x_k\}|$. In addition, let $\{F_i : 0 \leq i \leq k\}$ be a collection of closed subsets of $\mathbb{R}^n$ such that for every $\{i_0, \ldots, i_\ell\} \subseteq \{0, \ldots, k\}$ we have*
$$|\{x_{i_0}, \ldots, x_{i_\ell}\}| \subseteq F_{i_0} \cup \cdots \cup F_{i_\ell}.$$
*Then $\bigcap_{i=0}^{k} F_i \neq \emptyset$.*

**Proof.** If not, then $\tau = (\tau \setminus F_0) \cup \cdots \cup (\tau \setminus F_k)$. Let $\varepsilon > 0$ be a Lebesgue number for this covering (use that $\tau$ is compact and apply Lemma A.5.3). By Corollary 2.2.9 we can choose $m$ so large that $\text{mesh}(\text{sd}^{(m)}\mathcal{F}(\tau)) < \varepsilon$. Let $V$ be the set of vertices of elements of $\text{sd}^{(m)}\mathcal{F}(\tau)$. Let $v \in V$ and put
$$I_v = \{0 \leq i \leq k : \alpha_i(v) > 0\}$$
(here $\alpha_0(v), \ldots, \alpha_k(v)$ are the affine coordinates of $v$ with respect to $x_0, \ldots, x_k$ of course).

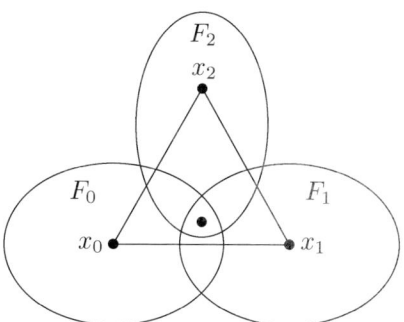

Figure 8.

Let $I_v = \{i_0, \ldots, i_\ell\}$. Then
$$v \in |\{x_{i_0}, \ldots, x_{i_\ell}\}| \subseteq F_{i_0} \cup \cdots \cup F_{i_\ell}.$$
So for $v \in V$ we can pick $h(v) \in \{0, \ldots, k\}$ such that
$$\alpha_{h(v)}(v) > 0, \quad v \in F_{h(v)}.$$
Then $h$ is an $m$-Sperner map for $\tau$. By Theorem 2.4.3 there exists a full simplex $\sigma \in \text{sd}^{(m)}\mathcal{F}(\tau)$ for $h$, say $\dot\sigma = \{v_0, \ldots, v_k\}$ with $h(v_i) = i$. Then clearly $v_i \in \sigma \cap F_i$ for every $0 \leq i \leq k$. But $\text{diam}(\sigma) < \varepsilon$ so there exists $i$ with $\sigma \subseteq \tau \setminus F_i$, i.e., $\sigma \cap F_i = \emptyset$. This a contradiction. □

We now come to the main result of this section.

**The Brouwer Fixed-Point Theorem 2.4.5.** *Let $f : \mathbb{I}^n \to \mathbb{I}^n$ be continuous. Then $f$ has a fixed-point.*

**Proof.** Let $\tau = |\{x_0, \ldots, x_n\}|$ be an $n$-simplex in $\mathbb{R}^n$. Then $\tau$ is homeomorphic to $\mathbb{I}^n$ (Exercise 1.1.24) and so it suffices to prove the Theorem for $\tau$. To this end, let $f \colon \tau \to \tau$ be continuous. For $i = 0, 1, \ldots, n$ let

$$F_i = \{x \in \tau : \alpha_i(f(x)) \leq \alpha_i(x)\}$$

(here $\alpha_0(x), \ldots, \alpha_n(x)$ are the affine coordinates of $x$ with respect to $x_0, \ldots, x_n$ of course).

Let $|\{x_{i_0}, \ldots, x_{i_\ell}\}|$ be a face of $\tau$ and $x \in |\{x_{i_0}, \ldots, x_{i_\ell}\}|$. Then for an element $i \notin \{i_0, \ldots, i_\ell\}$,

$$\alpha_i(x) = 0$$

so that

$$\alpha_{i_0}(x) + \cdots + \alpha_{i_\ell}(x) = 1 \geq \alpha_{i_0}(f(x)) + \cdots + \alpha_{i_\ell}(f(x))$$

and hence there must be $0 \leq j \leq \ell$ with

$$\alpha_{i_j}(x) \geq \alpha_{i_j}(f(x)).$$

Consequently, $x \in F_{i_j}$. We conclude that $|\{x_{i_0}, \ldots, x_{i_\ell}\}| \subseteq F_{i_0} \cup \cdots \cup F_{i_\ell}$.

Each $F_i$ is closed by continuity of $f$ and the functions $\alpha_i$ (Theorem 2.1.8). So by Lemma 2.4.4 there exists $x \in \bigcap_{i=1}^n F_i$. Then for every $0 \leq i \leq n$,

$$0 \leq \alpha_i(f(x)) \leq \alpha_i(x),$$

and

$$\sum_{i=1}^n \alpha_i(x) = 1 = \sum_{i=1}^n \alpha_i(f(x)).$$

This implies $\alpha_i(x) = \alpha_i(f(x))$ for every $0 \leq i \leq n$ and consequently, $x$ is equal to $f(x)$. $\square$

**Corollary 2.4.6.** *The Hilbert cube $Q$ has the fixed-point property.*

**Proof.** Let $f \colon Q \to Q$ be continuous. For every $n \in \mathbb{N}$ define

$$K_n = \{x \in Q : (x_1, \ldots, x_n) = (f(x)_1, \ldots, f(x)_n)\}.$$

It is clear that for every $n$ the set $K_n$ is closed in $Q$ and that $K_{n+1} \subseteq K_n$. Fix $n \in \mathbb{N}$, let $p_n \colon Q \to \mathbb{J}^n$ denote the projection and define a continuous function $f_n \colon \mathbb{J}^n \to \mathbb{J}^n$ by

$$f_n(x_1, \ldots, x_n) = (p_n \circ f)(x_1, \ldots, x_n, 0, 0, \ldots).$$

By Theorem 2.4.5, $f_n$ has a fixed-point, say $(x_1, \ldots, x_n)$, from which it follows that

$$(x_1, \ldots, x_n, 0, 0, \ldots) \in K_n.$$

We conclude that the $K_n$'s form a decreasing collection of nonempty closed subsets of $Q$ so that by compactness of $Q$,

$$K = \bigcap_{n=1}^{\infty} K_n \neq \emptyset.$$

It is clear that every point in $K$ is a fixed-point of $f$. □

**Corollary 2.4.7.** *Let $X$ be a compact* ANR. *If $f\colon X \to X$ is nullhomotopic then $f$ has a fixed-point. In particular, every compact* AR *has the fixed-point property.*

**Remark 2.4.8.** In view of this result, the following question naturally arises. Let $X$ be a compact ANR and let $f\colon X \to X$ be continuous. Assume that $f$ is homotopic to a continuous function $g\colon X \to X$ such that $g$ has a fixed-point. Does $f$ have a fixed-point? The answer to this question is in the negative, see Exercise 2.4.7. So in some sense, Corollary 2.4.7 is 'best possible'.

**Proof.** We may assume that $X$ is a subspace of $Q$ (Theorem A.4.4). Since every constant function $X \to X$ is extendable to a constant function $Q \to X$, by Corollary 1.4.3 it follows that $f$ can be extended to a continuous function

$$\bar{f}\colon Q \to X \hookrightarrow Q.$$

Since by Corollary 2.4.6, $Q$ has the fixed-point property, $\bar{f}$ has a fixed-point, say $x$. Since $\bar{f}[Q] \subseteq X$, $x$ belongs to $X$, and since $\bar{f}$ extends $f$, we conclude that $x$ is a fixed-point of $f$.

Since by Theorem A.12.4 every compact AR is contractible, the second part of the Corollary immediately follows from the first part and Proposition A.12.2. □

**Remark 2.4.9.** In §3.14 we will show by elementary but remote methods that Brouwer's Fixed-Point Theorem for $\mathbb{I}^3$, implies the Brouwer Fixed-Point Theorem for all dimensions. This seems to show that the 'real' power of the Brouwer Fixed-Point Theorem is already attained at dimension three. This is a rather curious phenomenon and it is not clear whether there is a direct route from Brouwer in $\mathbb{I}^3$ to Brouwer in all dimensions. This is of course only of interest for 'philosophical' reasons. The proof of Sperner's Lemma 2.4.3 is by induction, so it is equally complicated in all dimensions (except for the dimensions 1 and 2). Similar remarks can be made for the various other proofs that exist of Brouwer's Theorem in the literature. This seems to indicate that even if one finds a more direct route from Brouwer in $\mathbb{I}^3$ to Brouwer in all dimensions, this will not result in a simpler proof of Brouwer's Theorem. But this does not mean that clarifying the rather curious phenomenon that we will encounter in §3.14 would not be interesting.

We finish this section by presenting three applications of the techniques developed in this section.

**Application 1: The No-Retraction Theorem.** It is clear that the function $r\colon \mathbb{J}^n \to B^n$ defined by
$$r(x) = \begin{cases} \frac{x}{\|x\|} & (\|x\| \geq 1), \\ x & (\|x\| \leq 1). \end{cases}$$
is a retraction. Consequently, $B^n$ has the fixed-point property by Theorem 2.4.5 and Exercise A.12.10(2) (observe that in fact $\mathbb{I}^n$ and $B^n$ are homeomorphic, cf. Exercise 1.1.24).

**Theorem 2.4.10** (No-Retraction Theorem). *For every $n \in \mathbb{N}$, $\mathbb{S}^{n-1}$ is not a retract of $B^n$.*

**Proof.** To the contrary, suppose that $\mathbb{S}^{n-1}$ is a retract of $B^n$. As was remarked above, $B^n$ has the fixed-point property. Consequently, $\mathbb{S}^{n-1}$ has the fixed-point property by Exercise A.12.10(2). However, the antipodal mapping on $\mathbb{S}^{n-1}$ clearly demonstrates that $\mathbb{S}^{n-1}$ does not have the fixed-point property. Contradiction. $\square$

**Corollary 2.4.11.** *No $\mathbb{S}^n$ is contractible.*

**Proof.** Suppose that $\mathbb{S}^n$ is contractible. Then it is an AR by Corollaries 1.2.13 and 1.4.5. Consequently, there exists a retraction $r\colon B^n \to \mathbb{S}^{n-1}$, which contradicts Theorem 2.4.10. $\square$

**Application 2: The Theorem on Partitions.** The following result is the basis for dimension theory (see Chapter 3).

**The Theorem on Partitions 2.4.12.** *Consider $\mathbb{J}^n$ and for $i \leq n$ its opposite faces*
$$A_i = \{x \in \mathbb{J}^n : x_i = -1\}, \quad B_i = \{x \in \mathbb{J}^n : x_i = 1\}.$$
*If $C_i$ is a partition between $A_i$ and $B_i$ for $i \leq n$ then $\bigcap_{i=1}^n C_i \neq \emptyset$.*

**Proof.** To the contrary, assume that for every $i \leq n$, $C_i$ is a partition between $A_i$ and $B_i$ such that $\bigcap_{i=1}^n C_i = \emptyset$. There are closed subsets $E_i$ and $F_i$ in $\mathbb{J}^n$ for $i \leq n$ such that
$$A_i \subseteq E_i, \quad B_i \subseteq F_i, \quad E_i \cup F_i = \mathbb{J}^n, \quad E_i \cap F_i = C_i.$$
By Exercise A.4.9 there exist continuous functions $\xi_i \colon \mathbb{J}^n \to \mathbb{J}$ such that
$$\xi_i[A_i] = \{1\}, \quad \xi_i[B_i] = \{-1\}, \quad \xi_i^{-1}(0) = C_i.$$
Define $f\colon \mathbb{J}^n \to \mathbb{J}^n$ by $f(x) = (\xi_1(x), \ldots, \xi_n(x))$. Then $f$ is continuous and the point $\underline{0}$ does not belong to its range. For every $x \in \mathbb{J}^n \setminus \{\underline{0}\}$ the ray from $\underline{0}$ through $x$ intersects the 'boundary' $B = \bigcup_{i=1}^n A_i \cup \bigcup_{i=1}^n B_i$ of $\mathbb{J}^n$ in precisely one point, say $r(x)$. The function $r\colon \mathbb{J}^n \setminus \{\underline{0}\} \to B$ is easily seen to be continuous. The function $g = r \circ f \colon \mathbb{J}^n \to B$ has the following properties:
$$g\bigl[(-1,1)^n\bigr] \cap (-1,1)^n = \emptyset,$$

and for every $i \leq n$,
$$g[A_i] \subseteq B_i, \quad g[B_i] \subseteq A_i.$$
Therefore, $g$ has no fixed-point, which contradicts Theorem 2.4.5. □

**Corollary 2.4.13.** *Consider the Hilbert cube $Q$ and its opposite faces*
$$W_i^{-1} = \{x \in Q : x_i = -1\}, \quad W_i^1 = \{x \in Q : x_i = 1\}.$$
*If $C_i$ is a partition between $W_i^{-1}$ and $W_i^1$ for every $i$ then $\bigcap_{i=1}^n C_i \neq \emptyset$.*

**Proof.** For every $m$, define $f_m \colon \mathbb{J}^m \to Q$ by
$$f_m(x_1, \ldots, x_m) = (x_1, \ldots, x_m, 0, 0, \ldots).$$
Then $f_m$ is clearly an imbedding. It is easily seen that $f_m^{-1}[C_i]$ is a partition between $A_i$ and $B_i$ for every $i \leq m$. Consequently, by Theorem 2.4.12,
$$\bigcap_{i=1}^m f_m^{-1}[C_i] \neq \emptyset.$$
By compactness of $Q$ we therefore obtain that
$$\bigcap_{i=1}^\infty C_i \neq \emptyset,$$
which is as desired. □

**Application 3: The Non-Homogeneity Theorem.**

**Theorem 2.4.14.** *Let $n \in \mathbb{N}$. Then*

(1) *if $A \subseteq \mathbb{S}^{n-1}$ then $B^n \setminus A$ is contractible.*
(2) *if $A \subseteq B^n \setminus \mathbb{S}^{n-1}$ is nonempty then $B^n \setminus A$ is not contractible.*

**Proof.** (1) Define $H \colon (B^n \setminus A) \times \mathbb{I} \to B^n \setminus A$ by $H(x, t) = (1-t)x$. It is easily seen that $H$ is a contraction. (Alternatively, apply Exercise 1.2.5 and Theorem 1.4.4.)

(2) Without loss of generality, $\underline{0} \in A$. Assume to the contrary that
$$H \colon (B^n \setminus A) \times \mathbb{I} \to B^n \setminus A$$
is a contraction. Define $F \colon \mathbb{S}^{n-1} \times \mathbb{I} \to \mathbb{S}^{n-1}$ by
$$F(x, t) = \frac{H(x, t)}{|H(x, t)|}.$$
Then $F$ contracts $\mathbb{S}^{n-1}$ to a point which contradicts Corollary 2.4.11. □

Since $\mathbb{J}^n$ and $B^n$ are homeomorphic, cf. Exercise 1.1.24, this yields

**Corollary 2.4.15.** *Let $n \in \mathbb{N}$. Then $\mathbb{I}^n$ is not homogeneous.*

**Exercises for §2.4.** Let $X$ be a space. We say that a closed set $A$ in $X$ *separates* $X$ if $X \setminus A$ is not connected.

▶1. Make the 'it is geometrically obvious' part in the proof of Lemma 2.4.1 precise.

2. Prove that the 'No-Retraction Theorem', the 'Brouwer Fixed-Point Theorem' and the fact that no $\mathbb{S}^n$ is contractible are equivalent statements, in the sense that they are easily deduced from each other.

▶3. Let $C$ be a compact convex subset of a normed linear space $L$. Prove that for every $\varepsilon > 0$ there exists a map $f_\varepsilon \colon C \to C$ such that
   (1) $\hat{\varrho}(f_\varepsilon, 1) < \varepsilon$,
   (2) $f_\varepsilon[C]$ is contained in a finite dimensional linear space subspace of $L$.

4. Let $X$ be a space and let $f \colon X \to X$ be a map with the following properties:
   (1) the closure of $f[X]$ is compact,
   (2) for each $\varepsilon > 0$ there exists $x \in X$ with $\varrho(x, f(x)) < \varepsilon$.
   Prove that $f$ has a fixed-point.

5. Let $C$ be a compact convex subset of a normed linear space $L$. Prove that $C$ has the fixed-point property. (This is called the *Schauder Fixed-Point Theorem*.)

6. Prove that the antipodal map $\mathbb{S}^n \to \mathbb{S}^n$ is not nullhomotopic.

7. Give an example of a compact ANR $X$ having a homotopy $H \colon X \times \mathbb{I} \to X$ such that $H_0$ is the identity and $H_1$ is fixed-point free.

▶8. Let $A \subseteq \mathbb{R}^n$ be compact, where $n > 1$. Prove that if $A$ separates $\mathbb{R}^n$ then there is a continuous function $f \colon A \to \mathbb{S}^{n-1}$ which is not nullhomotopic.

9. Let $A \subseteq \mathbb{S}^n$ be compact, where $n \geq 1$. Prove that if $A$ separates $\mathbb{S}^n$ then there is a continuous function $f \colon A \to \mathbb{S}^{n-1}$ which cannot be extended over $\mathbb{S}^n$.

## 2.5. The Lusternik-Schnirelman-Borsuk theorem

The Brouwer Fixed-Point Theorem which was proved in the previous section, is one of the most central results in topology. We presented a detailed proof of it for two reasons. First of all, its proof uses subdivisions of triangulations of simplices. Getting used to this is important because that will be used extensively in ANR-theory later. Secondly, it is the basis for dimension theory.

A more powerful result is the so-called *Lusternik-Schnirelman-Borsuk Theorem* which can be proved along the same lines using a somewhat more complicated combinatorial lemma than the one of Sperner (Lemma 2.4.3). We will not present a proof of the Lusternik-Schnirelman-Borsuk Theorem here since it is used only twice in this book and for our purposes its proof does not give more insight than the proof of Sperner's Lemma.

## 2.5. THE LUSTERNIK-SCHNIRELMAN-BORSUK THEOREM

**The Lusternik-Schnirelman-Borsuk Theorem 2.5.1.** *If $\mathcal{M}$ is an arbitrary closed covering of $\mathbb{S}^n$ such that $|\mathcal{M}| = n+1$ then at least one element $M \in \mathcal{M}$ contains a pair of antipodal points.*

For a nice combinatorial proof, see DUGUNDJI and GRANAS [143, Theorem 4.4].

The Borsuk-Lusternik-Schnirelman Theorem implies many deep theorems about the topology of the Euclidean spaces $\mathbb{R}^n$.

**Exercises for §2.5.** A map $f \colon \mathbb{S}^m \to \mathbb{S}^n$ is called *antipodal-preserving* provided that $f(-x) = -f(x)$ for every $x \in \mathbb{S}^m$.

1. Prove that the Lusternik-Schnirelman-Borsuk Theorem is sharp by showing that there exists a closed covering $\mathcal{M}$ of $\mathbb{S}^n$ such that $|\mathcal{M}| = n+2$ while no $M \in \mathcal{M}$ contains a pair of antipodal points.

▶2. Prove that the following statements are equivalent in the sense that they are easily deduced from each other:
   (1) The Lusternik-Schnirelman-Borsuk Theorem for $\mathbb{S}^n$.
   (2) There is no antipodal-preserving map
   $$f \colon \mathbb{S}^n \to \mathbb{S}^{n-1}.$$
   (3) (Borsuk Antipodal Theorem) An antipodal-preserving map
   $$f \colon \mathbb{S}^{n-1} \to \mathbb{S}^{n-1}$$
   is not nullhomotopic.
   (4) (Borsuk-Ulam Theorem) Every continuous
   $$f \colon \mathbb{S}^n \to \mathbb{R}^n$$
   sends at least one pair of antipodal points to the same point.

3. Prove that the Lusternik-Schnirelman-Borsuk Theorem implies the Brouwer Fixed-Point Theorem.

CHAPTER 3

# Basic dimension theory

Dimension theory enables us to assign to every topological space $X$ a number, $\dim X$, in
$$\{-1, 0, 1, \ldots\} \cup \{\infty\}$$
having, among other things, the following properties:

(1) if $X$ and $Y$ are homeomorphic spaces then $\dim X = \dim Y$,
(2) $\dim \mathbb{R}^n = n$ for every $n \in \mathbb{N}$.

So $\dim X$ is a topological invariant of $X$, and by (2) it distinguishes between the euclidean spaces $\mathbb{R}^n$, $n \in \mathbb{N}$. A space $X$ for which $\dim X < \infty$ is called *finite dimensional*, and a space is *infinite-dimensional* if it is not finite dimensional.

The aim of this chapter is to present some basic results from dimension theory. Some of the presented results are classical: they mostly concern finite dimensional spaces. However, during the last decades significant contributions were made concerning the topology of infinite-dimensional spaces and the topology of hereditarily indecomposable continua. We shall also present some of these results in detail.

## 3.1. The covering dimension

Let $X$ be a space and let $\Gamma$ be an index set. The Theorem on Partitions 2.4.12 motivates the following definition: a family of pairs of disjoint closed sets $\tau = \{(A_i, B_i) : i \in \Gamma\}$ of $X$ is called *essential* if for every family
$$\{L_i : i \in \Gamma\},$$
where $L_i$ is an arbitrary partition between $A_i$ and $B_i$ for every $i$, we have
$$\bigcap_{i \in G} L_i \neq \emptyset;$$
if $\tau$ is not essential then it is called *inessential*. So Theorem 2.4.12 and Corollary 2.4.13 show that $\mathbb{I}^n$ has an essential family of $n$ pairs of disjoint closed sets and $Q$ has essential families of size $n$ for every $n \leq \aleph_0$. It follows easily that every subfamily of an essential family is again essential (Exercise 3.1.1).

**Theorem 3.1.1.** $\mathbb{I}^n$ has an essential family consisting of $n$ pairs of disjoint closed sets, but every family consisting of at least $n+1$ pairs of disjoint closed sets is inessential.

**Proof.** By the above remarks we need only to consider the second part of the theorem.

Let $A$ and $B$ be disjoint closed subsets of $\mathbb{I}^n$ and let $E \subseteq \mathbb{I}$ be dense.

**Claim 1.** There is a partition $D$ between $A$ and $B$ such that
$$D \subseteq \{x \in \mathbb{I}^n : (\exists i \leq n)(x_i \in E)\}.$$

*Proof.* This is easy. Every point $x \in A$ has a neighborhood of the form
$$\prod_{i=1}^{n}[a_i, b_i],$$
with $a_i, b_i \in E \cup \{0,1\}$ for every $i \leq n$, which misses $B$. There is a finite family $\mathcal{F}$ of these neighborhoods whose union covers $A$, and the boundary $D$ of this union is contained in the union of the boundaries of the elements of $\mathcal{F}$. We conclude that $D$ is the required partition between $A$ and $B$.  ◇

Now let $\tau = \{(A_i, B_i) : i \leq n+1\}$ be an arbitrary family consisting of $n+1$ pairs of disjoint closed subsets of $\mathbb{I}^n$. We will prove that $\tau$ is inessential. Indeed, there exist $n+1$ pairwise disjoint dense subsets of $\mathbb{I}$. One can take for example
$$E_1 = (\sqrt{2} + \mathbb{Q}) \cap \mathbb{I}, \quad E_2 = (\sqrt{3} + \mathbb{Q}) \cap \mathbb{I}, \quad E_3 = (\sqrt{5} + \mathbb{Q}) \cap \mathbb{I}, \quad \ldots,$$
respectively. By the above there exist partitions $D_i$ between $A_i$ and $B_i$ such that
$$D_i \subseteq \{x \in \mathbb{I}^n : (\exists j \leq n)(x_j \in E_i)\} \qquad (i \leq n+1).$$
Since the $E_i$ are pairwise disjoint, clearly $\bigcap_{i=1}^{n} D_i = \emptyset$.  □

Observe that in Theorem 3.1.1 we formulated a topological property of $\mathbb{I}^n$ shared by no $\mathbb{I}^m$ for $m \neq n$. In particular we obtain:

**Corollary 3.1.2.** Let $n, m \in \mathbb{N}$. If $n \neq m$ then $\mathbb{I}^n \not\approx \mathbb{I}^m$.

These remarks suggest the following definition: for a space $X$ define its *covering dimension*, $\dim X \in \{-1, 0, 1, \ldots\} \cup \{\infty\}$, by

$\dim X = -1 \quad \Leftrightarrow \quad X = \emptyset$,
$\dim X \leq n \quad \Leftrightarrow \quad$ every family of $n+1$ pairs of disjoint closed subsets of $X$ is inessential,
$\dim X = n \quad \Leftrightarrow \quad \dim X \leq n$ and $\dim X \not\leq n-1$,
$\dim X = \infty \quad \Leftrightarrow \quad \dim X \neq n$ for every $n \geq -1$.

A space $X$ with $\dim X < \infty$ is called *finite dimensional*; if it is not finite dimensional then it is called *infinite-dimensional*.

Observe that $\dim X \geq n$ if and only if there is an essential family consisting of $n$ pairs of disjoint closed subsets of $X$. So in a sense, there are $n$ different 'directions' in $X$. Also observe that it is not clear at all why we call $\dim X$ the covering dimension of $X$: the term *partition degree* seems to be more appropriate. We will explain our terminology later.

It is easy to see that if $X$ and $Y$ are homeomorphic spaces then $X$ and $Y$ have the same covering dimension.

In our new terminology, Theorem 3.1.1 reads as follows:

**Theorem 3.1.3.** $\dim \mathbb{I}^n = n$ for every $n$.

In §1.5 we called a space zero-dimensional if it is nonempty and has a base consisting of clopen sets. In Corollary 1.5.2 it was shown that a space $X$ is zero-dimensional if and only if $\emptyset$ is a partition between any pair of disjoint closed subsets of $X$. So, in our new terminology, a space $X$ is zero-dimensional in the sense of §1.5 if and only if $\dim X = 0$. It will turn out that zero-dimensional spaces are, in a sense, the 'building blocks' of all the other finite dimensional spaces.

Without too much trouble, the proof of Theorem 3.1.1 can be adapted to show that $\dim \mathbb{R}^n = n$ for every $n$. This equality however will turn out to follow trivially from Theorem 3.1.1 and results to be derived in §3.2. For that reason we will not verify it here.

For later use we shall now study some elementary properties of essential families of pairs of disjoint closed sets.

The following simple result is fundamental for dimension theory since it allows to extend partitions in subspaces to partitions in the whole space.

**Lemma 3.1.4.** *Let $Y$ be a subspace of a space $X$. In addition, let $A$ and $B$ be disjoint closed subsets of $X$. If $U$ and $V$ are open neighborhoods of $A$ and $B$, respectively, having disjoint closures, and if $S$ is a partition in $Y$ between $Y \cap \overline{U}$ and $Y \cap \overline{V}$, then there is a partition $T$ in $X$ between $A$ and $B$ such that $T \cap Y \subseteq S$.*

**Proof.** Write $Y \setminus S$ as the disjoint union of two open (in $Y$) sets $E$ and $F$ such that
$$Y \cap \overline{U} \subseteq E, \quad Y \cap \overline{V} \subseteq F.$$
Since $E \cap V = \emptyset$ we obtain $\overline{E} \cap B = \emptyset$, and similarly $\overline{F} \cap A = \emptyset$. From this it follows easily that $A \cup E$ and $B \cup F$ are separated. By Corollary A.8.2 there exist disjoint open neighborhoods $U'$ and $V'$ of $A \cup E$ and $B \cup F$, respectively. Clearly, $T = X \setminus (U' \cup V')$ is as required. □

For a closed subspace we can do a little better.

**Corollary 3.1.5.** *Let $Y$ be a closed subspace of a space $X$. In addition, let $A$ and $B$ be disjoint closed subsets of $X$. If $S$ is a partition in $Y$ between the sets $Y \cap A$ and $Y \cap B$, then there is a partition $T$ in $X$ between $A$ and $B$ such that $T \cap Y \subseteq S$.*

**Proof.** Write $Y \setminus S$ as the disjoint union of two open (in $Y$) sets $E$ and $F$ such that
$$Y \cap A \subseteq E, \quad Y \cap B \subseteq F.$$
Observe that $S \cup F \cup B$ is closed in $X$ and that $A \cap (S \cup F \cup B) = \emptyset$. By Corollary A.4.3 there is an open neighborhood $U$ of $A$ in $X$ such that
$$\overline{U} \cap (S \cup F \cup B) = \emptyset.$$
By a similar argument it is possible to find an open neighborhood $V$ of $B$ in $X$ such that
$$\overline{V} \cap (S \cup E \cup \overline{U}) = \emptyset.$$
By construction, $S$ is a partition between $\overline{U} \cap Y$ and $\overline{V} \cap Y$. Now apply Lemma 3.1.4. □

**Corollary 3.1.6.** *Let $Y$ be a zero-dimensional subspace of $X$. Then for all disjoint closed subsets $A$ and $B$ of $X$ there exists a partition $S$ between $A$ and $B$ in $X$ such that $S \cap Y = \emptyset$.*

**Proof.** Let $A$ and $B$ be disjoint closed subsets of $X$ and let $U$ and $V$ be disjoint neighborhoods of $A$ and $B$, respectively, such that $\overline{U} \cap \overline{V} = \emptyset$ (Corollary A.4.3). Since $Y$ is zero-dimensional, $\emptyset$ is in $Y$ a partition between the sets $\overline{U} \cap Y$ and $\overline{V} \cap Y$. Now apply Lemma 3.1.4. □

**Corollary 3.1.7.** *If a space $X$ is the union of at most $n+1$ zero-dimensional subspaces then $\dim X \leq n$.*

**Remark 3.1.8.** We will prove later in Corollary 3.3.9 that the converse of this result is also true; so a space is at most $n$-dimensional if and only if it is the union of a family of $n + 1$ (not necessarily pairwise distinct) zero-dimensional subspaces.

**Proof.** Let $X = \bigcup_{i=1}^{n+1} X_i$ with $\dim X_i \leq 0$ for $i \leq n+1$. Let
$$\tau = \{(A_i, B_i) : i \leq n+1\}$$
be an arbitrary family of $n+1$ pairs of disjoint closed subsets of $X$. By Corollary 3.1.6, for each $i \leq n+1$ there exists a partition $L_i$ between $A_i$ and $B_i$ in $X$ such that $L_i \cap X_i = \emptyset$. Then $\bigcap_{i=1}^{n+1} L_i = \emptyset$ so that $\tau$ is inessential. We conclude that $\dim X \leq n$. □

These results enable us to prove the following:

**Theorem 3.1.9.** *Let $X$ be a space, let $\{(A_i, B_i) : i \in \Gamma\}$ be essential in $X$, and let $\Gamma_0$ be a proper subset of $\Gamma$. If for every $i \in \Gamma_0$, $L_i$ is a partition between $A_i$ and $B_i$ and*
$$L = \bigcap_{i \in \Gamma_0} L_i$$
*then*
$$\{(L \cap A_i, L \cap B_i) : i \in \Gamma \setminus \Gamma_0\}$$
*is essential in $L$.*

**Proof.** For $i \notin \Gamma_0$ let $E_i$ be a partition in $L$ between $L \cap A_i$ and $L \cap B_i$. By Corollary 3.1.5, for $i \notin \Gamma_0$ we can find a partition $F_i$ in $X$ between $A_i$ and $B_i$ such that $F_i \cap L \subseteq E_i$. Then
$$\emptyset \neq \bigcap_{i \in \Gamma_0} L_i \cap \bigcap_{i \in \Gamma \setminus \Gamma_0} F_i \subseteq L \cap \bigcap_{i \in \Gamma \setminus \Gamma_0} E_i = \bigcap_{i \in \Gamma \setminus \Gamma_0} E_i,$$
which is as required. $\square$

**Corollary 3.1.10.** *Let $X$ be a space and let $n \geq 0$. If $\dim X \geq n$ then there exist disjoint closed subsets $A$ and $B$ of $X$ such that if $L$ is a partition between $A$ and $B$ then $\dim L \geq n - 1$.*

**Proof.** If $n = 0$ then there is nothing to prove. So assume that $n \geq 1$. Since $\dim X \geq n$ there exists an essential family
$$\tau = \{(A_i, B_i) : i \leq n\}$$
in $X$. We claim that $A_1$ and $B_1$ are as required. Indeed, let $L$ be a partition between $A_1$ and $B_1$. Observe that $L \neq \emptyset$ since $\tau$ is essential. We consider two cases. If $n = 1$ then since $L \neq \emptyset$, $\dim L \geq 0$ and so we are done. If $n > 1$ then by Theorem 3.1.9, $L$ has an essential family of cardinality $n - 1$, namely the collection
$$\{(L \cap A_i, L \cap B_i) : i = 2, \ldots, n\}.$$
And so $\dim L \geq n - 1$. $\square$

**Exercises for §3.1.** A space $X$ is called *countable dimensional* if it can be written as the union of countably many zero-dimensional subspaces (cf. Corollary 3.1.7). In addition, a space $X$ is called *strongly infinite-dimensional* if it has an infinite essential family. See e.g., §3.13 for more information on these notions.

Let $X$ be a space with disjoint closed subsets $A$ and $B$. We say that a closed subset $S$ of $X$ is an *irreducible partition* between $A$ and $B$ if $S$ is a partition between $A$ and $B$ while no proper closed subset of $S$ shares this property.

1. Prove every subfamily of an essential family is again essential.

2. Let $\tau = \{(A_i, B_i) : i \leq n\}$ be a family of pairs of disjoint closed subsets in the space $X$. Prove that if there exist different indices $i, j \leq n$ such that
$$(A_i, B_i) = (A_j, B_j)$$
then $\tau$ is inessential.

3. Let $\tau = \{(A_i, B_i) : i \leq n\}$ be a family of pairs of disjoint closed subsets in the space $X$. Prove that if $\tau$ is essential then $X$ can be mapped continuously onto $\mathbb{I}^n$.

4. Let $X$ be a compact space and let $\tau = \{(A_i, B_i) : i \in \Gamma\}$ be an essential family in $X$. Prove that $\Gamma$ is countable.

5. Prove that if $X$ is connected and contains more than one point then $X$ is at least one-dimensional.

6. Prove that if $X$ is strongly infinite-dimensional then $X$ is not countable dimensional. Give an example of a strongly infinite-dimensional space. Assuming that every finite dimensional space is countable dimensional, give an example of an infinite-dimensional countable dimensional compact space.

7. Let $X$ be a compact space, and let $\tau = \{(A_i, B_i) : i \in \Gamma\}$ be a family of pairs of disjoint closed subsets of $X$. Prove that $\tau$ is essential if and only if every finite subfamily is.

8. Let $X$ be a space. Assume that $\tau = \{(A_i, B_i) : i \in \Gamma\}$ is an inessential family of pairs of disjoint closed subsets of $X$. Prove that for every subspace $E \subseteq X$ the family
$$\{(A_i \cap E, B_i \cap E) : i \in \Gamma\}$$
is inessential in $E$.

▶9. Let $X$ be a compact space and let $\tau = \{(A_i, B_i) : i \in \Gamma\}$ be an essential family of pairs of disjoint closed subsets of $X$. Prove that there is a component $C$ of $X$ such that the family
$$\tau \restriction C = \{(A_i \cap C, B_i \cap C) : i \in \Gamma\}$$
is essential in $C$. Conclude that every strongly infinite-dimensional compact space has a strongly infinite-dimensional component.

10. Prove that $2^{\mathbb{I}}$, the hyperspace of $\mathbb{I}$, is infinite-dimensional.

11. Let $f \colon X \to Y$ be continuous, and let $\tau = \{(A_i, B_i) : i \in I\}$ be a collection of pairs of disjoint closed sets in $X$. Assume that the collection
$$\hat{\tau} = \{(\overline{f[A_i]}, \overline{f[B_i]}) : i \in I\}$$
consists of pairs of disjoint sets and is inessential. Prove that $\tau$ is inessential.

12. Let $f \colon X \to Y$ be continuous, and let $\tau = \{(A_i, B_i) : i \in I\}$ be an inessential collection of pairs of disjoint closed sets in $Y$. Prove that the collection
$$\hat{\tau} = \{(f^{-1}[A_i], f^{-1}[B_i]) : i \in I\}$$
is inessential in $X$.

13. Let $X$ be a space and let $\{(A_i, B_i) : i \in \Gamma\}$ be essential in $X$. Suppose that the set $Y \subseteq X$ is such that $Y \cap \bigcap_{i \in \Gamma} L_i \neq \emptyset$ for any choice of partitions $L_i$ of $A_i$ and $B_i$ in $X$. For each $i \in \Gamma$ let $U_i$ and $V_i$ be disjoint closed neighborhoods of $A_i$ and $B_i$, respectively. Prove that
$$\{(U_i \cap Y, V_i \cap Y) : i \in \Gamma\}$$
is essential in $Y$.

14. Let $X$ be a zero-dimensional space containing the pairwise disjoint closed subsets $A$, $B$ and $C$. Prove that $C$ is a partition between $A$ and $B$. (So in a zero-dimensional space every irreducible partition is empty.)

▶15. Give an example of a space $X$ containing two disjoint closed subsets $A$ and $B$ such that no partition between $A$ and $B$ is irreducible.

▶16. Prove that the Brouwer Fixed-Point Theorem and the fact that there is a space $X$ with $\dim X = \infty$ are equivalent statements, in the sense that they are easily deduced from each other. (This complements the list in Exercise 2.4.2 with an additional statement.)

## 3.2. Translation into open covers

We defined the covering dimension $\dim X$ in terms of properties of the family of all closed subsets of $X$. In this section we see that $\dim X$ is also describable in terms of properties of the family of all open covers of $X$. As an application of the results derived here we present a simple proof that $\mathbb{R}^n$ is $n$-dimensional for every $n$. We conclude from this that $\mathbb{R}^n$ and $\mathbb{R}^m$ are homeomorphic if and only if $n = m$.

Let $X$ be a space and let $\mathcal{A}$ and $\mathcal{B} = \{B(A) : A \in \mathcal{A}\}$ be families of subsets of $X$. We say that $\mathcal{B}$ is a *swelling* of $\mathcal{A}$ if

(1) for every $A \in \mathcal{A}$, $A \subseteq B(A)$,
(2) for every finite $\mathcal{F} \subseteq \mathcal{A}$, $\bigcap \mathcal{F} = \emptyset$ iff $\bigcap_{A \in \mathcal{F}} B(A) = \emptyset$.

Observe that if $B(A_0), B(A_1) \in \mathcal{B}$ are distinct then so are $A_0$ and $A_1$, but the converse need not hold.

Before we state the following proposition, let us recall that a locally finite collection is countable (see the remark on Page 486 following the definition of locally finite collection).

**Proposition 3.2.1.** *Let $\mathcal{F}$ be a locally finite collection of closed subsets of a space $X$. Then $\mathcal{F}$ has a swelling consisting of open subsets of $X$.*

**Proof.** We can enumerate $\mathcal{F}$ as $\{F_i : i \in \mathbb{N}\}$. Without loss of generality, assume that $F_1 = \emptyset$. By induction on $i \in \mathbb{N}$ we shall construct an open neighborhood $U_i$ of $F_i$ such that
$$\mathcal{F}_i = \{\overline{U}_1, \ldots, \overline{U}_i, F_{i+1}, F_{i+2}, \ldots\}$$

is a swelling of $\mathcal{F}$. Put $U_1 = \emptyset$, and assume that for some $i$ the sets $U_1, \ldots, U_i$ have been constructed. Observe that $\mathcal{F}_i$ is locally finite. Put
$$\mathcal{B} = \left\{ \bigcap \mathcal{E} : \mathcal{E} \text{ is a finite subcollection of } \mathcal{F}_i \text{ and } \left(\bigcap \mathcal{E}\right) \cap F_{i+1} = \emptyset \right\}.$$
Observe that $\mathcal{B}$ is locally finite as well and put $B = \bigcup \mathcal{B}$. Then $B$ is closed by Exercise A.7.1 and clearly $B \cap F_{i+1} = \emptyset$. By Corollary A.4.3 there consequently exists an open neighborhood $U_{i+1}$ of $F_{i+1}$ the closure of which misses $B$. It is clear that the set $U_{i+1}$ is as required.

We claim that $\{U_i : i \in \mathbb{N}\}$ is a swelling of $\mathcal{F}$. To this end, take arbitrary $i(1), \ldots, i(n) \in \mathbb{N}$ and assume that $\bigcap_{j=1}^n F_{i(j)} = \emptyset$. Let
$$m = \max\{i(1), \ldots, i(n)\}.$$
Then $\{U_1, \ldots, U_m\}$ is a swelling of $\{F_1, \ldots, F_m\}$ by our construction. We conclude that $\bigcap_{j=1}^n U_{i(j)} = \emptyset$. □

**Corollary 3.2.2.** *Let $\mathcal{F}$ be a locally finite family of closed subsets of a space $X$. Also, for every $F \in \mathcal{F}$ let $V(F)$ be a neighborhood of $F$. Then there exists a swelling $\{U(F) : F \in \mathcal{F}\}$ of $\mathcal{F}$ consisting of open subsets of $X$ such that for every $F \in \mathcal{F}$ we have $\overline{U(F)} \subseteq V(F)$.*

**Proof.** By Proposition 3.2.1 there exists an 'open' swelling $\{W(F) : F \in \mathcal{F}\}$ of $\mathcal{F}$. By Corollary A.4.3 there exists for every $F \in \mathcal{F}$ an open neighborhood $U(F)$ such that $\overline{U(F)} \subseteq V(F) \cap W(F)$. We claim that $\{U(F) : F \in \mathcal{F}\}$ is as required. To this end, let $\mathcal{G} \subseteq \mathcal{F}$ be finite such that $\bigcap \mathcal{G} = \emptyset$. Then clearly $\bigcap_{F \in \mathcal{G}} W(F) = \emptyset$ from which it follows that $\bigcap_{F \in \mathcal{G}} U(F) = \emptyset$. □

We now show how finite closed collections can be modified.

**Lemma 3.2.3.** *Let $A, A_1, \ldots, A_n$ be closed subsets of a space $X$. Moreover, let $V_1, \ldots, V_n$ be open subsets of $X$ with $A \cap A_i \subseteq V_i$ for every $i \leq n$. If*
$$\hat{A} = A \setminus \bigcup_{i=1}^n V_i, \quad \hat{A}_i = A_i \cup \overline{V_i}$$
*for $i \leq n$ then*
(1) $A \cup \bigcup_{i=1}^n A_i \subseteq \hat{A} \cup \bigcup_{i=1}^n \hat{A}_i$,
(2) $\hat{A} \cap \bigcap_{i=1}^n \hat{A}_i \subseteq \bigcap_{i=1}^n \operatorname{Fr} V_i$.

**Proof.** Observe that (1) is trivial. For (2), simply observe that for $i \leq n$,
$$\hat{A} \cap \hat{A}_i = (A \setminus V_i) \cap (A_i \cup \overline{V_i})$$
$$= [(A \cap A_i) \setminus V_i] \cup [A \cap (\overline{V_i} \setminus V_i)]$$
$$\subseteq \operatorname{Fr} V_i.$$
So we are done. □

One should think of the previous triviality in the following way. The closed sets $A$ are changed into the closed sets $\hat{A}$. But this is not done in an arbitrary way because (1) tells us that what is covered by the $A$'s is also covered by the $\hat{A}$'s, and (2) tells us where the intersection of the $\hat{A}$'s can be found.

The results obtained so far have nice applications in dimension theory.

**Corollary 3.2.4.** *Let $X$ be a space such that $\dim X \leq n < \infty$. Then for every countable open cover $\mathcal{U}$ of $X$ and for every subcollection $\mathcal{F} \subseteq \mathcal{U}$ of cardinality $n+2$ there exists an open shrinking $\mathcal{V} = \{V(U) : U \in \mathcal{U}\}$ of $\mathcal{U}$ having the following properties:*

(1) *for every $U \in \mathcal{U}$, $V(U) \subseteq \overline{V(U)} \subseteq U$,*
(2) *$\bigcap_{U \in \mathcal{F}} V(U) = \emptyset$.*

**Proof.** Enumerate $\mathcal{U}$ as $\{U_i : i \in \mathbb{N}\}$. Without loss of generality,
$$\mathcal{F} = \{U_1, \ldots, U_{n+2}\}.$$
By Proposition A.7.1 there exists a closed shrinking $\{B_i : i \in \mathbb{N}\}$ of $\mathcal{U}$. For every $i \leq n+1$ define $A_i = X \setminus U_i$. Then $\{(B_i, A_i) : i \leq n+1\}$ is a family of $n+1$ pairs of disjoint closed subsets of $X$. Since $\dim X \leq n$, there exist open sets $V_i \subseteq X$ for $i \leq n+1$ such that

(3) $B_i \subseteq V_i \subseteq \overline{V_i} \subseteq X \setminus A_i = U_i$,
(4) $\bigcap_{i=1}^{n+1} \operatorname{Fr} V_i = \emptyset$.

Now consider the closed sets $B_1, \ldots, B_{n+2}$. For $i \leq n+1$ put $\hat{B}_i = \overline{V_i}$ and let $\hat{B}_{n+2} = B_{n+2} \setminus \bigcup_{i=1}^{n+1} V_i$. Observe that the $\hat{B}$'s are the 'improved' sets we get from Lemma 3.2.3 by considering the collections of closed sets
$$B_{n+2}, B_1, \ldots, B_{n+1}$$
and open sets
$$V_1, \ldots, V_{n+1}.$$
So we get by (4) that

(5) $$\bigcap_{i=1}^{n+2} \hat{B}_i \subseteq \bigcap_{i=1}^{n+1} \operatorname{Fr} V_i = \emptyset.$$

Observe that
$$\hat{\mathcal{B}} = \{\hat{B}_1, \hat{B}_2, \ldots, \hat{B}_{n+2}, B_{n+3}, B_{n+4}, \ldots\}$$
covers $X$ since $\bigcup_{i=1}^{n+2} B_i \subseteq \bigcup_{i=1}^{n+2} \hat{B}_i$. In addition, the first $n+2$ members of $\hat{\mathcal{B}}$ have empty intersection by (5). Since $\hat{B}_i \subseteq U_i$ for every $i \leq n+2$, the desired result now follows from Corollary 3.2.2. □

Let $\mathcal{U}$ be a cover of a space $X$ and let $n \geq 0$ (we do not assume $\mathcal{U}$ to be open). We say that the *order* of $\mathcal{U}$ is at most $n$, $\text{ord}(\mathcal{U}) \leq n$, if for every $x \in X$,
$$|\{U \in \mathcal{U} : x \in U\}| \leq n+1.$$
We now come to the following central result.

**Theorem 3.2.5.** *Let $X$ be a nonempty space and let $n \geq 0$. The following statements are equivalent:*

(1) $\dim X \leq n$,
(2) *for every open cover $\mathcal{U}$ of $X$ there exists a locally finite closed refinement $\mathcal{V}$ of $\mathcal{U}$ such that $\text{ord}(\mathcal{V}) \leq n$,*
(3) *for every open cover $\mathcal{U}$ of $X$ there exists an open refinement $\mathcal{V}$ of $\mathcal{U}$ with $\text{ord}(\mathcal{V}) \leq n$,*
(4) *for every open cover $\mathcal{U}$ of $X$ there exists a closed shrinking $\mathcal{V}$ of $\mathcal{U}$ with $\text{ord}(\mathcal{V}) \leq n$,*
(5) *for every open cover $\mathcal{U}$ of $X$ there exists an open shrinking $\mathcal{V}$ of $\mathcal{U}$ with $\text{ord}(\mathcal{V}) \leq n$,*
(6) *for every finite open cover $\mathcal{U}$ of $X$ there exists a closed shrinking $\mathcal{V}$ of $\mathcal{U}$ such that $\text{ord}(\mathcal{V}) \leq n$,*
(7) *for every finite open cover $\mathcal{U}$ of $X$ there exists an open shrinking $\mathcal{V}$ of $\mathcal{U}$ such that $\text{ord}(\mathcal{V}) \leq n$.*

**Remark 3.2.6.** This theorem explains the earlier terminology *covering dimension*. The vague idea behind dimension theory is that one wishes to cover a given space with finitely many 'small' open sets in such a way that as few of the open sets as possible have points in common. If $\dim X \leq n$ then it is possible to do that in such a way that at most $n+1$ elements intersect. If $\dim X \geq n$ then no matter how the 'small' cover is chosen, there are always at least $n+1$ elements having a point in common. It is interesting and intriguing that these ideas can be used to distinguish between various spaces topologically.

**Remark 3.2.7.** Missing in the list of statements in Theorem 3.2.5 is:

(8) *for every open cover $\mathcal{U}$ of $X$ there exists a closed refinement $\mathcal{V}$ of $\mathcal{U}$ with $\text{ord}(\mathcal{V}) \leq n$.*

This is not a very interesting statement, since every open cover $\mathcal{U}$ of a space $X$ is refined by the closed cover $\{\{x\} : x \in X\}$ of $X$ which has order 0. The following statement seems to be more interesting:

(8') *for every open cover $\mathcal{U}$ of $X$ there exists a countable closed refinement $\mathcal{V}$ of $\mathcal{U}$ with $\text{ord}(\mathcal{V}) \leq n$.*

It is not true that (8') characterizes all at most $n$-dimensional spaces. We will show in Exercises 3.2.7 and 3.2.8 that Erdős' space described in Example 1.5.18 has the property that every open cover $\mathcal{U}$ of it has a countable *disjoint* closed refinement, yet $E$ is one-dimensional (as will be shown in Example 3.4.13). We do not know whether spaces that satisfy condition (8′) are at most $(n+1)$-dimensional.

**Proof.** *We prove* (1) ⇒ (2).

Since a refinement of a refinement is a refinement, by Corollary A.7.3 we may assume that $\mathcal{U}$ is locally finite. Enumerate $\mathcal{U}$ as $\{U_{0,i} : i \in \mathbb{N}\}$. In addition, let $\{F(j) : j \in \mathbb{N}\}$ enumerate the collection of all subsets of $\mathbb{N}$ of cardinality precisely $n+2$. By Corollary 3.2.4, for $j \in \mathbb{N}$ there exists an open cover $\mathcal{V}_j = \{U_{j,i} : i \in \mathbb{N}\}$ of $X$ having the following properties:

(8) for each $i$, $U_{j,i} \subseteq \overline{U}_{j,i} \subseteq U_{j-1,i}$,
(9) $\bigcap_{i \in F(j)} U_{j,i} = \emptyset$.

Now for each $i \in \mathbb{N}$ define

$$S_i = \bigcap_{j=1}^{\infty} \overline{U}_{j,i}.$$

**Claim 1.** The collection $\mathcal{V} = \{S_i : i \in \mathbb{N}\}$ covers $X$.

*Proof.* Take an arbitrary $x \in X$. Since $\mathcal{U}$ is locally finite, $x$ is contained in finitely many elements of $\mathcal{U}$ only. Pick $m \in \mathbb{N}$ such that $x \notin \bigcup_{i>m} U_{0,i}$. We conclude that for every $j \in \mathbb{N}$ there is an index $k(j) \leq m$ such that $x \in U_{j,k(j)}$. So there exists $k \leq m$ such that $x$ belongs to infinitely many of the $U_{j,k}$. By (8) this implies that $x \in S_k$, which is as required. ◇

We conclude that $\mathcal{V}$ is a closed shrinking of $\mathcal{U}$ and hence is locally finite. Moreover, ord($\mathcal{V}$) $\leq n$ by (8) and (9).

*We prove* (2) ⇒ (3).

By (2) there exists a locally finite closed refinement $\mathcal{S}$ of $\mathcal{U}$ such that ord($\mathcal{S}$) $\leq n$. For each $S \in \mathcal{S}$ pick $U(S) \in \mathcal{U}$ containing $S$. By Corollary 3.2.2 there exists an open swelling $\mathcal{V} = \{V(S) : S \in \mathcal{S}\}$ of $\mathcal{S}$ such that for every element $S \in \mathcal{S}$ we have $V(S) \subseteq U(S)$. Since ord($\mathcal{S}$) $\leq n$ we get ord($\mathcal{V}$) $\leq n$, and so $\mathcal{V}$ is as required.

*We prove* (3) ⇒ (5).

By (3) there exists an open refinement $\mathcal{W}$ of $\mathcal{U}$ such that ord($\mathcal{W}$) $\leq n$. For each $W \in \mathcal{W}$ pick $U(W) \in \mathcal{U}$ be such that $W \subseteq U(W)$. Now for each $U \in \mathcal{U}$ define

$$V(U) = \bigcup \{W \in \mathcal{W} : U(W) = U\}.$$

Clearly, $\mathcal{V} = \{V(U) : U \in \mathcal{U}\}$ is an open shrinking of $\mathcal{U}$ and it therefore suffices to prove that ord($\mathcal{V}$) $\leq n$. Take pairwise distinct

$$V(U_1), \ldots, V(U_{n+2}) \in \mathcal{V}$$

and assume that $x \in \bigcap_{i=1}^{n+2} V(U_i)$. For each $i \leq n+2$ there exists $W_i \in \mathcal{W}$ such that $U(W_i) = U_i$ and $x \in W_i$. Since the $U_i$ are pairwise distinct, this implies that the $W_i$ are pairwise distinct. But $\mathrm{ord}(\mathcal{W}) \leq n$, and so $x \in \bigcap_{i=1}^{n+2} W_i = \emptyset$. This is a contradiction.

*We prove* (5) $\Rightarrow$ (4).

Observe that if $\mathcal{A}$ is a cover of $X$ such that $\mathrm{ord}(\mathcal{A}) \leq n$ and $\mathcal{B}$ is a shrinking of $\mathcal{A}$ then $\mathrm{ord}(\mathcal{B}) \leq n$. So (4) follows directly from (5) and Corollary 3.2.4.

*We prove* (4) $\Rightarrow$ (6).

This is a triviality.

*We prove* (6) $\Rightarrow$ (7).

This follows immediately from Corollary 3.2.2.

*We prove* (7) $\Rightarrow$ (1).

Let $\{(A_i, B_i) : i \leq n+1\}$ be a family of $n+1$ pairs of disjoint closed subsets of $X$. Observe that

$$\bigcap_{i=1}^{n+1} A_i \cap \bigcup_{i=1}^{n+1} B_i = \emptyset$$

so that the collection

$$\left\{ X \setminus A_1, X \setminus A_2, \ldots, X \setminus A_{n+1}, X \setminus \bigcup_{i=1}^{n+1} B_i \right\}$$

covers $X$. By (7) there consequently exists an open cover $\mathcal{V} = \{V_i : i \leq n+2\}$ of $X$ such that

(10) $V_i \subseteq X \setminus A_i$ ($i \leq n+1$),
(11) $V_{n+2} \subseteq X \setminus \bigcup_{i=1}^{n+1} B_i$,
(12) $\mathrm{ord}(\mathcal{V}) \leq n$.

Let $\mathcal{F} = \{F_i : i \leq n+2\}$ be a closed shrinking of $\mathcal{V} = \{V_i : i \leq n+2\}$ (Proposition A.7.1). Now for $i \leq n+1$, define $\hat{A}_i$ and $\hat{B}_i$ by

$$\hat{A}_i = A_i \cup (F_{n+2} \setminus V_i),$$
$$\hat{B}_i = B_i \cup F_i.$$

**Claim 2.** For every $i \leq n+1$, $\hat{A}_i \cap \hat{B}_i = \emptyset$.

*Proof.* Simply observe that $A_i \cap B_i = \emptyset$, that $F_i \subseteq V_i$ from which it follows that $A_i \cap F_i = \emptyset$ and $(F_{n+2} \setminus V_i) \cap F_i = \emptyset$ and that $F_{n+2} \cap B_i = \emptyset$. ◇

**Claim 3.** $\bigcup_{i=1}^{n+1} (\hat{A}_i \cup \hat{B}_i) = X$.

*Proof.* Since $\mathcal{F}$ covers $X$, we are done if we show that $\bigcup_{i=1}^{n+1} F_i \subseteq \bigcup_{i=1}^{n+1} \hat{B}_i$ and $F_{n+2} \subseteq \bigcup_{i=1}^{n+1} \hat{A}_i$. The first inclusion is a triviality. For the second one observe that $F_{n+2} \subseteq V_{n+2}$ and that by (12), $V_{n+2} \cap \bigcap_{i=1}^{n+1} V_i = \emptyset$. We conclude that
$$F_{n+2} = F_{n+2} \setminus \bigcap_{i=1}^{n+1} V_i \subseteq \bigcup_{i=1}^{n+1} \hat{A}_i,$$
which proves the claim. ◇

By Corollary A.4.3 there exists for every $i \leq n+1$ a partition $L_i$ between $\hat{A}_i$ and $\hat{B}_i$. Since $A_i \subseteq \hat{A}_i$ and $B_i \subseteq \hat{B}_i$ for every $i$, $L_i$ is also a partition between the sets $A_i$ and $B_i$. By Claim 3 we obtain
$$\bigcap_{i=1}^{n+1} L_i \subseteq X \setminus \bigcup_{i=1}^{n+1} (\hat{A}_i \cup \hat{B}_i) = X \setminus X = \emptyset,$$
as required. □

Naturally, the reader wonders about the relation between the dimension of a space $X$ and the various dimensions of its subspaces. We finish this section by deriving two results about that relation.

**The Countable Closed Sum Theorem 3.2.8.** *If $X$ can be covered by countably many closed and at most $n$-dimensional sets then $X$ is at most $n$-dimensional.*

**Proof.** It is clear that without loss of generality we may assume that $n < \infty$. Enumerate the closed cover $\mathcal{F}$ as $\{F_i : i \in \mathbb{N}\}$, and put $F_0 = \emptyset$. Let $\mathcal{U}$ be a finite open cover of $X$. By induction on $i \geq 0$ we shall construct an open cover $\mathcal{U}(i)$ of $X$ such that the following conditions are satisfied:

(1) $\mathcal{U}(0) = \mathcal{U}$,
(2) if $0 \leq j < i$ then $\overline{\mathcal{U}(i)}$ is a shrinking of $\mathcal{U}(j)$,
(3) $\mathrm{ord}(\overline{\mathcal{U}(i)} \upharpoonright F_i) \leq n$.

Observe that conditions (2) and (3) are satisfied for $i = 0$. Now assume that we completed the construction for all $j$ with $0 \leq j < i$. Since $\dim F_i \leq n$, by assumption, the open cover $\mathcal{U}(i-1) \upharpoonright F_i$ of $F_i$ has an open shrinking
$$\mathcal{V} = \{V(U) : U \in \mathcal{U}(i-1)\}$$
of order at most $n$ (Theorem 3.2.5). For each $U \in \mathcal{U}(i-1)$ put
$$W(U) = (U \setminus F_i) \cup V(U).$$
It is clear that
$$\mathcal{W} = \{W(U) : U \in \mathcal{U}(i-1)\}$$
is an open shrinking of $\mathcal{U}(i-1)$ such that $\mathrm{ord}(\mathcal{W} \upharpoonright F_i) \leq n$. An easy application of Proposition A.7.1 and Corollary 3.2.2 now shows that there exists

an open cover $\mathcal{U}(i)$ of $X$ satisfying (2) and (3). This completes the inductive construction.

Let the cardinality of $\mathcal{U}$ be $k < \infty$. It is clear that for $i \in \mathbb{N}$ we can enumerate $\mathcal{U}(i)$ as $\{U_{m,i} : m \leq k\}$ such that for $0 \leq i < j$ and $m \leq k$ we have $\overline{U}_{m,j} \subseteq U_{m,i}$. Now for each $x \in X$ there exists $m(x) \leq k$ such that $x$ belongs to infinitely many of the $U_{m(x),i}$. So by construction we have

$$x \in \bigcap_{i=1}^{\infty} U_{m(x),i}.$$

By (2) and (3) this implies that

$$\left\{\bigcap_{i=1}^{\infty} U_{m,i} : m \leq k\right\} = \left\{\bigcap_{i=1}^{\infty} \overline{U}_{m,i} : m \leq k\right\}$$

is a closed shrinking of $\mathcal{U}$ and has order at most $n$. Consequently, $\dim X \leq n$ by Theorem 3.2.5. □

Since $\mathbb{R}^n$ is a countable union of homeomorphs of $\mathbb{I}^n$, Theorems 3.1.1 and 3.2.8 imply that $\dim \mathbb{R}^n \leq n$. In the remaining part of this section we shall present, among other things, two additional proofs of this inequality which are interesting in their own rights.

**The Subspace Theorem 3.2.9.** *Let $A$ be a subspace of a space $X$. Then $\dim A \leq \dim X$.*

**Proof.** If $\dim X = \infty$ then there is nothing to prove. So without loss of generality assume that $n = \dim X < \infty$.

First assume that $A$ is closed. Observe that if $\tau$ is an essential family in $A$ then it is also an essential family in $X$. From this it follows immediately that $\dim A \leq n$.

Next, assume that $A$ is open. Observe that $A$ is an $F_\sigma$-subset of $X$ by Exercise A.2.3. That $\dim A \leq n$ therefore follows from the above and the Countable Closed Sum Theorem 3.2.8.

Finally, assume that $A$ is an arbitrary subspace of $X$. In addition, let $\mathcal{U}$ be a cover of $A$ by sets that are open in $A$. For each $U \in \mathcal{U}$ pick an open subset $V(U)$ of $X$ with $V(U) \cap A = U$. Put $V = \bigcup_{U \in \mathcal{U}} V(U)$. Then $V$ is an open subset of $X$ and hence, by the above, $\dim V \leq n$. Since

$$\mathcal{V} = \{V(U) : U \in \mathcal{U}\}$$

covers $V$, Theorem 3.2.5 yields the existence of an open refinement $\mathcal{W}$ of $\mathcal{V}$ with $\operatorname{ord}(\mathcal{W}) \leq n$. Then $\mathcal{X} = \mathcal{W} \restriction A$ is an open refinement of $\mathcal{U}$ with order at most $n$. Another application of Theorem 3.2.5 now gives us that $\dim A \leq n$, as required. □

A different proof of Theorem 3.2.9 shall be given in Corollary 3.4.14(1)2.

Observe that since $\mathbb{R}^n$ is homeomorphic to $(0,1)^n$ we have by Theorems 3.1.1 and 3.2.9 that $\dim \mathbb{R}^n \leq n$.

The results in this section can also be used to prove that various spaces are zero-dimensional. We shall demonstrate this by the following example. For all $n \in \mathbb{N}$ and $0 \leq m \leq n$ let

$$\mathfrak{R}_{n,m} = \{x \in \mathbb{R}^n : \text{exactly } m \text{ coordinates of } x \text{ are rational}\}.$$

**Proposition 3.2.10.** *Let* $n \in \mathbb{N}$. *Then* $\mathbb{R}^n = \bigcup_{m=0}^n \mathfrak{R}_{n,m}$ *and* $\mathfrak{R}_{n,m}$ *is zero-dimensional for all* $0 \leq m \leq n$.

**Proof.** The first part of this proposition is trivial. For the second part, observe that if $m = 0$ then there is nothing to prove since $\mathfrak{R}_{n,0}$ is the product of $n$ copies of $\mathbb{P}$. So let $m \geq 1$ and let $A = \{i(1), \ldots, i(m)\}$ be a set of $m$ indices in $\{1, 2, \ldots, n\}$ and let $q_1, \ldots, q_m \in \mathbb{Q}$. Put

$$X = \{x \in \mathbb{R}^n : x_{i(j)} = q_j \text{ for every } j \leq m\}.$$

Observe that $X$ is a closed subspace of $\mathbb{R}^n$ and that

$$X \cap \mathfrak{R}_{n,m} = \{x \in \mathbb{R}^n : x_{i(j)} = q_j \text{ for every } j \leq m \text{ and } x_i \in \mathbb{P} \text{ for } i \notin A\}.$$

Consequently, $X \cap \mathfrak{R}_{n,m}$ is a closed subspace of $\mathfrak{R}_{n,m}$ which is homeomorphic to the product of $n - m$ copies of $\mathbb{P}$. We conclude that $X \cap \mathfrak{R}_{n,m}$ is zero-dimensional. Since $\mathfrak{R}_{n,m}$ is the union of countably many sets of the above form $X \cap \mathfrak{R}_{n,m}$, we conclude that $\dim \mathfrak{R}_{n,m} = 0$ by Theorem 3.2.8. $\square$

**Corollary 3.2.11.** *For all* $n \in \mathbb{N}$ *and* $0 \leq m \leq n$,

$$\dim\{x \in \mathbb{R}^n : \text{at most } m \text{ coordinates of } x \text{ are rational}\} \leq m.$$

**Proof.** Observe that the set of all $x \in \mathbb{R}^n$ for which at most $m$ coordinates are rational coincides with

$$\mathfrak{R}_{n,0} \cup \mathfrak{R}_{n,1} \cup \cdots \cup \mathfrak{R}_{n,m}.$$

So we are done by Proposition 3.2.10 and Corollary 3.1.7. $\square$

It therefore again follows that $\dim \mathbb{R}^n \leq n$.

We now come to what is sometimes called the 'Fundamental Theorem of Dimension Theory'.

**Theorem 3.2.12.** *For all* $n \in \mathbb{N}$,

$$\dim \mathbb{R}^n = \dim \mathbb{I}^n = \dim \mathbb{S}^n = n.$$

**Proof.** By Theorem 3.1.1, $\dim \mathbb{I}^n = n$. By Theorem 3.2.9 this implies that $\dim \mathbb{R}^n \geq n$. Since $\dim \mathbb{R}^n \leq n$, we obtain $\dim \mathbb{R}^n = n$. Also, $\mathbb{S}^n$ is the union of two homeomorphs of $\mathbb{I}^n$, so an appeal to Theorem 3.2.8 gives us that $\dim \mathbb{S}^n \leq n$. It now follows easily that $\dim \mathbb{S}^n = n$ as well. $\square$

Theorem 3.2.12 is of fundamental importance since it confirms our geometric intuition that $\mathbb{R}^n$, $\mathbb{I}^n$ and $\mathbb{S}^n$ are $n$-dimensional. Since the covering dimension is a topological notion, we also obtain the following

**Corollary 3.2.13.** *If $n \neq m$ then $\mathbb{R}^n$ and $\mathbb{R}^m$ are not homeomorphic.*

We finish this section with three characterizations of dimension, two of which are in the spirit of Theorem 3.4.4 and one of which is 'combinatorial'.

Let $\tau$ be an $n$-dimensional simplex in a linear space $L$. Exercise 1.1.24 gives us that $\tau$ is homeomorphic to $\mathbb{I}^n$, hence $\dim \tau = n$. Now let $X = |\mathcal{T}|$ be a polytope. Since $\mathcal{T}$ is countable, by the Countable Closed Sum Theorem 3.2.8 and the above remark we get

$$\dim X = \sup\{\dim \tau : \tau \in \mathcal{T}\}$$
$$= \sup\{k : \tau \text{ is a } k\text{-simplex}, \tau \in \mathcal{T}\}.$$

We conclude that the topological dimension of $X$ equals its 'combinatorial' dimension.

**Theorem 3.2.14.** *Let $X$ be a nonempty space and let $n \geq 0$. The following statements are equivalent:*

(1) *$\dim X \leq n$,*
(2) *for every open cover $\mathcal{U}$ of $X$ there exists a locally finite open refinement $\mathcal{V}$ of $\mathcal{U}$ such that such that $\operatorname{ord}(\mathcal{V}) \leq n$,*
(3) *for every open cover $\mathcal{U}$ of $X$ there exists a star-finite open refinement $\mathcal{V}$ of $\mathcal{U}$ such that such that $\operatorname{ord}(\mathcal{V}) \leq n$,*
(4) *for every open cover $\mathcal{U}$ of $X$ there exist a polytope $P$ such that $\dim P \leq n$ and a $\mathcal{U}$-mapping $f \colon X \to P$.*

**Proof.** We prove $(1) \Rightarrow (3)$.

Without loss of generality, $\mathcal{U}$ is star-finite (Theorem 2.3.5(1)). Now by Theorem 3.2.5 there exists an open shrinking $\mathcal{V}$ of $\mathcal{U}$ such that $\operatorname{ord}(\mathcal{V}) \leq n$. Since each shrinking of a star-finite cover is clearly star-finite, we are done.

We prove $(3) \Rightarrow (2) \Rightarrow (1)$.

That $(3) \Rightarrow (2)$ is a triviality and that $(2) \Rightarrow (1)$ follows from Theorem 3.2.5.

We prove $(3) \Rightarrow (4)$.

Let $\mathcal{U}$ be an open cover of $X$. By (3) there exists a (countable) star-finite open refinement $\mathcal{V}$ of $\mathcal{U}$ such that $\operatorname{ord}(\mathcal{V}) \leq n$. Let $P = |N(\mathcal{V})|$. Then

$$\kappa \colon X \to P$$

is a $\mathcal{V}$-mapping by Corollary 2.3.4, and hence a $\mathcal{U}$-mapping. Also, $P$ is a polytope by Theorem 2.3.5. Finally, since $\operatorname{ord}(\mathcal{V}) \leq n$, $N(\mathcal{V})$ has no simplices of dimension greater than $n$. By the Countable Closed Sum Theorem 3.2.8 it follows that $\dim P \leq n$.

We prove (4) $\Rightarrow$ (1).

Let $\mathcal{U}$ be an open cover of $X$. By (4) there exists a polytope $P$ such that $\dim P \leq n$ and a $\mathcal{U}$-mapping $f\colon X \to P$. For each $U \in \mathcal{U}$ let $E(U)$ be the set of all $y \in P$ with an open neighborhood $V_y$ such that $f^{-1}[V_y] \subseteq U$. Then the collection $\mathcal{E} = \{E(U) : U \in \mathcal{U}\}$ is an open cover of $P$ while moreover
$$f^{-1}[E(U)] \subseteq U$$
for every $U \in \mathcal{U}$. Since $\dim P \leq n$, Theorem 3.2.5 implies the existence of an open shrinking $\mathcal{V} = \{V(U) : U \in \mathcal{U}\}$ of $\mathcal{E} = \{E(U) : U \in \mathcal{U}\}$ such that $\operatorname{ord}(\mathcal{V}) \leq n$. It is clear that the collection $\mathcal{W} = \{f^{-1}[V(U)] : U \in \mathcal{U}\}$ is an open shrinking of $\mathcal{U}$ such that $\operatorname{ord}(\mathcal{W}) \leq n$. By Theorem 3.2.5 we therefore conclude that $\dim X \leq n$. $\square$

**Exercises for §3.2.** A space $X$ is called *almost zero-dimensional* if it has an open base $\mathcal{B}$ such that every $B \in \mathcal{B}$ has the property that $X \setminus \overline{B}$ is the union of clopen subsets of $X$.

1. Let $X$ be a space containing more than one point and let $p \in X$. Prove that $\dim(X \setminus \{p\}) = \dim X$. (So the dimension cannot be raised by the adjunction of a single point.)

2. Let $(X_n, f_n)_n$ be an inverse sequence of compact spaces such that
$$\dim X_n \leq m < \infty$$
for all $n$. Prove that the inverse limit of the sequence is at most $m$-dimensional. (The compactness condition is superfluous in this result. But we are not in the position yet to prove this. See Exercise 3.5.2 for more information.)

▶3. Let $X$ be a dense-in-itself space such that $\dim X \leq n < \infty$. Prove that for every cover $\{U_1, \ldots, U_m\}$ consisting of nonempty open sets there is a collection $\{F_1, \ldots, F_m\}$ consisting of closed sets having the following properties:
   (1) $\emptyset \neq F_i \subseteq U_i$ for every $i \leq m$,
   (2) $\bigcup_{i=1}^m F_i = X$,
   (3) if $i_1 < i_2 < \cdots < i_{n+2} \leq m$ are arbitrary then $\bigcap_{j=1}^{n+2} F_{i_j} = \emptyset$.

▶4. Prove that every $n$-dimensional compact space has a component which is also $n$-dimensional.

5. Let $X$ be a space. Prove that $X$ is almost zero-dimensional if and only if $X$ has a base $\mathcal{B}$ such that for each pair $G, H$ of elements of $\mathcal{B}$ with disjoint closures there is a clopen set $W$ in $X$ with $G \subseteq W \subseteq X \setminus H$.

6. Prove that every almost zero-dimensional space is totally disconnected.

▶7. Prove that Erdős' space $E$ is almost zero-dimensional.

8. Prove that if $X$ is almost zero-dimensional then every open cover of $X$ has a countable refinement by pairwise disjoint closed sets.

9. Prove that $\mathcal{C}(\mathbb{I}^2)$ is infinite-dimensional.

10. Let $X$ be a homogeneous space. Prove that every nonempty open subspace of $X$ has the same dimension as $X$.

## 3.3. The imbedding theorem

The aim of this section is to prove that every space $X$ with $\dim X \leq n$ can be imbedded in $\mathbb{R}^{2n+1}$. This is a result of fundamental importance.

For $n \geq 0$, define

$$\mathfrak{N}_n = \{x \in \mathbb{R}^{2n+1} : \text{at most } n \text{ coordinates of } x \text{ are rational}\}.$$

So

$$\mathfrak{N}_n = \bigcup_{k=0}^{n} \mathfrak{R}_{2n+1,k},$$

where the sets $\mathfrak{R}_{2n+1,k} \subseteq \mathbb{R}^{2n+1}$ are as in the previous section.

Observe that $\mathfrak{N}_0$ is the space of irrational numbers $\mathbb{P}$. The space $\mathfrak{N}_n$ is called *Nöbeling's universal n-dimensional space*. For later use, we do not aim at imbeddings in $\mathbb{R}^{2n+1}$ but at imbeddings in $\mathfrak{N}_n$. This does not require extra work and explains our terminology.

**Lemma 3.3.1.** *For $n \geq 0$, $\dim \mathfrak{N}_n = n$ and $\mathfrak{N}_n$ is the union of $n+1$ zero-dimensional subspaces.*

**Proof.** By Corollary 3.2.11 we get $\dim \mathfrak{N}_n \leq n$. To see that $\dim \mathfrak{N}_n \geq n$, we claim that $\mathfrak{N}_n$ contains a homeomorph of $\mathbb{I}^n$. This is easy. Define an imbedding $i\colon \mathbb{I}^n \to \mathfrak{N}_n$ by

$$i(x_1, \ldots, x_n) = (x_1, \ldots, x_n, \sqrt{2}, \sqrt{2}, \ldots, \sqrt{2}).$$

By Theorems 3.2.9 and 3.2.12 we now obtain $\dim \mathfrak{N}_n \geq n$.

Since, as observed above, $\mathfrak{N}_n = \mathfrak{R}_{2n+1,0} \cup \mathfrak{R}_{2n+1,1} \cup \cdots \cup \mathfrak{R}_{2n+1,n}$, an application of Proposition 3.2.10 gives us that $\mathfrak{N}_n$ is the union of $n+1$ zero-dimensional subspaces. $\square$

A finite subset $F = \{x_1, \ldots, x_n\}$ of $\mathbb{R}^m$ is said to be in *general position* if for every collection

$$i(0) < i(1) < \cdots < i(k)$$

of at most $m+1$ indices in $\{1, \ldots, n\}$, the set

$$\{x_{i(0)}, \ldots, x_{i(k)}\}$$

is geometrically independent, cf. §2.1.

**Lemma 3.3.2.** *Let $G = \{x_1, \ldots, x_n\}$ and $F = \{y_1, \ldots, y_k\}$ be subsets of $\mathbb{R}^m$ such that $G$ is in general position, and let $\varepsilon > 0$. Then for each $i \leq k$ there exists a point $x_{n+i} \in \mathbb{R}^m$ such that*

(1) $\|y_i - x_{n+i}\| < \varepsilon$,
(2) $\{x_1, \ldots, x_n, x_{n+1}, \ldots, x_{n+k}\}$ *is in general position.*

**Proof.** It is clear that the lemma follows by induction once we have established it for $k = 1$. So we assume without loss of generality that $k = 1$.

For every nonempty subset $F \subseteq \{1, \ldots, n\}$ of cardinality at most $m$, let
$$v(F) = \text{aff}(F),$$
the affine hull of $F$. Each $v(F)$ is a nowhere dense closed subset of $\mathbb{R}^m$ since $|F| \leq m$. Since $G = \{x_1, \ldots, x_n\}$ is finite, there consequently exists a point
$$x_{n+1} \in \mathbb{R}^m \setminus \bigcup \{v(F) : F \subseteq \{1, \ldots, n\}, |F| \leq m\}$$
such that $\varrho(y_1, x_{n+1}) < \varepsilon$. We claim that $\{x_1, \ldots, x_{n+1}\}$ is in general position. Since $G$ is, we need only prove that if $F \subseteq \{1, \ldots, n\}$ has cardinality at most $m$ then $\{x_i : i \in F\} \cup \{x_{n+1}\}$ is geometrically independent. This is clear however since by construction, $x_{n+1} \notin v(F)$ (Theorem 2.1.6). □

Now for the remaining part of this section, let $X$ be a fixed space such that $0 \leq \dim X \leq n$.

**Proposition 3.3.3.** *Let $f \in C(X, \mathbb{R}^{2n+1}; \varrho)$, let $\varepsilon > 0$, and let $H$ be an at most $n$-dimensional affine subspace of $\mathbb{R}^{2n+1}$. For every finite open cover $\mathcal{U}$ of $X$ there exists $g \in C(X, \mathbb{R}^{2n+1}; \varrho)$ such that*

(1) $\hat{\varrho}(f, g) < \varepsilon$,
(2) $\overline{g[X]} \cap H = \emptyset$,
(3) $g$ *is a $\mathcal{U}$-mapping.*

**Proof.** Since $f$ is bounded, $\overline{f[X]}$ is compact. There consequently exists a finite open cover $\mathcal{V}$ of it such that $\text{mesh}(\mathcal{V}) < \varepsilon/4$. Let $\mathcal{W}$ be the common refinement of $\mathcal{U}$ and $f^{-1}[\mathcal{V}]$. Observe that $\mathcal{W}$ is finite. Since $\dim X \leq n$, there exists by Theorem 3.2.5 a finite open shrinking $\mathcal{E}$ of $\mathcal{W}$ of order at most $n$. The essential properties of the cover $\mathcal{E}$ are the following ones:

(4) $\mathcal{E}$ has order at most $n$,
(5) $\mathcal{E} < \mathcal{U}$,
(6) for each $E \in \mathcal{E}$, $\text{diam}(f[E]) < \varepsilon/4$.

We assume without loss of generality that every member from $\mathcal{E}$ is nonempty.

Let $\mathcal{E} = \{E_1, \ldots, E_m\}$. Now for each $i \leq m$, pick an arbitrary point
$$z_i \in f[E_i].$$

There is a geometrically independent set $S = \{h_1, \ldots, h_k\} \subseteq \mathbb{R}^{2n+1}$ of cardinality at most $n+1$ such that $S$ spans $H$. By Lemma 3.3.2, for each $i \leq m$ there exists $y_i \in \mathbb{R}^{2n+1}$ such that

(7) $\varrho(y_i, z_i) < \varepsilon/4$,
(8) $\{h_1, \ldots, h_k, y_1, \ldots, y_m\}$ is in general position.

Let $\kappa_1, \ldots, \kappa_m \colon X \to \mathbb{I}$ denote the $\kappa$-functions with respect to the cover $\mathcal{E}$. Define $g \colon X \to \mathbb{R}^{2n+1}$ by

$$g(x) = \sum_{i=1}^{m} \kappa_i(x) \cdot y_i.$$

Then $g$ is clearly continuous and we claim that it is as required. First observe that $g$ is bounded because its range is contained in the simplex spanned by the vectors $\{y_1, \ldots, y_k\}$ which is compact by Lemma 1.1.1(2).

For $F \subseteq \{1, 2, \ldots, m\}$ we let $\vec{F}$ denote $\{y_i : i \in F\}$.

For every $x \in X$ let $F_x = \{i \leq m : x \in E_i\}$. Observe that $\kappa_i(x) > 0$ if and only if $i \in F_x$ so that $\sum_{i \in F_x} \kappa_i(x) = 1$.

It follows by (4) and (8) that $\vec{F}_x$ is geometrically independent. Consequently, the $\kappa_i(x)$, $i \in F_x$, are the affine coordinates of $g(x)$ with respect to $\vec{F}_x$. In particular, $g(x) \in \mathrm{aff}(\vec{F}_x)$.

**Claim 1.** $\hat{\varrho}(f, g) < \varepsilon$.

*Proof.* Fix an arbitrary $x \in X$. It follows by (6) and (7) that $\|f(x) - y_i\| < \varepsilon/2$ for every $i \in F_x$. Consequently,

$$\|f(x) - g(x)\| = \left\| \sum_{i \in F_x} \kappa_i(x) \cdot f(x) - \sum_{i \in F_x} \kappa_i(x) \cdot y_i \right\|$$
$$\leq \sum_{i \in F_x} \kappa_i(x) \|f(x) - y_i\|$$
$$< \sum_{i \in F_x} \kappa_i(x) \cdot \varepsilon/2$$
$$= \varepsilon/2.$$

We conclude that $\hat{\varrho}(f, g) \leq \varepsilon/2 < \varepsilon$, which is as required. ◇

**Claim 2.** $g[X] \subseteq \overline{g[X]} \subseteq \bigcup_{x \in X} \mathrm{aff}(\vec{F}_x) \subseteq \mathbb{R}^{2n+1} \setminus H$.

*Proof.* Let $x \in X$. By (4), $|F_x| \leq n+1$ and so if we put $E = \vec{F}_x \cup S$ then

$$|E| \leq (n+1) + (n+1) = 2n + 2.$$

By (8) this implies that $E$ is geometrically independent. Now observe that by Proposition 2.1.5 we have

$$\mathrm{aff}(\vec{F}_x) \cap H = \mathrm{aff}(\vec{F}_x) \cap \mathrm{aff}(S) = \mathrm{aff}(\vec{F}_x \cap S) = \mathrm{aff}(\emptyset) = \emptyset.$$

Since $g(x) \in \text{aff}(\vec{F}_x)$ and $x$ is arbitrary it therefore follows that
$$g[X] \subseteq \bigcup_{x \in X} \text{aff}(\vec{F}_x) \subseteq \mathbb{R}^{2n+1} \setminus H.$$
But since $\text{aff}(\vec{F}_x)$ is closed by Exercise 2.1.8 and there are only finitely many $F_x$'s, we even have that
$$g[X] \subseteq \overline{g[X]} \subseteq \bigcup_{x \in X} \text{aff}(\vec{F}_x) \subseteq \mathbb{R}^{2n+1} \setminus H,$$
as desired. $\diamond$

It remains to prove that $g$ is a $\mathcal{U}$-mapping.

Let $p \in \overline{g[X]}$. By Claim 2 there is a subset $G_p \subseteq \{1, 2, \ldots, m\}$ of cardinality at most $n+1$ such that $p \in \text{aff}(\vec{G}_p)$. We may assume that $G_p$ is minimal with respect to this property. Since $\vec{G}_p$ is geometrically independent, the affine coordinates of $p$ with respect to $\vec{G}_p$ are all non-zero.

In this way we associated to every element $p \in \overline{g[X]}$ a subset of indices $G_p \subseteq \{1, 2, \ldots, m\}$.

**Claim 3.** Let $p \in \overline{g[X]}$ and $i_0 \in G_p$. If $A \subseteq \{y_1, \ldots, y_m\} \setminus \{y_{i_0}\}$ has size at most $n+1$ then $p \notin \text{aff}(A)$.

*Proof.* Our reasoning is similar to the one in the previous claim. Striving for a contradiction, assume that $p \in \text{aff}(A)$. Observe that
$$|\vec{G}_p \cup A| \leq (n+1) + (n+1) = 2n+2.$$
By (8) this implies that $\vec{G}_p \cup A$ is geometrically independent. So by Proposition 2.1.5 we get $p \in \text{aff}(A) \cap \text{aff}(\vec{G}_p) = \text{aff}(A \cap \vec{G}_p)$. Since $i_0 \notin A$ and the affine coordinate of $p$ corresponding to $y_{i_0}$ is non-zero, by unicity of affine coordinates (Theorem 2.1.8), this is a contradiction. $\diamond$

Now let $B$ be the complement of
$$\bigcup \{\text{aff}(A) : A \subseteq \{y_1, \ldots, y_m\}, |A| \leq n+1$$
$$\text{and for some } i \in G_p, y_i \notin A\}.$$
By Claim 3, $B$ is an open neighborhood of $p$ in $\mathbb{R}^{2n+1}$.

**Claim 4.** $g^{-1}[B] \subseteq \bigcap_{i \in G_p} E_i$.

*Proof.* Pick an arbitrary $z \in g^{-1}[B]$ and let $i_0 \in G_p$. By the definition of $g$ we have $g(z) \in \text{aff}(\vec{F}_z)$. Since $|F_z| \leq n+1$ by (4), if $i_0 \notin F_z$ it would follow that
$$g(z) \in \text{aff}(\vec{F}_z) \subseteq \mathbb{R}^n \setminus B,$$
which is a contradiction. So $i_0 \in F_z$ or, equivalently, $z \in E_{i_0}$. $\diamond$

If $p \in U = \mathbb{R}^n \setminus \overline{g[X]}$ then $U$ is a neighborhood of $p$ and $g^{-1}[U] = \emptyset$.
Since by (5) we have $\mathcal{E} < \mathcal{U}$, this completes the proof. □

**Proposition 3.3.4.** Let $X$ be a space $0 \leq \dim X \leq n$. In addition, let
$$\{\mathcal{U}_i : i \in \mathbb{N}\}$$
be a sequence of finite open covers of $X$ such that
$$\mathrm{mesh}(\mathcal{U}_i) < 1/i$$
for every $i$. Then the function space $C(X, \mathbb{R}^{2n+1}; \varrho)$ contains a dense $G_\delta$-subset $\mathcal{C}$ having the following properties:

(1) Every $f \in \mathcal{C}$ is an imbedding.
(2) Every $f \in \mathcal{C}$ is a $\mathcal{U}_i$-map for every $i$.
(3) The range of every $f \in \mathcal{C}$ is contained in $\mathfrak{N}_n$.
(4) The range of every $f \in \mathcal{C}$ has compact closure in $\mathfrak{N}_n$.

**Proof.** Let $A = \{i(1), \ldots, i(n+1)\} \subseteq \{1, 2, \ldots, 2n+1\}$ and $q_1, \ldots, q_{n+1} \in \mathbb{Q}$ be arbitrary. Put
$$H = \{x \in \mathbb{R}^{2n+1} : x_{i(j)} = q_j \text{ for every } j \leq n+1\}.$$
Observe that $H$ is an $n$-dimensional affine subspace of $\mathbb{R}^{2n+1}$ and that
$$H \cap \mathfrak{N}_n = \emptyset.$$
In addition, each point of $\mathbb{R}^{2n+1} \setminus \mathfrak{N}_n$ has at least $n+1$ rational coordinates and is therefore contained in an affine subspace of the form $H$. We conclude that the complement of $\mathfrak{N}_n$ is the union of countably many $n$-dimensional affine subspaces, say $\{L_i : i \in \mathbb{N}\}$.

Now consider the function space $C(X, \mathbb{R}^{2n+1}; \varrho)$. For every $i \in \mathbb{N}$, put
$$\mathcal{C}_i = \{f \in C(X, \mathbb{R}^{2n+1}; \varrho) : f \text{ is a } \mathcal{U}_i\text{-map and } \overline{f[X]} \cap L_i = \emptyset\}.$$
It follows from Proposition 3.3.3 that $\mathcal{C}_i$ is dense in $C(X, \mathbb{R}^{2n+1}; \varrho)$ for every $i$.

**Claim 1.** $\mathcal{C}_i$ is open in $C(X, \mathbb{R}^{2n+1}; \varrho)$ for every $i$.

*Proof.* Take an arbitrary $f \in \mathcal{C}_i$. Since $f$ is bounded, $\overline{f[X]}$ is compact. Therefore, since $\overline{f[X]} \cap L_i = \emptyset$, there exists $\varepsilon > 0$ such that

(5) $$\{y \in \mathbb{R}^{2n+1} : \varrho(y, \overline{f[X]}) \leq \varepsilon\} \cap L_i = \emptyset$$

(Corollary A.5.4). Now since $f$ is a $\mathcal{U}_i$-map, every $y$ in the compact set $\overline{f[X]}$ has a neighborhood $V_y$ in $\mathbb{R}^{2n+1}$ such that $f^{-1}[V_y]$ is contained in an element of $\mathcal{U}_i$. Put $\mathcal{V} = \{V_y : y \in \overline{f[X]}\}$. By Lemma A.5.3 there exists $\delta > 0$ such that every $A \subseteq \mathbb{R}^{2n+1}$ with $\mathrm{diam}(A) < \delta$ and which moreover intersects $\overline{f[X]}$ is contained in an element of $\mathcal{V}$. Let $\gamma = \min\{\delta/3, \varepsilon\}$. We claim that the open ball about $f$ with radius $\gamma$ is contained in $\mathcal{C}_i$. To this end, take an arbitrary function $g \in C(X, \mathbb{R}^{2n+1}; \varrho)$ such that $\hat{\varrho}(f, g) < \gamma$. We shall first prove that

$$\overline{g[X]} \cap L_i = \emptyset.$$

This is easy. Take an arbitrary $x \in X$. Then $\varrho(f(x), g(x)) < \varepsilon$, which implies that $g(x)$ is contained in the compact set

$$\{y \in \mathbb{R}^{2n+1} : \varrho(y, \overline{f[X]}) \leq \varepsilon\}.$$

So by (5), $\overline{g[X]} \cap L_i = \emptyset$. We next prove that $g$ is a $\mathcal{U}_i$-map. Take an arbitrary element $p \in \overline{g[X]}$. There exists $x \in X$ such that $g(x) \in B(p, \delta/6)$. Consider the open neighborhood

$$B = g^{-1}[B(g(x), \delta/3)]$$

of $x$. We claim that $B$ is contained in an element of $\mathcal{U}_i$. If $z \in B$ then clearly

$$\varrho(g(x), g(z)) < \delta/3.$$

Since $\hat{\varrho}(f, g) < \delta/3$, we also get

$$\varrho(g(x), f(x)) < \delta/3, \quad \varrho(g(z), f(z)) < \delta/3.$$

From these inequalities it follows that

$$\varrho(f(z), f(x)) \leq \varrho(f(z), g(z)) + \varrho(g(z), g(x)) + \varrho(g(x), f(x))$$
$$< \delta/3 + \delta/3 + \delta/3$$
$$= \delta$$

and so

(6) $$g^{-1}[B(g(x), \delta/3)] \subseteq f^{-1}[B(f(x), \delta))].$$

Since by the special choice of $\delta$ the set $B(f(x), \delta)$ is contained in an element of $\mathcal{V}$, $f^{-1}[B(f(x), \delta)]$ is contained in an element of $\mathcal{U}_i$. By (6) it therefore follows that $g^{-1}[B(g(x), \delta/3)]$ is contained in an element of $\mathcal{U}_i$. But this is as desired since $B(p, \delta/6) \subseteq B(g(x), \delta/3)$. ◇

By Corollary 1.3.6, $C(X, \mathbb{R}^{2n+1}; \varrho)$ is a topologically complete space, and is therefore a Baire space (Theorem A.6.6). Consequently, Claim 1 implies that

$$\mathcal{C} = \bigcap_{i=1}^{\infty} \mathcal{C}_i$$

is a dense $G_\delta$-subset of $C(X, \mathbb{R}^{2n+1}; \varrho)$. We shall prove that $\mathcal{C}$ is as desired. Observe that (2) follows by construction.

Take any function $f \in \mathcal{C}$. Then $\overline{f[X]} \cap \bigcup_{i=1}^{\infty} L_i = \emptyset$, i.e., $\overline{f[X]} \subseteq \mathfrak{N}_n$. This proves (3) and (4) for $f$. So it suffices to prove that $f$ is an imbedding.

We shall prove that for every $x \in X$ and every neighborhood $W$ of $x$ there exists a neighborhood $V$ of $f(x)$ such that $f^{-1}[V] \subseteq W$. From this it follows that $f$ is one-to-one and that $f: X \to f[X]$ is open, i.e., $f$ is an imbedding. Take an arbitrary $x \in X$ and a neighborhood $W$ of $x$. Let $\varepsilon > 0$ be such that $B(x,\varepsilon) \subseteq W$. There exists $i \in \mathbb{N}$ with $1/i < \varepsilon$. Since $f$ is a $\mathcal{U}_i$-map, there exists a neighborhood $V$ of $f(x)$ such that $f^{-1}[V]$ is contained in an element $U$ of $\mathcal{U}_i$. So $x \in U$ and since $\operatorname{diam}(U) < 1/i < \varepsilon$, we conclude that $f^{-1}[V] \subseteq U \subseteq B(x,\varepsilon) \subseteq W$, as required. □

We now come to the main result in this section.

**The Imbedding Theorem 3.3.5.** *Let $X$ be a space with $0 \leq \dim X \leq n$. Then there exists an imbedding $i \colon X \to \mathfrak{N}_n$ such that $i[X]$ has compact closure in $\mathfrak{N}_n$.*

**Proof.** By Exercise A.6.5, $X$ has an admissible metric $\varrho$ for which there exists a sequence of finite open covers $\{\mathcal{U}_i : i \in \mathbb{N}\}$ such that for every $i$, $\operatorname{mesh}(\mathcal{U}_i) < 1/i$ ($i \in \mathbb{N}$). So we are in a position to apply Proposition 3.3.4 to get what we want. □

**Remark 3.3.6.** Notice that we proved in fact the stronger statement that for a space $X$ with $0 \leq \dim X \leq n$, any bounded map $f \colon X \to \mathbb{R}^{2n+1}$ can be approximated arbitrarily closely by an imbedding in $\mathfrak{N}_n$ the range of which has compact closure in $\mathfrak{N}_n$.

**Remark 3.3.7.** Theorem 3.3.5 is 'best possible' for there exist $n$-dimensional spaces that cannot be imbedded in $\mathbb{R}^{2n}$. An example of such a space is the union $X$ of all $n$-faces of a $(2n+2)$-dimensional simplex. This $X$ is obviously $n$-dimensional and that it cannot be imbedded in $\mathbb{R}^{2n}$ is due to FLORES [164]. The proof of this fact for $n = 1$ is simple and is based on the Jordan Curve Theorem for $\mathbb{R}^2$. For $n > 1$, the proof is more complicated. Its main ingredient is the Borsuk-Ulam Antipodal Theorem in Exercise 2.5.2(4), which is equivalent to the Lusternik-Schnirelman-Borsuk Theorem 2.5.1. See ENGELKING [155, p. 106] for details and additional information.

**Some applications.** We now present some applications of the results in this section. The proofs are based on Theorem 3.3.5 which explains our interest in imbeddings in $\mathfrak{N}_n$ instead of imbeddings in $\mathbb{R}^{2n+1}$.

**Corollary 3.3.8.** *If $\dim X \leq n$, $n \geq 0$, then $X$ has a compactification $\gamma X$ such that $\dim \gamma X \leq n$.*

**Proof.** Let $i \colon X \to \mathfrak{N}_n$ be an imbedding such that $\gamma X = \overline{i[X]}$ is a compact subset of $\mathfrak{N}_n$ (Theorem 3.3.5). Then $\dim \gamma X \leq n$ by Lemma 3.3.1 and the Subspace Theorem 3.2.9. □

**Corollary 3.3.9.** *Let $X$ be a nonempty space and let $n \geq 0$. The following statements are equivalent:*

(1) $\dim X \leq n$,
(2) $X$ is the union of at most $n+1$ zero-dimensional subspaces.

**Proof.** Suppose that $\dim X \leq n$. By Theorem 3.3.5 we may assume that $X$ is contained in $\mathfrak{N}_n$. By Lemma 3.3.1, $\mathfrak{N}_n$ is the union of $n+1$ zero-dimensional subspaces. Now apply the Subspace Theorem 3.2.9.

Conversely, assume that $X$ is the union of at most $n+1$ zero-dimensional subspaces. Then apply Corollary 3.1.7 to conclude that $\dim X \leq n$. □

We now prove a generalization of Corollary 3.1.6 to $n$-dimensional spaces.

**Corollary 3.3.10.** *Let $0 \leq n < \infty$ and let $X$ be an $n$-dimensional subspace of the space $Y$. Then for all disjoint closed subsets $A$ and $B$ of $Y$ there exists a partition $S$ between $A$ and $B$ in $Y$ such that $\dim(S \cap X) \leq n - 1$.*

**Proof.** Write $X = \bigcup_{i=1}^{n+1} X_i$, where $\dim X_i \leq 0$ for every $i$ (Corollary 3.3.9). Let $A$ and $B$ be disjoint closed subsets of $X$. By Corollary 3.1.6 there is a partition $S$ between $A$ and $B$ such that $S \cap X_1 = \emptyset$. But then $S \cap X$ is contained in $\bigcup_{i=2}^{n+1} X_i$ and hence is at most $(n-1)$-dimensional by Corollary 3.3.9. □

By a repeated application of the previous corollary, we get the following result.

**Corollary 3.3.11.** *Let $X$ be a space with subspace $Y$ such that $\dim Y \leq n$, where $0 \leq n < \infty$. For each collection $\tau = \{(F_i, G_i) : i \leq n+1\}$ of pairs of disjoint closed subsets of $X$ there exists for every $i \leq n+1$ a partition $S_i$ between $F_i$ and $G_i$ such that*

$$Y \cap \bigcap_{i=1}^{n+1} S_i = \emptyset.$$

The following result is known as the $G_\delta$-*Enlargement Theorem*.

**Corollary 3.3.12.** *Let $M$ be a subspace of a space $X$. If $\dim M \leq n$ then there exists a $G_\delta$-subset $S \subseteq X$ with $M \subseteq S$ and $\dim S \leq n$.*

**Proof.** We first prove the corollary for the special case $\dim M = 0$. Let $\mathcal{E}$ be a countable open base for $\overline{M}$. For every pair $(E_0, E_1)$ of elements of $\mathcal{E}$ such that $\overline{E_0} \subseteq E_1$ there exists by Corollary 3.1.6 an open set $B \subseteq \overline{M}$ such that

$$\overline{E_0} \subseteq B \subseteq \overline{B} \subseteq E_1$$

while moreover $M \cap \operatorname{Fr} B = \emptyset$. The collection $\mathcal{B}$ of $B$'s obtained in this way is clearly a countable base for $\overline{M}$. In addition,

$$F = \bigcup_{B \in \mathcal{B}} \operatorname{Fr} B$$

is an $F_\sigma$-subset of $\overline{M}$ which misses $M$. Let $S = \overline{M} \setminus F$. Then $S$ is a $G_\delta$-subset of the $G_\delta$-subset $\overline{M}$ and hence is a $G_\delta$-subset of $X$. It is clear that $\mathcal{B} \upharpoonright S$ consists of (relative) clopen sets, and so $S$ is zero-dimensional.

We will now use the just proved special case to prove the general case. To this end, write
$$M = M_0 \cup \cdots \cup M_n,$$
where $M_i$ is zero-dimensional for every $0 \leq i \leq n$ (Corollary 3.3.9). By the above there exists for every $0 \leq i \leq n$ a zero-dimensional $G_\delta$-subset $M_i^*$ of $X$ with $M_i \subseteq M_i^*$. The union of the $M_i^*$'s is clearly a $G_\delta$-subset of $X$ and is at most $n$-dimensional by Corollary 3.3.9. □

**Exercises for §3.3.**

1. Give an example of a space $X$, an open cover $\mathcal{U}$ of $X$, and an imbedding $f \colon X \to X$ such that $f$ is not a $\mathcal{U}$-mapping.
2. Let $X$ be a space. Prove that every continuous function $f \colon X \to Q$ can be approximated arbitrarily closely by an imbedding in $Q$.
3. Prove that for every $n \geq 0$ there exists a compact space $X_n$ such that
   (1) $\dim X_n = n$,
   (2) every space $Y$ with $\dim Y \leq n$ can be imbedded in $X_n$.
▶4. Let $X$ be a compact space without isolated points such that
$$0 \leq \dim X \leq n < \infty.$$
Prove that there exists a continuous surjection $f \colon \mathsf{C} \to X$ such that each fiber of $f$ has cardinality at most $n + 1$
5. Let $X$ be a compact space and let $f \colon \mathsf{C} \to X$ be a continuous surjection such that each fiber of $f$ has cardinality at most $n + 1$. Prove that $X$ is at most $n$-dimensional.
6. Let $X$ be a compact space with $0 \leq n = \dim X < \infty$. Prove that every open cover $\mathcal{U}$ of $X$ has a closed refinement $\mathcal{F}$ such that for every $F \in \mathcal{F}$ the collection
$$\{F' \in \mathcal{F} : F' \cap F \neq \emptyset\}$$
has cardinality at most $3^{2n+1}$.
▶7. Prove that every $(n+1)$-dimensional compactum contains an $n$-dimensional compactum (and hence an $n$-dimensional continuum by Exercise 3.2.4).

## 3.4. The inductive dimension functions ind and Ind

There are two additional dimension functions that are important in dimension theory, namely, the small and the large inductive dimension function, abbreviated ind and Ind, respectively. It turns out that for a given space these functions and the dimension function dim take the same value. The functions ind and Ind are important because in certain situations it is easier to deal with them than with dim. For example, it is a triviality to verify that for all $X$ and $Y$,
$$\operatorname{ind}(X \times Y) \leq \operatorname{ind} X + \operatorname{ind} Y;$$

by equality of ind and dim it therefore follows that
$$\dim(X \times Y) \leq \dim X + \dim Y.$$
However, to verify this straight from the definition of dim is unpleasant. The aim of this section is to study basic properties of the dimension functions ind and Ind and to prove equality of all three dimension functions.

We shall now give the definition of ind, which differs from the definition of dim in the sense that it is an inductive definition. Indeed, for a space $X$ define its *small inductive dimension* ind $X \in \{-1, 0, 1, \ldots\} \cup \{\infty\}$ as follows:

ind $X = -1$    $\Leftrightarrow$    $X = \emptyset$,
ind $X \leq n$    $\Leftrightarrow$    for every $x \in X$ and every closed subset $A$ of $X$ not containing $x$ there exists a partition $L$ between $\{x\}$ and $A$ such that ind $L \leq n - 1$,
ind $X = n$    $\Leftrightarrow$    ind $X \leq n$ and ind $X \not\leq n - 1$,
ind $X = \infty$    $\Leftrightarrow$    ind $X \neq n$ for every $n \geq -1$.

It is clear that if $X$ and $Y$ are homeomorphic spaces then ind $X = $ ind $Y$.

The property of having small inductive dimension at most $n$ can be expressed in terms of a special countable base.

**Lemma 3.4.1.** *A space $X$ satisfies $0 \leq $ ind $X \leq n$ if and only if $X$ has a countable base $\mathcal{B}$ such that ind Fr $B \leq n - 1$ for all $B \in \mathcal{B}$.*

**Proof.** If $\in X \leq n$ then $X$ has clearly a base $\mathcal{U}$ such that ind Fr $U \leq n - 1$ for every $U \in \mathcal{U}$. This base has a countable subcollection $\mathcal{B}$ which is still a base (Exercise A.2.12). So $\mathcal{B}$ is as required.

The converse of the lemma is trivial. □

A consequence of this lemma is that for a space $X$ we have ind $X = 0$ if and only if $X$ is nonempty and has a base consisting of clopen sets. This means that $X$ is zero-dimensional if and only if ind $X = 0$.

We shall now derive a few additional properties of the dimension function ind. The following triviality shall be used several times in the forthcoming: if $X$ is a space, $Y \subseteq X$, $y \in Y$ and $A$ is a (relatively) closed subset of $Y$ not containing $y$, then $y$ does not belong to the closure of $A$ in $X$.

**Proposition 3.4.2.** *Let $X$ be a space. Then $\dim X = 0$ iff ind $X = 0$.*

**Proof.** This is a triviality since, as observed above, ind $X = 0$ iff the clopen subsets of $X$ form a base iff $\dim X = 0$ (see Page 41). □

**Lemma 3.4.3.** *Let $X$ be a space, and let $A$ be a subspace of $X$. Then ind $A \leq $ ind $X$.*

**Proof.** There is nothing to prove if ind $X = \infty$, so we assume that ind $X$ is finite. Again, there is nothing to prove if ind $X = -1$. So assume the lemma to be true for all spaces $Y$ with $0 \leq \operatorname{ind} Y \leq n-1$, and assume that $\operatorname{ind} X \leq n$. Take $x \in A$ and let $C$ be a closed subset of $A$ not containing $x$. Since $x \notin \overline{C}$ there is a partition $L$ in $X$ between $x$ and $\overline{C}$ such that $\operatorname{ind} L \leq n - 1$. Then $L \cap A$ is a partition between $x$ and $C$ in $A$ so that, by our inductive assumption, $\operatorname{ind}(L \cap A) \leq n - 1$. We conclude that $\operatorname{ind} A \leq n$. □

This result enables us to prove the following:

**The Addition Theorem 3.4.4.** *If $A$ and $B$ are subspaces of a space $X$ then*
$$\operatorname{ind}(A \cup B) \leq \operatorname{ind} A + \operatorname{ind} B + 1.$$

**Proof.** If $\operatorname{ind} A = \infty$ or $\operatorname{ind} B = \infty$ then there is nothing to prove. So we assume that $\operatorname{ind} A$ and $\operatorname{ind} B$ are both finite. We induct on $\operatorname{ind} A + \operatorname{ind} B$. If $A$ and $B$ are both empty, i.e., if $\operatorname{ind} A = \operatorname{ind} B = -1$, then there is again nothing to prove. So assume that the theorem is true for all subspaces $E$ and $F$ of $X$ with
$$\operatorname{ind} E + \operatorname{ind} F \leq n - 1,$$
where $n \geq -1$. Let $\alpha = \operatorname{ind} A$ and $\beta = \operatorname{ind} B$ and assume that $\alpha + \beta = n$. We shall prove that $\operatorname{ind}(A \cup B) \leq n+1$. Take an arbitrary $x \in A \cup B$, say $x \in A$, and let $C$ be a closed subset of $A \cup B$ not containing $x$. Since $x \notin \overline{C}$, there is a closed neighborhood $Z$ of $\overline{C}$ such that $x \notin Z$. Let $L$ be a partition in $A$ between $\{x\}$ and $Z \cap A$ such that $\operatorname{ind} L \leq \alpha - 1$. Write $A \setminus L$ as the union of two relatively open disjoint sets $A_0$ and $A_1$ such that $x \in A_0$ and $Z \cap A \subseteq A_1$. There exists an open neighborhood $W$ of $x$ in $X$ such that $\overline{W} \cap A \subseteq A_0$ and $\overline{W} \cap Z = \emptyset$. Now observe that $L$ is a partition between $\overline{W} \cap A$ and $Z \cap A$. By Lemma 3.1.4, there exists a partition $S$ in $X$ between $\{x\}$ and $\overline{C}$ such that $S \cap A \subseteq L$. Now observe that by Lemma 3.4.3, $\operatorname{ind}(S \cap A) \leq \alpha - 1$. Similarly, $\operatorname{ind}(S \cap B) \leq \beta$ since $\operatorname{ind} B = \beta$. Consequently, we obtain by our inductive assumption,
$$\operatorname{ind}\bigl(S \cap (A \cup B)\bigr) \leq \alpha - 1 + \beta + 1 = \alpha + \beta = n.$$
Since $S \cap (A \cup B)$ is a partition in $A \cup B$ between $\{x\}$ and $C$, this shows that indeed $\operatorname{ind}(A \cup B) \leq n+1$. □

**Remark 3.4.5.** If $X = \mathbb{R}$, $A = \mathbb{Q}$ and $Y = \mathbb{P}$ then $\operatorname{ind} X = 1$ and
$$\operatorname{ind} A = 0 = \operatorname{ind} B.$$
This shows that Theorem 3.4.4 is 'best possible'.

**Corollary 3.4.6.** *If $X$ can be written as the union of $n+1$ zero-dimensional subspaces then $\operatorname{ind} X \leq n$.*

**Proof.** Since for every space $Y$, $\dim Y = 0$ iff $\operatorname{ind} Y = 0$ (Proposition 3.4.2) the result follows immediately from Theorem 3.4.4. □

We want to prove the equality of dim and ind. Before being able to do that, we need to derive a preliminary lemma.

**Lemma 3.4.7.** *Let $X$ be space and let $n \geq 0$. The following two statements are equivalent:*

(1) *For every $x \in X$ and for every open neighborhood $U$ of $x$ there exists a partition $L$ between $\{x\}$ and $X \setminus U$ such that $\dim L \leq n - 1$.*
(2) *For every pair $A$ and $B$ of disjoint closed subsets of $X$ there exists a partition $L$ between $A$ and $B$ such that $\dim L \leq n - 1$.*

**Proof.** First observe that (2) $\Rightarrow$ (1) is a triviality.

For (1) $\Rightarrow$ (2), let $A$ and $B$ be disjoint closed subsets of $X$. By assumption, for each $x \in X$ there exist open sets $U(x)$ and $V(x)$ such that

(3) $x \in U(x)$, $U(x) \cap V(x) = \emptyset$,
(4) if $L(x) = X \setminus (U(x) \cup V(x))$ then $\dim L(x) \leq n - 1$,
(5) $\overline{U(x)} \cap A = \emptyset$ or $\overline{U(x)} \cap B = \emptyset$.

The cover $\{U(x) : x \in X\}$ has a countable subcover, say $\mathcal{U}$ (Corollary A.2.3). Put $L = \bigcup\{L(x) : U(x) \in \mathcal{U}\}$. Observe that by (4) and the Countable Closed Sum Theorem 3.2.8 we have $\dim L \leq n - 1$. Let $\mathcal{U}_0 = \{U \in \mathcal{U} : \overline{U} \cap A \neq \emptyset\}$ and $\mathcal{U}_1 = \mathcal{U} \setminus \mathcal{U}_0$. Enumerate $\mathcal{U}_0$ as $\{U_{0,i} : i \in \mathbb{N}\}$ and $\mathcal{U}_1$ as $\{U_{1,i} : i \in \mathbb{N}\}$, respectively. For each $i \in \mathbb{N}$ put

(6) $E_i = U_{0,i} \setminus \bigcup_{j<i} \overline{U}_{1,j}$ and $F_i = U_{1,i} \setminus \bigcup_{j \leq i} \overline{U}_{0,j}$.

By (5) and (6) it follows easily that $E = \bigcup_{i=1}^{\infty} E_i$ and $F = \bigcup_{i=1}^{\infty} F_i$ are disjoint open sets with $A \subseteq E$ and $B \subseteq F$. Consequently, $G = X \setminus (E \cup F)$ is a partition between $A$ and $B$. We will prove that $\dim G \leq n - 1$ by proving that $G \subseteq L$ and applying the Subspace Theorem 3.2.9. Therefore, $G$ will turn out to be the required partition between $A$ and $B$.

Take an arbitrary point $y \in G$. Let $i$ be the first natural number with the property that $y \in \overline{U}_{0,i} \cup \overline{U}_{1,i}$. Observe that such an $i$ exists. Suppose first that $y \in \overline{U}_{0,i}$. If $y \in U_{0,i}$ then $y \in E_i$, which is not the case. So $y \notin U_{0,i}$, hence $y \in \overline{U}_{0,i} \setminus U_{0,i} \subseteq L$, and we are done. Suppose therefore that $y \notin \overline{U}_{0,i}$. If $y \in U_{1,i}$ then $y \in F_i$, which is also not the case. Consequently, $y \notin U_{1,i}$, which implies that in this case $y \in L$ as well. □

**Theorem 3.4.8.** *For every space $X$, $\dim X = \operatorname{ind} X$.*

**Proof.** We shall first prove that $\dim X \leq \operatorname{ind} X$. If $\operatorname{ind} X = \infty$ then there is nothing to prove. So assume that $\operatorname{ind} X < \infty$. We induct on $\operatorname{ind} X$. If $\operatorname{ind} X = 0$ then apply Proposition 3.4.2. Now assume that the inequality holds for all spaces $Y$ with $\operatorname{ind} Y \leq n - 1$, $n \geq 1$, and let $\operatorname{ind} X = n$. Then for each point $x \in X$ and for every open neighborhood $U$ of $x$ there exists a partition $L$ between $\{x\}$ and $X \setminus U$ such that $\operatorname{ind} L \leq n - 1$. By our inductive hypothesis, all these partitions have covering dimension at most $n-1$. We conclude that $\dim X \leq n$ by Lemma 3.4.7 and Corollary 3.1.10.

We shall now prove that $\operatorname{ind} X \leq \dim X$. If $\dim X = -1$ or $\dim X = \infty$ then this is a triviality. So assume that $0 \leq \dim X = n$. By Corollary 3.3.9, $X$ is the union of at most $n+1$ zero-dimensional subspaces. So by Corollary 3.4.6 we get $\operatorname{ind} X \leq n = \dim X$, which is what we want. □

Motivated by Lemma 3.4.7, we shall now present the definition of the *large inductive dimension* Ind. Indeed, for a space $X$ put:

$\operatorname{Ind} X = -1 \quad \Leftrightarrow \quad X = \emptyset$,

$0 \leq \operatorname{Ind} X \leq n \quad \Leftrightarrow \quad$ for every pair of disjoint closed subsets $A$ and $B$ of $X$ there exists a partition $L$ between $A$ and $B$ such that $\operatorname{Ind} L \leq n-1$,

$\operatorname{Ind} X = n \quad \Leftrightarrow \quad \operatorname{Ind} X \leq n$ and $\operatorname{Ind} X \not\leq n-1$,

$\operatorname{Ind} X = \infty \quad \Leftrightarrow \quad \operatorname{Ind} X \neq n$ for every $n \geq -1$.

As in the case of ind, it is easy to see that if $X$ and $Y$ are homeomorphic spaces then $\operatorname{Ind} X = \operatorname{Ind} Y$. We shall prove that Ind = dim, thereby establishing the announced equality.

**Lemma 3.4.9.** *Let $X$ be a space. Then* $\operatorname{ind} X \leq \operatorname{Ind} X$.

**Proof.** If $\operatorname{Ind} X = \infty$ or $\operatorname{Ind} X = -1$ then there is nothing to prove, so assume that the lemma is true for all spaces $Y$ with $-1 \leq \operatorname{Ind} Y \leq n-1$ and $0 \leq n < \infty$, and assume that $\operatorname{Ind} X = n$. Let $x \in X$ and let $U$ be an open neighborhood of $x$. Since $\{x\}$ and $X \setminus U$ are disjoint closed sets in $X$, there is a partition $L$ between them such that $\operatorname{Ind} L \leq n-1$. By our inductive assumption we have $\operatorname{ind} L \leq \operatorname{Ind} L \leq n-1$. From this we consequently conclude that $\operatorname{ind} X \leq n = \operatorname{Ind} X$. □

We now come to the announced:

**The Coincidence Theorem 3.4.10.** *For every space $X$ we have*

$$\dim X = \operatorname{ind} X = \operatorname{Ind} X.$$

**Proof.** By Theorem 3.4.8 and Lemma 3.4.9 we need only prove that

$$\operatorname{Ind} X \leq \dim X.$$

If $\dim X = \infty$ or $\dim X = -1$ then there is nothing to prove. Assume therefore that the theorem is true for all spaces $Y$ with $\dim Y \leq n-1$, where $0 \leq n < \infty$. Assume that $\dim X = n$. Since $\operatorname{ind} X = \dim X$ (Theorem 3.4.8), for every $x \in X$ and every closed set $A$ in $X$ with $x \notin A$, there exists a partition $L$ between $\{x\}$ and $A$ such that $\operatorname{ind} L \leq n-1$. By an application of Theorem 3.4.8 it follows that all these partitions have covering dimension at most $n-1$. Consequently, by Lemma 3.4.7, for every pair of disjoint closed subsets $A$ and $B$ of $X$ there exists a partition $L$ between $A$ and $B$ such that $\dim L \leq n - 1$. By our inductive hypothesis it follows that these partitions all have large inductive dimension at most $n - 1$. From this we conclude that indeed $\operatorname{Ind} X \leq n = \dim X$. □

**Remark 3.4.11.** From now on we will no longer formally distinguish between dim, ind and Ind. Theorems proved for dim will be used for ind and vice versa, etc.

Let $X$ and $Y$ be spaces. It is natural to wonder about the relation between $\dim X$, $\dim Y$ and $\dim(X \times Y)$. Since $\dim \mathbb{R}^n = n$ for every $n$, one would expect the relation $\dim(X \times Y) = \dim X + \dim Y$ to hold for all nonempty $X$ and $Y$. Unfortunately, this is not true, see Example 3.4.13. We shall now prove 'half' of the expected equality.

**Theorem 3.4.12.** *Let $X$ and $Y$ be nonempty spaces. Then*
$$\dim(X \times Y) \leq \dim X + \dim Y.$$

**Proof.** By the Coincidence Theorem 3.4.10, it suffices to prove that
$$\operatorname{ind}(X \times Y) \leq \operatorname{ind} X + \operatorname{ind} Y.$$
First observe that without loss of generality, $0 \leq \operatorname{ind} X, \operatorname{ind} Y < \infty$. We shall prove the theorem by induction on $\operatorname{ind} X + \operatorname{ind} Y$. If $\operatorname{ind} X + \operatorname{ind} Y = 0$ then both $X$ and $Y$ are zero-dimensional, and so is $X \times Y$. Assume that the theorem is true for all spaces $X$ and $Y$ with
$$\operatorname{ind} X + \operatorname{ind} Y \leq n - 1,$$
where $n \geq 1$, and let $X$ and $Y$ be spaces such that $\operatorname{ind} X = \alpha$, $\operatorname{ind} Y = \beta$ and $\alpha + \beta = n$. By Lemma 3.4.1 there is a countable base $\mathcal{B}$ for $X$ such that $\dim \operatorname{Fr} B \leq \alpha - 1$ for every $B \in \mathcal{B}$. Put $B' = \bigcup_{B \in \mathcal{B}} \operatorname{Fr} B$. Then $B'$ is an $F_\sigma$-subset of $X$ and hence by the Countable Closed Sum Theorem 3.2.8 it follows that $\dim B' \leq \alpha - 1$. Since $\operatorname{Fr} B \cap (X \setminus B') = \emptyset$ for every $B \in \mathcal{B}$ it also follows that $\mathcal{B} \restriction (X \setminus B')$ consists of relatively clopen sets, i.e., $\dim(X \setminus B) \leq 0$. Similarly, construct an $F_\sigma$-subset $E \subseteq Y$ such that $\dim E \leq \beta - 1$ and $\dim(Y \setminus E) \leq 0$. Now observe that by our inductive hypothesis and the Countable Closed Sum Theorem 3.2.8, $B' \times Y$ is an at most $(\alpha - 1) + \beta$-dimensional $F_\sigma$-subset of $X \times Y$. It follows similarly that $X \times E$ is an at most $\alpha + (\beta - 1)$-dimensional $F_\sigma$-subset of $X \times Y$. Again by the Countable Closed Sum Theorem 3.2.8 we get
$$\dim\left((B' \times Y) \cup (X \times E)\right) \leq \alpha + \beta - 1.$$
Since the complement of $(B' \times Y) \cup (X \times E)$ is zero-dimensional, being the product $(X \setminus B') \times (Y \setminus E)$, we get by Theorem 3.4.4 that
$$\dim(X \times Y) \leq (\alpha + \beta - 1) + 0 + 1 = \alpha + \beta,$$
as required. □

The following example shows that the other 'half' of the expected equality $\dim(X \times Y) = \dim X + \dim Y$ need not hold. See Theorem 3.9.5 for a (strong) generalization of this result to all dimensions.

**Example 3.4.13.** There exists a one-dimensional space $E$ such that $E \times E$ is one-dimensional as well.

We will show that the space $E$ in Example 1.5.18, i.e., Erdős' space, is the required example. From Exercise 1.5.14 we know that $E \approx E \times E$. So it suffices to prove that $\dim E = 1$. In Example 1.5.18 we showed that $\dim E \geq 1$. We shall prove here that $\operatorname{ind} E \leq 1$.

First observe that $E$ is a subgroup of $\ell^2$, and hence is a *topological group*. By homogeneity, it therefore suffices to prove that $\operatorname{ind} E \leq 1$ at the zero-element of $E$. We claim that for every $\varepsilon > 0$, the sphere $S_\varepsilon$ consisting of all points in $E$ with norm $\varepsilon$, is zero-dimensional. Indeed, since all elements of $S_\varepsilon$ have the same norm, Lemma 1.1.12 shows that $S_\varepsilon$ is homeomorphic to a subspace of the countably infinite product of rational numbers, which is zero-dimensional by Proposition 1.5.3 and the trivial observation that products of zero-dimensional spaces are again zero-dimensional. So $S_\varepsilon$ is indeed zero-dimensional, being a subspace of a zero-dimensional space.

We now summarize the results obtained so far in the following

**Corollary 3.4.14.**

(1) Let $X$ be a nonempty space and let $n \geq 0$. The following statements are equivalent:
   1. $\dim X \leq n$,
   2. for every subspace $A$ of $X$, $\dim A \leq n$,
   3. $X$ has a compactification $\gamma X$ such that $\dim \gamma X \leq n$,
   4. for every pair of disjoint closed subsets $A$ and $B$ of $X$ there exists a partition $L$ between $A$ and $B$ such that $\dim L \leq n-1$,
   5. $X$ is the union of at most $n+1$ zero-dimensional subspaces.

(2) If $X$ is a space and if $A$ and $B$ are subspaces of $X$ then
$$\dim(A \cup B) \leq \dim A + \dim B + 1,$$
$$\dim(A \times B) \leq \dim A + \dim B.$$

**Exercises for §3.4.** Let $X$ be a space. Define the *compactness degree* $\operatorname{cmp} X$ of $X$, as follows:

$\operatorname{cmp} X = -1 \quad \Leftrightarrow \quad X$ is compact,

$\operatorname{cmp} X \leq n \quad \Leftrightarrow \quad$ for every $x \in X$ and every closed subset $A$ of $X$ not containing $x$ there exists a partition $L$ between $\{x\}$ and $A$ such that $\operatorname{cmp} L \leq n-1$,

$\operatorname{cmp} X = n \quad \Leftrightarrow \quad \operatorname{cmp} X \leq n$ and $\operatorname{cmp} X \not\leq n-1$,

$\operatorname{cmp} X = \infty \quad \Leftrightarrow \quad \operatorname{cmp} X \neq n$ for every $n \geq -1$.

So the compactness degree of a space is intuitively its small inductive dimension modulo the class of all compact spaces. It is clear that if $X$ and $Y$ are homeomorphic spaces then they have the same compactness degree.

Let $\gamma X$ be a compactification of $X$. We call the space $\gamma X \setminus X$ the *remainder* of the compactification $\gamma X$ of $X$.

Let the *compactness deficiency* be the least integer $n \in [-1, \infty]$ such that $X$ has a compactification with remainder of dimension $n$. Observe that homeomorphic spaces have the same compactness deficiency.

Given a space $X$ and $n \in \mathbb{N}$, we shall denote by $X_{(n)}$ the set of all points in $X$ that have arbitrarily small neighborhoods with at most $(n-1)$-dimensional boundaries.

1. Prove that Erdős' space $E$ is not topologically complete. Give an example of a one-dimensional totally disconnected topologically complete space $F$ such that $F$ and $F \times F$ are one-dimensional
2. Let $X$ be a space. Prove that $\operatorname{cmp} X \leq \operatorname{def} X \leq \dim X$.
3. Let $X$ be a totally disconnected space which is not compact and let $\gamma X$ be a compactification of $X$. Prove that $\dim \gamma X \setminus X \geq \dim X$, i.e., $X$ cannot be compactified by adding a set of smaller dimension than the dimension of $X$.
▶4. Let $X$ be a compact space such that $0 \leq \dim X = n < \infty$, and let $Y$ be the product $\mathbb{Q} \times X$.
    (1) Prove that if $S$ is a topologically complete space containing $Y$ then $S \setminus Y$ contains a homeomorph of $X$. As a consequence, the dimension of $S \setminus Y$ is at least $n$.
    (2) Prove that $\operatorname{def} Y = n$.
5. Let $X$ be a space and $n \in \mathbb{N}$. Prove that there is a countable family open sets $\mathcal{U}$ such that
    (1) $\mathcal{U}$ is a local base at every point of $X_{(n)}$,
    (2) $\dim(\overline{U} \setminus U) \leq n - 1$ for every $U \in \mathcal{U}$.
   Prove that $\dim X_{(n)} \leq n$.

## 3.5. Dimensional properties of compactifications

In this section we collect some results on compactifications preserving certain dimensional properties.

We saw in Corollary 3.3.8 that every $m$-dimensional space $X$ has an $m$-dimensional compactification. Such a compactification is called *dimension preserving*. The following result generalizes this and is central in this section.

**Theorem 3.5.1.** *Let $X$ be a space, and let $\{(F_n, G_n) : n \in \mathbb{N}\}$ be a countable collection of pairs of disjoint closed subsets of $X$. Then there is a dimension preserving compactification $aX$ of $X$ such that for every $n$ the sets $F_n$ and $G_n$ have disjoint closures in $aX$.*

**Proof.** We will prove this in the case of finite dimensional $X$ only, the infinite-dimensional case being simpler.

Let $\varrho$ be an admissible metric on $X$ for which there exists a sequence
$$\{\mathcal{U}_n : n \in \mathbb{N}\}$$
of finite open covers of $X$ such that $\mathrm{mesh}(\mathcal{U}_n) < 1/n$ for every $n$ (Exercise A.6.5). Since $\mathcal{A}_n = \{X \setminus F_n, X \setminus G_n\}$ is an open cover of $X$, we can replace $\mathcal{U}_n$ by the common refinement of $\mathcal{U}_n$ and $\mathcal{A}_n$. So we may assume without loss of generality that for every $n$ and $U \in \mathcal{U}_n$ we have that $U \cap F_n = \emptyset$ or $U \cap G_n = \emptyset$.

Let $m = \dim X$. By Proposition 3.3.4 there is an imbedding $i\colon X \to \mathfrak{N}_m$ such that

(1) $i[X]$ has compact closure in $\mathfrak{N}_m$.
(2) $i$ is a $\mathcal{U}_n$-map for every $n$.

Put $aX = \overline{i[X]}$. Then $aX$ is a dimension preserving compactification of $X$ (cf. Corollary 3.3.8).

Fix $n \in \mathbb{N}$. We claim that $i[F_n]$ and $i[G_n]$ have disjoint closures in $aX$. Since $i$ is a $\mathcal{U}_n$-map, there is an open cover $\mathcal{G}$ of $\mathbb{R}^{2m+1}$ such that
$$\{i^{-1}[G] : G \in \mathcal{G}\}$$
refines $\mathcal{U}_n$. Striving for a contradiction, assume that
$$p \in \overline{i[F_n]} \cap \overline{i[G_n]}.$$
There is an element $G \in \mathcal{G}$ such that $p \in G$. Pick an element $U \in \mathcal{U}_n$ such that $i^{-1}[G] \subseteq U$. We may assume without loss of generality that $U \cap F_n = \emptyset$. But then $i^{-1}[G] \cap F_n = \emptyset$ and so $G \cap i[F_n] = \emptyset$. Since $\mathbb{R}^{2m+1} \setminus G$ is closed, this implies that $p \notin \overline{i[F_n]}$, which is a contradiction. □

**Corollary 3.5.2.** *Let $X$ be a space, and let $bX$ be a compactification of $X$. Then there is a compactification $aX$ of $X$ which is dimension preserving while moreover $aX \geq bX$.*

**Proof.** By Lemma A.5.6 there is a collection $\{(A'_n, B'_n) : n \in \mathbb{N}\}$ of pairs of disjoint closed subsets of $bX$ such that for any pair $(E, F)$ of disjoint closed subsets in $bX$ there is an $m \in \mathbb{N}$ such that $E \subseteq A'_m$ and $F \subseteq B'_m$. For every $n$, let $A_n = A'_n \cap X$ and $B_n = B'_n \cap X$, respectively. By Theorem 3.5.1 there is a compactification $aX$ of $X$ such that for every $n$ the closures in $aX$ of $A_n$ and $B_n$ are disjoint, while moreover $\dim aX = \dim X$. But this implies by Lemma A.8.3 that the identity function $X \to X$ can be extended to a continuous function $aX \to bX$, i.e., $aX \geq bX$. □

We now show that compactifications can also preserve the dimensions of countably many closures of closed sets.

## 3.5. DIMENSIONAL PROPERTIES OF COMPACTIFICATIONS

**Theorem 3.5.3.** *Let $X$ be a space, and let $\mathcal{H}$ be a countable family closed subsets of $X$. Then there is a Wallman base $\mathcal{F}$ of $X$ such that*

(1) $\mathcal{H} \subseteq \mathcal{F}$,
(2) $\dim \overline{H} = \dim H$ *for every* $H \in \mathcal{H}$ *(here closure means closure in $\omega(X, \mathcal{F})$).*

**Proof.** List $\mathcal{H}$ as $\{H_i : i \in \mathbb{N}\}$ such that every $H \in \mathcal{H}$ is listed infinitely often. For every $H \in \mathcal{H}$ put $n_H = \dim H$, and if $H = H_i$ then $n_i = n_H$.

Let $\mathcal{F}_1$ be an arbitrary Wallman base for $X$ (Lemma A.9.1) and let $a_1 X$ be the compactification $\omega(X, \mathcal{F}_1)$ of $X$.

Consider the set $H_1$. The closure of $H_1$ in $a_1 X$ is a compactification of $H_1$, say $\beta_1 H_1$. By Corollary 3.5.2 there is a compactification $\beta_2 H_1 \geq \beta_1 H_1$ such that $\dim \beta_2 H_1 = n_1$. By an application of Exercise A.11.8, there is a compactification $t_2 X$ of $X$ for which there is a continuous function

$$f_1 : t_2 X \to a_1 X$$

which extends the identity on $X$ while moreover the closure of $H_1$ in $t_2 X$ is $\beta_2 H_1$. Hence the closure of $H_1$ in $t_2 X$ has dimension $n_1$. By Corollary A.9.8 there is a Wallman base $\mathcal{T}_1$ of $X$ such that $t_1 X$ and $\omega(X, \mathcal{T}_1)$ are equivalent compactifications.

Let $\mathcal{F}_2$ be a Wallman base of $X$ containing the collection

$$\mathcal{F}_1 \cup \mathcal{T}_1 \cup \{H_1\}$$

(Lemma A.9.1), and let $a_2 X$ be $\omega(X, \mathcal{F}_2)$. By Exercise A.9.4 there is a continuous function $g_1 : a_2 X \to t_1 X$ which extends the identity on $X$.

Observe that at this stage we have the following compactifications and continuous functions:

$$a_1 X = \omega(X, \mathcal{F}_1) \xleftarrow{f_1} t_1 X \xleftarrow{g_1} \omega(X, \mathcal{F}_2) = a_2 X.$$

The Wallman bases $\mathcal{F}_1$ and $\mathcal{F}_2$ have the property that $\mathcal{F}_1 \subseteq \mathcal{F}_2$, and the compactification $t_1 X$ has the property that the closure of $H_1$ in it has the 'right' dimension, namely $n_1$.

This is the first step in an inductive process.

We now proceed recursively as above, dealing with the set $H_i$ in the $i$-th step of our process. We thus obtain an inverse sequence

$$a_1 X \xleftarrow{f_1} t_1 X \xleftarrow{g_1} a_2 X \xleftarrow{f_2} t_2 X \xleftarrow{g_2} a_3 X \xleftarrow{f_3} \cdots$$

of compactifications and Wallman bases $\mathcal{F}_i$ of $X$ having for every $i$ the following properties:

(3) $\mathcal{F}_{i-1} \subseteq \mathcal{F}_i$,
(4) $H_i \in \mathcal{F}_i$,
(5) $\omega(X, \mathcal{F}_i) = a_i X$,

(6) $f_i \colon t_i X \to a_i X$ extends the identity on $X$,
(7) $g_i \colon a_{i+1} X \to t_i X$ extends the identity on $X$,
(8) the closure of $H_i$ in $t_i X$ has dimension $n_i$.

Put $\mathcal{F} = \bigcup_{i=1}^{\infty} \mathcal{F}_i$. We claim that $\mathcal{F}$ is the desired Wallman base of $X$ (that it is a Wallman base follows from Exercise 1.10.11). Let $a_\infty X$ be the inverse limit of our inverse sequence. We already know by Exercise 1.10.11 that

$$a_\infty X = \omega(X, \mathcal{F}).$$

So $a_\infty X$ is a compactification of $X$ and every projection restricts to the identity on $X$.

Now let $H \in \mathcal{H}$ be arbitrary. The closure of $H$ in $t_i X$ is mapped by $f_i$ onto the closure of $H$ in $a_i X$. Similarly, the closure of $H$ in $a_{i+1} X$ is mapped by $g_i$ onto the closure of $H$ in $t_i X$. Also, the closure of $H$ in $a_\infty X$ is mapped by the various projections onto the closures of $H$ in the corresponding compactifications. This is all true because of the compactness of the spaces involved and the fact that every function in our inverse sequence restricts to the identity on $X$. So all the closures of $H$ in the various compactifications form an inverse system of which the limit is the closure of $H$ in $a_\infty X$ (Exercise 1.10.7). Since $H$ is listed infinitely often, by (8) this sequence has a subsequence all elements of which have dimension $n_H$. But then, since every subsequence of an inverse system has the same limit (Lemma 1.10.6), by Exercise 3.2.2 it consequently follows that the closure of $H$ in $a_\infty X$ has dimension at most $n_H$. But since $H$ is a subspace of its own closure in $a_\infty X$, by the Subspace Theorem 3.2.9 we get that this closure is at least $n_H$-dimensional. So we are done. □

**Corollary 3.5.4.** *Let $X$ be a space, and let $\mathcal{F}$ be a countable family closed subsets of $X$. Then there is a compactification $aX$ of $X$ such that*

$$\dim \overline{F} = \dim F$$

*for every $F \in \mathcal{F}$.*

**Remark 3.5.5.** An inspection of the proof of Theorem 3.5.3 shows that for this corollary it is not necessary to deal with Wallman bases. The proof of it can be completed entirely within the framework of inverse limits.

The following corollary to Theorem 3.5.3, which will be important in §3.11, does however not follow from Corollary 3.5.4. An inspection of its proof will show that we need to control the dimensions of various partitions between points and closed sets, and for that control over closures alone does not suffice.

It turns out that via the use of Wallman bases, one gets the desired control automatically. One simply has to put the right collection of closed sets in a suitable Wallman base, and the Wallman compactification theory does the rest.

## 3.5. DIMENSIONAL PROPERTIES OF COMPACTIFICATIONS

**Corollary 3.5.6.** *Let $X$ be a space with countable family closed subsets $\mathcal{G}$. Then $X$ has a compactification $aX$ having the following properties:*

(1) $\dim \overline{G} = \dim G$ for all $G \in \mathcal{G}$,
(2) $X_{(n)} \subseteq (aX)_{(n)}$ for every $n \geq 0$.

**Proof.** For $n \geq 0$ there is by Exercise 3.4.5 a countable collection of open sets $\mathcal{U}_n$ in $X$ such that

(3) $\mathcal{U}_n$ is a local base at every point of $X_{(n)}$,
(4) $\dim(\overline{U} \setminus U) \leq n - 1$ for every $U \in \mathcal{U}_n$.

Let $\mathcal{H}$ be the collection

$$\{\overline{U} : U \in \mathcal{U}_n, n \geq 0\} \cup \{X \setminus U : U \in \mathcal{U}_n, n \geq 0\} \cup$$
$$\cup \{\overline{U} \setminus U : U \in \mathcal{U}_n, n \geq 0\} \cup \mathcal{G}.$$

Then $\mathcal{H}$ is a countable collection closed subsets of $X$. By Theorem 3.5.3 there is a Wallman base $\mathcal{F}$ of $X$ such that

(5) $\mathcal{H} \subseteq \mathcal{F}$,
(6) $\dim \overline{H} = \dim H$ for every $H \in \mathcal{H}$ (here closure means closure in $\omega(X, \mathcal{F})$).

Put $aX = \omega(X, \mathcal{F})$. Observe that (1) follows from (6) since $\mathcal{G} \subseteq \mathcal{H}$. So we need only worry about (2). To this end, let $x \in X_{(n)}$ for certain $n \geq 0$ and let $V$ be an arbitrary open neighborhood of $x$ in $aX$. Let $W$ be a closed neighborhood of $x$ in $aX$ such that $W \subseteq V$. There is an element $U \in \mathcal{U}_n$ such that

$$x \in U \subseteq \overline{U} \subseteq W \cap X$$

(closure means closure in $X$ here). Observe that

$$\{\overline{U}, \overline{U} \setminus U, X \setminus U\} \subseteq \mathcal{F}.$$

This means that

$$\overline{U}^* \cup (X \setminus U)^* = \omega(X, \mathcal{F})$$

and

$$\overline{U}^* \cap (X \setminus U)^* = (\overline{U} \setminus U)^*.$$

(Lemma A.9.4). Observe that $\overline{U}^*$ is the closure of $\overline{U}$ in $\omega(X, \mathcal{F})$ (Exercise A.9.1). Similarly, $(\overline{U} \setminus U)^*$ is the closure of $\overline{U} \setminus U$ in $\omega(X, \mathcal{F})$. Since $W$ is a closed neighborhood of $x$, it consequently contains $\overline{U}^*$. Hence $(\overline{U} \setminus U)^*$ is a partition between $\{x\}$ and $aX \setminus W$ which is at most $(n-1)$-dimensional by (4) and (6). Since $V$ was arbitrary, this proves that $x \in (aX)_{(n)}$, as required. $\square$

Let $X$ be a space, and let $f \colon X \to X$ be continuous. A set $A \subseteq X$ such that $f[A] \cap A = \emptyset$ or, equivalently, $A \cap f^{-1}[A] = \emptyset$, is called a *color* of $f$. A *coloring* of $f$ is a finite collection of colors of $f$ which covers $X$. We say that $f$ is *colorable* if there is a finite cover of $X$ consisting of closed colors of $f$.

Observe that every colorable map is fixed-point free. There is an easy to describe example of a fixed-point free map which is not colorable. The verification of this is difficult. See §3.12 for details.

Let $f\colon X \to X$ be fixed-point free. Then every $x \in X$ has a closed neighborhood $V_x$ such that $V_x \cap f[V_x] = \emptyset$. By Corollary A.2.3, the cover
$$\{V_x : x \in X\}$$
has a countable subcover. There consequently exists for every fixed-point free map $f\colon X \to X$ a *countable* and potentially infinite closed cover $\mathcal{V}$ of $X$ such that every $V \in \mathcal{V}$ misses its own image. So a fixed-point free map on $X$ is colorable if such a cover can be chosen to be *finite*.

**Corollary 3.5.7.** *Let $X$ be a space and let $\mathcal{F}$ be a countable collection of continuous functions from $X$ to $X$. Then there is a compactification $aX$ of $X$ such that*

(1) $\dim X = \dim aX$.
(2) *Every $f \in \mathcal{F}$ can be extended to a continuous function $\bar{f}\colon aX \to aX$. If $f$ is a homeomorphism then so is $\bar{f}$. In addition, if $f \in \mathcal{F}$ is colorable then $\bar{f}$ is fixed-point free.*

**Proof.** We assume without loss of generality that the identity on $X$ belongs to $\mathcal{F}$ and that for each homeomorphism $f \in \mathcal{F}$ we also have that $f^{-1} \in \mathcal{F}$.

Assume that $f \in \mathcal{F}$ is colorable. Then there is a finite closed cover $\mathcal{G}_f$ of $X$ such that $G \cap f^{-1}[G] = \emptyset$ for every $G \in \mathcal{G}_f$. Put $\mathcal{H}_f = \mathcal{G}_f \cup f^{-1}[\mathcal{G}_f]$. Let
$$\mathcal{H} = \bigcup\{\mathcal{H}_f : f \in \mathcal{F}, f \text{ colorable}\} \cup \{X\}.$$
Then $\mathcal{H}$ is a countable collection closed subsets of $X$. Let $\mathcal{A}_1$ be the Wallman base of $X$ given by Theorem 3.5.3 for $\mathcal{H}$. Continuing in this way, let $\mathcal{A}_{n+1}$ be the Wallman base of $X$ given by Theorem 3.5.3 for
$$\bigcup_{f \in \mathcal{F}} f^{-1}[\mathcal{A}_n].$$
Observe that since the identity belongs to $\mathcal{F}$ we have $\mathcal{A}_n \subseteq \mathcal{A}_{n+1}$. In addition, since $X \in \mathcal{A}_n$ we have $\dim \omega(X, \mathcal{A}_n) = \dim X$.

As in the proof of Theorem 3.5.1,
$$\mathcal{A} = \bigcup_{n=1}^{\infty} \mathcal{A}_n$$
is a Wallman base such that $aX = \omega(X, \mathcal{A})$ is the inverse limit of an inverse sequence of compact spaces having the same dimension as $X$, hence $aX$ is dimension preserving.

Since
$$f^{-1}[A] \subseteq \mathcal{A}$$
for every $f \in \mathcal{F}$ we have that $f$ can be extended to a continuous function $\bar{f}: aX \to aX$ by Exercise A.9.4.

If $f$ is a homeomorphism then so is $\bar{f}$ because $f^{-1} \in \mathcal{F}$ can be extended as well and this extension is the inverse of $\bar{f}$.

Now assume that there exists an $f \in \mathcal{F}$ which is colorable. The collection
$$\mathcal{H}_f = \mathcal{G}_f \cup f^{-1}[\mathcal{G}_f]$$
is a subcollection of $\mathcal{A}$.

We claim that $\bar{f}: aX \to aX$ is fixed-point free. To see this, observe that the finite collection $\mathcal{G}$ covers $X$ and so does $\{f^{-1}[G] : G \in \mathcal{G}\}$. For a given $p \in aX$ there consequently exists $G \in \mathcal{G}_f$ such that $p \in \overline{f^{-1}[G]}$. The continuity of $\bar{f}: aX \to aX$ therefore implies that
$$\bar{f}(p) \in \bar{f}\left[\overline{f^{-1}[G]}\right] \subseteq \overline{\bar{f}[f^{-1}[G]]} \subseteq \overline{G}.$$
Since $G$ and $f^{-1}[G]$ are disjoint and both belong to $\mathcal{A}$, they have disjoint closures in $aX$ (Lemma A.9.4(2)). So $\bar{f}(p) \neq p$. □

**The dimension of almost zero-dimensional spaces.** We will use compactifications to prove that every almost zero-dimensional space is at most one-dimensional.

A subset $X$ of a compactum $K$ is *L-imbedded* in $K$ if for every open cover $\mathcal{U}$ of $K$ there is a neighborhood $V$ of $X$ in $K$ such that every subcontinuum of $K$ which is a subset of $V$ is contained in an element of $\mathcal{U}$.

Let $A$ and $B$ be disjoint closed sets in a space $X$, and let $\varepsilon > 0$. A triple
$$(G, H, \mathcal{F})$$
is an $L_\varepsilon$-*enlargement* for the pair $(A, B)$ if

(1) $G$ and $H$ are open neighborhoods of $A$ and $B$, respectively,
(2) $\overline{G} \cap \overline{H} = \emptyset$,
(3) $\mathcal{F}$ is a discrete collection of closed sets of diameter at most $\varepsilon$ such that $X \setminus (G \cup H) = \bigcup \mathcal{F}$.

**Theorem 3.5.8.** *Let $X$ be an L-imbedded subspace of a compact space $(K, \varrho)$. If $A$ and $B$ are disjoint closed subsets of $X$ having disjoint closures in $K$ then for each $\varepsilon > 0$ there is in $X$ an $L_\varepsilon$-enlargement $(G, H, \mathcal{F})$ for the pair $A, B$ such that for every $F \in \mathcal{F}$ the sets*
$$\overline{G} \cap F, \quad \overline{H} \cap F$$
*have disjoint closures in $K$ (here closure means closure in $K$).*

**Proof.** Let
$$\delta = \varrho(\overline{A}, \overline{B}).$$
Our assumptions give us that $\delta > 0$.

Pick an arbitrary $\varepsilon > 0$. We will construct in $X$ an $L_\varepsilon$-enlargement for the pair $(A, B)$. There is by assumption an open neighborhood $U$ of $X$ in $K$ such that all continua in $U$ have diameter at most
$$\tfrac{1}{2}\min\{\varepsilon, \delta\}.$$
Let $\varphi \colon K \to \mathbb{I}$ be continuous such that $U = \varphi^{-1}\big[(0,1]\big]$ (Lemma A.4.1). For each $n$, let
$$K_n = \varphi^{-1}\big[[\tfrac{1}{(n+1)}, \tfrac{1}{n}]\big].$$
Observe that every $K_n$ is compact. Fix $n$ for a moment. If $C$ is a component of $K_n$ then $C$ is not a continuum from $\overline{A} \cap K_n$ to $\overline{B} \cap K_n$. This is so because $\mathrm{diam}(C) \leq \tfrac{1}{2}\delta$ since $C \subseteq U$ and $\varrho(\overline{A}, \overline{B}) = \delta$. Hence $K_n$ can be split into two disjoint closed sets $P_n$ and $Q_n$ containing $\overline{A} \cap K_n$ and $\overline{B} \cap K_n$, respectively (Exercise A.11.5).

Observe that the sets
$$S = \bigcup_{n=1}^{\infty} P_{2n} \cup (\overline{A} \cap U), \quad T = \bigcup_{n=1}^{\infty} Q_{2n} \cup (\overline{B} \cap U)$$
are disjoint closed subsets of $U$. Let $\alpha \colon U \to \mathbb{I}$ be a Urysohn function such that $\alpha[S] \subseteq \{0\}$ and $\alpha[T] \subseteq \{1\}$. Put
$$V = \alpha^{-1}\big[[0, \tfrac{1}{4})\big], \quad W = \alpha^{-1}\big[(\tfrac{3}{4}, 1]\big].$$
Clearly, $\overline{V} \cap U$ and $\overline{W} \cap U$ are disjoint. Observe that $U \setminus (V \cup W)$ is contained in the union of the discrete collection
$$\mathcal{K} = \{K_{2n-1} : n \in \mathbb{N}\}$$
in $U$. Each $K_{2n-1}$ is a compactum whose components have all diameter at most $\tfrac{1}{2}\varepsilon$. So by Corollary A.10.2 we can split $K_{2n-1}$ into a finite pairwise disjoint collection compacta $\mathcal{F}_{2n-1}$ of diameter at most $\varepsilon$. Observe that the traces of $\overline{V}$ and $\overline{W}$ on any of these compact pieces are disjoint. We conclude that the triple
$$(V \cap X, W \cap X, \bigcup_{n=1}^{\infty} \mathcal{F}_{2n-1} \upharpoonright X)$$
is the required $L_\varepsilon$-enlargement for $(A, B)$. $\square$

**Corollary 3.5.9.** *Let $X$ be an L-imbedded subspace of a compact space $K$. Then $\dim X \leq 1$.*

**Proof.** In this proof $\overline{A}$ stands for the closure of $A$ in $K$.

Let $F$ be a closed subset of $X$, and let $p \in X \setminus F$. We will prove that there are disjoint open subsets $V, W$ in $X$ with $p \in V$, $F \subseteq W$ and $\dim\bigl((X \setminus (V \cup W))\bigr) \leq 0$. By Theorem 3.4.10, this suffices.

We shall construct two increasing sequences of open sets
$$p \in G_1 \subseteq G_2 \subseteq \cdots, \quad F \subseteq H_1 \subseteq H_2 \subseteq \cdots$$
and for every $n \in \mathbb{N}$ a collection $\mathcal{F}_n$ of closed subsets of $X$ such that

(1) $\overline{G}_n \cap \overline{H}_n \cap X = \emptyset$,
(2) the triple $(G_{n+1}, H_{n+1}, \mathcal{F}_n)$ is an $L_{1/n}$-enlargement for the pair
$$(\overline{G}_n \cap X, \overline{H}_n \cap X),$$
(3) if $F \in \mathcal{F}_n$ is arbitrary then
$$\overline{F \cap \overline{G}_n} \cap \overline{F \cap \overline{H}_n} = \emptyset.$$

Let $G_1, H_1$ be any pair of open sets containing $p$ and $F$, respectively, having disjoint closures in $K$. Suppose $G_n, H_n$ have been defined. For $n = 1$, we get the triple $(G_2, H_2, \mathcal{F}_1)$ directly from Theorem 3.5.8.

Assume $n \geq 2$. For each $F \in \mathcal{F}_{n-1}$ consider the disjoint closed sets
$$A(F) = \overline{G}_n \cap F, \quad B(F) = \overline{H}_n \cap F$$
in $X$. These sets have disjoint closures in $K$ by (3). By Theorem 3.5.8 there is an $L_{1/n}$-enlargement
$$(G(F), H(F), \mathcal{R}_F)$$
of the pair $\bigl(A(F), B(F)\bigr)$, such that for every $R \in \mathcal{R}_F$ the sets
(4) $$\overline{G(F)} \cap R, \quad \overline{H(F)} \cap R$$
have disjoint closures in $K$. For each $F \in \mathcal{F}_{n-1}$ let $U(F) \supseteq F$ be an open set in $X$ such that the collection
(5) $$\{\overline{U(F)} \cap X : F \in \mathcal{F}_{n-1}\}$$
is discrete in $X$ (Exercise A.7.3). Put
$$G_{n+1} = G_n \cup \bigcup\{G(F) \cap U(F) : F \in \mathcal{F}_{n-1}\},$$
$$H_{n+1} = H_n \cup \bigcup\{H(F) \cap U(F) : F \in \mathcal{F}_{n-1}\}.$$
It is clear that $G_{n+1}$ and $H_{n+1}$ are open, and by (5) and (1) it follows that
$$\overline{G}_{n+1} \cap \overline{H}_{n+1} \cap X = \emptyset.$$
Put
$$\mathcal{F}_n = \{F \cap R : F \in \mathcal{F}, R \in \mathcal{R}_F\}.$$
Then $\mathcal{F}_n$ is a discrete collection of closed sets with diameter at most $1/n$.

It is easy to see that the triple $(G_{n+1}, H_{n+1}, \mathcal{F}_n)$ is an $L_{1/n}$-enlargement for the pair
$$(\overline{G}_n \cap X, \overline{H}_n \cap X).$$

This completes the inductive construction.

Now put
$$V = \bigcup_{n=1}^{\infty} G_n, \quad W = \bigcup_{n=1}^{\infty} H_n.$$
Observe that for each $\varepsilon > 0$ the complement $X \setminus (V \cup W)$ is the union of a discrete collection of closed sets of diameter at most $\varepsilon$. As a consequence, the partition $X \setminus (V \cup W)$ is at most zero-dimensional (Exercise 1.5.3). □

We are now in a position to present the proof of the following interesting result.

**Theorem 3.5.10.** *If $X$ is almost zero-dimensional then $\dim X \leq 1$.*

**Proof.** Let $X$ be nonempty and almost zero-dimensional. In addition, let $\mathcal{B}$ be a base for $X$ with the properties stated in Exercise 3.2.5. We may assume that $\mathcal{B}$ is countable (Exercise A.2.12). For every pair $G, H \in \mathcal{B}$ such that
$$\overline{G} \cap \overline{H} = \emptyset,$$
pick a clopen set $C$ in $X$ with
$$G \subseteq C \subseteq H.$$
Since $\mathcal{B}$ is countable, the collection $\mathcal{C}$ of clopen sets picked in this way is countable as well.

There is a compactification $\gamma X$ of $X$ such that for every $C \in \mathcal{C}$ we have that $C$ and $X \setminus C$ have disjoint closures in $\gamma X$ (Exercise A.9.3). That is, the closure of $C$ in $\gamma X$ is clopen in $\gamma X$.

For every $B \in \mathcal{B}$ let $B^*$ be an open subset of $\gamma X$ such that $B^* \cap X = B$.

We claim that $X$ is $L$-imbedded in $\gamma X$. To this end, let $\varepsilon > 0$, and put
$$\mathcal{H} = \{B \in \mathcal{B} : \operatorname{diam} B^* < {}^1\!/_3 \varepsilon\},$$
and
$$U = \bigcup \mathcal{H},$$
respectively. If $x \in X$ then there is an element $B \in \mathcal{B}$ such that
$$B \subseteq B(x, {}^1\!/_7 \varepsilon) \cap X.$$
Since $B$ is dense in $B^*$, the set $B^*$ is contained in the closed ball $D(x, {}^1\!/_7 \varepsilon)$, hence $B^* \subseteq U$. We conclude that $U$ is an open neighborhood of $X$ in $\gamma X$.

To conclude the proof, let $K \subseteq U$ be an arbitrary continuum. Pick two elements $B_0, B_1 \in \mathcal{H}$ such that
(1) $$B_0^* \cap K \neq \emptyset \neq K \cap B_1^*.$$
If $B_0$ and $B_1$ have disjoint closures in $X$ then there is an element $C \in \mathcal{C}$ such that
$$B_0 \subseteq C \subseteq B_1.$$

The closure of $C$ in $\gamma X$ is clopen. By (1), $K$ intersects this closure, as well as its complement. But this contradicts the connectivity of $K$. So the closures of $B_0$ and $B_1$ intersect. Since there is a subfamily $\mathcal{E}$ of $\mathcal{H}$ such that
$$K \subseteq \bigcup \{B^* : B \in \mathcal{E}\}$$
this proves that the diameter of $K$ is at most $\varepsilon$.

So an appeal to Theorem 3.5.9 finishes the job. $\square$

**Exercises for §3.5.**

1. Let $X$ be a space and let $Y$ be a compact space for which there is a continuous function $f \colon X \to Y$. Prove that there is a dimension preserving compactification $\gamma X$ of $X$ such that the function $f$ can be extended to a continuous function $g \colon \gamma X \to Y$.

▶2. Let $(X_n, f_n)_n$ be an inverse sequence consisting of at most $m$-dimensional spaces. Prove that its inverse limit is at most $m$-dimensional.

3. Let $X$ be a space which is not compact. Prove that $X$ has a compactification $\gamma X$ such that $\dim \gamma X = \infty$.

▶4. Give an example of a compact space $X$ containing two disjoint closed subsets $A$ and $B$ such that no partition between $A$ and $B$ is irreducible.

5. Let $X$ be a subspace of the compact space $Y$. Prove that the following statements are equivalent:
   (1) $X$ is $L$-imbedded in $Y$.
   (2) For every $\varepsilon > 0$ there is a neighborhood $U$ of $Y$ in $X$ such that every continuum in $U$ has diameter less than $\varepsilon$.

▶6. Let $A$ and $B$ be $L$-imbedded subspaces of the compacta $X$ and $Y$, respectively. Prove that $A \times B$ is $L$-imbedded in $X \times Y$. Conclude that $A \times B$ is at most one-dimensional.

## 3.6. Mappings into spheres

In the previous sections we presented several characterizations of dimension, some 'internal' (e.g., Theorem 3.2.14(2) and (3)) and some 'external' (e.g., Theorem 3.2.14(4)). In this section we shall prove that a space $X$ is at most $n$-dimensional if and only if every continuous function $f \colon A \to \mathbb{S}^n$, where $A \subseteq X$ is closed, can be extended over $X$, thereby deducing another important 'external' characterization of dimension. As an application of this result we shall present a proof of the Brouwer Invariance of Domain Theorem.

We first formulate and prove two lemmas that are needed in the proof of the main result in this section.

**Lemma 3.6.1.** *Let $X$ be a space and let $A_1$ and $A_2$ be disjoint closed subsets of $X$ such that $0 \leq \dim \left( X \setminus (A_1 \cup A_2) \right) \leq n$. Then there exists a partition $L$ between $A_1$ and $A_2$ in $X$ such that $\dim L \leq n - 1$.*

**Proof.** Put $A = A_1 \cup A_2$. By Corollary A.4.3 there exist open subsets $U$ and $V$ of $X$ such that $A_1 \subseteq U$, $A_2 \subseteq V$ and $\overline{U} \cap \overline{V} = \emptyset$. Since $\dim(X \setminus A) \leq n$ and $\overline{U} \cap (X \setminus A)$ and $\overline{V} \cap (X \setminus A)$ are disjoint closed subsets of $X \setminus A$, by Corollary 3.4.14 there exist disjoint open sets $E$ and $F$ in $X \setminus A$ such that

(1) $\overline{U} \cap (X \setminus A) \subseteq E$,
(2) $\overline{V} \cap (X \setminus A) \subseteq F$,
(3) if $L = (X \setminus A) \setminus (E \cup F)$ then $\dim L \leq n - 1$.

Since $A_1 \cup E = U \cup E$, and $E$ is open in $X$ being open in the subspace $X \setminus A$, we conclude that $A_1 \cup E$ is open in $X$. Similarly, $A_2 \cup F$ is open in $X$. Consequently, $L$ is closed in $X$ and is a partition between $A_1$ and $A_2$ in $X$ with $\dim L \leq n - 1$ by (3). So $L$ is as required. $\square$

**Corollary 3.6.2.** *Let $X$ be a space and let $A_1$ and $A_2$ be closed subspaces of $X$ such that $0 \leq \dim(X \setminus (A_1 \cup A_2)) \leq n$. Then there exist closed subspaces $X_1$ and $X_2$ of $X$ such that*

(1) $A_1 = X_1 \cap (A_1 \cup A_2)$,
(2) $A_2 = X_2 \cap (A_1 \cup A_2)$,
(3) $\dim((X_1 \cap X_2) \setminus (A_1 \cup A_2)) \leq n - 1$,
(4) $X_1 \cup X_2 = X$.

**Proof.** Put $Y = X \setminus (A_1 \cap A_2)$. By Lemma 3.6.1 there exists a closed set $L$ in $Y$ with $\dim L \leq n - 1$, such that $Y \setminus L$ can be written as the disjoint union of two open sets $E$ and $F$ such that $A_1 \cap Y \subseteq E$ and $A_2 \cap Y \subseteq F$. It is easy to see that $X_1 = A_1 \cup E \cup L$ and $X_2 = A_2 \cup F \cup L$ are as required. $\square$

These simple results enable us to derive the following:

**Theorem 3.6.3.** *Let $X$ be a space and let $A$ be a closed subspace of $X$ such that $0 \leq \dim(X \setminus A) \leq n$. Then every continuous function $g \colon A \to \mathbb{S}^n$ can be extended to a continuous function $\bar{g} \colon X \to \mathbb{S}^n$.*

**Proof.** We shall prove the theorem by induction on $n \geq 0$. Suppose first that $n = 0$ and recall that $\mathbb{S}^0 = \{-1, 1\}$. By Corollary 1.5.2 there exists a clopen set $C \subseteq X$ such that $g^{-1}(-1) \subseteq C$ and $g^{-1}(1) \subseteq X \setminus C$. Define the function $\bar{g} \colon X \to \mathbb{S}^0$ by

$$\bar{g}(x) = \begin{cases} -1 & (x \in C), \\ 1 & (x \notin C). \end{cases}$$

An easy check shows that $\bar{g}$ is as required.

Now assume that the theorem is true for $n - 1$, $n \geq 1$, and let $g \colon A \to \mathbb{S}^n$. Define $\mathbb{S}_1^n = \{x \in \mathbb{S}^n : x_1 \geq 0\}$ and $\mathbb{S}_2^n = \{x \in \mathbb{S}^n : x_1 \leq 0\}$, respectively. Observe that $\mathbb{S}_1^n$ and $\mathbb{S}_2^n$ are both homeomorphic to $\mathbb{I}^n$ and that $\mathbb{S}_1^n \cap \mathbb{S}_2^n$ is homeomorphic to $\mathbb{S}^{n-1}$. Now put $A_1 = g^{-1}[\mathbb{S}_1^n]$ and $A_2 = g^{-1}[\mathbb{S}_2^n]$, respectively. Let $X_1$ and $X_2$ be such as in Corollary 3.6.2 for $A_1$ and $A_2$. By our inductive assumption, we can extend $g \upharpoonright A_1 \cap A_2$ to a continuous function

$$h \colon X_1 \cap X_2 \to \mathbb{S}_1^n \cap \mathbb{S}_2^n.$$

Now for $i = 1, 2$ define $g_i \colon A_i \cup (X_1 \cap X_2) \to \mathbb{S}_i^n$ by $g_i = (g \upharpoonright A_i) \cup h$. Since $\mathbb{S}_i^n$ is homeomorphic to $\mathbb{I}^n$ it is an AR (Corollary 1.2.12) and so we can extend $g_i$ for $i = 1, 2$ to a continuous function $\bar{g}_i \colon X_i \to \mathbb{S}_i^n$. Clearly,

$$\bar{g} = \bar{g}_1 \cup \bar{g}_2 \colon X \to \mathbb{S}^n$$

is a continuous extension of $g$. □

**Corollary 3.6.4.** *Let $X$ be a space and let $f, g \colon X \to \mathbb{S}^n$ be continuous. If*

$$D(f, g) = \{x \in X : f(x) \neq g(x)\}$$

*is at most $(n-1)$-dimensional then $f$ and $g$ are homotopic.*

**Proof.** Consider the product $X \times \mathbb{I}$ and put

$$A = \bigl(X \times \{0, 1\}\bigr) \cup \bigl((X \setminus D(f, g)) \times \mathbb{I}\bigr).$$

Then $A$ is closed in $X \times \mathbb{I}$ (Exercise A.1.9) and the function $\xi \colon A \to \mathbb{S}^n$ defined by

$$\xi(x, t) = \begin{cases} f(x) & (t = 0), \\ f(x) = g(x) & (x \in X \setminus D(f, g)), \\ g(x) & (t = 1), \end{cases}$$

is clearly continuous. In addition, $(X \times \mathbb{I}) \setminus A$ is contained in

$$D(f, g) \times \mathbb{I}$$

and hence is at most $\bigl((n-1) + 1\bigr)$-dimensional by Theorem 3.4.12. So Theorem 3.6.3 implies that $\xi$ can be extended to a continuous function

$$\bar{\xi} \colon X \times \mathbb{I} \to \mathbb{S}^n,$$

which means that $f$ and $g$ are homotopic. □

**Theorem 3.6.5.** *Let $X$ be a space and let $n \geq 0$. The following statements are equivalent:*

(1) $\dim X \leq n$,
(2) *for every closed subset $A$ of $X$, every continuous function $f \colon A \to \mathbb{S}^n$ can be continuously extended over $X$.*

**Proof.** For (1) $\Rightarrow$ (2) observe that $\dim A \leq n$ and apply Theorem 3.6.3.

For (2) $\Rightarrow$ (1), let $\tau = \{(A_i, B_i) : i \leq n+1\}$ be a family of pairs of disjoint closed subsets of $X$. We shall prove that $\tau$ is inessential. To this end, for every $i \leq n+1$ let $\alpha_i \colon X \to \mathbb{I}$ be a Urysohn function such that $\alpha_i \restriction A_i \equiv 0$ and $\alpha_i \restriction B_i \equiv 1$ (Corollary A.4.1). Define $\alpha \colon X \to \mathbb{I}^{n+1}$ by

$$\alpha(x) = \big(\alpha_1(x), \ldots, \alpha_{n+1}(x)\big).$$

Put $A = \bigcup_{i=1}^{n+1}(A_i \cup B_i)$ and $\beta = \alpha \restriction A$. Then $\beta[A]$ is contained in the boundary $B$ of $\mathbb{I}^{n+1}$ which is homeomorphic to $\mathbb{S}^n$ by Exercise 1.1.24. By assumption, there exists a continuous extension $\gamma \colon X \to B$ of $\beta$. Now for every $i \leq n+1$ put $E_i = (\pi_i \circ \gamma)^{-1}(1/2)$. Then clearly $E_i$ is a partition between $A_i$ and $B_i$ for every $i$ such that $\bigcap_{i=1}^{n+1} E_i = \emptyset$. Consequently, $\tau$ is inessential. $\square$

**Application 1: The Brouwer Invariance of Domain Theorem.**
To begin with, we shall first present an 'internal' characterization of the boundary points of an arbitrary closed subset of a fixed $\mathbb{R}^n$.

**Proposition 3.6.6.** *Let $n \in \mathbb{N}$ and let $X$ be a closed subspace of $\mathbb{R}^n$. Then a point $x$ in $X$ belongs to the boundary $\mathrm{Fr}(X)$ of $X$ in $\mathbb{R}^n$ if and only if $x$ has arbitrarily small neighborhoods $U$ in $X$ such that every continuous function $g \colon X \setminus U \to \mathbb{S}^{n-1}$ can be extended continuously over $X$.*

**Proof.** First assume that $x \in \mathrm{Fr}(X)$. Without loss of generality, $x = \underline{0}$. If $V$ is a neighborhood of $\underline{0}$ in $X$ then there is $\varepsilon > 0$ such that

$$B(x, \varepsilon) \cap X \subseteq V.$$

Consequently, it suffices to consider for some $\varepsilon > 0$ a 'spherical' neighborhood $B(\underline{0}, \varepsilon)$ of $\underline{0}$ and a continuous function $g \colon X \setminus B(\underline{0}, \varepsilon) \to \mathbb{S}^{n-1}$. Without loss of generality assume that $\varepsilon = 1$ which means that the boundary of $B(\underline{0}, \varepsilon)$ is $\mathbb{S}^{n-1}$.

Since $\underline{0}$ belongs to the boundary of $X$, there is a point $q \in B(\underline{0}, 1) \setminus X$. For each $p \in B^n \setminus \{q\}$ let $\tau(p)$ denote the 'projection' of $p$ on $\mathbb{S}^{n-1}$ from $q$.

Put $X_0 = X \cap B^n$ and $X_1 = X \setminus B(\underline{0}, 1)$, respectively. Then $X_0$ and $X_1$ are closed in $X$, cover $X$, and have the following additional property:

$$X_0 \cap \mathbb{S}^{n-1} = X_1 \cap \mathbb{S}^{n-1}.$$

Put $Y = X_0 \cap X_1$.

Since $\dim \mathbb{S}^{n-1} = n-1$ (Theorem 3.2.12), the function $g \restriction Y$ can be extended to a continuous function $\overline{g \restriction Y} \colon \mathbb{S}^{n-1} \to \mathbb{S}^{n-1}$ by Theorem 3.6.5. In addition, define $h \colon X_0 \to \mathbb{S}^{n-1}$ by $h = \overline{g \restriction Y} \circ \tau$. Observe that $\tau(y) = y$ for $y \in Y$ so that

$$h(y) = \overline{g \restriction Y}(y) = g(y).$$

This means that the functions $g$ and $h$ agree on $Y$ from which it follows that the function

$$\bar{g} = h \cup g$$

is the desired continuous extension of $g$.

Conversely, let $x \in X$ and assume that $x$ is an interior point of $X$. There is $\varepsilon > 0$ such that $B = D(x,\varepsilon) \subseteq X$. Let $U$ be an open neighborhood of $x$ in $X$ such that $U \subseteq B(x,\varepsilon)$. We identify $\mathbb{S}^{n-1}$ and the boundary of $B$. For each $p \in X \setminus U$ let $\tau(p)$ denote the 'projection' of $p$ on $\mathbb{S}^{n-1}$ from $x$. Then $\tau$ cannot be extended over $X$ since this would yield a retraction from $B$ onto its boundary, which is impossible by Theorem 2.4.10. $\square$

**Corollary 3.6.7.** *Let $n \in \mathbb{N}$ and let $X$ and $Y$ be closed subspaces of $\mathbb{R}^n$. If $f \colon X \to Y$ is a homeomorphism then $f[\mathrm{Fr}(X)] = \mathrm{Fr}(Y)$.*

We are now in a position to present a proof of the following interesting:

**Brouwer Invariance of Domain Theorem 3.6.8.** *Let $n \in \mathbb{N}$ and let $U$ be an open subset of $\mathbb{R}^n$. If $f \colon U \to \mathbb{R}^n$ is injective and continuous then the following statements hold:*

(1) *$f[U]$ is open in $\mathbb{R}^n$,*
(2) *$f \colon U \to f[U]$ is a homeomorphism.*

**Proof.** We shall first prove (1). Take an arbitrary $x \in U$. We shall prove that $f(x)$ belongs to the interior of $f[U]$. There exists $\varepsilon > 0$ such that
$$D(x,\varepsilon) = \{y \in \mathbb{R}^n : \|x - y\| \leq \varepsilon\} \subseteq U.$$
Since $B$ is compact, by Exercise A.5.9, $f \restriction B$ is a homeomorphism. By Corollary 3.6.7 we conclude that $f(x)$ belongs to the interior of $f[B]$ and hence to the interior of $f[U]$.

We shall now prove (2). This is easy. If $V$ is an open subset of $U$ then (1) applied to $f \restriction V$ shows that $f[V]$ is open in $\mathbb{R}^n$ and hence in $f[U]$. We conclude that $f \colon U \to f[U]$ is an open mapping and therefore by injectivity is a homeomorphism. $\square$

The question naturally arises whether something like Theorem 3.6.8 can also be derived for the class consisting of all *infinite-dimensional* linear spaces. As to be expected, this is not possible. For the first part of the theorem, this can be demonstrated quite easily by considering Hilbert space $\ell^2$. The function $f \colon \ell^2 \to \ell^2$ defined by
$$f(x_1, x_2, \dots) = (0, x_1, x_2, \dots),$$
is an imbedding with nowhere dense range.

We now turn to the second part of Theorem 3.6.8.

**Theorem 3.6.9.** *For an infinite-dimensional normed linear space $L$ there exists a bijective continuous function $f \colon L \to L$ such that $f$ is not a homeomorphism.*

**Proof.** Let $S$ be the unit sphere in $L$. Since $L$ is infinite-dimensional, $S$ is not compact by Exercise 1.1.22. Consequently, by Exercise A.5.13 there exists a continuous function $\lambda\colon S \to (0,1]$ such that $\inf \lambda(S) = 0$. Now define $f\colon L \to L$ by the formula
$$\begin{cases} f(y) = \lambda\bigl(\tfrac{y}{\|y\|}\bigr) \cdot y & (y \neq \underline{0}), \\ f(\underline{0}) = \underline{0}. \end{cases}$$

**Claim 1.** If $x \in L$ and $\alpha \in [0, \infty)$ then $f(\alpha x) = \alpha f(x)$.

*Proof.* This is a triviality. If $x = \underline{0}$ or $\alpha = 0$ then there is nothing to prove. In addition, if $x \neq \underline{0}$ and $\alpha \neq 0$ then
$$f(\alpha x) = \lambda\Bigl(\frac{\alpha x}{\|\alpha x\|}\Bigr) \cdot \alpha x = \alpha f(x),$$
as required. $\diamond$

We shall now prove that $f$ is continuous. It is clear that this need only be checked at the origin. To this end, let $(x_n)_n$ be a sequence in $L \setminus \{\underline{0}\}$ such that $\lim_{n \to \infty} x_n = \underline{0}$. Then
$$\|f(x_n)\| = \lambda\Bigl(\frac{x_n}{\|x_n\|}\Bigr) \cdot \|x_n\| \leq 1 \cdot \|x_n\| \xrightarrow{n \to \infty} 0.$$

We conclude that $\lim_{n \to \infty} f(x_n) = \underline{0}$.

We shall next prove that $f$ is injective. To this end, take $x, y \in L$ and assume that $f(x) = f(y)$. Without loss of generality, $x \neq \underline{0}$. Then $f(x) \neq \underline{0}$ so that $f(y) \neq \underline{0}$ from which it follows that $y \neq \underline{0}$. So $f(x) = f(y)$ gives us that
$$\lambda\Bigl(\frac{x}{\|x\|}\Bigr) \cdot x = \lambda\Bigl(\frac{y}{\|y\|}\Bigr) \cdot y,$$
and since $\lambda(S) \subseteq (0,1]$ this implies that there exists $\alpha > 0$ such that $x = \alpha y$. By the claim, this yields $f(x) = \alpha f(y)$, i.e., $\alpha = 1$. We conclude that $x = y$.

Now take $y \in L \setminus \{\underline{0}\}$. Put
$$x = \frac{y}{\lambda(\frac{y}{\|y\|})}.$$

An easy application of the claim gives us that $f(x) = y$.

We shall now prove that $f$ is not a homeomorphism by showing that $f^{-1}$ is not continuous. Since $\inf \lambda(S) = 0$, we can choose a sequence $(x_n)_n$ in $S$ such that
$$\lim_{n \to \infty} \lambda(x_n) = 0.$$

It follows that
$$\|f(x_n)\| = \lambda(x_n) \cdot \|x_n\| \to 0 \cdot 1 = 0,$$
so $f(x_n) \to \underline{0}$. However, $\|x_n\| = 1$ for every $n$, so $x_n \not\to \underline{0}$. $\square$

**Application 2: Dimension and mappings.** We shall now discuss the question how continuous mappings can lower or raise dimension. The results obtained earlier in this section are important in the proof of Theorem 3.6.10 below. They are not used in the proof of the subsequent Theorem 3.6.14. However, Theorem 3.6.14 is motivated by Theorem 3.6.10 and so it is natural to discuss it here.

We already know that every compact space is a continuous image of C (Theorem 1.5.10). So continuous closed surjections, can arbitrarily raise dimension. In general it is therefore not possible to conclude anything. Certain conditions have to be imposed.

It is natural to look at the fibres of the map under consideration. If something can be said about their dimensions, then our intuition tells us that there should be a relation between the dimensions of the domain and the range of the map. This is what the next result is about.

**Theorem 3.6.10.** *Let $X$ and $Y$ be spaces and let $f\colon X \to Y$ be a continuous closed surjection. If there is an integer $k \geq 0$ such that $\dim f^{-1}(y) \leq k$ for every $y \in Y$ then*
$$\dim X \leq \dim Y + k.$$

**Proof.** We may assume without loss of generality that $\dim Y < \infty$. We shall prove the theorem by induction on $\dim Y$. If $\dim Y = -1$ then $X$ and $Y$ are empty and so the theorem trivially holds. Now assume the theorem holds for all spaces $X$ and $Y$ and all continuous closed surjections $f\colon X \to Y$ such that $\dim Y < n$, where $n \geq 0$, and assume that we are in the situation that $Y$ is a space such that $\dim Y = n$.

**Claim 1.** If $y \in Y$ and $U$ is an open neighborhood of $f^{-1}(y)$ in $X$ then there is an open neighborhood $V$ of $f^{-1}(y)$ such that $\overline{V} \subseteq U$ while moreover
$$\dim \operatorname{Fr} V \leq n - 1 + k.$$

*Proof.* Put $W = Y \setminus f[X \setminus U]$. Since the map $f$ is closed, $W$ is an open neighborhood of $y$. Let $E$ be an open neighborhood of $y$ in $Y$ such that
$$E \subseteq \overline{E} \subseteq W$$
while moreover $\dim \operatorname{Fr} E \leq n - 1$ (Theorem 3.4.10). Then $V = f^{-1}[E]$ is as required since clearly $\overline{V} \subseteq U$ and
$$\operatorname{Fr} V = \operatorname{Fr} f^{-1}[E] = \overline{f^{-1}[E]} \setminus f^{-1}[E] \subseteq f^{-1}[\overline{E}] \setminus f^{-1}[E] = f^{-1}[\operatorname{Fr} E]$$
and so $\dim \operatorname{Fr} V \leq n-1+k$ by the Subspace Theorem 3.2.9 and our inductive hypothesis applied to the closed map $f \restriction f^{-1}[\operatorname{Fr} E]\colon f^{-1}[\operatorname{Fr} E] \to \operatorname{Fr} E$. ◊

Put $m = n + k$. Now let $A$ be an arbitrary closed subset of $X$, and let $h\colon A \to \mathbb{S}^m$ be continuous. We will prove that $h$ can be extended to a continuous function $\bar{h}\colon X \to \mathbb{S}^m$. It will then follow that $\dim X \leq m = n+k$ by Theorem 3.6.5.

Let $y \in Y$ be arbitrary. Then $\dim f^{-1}(y) \leq k \leq n+k$ and so we can extend $h$ continuously over $A \cup f^{-1}(y)$ (Theorem 3.6.3). Since $\mathbb{S}^m$ is an ANR, this function can be extended over a neighborhood $U_y$ of $A \cup f^{-1}(y)$ (Corollary 1.2.13). By Claim 1 there is an open neighborhood $V_y$ of $f^{-1}(y)$ such that $\overline{V}_y \subseteq U_y$ while moreover $\dim \operatorname{Fr} V_y \leq n-1+k$. So $h$ can evidently be extended over $A \cup \overline{V}_y$.

The open cover $\mathcal{V} = \{V_y : y \in Y\}$ has a countable subcover by Corollary A.2.3, say $\{V_{y_i} : i \in \mathbb{N}\}$. We shall inductively define a sequence of continuous functions $h_i \colon A \cup \bigcup_{j=1}^{i} \overline{V}_{y_j} \to \mathbb{S}^m$ such that

$$(*) \qquad h_i \upharpoonright A \cup \bigcup_{j=1}^{i-1} \overline{V}_{y_j} = h_{i-1},$$

for every $i > 1$. Let $h_1$ be an arbitrary continuous extension of $h$ over $A \cup \overline{V}_{y_1}$. Assume that the functions $h_1, \ldots, h_{i-1}$ have been defined satisfying $(*)$. Put

$$K = A \cup \bigcup_{j=1}^{i-1} \overline{V}_{y_j}, \quad L = A \cup \left( \overline{V}_{y_k} \setminus \bigcup_{j=1}^{i-1} V_{y_j} \right),$$

respectively. Observe that $K \cup L = A \cup \bigcup_{j=1}^{i} \overline{V}_{y_j}$. By the above we can extend the function $f$ to a continuous function $g \colon L \to \mathbb{S}^m$. Observe that

$$D = \{x \in L \cap K : h_{i-1}(x) \neq g(x)\} \subseteq P = \bigcup_{j=1}^{i-1} \overline{V}_{y_j} \setminus \bigcup_{j=1}^{i-1} V_{y_j} = \bigcup_{j=1}^{i-1} \operatorname{Fr} V_{y_j}.$$

Since $\dim P \leq n-1+k = m-1$ by the Countable Closed Sum Theorem 3.2.8, we have that $\dim D \leq m-1$ by the Subspace Theorem 3.2.9. We conclude that the functions $h_{i-1} \upharpoonright K \cap L$ and $g \upharpoonright K \cap L$ are homotopic by Corollary 3.6.4. The Borsuk Homotopy Theorem 1.4.2 implies that the function $h_{k-1} \upharpoonright K \cap L$ can be extended to a continuous function $\tilde{g} \colon L \to \mathbb{S}^m$. Since $h_{k-1}$ and $\tilde{g}$ agree on the closed set $K \cap L$ it follows that their union is the desired function $h_k$. This completes the inductive construction.

Now define $\bar{h} \colon X \to \mathbb{S}^m$ by

$$\bar{h}(x) = h_i(x) \qquad (x \in V_{y_i}).$$

Then $\bar{h}$ is clearly a continuous function extending $h$. $\square$

As we saw above, closed continuous surjections can arbitrarily raise dimension. It is natural to ask whether the same can happen for open continuous surjections. But things are slightly different here. For observe that if $X$ is zero-dimensional, and $f \colon X \to Y$ is an open and closed continuous surjection, then $Y$ is zero-dimensional (Exercise 1.5.5). So open maps defined on

**Proposition 3.6.11.** For each $n \geq 1$ there exists a continuous function
$$f_n \colon B^{n+1} \to 2^{\mathbb{S}^n}$$
such that for every $x \in \mathbb{S}^n$, $f_n(x) = \{x\}$.

**Proof.** Define $f_n \colon B^{n+1} \to 2^{\mathbb{S}^n}$ by
$$f_n(x) = \begin{cases} \{\mathbb{S}^n\} & (x = 0), \\ f_n(x) = \left\{ z \in \mathbb{S}^n : \left\| z - \frac{x}{\|x\|} \right\| \leq 2 - 2 \cdot \|x\| \right\} & (x \neq 0). \end{cases}$$
It is easy to visualize this function e.g., for $n = 1$. It is clear that $f_n$ is continuous and that $f_n(x) = \{x\}$ for every $x \in \mathbb{S}^n$. □

**Lemma 3.6.12.** Let $X$ be a compact space with closed sets $A, B, E$ and $F$ such that $A \cap B = \emptyset$ and $E \cap F = \emptyset$. Then there exist a compact space $Y$ and an open surjection $f \colon Y \to X$ such that the collection of pairs
$$\{(f^{-1}[A], f^{-1}[B]), (f^{-1}[E], f^{-1}[F])\}$$
is inessential in $Y$.

**Proof.** Put $Z = (A \cup B) \cup (E \cup F)$. In addition, pick Urysohn functions $\xi, \eta \colon Z \to \mathbb{I}$ such that $\xi[A] = 0, \xi[B] = 1, \eta[E] = 0$ and $\eta[F] = 1$ (Lemma A.4.1). The function
$$h(x) = \big(\xi(x), \eta(x)\big) \qquad (x \in Z)$$
maps $Z$ into the boundary of $\mathbb{I}^2$. Let $\bar{h} \colon X \to \mathbb{I}^2$ be a continuous extension of $h$ (Corollary 1.2.12). By the fact that $\mathbb{I}^2 \approx B^2$ under a homeomorphism which (necessarily) maps its boundary $W = \partial \mathbb{I}^2$ onto $\mathbb{S}^1$ (Exercise 1.1.24), there exists by Proposition 3.6.11 a continuous function $g \colon \mathbb{I}^2 \to 2^W$ such that $g(x) = \{x\}$ for every $x \in W$. Now define
$$Y = \bigcup_{x \in X} \big(\{x\} \times g(\bar{h}(x))\big) \subseteq X \times W.$$
This is the space $Y$ we are looking for. We first claim that $Y$ is closed in $X \times W$. To this end, suppose that $(x_n, y_n)_n$ is a sequence in $Y$ converging to an element $(x, y) \in X \times W$. Then $x_n \to x$ and so $g \circ \bar{h}(x_n) \to g \circ \bar{h}(x)$. Since
$$y_n \in g \circ \bar{h}(x_n)$$
for every $n$, and $y_n \to y$, it therefore follows that
$$y \in g \circ \bar{h}(x)$$
(Lemma 1.11.2(2)). This proves that $Y$ is closed, whence compact.

Now let $f = \pi_1 \upharpoonright Y \to X$, where $\pi_1 \colon X \times W \to X$ is the projection. This is the map $f$ we are looking for. The function $F \colon X \to 2^{X \times W}$ defined by

$$F(x) = \{x\} \times [g \circ \bar{h}(x)]$$

is continuous because $g$ and $\bar{h}$ are. As a consequence, $f$ is open by Exercise 1.11.16.

Let $\pi_2 \colon X \times W \to W$ denote the projection.

Now take arbitrary elements $(x, y) \in f^{-1}[A]$ and $(p, q) \in f^{-1}[B]$. Then obviously $x \in A$ and $y \in g(\bar{h}(x))$. But since $x \in Z$, it follows that

$$\bar{h}(x) = h(x) \in W$$

and so

$$g(\bar{h}(x)) = \{h(x)\}.$$

So this implies that $(x, y) = (x, h(x))$. Similarly it follows that $p \in B$ and that $(p, q) = (p, h(p))$. Since $h(x) \neq h(p)$ and $(x, y)$ and $(p, q)$ were arbitrarily chosen, this implies $\pi_2[f^{-1}[A]] \cap \pi_2[f^{-1}[B]] = \emptyset$. It follows similarly that

$$\pi_2[f^{-1}[E]] \cap \pi_2[f^{-1}[F]] = \emptyset.$$

This implies that the collection of pairs

$$\{(f^{-1}[A], f^{-1}[B]), (f^{-1}[E], f^{-1}[F])\}$$

is by Exercise 3.1.11 inessential since $\dim W = 1$. $\square$

**Corollary 3.6.13.** *Let $X$ be a compact space. Then there exist a compact space $Y$ and an open surjective map $f \colon Y \to X$ such that if $A, B, E$ and $F$ are arbitrary closed subsets of $X$ such that $A \cap B = \emptyset$ and $E \cap F = \emptyset$ then the pairs*

$$\{(f^{-1}[A], f^{-1}[B]), (f^{-1}[E], f^{-1}[F])\}$$

*are inessential.*

**Proof.** Let $\mathcal{E} = \{(E_n, F_n) : n \in \mathbb{N}\}$ be a collection of pairs of disjoint closed sets for $X$ such as in Lemma A.5.6. That is, if $(A, B)$ is any pair of disjoint closed subsets of $X$ then there exists $n \in \mathbb{N}$ such that $A \subseteq E_n$ and $F \subseteq B_n$.

Let $\{(n_i, m_i) : i \in \mathbb{N}\}$ be an enumeration of all pairs of natural numbers.

We aim at applying Lemma 3.6.12 repeatedly. In the first step of the construction we find a compact space $Y_1$ and an open continuous surjection $f_0 \colon Y_1 \to X$ such that the collection of pairs

(**) $$\{(f_0^{-1}[E_{n_1}], f_0^{-1}[F_{n_1}]), (f_0^{-1}[E_{m_1}], f_0^{-1}[F_{m_1}])\}$$

is inessential. Then we again apply Lemma 3.6.12 but now with $Y_1$ and the collection of pairs

$$\{(f_0^{-1}[E_{n_2}], f_0^{-1}[F_{n_2}]), (f_0^{-1}[E_{m_2}], f_0^{-1}[F_{m_2}])\},$$

obtaining an open surjection $f_1\colon Y_2 \to Y_1$ such that the collection of pairs
$$\{(f_1^{-1}[f_0^{-1}[E_{n_2}]], f_1^{-1}[f_0^{-1}[F_{n_2}]]), (f_1^{-1}[f_0^{-1}[E_{m_2}]], f_1^{-1}[f_0^{-1}[F_{m_2}]])\},$$
is inessential. Observe that by (∗∗) and Exercise 3.1.12 it follows that the collection of pairs
$$\{(f_1^{-1}[f_0^{-1}[E_{n_1}]], f_1^{-1}[f_0^{-1}[F_{n_1}]]), (f_1^{-1}[f_0^{-1}[E_{m_1}]], f_1^{-1}[f_0^{-1}[F_{m_1}]])\},$$
is inessential as well.

So now it is clear how to proceed. In an infinite process we can deal with all pairs in the collection $\mathcal{E}$, thus obtaining an inverse sequence
$$X \xleftarrow{f_0} Y_1 \xleftarrow{f_1} Y_2 \leftarrow \cdots \leftarrow Y_n \xleftarrow{f_n} Y_{n+1} \leftarrow \cdots$$
with continuous, surjective, open bonding mappings. If $Y_\infty$ denotes its inverse limit, then the projection $f_\infty\colon Y_\infty \to X$ is an open continuous surjection by Exercise 1.10.5. In addition, if $A, B, E$ and $F$ are closed subsets of $X$ such that $A \cap B = \emptyset$ and $E \cap F = \emptyset$ then there exist $n, m \in \mathbb{N}$ such that
$$A \subseteq E_n, \quad B \subseteq F_n, \quad E \subseteq E_m, \quad F \subseteq F_m.$$
Since by construction the pairs
$$\{(f_\infty^{-1}[E_n], f_\infty^{-1}[F_n]), (f_\infty^{-1}[E_m], f_\infty^{-1}[F_m])\}$$
are inessential, the same holds for the pairs
$$\{(f_\infty^{-1}[A], f_\infty^{-1}[B]), (f_\infty^{-1}[E], f_\infty^{-1}[F])\}$$
So the space $Y_\infty$ and the map $f_\infty$ are as required. □

This leads us to the following interesting result.

**Theorem 3.6.14.** *Let $X$ be a compactum. Then there exists an at most one-dimensional compactum $Y$ which can be mapped onto $X$ by an open map.*

**Proof.** By applying Corollary 3.6.13 we obtain an inverse sequence
$$Y_0 = X \xleftarrow{f_0} Y_1 \xleftarrow{f_1} Y_2 \leftarrow \cdots \leftarrow Y_n \xleftarrow{f_n} \cdots$$
of compact spaces, with continuous, open and surjective bonding mappings, having the following additional property: if for some $n \geq 0$, $A, B, E$ and $F$ are closed sets in $Y_n$ such that $A \cap B = \emptyset$ and $E \cap F = \emptyset$ then the collection of pairs
$$\{(f_n^{-1}[A], f_n^{-1}[B]), (f_n^{-1}[E], f_n^{-1}[F])\}$$
is inessential in $Y_{n+1}$. Let $Y_\infty$ denote the inverse limit of the sequence. By similar arguments as the ones in the proof of Corollary 3.6.13 it follows that for every $n \geq 0$ the canonical projection $f_\infty^n$ from $Y_\infty$ onto $Y_n$ is an open surjection. So we are done if we can prove that $\dim Y_\infty \leq 1$. But this follows by a compactness argument. For let $A, B, E$ and $F$ be closed subsets of $Y_\infty$ such that $A \cap B = \emptyset$ and $E \cap F = \emptyset$. By Exercise 1.10.12 it follows that

there exists $n \in \mathbb{N}$ such that $f^n_\infty[A] \cap f^n_\infty[B] = \emptyset$ and $f^n_\infty[E] \cap f^n_\infty[F] = \emptyset$. By construction we have that the pairs

$$\{(f_n^{-1}[f^n_\infty[A]], f_n^{-1}[f^n_\infty[B]]), (f_n^{-1}[f^n_\infty[E]], f_n^{-1}[f^n_\infty[F]])\}$$

are inessential. So now we are done by using Exercise 3.1.11 and by observing that $f^{n+1}_\infty[A] \subseteq f_n^{-1}[f^n_\infty[A]]$, etc. □

**Exercises for §3.6.**

▶1. Let $n \geq 0$ and let $X$ be a compact space which can be written as the union of closed sets $F_i$, $i \in \mathbb{N}$, such that for all distinct $i, j \in \mathbb{N}$,
$$\dim(F_i \cap F_j) < n.$$
Prove that every continuous function from $F_1$ into $\mathbb{S}^n$ is continuously extendable over $X$.

2. Observe that the previous exercise is a generalization of Theorem A.10.6.

3. Let $X$ be a compact space and let $H$ be the union of all components of $X$ of dimension at most $n$. Prove that $\dim H \leq n$.

▶4. Let $A$ be a closed subspace of a space $X$ such that $\dim(X \setminus A) \leq n$, let $0 \leq k \leq n$, and let $f\colon A \to \mathbb{S}^k$ be continuous. Prove that there exists a closed subspace $B$ of $X$ such that $A \cap B = \emptyset$ and $\dim B \leq n - k - 1$, while moreover the function $f$ can be extended over $X \setminus B$.

5. Let $n \geq 2$ and consider the $n$-dimensional cube $\mathbb{I}^n$. Define
$$E^{n-1} = \{(x_1, \ldots, x_{n-1}, 1) : 0 < x_i < 1, i = 1, \ldots, n-1\}.$$
Prove that if $X = \mathbb{I}^n \setminus E^{n-1}$ then $\operatorname{def} X = n - 1$.

6. Let $X$ be an at most $(n-1)$-dimensional space. Prove that every continuous function $f\colon X \to \mathbb{S}^n$ is nullhomotopic.

7. Prove that $\mathbb{S}^n$ is unicoherent for every $n \geq 2$.

## 3.7. Dimension of subsets of $\mathbb{R}^n$ and certain generalizations

In this section we prove some interesting dimensional properties of subsets of the Euclidean spaces $\mathbb{R}^n$, $n \in \mathbb{N}$.

**Theorem 3.7.1.** *Let $X \subseteq \mathbb{R}^n$. Then*
$$\dim X = n \iff \operatorname{Int} X \neq \emptyset.$$

**Proof.** First, if $\operatorname{Int} X \neq \emptyset$ then $X$ contains a homeomorph of $\mathbb{I}^n$ and is therefore $n$-dimensional by Theorems 3.2.12 and Corollary 3.4.14.

Next, assume that $\operatorname{Int} X = \emptyset$. There consequently exists a countable set $D \subseteq \mathbb{R}^n \setminus X$ which is dense in $\mathbb{R}^n$.

## 3.7. DIMENSION OF SUBSETS OF $\mathbb{R}^n$ AND CERTAIN GENERALIZATIONS

By Proposition 3.2.10, $\mathbb{R}^n = \mathfrak{R}_{n,0} \cup \mathfrak{R}_{n,1} \cup \cdots \cup \mathfrak{R}_{n,n}$. Put

$$Y = \mathbb{R}^n \setminus \mathfrak{R}_{n,n} = \bigcup_{i=0}^{n-1} \mathfrak{R}_{n,i}.$$

Notice that $\mathfrak{R}_{n,n}$ consists of all points of $\mathbb{R}^n$ having rational coordinates only, and hence is a countable dense set. By Proposition 3.2.10 and Corollary 3.1.7 we get that $\dim Y \leq n-1$.

By the countable dense homogeneity of $\mathbb{R}^n$ (Corollary 1.6.10), there is an element $h \in \mathcal{H}(\mathbb{R}^n)$ such that $h[D] = \mathfrak{R}_{n,n}$. This implies that $h[X] \subseteq Y$, and hence by Theorem 3.2.9 that $\dim X = \dim h[X] \leq n-1$, as required. □

**Corollary 3.7.2.** *Let $U \subseteq \mathbb{R}^n$ be nonempty and open. If $U$ is not dense then $\dim \operatorname{Fr} U = n-1$.*

**Proof.** Put $F = \operatorname{Fr} U$. Then by Theorem 3.7.1 it follows that $\dim F \leq n-1$. Pick an arbitrary point $m \in \mathbb{R}^n \setminus \overline{U}$.

Striving for a contradiction, assume that $\dim F \leq n-2$.

We will first prove that we may assume without loss of generality that $\overline{U}$ is compact. For this we will use a trick on 'exchanging' points at infinity. The one-point compactification $X = \mathbb{R}^n \cup \{\infty\}$ is homeomorphic to $\mathbb{S}^n$ (Exercise A.4.8). The boundary of $U$ in $X$ is $F$ if it is compact and $F \cup \{\infty\}$ otherwise. Since $\dim F = \dim(F \cup \{\infty\})$ (Exercise 3.2.1) we arrive at the conclusion that the boundary of $U$ in $X$ is at most $(n-2)$-dimensional. Now observe that $Y = X \setminus \{m\}$ is homeomorphic to $\mathbb{R}^n$, that the closure of $U$ in $Y$ is compact, and that its boundary is at most $(n-2)$-dimensional.

So from now on assume that $\overline{U}$ is compact in $\mathbb{R}^n$ and that $\operatorname{Fr} U$ is at most $(n-2)$-dimensional. We may assume without loss of generality that the origin $\underline{0}$ of $\mathbb{R}^n$ belongs to $U$. By compactness of $\overline{U}$, the collection

$$\{r \cdot \overline{U} : r \in (0,1]\}$$

is a local base at $\underline{0}$ consisting of compact neighborhoods all of whose boundaries are at most $(n-2)$-dimensional. But since $\mathbb{R}^n$ is homogeneous, the same is true for all points of $\mathbb{R}^n$. By Theorem 3.4.10 (applied twice) this implies that $\dim \mathbb{R}^n \leq n-1$, which contradicts Theorem 3.2.12. □

**Corollary 3.7.3.** *Let $L$ be a closed subset of $\mathbb{R}^n$ with $\dim L \leq n-2$. Then $L$ does not separate $\mathbb{R}^n$.*

**Proof.** Suppose that $\mathbb{R}^n \setminus L = U \cup V$, where $U$ and $V$ are disjoint nonempty open sets. Then $\operatorname{Fr} U \subseteq L$ and so $\dim \operatorname{Fr} L \leq n-2$ which contradicts Corollary 3.7.2. □

We now aim at proving in Theorem 3.7.6 a much stronger version of this corollary. First we need to do derive some elementary results about essential families and continua.

**Proposition 3.7.4.** *Let $X$ be a compact space, let $\{(A_i, B_i) : i \in \Gamma\}$ be an essential family of pairs of disjoint closed subsets of $X$ and let $\Gamma(0) \subseteq \Gamma$. If for each $i \in \Gamma(0)$, $L_i$ is a partition between $A_i$ and $B_i$ in $X$ and $n \in \Gamma \setminus \Gamma(0)$ then $L = \bigcap_{i \in \Gamma(0)} L_i$ contains a continuum from $A_n$ to $B_n$.*

**Proof.** Suppose that $L$ does not contain any continuum from $A_n$ to $B_n$. Let
$$H = A_n \cap L, \quad G = B_n \cap L,$$
and put
$$\mathcal{E} = \{C : C \text{ is a component of } L \text{ and } C \cap H \neq \emptyset\}, \quad E = \bigcup \mathcal{E},$$
and
$$\mathcal{F} = \{C : C \text{ is a component of } L \text{ and } C \cap G \neq \emptyset\}, \quad F = \bigcup \mathcal{F},$$
respectively. From Proposition A.10.3 it follows that both $E$ and $F$ are closed. Also, by assumption, $E \cap F = \emptyset$. Take an arbitrary $C \in \mathcal{E}$. Since $L \setminus F$ is an open neighborhood of $C$ in $L$, by Lemma A.10.1 there exists a clopen (in $L$) set $U_C$ which contains $C$ but misses $F$. By compactness, finitely many $U_C$'s cover $E$. We conclude that $\emptyset$ is a partition in $L$ between $H$ and $G$. By Corollary 3.1.5 there is a partition $T$ in $X$ between $A_n$ and $B_n$ such that $T \cap L = \emptyset$. But this contradicts the fact that
$$\{(A_i, B_i) : i \in \Gamma(0)\} \cup \{(A_n, B_n)\}$$
is essential (Exercise 3.1.1). □

**Corollary 3.7.5.** *Let $X$ be a compact space, let*
$$\{(A_i, B_i) : i \in \Gamma\}$$
*be an essential family of pairs of disjoint closed subsets of $X$ and let $n \in \Gamma$. Suppose that $Y \subseteq X$ is such that $Y$ meets every continuum from $A_n$ to $B_n$. For each $i \in \Gamma \setminus \{n\}$ let $U_i$ and $V_i$ be disjoint closed neighborhoods of $A_i$ and $B_i$, respectively. Then*
$$\{(U_i \cap Y, V_i \cap Y) : i \in \Gamma \setminus \{n\}\}$$
*is essential in $Y$.*

**Proof.** By Proposition 3.7.4, if for every $i \in \Gamma \setminus \{n\}$ the set $L_i$ is a partition between $A_i$ and $B_i$ then $Y \cap \bigcap_{i \in \Gamma \setminus \{n\}} L_i \neq \emptyset$. Now apply Exercise 3.1.13. □

A space $X$ is called *continuum-connected* provided that for any pair of points $x, y \in X$ there is a subcontinuum $C$ of $X$ containing both $x$ and $y$. A continuum-connected space is obviously connected, but the converse need not be true (Exercise 3.7.1).

**Theorem 3.7.6.** *Let $M \subseteq G$, where $G \subseteq \mathbb{R}^n$ is open and connected. If $M$ is at most $(n-2)$-dimensional then $G \setminus M$ is continuum-connected.*

**Proof.** We first prove the special case that $G = \mathbb{R}^n$. Take two arbitrary distinct points $x, y \in \mathbb{R}^n \setminus M$ and let $K$ denote the closed ball in $\mathbb{R}^n$ with center $\frac{1}{2}(x+y)$ and radius $\frac{1}{2}\|x-y\|$. Let $f \colon \mathbb{I}^n \to K$ be a continuous surjection which maps a pair of opposite faces $A$ and $B$ of $\mathbb{I}^n$ to $x$ and $y$, respectively, and has the property that the restriction $g = f \restriction \mathbb{I}^n \setminus (A \cup B)$ is a homeomorphism from $\mathbb{I}^n \setminus (A \cup B)$ onto $K \setminus \{x, y\}$. Such a function can easily be constructed with the technique used in the solution of Exercise 1.1.24. The set $f^{-1}[M \cap K]$ is at most $(n-2)$-dimensional. Hence by Corollary 3.7.5 there is a continuum $C$ from $A$ to $B$ which misses $f^{-1}[M \cap K]$. But then $f[C]$ is a continuum containing $x$ and $y$ but missing $M$.

Now consider an arbitrary open and connected set $G \subseteq \mathbb{R}^n$ and arbitrary points $x, y \in G \setminus M$. By Lemma 1.5.21 there exist open balls $B_1, B_2, \ldots, B_k$ in $G$ such that $x \in B_1$, $y \in B_k$ and $B_i \cap B_{i+1} \neq \emptyset$ for all $i \leq k-1$ (here we use that $G$ is open and connected). Since $M$ has empty interior by Theorem 3.7.1 there exists for $i \leq k-1$ a point $z_i \in (B_i \cap B_{i+1}) \setminus M$. Put $z_0 = x$ and $z_k = y$. By the above for $i \leq k$ there is a continuum $C_i \subseteq B_i \setminus M$ containing both $z_{i-1}$ and $z_i$. Then $C = \bigcup_{i=1}^{k} C_i$ is the desired continuum in $G \setminus M$ connecting $x$ and $y$. □

It is easy to see that Corollary 3.7.3 follows from Theorem 3.7.6.

Corollary 3.7.3 leads to the definition of a so-called Cantor-manifold. A compact space $X$ with $\dim X = n \geq 1$ is called an *n-dimensional Cantor-manifold* if for each closed subset $L \subseteq X$ with

$$\dim L \leq n - 2$$

the complement $X \setminus L$ is connected. The examples of $n$-dimensional totally disconnected spaces for all $n$ that will be constructed in §3.9 show that a generalization of this concept to noncompact spaces makes no sense; simply observe that the singletons are the only connected subsets of a totally disconnected space. Observe that each Cantor-manifold is a continuum and that the one-dimensional Cantor-manifolds are precisely the one-dimensional continua.

**Proposition 3.7.7.** *Every $n$-dimensional compactification of $\mathbb{R}^n$ is an $n$-dimensional Cantor-manifold. In particular, $\mathbb{S}^n$ and $\mathbb{I}^n$ are $n$-dimensional Cantor-manifolds.*

**Proof.** Let $\gamma \mathbb{R}^n$ be an $n$-dimensional compactification of $\mathbb{R}^n$, and let $A$ in $\gamma X$ be at most $(n-2)$-dimensional. Then $B = \mathbb{R}^n \setminus A$ is connected by Corollary 3.7.3. In addition, $A \cap \mathbb{R}^n$ has empty interior in $\mathbb{R}^n$ by Theorem 3.7.1

which implies that $B$ is dense. We conclude that $\gamma \mathbb{R}^n \setminus A$ contains the dense connected set $B$ and is therefore connected itself. □

We will now show that there are 'many' $n$-dimensional Cantor-manifolds.

**Theorem 3.7.8.** *Let $n \geq 1$. Every compact $n$-dimensional space $X$ contains an $n$-dimensional Cantor-manifold.*

**Proof.** By Theorem 3.6.5 there exists a closed subset $A \subseteq X$ and a continuous function $f\colon A \to \mathbb{S}^{n-1}$ such that $f$ admits no continuous extension over $X$. Let $\mathcal{B} = \{B_i : i \in \mathbb{N}\}$ be a countable open base for $X$.

We define a decreasing sequence $(X_i)_{i \geq 0}$ of closed subsets of $X$, as follows. First, let $X_0 = X$. If $X_i$ has been defined then we consider two cases. If $f$ does not admit a continuous extension over $A \cup (X_i \setminus B_{i+1})$ then put
$$X_{i+1} = X_i \setminus B_{i+1},$$
otherwise put
$$X_{i+1} = X_i.$$
We claim that $M = \bigcap_{i=0}^{\infty} X_i$ is an $n$-dimensional Cantor-manifold.

Observe that by construction, for no $i$ the function $f$ admits a continuous extension over $A \cup X_i$.

**Claim 1.** *$f$ does not admits a continuous extension over $A \cup M$.*

*Proof.* Striving for a contradiction, assume that $\bar{f}\colon A \cup M \to \mathbb{S}^{n-1}$ is a continuous extension of $f$. By Corollary 1.2.13 it follows that there is an open neighborhood $U$ of $M$ such that $\bar{f}$ can be extended to a continuous function $\tilde{f}\colon A \cup U \to \mathbb{S}^{n-1}$. Pick $N$ so large that $M \subseteq X_N \subseteq U$ (here we use that $X$ is compact, see Exercise A.5.17). Then $\tilde{f} \restriction (A \cup X_N)$ extends $f$ which is impossible by construction. ◇

So by Theorem 3.6.3 it follows that $\dim M \geq n$. Since $\dim M \leq n$ being a subspace of $X$ we conclude that $\dim M = n$.

To complete the proof it clearly suffices to prove that if $M_1$ and $M_2$ are proper closed subsets of $M$ with $M = M_1 \cup M_2$ then $\dim(M_1 \cap M_2) \geq n-1$. Striving for a contradiction, assume that there exist proper closed subsets $M_1$ and $M_2$ of $M$ with
$$M = M_1 \cup M_2, \quad \dim(M_1 \cap M_2) \leq n-2.$$
Put $A_1 = A \cup M_1$ and $A_2 = A \cup M_2$, respectively.

**Claim 2.** *$f$ admits continuous extensions over $A_1$ and $A_2$, say to $f_1$ and $f_2$, respectively.*

*Proof.* It suffices to prove this for $A_1$. To this end, pick $x \in M_2 \setminus M_1$ (here we use the fact that $M_1$ and $M_2$ are proper subsets of $M$). Pick $i \in \mathbb{N}$ so large that $x \in B_{i+1} \subseteq X \setminus M_1$. Consider the set $X_{i+1}$. It contains $x$ and hence it intersects $B_{i+1}$. By construction this can only happen when $f$ can be extended continuously over $A \cup (X_i \setminus B_{i+1})$. Since $A_1 \subseteq A \cup (X_i \setminus B_{i+1})$, we are done. $\diamondsuit$

Put $B = A \cup (M_1 \cap M_2)$. The functions $f_1 \restriction B$ and $f_2 \restriction B$ only differ in a subset of $M_1 \cap M_2$, and hence on a set of dimension $n - 2$ at most. We conclude that $f_1 \restriction B$ and $f_2 \restriction B$ are homotopic by Corollary 3.6.4. Since $f_2 \restriction B$ admits a continuous extension over $A \cup M_2$, the Borsuk Homotopy Extension Theorem 1.4.2 implies that the same is true for $f_1 \restriction B$, say to the function $f_1' \colon A \cup M_2 \to \mathbb{S}^{n-1}$. Since
$$B = (A \cup M_1) \cap (A \cup M_2),$$
it follows that the formula
$$\xi(x) = \begin{cases} f_1(x) & (x \in A \cup M_1), \\ f_1'(x) & (x \in A \cup M_2), \end{cases}$$
defines a continuous extension of $f$ over $A \cup M$. This is a contradiction. $\square$

This yields another solution to Exercise 3.2.4.

**Corollary 3.7.9.** *If $X$ is compact and $\dim X = n$ then one of its components is $n$-dimensional.*

**Exercises for §3.7.**

1. Give an example of a connected space which is not continuum-connected.

2. Give an example of a space $X$ having the property that for every component $C \subseteq X$ we have $\dim C < \dim X$.

3. Let $X$ be compact and let $A$ and $B$ be disjoint closed subsets of $X$. In addition, let $L$ be a closed subset of $X$ such that no continuum in $L$ intersects both $A$ and $B$. Prove that there is an open neighborhood $U$ of $L$ such that no continuum in $\overline{U}$ intersects both $A$ and $B$.

▶4. Let $X$ be a compact space, let $\{(A_i, B_i) : i \in \Gamma\}$ be an essential family of pairs of disjoint closed subsets of $X$ and let $\Gamma(0) \subseteq \Gamma$. Suppose that for each $i \in \Gamma(0)$, $L_i$ is a partition between $A_i$ and $B_i$ in $X$ and let
$$L = \bigcap_{i \in \Gamma(0)} L_i.$$
Finally, let $n_1, n_2 \in \Gamma \setminus \Gamma(0)$ be distinct.

Call a subset $T \subseteq X$ *small* if no continuum in $T$ meets both $A_{n_1}$ and $B_{n_1}$.

Prove that if $\mathcal{N}$ is a finite pairwise disjoint collection of small closed subsets of $X$ then there is a continuum in

$$L \setminus \bigcup \mathcal{N}$$

from $A_{n_2}$ to $B_{n_2}$.

5. Let $X$ be a compact space such that $\dim X \geq 2$. Prove that there are real numbers $\delta > 0$ and $\varepsilon > 0$ such that if $\mathcal{N}$ is any finite collection of pairwise disjoint closed sets with $\operatorname{mesh}(\mathcal{N}) < \delta$ then there is a continuum $C \subseteq X \setminus \bigcup \mathcal{N}$ with diameter at least $\varepsilon$.

▶6. Let $X$ be a compact space with $\dim X \leq 1$. Prove that for every real number $\varepsilon > 0$ there is a family $\mathcal{N}$ of pairwise disjoint closed subsets of $X$ with $\operatorname{mesh}(\mathcal{N}) < \varepsilon$ such that each continuum $C \subseteq X \setminus \bigcup \mathcal{N}$ has diameter less than $\varepsilon$.

## 3.8. Higher-dimensional hereditarily indecomposable continua

Indecomposable continua are strange objects. We proved their existence in §1.10. In this section we go one step further, and prove the existence of 'many' hereditarily indecomposable continua of arbitrary large dimension. Higher-dimensional hereditarily indecomposable continua were first constructed in 1951. But only the last decade it was realized that they are fundamental objects that can be used for the solution of various open problems in dimension theory. See the notes for more information.

The following result is the heart of the construction of hereditarily indecomposable continua.

**Theorem 3.8.1.** *Let $X$ be a compactum and let $F_0, F_1$ be disjoint closed sets in $X$. Then there are disjoint open neighborhoods $W_0$ and $W_1$ of $F_0$ and $F_1$, respectively, such that any two intersecting continua $K_0, K_1$ in*

$$X \setminus (W_0 \cup W_1)$$

*are comparable, i.e., $K_0 \subseteq K_1$ or $K_1 \subseteq K_0$.*

**Remark 3.8.2.** A compact space $X$ is called *Bing* provided that for all subcontinua $A$ and $B$ of $X$ we have $A \cap B = \emptyset$ or $A \subseteq B$ or $B \subseteq A$. So a zero-dimensional compact space is Bing. Theorem 3.8.1 shows that disjoint closed sets in a compact space can always be partitioned by a Bing compactum.

**Proof.** Let $\varrho$ be an admissible metric on $X$. The Euclidean metric on $\mathbb{I}$ will be denoted by $d$.

Let $P$ be the union of the five open rectangles in $\mathbb{I}^2$, described in Figure 9. The set $\mathbb{I}^2 \setminus \overline{P}$ is the union of the disjoint open sets $M_0$ and $M_1$ in the picture. Clearly, $\overline{M_0} \cap \overline{M_1} = \emptyset$. Observe that

(1) $\qquad\qquad \{0\} \times \mathbb{I} \subseteq M_0, \quad \{1\} \times \mathbb{I} \subseteq M_1.$

## 3.8. HIGHER-DIMENSIONAL HEREDITARILY INDECOMPOSABLE CONTINUA

Hence $\overline{P}$ is a partition between $\{0\} \times \mathbb{I}$ and $\{1\} \times \mathbb{I}$ in $\mathbb{I}^2$. Let $W_{0,0}$ and $W_{1,0}$ be open neighborhoods of $F_0$ and $F_1$, respectively, such that

(2) $$\overline{W}_{0,0} \cap \overline{W}_{1,0} = \emptyset.$$

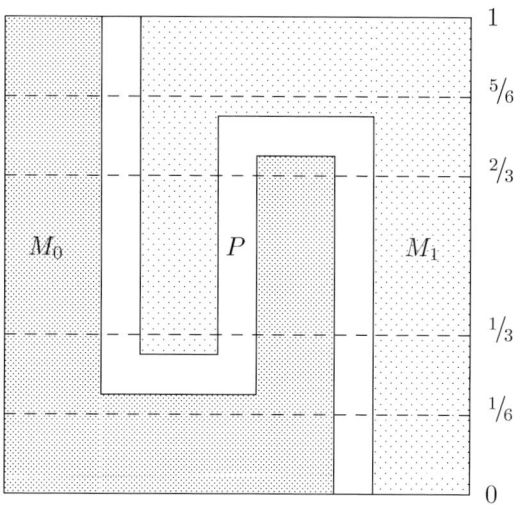

Figure 9.

In addition, let $F = \{f_i : i \in \mathbb{N}\}$ be a dense subset of $C(X, \mathbb{I})$ (Proposition 1.3.3). Recursively, determine for every $i \geq 1$ open sets $W_{0,i}$ and $W_{1,i}$ in $X$ and a continuous function $u_i \colon X \to \mathbb{I}^2$, by the following formulas:

(3) $$u_i(x) = \left( \frac{\varrho(x, W_{0,i-1})}{\varrho(x, W_{0,i-1}) + \varrho(x, W_{1,i-1})}, f_i(x) \right),$$

(4) $$W_{0,i} = u_i^{-1}[M_0], \quad W_{1,i} = u_i^{-1}[M_1].$$

Since $\overline{W}_{0,0} \cap \overline{W}_{0,1} = \emptyset$, it follows that $u_1$ is well-defined and continuous. Notice that by (1) and the fact that the first component of $u_1$ is an appropriate Urysohn function, we have

$$u_1[W_{0,0}] \subseteq \{0\} \times \mathbb{I} \subseteq M_0, \quad u_1[W_{1,0}] \subseteq \{1\} \times \mathbb{I} \subseteq M_1.$$

By (4) and the fact that $\overline{M}_0 \cap \overline{M}_1 = \emptyset$, we consequently obtain that

$$\overline{W}_{0,0} \subseteq W_{0,1}, \quad \overline{W}_{1,0} \subseteq W_{1,1}, \quad \overline{W}_{0,1} \cap \overline{W}_{1,1} = \emptyset.$$

By induction on $i$ this reasoning can easily be extended. So

$$W_0 = \bigcup_{i=0}^{\infty} W_{0,i}, \quad W_1 = \bigcup_{i=0}^{\infty} W_{1,i},$$

are disjoint open neighborhoods of $F_0$ and $F_1$, respectively. We claim that these neighborhoods are as desired.

Let $K_0$ and $K_1$ be arbitrary intersecting continua of $X$ such that
$$K_0 \cup K_1 \subseteq X \setminus (W_0 \cup W_1).$$
Striving for a contradiction, assume that there exist points
$$a_0 \in K_0 \setminus K_1, \quad a_1 \in K_1 \setminus K_0.$$
There is by Lemma A.4.2 a Urysohn function $u \colon X \to \mathbb{I}$ such that
$$u(a_0) = 0, \quad u(a_1) = 1, \quad u[K_0] \subseteq [0, 1/2], \quad u[K_1] \subseteq [1/2, 1].$$
Since $F$ is dense in $C(X, \mathbb{I})$, there exists $i$ such that $\hat{d}(u, f_i) < 1/6$. So we have

(5) $\quad f_i(a_0) \in [0, 1/6), \ f_i(a_1) \in (5/6, 1], \ f_i[K_0] \subseteq [0, 2/3], \ f_i[K_1] \subseteq [1/3, 1],$

hence

(6) $\quad\quad\quad K_0 \cap f_i^{-1}(2/3, 1] = \emptyset, \quad K_1 \cap f_i^{-1}[0, 1/3) = \emptyset.$

This shows that the continua
$$T_0 = u_i[K_0], \quad T_1 = u_i[K_1],$$
have the property that

(7) $\quad\quad\quad T_0 \subseteq \mathbb{I} \times [0, 2/3], \quad T_0 \subseteq \mathbb{I} \times [1/3, 1].$

Since $K_0$ and $K_1$ are disjoint from $W_{0,i} \cup W_{1,i}$, the continua $T_0$ and $T_1$ are contained in $\overline{P}$, cf. (4). By (5), $T_0$ intersects the strip $\mathbb{I} \times [0, 1/6)$ and, similarly, $T_1$ intersects $\mathbb{I} \times (5/6, 1]$. Since $T_0$ and $T_1$ are continua, the geometry of $\overline{P}$ and (7) imply that they are disjoint. But this is a contradiction since $K_0$ and $K_1$ intersect, hence $T_0 \cap T_1 \neq \emptyset$. $\square$

**Corollary 3.8.3.** *Every $(n+1)$-dimensional compactum $X$ contains an $n$-dimensional hereditarily indecomposable subcontinuum.*

**Proof.** Let $X$ be an $(n+1)$-dimensional compact space with essential family $\{(A_i, B_i) : i \leq n+1\}$. An arbitrary partition $T$ between $A_1$ and $B_1$ must be at least $n$-dimensional by Theorem 3.1.9. Theorem 3.8.1 shows that for $T$ we may take a closed set having the property that each of its components is hereditarily indecomposable (Exercise 1.10.21). One of those components, say $C$, must have the same dimension as $T$ by Exercise 3.2.4 or Corollary 3.7.9. So $C$ is a hereditarily indecomposable continuum of dimension $n$ or $n+1$. But every $(n+1)$-dimensional compactum contains an $n$-dimensional compactum by Exercise 3.3.7, and hence an $n$-dimensional continuum by Exercise 3.2.4. So $X$ contains an $n$-dimensional hereditarily indecomposable continuum. $\square$

**Corollary 3.8.4.** *Every strongly infinite-dimensional compact space contains a strongly infinite-dimensional hereditarily indecomposable continuum.*

**Proof.** If $X$ is a strongly infinite-dimensional compactum with essential family $\{(A_i, B_i) : i \in \mathbb{N}\}$ then an arbitrary partition between $A_1$ and $B_1$ must be strongly infinite-dimensional by Theorem 3.1.9. By using Exercise 3.1.9 one of the components of such a partition $T$ must be strongly infinite-dimensional as well. We can therefore use the same method as in the proof of Corollary 3.8.3. □

**Remark 3.8.5.** In view of Corollary 3.8.3 and its proof, the question naturally arises whether there exists for $n \geq 2$ a hereditarily indecomposable continuum separating the top from the bottom in $\mathbb{I}^{n+1}$. The answer to this question is in the affirmative. Simply observe that $\mathbb{I}^{n+1}$ is locally connected and by contractibility is unicoherent (Corollary A.12.10). As a consequence, the Bing partition between the top and the bottom of $\mathbb{I}^3$ considered in the proof of Corollary 3.8.3 contains a connected partition by Exercises 3.10.2 and 3.10.3. This partition is clearly a hereditarily indecomposable continuum.

**Henderson compacta.** A compactum $X$ is called *Henderson* if it is infinite-dimensional but every finite dimensional closed subspace is zero-dimensional. It is not clear at all that such a compactum exists. By using Corollary 3.8.4 we will present a surprisingly simple construction of such a space (for a stronger result, see Theorem 3.13.10).

Let $X$ be a compact space. For $t > 0$ and $n \geq 0$ put
$$X_{n,t} = \bigcup \{C \subseteq X : C \text{ is a continuum with } \dim C \leq n \text{ and } \operatorname{diam} C \geq t\}.$$

**Lemma 3.8.6.** *Let $X$ be a Bing compactum. Then for every $t > 0$ and $n \geq 0$ we have $\dim X_{n,t} \leq n$.*

**Proof.** Let $F$ be an arbitrary closed subset of $X$ with $\operatorname{diam} F < t$.

**Claim 1.** *The components of $F$ of dimension at most $n$ cover $X_{n,t} \cap F$. As a consequence,*
$$\dim(X_{n,t} \cap F) \leq n.$$

*Proof.* Indeed, let $x \in X_{n,t} \cap F$. There is a continuum $A$ contained in $X_{n,t}$ with $\dim A \leq n$ and $\operatorname{diam} A \geq t$ such that $x \in A$. Let $B$ be the component of $x$ in $F$. Then $\operatorname{diam} B < t$ and $B \cap A \neq \emptyset$. So $A \not\subseteq B$ and hence since $X$ is Bing it follows that $B \subseteq A$. By the Subspace Theorem 3.2.9 we conclude that $\dim B \leq n$. Hence $x$ is contained in an at most $n$-dimensional component of $F$.

Let $H$ be the union of all components of $F$ with dimension at most $n$. Then $\dim H \leq n$ by Exercise 3.6.3. We just observed that $H \supseteq X_{n,t} \cap F$. So we are done by the Subspace Theorem 3.2.9. ◇

Since $X$ is compact, there is a finite closed cover $\mathcal{F}$ of $X$ with mesh less than $t$. By the claim, $X_{n,t} \cap F$ is at most $n$-dimensional for every $F$. Hence $\dim X_{n,t} \leq n$ by the Countable Closed Sum Theorem 3.2.8. □

This leads us to the following unexpected result.

**Theorem 3.8.7.** *Let $X$ be a Bing compactum. The union of all nondegenerate finite dimensional subcontinua of $X$ is countable dimensional.*

**Proof.** Every nondegenerate subcontinuum of $X$ has positive diameter. So the set we are interested in is a subset of

$$\bigcup_{n=1}^{\infty} X_{n,1/n}.$$

This means that we are done by Lemma 3.8.6, Corollary 3.3.9 and Theorem 3.2.9. □

This is all we need for the construction of a Henderson compactum.

**Corollary 3.8.8.** *Every strongly infinite-dimensional compactum contains a strongly infinite-dimensional Henderson compactum.*

**Proof.** By Corollary 3.8.4 we may assume without loss of generality that $X$ is a hereditarily indecomposable continuum. Let

$$\{(A_i, B_i) : i \in \mathbb{N}\}$$

be essential in $X$. By Theorem 3.8.7 there is a countable family

$$\{N_{2i} : i \in \mathbb{N}\}$$

of zero-dimensional subspace of $X$ covering all finite dimensional nondegenerate subcontinua of $X$. For every $i \in \mathbb{N}$ let $T_i$ be a partition between the sets $A_{2i}$ and $B_{2i}$ such that $T_i \cap N_{2i} = \emptyset$. By Theorem 3.1.9 it follows that the compactum

$$T = \bigcap_{i=1}^{\infty} T_i$$

is strongly infinite-dimensional. In addition, $T$ contains no nondegenerate finite dimensional subcontinua. For such a subcontinuum is contained in $\bigcup_{i=1}^{\infty} N_{2i}$ and hence misses $T$.

We claim that $T$ is Henderson. For if $A$ is a finite-dimensional closed subspace of $T$ which is not zero-dimensional, then one of its components is not zero-dimensional (Exercise 3.2.4) and hence by construction misses $T$. So this is impossible. □

## 3.8. HIGHER-DIMENSIONAL HEREDITARILY INDECOMPOSABLE CONTINUA

**Homogeneity.** As we observed in §1.10, the dyadic solenoid $\Sigma_2$ is a topological group and hence is homogeneous. So there is a homogeneous indecomposable continuum. Is there an example of a homogeneous *hereditarily* indecomposable continuum? Surprisingly, there is such a space which is even planar. It is the so-called *pseudo-arc* which can be constructed by a slightly more complicated procedure than the one used for the indecomposable continuum $P$ in Figure 5 on Page 95. So the construction of the pseudo-arc is not difficult. The verification that it is hereditarily indecomposable is not difficult as well. But the proof of its homogeneity is not simple. This was established by BING in [55].

The pseudo-arc is clearly one-dimensional. So in view of the existence of higher-dimensional hereditarily indecomposable continua (Corollary 3.8.3), the question arises whether such a space can be homogeneous. The answer to this question is in the negative.

**Proposition 3.8.9.** *If $X$ is a Bing compactum such that $2 \leq \dim X < \infty$ then $X$ is not homogeneous.*

**Proof.** Let $\dim X = n$. Let $\mathcal{U}$ be a finite open cover of $X$ of mesh less than 1. Since every element of $\mathcal{U}$ is an $F_\sigma$-subset of $X$ (Exercise A.2.3), the Countable closed Sum Theorem 3.2.8 shows that there are an element $U \in \mathcal{U}$ and a compact subset $F \subseteq U$ such that $\dim F = n$. Corollary 3.7.9 shows that we may assume without loss of generality that $F$ is a continuum. So $X$ contains an $n$-dimensional continuum of diameter less than 1.

Observe that besides compactness, we did not use any of the properties of $X$. So by repeating this reasoning, we can construct by induction a decreasing sequence $(F_i)_i$ of $n$-dimensional subcontinua of $X$ such that $\operatorname{diam} F_i < 1/i$ for every $i$.

By compactness, the intersection $\bigcap_{i=1}^{\infty} F_i \neq \emptyset$. In fact, it is a single point since $\operatorname{diam} F_i < 1/i$ for every $i$. Let $x$ be the unique point in that intersection.

**Claim 1.** *If $C$ is a nondegenerate subcontinuum of $X$ containing $x$ then $\dim C = n$.*

*Proof.* For every $i$, $x \in F_i \cap C$. Since $X$ is hereditarily indecomposable, this implies that $F_i \subseteq C$ or $C \subseteq F_i$ (Exercise 1.10.21). So if there exists $i$ such that $F_i \subseteq C$ then we are done since by the Subspace Theorem 3.2.9 (applied twice) we then have
$$n = \dim F_i \leq \dim C \leq \dim X = n.$$
If there does not exist such $i$ then $C \subseteq F_i$ for every $i$ and hence
$$C \subseteq \bigcap_{i=1}^{\infty} F_i = \{x\}.$$
But this contradicts the fact that $C$ is nondegenerate.

By equality of dim and ind (Theorem 3.4.8), there is a closed subspace $G$ of $X$ such that $\dim G = n - 1$. Corollary 3.7.9 shows that we may assume that $G$ is a continuum. Since $n - 1 \geq 1$, the set $G$ is a nondegenerate subcontinuum. Let $y$ be an arbitrary element in $G$. Then by Claim 1 there does not exist a homeomorphism of $X$ which maps $x$ onto $y$. Hence $X$ is not homogeneous. □

**Remark 3.8.10.** The proof of Proposition 3.8.9 clearly breaks down if $X$ is infinite-dimensional. It would work if every infinite-dimensional compact space would contain a nondegenerate finite dimensional continuum. But this is not true, as was shown in Corollary 3.8.8 (see also §3.13). By a different technique, ROGERS [354] proved that there is no infinite-dimensional hereditarily indecomposable homogeneous continuum. It is unknown however whether there is an infinite-dimensional homogeneous indecomposable continuum (this is a question of ROGERS).

**Exercise for §3.8.**

1. Let $X$ be a space and let $F_0, F_1$ be disjoint closed sets in $X$. Prove that there are disjoint open neighborhoods $W_0$ and $W_1$ of $F_0$ and $F_1$, respectively, such that any two intersecting continua $K_0, K_1$ in $X \setminus (W_0 \cup W_1)$ are comparable, i.e., $K_0 \subseteq K_1$ or $K_1 \subseteq K_0$.

## 3.9. Totally disconnected spaces

In this section we shall prove that totally disconnected spaces of every dimension exist, cf. Examples 1.5.18 and 3.4.13 where we constructed an example of a one-dimensional totally disconnected space. Our main tool is the concept of a so-called '$G_\delta$-selection'. After some preparatory work, the construction of the examples is surprisingly simple.

**The lexicographic order.** Suppose that $X$ is compact and that

$$f \colon X \to Y$$

is a continuous surjection. The set-valued function

$$F \colon Y \Rightarrow X$$

defined by $F(y) = f^{-1}(y)$ in general unfortunately does not admit a continuous selection, see Page 13. We shall show however that there always exists a selection $s$ for $F$ such that the range of $s$ is a $G_\delta$-subset of $X$.

Let $n \in \mathbb{N} \cup \{\infty\}$ and consider the product $\mathbb{J}^n$. It can be ordered by the so-called *lexicographic order* $\preceq$, as follows:

$$x \preceq y \iff (\exists n \in \mathbb{N})(x_n \leq y_n \text{ and } x_i = y_i \text{ if } i < n).$$

A straightforward verification shows that $\preceq$ is a linear order on $\mathbb{J}^n$.

## 3.9. TOTALLY DISCONNECTED SPACES

**Lemma 3.9.1.** *Let $n \in \mathbb{N} \cup \{\infty\}$ and let $X \subseteq \mathbb{J}^n$ be compact. Then $X$ has a smallest element with respect to $\preceq$.*

**Proof.** Since $X$ is compact and $\pi_1$ is continuous, $\pi_1[X]$ is compact as well and hence $x_1 = \min \pi_1[X]$ exists. Observe that $X \cap \pi_1^{-1}(x_1)$ is a nonempty compact subset of $X$. So $x_2 = \min \pi_2[X \cap \pi_1^{-1}(x_1)]$ exists as well. Observe that the set
$$X \cap \pi_1^{-1}(x_1) \cap \pi_2^{-1}(x_2)$$
is again a nonempty compact subset of $X$. So the minimum of its projection onto the third factor exists. Etc. It is clear that the point $x = (x_i)_i$ defined in this way is the smallest element of $X$ with respect to $\preceq$ (observe that by compactness of $X$ it belongs to $X$, even if $n = \infty$). $\square$

**Theorem 3.9.2.** *Let $X$ be a compact space and let $f \colon X \to Y$ be a continuous surjection. There exists a $G_\delta$-subset $S$ of $X$ which intersects each fiber of $f$ in precisely one point.*

**Proof.** By Corollary A.4.4 we may assume that $X$ is a subspace of $Q$. Since $f$ is a surjection, for every $y \in Y$ the fiber $f^{-1}(y)$ is a nonempty compact subset of $X$. Hence it has a smallest element with respect to the lexicographic ordering $\preceq$ on $Q$ (Lemma 3.9.1), say $p(y)$. We claim that the set
$$S = \{p(y) : y \in Y\}$$
is a $G_\delta$-subset of $X$ and is therefore as required.

Define for $n, m \in \mathbb{N}$ the following sets:
$$U_{n,m} = \{x \in X : \text{for all } z \in X \text{ such that}$$
$$x_i = z_i \text{ for } i \leq n-1 \text{ and } z_n \leq x_n - 1/m$$
$$\text{we have } f(x) \neq f(z)\}.$$

We first prove that the $U_{n,m}$'s are open by establishing that their complements are closed. To this end, for each $j \in \mathbb{N}$ take $x(j) \in X \setminus U_{n,m}$ and assume that the sequence $\big(x(j)\big)_j$ converges to a point $x \in X$. For each $j$ there exists an element $z(j)$ in $X$ such that $x(j)_i = z(j)_i$ for every $i \leq n-1$ while moreover $z(j)_n \leq x(j)_n - 1/m$ and $f\big(x(j)\big) = f\big(z(j)\big)$. By compactness, without loss of generality we assume that $\big(z(j)\big)_j$ converges to a point $z$ in $X$. By continuity of the projections and $f$ we get
$$x_i = z_i \text{ for every } i \leq n-1, z_n \leq x_n - 1/m \text{ and } f(x) = f(z),$$
i.e., $x \in X \setminus U_{n,m}$. We conclude that $X \setminus U_{n,m}$ is closed.

We next claim that for all $n$ and $m$, $S \subseteq U_{n,m}$. To this end, take an arbitrary $y \in Y$ and consider the point $p(y)$. Fix arbitrary $n, m \in \mathbb{N}$. Take a point $z \in X$ such that $p(y)_i = z_i$ for all $i \leq n-1$ and $z_n \leq p(y)_n - 1/m$. If $z \in f^{-1}(y)$ then by construction $p(y)_n \leq z_n$. But this is impossible and so $z \notin f^{-1}(y)$, as required.

We finally claim that the intersection $U$ of all the $U_{n,m}$'s is equal to $S$. Since $S$ intersects every fiber of $f$ in precisely one point, and as we just observed, $S \subseteq U$, it suffices to prove that $U$ intersects every fiber of $f$ in at most one point. To this end, take an arbitrary $y \in Y$ and assume that there are distinct points $x(1), x(2) \in U \cap f^{-1}(y)$. Let $n \in \mathbb{N}$ be the first index such that $x(1)_n \neq x(2)_n$. We may assume without loss of generality that

$$x(1)_n < x(2)_n.$$

There exists $i \in \mathbb{N}$ such that $x(1)_n \leq x(2)_n - 1/i$. Since $x(2) \in U_{n,i}$ and $x(1)_m = x(2)_m$ for all $m \leq n - 1$, we get $f(x(1)) \neq f(x(2))$. This is a contradiction. □

**The existence of $n$-dimensional totally disconnected spaces.** We are now in a position to prove the main result of this section.

**Theorem 3.9.3.** *For each $n \geq 0$ there exists an $n$-dimensional topologically complete totally disconnected space. There also exists a strongly infinite-dimensional topologically complete totally disconnected space.*

**Remark 3.9.4.** The existence of a strongly infinite-dimensional topologically complete totally disconnected space turns out to have a surprising application: it is the main ingredient in the solution of a famous problem in dimension theory that remained open for more than 50 years. See Theorem 3.13.8 for details.

**Proof.** We shall construct the examples simultaneously. Consider

$$X = \mathbb{J} \times \mathbb{J}^n,$$

where $n \in \mathbb{N} \cup \{\infty\}$. Let $\pi \colon X \to \mathbb{J}$ be the projection onto its first factor and put $A = \pi^{-1}(-1)$ and $B = \pi^{-1}(1)$, respectively. Define

$$\mathcal{C} = \{C \in 2^X : C \text{ is a continuum from } A \text{ to } B\}.$$

By Exercise 1.11.21 it follows that $\mathcal{C}$ is a closed subspace of $2^X$. Since $2^X$ is compact by Proposition 1.11.4, this proves that $\mathcal{C}$ is compact.

Let $\mathsf{C}$ be a Cantor set in $\mathbb{J}$ and since $\mathcal{C} \neq \emptyset$ there is a continuous surjection $\alpha \colon \mathsf{C} \to \mathcal{C}$ (Theorem 1.5.10). Now put

$$Y = \bigcup \{\pi^{-1}(t) \cap \alpha(t) : t \in \mathsf{C}\}.$$

**Claim 1.** *$Y$ is compact and $\pi[Y] = \mathsf{C}$.*

*Proof.* We shall first prove that $\pi[Y] = \mathsf{C}$. Clearly, $\pi[Y] \subseteq \mathsf{C}$. Take an arbitrary $t \in \mathsf{C}$. Then $\alpha(t)$ is a continuum from $A$ to $B$. Consequently, $\pi[\alpha(t)]$ is a continuum in $\mathbb{J}$ from $-1$ to $1$ and therefore is equal to $\mathbb{J}$. So there exists an element $y \in \alpha(t)$ with $\pi(y) = t$. Since this $y$ belongs to $Y$, this proves that $\mathsf{C} \subseteq \pi[Y]$.

We shall next show that $Y$ is compact. To this end, let $(y_i)_i$ be a sequence in $Y$ converging to a point $y \in X$. Then $t_i = \pi(y_i)$ belongs to $\mathsf{C}$ for every $i$ and $(t_i)_i$ converges to $t = \pi(y)$. Consequently, $t$ belongs to $\mathsf{C}$. Since $\alpha$ is continuous, the sequence of continua $\bigl(\alpha(t_i)\bigr)_i$ converges to $\alpha(t)$ in $\mathfrak{C}$. We claim that $y \in \alpha(t)$. If this is not the case then we can find $\varepsilon > 0$ such that
$$y \notin D(\alpha(t), \varepsilon).$$
Since $\bigl(\alpha(t_i)\bigr)_i$ converges to $\alpha(t)$, for all but finitely many $i$,
$$y_i \in \alpha(t_i) \subseteq D(\alpha(t), \varepsilon)$$
which implies that $y \in D(\alpha(t), \varepsilon)$. This is a contradiction. We conclude that $y \in Y$. ◇

Since $\pi$ is continuous and $Y$ is compact (Claim 1), by Theorem 3.9.2 there exists a $G_\delta$-subset $S$ of $Y$ which intersects each fiber of $\pi \restriction Y$ in precisely one point. The compactness of $Y$ implies that $S$ is topologically complete (Theorem A.6.3).

**Claim 2.** $S$ intersects each continuum from $A$ to $B$.

*Proof.* Take an arbitrary $\hat{C} \in \mathfrak{C}$. Since $\alpha$ is surjective, there exists $t \in \mathsf{C}$ such that $\alpha(t) = \hat{C}$. Observe that
$$(\pi \restriction Y)^{-1}(t) = \pi^{-1}(t) \cap Y = \pi^{-1}(t) \cap \alpha(t) = \pi^{-1}(t) \cap \hat{C} \subseteq \hat{C}.$$
Since $S$ intersects each fiber of $\pi \restriction Y$ we consequently get $S \cap \hat{C} \neq \emptyset$. ◇

Observe that $\pi \restriction S$ is a one-to-one continuous function from $S$ onto the zero-dimensional space $\mathsf{C}$. This implies that $S$ is totally disconnected, cf. Example 1.5.18.

Now assume that $n$ is finite. Then $\dim S \geq n$ by Theorem 2.4.12 and Corollary 3.7.5. Also $\dim S \leq n$ since $S$ is contained in $\mathsf{C} \times \mathbb{J}^n$, which is $n$-dimensional by Theorems 3.2.12 and 3.4.12. We conclude that $\dim S = n$.

Now if $n = \infty$ then $S$ is strongly infinite-dimensional by Corollaries 2.4.13 and 3.7.5. □

**Application: The dimension of infinite products.** As an unexpected application of the results obtained here we will show that there exists for every $n \geq 0$ a topologically complete $n$-dimensional space whose infinite power is also $n$-dimensional. This shows that there exists an example such as Example 3.4.13 in every dimension.

We will in fact show that the spaces constructed in the proof of Theorem 3.9.3 are as required. We have to be a little careful though. In the proof of Theorem 3.9.3 the precise nature of the '$G_\delta$-selection' $S$ is irrelevant. We used Theorem 3.9.2 to prove its existence, but how it was created in the proof of Theorem 3.9.2 did play no role. But this *is* important here. So we assume in the proof of Theorem 3.9.5 that the '$G_\delta$-selection' $S$ was constructed via the method used in the proof of Theorem 3.9.2, i.e., via the lexicographic order on $\mathbb{J}^n$.

**Theorem 3.9.5.** *For every $n \geq 0$ there exists an $n$-dimensional topologically complete space whose infinite power is also $n$-dimensional.*

**Proof.** Let $S$ be the space constructed in the proof of Theorem 3.9.3. We claim that $S$ is as required. Observe that $S \subseteq \mathbb{J} \times \mathbb{J}^n$ and that the projection

$$\pi\colon \mathbb{J} \times \mathbb{J}^n \to \mathbb{J}$$

maps $S$ in a one-to-one way onto the Cantor set $\mathsf{C}$. Define a function

$$f\colon S \times S \to \mathsf{C}^2 \times \mathbb{R}^n$$

as follows:

$$f\big((x,z),(x',z')\big) = \big((x,x'), z+z'\big).$$

Since $\dim(\mathsf{C}^2 \times \mathbb{R}^n) \leq n$ (Theorems 3.2.12 and 3.4.12) it will follow that $\dim S^2 \leq n$ if we can prove that $f$ is an imbedding. It is clear that $f$ is one-to-one and continuous. The potential problem is the continuity of the inverse. To check this, let $\big(f((x_i,z_i),(x'_i,z'_i))\big)_i$ be a sequence in $f[S^2]$ converging to an element $f\big((x,z),(x',z')\big)$ in $f[S^2]$. By the definition of $f$ this implies that

$$x_i \to x, \quad x'_i \to x', \quad z_i + z'_i \to z + z'.$$

By compactness we may assume without loss of generality that

$$z_i \to a, \quad z'_i \to b,$$

where $a, b \in \mathbb{J}^n$. So $a + b = z + z'$. Since $(x_i, z_i) \in \alpha(x_i)$ for every $i$ and $\alpha(x_i) \to \alpha(x)$ by continuity of $\alpha$ it follows easily that $(x, a) \in \alpha(x)$ (Lemma 1.11.2(b)). But this implies by construction that $z \preceq a$. It follows similarly that $z' \preceq b$. Since $z + z' = a + b$ we get $z = a$ and $z' = b$. This implies that

$$\big((x_i,z_i),(x'_i,z'_i)\big) \to \big((x,z),(x',z')\big),$$

as required.

So $\dim S^2 \leq n$ from which it follows that $\dim S^2 = n$ since $S^2$ contains a (closed) copy of the $n$-dimensional space $S$ (Theorem 3.2.9).

The same method can be be used to show that the dimension of each finite power of $S$ is equal to $n$. Since the countably infinite power of $S$ is the inverse limit of the inverse sequence consisting of all of its finite powers (Example 1.10.3), we are done by an appeal to Exercise 3.5.2. □

**Remark 3.9.6.** It was shown by POL [345] that if $X$ is a compact space which is $(n+1)$-dimensional then it contains an $n$-dimensional $G_\delta$-set $S$ such that $\dim S^\infty = n$. The methods used in this section show that this is true for the special case $X = \mathbb{J}^{n+1}$.

**Exercises for §3.9.**

1. Let $X$ be a compact space such that $\dim X \geq n+1$. Prove that $X$ contains a totally disconnected $G_\delta$-subset $S$ such that $\dim S \geq n$.

2. Let $X$ and $Y$ be totally disconnected. Prove that $X \times Y$ is totally disconnected.

3. Prove that there exists a countable dimensional infinite-dimensional topologically complete totally disconnected space.

## 3.10. The origins of dimension theory

The first definition of a dimension function is due to BROUWER [77]. He seems to have been put on the right track by POINCARÉ [337, 338] who had proposed to call a continuum $n$-dimensional, if it can be dissected in separate pieces by one or more $(n-1)$-dimensional continua. Brouwer refined this idea and defined a dimensional invariant intended for topologically complete spaces without isolated points (in the terminology of his days, normal sets in the sense of Fréchet) which he called *Dimensionsgrad*. There is no reason for the restriction to topologically complete spaces without isolated points, so we will state his definition for arbitrary spaces instead. Indeed, define the *Dimensionsgrad* $\operatorname{Dg} X$ of a space $X$, as follows:

$\operatorname{Dg} X = -1 \quad \Leftrightarrow \quad X = \emptyset$,

$0 \leq \operatorname{Dg} X \leq n \quad \Leftrightarrow \quad X$ is nonempty and for every pair $A, B$ of disjoint closed subsets of $X$ there exists a closed set $T \subseteq X$ which is a cut between $A$ and $B$ such that $\operatorname{Dg} T$ is at most $n-1$,

$\operatorname{Dg} X = n \quad \Leftrightarrow \quad \operatorname{Dg} X \leq n$ and $\operatorname{Dg} X$ is not smaller than $n$,

$\operatorname{Dg} X = \infty \quad \Leftrightarrow \quad \operatorname{Dg} X \neq n$ for every $n \geq -1$.

So this invariant is very similar to the large inductive dimension, which was defined much later.

It is easy to see that if $X$ and $Y$ are homeomorphic spaces then $\operatorname{Dg} X$ is equal to $\operatorname{Dg} Y$. Hence Dg is a topological invariant, as are all dimension functions considered in this chapter.

It is clear that $\operatorname{Dg} X = 0$ if and only if $X \neq \emptyset$ and $X$ contains no continuum of cardinality larger than 1. Such spaces are called *punctiform*.

There is something very peculiar about the definition of Dimensionsgrad, namely, a cut between two closed subsets of an arbitrary space is not necessarily a partition, even in compact spaces. In the $\sin(1/x)$-continuum in the plane

$$\{(x, \sin(1/x)) : 0 < x \leq 1/\pi\} \cup \{(0, y) : -1 \leq y \leq 1\}$$

the point $(0, 1)$ cuts between the points $(1/\pi, 0)$ and $(0, -1)$. But the complement of $(0, 1)$ is connected! (See also Exercise A.10.7.) This was pointed out to Brouwer by Urysohn. As a result of this, Brouwer issued a flood of notes and papers correcting his 'mistake'. For a description of the troubles this caused in the early days of dimension theory, see e.g., JOHNSON [213].

**Lemma 3.10.1.** *For every space $X$ we have $\operatorname{Dg} X \leq \dim X$.*

**Proof.** This is easy. If $\dim X = -1$ then $X = \emptyset$ and so $\operatorname{Dg} X = -1$. Suppose that the inequality is true for all spaces $X$ with $\dim X \leq n-1$, where $n \geq 0$. By the Coincidence Theorem 3.4.10 it follows that for every pair of disjoint closed sets $A, B \subseteq X$ there exists a partition $T$ between $A$ and $B$ such that $\dim T \leq n-1$. By our inductive assumption, $\operatorname{Dg} T \leq \dim T \leq n-1$. Since a partition is a cut, it follows that $\operatorname{Dg} X \leq n$, as required. □

The proof of Lemma 3.10.1 and Urysohn's Example suggest the question whether every cut is a partition in some well-behaved class of spaces.

**Lemma 3.10.2.** *Let $X$ be a space which is both topologically complete and locally connected. If $A$ and $B$ are closed subsets of $X$ and $T$ cuts between $A$ and $B$ then $T$ is a partition between $A$ and $B$.*

**Proof.** Let $\mathcal{U}$ be the family of all components of $X \setminus T$. Since $X$ is locally connected, every member of $\mathcal{U}$ is open (Exercise A.2.8). Assume that some $U \in \mathcal{U}$ intersects both $A$ and $B$. Then $U$ is a connected, locally connected and topologically complete space (Theorem A.6.3). But then by the Mazurkiewicz Theorem 1.5.22 it follows that there is an imbedding $\Phi \colon \mathbb{I} \to U$ such that $\Phi[\mathbb{I}]$ meets both $A$ and $B$. But this contradicts the fact that $T$ cuts between $A$ and $B$. Now let $U$ be the union of all elements of $\mathcal{U}$ that meet $A$ and let $V$ be the union of all the other elements of $\mathcal{U}$. Then $U$ is an open neighborhood of $A$, $V$ is an open neighborhood of $B$ and $U \cup V = X \setminus T$. Hence $T$ is a partition between $A$ and $B$. □

So from Lemma 3.10.2 one would hope to be able to prove that Dg and dim take the same values on locally connected, topologically complete spaces. This was indeed widely believed to be true for a long time. In the book by HUREWICZ and WALLMAN [208], one can read on Page 4 that Dimensionsgrad and dimension agree even in the class of all locally connected spaces. A proof of this assertion was not presented. Similar statements for various classes of locally connected spaces were subsequently made in various other books on dimension theory. None of these claims was supported by a proof.

That there are troubles, even for *compact* locally connected spaces, is easily demonstrated. Let $X$ be a compact space of Dimensionsgrad 0. Then it contains no nontrivial continuum, and hence is zero-dimensional (Corollary 3.7.9). This triviality will be used in the proof of Theorem 3.10.3 below. If moreover $X$ is locally connected and $\operatorname{Dg} X \leq 1$ then by Lemma 3.10.2 every two disjoint closed sets can be partitioned by a zero-dimensional set, i.e., the large inductive dimension of $X$ is at most 1. But if $X$ is compact and locally connected and has Dimensionsgrad at most 2 then all there can be concluded from the above is that every pair of disjoint closed subsets can be partitioned by a closed subset with Dimensionsgrad at most 1. But that closed set does not need to be locally connected, so we are at a dead end.

Brouwer always claimed that his Dimensionsgrad could serve as the basis of an equally important theory of dimension as ordinary dimension. But one would guess from the above that for a sufficiently 'bad' compact space dim and Dg can differ. The following surprising result shows that this is not true and that Brouwer, as usual, was right after all with his claim – at least for compact spaces.

**Theorem 3.10.3.** *Let $X$ be a compact space. Then* $\dim X = \mathrm{Dg}\, X$.

**Proof.** By Lemma 3.10.1 we only need to prove that $\dim X \leq \mathrm{Dg}\, X$ for all compact spaces $X$. If $\mathrm{Dg}\, X = 0$ then by the above remarks we have that $\dim X = 0$. So now suppose that for every compact space $X$ such that
$$0 \leq \mathrm{Dg}\, X \leq n-1$$
we have
$$\dim X \leq \mathrm{Dg}\, X.$$
Assume that $X$ is a compact space such that $\mathrm{Dg}\, X = n$. Let $\mathcal{B}$ be a countable base for $X$. For every pair $(B_0, B_1)$ of elements of $\mathcal{B}$ such that $\overline{B_0} \subseteq B_1$ there is a cut between $\overline{B_0}$ and $X \setminus B_1$ with Dimensionsgrad at most $n-1$. So by our inductive assumption, every such cut has dimension at most $n-1$. By the Countable Closed Sum Theorem 3.2.8, the union of these cuts is at most $(n-1)$-dimensional. By Corollary 3.3.12 this union is contained in an at most $(n-1)$-dimensional $G_\delta$-subset $S$ of $X$. Then $T = X \setminus S$ is an $F_\sigma$-subset of $X$. We claim that $T$ contains no nontrivial continuum. Striving for a contradiction, assume that $C$ is a nontrivial continuum contained in $T$, and pick arbitrary distinct points $a, b \in C$. There exist elements $B_0, B_1 \in \mathcal{B}$ such that
$$a \in B_0 \subseteq \overline{B_0} \subseteq B_1 \setminus \{b\}.$$
By construction, $S$ contains a cut between $\overline{B_0}$ and $X \setminus B_1$. But $C$ is a continuum meeting both $\overline{B_0}$ and $X \setminus B_1$ but missing $S$. This is a contradiction. So we conclude that $T$ contains no nontrivial continua. But then every compact subset of $T$ is zero-dimensional by the same reasons as before. So $T$ is a countable union of compact zero-dimensional subspaces, and hence is zero-dimensional itself (again by the Countable Closed Sum Theorem 3.2.8). So by Theorem 3.4.4 we conclude that
$$\dim X = \dim(S \cup T) \leq \dim S + \dim T + 1 = (n-1) + 0 + 1 = n,$$
as required. $\square$

It is clear that there are spaces $X$ such that $\mathrm{Dg}\, X \neq \dim X$. For let $X$ be a totally disconnected $n$-dimensional space (Theorem 3.9.3). Then $X$ is punctiform and so $\mathrm{Dg}\, X = 0$. The gap between Dimensionsgrad and dimension can therefore be arbitrarily large. In view of the claim in the book by HUREWICZ and WALLMAN [208], the question naturally arises whether there are similar examples within the class of locally connected spaces. Observe that Lemma 3.10.2 draws specific attention to the class of all topologically complete locally connected spaces, for there the notions of a cut and a partition coincide.

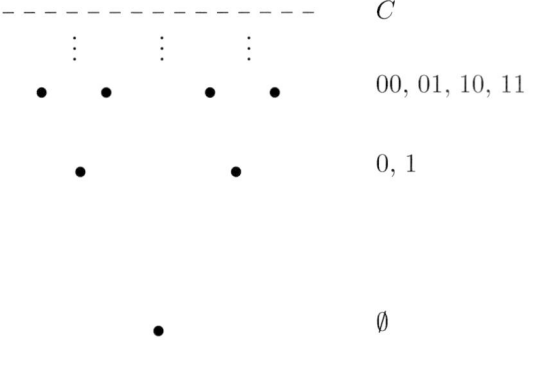

Figure 10.

**Theorem 3.10.4.** *For each $n = 2, 3, \ldots, \infty$ there exists a locally connected topologically complete space $X_n$ such that $\mathrm{Dg}\, X_n = 1 < \dim X_n = n$.*

**Proof.** Represent the set $\Sigma$, the set consisting of all finite sequences consisting of 0's and 1's (including the empty sequence), by dots in the plane as in Figure 10. These dots are compactified in a natural way by a copy $C$ of the Cantor set.

We now connect every dot with its two 'successor' dots by an interval (see Figure 11). In this way we obtain a connected space which we call $T$. By $U$ we denote one of the basic neighborhoods of a point of $C$. The set $U$ is clearly connected, so $T$ is locally connected. But it has a stronger property which we will now describe.

**Claim 1.** *For an arbitrary subset $C_0 \subseteq C$ the space $T_0 \cup C_0$ is locally connected.*

*Proof.* In fact, a basic neighborhood of an arbitrary point $c \in C_0 \subseteq T_0 \cup C_0$ has the form
$$(U \cap T_0) \cup (U \cap C_0).$$
This set is connected, since it contains the dense connected set $U \cap T_0$. ◇

## 3.10. THE ORIGINS OF DIMENSION THEORY

Now let $f\colon C \to Q$ be a continuous surjection from $C$ onto the Hilbert cube $Q$ (Theorem 1.5.10). Let $Z = T \cup_f Q$ and let $g\colon T \to Z$ be the quotient map. So $Z$ is the disjoint union of $T_0$ and the Hilbert cube $Q$.

For every $n$ let $Y_n \subseteq Q$ be an $n$-dimensional totally disconnected $G_\delta$-set (Theorem 3.9.3) and let $X_n = T_0 \cup Y_n$.

**Claim 2.** $X_n$ is a topologically complete space.

*Proof.* This is clear since the complement of $X_n$ in $Q$ is an $F_\sigma$-subset of $Q$ by Theorem A.6.3. So the complement of $X_n$ in the compact space $Z$ is an $F_\sigma$-subset of $Z$ as well. ◇

**Claim 3.** $X_n$ is locally connected.

*Proof.* In fact, $X_n = f[f^{-1}[X_n]]$. The set $f^{-1}[X_n]$ is locally connected by Claim 1. Hence $X_n$ is locally connected by Exercise A.2.11. ◇

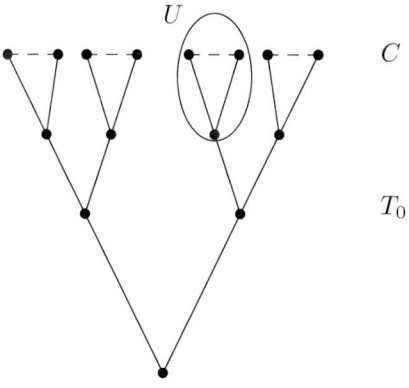

Figure 11.

**Claim 4.** $\dim X_n = n$.

*Proof.* It is clear that $Y_n$ is a closed $n$-dimensional subset of $X_n$. So $X_n$ is at least $n$-dimensional by the Subspace Theorem 3.2.9. On the other hand, $X_n \setminus Y_n$ is a countable union of one-dimensional compacta. Hence $\dim X_n \leq n$ by the Countable Closed Sum Theorem 3.2.8. ◇

**Claim 5.** $\mathrm{Dg}\, X_n = 1$.

*Proof.* Let $A$ and $B$ be disjoint closed subsets of $X_n$. Take an arbitrary partition $S_0$ in $Y_n$ between $A \cap Y_n$ and $B \cap Y_n$. Corollary 3.3.10 gives us a partition $S$ in $X_n$ between $A$ and $B$ such that $S \cap Y_n \subseteq S_0$ and $\dim(S \setminus S_0) \leq 0$.

We claim that $S$ is totally disconnected. So we must show that $\emptyset$ is a partition between any pair of distinct points $a$ and $b$ of $S$. Since $S \setminus Y_n$ is an open zero-dimensional subspace of $S$, a moments reflection shows that we may assume without loss of generality that both $a$ and $b$ belong to the totally disconnected closed space $Y_n$ of $X_n$. So $\emptyset$ is a partition between $a$ and $b$ in $S \cap Y_n$. By another appeal to Corollary 3.3.10 it follows that $\emptyset$ can be extended to a partition between $a$ and $b$ in $S$ such that the dimension of its intersection with $S \setminus Y_n$ is $-1$. We conclude that $\emptyset$ is a partition in $S$ between $a$ and $b$.

Since a totally disconnected space is punctiform, it has Dimensionsgrad 0. We conclude that $\operatorname{Dg} X_n \leq 1$. That

$$\operatorname{Dg} X_n \neq 0$$

is clear since $X_n$ contains copies of the unit interval. $\diamond$

This concludes the proof of the theorem. $\square$

Let us remark that there also is an infinite-dimensional locally connected topologically complete space with Dimensionsgrad 1. It suffices to take the topological sum of the spaces $X_n$ in Theorem 3.10.4. This space is countable dimensional. There is also a strongly infinite-dimensional example since there is a strongly infinite-dimensional totally disconnected $G_\delta$-subset in $Q$ (Theorem 3.9.3).

**Exercises for §3.10.**

1. Let $X$ be a $\sigma$-compact space. Prove that $\operatorname{Dg} X = \dim X$.
▶2. Let $X$ be a space which is both topologically complete and locally connected. Suppose that $A$ and $B$ are disjoint closed subsets of $X$. Prove that every partition $S$ between $A$ and $B$ contains an irreducible partition $T$ between $A$ and $B$.
▶3. Let $X$ be a unicoherent Peano continuum containing two disjoint continua $A$ and $B$. Prove that if $S$ is an irreducible partition between $A$ and $B$ then $S$ is connected.
4. Show that the fact that $A$ and $B$ are continua in the previous exercise is essential.
▶5. Prove that $\mathbb{S}^n$ for $n \geq 2$ cannot be imbedded in a product of the form

$$X_1 \times \cdots \times X_n,$$

such that every $X_i$ is at most one-dimensional.
6. Prove that the local connectivity assumption is essential in Exercise 3.10.2, even for compact spaces.

## 3.11. The dimensional kernel of a space

For a space $X$ and a point $x \in X$, $\operatorname{ind}_x X$ denotes the *dimension of $X$ at the point $x$*, and is defined as follows:

$\operatorname{ind}_x X = -1 \Leftrightarrow X = \emptyset$,
$0 \le \operatorname{ind}_x X \le n \Leftrightarrow$ for every closed subset $A$ of $X$ not containing $x$ there exists a partition $L$ between $\{x\}$ and $A$ such that $\dim L \le n-1$,
$\operatorname{ind}_x X = n \Leftrightarrow \operatorname{ind}_x X \le n$ and $\operatorname{ind}_x X \not\le n-1$,
$\operatorname{ind}_x X = \infty \Leftrightarrow \operatorname{ind}_x X \ne n$ for every $n \ge -1$.

Observe that $\dim X = \operatorname{ind} X \le n$ if and only if $\operatorname{ind}_x X \le n$ for every $x \in X$.

If $X$ is $n$-dimensional then its *dimensional kernel* $\Lambda(X)$ is the set

$$\{x \in X : \operatorname{ind}_x X = n\}.$$

**Lemma 3.11.1.** *Let $X$ be an $n$-dimensional space, $n \ge 1$.*

(1) *The dimensional kernel of $X$ is an $F_\sigma$-subset of $X$.*
(2) *The dimensional kernel of $X$ has dimension $\ge n-1$.*

**Proof.** Let $E$ denote the complement of the dimensional kernel of $X$. For every $i \in \mathbb{N}$ and for every $x \in E$ there is an open neighborhood $U_x^i$ of $x$ such that

$$\operatorname{ind} \operatorname{Fr}(U_x^i) \le n-2, \quad \operatorname{diam}(U_x^i) \le 1/i.$$

The cover $\{U_x^i : x \in E\}$ of $E$ has a countable subcover, say $\mathcal{U}_i$ (Corollary A.2.3); put $U_i = \bigcup \mathcal{U}_i$. Then $U_i$ is open, and

$$E \subseteq U = \bigcap_{i=1}^{\infty} U_i.$$

We claim that $U \subseteq E$ as well. Pick arbitrary $x \in U$ and $\varepsilon > 0$. Let $i$ be so large that $1/i < \varepsilon$. Let $V \in \mathcal{U}_i$ be such that $x \in V$. Then

$$\operatorname{diam}(V) = \operatorname{diam} \overline{V} \le 1/i < \varepsilon$$

which implies that $\overline{V} \subseteq B(x, \varepsilon)$. Since $\operatorname{ind} \operatorname{Fr}(V) \le n-2$ and $\varepsilon > 0$ was arbitrarily chosen, this implies that $\operatorname{ind}_x X \le n-1$, i.e., $x \in E$. We conclude that $U \subseteq E$ and hence that $E = U$. As a consequence, $E$ is a $G_\delta$-subset of $X$, which proves (1).

For (2), let

$$F = \bigcup_{i=1}^{\infty} \bigcup_{U \in \mathcal{U}_i} \operatorname{Fr}(U).$$

Then $F$ is the union of countably many closed subsets of $X$ with dimension at most $n-2$. By the above, the collection $\mathcal{U} = \bigcup_{i=1}^{\infty} \mathcal{U}_i$ is a local base at every point of $E$ (cf. Exercise A.2.15). So it follows that the collection $\mathcal{U} \upharpoonright (E \setminus F)$

consists of open and closed subsets of $E \setminus F$ and is in $E \setminus F$ a local base at any point. From this we see that $\mathrm{ind}(E \setminus F) \leq 0$.

Now let $K = X \setminus E$ be the dimensional kernel of $X$, and suppose that $\mathrm{ind}\, K \leq n - 2$. We will derive a contradiction. By (1) and the above, $K \cup F$ is an $F_\sigma$-subset of $X$ and by Theorems 3.4.10 and 3.2.8 it follows that

$$\mathrm{ind}(K \cup F) \leq n - 2.$$

But then by Theorem 3.4.4,

$$\mathrm{ind}\, X = \mathrm{ind}\,((K \cup F) \cup (E \setminus F)) \leq n - 2 + 0 + 1 = n - 1,$$

which is a contradiction. $\square$

For compact spaces, we can do a little better.

**Theorem 3.11.2.** *Let $X$ be an $n$-dimensional compact space. Then the dimensional kernel of $M$ is $n$-dimensional and $\sigma$-compact.*

**Proof.** The dimensional kernel is $\sigma$-compact by Lemma 3.11.1(1). To prove that it is $n$-dimensional, first observe that if $n = 0$ then there is nothing to prove. So let $n \geq 1$. By Theorem 3.7.8, $X$ contains an $n$-dimensional Cantor-manifold $Y$. We claim that $Y$ is contained in the dimensional kernel of $X$. If so then we are done. To this end, let $y \in Y$ be arbitrary. Since $\dim Y \geq 1$ there exists a point $z \in Y \setminus \{y\}$. If $\mathrm{ind}_y X < n$ then there is a partition $F$ between $\{y\}$ and $\{z\}$ such that $\dim F \leq n - 2$. But then $F \cap Y$ separates $Y$ and is at most $(n-2)$-dimensional, which is a contradiction since $Y$ is an $n$-dimensional Cantor-manifold. $\square$

Let $X$ be zero-dimensional, and let $x \in X$. It is clear that $\mathrm{ind}_x X = 0$. So the dimensional kernel of $X$ is $X$ itself. This explains why in the following definition we require $X$ to be at least one-dimensional.

A space is called *weakly $n$-dimensional*, where $n \geq 1$, if it is $n$-dimensional, but its dimensional kernel is of dimension $n - 1$. By Theorem 3.11.2, no compact subspace of a weakly $n$-dimensional space is $n$-dimensional.

It is not clear at all whether weakly $n$-dimensional spaces exist. But in Theorem 3.11.8 we will construct for every $n \geq 1$ such a space. We have to so some preparatory work first.

**Lemma 3.11.3.** *Let $X$ be weakly $n$-dimensional with subspace $F$. Then $F$ is at most $(n-1)$-dimensional or weakly $n$-dimensional.*

**Proof.** Assume that $\dim F = n$. We claim that if $x$ belongs to the dimensional kernel of $F$ then $x$ belongs to the dimensional kernel of $X$. For assume that $x \in F$ does not belong to the dimensional kernel of $X$. Then $x$ has arbitrarily small neighborhoods $U$ in $X$ such that $\mathrm{ind}\,\mathrm{Fr}\, U \leq n - 2$. By Lemma 3.4.3 and Exercise A.1.16 it follows that the boundary of $U \cap F$ in $F$

has dimension at most $n-2$. So this shows that $x$ does not belong to the dimensional kernel of $F$. So the dimensional kernel of $F$ is a subspace of the dimensional kernel of $X$ and hence is at most $(n-1)$-dimensional by the Subspace Theorem 3.2.9 (or by Lemma 3.4.3). □

**Lemma 3.11.4.** *Let $n \geq 1$. In addition, let $X$ be compact such that there exists a continuous function $f \colon X \to \mathbb{I}$ such that for every $q \in \mathbb{Q} \cap (0,1)$ we have $\dim f^{-1}(q) \leq n$. Then there is a $G_\delta$-subset $S \subseteq (0,1)$ with $\mathbb{Q} \cap (0,1) \subseteq S$ such that $\dim f^{-1}(z) \leq n$ for every $z \in S$.*

**Proof.** By the Countable Closed Sum Theorem 3.2.8 it follows that

$$\bigcup_{q \in \mathbb{Q}} f^{-1}(q)$$

is at most $n$-dimensional and hence can be enlarged to a $G_\delta$-subset $S$ of $X$ with $\dim S \leq n$ (Corollary 3.3.12). Then $T = f[X \setminus S]$ is by compactness of $X$ an $F_\sigma$-subset of $\mathbb{I}$ which misses $\mathbb{Q}$. It is clear that for every $z$ in the $G_\delta$-subset $\mathbb{I} \setminus T$ of $\mathbb{I}$ we have that $f^{-1}(z) \subseteq S$. Hence $f^{-1}(z)$ is at most $n$-dimensional. □

**Lemma 3.11.5.** *Let $K$ be an $(n+1)$-dimensional compact space with disjoint closed subsets $A$ and $B$. There exist a continuous function $f \colon K \to \mathbb{I}$ and a Cantor set $\triangle$ in $[1/3, 2/3]$ such that $f[A] \subseteq \{0\}$, $f[B] \subseteq \{1\}$ and $\dim\bigl(f^{-1}(t)\bigr) \leq n$ for each $t \in \triangle$.*

**Proof.** Let $\alpha \colon K \to \mathbb{I}$ be a Urysohn function such that

$$\alpha[A] \subseteq \{0\}, \quad \alpha[B] \subseteq \{1\}$$

(Lemma A.4.1). Identify $\mathbb{I}$ and the set

$$\mathbb{I} \times \underbrace{\{0\} \times \cdots \times \{0\}}_{2n+2}$$

in $\mathbb{R}^{2n+3}$. By Remark 3.3.6 there is an imbedding $g \colon K \to \mathfrak{N}_{n+1}$ such that $\hat{\varrho}(\alpha, g) < 1/3$. It follows that we may think of $K$ as subspace of $\mathfrak{N}_{n+1}$ having the additional property that $\pi_1[A] \subseteq (-\infty, 1/3)$ and $\pi_2[B] \subseteq (2/3, \infty)$. Now consider an arbitrary element $q \in \mathbb{Q} \cap [1/3, 2/3]$. The set

$$S = \{x \in \mathbb{R}^{2n+3} : x_1 = q\}$$

can be identified with $\mathbb{R}^{2n+2}$, and the set $K \cap S$ with a subspace of

$$\{x \in \mathbb{R}^{2n+2} : \text{at most } n \text{ coordinates of } x \text{ are rational}\}.$$

But this implies that $\dim(K \cap S) \leq n$ by Corollary 3.2.11. So the function

$$\xi = \pi_1 \restriction K \colon K \to \mathbb{R}$$

has the property that for every $q \in \mathbb{Q} \cap [1/3, 2/3]$ we have $\dim \xi^{-1}(q) \leq n$. By Lemma 3.11.4 it follows that we may enlarge $\mathbb{Q} \cap [1/3, 2/3]$ to a $G_\delta$-subset $S$ having the same property. Observe that by the Baire Category Theorem A.6.6

and the fact that $\mathbb{Q} \cap [1/3, 2/3]$ is dense in $[1/3, 2/3]$, it follows that $S$ is uncountable. So $S$ contains a Cantor set $\triangle$ by Theorem 1.5.12, and all we need to do is to adjust $\xi$ slightly so as to satisfy the additional property that it sends $A$ to 0 and $B$ to 1. This adjustment is left as Exercise 3.11.1 to the reader. $\square$

**Lemma 3.11.6.** If $X$ is an $n$-dimensional space, where $n \geq 0$, then there exists a zero-dimensional $F_\sigma$-subset $N \subseteq X$ such that $\dim(X \setminus N) \leq n - 1$.

**Proof.** If $\dim X = 0$ then $N = X$ is as required. Suppose therefore that the lemma is true for all spaces $Z$ with $\dim Z \leq n-1$, where $n \geq 1$. Assume that $\dim X = n$. Let $\mathcal{B}$ be a countable base for $X$ such that $\dim \mathrm{Fr}(B) \leq n - 1$ for every $B \in \mathcal{B}$ (Lemma 3.4.1). Put

$$Y = \bigcup_{B \in \mathcal{B}} \mathrm{Fr}(B).$$

Then $Y$ is an $F_\sigma$-subset of $X$ since every $\mathrm{Fr}(B)$ is closed in $X$, and so $Y$ is at most $(n-1)$-dimensional by the Countable Closed Sum Theorem 3.2.8. In addition, $\dim(X \setminus Y) \leq 0$ since $\mathcal{B} \restriction X \setminus Y$ is a clopen base for $X \setminus Y$. By our inductive assumption there exists a zero-dimensional $F_\sigma$-subset $N \subseteq Y$ such that $\dim(Y \setminus N) \leq n - 2$. Observe that $N$ is an $F_\sigma$-subset of $X$ as well. Since $\dim(X \setminus Y) \leq 0$, it follows by Theorem 3.4.4 that

$$\begin{aligned}\dim(X \setminus N) &\leq \dim(Y \setminus N) + \dim(X \setminus Y) + 1 \\ &\leq (n-2) + 0 + 1 \\ &= n - 1,\end{aligned}$$

as required. $\square$

**Lemma 3.11.7.** Let $X$ and $Y$ be compact spaces, and let $f \colon X \to Y$ be a continuous surjection. If $Y$ is zero-dimensional at $y \in Y$ and if $f^{-1}(y)$ is zero-dimensional then $X$ is zero-dimensional at every point of $f^{-1}(y)$.

**Proof.** Pick an arbitrary $x \in f^{-1}(y)$, and let $U$ be a neighborhood of $x$ in $X$. Since $\dim f^{-1}(y) = 0$ there is a clopen subset $C$ of $f^{-1}(y)$ such that $x \in C \subseteq U$. Pick disjoint open subsets $E$ and $F$ in $X$ such that

$$E \cap f^{-1}(y) = C, \quad F \cap f^{-1}(y) = f^{-1}(y) \setminus C.$$

It is clear that without loss of generality we may assume that $E \subseteq U$. Observe that $E \cup F$ is a neighborhood of $f^{-1}(y)$ in $X$. Consequently, since $f$ is a closed map, there is a neighborhood $V$ of $y$ in $Y$ such that $f^{-1}[V] \subseteq E \cup F$ (Exercise A.1.15). Since $Y$ is zero-dimensional at $y$, we may assume without loss of generality that $V$ is clopen. Then $f^{-1}[V] \cap E$ is a clopen neighborhood of $x$ in $X$ which is contained in $U$. $\square$

We now come to the main result in this section.

## 3.11. THE DIMENSIONAL KERNEL OF A SPACE

**Theorem 3.11.8.** *Let $n \geq 1$ and let $K$ be an $(n+1)$-dimensional compactum. Then $K$ contains a weakly $n$-dimensional $G_\delta$-subset $X$ (hence $X$ is topologically complete).*

**Proof.** Let $\tau = \{(A_0, B_0), \ldots, (A_n, B_n)\}$ be an essential family in $K$. By Lemma 3.11.5 there exist a continuous function $f \colon K \to \mathbb{I}$ and a Cantor set $\triangle \subseteq (0,1)$ such that $f[A_0] = \{0\}$, $f[B_0] = \{1\}$ and $\dim\bigl(f^{-1}(t)\bigr) \leq n$ for every $t \in \triangle$.

We now closely follow the construction in Theorem 3.9.3. Put

$$\mathcal{C} = \{C \in 2^K : C \text{ is a continuum from } A_0 \text{ to } B_0\}.$$

Then $\mathcal{C}$ is closed in $2^K$ and hence is a compact space (see the proof of Claim 1 of Theorem 3.9.3). There consequently is a continuous surjection $\phi \colon \triangle \to \mathcal{C}$ (Theorem 1.5.10). Put

$$Z = \bigcup\{f^{-1}(t) \cap \phi(t) : t \in \triangle\}.$$

Then $Z$ is compact, and $f[Z] = \triangle$ (see the proof of Claim 2 of Theorem 3.9.3). Now $\dim Z \geq n$ because $Z$ intersects every continuum from $A_0$ to $B_0$ (Corollary 3.7.5). In addition, the fibers of $g = f \restriction Z$ are at most $n$-dimensional and $\triangle$ is zero-dimensional. We conclude that $\dim Z \leq n$ by Theorem 3.6.10.

Now let $\Lambda$ denote the dimensional kernel of $Z$. Then because $Z$ is compact, $\Lambda$ is an $n$-dimensional $\sigma$-compact subset of $Z$ (Theorem 3.11.2). There consequently exists a $\sigma$-compact zero-dimensional subset $N$ of $\Lambda$ such that $\dim(\Lambda \setminus N) \leq n-1$ (Lemma 3.11.6). Now put $X = Z \setminus N$. Then $X$ is clearly a $G_\delta$-subset of $K$ and we first claim that it is $n$-dimensional.

Let $C \in \mathcal{C}$. We will prove that $C$ meets $X$. Pick $t \in \triangle$ such that $\phi(t) = C$. Observe that $f^{-1}(t) \cap C \subseteq Z$. If $\dim f^{-1}(t) \cap C > 0$ then $f^{-1}(t) \cap C$ intersects $X$ because the complement of $X$ in $Z$ is zero-dimensional. We may therefore assume that $\dim f^{-1}(t) \cap C = 0$. But since $\triangle$ is zero-dimensional, we get that $Z$ is zero-dimensional at all points of $f^{-1}(t) \cap C$ (Lemma 3.11.7). Consequently, $f^{-1}(t) \cap C \cap \Lambda = \emptyset$, i.e., $\emptyset \neq f^{-1}(t) \cap C \subseteq X$. So we conclude from Corollary 3.7.5, that $\dim X \geq n$. However, because $X$ is a subspace of $Z$, we also have $\dim X \leq n$. Consequently, $\dim X = n$, as required.

We finally claim that $X$ is weakly $n$-dimensional. This is however a triviality (cf. the proof of Lemma 3.11.3). Simply observe that if $x$ is a point of $X$ at which $X$ is $n$-dimensional, then $Z$ is $n$-dimensional at $x$, which implies that $x \in X \cap \Lambda$. Since by construction, $\dim(X \cap \Lambda) \leq n-1$, we are done. $\square$

**Remark 3.11.9.** If one takes $K = \mathbb{I}^{n+1}$ in the above theorem then there is no need to use Lemma 3.11.5 in the proof because we can use for $f$ the projection onto the first coordinate. It seems that this gives the easiest known examples of weakly $n$-dimensional spaces.

**A universal weakly $n$-dimensional space.** In §3.3 we showed that for every $n$ there is a universal space for $n$-dimensional spaces. That is, there is an $n$-dimensional space $X$ which contains a homeomorphic copy of any $n$-dimensional space. We will prove now that a similar result holds for weakly $n$-dimensional spaces.

**Theorem 3.11.10.** *For each $n \geq 1$ there exists a weakly $n$-dimensional space $E$ such that any weakly $n$-dimensional space imbeds in $E$.*

**Proof.** Let $p\colon \{0,1\}^\infty \times Q \to \{0,1\}^\infty$ be the projection. It will be convenient in this proof to denote the hyperspace of $Q$, i.e., the space $2^Q$, by $\mathcal{K}$.

For a fixed natural number $n$, let $\mathcal{M}$ be the subspace of the product
$$\mathcal{K} \times \mathcal{K}^\infty \times (\mathcal{K} \times \mathcal{K})^\infty,$$
consisting of all elements of the form

(1) $\qquad\qquad (K, (L_i)_i, (C_i, D_i)_i)$

satisfying the following conditions:

(2) $\qquad L_1 \subseteq L_2 \subseteq \cdots \subseteq K, \quad \dim L_i \leq n-1 \qquad (i \in \mathbb{N}),$

(3) $\qquad K = C_i \cup D_i, \quad \dim(C_i \cap D_i) \leq n-2 \qquad (i \in \mathbb{N}).$

There are a subspace $T$ of $\{0,1\}^\infty$ and a continuous mapping $\varphi$ from $T$ onto the space $\mathcal{M}$ (use Exercise 1.5.10 and Corollary 1.5.7). It will be convenient to denote $\varphi$ as follows:

(4) $\qquad\qquad t \mapsto (K(t), (L_i(t))_i, (C_i(t), D_i(t))_i).$

For any $t \in T$, let

(5) $H(t) = \{z \in K(t) \setminus \bigcup_{i=1}^\infty L_i(t) :$ for any neighborhood $U$ of $z$ in $Q$
$\qquad\qquad\qquad\qquad$ there is an $i \in \mathbb{N}$ with $z \in C_i(t) \subseteq U$
$\qquad\qquad\qquad\qquad$ and $z \notin D_i(t)\}.$

For $t \in T$, put

(6) $\qquad\qquad E(t) = H(t) \cup \bigcup_{i=1}^\infty L_i(t).$

Finally, let $E \subseteq \{0,1\}^\infty \times Q$ be the subspace of all points of the form $(t,z)$, where $t \in T$ and $z \in E(t)$.

We shall show that

(7) $\qquad\qquad E$ is weakly $n$-dimensional,

and

(8) $\qquad\qquad$ any weakly $n$-dimensional space imbeds in $E$.

## 3.11. THE DIMENSIONAL KERNEL OF A SPACE

To this end let, us consider the sets

$$K^* = \{(t, z) : t \in T, z \in K(t)\}, \quad L_i^* = \{(t, z) : t \in T, z \in L_i(t)\},$$
$$C_i^* = \{(t, z) : t \in T, z \in C_i(t)\}, \quad D_i^* = \{(t, z) : t \in T, z \in D_i(t)\}.$$

**Claim 1.** For every $i \in \mathbb{N}$ we have

(9) $\qquad \dim L_i^* \leq n - 1, \quad E \subseteq C_i^* \cup D_i^*, \quad \dim(C_i^* \cap D_i^*) \leq n - 2.$

*Proof.* It follows from Exercise 1.11.26 that the projection $\pi = p \restriction L_i^*$ is closed. Observe that the fibers of $\pi$ are of the form

$$\{t\} \times L_i(t), \quad (t \in T).$$

Hence by (2) the fibers of $\pi$ are at most $(n-1)$-dimensional. Since $\dim T = 0$, the desired result follows from Theorem 3.6.10.

By (2) and (3) the other statements follow by similar considerations. ◇

By (9) and Theorem 3.2.8, $\dim(\bigcup_{i=1}^{\infty} L_i^*) \leq n - 1$.

**Claim 2.**

(10) $$E \setminus \bigcup_{i=1}^{\infty} L_i^* \subseteq E_{(n-1)}.$$

This implies among other things that the dimension of the set $E \setminus E_{(n-1)}$ is at most $n - 1$. So if $\dim E > n$ then we contradict Lemma 3.11.1. This implies that $\dim E \leq n$ and also by the same argument that $E$ is weakly $n$-dimensional provided that $E$ is $n$-dimensional. The latter fact will follow once we established (8). Simply observe that there are weakly $n$-dimensional spaces (Theorem 3.11.8) which imbed into $E$ by (8). Hence $E$ is at least $n$-dimensional since it contains $n$-dimensional subsets (Theorem 3.2.9).

*Proof.* Let us consider any point $a = (t, z)$ from the set on the left hand side of the inclusion (10), i.e., $z \in H(t)$ by (6). Let $W$ be a neighborhood of $a$ in $\{0,1\}^\infty \times Q$, and let

$$U = \{x \in Q : (t, x) \in W\}.$$

Then $U$ is a neighborhood of $z$; let $i \in \mathbb{N}$ be an index given by (5). The projection $p[C_i^* \setminus W]$ is closed in $T$ (cf. the proof of Claim 1). It also misses $t$ since $C_i(t) \subseteq U$. Let $V$ be a clopen neighborhood of $t$ in $T$, disjoint from $p[C_i^* \setminus W]$. Then $C_i^* \cap p^{-1}[V] \subseteq W$ and therefore

$$C_i^* \cap D_i^* \cap p^{-1}[V]$$

is an $(n-2)$-dimensional partition between the point $a$ and the set $E \setminus W$, cf. (9). ◇

Let us consider now an arbitrary weakly $n$-dimensional space $X$. We shall find an element $t \in T$ such that $X$ embeds in $E(t)$. The dimensional kernel of $X$ is an $F_\sigma$-set in $X$ (Lemma 3.11.1). Hence there exist closed sets

$$Z_1 \subseteq Z_2 \subseteq \cdots$$

in $X$ such that $\dim Z_i \leq n-1$ for every $i$ and $X \setminus \bigcup_{i=1}^\infty Z_i \subseteq X_{(n-1)}$.

Corollary 3.5.6 and the fact that every compact space imbeds in $Q$ (Corollary A.4.4), provide an imbedding $h\colon X \to Q$ such that $\dim \overline{h[Z_i]} \leq n-1$, for $i = 1, 2, \ldots$, and $h(x) \in (\overline{h[X]})_{(n-1)}$ for any $x \in X \setminus \bigcup_{i=1}^\infty Z_i$. Let

(11) $$K = \overline{h[X]}, \quad L_i = \overline{h[Z_i]}.$$

Then

(12) $$h[X] \setminus \bigcup_{i=1}^\infty L_i \subseteq K_{(n-1)}.$$

By (12) there exist pairs $(C_i, D_i)$, $i \in \mathbb{N}$, of compact sets in $K$, with

$$K = C_i \cup D_i, \quad \dim(C_i \cap D_i) \leq n-2,$$

such that for any $z \in h[X] \setminus \bigcup_{i=1}^\infty L_i$ and any neighborhood $U$ of $z$ in $Q$, there is an $i \in \mathbb{N}$ with $z \in C_i \subseteq U$ and $z \notin D_i$. The sets $K$ and $L_i$ in (11), together with the pairs $(C_i, D_i)$ determine a point in the set $\mathfrak{M}$, cf. (1), (2) and (3). So there exists $t \in T$ such that

(13) $$K = K(t), \quad L_i = L_i(t), \quad C_i = C_i(t), \quad D_i = D_i(t).$$

Let us consider the section $E(t)$ of the set $E$, cf. (6). According to formula (5),

$$h[X] \setminus \bigcup_{i=1}^\infty L_i \subseteq H(t),$$

and hence by (13), $h[X] \subseteq E(t)$.

This completes the proof of (8) and ends the proof of the theorem. $\square$

**Dimension of products of weakly one-dimensional spaces.** In his paper [284], MENGER asked whether there exists a weakly one-dimensional space $X$ such that $\dim X^n = n$ for every $n$. We will prove here among other things that the product of an arbitrary family weakly one-dimensional spaces is at most one-dimensional, thereby giving a strong negative answer to MENGER's question.

**Theorem 3.11.11.** *If $X$ is weakly one-dimensional then $X$ has a compactification $\gamma X$ in which it is $L$-imbedded.*

**Proof.** We put $K = \Lambda(X)$. In addition, let $\mathcal{C}$ be a countable collection clopen subsets of $X$ having the property that it is a local base at every point of $X \setminus K$ (Exercise A.2.12).

There is a compactification $\gamma X$ of $X$ such that for every $C \in \mathcal{C}$ we have that $C$ and $X \setminus C$ have disjoint closures in $\gamma X$ (Exercise A.9.3). That is, the closure of $C$ in $\gamma X$ is clopen in $\gamma X$. We claim that $X$ is $L$-imbedded in $\gamma X$.

To this end, let $\mathcal{U}$ be a finite open cover of $X$. For every point $x \in X \setminus K$, pick elements $C_x \in \mathcal{C}$ and $U_x \in \mathcal{U}$ such that

(1) $$x \in C_x \subseteq \overline{C}_x \subseteq U_x.$$

Put $A = X \setminus \bigcup_{x \in X \setminus K} C_x$. Then $A$ is closed in $X$ and a subspace of its dimensional kernel. As a consequence, $A$ is zero-dimensional. The finite open cover $\mathcal{U} \restriction A$ of $A$ has a shrinking $\mathcal{A}$ consisting of pairwise disjoint closed sets (Theorem 3.2.5). Observe that $\mathcal{A}$ is finite. For every $A \in \mathcal{A}$ pick an open set $V_A$ in $X$ and an element $U_A \in \mathcal{U}$ such that

(2) $A \subseteq V_A \subseteq U_A$,
(3) the collection $\{V_A : A \in \mathcal{A}\}$ is pairwise disjoint.

Observe that (2) is easily taken care of because $\mathcal{A}$ is a finite collection of pairwise disjoint closed subsets of $X$. For every $A \in \mathcal{A}$ let $W_A \subseteq \gamma X$ be open such that $W_A \subseteq U_A$ and $W_A \cap X = V_A$. Observe that the collection

$$\{W_A : A \in \mathcal{A}\}$$

is pairwise disjoint by (3) and the fact that $X$ is dense in $\gamma X$.

Now put

$$E = \bigcup_{x \in X \setminus K} \overline{C}_x \cup \bigcup_{A \in \mathcal{A}} W_A.$$

Then $E$ is an open neighborhood of $X$ in $\gamma X$. Let $Z \subseteq E$ be an arbitrary continuum. If $Z \cap \overline{C}_x \neq \emptyset$ for some $x \in X \setminus K$ then $Z \subseteq \overline{C}_x \subseteq U_x$ by connectivity of $Z$ and (1). So we may assume without loss of generality that $Z \cap \overline{C}_x = \emptyset$ for every $x \in X \setminus K$. But then

$$Z \subseteq \bigcup_{A \in \mathcal{A}} W_A$$

so that by (3) and the connectivity of $Z$ there is an $A \in \mathcal{A}$ such that $Z \subseteq W_A$. So this shows that $Z \subseteq U_A$. □

**Corollary 3.11.12.** *The product of a finite family of weakly one-dimensional spaces is one-dimensional.*

**Proof.** Such a product if $L$-imbedded in one of its compactifications by Exercise 3.5.6 and Theorem 3.11.11. So the desired result follows by an application of Corollary 3.5.9. □

In the light of this result, the question arises whether the dimension of products of weakly $n$-dimensional spaces can be estimated more accurately than the bounds we get from Theorem 3.4.12. It is unclear whether this is possible.

**Exercises for §3.11.**

1. Let $X$ be a space and let $A$ and $B$ be disjoint closed subsets of $X$. In addition, let $f\colon X \to \mathbb{I}$ be continuous such that
$$f[A] \subseteq [0, \tfrac{1}{3}), \quad g[B] \subseteq (\tfrac{2}{3}, 1].$$
Prove that there is a continuous function $\hat{f}\colon X \to \mathbb{I}$ having the following properties:
   (1) $\hat{f}(x) = f(x)$ if $x \in f^{-1}\big[[\tfrac{1}{3}, \tfrac{2}{3}]\big]$,
   (2) $\hat{f}[A] \subseteq \{0\}$ and $\hat{f}[B] \subseteq \{1\}$.

2. For every $n \in \mathbb{N}$ let $X_n$ be weakly one-dimensional. Prove that $\prod_{n=1}^{\infty} X_n$ is one-dimensional.

We now present a different proof of the fact that the product of two weakly one-dimensional spaces is one-dimensional.

Let $X$ and $Y$ be weakly one-dimensional.

Pick an arbitrary point $(x, y) \in (X \times Y) \setminus \big(\Lambda(X) \times \Lambda(Y)\big)$. We claim that $X \times Y$ is at most one-dimensional at $(x, y)$. We first show that this suffices. Striving for a contradiction, assume that $X \times Y$ is two-dimensional. Then by our claim,
$$\Lambda(X \times Y) \subseteq \Lambda(X) \times \Lambda(Y)$$
hence $\dim\big(\Lambda(X) \times \Lambda(Y)\big) \leq 0$. This contradicts the fact that $\Lambda(X \times Y)$ is at least one-dimensional (Lemma 3.11.1(2)).

We assume without loss of generality that $y \notin \Lambda(Y)$. If in addition $x \notin \Lambda(X)$ then $X \times Y$ is zero-dimensional at $(x, y)$. So we assume further that $x \in \Lambda(X)$. Let $U$ and $V$ be arbitrary open subsets of $X$ and $Y$, respectively, such that $x \in U$ and $y \in V$. We will construct an open subset $E \subseteq X \times Y$ such that
$$(x, y) \in E \subseteq U \times V$$
while moreover $\operatorname{ind}\operatorname{Fr} E \leq 0$. Since $y \notin \Lambda(Y)$ we may assume without loss of generality that $V$ is clopen. Let $U'$ be an open neighborhood of $x$ such that
$$\overline{U'} \subseteq U, \quad \operatorname{Fr} U' \subseteq X \setminus \Lambda(X).$$
It is possible to pick $U'$ since $\operatorname{ind}\big(X \setminus \Lambda(X)\big) \leq 0$ (Corollary 3.1.6). If $\dim V \leq 0$ then
$$\operatorname{Fr}(U' \times V) = \operatorname{Fr}(U') \times V$$
is at most zero-dimensional and so we are done. Assume therefore that $\dim V = 1$. Then $V$ is weakly one-dimensional (Lemma 3.11.3), and so we may assume without loss of generality that $V = Y$. Put $A = \operatorname{Fr} U'$ and let $U''$ be an open subset of $X$ such that $A \subseteq U'' \subseteq \overline{U''} \subseteq U$.

3. Prove that for every $n \in \mathbb{N}$ there exist pairwise disjoint clopen subsets $U_{1n}, U_{2n}, \ldots$ of $X$ such that
   (1) $U_{in} \cap A \neq \emptyset$ for every $i$,

(2) $\operatorname{diam} U_{in} < 1/n$ for every $i$,
(3) $A \subseteq \bigcup_{i=1}^{\infty} U_{in} \subseteq \overline{\bigcup_{i=1}^{\infty} U_{in}} \subseteq U''$,
(4) $\operatorname{Fr}(\bigcup_{i=1}^{\infty} U_{in}) \subseteq \Lambda(X)$.

Now for every $n \in \mathbb{N}$ let $\mathcal{E}_n$ denote the collection of all clopen subsets of $Y$ of diameter at most $1/n$ and put

$$B_n = Y \setminus \bigcup \mathcal{E}_n.$$

Observe that if $b \in Y$ is such that for every $n$ there exists a clopen subset $C_n$ of $Y$ containing $b$ and of diameter at most $1/n$ then $Y$ is zero-dimensional at $b$. So if $b \in \Lambda(Y)$ then there exists $n \in \mathbb{N}$ such that $b \in B_n$.

4. Prove that for every $n$ there is a decreasing sequence $V_{in}$, $i \in \mathbb{N}$, of clopen subsets of $Y$ such that $\bigcap_{i=1}^{\infty} V_{in} = B_n$.

5. Let
$$W = \bigcup_{n=1}^{\infty} \bigcup_{i=1}^{\infty} (U_{in} \times V_{in}).$$
Prove that $A \times \Lambda(Y) \subseteq W$.

6. Prove that $\operatorname{Fr} W \subseteq \bigl(\Lambda(X) \times \Lambda(Y)\bigr) \cup (A \times Y)$.

Put
$$E = W \cup (U' \times Y).$$
Then $E$ is an open neighborhood of $(x,y)$ and by Exercise 3.11.2(3) it follows that
$$\overline{E} \subseteq U \times Y.$$
We claim that $\operatorname{Fr} E$ is zero-dimensional. First observe that

(∗) $\qquad \operatorname{Fr} E \subseteq \operatorname{Fr} W \cup (\operatorname{Fr} U' \times Y) = \operatorname{Fr} W \cup (A \times Y).$

Put $B_0 = \operatorname{Fr} E \cap (A \times Y)$ and $B_1 = \operatorname{Fr} E \setminus B_0$, respectively. Since $A \times \Lambda(Y) \subseteq W$ (Exercise 3.12.4) and $\operatorname{Fr} E \cap W = \emptyset$, it follows that
$$B_0 \subseteq A \times \bigl(Y \setminus \Lambda(Y)\bigr),$$
which is zero-dimensional. We conclude that $B_0$ is a zero-dimensional closed subspace of $\operatorname{Fr} E$. In addition, Exercise 3.12.5 and (∗) imply that
$$B_1 = \operatorname{Fr} E \setminus B_0 \subseteq \operatorname{Fr} E \setminus (A \times Y) \subseteq \operatorname{Fr} W \setminus (A \times Y) \subseteq \Lambda(X) \times \Lambda(Y),$$
which is also zero-dimensional. So $B_1$ is zero-dimensional as well. Since $B_0$ is closed, the Countable Closed Sum Theorem 3.2.8 now easily gives us that $\dim E \leq 0$, as desired.

## 3.12. Colorings of maps

Let $X$ be a space and let $f \colon X \to X$ be a fixed-point free continuous function. On Page 188 we defined $f$ to be colorable if there is a finite closed cover $\mathcal{A}$ of $X$ such that $A \cap f[A] = \emptyset$ for every $A \in \mathcal{A}$. Since finite open covers can be shrunk to closed covers, and finite closed covers can be swelled to open covers, finite open covers do equally well (see §3.2).

There are two natural questions we consider here. First of all the question whether every fixed-point free continuous function can be colored. And secondly, whether something can be said about the number of colors needed for a colorable function.

**Negative results.**

**Theorem 3.12.1.** *There is a space $X$ with a fixed-point free involution which admits no coloring.*

**Proof.** If $\alpha_n \colon \mathbb{S}^n \to \mathbb{S}^n$ is the antipodal map, then every closed cover $\mathcal{F}$ of $\mathbb{S}^n$ such that $F \cap \alpha[F] = \emptyset$ for every $F \in \mathcal{F}$ has at least $n+2$ elements by Theorem 2.5.1. Let $X$ be the topological sum of the $\mathbb{S}^n$ ($n \in \mathbb{N}$), and let $h$ be the involution on $X$ which is defined by the requirement that $h \restriction \mathbb{S}^n = \alpha_n$ for every $n$.

Now assume that there is a finite closed cover $\mathcal{F}$ such that $F \cap h[F] = \emptyset$ for every $F \in \mathcal{F}$. Pick $m$ so large that $|\mathcal{F}| - 2 < m$. Then $\mathcal{G} = \mathcal{F} \restriction \mathbb{S}^m$ is a closed cover of $\mathbb{S}^m$ no element of which contains an antipodal pair. This implies that $m + 2 \leq |\mathcal{G}| \leq |\mathcal{F}|$, which is a contradiction. $\square$

Observe that the space $X$ in this theorem is infinite-dimensional. It is therefore natural to ask whether there is also a finite dimensional example. The answer to this question is quite interesting. For continuous functions there is such an example, but for homeomorphisms this is not true. Before we can answer it, we need to make a very interesting detour which will lead us via topological Ramsey theory to the desired result.

We consider the Cantor set $\{0,1\}^\infty$ with its usual Tychonoff product topology. As was remarked on Page 458, the function $f \colon \mathcal{P}(\mathbb{N}) \to \{0,1\}^\infty$ defined by $f(A) = \chi_A$, the characteristic function of $A$, is a bijection. The product topology on $\{0,1\}^\infty$ therefore induces a topology on $\mathcal{P}(\mathbb{N})$. With this topology, $\mathcal{P}(\mathbb{N})$ is homeomorphic to $\{0,1\}^\infty$ and hence is a Cantor set. The collection of all sets of the form

$$\langle \sigma, \tau \rangle = \{A \in \mathcal{P}(\mathbb{N}) : (\sigma \subseteq A) \wedge (A \cap \tau = \emptyset)\},$$

where $\sigma$ and $\tau$ are arbitrary *finite* subsets of $\mathbb{N}$, is a clopen base for $\mathcal{P}(\mathbb{N})$. For this simply observe that if $\sigma$ and $\tau$ are disjoint finite subsets of $\mathbb{N}$ then the basic open subset

$$\{x \in \{0,1\}^\infty : (i \in \sigma \Rightarrow x_i = 1) \text{ and } (i \in \tau \Rightarrow x_i = 0)\}$$

of $\{0,1\}^\infty$ corresponds under this function to $\langle \sigma, \tau \rangle$. So the finite subsets of $\mathbb{N}$ form a countable dense subset of $\mathcal{P}(\mathbb{N})$. The complement, i.e., the collection of all infinite subsets of $\mathbb{N}$, is a $G_\delta$-subset which is also dense by the Baire Category Theorem A.6.6.

If $H \subseteq \mathbb{N}$ is infinite then $[H]^\omega$ denotes the collection of all infinite subsets of $H$.

The *usual topology* on $[\mathbb{N}]^\omega$ is the subspace topology that $[\mathbb{N}]^\omega$ inherits from $\mathcal{P}(\mathbb{N})$ (with the just defined topology).

If $A, B \subseteq \mathbb{N}$ then we write $A < B$ provided that $\max(A) < \min(B)$. So this makes sense only if $A$ is finite. For $a < A$, where $a \subseteq \mathbb{N}$ is (necessarily) finite and $A \subseteq \mathbb{N}$ is infinite, let

(*) $$[a, A] = \{S \in [\mathbb{N}]^\omega : a \subseteq S \subseteq a \cup A\}.$$

Observe that $[\emptyset, A]$ is simply $[A]^\omega$.

The *Ellentuck topology* on $[\mathbb{N}]^\omega$ has as basic open sets all sets of the form (*). It is easy to verify that they satisfy the axioms for a base for a topology. There is a problem though with the Ellentuck topology since it does not have a countable base. So it violates our assumption that all topological spaces are separable and metrizable. This causes no problem, since the Ellentuck topology is used merely as a tool here to prove interesting theorems on $[\mathbb{N}]^\omega$ endowed with its usual topology, and the language of topology which we are used to is convenient for that.

**Lemma 3.12.2.** *If $H \subseteq \mathbb{N}$ is infinite then the subspace $[H]^\omega$ of $[\mathbb{N}]^\omega$ with the Ellentuck topology is homeomorphic to $[\mathbb{N}]^\omega$.*

**Proof.** Let $f\colon H \to \mathbb{N}$ be any bijection. It is easy to see that the function $F\colon \mathcal{P}(H) \to \mathcal{P}(\mathbb{N})$ defined by

$$F(A) = f[A]$$

is a homeomorphism. □

**Lemma 3.12.3.** *The Ellentuck topology on $[\mathbb{N}]^\omega$ contains the usual topology on $[\mathbb{N}]^\omega$.*

**Proof.** It suffices to prove that if $\sigma$ and $\tau$ are arbitrary finite subsets of $\mathbb{N}$ then

$$\langle \sigma, \tau \rangle \cap [\mathbb{N}]^\omega$$

is open in the Ellentuck topology. To this end, let $A \subseteq \mathbb{N}$ be infinite such that $A \in \langle \sigma, \tau \rangle$. Put $B = A \setminus \sigma$. Then $A \in [\sigma, B] \subseteq \langle \sigma, \tau \rangle$, as required. □

A set $X \subseteq [\mathbb{N}]^\omega$ is called *Ramsey* if there is an $A \in [\mathbb{N}]^\omega$ with

$$[\emptyset, A] \subseteq X \quad \text{or} \quad [\emptyset, A] \subseteq [\mathbb{N}]^\omega \setminus X.$$

It is called *completely Ramsey* if for every $a, A \subseteq \mathbb{N}$ with $a$ finite and $A$ infinite and $a < A$ there is an infinite set $B \subseteq A$ with

$$[a, B] \subseteq X \quad \text{or} \quad [a, B] \subseteq [\mathbb{N}]^\omega \setminus X.$$

We introduce some additional notation. Let $U \subseteq [\mathbb{N}]^\omega$ be open in the Ellentuck topology. Call $[a, A]$ *good* if for some $B \subseteq A$ we have $[a, B] \subseteq U$. If $[a, A]$ is not good then it is called *bad*. We shall call $[a, A]$ *very bad* if it is bad and for every $n \in A$, $[a \cup \{n\}, A/n]$ is bad, where $A/n = \{m \in A : m > n\}$. Notice that if $[a, A]$ is (very) bad and $B \subseteq A$ is infinite then $[a, B]$ is (very) bad.

**Lemma 3.12.4.** *If $[a, A]$ is bad then there is an infinite subset $B \subseteq A$ such that $[a, B]$ is very bad.*

**Proof.** Indeed, suppose that this is not true. Then $[a, A]$ is bad but not very bad and so there exists $n_0 \in A$ such that $[a \cup \{n_0\}, A/n_0]$ is good. There consequently exists an infinite $B_0 \subseteq A/n_0$ such that $[a \cup \{n_0\}, B_0] \subseteq U$. Since $[a, B_0]$ is not very bad, there exists $n_1 \in B$ such that $[a \cup \{n_1\}, B_0/n_1]$ is good. Observe that $n_0 < n_1$. So there exists an infinite $B_1 \subseteq B_0/n_1$ with $[a \cup \{n_1\}, B_1] \subseteq U$, etc. Let $B = \{n_0, n_1, \dots\}$. Pick an arbitrary element $C \in [a, B]$. Then $a \subseteq C \subseteq a \cup B$. Let $n_i$ be the first integer in $(C \cap B) \setminus a$. Since $[a \cup \{n_i\}, B_i] \subseteq U$ and $C \in [a \cup \{n_i\}, B_i]$ we conclude that $C \in U$. Since $C$ was arbitrary, it therefore follows that $[a, B] \subseteq U$, so $[a, A]$ is good, which is a contradiction. □

**Proposition 3.12.5.** *Let $U \subseteq [\mathbb{N}]^\omega$ be open in the Ellentuck topology. Then $U$ is completely Ramsey.*

**Proof.** Suppose that $[a, A]$ is given. We may assume without loss of generality that it is bad.

**Claim 1.** There is a decreasing sequence
$$A \supseteq B_0 \supseteq B_1 \supseteq \cdots,$$
with $n_i = \min(B_i)$ strictly increasing, such that for any $b \subseteq \{n_0, \dots, n_{i-1}\}$ we have that $[a \cup b, B_i]$ is very bad.

*Proof.* By Lemma 3.12.4 there exists an infinite $B_0 \subseteq A$ such that $[a, B_0]$ is very bad. Let $n_0 = \min(B)$.

Since $[a \cup \{n_0\}, B/\{n_0\}]$ is bad, again by Lemma 3.12.4 there exists an infinite $B_1 \subseteq B/n_0$ such that $[a \cup \{n_0\}, B_1]$ is very bad. Then $[a, B_1]$ is very bad since $B_1 \subseteq B$ and $[a \cup \{n_0\}, B_1]$ is very bad by construction.

Now assume that for $i \geq 1$
$$A \supseteq B_0 \supseteq B_1 \supseteq \cdots \supseteq B_i,$$
are as we want. So $n_{i-1} < n_i = \min(B_i)$ and if $b \subseteq \{n_0, \dots, n_{i-1}\}$ is arbitrary then $[a \cup b, B_i]$ is very bad. So $[a \cup b \cup \{n_i\}, B/n_i]$ is bad and we can use Lemma 3.12.4 repeatedly (finitely often) to get an infinite set $B_{i+1} \subseteq B_i$ such that for every $b \subseteq \{n_0, \dots, n_{i-1}\}$ we have that
$$[a \cup b \cup \{n_i\}, B_{i+1}]$$

is very bad. Then $B_{i+1}$ is clearly as required. ◊

Let $B = \{n_0, n_1, \ldots\}$. We claim that $[a, B] \subseteq [\mathbb{N}]^\omega \setminus U$. If not then
$$[a, B] \cap U \neq \emptyset$$
and so since $U$ is open there is $[a', B'] \subseteq [a, B]$ such that $[a', B'] \subseteq U$. By Exercise 3.12.1 for some $i$, $a' = a \cup b$ with $b \subseteq \{n_0, \ldots, n_i\}$ and $B'/n_i \subseteq B_i/n_i$. So since $[a \cup b, B'/n_i] \subseteq U$, we have that $[a \cup b, B_i/n]$ is good. This is a contradiction. □

**Corollary 3.12.6.** *Let $[\mathbb{N}]^\omega = P_0 \cup \cdots \cup P_k$, where each $P_i$ is open in the natural topology. Then there are an infinite $H \subseteq \mathbb{N}$ and an $i \leq k$ such that every infinite subset of $H$ is contained in $P_i$.*

**Proof.** It is clear that we may restrict ourselves to the case that $k = 1$ (Lemma 3.12.2). Since $P_0$ is open in the Ellentuck topology by Lemma 3.12.3 it is completely Ramsey by Proposition 3.12.5. So we are done. □

We are now ready to present our second example of a map which cannot be colored.

**Theorem 3.12.7.** *There are a zero-dimensional space $X$ and a fixed-point free continuous function $f: X \to X$ such that*

(1) *$f$ is closed,*
(2) *$f$ is finite-to-one,*
(3) *$f$ cannot be colored.*

**Proof.** Put $X = [\mathbb{N}]^\omega$ endowed with its natural topology. Define $f: X \to X$ by
$$f(A) = A \setminus \{\min A\} \qquad (A \in X).$$
It is clear that $f$ is fixed-point free. We claim that it is continuous. To this end, let $(A_n)_n$ be a sequence in $X$ converging to an element $A \in X$. Let
$$p = \min A, \quad \alpha = \{p\}, \quad \beta = \{1, 2, \ldots, p-1\},$$
respectively. Then $A \in \langle \alpha, \beta \rangle$ and since $A_n \to A$ we may therefore assume without loss of generality that $\min A_n = \min A$ for all $n$. Now assume that
$$A \setminus \{\min A\} \in \langle \sigma, \tau \rangle$$
for certain finite $\sigma, \tau \subseteq \mathbb{N}$. Then if $\sigma' = \sigma \cup \{\min A\}$ and $\tau' = \tau \setminus \{\min A\}$, it follows that $A \in U$, where $U = \langle \sigma', \tau' \rangle$. There exists $N \in \mathbb{N}$ such that $A_n \in U$ for all $n \geq N$. But since $\min A_n = \min A$ it follows that $\sigma \subseteq A_n \setminus \{\min A_n\}$ and $\tau \cap (A_n \setminus \{\min A_n\}) = \emptyset$. We conclude that $f(A_n) \in \langle \sigma, \tau \rangle$ for all $n \geq N$.

To prove that $f$ is closed, let $A \subseteq X$ be closed, and assume that $B \notin f[A]$. Let $p = \min B$. Observe that
$$f^{-1}[B] = \{B \cup \{i\} : i \leq p - 1\}.$$

Since $\mathcal{A}$ is closed, for every $i \leq p-1$ there exist finite sets $\sigma_i, \tau_i \subseteq \mathbb{N}$ such that

(∗∗) $\qquad B \cup \{i\} \in \langle \sigma_i, \tau_i \rangle \subseteq [\mathbb{N}]^\omega \setminus \mathcal{A}.$

Put $\sigma = \{p\} \cup (\bigcup_{i=1}^{p-1} \sigma_i \setminus \{i\})$ and $\tau = \bigcup_{i=1}^{p-1} \tau_i$, respectively, and let $V = \langle \sigma, \tau \rangle$. Then $B \in V$. We claim that $V \cap f[\mathcal{A}] = \emptyset$. Striving for a contradiction, assume that there exists $A \in \mathcal{A}$ such that $A \setminus \{\min A\} \in V$. Then since

$$\sigma \subseteq A \setminus \{\min A\}$$

it follows that $p \in A \setminus \{\min A\}$ and so $\min A < p$. Put $i = \min A$. Observe that $\tau_i \cap (A \setminus \{i\}) = \emptyset$ and that $i \notin \tau_i$. So $A \in \langle \sigma_i, \tau_i \rangle$ which contradicts (∗∗).

We already implicitly proved that $f$ is finite-to-one. To show that it cannot be colored, let $\mathcal{U}$ be an arbitrary finite open cover of $X$. By Corollary 3.12.6 there are an element $U \in \mathcal{U}$ and an infinite subset $H \subseteq \mathbb{N}$ such that $[H]^\omega \subseteq U$. But then $f(H) \in U$ so that $f[U] \cap U \neq \emptyset$, i.e., $\mathcal{U}$ does not color $f$. □

**Positive results.** We will now present the main results on the existence of colorings.

If $f \colon X \to X$ is a homeomorphism then $A \subseteq X$ is called $f$-*invariant* if $f[A] = A$.

**Theorem 3.12.8.** *Let $X$ be finite dimensional, and let $f$ be a fixed-point free homeomorphism. Then $f$ can be colored.*

**Proof.** We will first prove that if $\dim X = 0$ then $f$ can be colored with at most three pairwise disjoint colors. By the remarks on Page 188 there exists an open cover $\mathcal{U}$ of $X$ such that $f[U] \cap U = \emptyset$ for all $U \in \mathcal{U}$. By Proposition 1.5.1 we may assume without loss of generality that $\mathcal{U}$ is a clopen partition. So $\mathcal{U}$ is countable and we may list it as $\{U_n : n \in \mathbb{N}\}$, where $U_1 = \emptyset$. By induction on $n$ we shall now construct a clopen partition $\{V_1^n, V_2^n, V_3^n\}$ of $U_n$ such that

(1) $\qquad f\left[\bigcup_{i=1}^n V_j^i\right] \cap \bigcup_{i=1}^n V_j^i = \emptyset$

for $j \leq 3$. Put $V_j^1 = \emptyset$ for $j \leq 3$, and assume that the $V_j^i$'s have been constructed for $i \leq n$ and $j \leq 3$. For $j \leq 3$ put

$$E_j = \bigcup_{i=1}^n V_j^i.$$

**Claim 1.**

$$\bigcap_{j=1}^3 \left( f^{-1}[E_j] \cup f[E_j] \right) = \emptyset.$$

*Proof.* First observe that $\bigcap_{j=1}^{3} E_j = \emptyset$ from which it follows that both the sets $f[\bigcap_{j=1}^{3} E_j]$ and $f^{-1}[\bigcap_{j=1}^{3} E_j]$ are empty since $f$ is one-to-one. So if the claim is not true then we may assume without loss of generality that e.g.,
$$f^{-1}[E_1] \cap f^{-1}[E_2] \cap f[E_3] \neq \emptyset.$$
But this is impossible since $E_1 \cap E_2 = \emptyset$. ◇

So the claim implies that
$$\{U_n \setminus (f^{-1}[E_j] \cup f[E_j]) : j \leq 3\}$$
is an open cover of $U_n$, which consequently can be refined by a clopen partition (Proposition 1.5.1). So there exists a clopen partition
$$\{V_1^{n+1}, V_2^{n+1}, V_3^{n+1}\}$$
of $U_{n+1}$ such that
(2) $$V_j^{n+1} \cap (f^{-1}[E_j] \cup f[E_j]) = \emptyset$$
for $j \leq 3$.

**Claim 2.** *The $V_j^{n+1}$'s are as required.*

*Proof.* If (1) fails for $n+1$ then since it holds for $n$ for some $j \leq 3$ either
$$f[V_j^{n+1}] \cap \bigcup_{i=1}^{n} V_j^i = f[V_j^{n+1}] \cap E_j \neq \emptyset$$
which is impossible by (2), or
$$f[V_j^{n+1}] \cap V_j^{n+1} \neq \emptyset$$
which is impossible since $f[U_{n+1}] \cap U_{n+1} = \emptyset$, or
$$f\left[\bigcup_{i=1}^{n} V_j^i\right] \cap V_j^{n+1} = f[E_j] \cap V_j^{n+1} \neq \emptyset$$
which is also impossible by (2). ◇

This completes the inductive construction. So $V_j = \bigcup_{i=1}^{\infty} V_j^i$ for $j \leq 3$ is the desired coloring of $f$ with three colors.

Now assume that $\dim X = n+1$ and that the theorem has been proved for all spaces of dimension at most $n$, where $n \geq 0$. Let $D$ be a countable dense subset of $X$ and let $\mathcal{B}$ be a countable base for $X$ such that $\dim \operatorname{Fr} B \leq n$ for all $B \in \mathcal{B}$. Put $S = D \cup \bigcup_{B \in \mathcal{B}} \operatorname{Fr} B$, and
$$T = \bigcup_{n \in \mathbb{Z}} f^n[S],$$
respectively. It is clear that $T$ is a dense $f$-invariant $F_\sigma$-subset of $X$ which by the Countable Closed Sum Theorem 3.2.8 has dimension at most $n$. In

addition, $Z = X \setminus T$ is also $f$-invariant and is of dimension at most 0 since $\operatorname{Fr} B \cap Z = \emptyset$ for all $B \in \mathcal{B}$ (so $\mathcal{B} \upharpoonright Z$ is a countable base for $Z$ consisting of relatively clopen sets). There exists by our inductive assumption a finite cover $\mathcal{U}$ of $T$ consisting of relatively open sets such that $f[U] \cap U = \emptyset$ for all $U \in \mathcal{U}$. For every $U \in \mathcal{U}$ let $U' \subseteq X$ be open such that $U' \cap T = U$. Then $f[U'] \cap U' = \emptyset$ since $f$ is a homeomorphism and $T$ is dense in $X$. We conclude that $\mathcal{U}' = \{U' : U \in \mathcal{U}\}$ is a finite collection of open subsets of $X$ which 'colors' $T$. Now consider the zero-dimensional $f$-invariant set $Z$. We would like to apply the same technique as the one we just used. But $Z$ need not be dense in $X$. But we know that a strong form of the theorem is true for zero-dimensional spaces, so we will proceed from there. There exist relatively open pairwise disjoint sets $V_1, V_2$ and $V_3$ in $Z$ that 'color' $Z$. Observe that if e.g., $x \in V_1$ then $f(x) \in V_2 \cup V_3$. By Lemma A.8.1 there are pairwise disjoint open sets $V_1', V_2'$ and $V_3'$ in $X$ such that $V_i' \cap Z = V_i$ for $i \leq 3$. Now put $V_1'' = V_1' \cap f^{-1}[V_2' \cup V_3']$. Then $V_1''$ is an open subset of $X$ and clearly $f[V_1''] \cap V_1'' = \emptyset$. We conclude by proving that $V_1 \subseteq V_1''$. Indeed, take an arbitrary element $x \in V_1$. Then $f(x) \in V_2 \cup V_3$ from which it follows that $x \in f^{-1}[V_2 \cup V_3] \subseteq f^{-1}[V_2' \cup V_3']$. We conclude that $x \in V_1''$. In other words, $V_1''$ 'colors' all the points of $V_1$. Define the sets $V_2''$ and $V_3''$ similarly. So $\mathcal{U} \cup \{V_0'', V_1'', V_3''\}$ is the desired coloring of $X$. □

**Remark 3.12.9.** Observe that the proof of Theorem 3.12.8 shows that if $X$ is $n$-dimensional and $f$ is a fixed-point free homeomorphism of $X$ then $f$ can be colored with at most $3n + 3$ colors.

**The number of colors.** It is a natural to try to reduce the number of colors needed to color a given function. This is sometimes possible and we will come to that after some preparatory work.

Our main tool is the following theorem, which is interesting in its own right.

**Theorem 3.12.10.** *Let $X$ be a space with $\dim X \leq n$ and let $\mathcal{U}$ be a finite open cover of $X$ with $|\mathcal{U}| \geq n + 3$. In addition, let $f : X \to X$ be a homeomorphism. Then there is an open shrinking $\mathcal{V} = \{V_U : U \in \mathcal{U}\}$ of $\mathcal{U}$ such that for any $\mathcal{F} \subseteq \mathcal{U}$ with $|\mathcal{F}| = n + 3$ we have $\bigcap_{U \in \mathcal{F}}(V_U \cup f^{-1}[V_U]) = \emptyset$.*

It is tempting to think of the shrinking $\mathcal{V}$ in this result as a cover for which

(∗) $\qquad \operatorname{ord}\{V_U \cup f^{-1}[V_U] : U \in \mathcal{U}\} \leq n + 1$.

But since simple examples show that for different $U$'s the sets $V_U \cup f^{-1}[V_U]$ can be the same, the property of $\mathcal{V}$ stated in the theorem is stronger than the one in (∗). The following is such an example.

**Example 3.12.11.** Define $f : \{0, 1, 2, 3, 4\} \to \{0, 1, 2, 3, 4\}$ as follows

$$f(0) = 1, f(1) = 0, f(2) = 3, f(3) = 2, f(4) = 4.$$

In addition, let $\mathcal{V} = \{\{0,4\}, \{1,4\}, \{2,4\}, \{3,4\}\}$. Then the collection
$$\mathcal{W} = \{V \cup f^{-1}[V] : V \in \mathcal{V}\}$$
has precisely two elements, namely, the sets $\{0,1,4\}$ and $\{2,3,4\}$. So the order of $\mathcal{W}$ is 1. But since $\bigcap \mathcal{W} \neq \emptyset$, the cover $\mathcal{V}$ does not have the property stated in Theorem 3.12.10.

Before presenting the proof of Theorem 3.12.10, we will first show how we can use it to reduce the number of colors of a homeomorphism.

**Theorem 3.12.12.** *Let $X$ be a space with $\dim X \leq n$, and let $f: X \to X$ be a fixed-point free homeomorphism. If $f$ is colorable, then it can be colored with $n + 3$ colors.*

**Remark 3.12.13.** The space $X = \{0, 1, 2\}$ and the homeomorphism
$$f = \{(0,1), (1,2), (2,0)\}$$
of $X$ show that Theorem 3.12.12 is sharp for $n = 0$. By results of STEINLEIN [380, 379] it follows that Theorem 3.12.12 is sharp for every $n$.

**Proof.** Let $k$ be the minimum cardinality of a coloring of $f$, and let
$$\mathcal{U} = \{U_1, \ldots, U_k\}$$
be a coloring. Assume that $k > n + 3$. We will derive a contradiction. Observe that $U_i \neq U_j$ if $i \neq j$. By Theorem 3.12.10 we may without loss of generality assume that for any $\mathcal{F} \subseteq \mathcal{U}$ of cardinality $n + 3$ we have
$$\bigcap_{U \in \mathcal{F}} (U \cup f^{-2}[U]) = \emptyset.$$
Since $f$ is a homeomorphism, this means that for any $\mathcal{F} \subseteq \mathcal{U}$ of cardinality $n + 3$ we have

(1) $$\bigcap_{U \in \mathcal{F}} (f[U] \cup f^{-1}[U]) = \emptyset.$$

Let $\mathcal{F} = \{F_1, \ldots, F_k\}$ be a closed shrinking of $\mathcal{U}$ (Proposition A.7.1). For every $i \leq k - 1$, define
$$B_i = F_i \cup \left( F_k \cap f[X \setminus U_i] \cap f^{-1}[X \setminus U_i] \right).$$
We claim that $\mathcal{B} = \{B_1, \ldots, B_{k-1}\}$ is a coloring of $f$. To see that
$$f[B_i] \cap B_i = \emptyset$$
observe that
$$f[B_i] = f[F_i] \cup \left( f[F_k] \cap f^2[X \setminus U_i] \cap (X \setminus U_i) \right).$$
This gives us what we want since
$$F_i \cap f[F_i] = \emptyset, \quad F_i \cap (X \setminus U_i) = \emptyset, \quad F_k \cap f[F_k] = \emptyset.$$

To see that $\mathcal{B}$ covers, first note that $\bigcup_{i=1}^{k-1} F_i \subseteq \bigcup_{i=1}^{k-1} B_i$. Hence it suffices to show that $F_k$ is covered. To this end, pick an arbitrary $x \in F_k$ and consider the collection

$$\{f[U_i] \cup f^{-1}[U_i] : i \leq k-1\}.$$

Since $k - 1 > n + 2$, (1) implies that there exists $i \leq k - 1$ such that

$$x \notin f[U_i] \cup f^{-1}[U_i].$$

But then $x \in B_i$. This contradiction completes the proof of the theorem. □

Before turning to the proof of Theorem 3.12.10, we make the following remarks.

Recall that we wish to shrink an open cover $\mathcal{U}$ to an open cover

$$\mathcal{V} = \{V_U : U \in \mathcal{U}\}$$

such that for any subfamily $\mathcal{W} \subseteq \mathcal{U}$ of cardinality $n + 3$ we have

$$\bigcap_{U \in \mathcal{W}} (U \cup f^{-1}[U]) = \emptyset.$$

A moments reflection shows that our task is to construct $\mathcal{V}$ in such a way that for any subfamily $\mathcal{W}$ of $\mathcal{V}$ of cardinality $n + 3$ and any partition $\mathcal{F} \cup \mathcal{G}$ of $\mathcal{W}$ we have

$$\bigcap_{F \in \mathcal{F}} F \cap \bigcap_{G \in \mathcal{G}} f^{-1}[G] = \emptyset.$$

**Lemma 3.12.14.** *Let $X$ be a space with $\dim X \leq n$. Let $F_i \subseteq U_i$ for $i \leq m$, with $F_i$ closed and $U_i$ open. In addition, Let $G_j \subseteq U'_j$ for $j \leq k$, with $G_j$ closed and $U'_j$ open. If $m + k \geq n + 3$ then there exist closed subsets*

$$\hat{F}_1, \ldots, \hat{F}_m, \quad \hat{G}_1, \ldots, \hat{G}_k$$

*such that*

(1) $\hat{F}_i \subseteq U_i$ for $i \leq m$, and $\hat{G}_j \subseteq U'_j$ for $j \leq k$,
(2) $\bigcup_{i=1}^m F_i \subseteq \bigcup_{i=1}^m \hat{F}_i$, and $\bigcup_{j=1}^k G_j \subseteq \bigcup_{j=1}^k \hat{G}_j$,
(3) $\bigcap_{i=1}^m \hat{F}_i \cap \bigcap_{j=1}^k \hat{G}_j = \emptyset$.

**Proof.** We assume without loss of generality that $m + k = n + 3$. Suppose first that $m \geq n + 2$. Since $\dim X \leq n$, there exist open sets $V_1, \ldots, V_{n+1}$ such that

(4) $F_i \subseteq V_i \subseteq \overline{V_i} \subseteq U_i$ for $i \leq n + 1$,
(5) $\bigcap_{i=1}^{n+1} \operatorname{Fr} V_i = \emptyset$.

So we get what we want by a direct application of Lemma 3.2.3 (let $A_i = F_i$ for $i \leq n + 1$ and $A = F_{n+2}$). We may therefore assume without loss of generality that $m \leq n+1$ and, similarly, that $k \leq n+1$. Since $m+k = n+3$, this implies that $k, m \geq 2$.

Observe that for $i \leq m - 1$ and $j \leq k - 1$ we have
$$F_i \cap F_m \subseteq U_i, \quad G_j \cap G_k \subseteq U'_j.$$
Since $\dim X \leq n$ and $(m-1) + (k-1) = n + 1$, there exist open sets
$$V_1, \ldots, V_{m-1}, \quad W_1, \ldots, W_{k-1}$$
such that

(6) $F_i \cap F_m \subseteq V_i \subseteq \overline{V_i} \subseteq U_i$ for $i \leq m - 1$,
(7) $G_j \cap G_k \subseteq W_j \subseteq \overline{W_j} \subseteq U'_j$ for $j \leq k - 1$,
(8) $\bigcap_{i=1}^{m-1} \operatorname{Fr} V_i \cap \bigcap_{j=1}^{k-1} \operatorname{Fr} W_j = \emptyset$.

So we again get what we want by a direct application of Lemma 3.2.3. □

**Proof of Theorem 3.12.10.** Let $\mathcal{F}$ be a closed shrinking of $\mathcal{U}$ (Proposition A.7.1). Fix $n + 3$ different elements of $\mathcal{U}$, say,
$$\mathcal{G} = \{U_1, \ldots, U_m, U'_1, \ldots, U'_k\},$$
where $m + k = n + 3$. Let
$$\mathcal{H} = \{F_1, \ldots, F_m, G_1, \ldots, G_k\},$$
be the elements of $\mathcal{F}$ corresponding to the elements in $\mathcal{G}$. So $F_i \subseteq U_i$ for $i \leq m$ and $f^{-1}[G_j] \subseteq f^{-1}[U'_j]$ for $j \leq k$. Since $m + k = n + 3$, by Lemma 3.12.14, there exist closed sets $A_1, \ldots, A_m, B_1, \ldots, B_k$ such that

(1) $A_i \subseteq U_i$ for $i \leq m$, and $B_j \subseteq f^{-1}[U'_j]$ for $j \leq k$,
(2) $\bigcup_{i=1}^m F_i \subseteq \bigcup_{i=1}^m A_i$, and $\bigcup_{j=1}^k f^{-1}[G_j] \subseteq \bigcup_{j=1}^k B_j$,
(3) $\bigcap_{i=1}^m A_i \cap \bigcap_{j=1}^k B_j = \emptyset$.

Put $A'_j = f[B_j]$ for $j \leq k$. In $\mathcal{F}$ replace $F_i$ by $A_i$ for $i \leq m$ and $G_j$ by $A'_j$ for $j \leq k$. The other elements of $\mathcal{F}$ are not being replaced. We claim that the collection $\mathcal{F}'$ obtained in this way covers $X$, and hence is a closed shrinking of $\mathcal{U}$. To see that it covers, pick an arbitrary $x \in X$. If $x \in \bigcup_{i=1}^m F_i$ then we are done by the first part of (2). If $x \in \bigcup_{j=1}^k G_j$ then $f^{-1}(x) \in \bigcup_{j=1}^k f^{-1}[G_j] \subseteq \bigcup_{j=1}^k B_j$ by the second part of (2), As a consequence, for some $j \leq k$, $x \in A'_j$. Finally, if $x$ does not belong to any element of $\mathcal{H}$ then the element of $\mathcal{F}$ that contains $x$ belongs to $\mathcal{F}'$. Observe next that $\bigcap_{i=1}^m A_i \cap \bigcap_{j=1}^k f^{-1}[A'_j] = \emptyset$.

Since $f$ is a homeomorphism, there exists an open swelling $\mathcal{V}$ of $\mathcal{F}'$ which is simultaneously a shrinking of $\mathcal{U}$ (Corollary 3.2.2) and which is as required.

So we arrive at the conclusion that $\mathcal{U}$ has an open shrinking $\mathcal{V}$ such that the $V$'s and $f^{-1}[V]$'s corresponding to $\mathcal{G}$ have empty intersection. The same procedure can be repeated with any subfamily of $\mathcal{V}$ of cardinality $n + 3$. So after finitely many steps, we arrive at the required shrinking of $\mathcal{U}$. □

**Corollary 3.12.15.** Let $X$ be a space with $\dim X \leq n$ and let $f$ be a fixed-point free homeomorphism of $X$. Then $f$ can be colored with at most $n+3$ colors.

**Proof.** By Theorem 3.12.8, $f$ is colorable, and by Theorem 3.12.12 the number of colors can be reduced to $n+3$. □

**Coloring continuous functions.** An inspection of the proof of Theorem 3.12.12 shows that it very strongly depends on the fact that we are dealing with homeomorphisms. In fact, as Theorem 3.12.7 shows, homeomorphisms are essential for the basic coloring result Theorem 3.12.8 and they are also essential for the process of reducing the number of colors in Theorem 3.12.12. We will investigate now the natural question whether the results obtained so far can be generalized. In the light of Theorem 3.12.7, the only reasonable question is whether colorable maps can be colored by a color of small size. So we are aiming at a generalization of Theorem 3.12.12.

The only place in the proof of Theorem 3.12.12 where we used that the map under consideration is a homeomorphism, is in the proof of Theorem 3.12.10. So we will first think about the question whether Theorem 3.12.10 is also true for continuous maps instead of homeomorphisms.

For $n = 0$ there are no problems. To see this, assume that $X$ is zero-dimensional, $\mathcal{U}$ is a finite open cover of $X$, and $f\colon X \to X$ is continuous. Since $\dim X = 0$, there is an open shrinking $\mathcal{V} = \{V_U : U \in \mathcal{U}\}$ of $\mathcal{U}$ with $\operatorname{ord}(\mathcal{V}) \leq 0$, i.e., $\mathcal{V}$ is pairwise disjoint (Theorem 3.4.4). But then $\mathcal{V}$ is as required. For let $x \in X$, and observe that there is precisely one element of $\mathcal{V}$ that contains $x$. Similarly, there is precisely one element of $f^{-1}[\mathcal{V}]$ that contains $x$.

But for larger $n$, Theorem 3.12.10 does not hold for continuous functions, as the following simple example shows.

**Example 3.12.16.** Let $X$ be the topological sum of the spaces $\mathbb{I}_n$, $n \in \mathbb{N}$, where each $\mathbb{I}_n$ is a copy of $\mathbb{I}$. In addition, let $\{d_i : i \geq 2\}$ be a countable dense subset of $\mathbb{I}_1$. Now define $f\colon X \to X$ as follows. $f \restriction \mathbb{I}_1$ is a homeomorphism from $\mathbb{I}_1$ onto $\mathbb{I}_2$. In addition, $f \restriction \mathbb{I}_i$ is the function with constant value $d_i$ for every $i \geq 2$. Then $f$ is clearly continuous (and has no fixed-point). For every $i \in \mathbb{N}$ let $E_i$ and $F_i$ be open subsets of $\mathbb{I}_i$ such that

$$E_i \cup F_i = \mathbb{I}_i$$

and

$$\mathbb{I}_i \setminus E_i \neq \emptyset \neq \mathbb{I}_i \setminus F_i.$$

Define

$$U_1 = E_1, \quad U_2 = F_1, \quad U_3 = \bigcup_{i \geq 2} E_i, \quad U_4 = \bigcup_{i \geq 2} F_i.$$

Then $\mathcal{U} = \{U_1, \ldots, U_4\}$ is an open cover of $X$. Let $\mathcal{V} = \{V_1, \ldots, V_4\}$ be an arbitrary open shrinking of $\mathcal{U}$. Then $V_1$ and $V_2$ cover $\mathbb{I}_1$, and are clearly proper subsets of $\mathbb{I}_1$. So by connectivity of $\mathbb{I}_1$, $V_1 \cap V_2 \neq \emptyset$. Pick $i \geq 2$ with $d_i \in V_1 \cap V_2$. Since $V_3$ and $V_4$ cover $\bigcup_{i \geq 2} \mathbb{I}_i$, the same argument shows that $V_3 \cap V_4 \cap \mathbb{I}_i \neq \emptyset$. We conclude that

$$f^{-1}[V_1] \cap f^{-1}[V_2] \cap V_3 \cap V_4 \neq \emptyset,$$

as required.

This example shows that if one wishes to color continuous functions, the method used so far does not work. Fortunately, there is another way to obtain what we want.

**Theorem 3.12.17.** *Let $X$ be a space and let $f \colon X \to X$ be fixed-point free and continuous. If $f$ is colorable and $\dim X \leq n$, then $f$ can be colored with at most $n + 3$ colors.*

**Proof.** We claim that by Corollary 3.5.7 we may assume without loss of generality that $X$ is compact. For let $aX$ be an at most $n$-dimensional compactification of $X$ such that $f$ can be extended to a fixed-point free continuous function $\bar{f} \colon aX \to aX$. By compactness of $aX$ the function $\bar{f}$ can be colored. So if the theorem for compact spaces is true then $\bar{f}$ can be colored with at most $n + 3$ colors. It suffices therefore to observe that if $\mathcal{F}$ colors $\bar{f}$ then $\mathcal{F} \upharpoonright X$ colors $f$.

So assume that $X$ is compact. We also assume for the time being that $f$ is surjective. The general case will be derived from this special case and will be dealt with later.

Consider the dynamical system $(X, f)$ and the inverse system

$$X \xleftarrow{f} X \xleftarrow{f} \cdots \xleftarrow{f} X \xleftarrow{f} \cdots$$

It was shown on Page 92 that the shift function $\sigma \colon X_\infty \to X_\infty$ defined by

$$\sigma(x_1, x_2, \ldots) = (f(x_1), x_1, x_2, \ldots).$$

is a homeomorphism. Moreover, for each $n$ the diagram

$$\begin{array}{ccc} X & \xleftarrow{f_n^\infty} & X_\infty \\ {\scriptstyle f}\downarrow & & \downarrow{\scriptstyle \sigma} \\ X & \xleftarrow{f_n^\infty} & X_\infty \end{array}$$

commutes and $\sigma$ is fixed-point free (Lemma 1.10.24).

Now observe that $\dim X_\infty \leq n$ by Exercise 3.2.2. So by Theorem 3.12.10 there is a closed cover $\mathcal{F}$ of $X_\infty$ such that $\sigma[F] \cap F = \emptyset$ for every $F \in \mathcal{F}$

while moreover $|\mathcal{F}| \leq n+3$. Now for each $F \in \mathcal{F}$ by compactness there exists an $N_F \in \mathbb{N}$ such that $f_m^\infty[F] \cap f_m^\infty[\sigma[F]] = \emptyset$ for every $m \geq N_F$ (Exercise 1.10.12). So if $n = \max\{N_F : F \in \mathcal{F}\}$ then $f_n^\infty[F] \cap f_n^\infty[\sigma[F]] = \emptyset$ for every $F \in \mathcal{F}$. So the commutativity of the above diagram implies that the closed collection $\mathcal{B} = \{f_n^\infty[F] : F \in \mathcal{F}\}$ covers $f_n^\infty[X_\infty]$ and has the property that every $B \in \mathcal{B}$ misses its image under $f$. The problem is that $\mathcal{B}$ in general does not need to be a cover of $X$. But since $f$ is surjective by assumption, $f_n^\infty$ is surjective by Lemma 1.10.8. So we conclude that $\mathcal{B}$ is the required coloring of $f$.

We now consider the general case that $f$ is not necessarily surjective, and put

$$K = \bigcap_{i=1}^\infty f^i[X].$$

Then $f[K] = K$ by Exercise 1.10.18. By the result just proved there is a cover $\mathcal{C} = \{C_i : i \leq n+3\}$ of $K$ consisting of relatively open sets such that $f[C_i] \cap C_i = \emptyset$ for every $i$. Let $\{D_i : i \leq n+3\}$ be a closed shrinking of $\mathcal{C}$ (Proposition A.7.1). Then $f[D_i] \cap D_i = \emptyset$ for every $i$. Let $U_i \subseteq X$ be an open neighborhood of $D_i$ for $i \leq n+3$ such that $f[U_i] \cap U_i = \emptyset$. The union $U$ of the $U_i$'s is an open neighborhood of $K$. By compactness, for a sufficiently large $k$ we have $f^k[X] \subseteq U$. So

$$\{f^{-k}[U_i] : i \leq n+3\}$$

is an open cover of $X$. We claim that it is a coloring of $f$. To prove this, take an arbitrary $i \leq n$ and assume that there exists

$$x \in f^{-k}[U_i] \cap f[f^{-k}[U_i]].$$

Then $f^k(x) \in U_i$ and there exists an element $z \in f^{-k}[U_i]$ such that $f(z) = x$. Then on the one hand $f^k(x) \in U_i$ while on the other hand

$$f^k(x) = f^k(f(z)) = f(f^k(z)) \in f[U_i].$$

This shows that $f^k(x) \in U_i \cap f[U_i]$, which is a contradiction. □

**Exercises for §3.12.** Let $f : X \to X$ be a fixed-point free and continuous. If $f$ is colorable then its *color number* $C(f)$ is the minimal cardinality of a coloring of $f$. If $f$ cannot be colored then we put $C(f) = \infty$.

Suppose that $f : X \to X$ is a fixed-point free and continuous. We say that a subset $A$ of $X$ is *invariant* if $f[A] \subseteq A$. Observe that a color of $f \restriction A$ is a subset of $A$.

1. Let $a, b, A, B \subseteq \mathbb{N}$ be such that $a$ and $b$ are finite, while $A$ and $B$ are infinite. Prove that $[a, A] \subseteq [b, B]$ iff $a \supseteq b$, $a \setminus b \subseteq B$ and $A \subseteq B$.

2. Prove that the collection in (∗) on Page 239 is a base for a topology on $[\mathbb{N}]^\omega$.

3. Let $\mathcal{A}$ be a countable collection infinite subsets of $\mathbb{N}$. Prove that there is an infinite subset $X \subseteq \mathbb{N}$ such that for every $A \in \mathcal{A}$ we have
$$X \cap A \neq \emptyset \neq A \cap (\mathbb{N} \setminus X).$$

4. Prove that the Ellentuck topology on $[\mathbb{N}]^\omega$ does not have a countable base.

We now present an interesting and unexpected addition theorem for the color number of a fixed-point free map.

**Theorem 3.12.18.** *Let $X$ be a space and let $f\colon X \to X$ be a continuous fixed-point free function such that $X = A \cup B$ for invariant subsets $A$ and $B$ of $X$. Then*
$$C(f) \leq C(f \restriction A) + C(f \restriction B) - 1.$$

The following two simple exercises will be useful.

5. Let $f\colon X \to X$ be a fixed-point free map with invariant subspace $A$. Suppose that the subset $C \subseteq A$ is a closed color of $f \restriction A$. Prove that there exists an open subset $W$ of $X$ such that $(f \restriction A)^{-1}[C] \subseteq W$ and $W$ is a color of $f$.

6. Suppose that $f\colon X \to X$ is a fixed-point free map and that $\mathcal{C}$ is a collection of $n$ colors of $f$ each of which is open or closed. Prove that there exists an open coloring of $f$ of size at most $n$.

From Exercise 3.13.6 it follows that the color number of a map $f\colon X \to X$ is the minimal cardinality of a cover of $X$ consisting of colors that are open or closed. The following exercise relates an open coloring of an invariant subspace to a collection of open colors of the ambient space.

7. Suppose that $f\colon X \to X$ is a fixed-point free map and that the subset $A$ of $X$ is invariant. Let $\mathcal{U}$ be an open coloring of $f \restriction A$ with $|\mathcal{U}| = m < \infty$. Prove that there exists a collection $\mathcal{V}$ consisting of at most $m$ open colors of $f$ such that $A \subseteq \bigcup \mathcal{V}$.

We are now in a position to finish the proof of Theorem 3.12.18.

8. Prove Theorem 3.12.18.

## 3.13. Various kinds of infinite-dimensionality

Recall that a space $X$ is *strongly infinite-dimensional* if it has an infinite essential family of pairs of disjoint closed sets. By Corollary 2.4.13, the Hilbert cube $Q$ is an example of a strongly infinite-dimensional space. A space $X$ is called *weakly infinite-dimensional* if it is not strongly infinite-dimensional, i.e., if for every family
$$\{(A_i, B_i) : i \in \mathbb{N}\}$$
of pairs of disjoint closed subsets of $X$ there exist partitions $D_i$ between $A_i$ and $B_i$ such that $\bigcap_{i=1}^{\infty} D_i = \emptyset$.

**Proposition 3.13.1.** *Let $X \subseteq Q$ be weakly infinite-dimensional. Then for every sequence $\{(A_i, B_i) : i \in \mathbb{N}\}$ of pairs of disjoint closed subsets of $Q$ there exist partitions $D_i$ between $A_i$ and $B_i$ in $Q$ such that*

$$\bigcap_{i=1}^{\infty} D_i \cap X = \emptyset.$$

**Proof.** For every $i \in \mathbb{N}$ let $U_i$ and $V_i$ be disjoint closed neighborhoods of $A_i$ and $B_i$, respectively (Corollary A.4.3). Since $X$ is weakly infinite-dimensional, the sequence $\{(U_i \cap X, V_i \cap X), i \in \mathbb{N}\}$ is inessential. Consequently, there exist partitions $S_i$ between $U_i \cap X$ and $V_i \cap X$ in $X$ such that the intersection of the $S_i$'s is empty. By Lemma 3.1.4 there exists for every $i$ a partition $D_i$ between $A_i$ and $B_i$ in $Q$ such that $D_i \cap X \subseteq S_i$. Then clearly

$$\bigcap_{i=1}^{\infty} D_i \cap X \subseteq \bigcap_{i=1}^{\infty} S_i = \emptyset,$$

as required. □

Finally, recall that a space X is called *countable dimensional* if it can be written as the union of countably many zero-dimensional subspaces. So every finite dimensional space is countable dimensional by Corollary 3.3.9, but not conversely. An important example of a countable dimensional space which is infinite-dimensional is given in the following example.

**Example 3.13.2.** As in §1.1, let

$$\sigma = \{x \in s : (\exists n \in \mathbb{N})(\forall m \geq n)(x_m = 0)\}.$$

For each $n \in \mathbb{N}$ put

$$\sigma_n = \{x \in s : (\forall m \geq n)(x_m = 0)\}.$$

It is clear that $\sigma_n$ is homeomorphic to $\mathbb{R}^n$, $n \in \mathbb{N}$, and hence is the union of $n+1$ zero-dimensional subspaces (Corollary 3.3.9). Since $\sigma$ is equal to the union of the $\sigma_n$'s, it follows that $\sigma$ is countable dimensional. Observe that $\sigma$ is infinite-dimensional by Theorem 3.2.12 and Corollary 3.4.14(1).

Observe that every subspace of a countable dimensional space is again countable dimensional.

**Example 3.13.3.** There exists an infinite-dimensional compact space which is countable dimensional.

This is a triviality. Let $X$ be the one-point compactification of the topological sum of the spaces $\mathbb{I}$, $\mathbb{I}^2$, $\mathbb{I}^3$, $\cdots$. Then $X$ is infinite-dimensional for the same reason $\sigma$ is. In addition, $X$ is countable dimensional by Corollary 3.4.14(1).

**Proposition 3.13.4.** *Every countable dimensional space is weakly infinite-dimensional.*

**Proof.** Let $X$ be countable dimensional. Then $X$ is the union of countably many zero-dimensional subspaces, say $X_i$, $i \in \mathbb{N}$. Let $\{(A_i, B_i) : i \in \mathbb{N}\}$ be a family of pairs of disjoint closed subsets of $X$. By Corollary 3.1.6 for every $i$ there exists a partition $D_i$ between $A_i$ and $B_i$ such that $D_i \cap X_i = \emptyset$. Since the union of the $X_i$'s equals $X$, the intersection of the $D_i$'s is empty, which is as required. □

**Corollary 3.13.5.** *No strongly infinite-dimensional space is countable dimensional.*

**Corollary 3.13.6.** *The Hilbert cube cannot be written as the union of countably many zero-dimensional subspaces.*

In view of the above results, the question naturally arises whether the converse to Proposition 3.13.4 holds. This question was known as *Alexandroff's problem* for decades, [8, §4, Hypothesis]. We shall answer this problem in the negative. However, we need to derive the following result first.

**Theorem 3.13.7.** *Every topologically complete space $X$ has a compactification $\gamma X$ such that the remainder $\gamma X \setminus X$ is countable dimensional.*

**Proof.** We may assume that $X$ is a subspace of the Hilbert cube $Q$ (Corollary A.4.4). Since $X$ is topologically complete, there is a decreasing sequence $(U_n)_n$ of open sets in $Q$ such that

$$X = \bigcap_{n=1}^{\infty} U_n$$

(Theorem A.6.3). Since $Q$ is compact, for each $\varepsilon > 0$ there is a finite open cover of $Q$ with mesh less than $\varepsilon$. Consequently, for each $n$ there exist finitely many open sets $U_{n,1}, \ldots, U_{n,m_n} \subseteq Q$ such that

(1) $U_{n,1} \cup \cdots \cup U_{n,m_n} = U_n$,
(2) $\mathrm{mesh}(\{U_{n,1}, \ldots, U_{n,m_n}\}) < 1/n$.

For each $n \in \mathbb{N}$ and $m \leq m_n$, define $f_{n,m} \colon Q \to \mathbb{I}$ by

$$f_{n,m}(x) = 1/2 \varrho(x, Q \setminus U_{n,m})$$

and define $f \colon Q \to Q$ coordinatewise as follows:

$$f(x) = (f_{1,1}(x), \ldots, f_{1,m_1}(x), f_{2,1}(x), \ldots, f_{2,m_2}(x), f_{3,1}(x), \ldots, \ldots).$$

By (2) it follows easily that $f \restriction X$ is an imbedding. We claim that

$$f[Q] \setminus f[X] \subseteq \sigma.$$

Take an arbitrary point $f(q) \in f[Q] \setminus f[X]$. Since $q \notin X$, by (1) there exists an $N \in \mathbb{N}$ such that $q \notin U_N$. Consequently, since the $U_n$'s decrease, for every $n \geq N$ and $m \leq m_n$ we have

$$f_{n,m}(q) = 0,$$

as required. Now let $\gamma X$ be the closure of $f[X]$ in $f[Q]$. Then $\gamma X$ is a compactification of $X$ the remainder $\gamma X \setminus X$ of which is contained in the countable dimensional space $\sigma$ (Example 3.13.2). So $\gamma X \setminus X$ is countable dimensional as well. □

The topological completeness of the spaces under discussion in the above theorem is essential. It can be shown that $\sigma$ is an example of a countable dimensional space the remainder of each compactification of which is strongly infinite-dimensional, see Exercise 3.13.4.

We shall now present the solution to Alexandroff's problem.

**Theorem 3.13.8.** *There exists a weakly infinite-dimensional compact space which is not countable dimensional.*

**Proof.** By Theorem 3.9.3 there exists a strongly infinite-dimensional totally disconnected and topologically complete space, say $S$. Let $X = \gamma S$ be a compactification of $S$ with countable dimensional remainder (Theorem 3.13.7). We claim that $X$ is the required example. To this end, first observe that $X$ is not countable dimensional since by Corollary 3.13.5 its subspace $S$ is not countable dimensional. Now let $\tau = \{(A_i, B_i) : i \geq 0\}$ be a sequence of pairs of disjoint closed subsets of $X$. Write $X \setminus S$ as the union of countably many zero-dimensional subspaces, say $X_i$, $i \in \mathbb{N}$. By Corollary 3.1.6 for each $i \in \mathbb{N}$ there exists a partition $D_i$ between $A_i$ and $B_i$ such that $D_i \cap X_i = \emptyset$. We conclude that the compact set $D = \bigcap_{i=1}^{\infty} D_i$ is contained in $S$, and $S$ being totally disconnected, therefore has to be zero-dimensional (Exercise 1.5.12). By another appeal to Corollary 3.13.5 there exists a partition $D_0$ between $A_0$ and $B_0$ such that $D_0 \cap D = \emptyset$. We conclude that $\bigcap_{i=0}^{\infty} D_i = \emptyset$. □

An infinite-dimensional space $X$ is called *hereditarily infinite-dimensional* if for every nonempty subspace $A$ of $X$ either $\dim A = 0$ or $\dim A = \infty$. A hereditarily infinite-dimensional compact space is clearly Henderson, but the converse need not be true. See the notes for more information.

We finish this section by presenting an example of a hereditarily infinite-dimensional space.

**Lemma 3.13.9.** *Let $K \subseteq \mathbb{J}$ be a Cantor set, let $\xi \in \mathbb{N}$ and let $\kappa \subseteq \mathbb{N}\setminus\{\xi\}$ be an infinite set the complement of which is also infinite. For each $j \in \kappa$ there exists a partition $S_j$ between the opposite faces $W_j^{-1}$ and $W_j^1$ of $Q$ such that each subset $X \subseteq \bigcap_{j \in \kappa} S_j$ with $K \subseteq \pi_\xi[X]$ is infinite-dimensional.*

**Proof.** Without loss of generality we may assume that $\xi = 1$ and $\kappa$ is the set of even natural numbers. For each $i \in \mathbb{N}$ let

$$C_i = \pi_i^{-1}[-1, -1/2] \text{ and } D_i = \pi_i^{-1}[1/2, 1],$$

respectively. Now put
$$\Lambda = \{f \in C(Q,Q) : (\forall i \in \mathbb{N})(f^{-1}[W_i^{-1}] = C_{2i} \text{ and } f^{-1}[W_i^1] = D_{2i})\}.$$
Then $\Lambda$ is a separable metric space since by Proposition 1.3.3 $C(Q,Q)$ is. By Exercise 1.5.10 there exists a subspace $T$ of $K$ such that $T$ can be mapped onto $\Lambda$, say by the map $\tau$. Let
$$E = T \times \mathbb{J}_2 \times \mathbb{J}_3 \times \cdots \subseteq Q$$
and define $\Phi \colon E \to Q$ by
$$\Phi(x) = \tau(x_1)(x).$$
Then $\Phi$ is clearly continuous.

**Claim 1.** For each $i$, $(\pi_i \circ \Phi)^{-1}(0)$ is a partition in $E$ between the sets $C_{2i} \cap E$ and $D_{2i} \cap E$.

*Proof.* Fix $i \in \mathbb{N}$. Then
$$\begin{aligned}\Phi^{-1}(W_i^{-1}) &= \{x \in E : \Phi(x) \in W_i^{-1}\} \\ &= \{x \in E : \tau(x_1)(x) \in W_i^{-1}\} \\ &= \{x \in E : x \in \tau(x_1)^{-1}[W_i^{-1}]\} \\ &= E \cap C_{2i},\end{aligned}$$
and similarly, $\Phi^{-1}[W_i^1] = E \cap D_{2i}$. Now since $\pi_i^{-1}(0)$ is a partition between $W_i^{-1}$ and $W_i^1$, the claim follows. $\diamond$

Since $W_{2i}^{-1}$ and $W_{2i}^1$ are in the interior of $C_{2i}$ and $D_{2i}$, respectively, there exists by Lemma 3.1.4 a partition $S_{2i}$ between $W_{2i}^{-1}$ and $W_{2i}^1$ in $Q$ such that
$$(*) \qquad S_{2i} \cap E \subseteq (\pi_i \circ \Phi)^{-1}(0).$$
We claim that
$$S = \bigcap_{i=1}^{\infty} S_{2i}$$
is as required.

**Claim 2.** $S \cap E \subseteq \Phi^{-1}(0)$.

*Proof.* This immediately follows from $(*)$. $\diamond$

Now let $M$ be a subspace of $S$ such that $K \subseteq \pi_1[M]$. We shall prove that $M$ is strongly infinite-dimensional. Striving for a contradiction, suppose that this is not the case, i.e., that $M$ is weakly infinite-dimensional. By Proposition 3.13.1 for $i \in \mathbb{N}$ there exist partitions $M_i$ in $Q$ between $C_{2i}$ and $D_{2i}$ such that $\bigcap_{i=1}^{\infty} M_i \cap M = \emptyset$. For $i \in \mathbb{N}$ there exists a continuous function $f_i \colon Q \to \mathbb{J}_i$ such that
$$f_i^{-1}(-1) = C_{2i}, \quad f_i^{-1}(1) = D_{2i}, \quad f_i^{-1}(0) = M_i$$

(Exercise A.4.9). Define $f \colon Q \to Q$ by
$$f(x) = (f_1(x), f_2(x), \dots).$$

**Claim 3.** $f \in \Lambda$ and $M \cap f^{-1}(\mathbf{0}) = \emptyset$.

*Proof.* That $f$ is continuous is trivial. Now take $i \in \mathbb{N}$. Then
$$f^{-1}[W_i^{-1}] = \{x \in Q : f(x)_i = -1\} = C_{2i}$$
by the definition of $f_i$. Similarly one proves that $f^{-1}[W_i^1] = D_{2i}$. We conclude that $f \in \Lambda$. Finally,
$$f^{-1}(\mathbf{0}) \cap M = \bigcap_{i=1}^{\infty} M_i \cap M = \emptyset,$$
as required. ◇

Now since $\tau$ is onto, there exists $p \in T$ such that $\tau(p) = f$. Since $T \subseteq \pi_1[M]$, there exists $x \in M$ such that $x_1 = p$. Then $x \in S \cap E$ from which it follows by Claim 2 that
$$f(x) = \tau(p)(x) = \tau(x_1)(x) = \Phi(x) = \mathbf{0}.$$
But this contradicts Claim 3. □

This result enables us to construct our final example.

**Theorem 3.13.10.** *There exists a hereditarily infinite-dimensional compact space.*

**Proof.** By Exercise 3.13.6 there exists a collection $\{K_i : i \in \mathbb{N}\}$ of pairwise disjoint Cantor sets in $\mathbb{J}$ such that each nondegenerate subinterval of $\mathbb{J}$ contains one of the $K_i$. So for each $i$ choose pairwise disjoint such Cantor sets $K_{i1}, K_{i2}, \dots$ in $\mathbb{J}_i = [-1, 1]_i$. In addition, let $\kappa_{ik}$ ($i, k \in \mathbb{N}$) be a collection of pairwise disjoint infinite subsets of $\mathbb{N} \setminus \{1\}$ such that
$$i \notin \kappa_{ik} \text{ for all } i \text{ and } k.$$
Fix $i, k \in \mathbb{N}$. By Lemma 3.13.9, for every $j \in \kappa_{ik}$ we can choose a partition $S_j$ in $Q$ between $W_j^{-1}$ and $W_j^1$ such that each subset in the intersection
$$S_{ik} = \bigcap \{S_j : j \in \kappa_{ik}\}$$
whose projection onto the $i$-th axis contains $K_{ik}$, is infinite-dimensional. Now put
$$S = \bigcap_{i,k} S_{ik}.$$
We claim that $S$ is hereditarily infinite-dimensional. Clearly $S$ is a compact subspace of $Q$. Since $1 \notin \bigcup_{i,k} \kappa_{ik}$, by Corollary 2.4.13 $S$ intersects every partition between $W_1^{-1}$ and $W_1^1$. By Exercise 3.1.13 it therefore follows

that $S$ is not zero-dimensional (for otherwise there would exist a partition $P$ between $W_1^{-1}$ and $W_1^1$ such that $P \cap S = \emptyset$, contradicting the fact that the family of opposite faces of $Q$ is essential). Now let $M$ be an arbitrary nonempty subspace of $S$. There are two cases to consider. First assume that for each $i \in \mathbb{N}$, $\pi_i[M]$ is zero-dimensional. Then $M$ is contained in the set

$$\prod_{i=1}^{\infty} \pi_i[M],$$

which is clearly zero-dimensional. So then $\dim M = 0$ as well. Next assume that there exists $i \in \mathbb{N}$ such that $\pi_i[M]$ is not zero-dimensional. Then $\pi_i[M]$ contains a nondegenerate subinterval of $\mathbb{J}_i$ by Proposition 1.5.3. Consequently, by construction, $\pi_i[M]$ contains some $K_{ij}$ which implies that $M$ is (strongly) infinite-dimensional because $M$ is contained in $S_{ij}$. □

**Exercises for §3.13.** A real-valued function $f$ on a space $X$ is *countably continuous* provided that $X$ can be partitioned into countably many sets $E_1, E_2, \ldots$ such that for every $i$ the restriction $f \restriction E_i$ is continuous.

1. Prove that if a space $X$ is the union of countably many weakly infinite-dimensional subspaces then $X$ is weakly infinite-dimensional. Conclude that the Hilbert cube is not the union of countably many weakly infinite-dimensional subspaces.

2. Let $X$ be a strongly infinite-dimensional compact space. Prove that $X$ contains a hereditarily infinite-dimensional closed subspace.

▶3. Prove that if $G$ is a $G_\delta$-subset of $s$ containing $\sigma$ then $G \setminus \sigma$ contains a copy of the Hilbert cube.

4. Prove that if $\gamma\sigma$ is a compactification of $\sigma$ then the remainder $\gamma\sigma \setminus \sigma$ contains a copy of the Hilbert cube and is therefore strongly infinite-dimensional.

▶5. Give an 'explicit' example of an upper semi-continuous function that is not countably continuous.

6. Prove that there exists a collection $\{K_i : i \in \mathbb{N}\}$ of pairwise disjoint Cantor sets in $\mathbb{J}$ such that each nondegenerate subinterval of $\mathbb{J}$ contains one of the $K_i$.

## 3.14. The Brouwer fixed-point theorem revisited

In this section we show that the existence of a two-dimensional hereditarily indecomposable continuum, which uses Brouwer's Fixed-Point Theorem for $\mathbb{I}^3$ only, implies the Brouwer Fixed-Point Theorem for all dimensions. This seems to show that the 'real' power of the Brouwer Fixed-Point Theorem is already attained at dimension three.

**Kelley's construction.** Let $X$ be a hereditarily indecomposable continuum with $\dim X \geq 2$. The existence from such a space follows from Corollary 3.8.3. An inspection of the proof shows that for the construction of $X$ one needs Brouwer's Fixed-Point Theorem for $\mathbb{I}^3$ only.

We will show that from $X$ we can build an infinite-dimensional space. Our strategy will be to 'blow up' the points of $X$ and to topologize them in such a way that we get an infinite-dimensional hereditarily indecomposable continuum.

Since $\dim X \geq 2$, by Exercise 3.7.5 the following holds for $X$:

(1) there are real numbers $\delta > 0$ and $\varepsilon > 0$ such that if $\mathcal{N}$ is any finite collection of pairwise disjoint closed sets with $\operatorname{mesh}(\mathcal{N}) \leq 2\delta$ then there is a continuum $C \subseteq X \setminus \bigcup \mathcal{N}$ with diameter at least $\varepsilon$.

Let $\omega \colon 2^X \to \mathbb{I}$ be a Whitney map (Proposition 1.11.15). By Exercise 1.11.25 there is an element $t \in (0, \infty)$ such that if $A \in 2^X$ is such that $\omega(A) \leq t$ then $\operatorname{diam}(A) < \varepsilon$. Consider the Whitney level

$$\mathcal{E} = \{A \in \mathcal{C}(X) : \omega(A) = t\}.$$

Then $\mathcal{E}$ is an upper semi-continuous decomposition of $X$ by Corollary 1.11.17. Here we use for the first time that $X$ is hereditarily indecomposable.

By Exercise 1.11.24 we find that

(2) $$\gamma = \inf\{\operatorname{diam}(E) : E \in \mathcal{E}\} > 0.$$

This property of the collection $\mathcal{E}$ is crucial.

So for the diameter of an arbitrary element $E \in \mathcal{E}$ we have

(3) $$0 < \gamma \leq \operatorname{diam} E < \varepsilon.$$

Let $Y$ be the space $X/\mathcal{E}$, and let $\pi \colon X \to Y$ be the natural quotient map. Since $\pi$ is monotone, we already know from general considerations that $Y$ is a hereditarily indecomposable continuum (Exercise 1.10.22).

Observe that $\pi$ is open by Exercise 1.11.23. So $Y$ is homeomorphic to the subspace $\mathcal{E}$ of $2^X$, and the formula

$$d(y_1, y_2) = \varrho_H\left(f^{-1}(y_1), f^{-1}(y_2)\right)$$

defines an admissible metric on $Y$ (Exercise 1.11.17). This is the metric on $Y$ that we will use in the sequel.

Observe that since $\mathcal{E}$ consists of continua, the space $Y$ is in fact homeomorphic to a subspace of $\mathcal{C}(X)$. This observation will be used in the solution to Exercise 3.14.1.

Having found the collection $\mathcal{E}$ of "blown up" points, we fix any $n \geq 2$ and set

(4) $$\delta_n = \min\left\{\frac{\gamma}{4n}, \delta\right\}.$$

Observe that by (1) and (2), $\delta_n > 0$.

**Proposition 3.14.1.** *For any finite closed cover $\mathcal{F}$ of $Y$ of mesh less than $\delta_n$ there is an element $A \in \mathcal{F}$ intersecting at least $n$ elements of $\mathcal{F} \setminus \{A\}$.*

**Proof.** It will be convenient to introduce the following notation. For $A \subseteq Y$ we put
$$\mathcal{E}(A) = \{\pi^{-1}(y) : y \in A\}.$$
Striving for a contradiction, let $\mathcal{F}$ be a finite closed cover of $Y$ with mesh less than $\delta_n$ such that each element of $A \in \mathcal{F}$ meets at most $n-1$ elements of $\mathcal{F} \setminus \{A\}$. We shall associate to each element $A \in \mathcal{F}$ a compact set

(5) $$\varphi(A) \subseteq \pi^{-1}[A]$$

such that

(6) $\varphi(A) \cap E \neq \emptyset$ for any $E \in \mathcal{E}(A)$,
(7) $\operatorname{diam} \varphi(A) \leq 2\delta_n$,
(8) if $A, B \in \mathcal{F}$ are distinct then $\varphi(A) \cap \varphi(B) = \emptyset$.

Assume that $\varphi$ is already defined on $\mathcal{G} \subseteq \mathcal{F}$, and let $A \in \mathcal{F} \setminus \mathcal{G}$ be arbitrary. The set $\mathcal{H}$ of all elements of $\mathcal{G}$ hitting $A$ has by assumption cardinality less than or equal to $n-1$. For each $D \in \mathcal{G}$, let $B(D)$ be the closed $\delta_n$-ball about $\varphi(D)$. Observe that by (7) we have

(9) $$\operatorname{diam} B(D) \leq 4\delta_n.$$

Let us fix an arbitrary element $E(A) \in \mathcal{E}(A)$. We shall show that

(10) $$E(A) \setminus \bigcup \{B(D) : D \in \mathcal{H}\} \neq \emptyset.$$

Otherwise, $E(A)$ is covered by the collection
$$\{B(D) : D \in \mathcal{H}\}$$
and hence, $E(A)$ being a continuum, by Exercise A.10.3 and (9) we get
$$\operatorname{diam} E(A) \leq |\mathcal{H}| \cdot \max_{D \in \mathcal{H}} \operatorname{diam} B(D) \leq (n-1) \cdot 4\delta_n,$$
which contradicts (4). Let $a$ be any element given by (10) and let $B$ be the closed $\delta_n$-ball about $a$. We set

(11) $$\varphi(A) = B \cap \pi^{-1}[A].$$

We claim that $\varphi(A)$ is as required. It is clear that (7) is satisfied. Let $E$ be an arbitrary element of $\mathcal{E}(A)$. The singleton sets $\pi(E(A))$ and $\pi(E)$ are less than $\delta_n$ apart since they both belong to the set $A \in \mathcal{F}$ and $\mathcal{F}$ has mesh less than $\delta_n$. The Hausdorff $\varrho$-distance of $E(A)$ and $E$ is therefore less than $\delta_n$ and so $\varrho(a, E) < \delta_n$ (Exercise 1.11.13). This shows that $B$ meets $E$ and hence $\varphi(A)$ meets $E$. We conclude that (6) is satisfied.

If $D \in \mathcal{H}$ then $a$ does not belong to the closed $\delta_n$-ball about $\varphi(D)$. So the closed $\delta_n$-ball about $a$ misses $\varphi(D)$, i.e., $\varphi(A) \cap \varphi(D) = \emptyset$. Moreover, if $D$ belongs to $\mathcal{G} \setminus \mathcal{H}$ then $A \cap D = \emptyset$ and hence $\varphi(A) \cap \varphi(D) = \emptyset$ since

$$\varphi(A) \subseteq \pi^{-1}[A], \quad \varphi(D) \subseteq \pi^{-1}[D], \quad \pi^{-1}[A] \cap \pi^{-1}[D] = \emptyset.$$

So (8) is also satisfied (for $\mathcal{G} \cup \{A\}$).

We conclude that we can extend $\varphi$ over $\mathcal{G} \cup \{A\}$ without creating problems. By the fact that $\mathcal{F}$ is finite, we may therefore assume that $\varphi$ is defined on all of $\mathcal{F}$.

The collection

(12) $$\mathcal{N} = \{\varphi(A) : A \in \mathcal{F}\}$$

is disjoint and has by (7) mesh less than or equal to $2\delta_n \leq 2\delta$. So it can serve as input for (1).

To reach a contradiction, let $C$ be an arbitrary continuum in $X \setminus \bigcup \mathcal{N}$. Let $E \in \mathcal{E}$ intersect $C$ and pick $A \in \mathcal{F}$ such that $E \in \mathcal{E}(A)$. By (6),

$$E \cap \bigcup \mathcal{N} \neq \emptyset.$$

Since $C$ does not intersect $\bigcup \mathcal{N}$, this shows that $E$ is not a subset of $C$. But then since $X$ is hereditarily indecomposable, it must follows that $C$ is a subset of $E$. This implies by (3) that

$$\operatorname{diam} C \leq \operatorname{diam} E < \varepsilon.$$

By (1) there is such a continuum $C$ with diameter at least $\varepsilon$. This contradiction establishes the proof. □

By Exercise 3.3.6, we now come to the following surprising conclusion.

**Corollary 3.14.2.** *The existence of an at least two-dimensional hereditarily indecomposable continuum implies the existence of an hereditarily indecomposable infinite-dimensional continuum.*

**Dimension theory.** By Exercise 3.1.16 we know that the Brouwer Fixed-Point Theorem and the fact that there is a space $X$ with $\dim X = \infty$ are equivalent statements, in the sense that they are easily deduced from each other. So from Brouwer's Fixed-Point Theorem for $\mathbb{I}^3$, implying the existence of a two-dimensional hereditarily indecomposable continuum, we derived the Brouwer Fixed-Point Theorem for all dimensions. This is a rather curious phenomenon and it is not clear whether there is a more direct route from Brouwer in $\mathbb{I}^3$ to Brouwer in all dimensions. This is of course only of interest for 'philosophical' reasons. The proof of Sperner's Lemma 2.4.3 is by induction, so it is equally complicated in all dimensions (except for the dimensions one and two). Similar remarks can be made for the various other known proofs of Brouwer's Theorem. This seems to indicate that even if one finds a more direct route from Brouwer in $\mathbb{I}^3$ to Brouwer in all dimensions, this will not result in a simpler proof of Brouwer's Theorem. But this does not mean that clarifying the rather curious phenomenon that we encountered in this section would not be interesting.

**Exercises for §3.14.**

1. Let $Z$ be a compact space with $\dim Z \geq 3$. Prove that $\dim \mathcal{C}(Z) = \infty$.
2. Give an example of a continuum $X$ such that $\dim \mathcal{C}(X) = 2$.

CHAPTER 4

# Basic ANR theory

In this chapter we shall present several basic results from ANR theory. We introduce and characterize ANR-pairs. They enable us to prove that the hyperspace of a Peano continuum is an AR, which is an important step in the proof of Curtis-Schori-West Hyperspace Theorem that $2^X \approx Q$ if and only if $X$ is a Peano continuum, and that there are deformations of hyperspaces through various interesting subspaces. A curious important result that we will prove is that for a Peano continuum $X$ there are arbitrarily close to the identity maps from $2^X$ into the finite subsets of $X$. Our concept of an ANR-pair, which was motivated by TORUŃCZYK's concept of a locally homotopy negligible set from [390], turns out to be a convenient framework for obtaining such results.

## 4.1. Some properties of ANR's

In this section we shall derive a few useful properties of ANR's. We have to introduce some terminology first.

Let $Y$ be a space and let $\mathcal{U}$ be an open cover of $Y$. Two continuous functions $f, g \colon X \to Y$ are called $\mathcal{U}$-close if for every $x \in X$ there exists an element $U \in \mathcal{U}$ such that $\{f(x), g(x)\} \subseteq U$. A homotopy $H \colon X \times \mathbb{I} \to Y$ is said to be *limited by* $\mathcal{U}$ provided that for any $x \in X$ there exists $U \in \mathcal{U}$ such that

$$H[\{x\} \times \mathbb{I}] \subseteq U.$$

Two continuous functions $f, g \colon X \to Y$ are $\mathcal{U}$-*homotopic*, if there is a homotopy $H \colon X \times \mathbb{I} \to Y$ which is limited by $\mathcal{U}$ and connects $f$ and $g$, i.e., $H_0 = f$ and $H_1 = g$.

**Small functions and small homotopies.** Here is our first property of ANR's.

**Theorem 4.1.1.** *Let $X$ be an ANR. Then for every open cover $\mathcal{U}$ of $X$ there exists an open refinement $\mathcal{V}$ of $\mathcal{U}$ such that for every space $Y$, any two $\mathcal{V}$-close maps $f, g \colon Y \to X$ are $\mathcal{U}$-homotopic.*

**Proof.** By Corollary 1.1.8 we may assume that $X$ is a closed subspace of a normed linear space $L$. Observe that $L$ is locally convex (Exercise 1.1.6). Since $X$ is an ANR, there are a neighborhood $V$ of $X$ in $L$ and a retraction $r\colon V \to X$. Put $\mathcal{E} = \{r^{-1}[U] : U \in \mathcal{U}\}$. Then $\mathcal{E}$ is an open cover of $V$, and since $V$ is open in $L$, there is an open cover $\mathcal{F}$ of $V$ having the following properties:

(1) $\mathcal{F} < \mathcal{E}$,
(2) $\mathcal{F}$ consists of convex sets.

Let $\mathcal{V} = \mathcal{F} \upharpoonright X$. We claim that $\mathcal{V}$ is the required open cover of $X$. To this end, let $Y$ be a space, and assume that $f, g\colon Y \to X$ are continuous and $\mathcal{V}$-close. Define a homotopy $G\colon Y \times \mathbb{I} \to L$ by

$$G(y,t) = tg(y) + (1-t)f(y).$$

Then $G$ is continuous and clearly connects $f$ and $g$. Since by (2) the elements of $\mathcal{F}$ are convex, by the special definition of $G$ it follows that for every $y \in Y$ there exists $F \in \mathcal{F}$ such that

(3) $$G[\{y\} \times \mathbb{I}] \subseteq F.$$

We conclude that $G$ can be considered to be a mapping from $Y \times \mathbb{I}$ into $V$ and that moreover $G$ is an $\mathcal{F}$-homotopy. Now define $H\colon Y \times \mathbb{I} \to X$ by

$$H = r \circ G.$$

Then $H$ connects $f$ and $g$ since $r$ is a retraction. Moreover, by (3) and (1) it follows easily that $H$ is limited by $\mathcal{U}$. □

**Remark 4.1.2.** It is a natural problem whether the property of ANR's stated in Theorem 4.1.1 in fact characterizes the class of all ANR's. This was a difficult and fundamental problem which remained unanswered for decades. It was finally solved in 1994 by CAUTY [87] in the negative by his example of linear space which is not an ANR (Remark 1.2.4). To see that Cauty's Example really solves the problem, let $L$ be a linear space. In addition, let $\mathcal{U}$ be an open cover of $L$. The function $\lambda\colon L \times L \times \mathbb{I} \to L$ defined by

$$\lambda(x,y,t) = (1-t) \cdot x + t \cdot y,$$

is defined in terms of the algebraic operations on $L$ and is therefore continuous. For every $x \in L$ pick an element $U_x \in \mathcal{U}$ containing $x$. Since $\lambda$ is continuous and $\lambda[\{x\} \times \{x\} \times \mathbb{I}] = \{x\}$, by compactness of $\mathbb{I}$ there exists for every $x \in X$ a neighborhood $V_x$ of $x$ such that $\lambda[V_x \times V_x \times \mathbb{I}] \subseteq U_x$ (Exercise A.5.7). Put $\mathcal{V} = \{V_x : x \in L\}$. We claim that $\mathcal{V}$ is as required. To this end, let $X$ be a space and let $f, g\colon X \to L$ be continuous $\mathcal{V}$-close functions. Define a homotopy $H\colon X \times \mathbb{I} \to L$ in the obvious way by the formula

$$H(x,t) = (1-t) \cdot f(x) + t \cdot g(x).$$

Then clearly $H_0 = f$ and $H_1 = g$. Fix an arbitrary $x \in X$. Since $f$ and $g$ are $\mathcal{V}$-close, there exists an element $p \in L$ such that $f(x), g(x) \in V_p$. But then
$$\{f(x)\} \times \{g(x)\} \times \mathbb{I} \subseteq V_p \times V_p \times \mathbb{I}$$
from which it follows that $H(x,t) = \lambda(f(x), g(x), t) \in U_p$ for every $t \in \mathbb{I}$. So this indeed proves that $f$ and $g$ are $\mathcal{U}$-homotopic.

**Homotopies with control.** In many applications one needs homotopies 'with control'. We will therefore present a 'controlled' version of the Borsuk Homotopy Extension Theorem, cf. Theorem 1.4.2.

**Theorem 4.1.3.** *Let $X$ be an ANR and let $\mathcal{U}$ be an open cover of $X$. Then for every closed subset $A$ of a space $Y$ and for every homotopy $H: A \times \mathbb{I} \to X$ such that*

(1) *$H$ is limited by $\mathcal{U}$,*
(2) *$H_0$ can be extended to a continuous function $h_0: Y \to X$*

*there exists a homotopy $\bar{H}: Y \times \mathbb{I} \to X$ such that*

(3) *$\bar{H}$ is limited by $\mathcal{U}$,*
(4) *$\bar{H}_0 = h_0$ and $\bar{H} \restriction (A \times \mathbb{I}) = H$.*

**Proof.** Let $H, A$ and $Y$ be as in the formulation of the theorem. By Theorem 1.4.2 there exists a homotopy $F: Y \times \mathbb{I} \to X$ such that $F_0 = h_0$ and
$$F \restriction (A \times \mathbb{I}) = H.$$
For each $a \in A$ there exists $U_a \in \mathcal{U}$ containing $H[\{a\} \times \mathbb{I}]$. There consequently exists by compactness of $\mathbb{I}$ and continuity of $H$ a neighborhood $E_a$ of $a$ in $Y$ such that
$$F[E_a \times \mathbb{I}] \subseteq U_a$$
(Exercise A.5.7). Put $E = \bigcup_{a \in A} E_a$. Then $E$ is a neighborhood of $A$ in $Y$. By Corollary A.4.1 there is a Urysohn function $\lambda: Y \to \mathbb{I}$ such that $\lambda \restriction A \equiv 1$ and $\lambda \restriction Y \setminus E \equiv 0$. Now define $\bar{H}: Y \times \mathbb{I} \to X$ as follows:
$$\bar{H}(y, t) = F(y, \lambda(y) \cdot t).$$
Then $\bar{H}$ is clearly continuous and $\bar{H} \restriction (A \times \mathbb{I}) = H$. In addition,
$$\bar{H}(y, 0) = F(y, 0) = h_0(y)$$
for every $y \in Y$ so $\bar{H}_0 = h_0$. It remains to prove that $\bar{H}$ is limited by $\mathcal{U}$. This is a triviality. Take an arbitrary $y \in Y$. If $y \in E$ then there exists $a \in A$ such that $y \in E_a$. Consequently,
$$\bar{H}(\{y\} \times \mathbb{I}) \subseteq F[\{y\} \times \mathbb{I}] \subseteq U_a.$$
Now if $y \notin E$ then $\lambda(y) = 0$ from which it follows that
$$\bar{H}(\{y\} \times \mathbb{I}) = \{F(y, 0)\} = \{h_0(y)\}$$

consists of precisely one point and is therefore also contained in an element of $\mathcal{U}$. □

**ANR-pairs.** We shall now derive another important property of ANR's. For later use, we will state it in terms of ANR-pairs. Let $X$ be a space with subspace $Y$. We say that the pair $(X,Y)$ is an ANR-*pair* provided that for every space $Z$ and every closed set $A \subseteq Z$ and every continuous function $f: A \to X$ there exist a neighborhood $U$ of $A$ in $Z$ and a continuous extension $\bar{f}: U \to X$ of $f$ having the additional property that $\bar{f}[U \setminus A] \subseteq Y$.

Observe that if $(X,Y)$ is an ANR-pair then so is $(E,Y)$, where $E$ is any subspace of $X$ containing $Y$.

Define the notion of an AR-*pair* similarly.

Observe that if $(X,Y)$ is an A(N)R-pair then both $X$ and $Y$ are A(N)R's (see Exercise 4.1.7). Moreover, if $X$ is an ANR then $(X,X)$ is an ANR-pair.

Also, $Y$ is dense in $X$. For if $Y$ is not dense in $X$ then there exists a point $x \in X$ such that for some $\varepsilon > 0$ we have that $B(x,\varepsilon) \cap Y = \emptyset$. Now let $Z = \mathbb{I}$, $A = \{0\}$ and $f: A \to X$ the constant function with value $x$. Then obviously $f$ cannot be extended over a neighborhood of $A$ with 'new' values in $Y$.

Let $X$ be a space, and let $Y \subseteq X$. A *deformation of $X$ through $Y$* is a homotopy $H: X \times \mathbb{I} \to X$ such that $H_0$ is the identity on $X$, and $H_t[X] \subseteq Y$ for every $t \in (0,1]$.

**Lemma 4.1.4.** *Let $X$ be a space and let $U$ be a neighborhood of $X \times \{0\}$ in $X \times \mathbb{I}$. Then there is a continuous function $\alpha: X \to (0,1]$ such that*

$$\{(x,t) : t \leq \alpha(x)\} \subseteq U.$$

**Proof.** Let $\varrho$ be an admissible metric on $X$ which is bounded by 1 (Exercise A.1.6). The function

$$d\big((x,t),(x',t')\big) = \varrho(x,x') + |t - t'|$$

defines an admissible metric on $X \times \mathbb{I}$ (cf. Exercise A.1.10). As a consequence, the function $\alpha: X \to (0,1]$ defined by

$$\alpha(x) = \tfrac{1}{2}d\big((x,0),(X \times \mathbb{I}) \setminus U\big)$$

is continuous (and clearly well-defined since $\varrho$ is bounded by 1). We claim that it is as required. To this end, consider for some $x \in X$ a point of the form $(x,t)$, where $t \leq \alpha(x)$. If $(x,t) \in (X \times \mathbb{I}) \setminus U$ then

$$\alpha(x) = \tfrac{1}{2}d\big((x,0),(X \times \mathbb{I}) \setminus U\big) \leq \tfrac{1}{2}d\big((x,0),(x,t)\big) = \tfrac{1}{2}|t| \leq \tfrac{1}{2}\alpha(x).$$

So we conclude that $\alpha(x) = 0$, which is a contradiction. □

**Corollary 4.1.5.** *Let $X$ be a space with subspace $Y$. If $X$ can be deformed through $Y$ then for every open cover $\mathcal{U}$ of $X$ there exists a homotopy*
$$H\colon X \times \mathbb{I} \to X$$
*such that*

(1) *$H$ deforms $X$ through $Y$,*
(2) *$H$ is limited by $\mathcal{U}$.*

**Proof.** Let $F\colon X \times \mathbb{I} \to X$ be a deformation through $Y$. For every $x \in X$ pick $U_x \in \mathcal{U}$ containing $x$. Since $F^{-1}[U_x]$ is a neighborhood of $(x,0)$ in $X \times \mathbb{I}$, there are an open neighborhood $V_x$ of $x$ and $0 < t_x \leq 1$ such that
$$F[V_x \times [0, t_x]] \subseteq U_x.$$
Put
$$V = \bigcup_{x \in X} V_x \times [0, t_x).$$
By Lemma 4.1.4 there is a continuous function $\alpha\colon X \to (0, 1]$ such that
$$\{(x, t) : x \in X, t \leq \alpha(x)\} \subseteq V.$$
Now define $H\colon X \times \mathbb{I} \to X$ by
$$H(x, t) = F(x, t \cdot \alpha(x)).$$
Since $\alpha(x) > 0$ for every $x$, we clearly have that $H$ deforms $X$ through $Y$. Now if $x \in X$ then $(x, \alpha(x)) \in V$. So there exists $x' \in X$ such that
$$(x, \alpha(x)) \in V_{x'} \times [0, t_{x'}).$$
But then
$$\{x\} \times [0, \alpha(x)] \subseteq V_{x'} \times [0, t_{x'})$$
and so
$$H[\{x\} \times \mathbb{I}] \subseteq F[\{x\} \times [0, \alpha(x)]] \subseteq U_{x'},$$
as required. $\square$

These results enable us to characterize A(N)R-pairs, as follows:

**Theorem 4.1.6.** *Let $X$ be a space, with subspace $Y$. The following statements are equivalent:*

(1) *$(X, Y)$ is an A(N)R-pair.*
(2) *$X$ is an A(N)R and there exists a deformation from $X$ through $Y$.*

**Proof.** We will show this for ANR-pairs only, the proof for AR-pairs being entirely similar. For (1) $\Rightarrow$ (2), consider $X \times \mathbb{I}$, its subspace $X \times \{0\}$ and the function $f\colon X \times \{0\} \to X$ defined by $f(x, 0) = x$. There exist a neighborhood $U$ of $X \times \{0\}$ in $X \times \mathbb{I}$ and a continuous extension $\bar{f}\colon U \to X$ of $f$ such

that $\bar{f}[U \setminus (X \times \{0\})] \subseteq Y$. Let $\alpha\colon X \to (0,1]$ be the function in Lemma 4.1.4 for $U$. Now define $H\colon X \times \mathbb{I} \to X$ by the formula
$$H(x,t) = \bar{f}(x, t \cdot \alpha(x)).$$
Then $H$ is a deformation through $Y$. Since, as observed earlier, $X$ is an ANR, we are done.

For (2) $\Rightarrow$ (1), let $X$ be an ANR and $H\colon X \times \mathbb{I} \to X$ be a deformation through $Y$. In addition, let $Z$ be a space, $A \subseteq Z$ be closed and $f\colon A \to X$ continuous. Since $X$ is an ANR, there are a neighborhood $U$ of $A$ in $Z$ and a continuous extension $\bar{f}\colon U \to X$. Let $\varrho$ be an admissible metric for $Z$ which is bounded by 1 (Exercise A.1.6), and define $g\colon Z \to X$ by the formula
$$g(z) = H(\bar{f}(z), \varrho(z, A)).$$
Then $g$ is clearly as required. □

As we saw, it is sometimes possible to replace a homotopy by one with better properties. Another example of such a phenomenon is the following result, which shall be used frequently in the sequel.

**Proposition 4.1.7.** *Let $(X, \varrho)$ be a space with subspace $Y$ and let $H$ be a deformation of $X$ through $Y$. Then $H$ can be replaced by a deformation $F$ of $X$ through $Y$ such that*
$$\varrho(F(x,t), x) \leq t$$
*for all $x \in X$ and $t \in \mathbb{I}$.*

**Proof.** For every $x \in X$ and $t \in \mathbb{I}$ put
$$G(x,t) = \{s \in \mathbb{I} : \operatorname{diam}(H[\{x\} \times [0,s]]) \geq t\}.$$
Observe that possibly $G(x,t) = \emptyset$.

**Claim 1.** $G(x,t)$ is closed in $\mathbb{I}$ for every $x$ and $t$.

*Proof.* This follows easily from the continuity of the functions $H$ and diam (Exercise 1.11.12). ◇

Now define $\xi\colon X \times \mathbb{I} \to \mathbb{I}$ by
$$\xi(x,t) = \begin{cases} \min G(x,t) & (G(x,t) \neq \emptyset), \\ 1 & (G(x,t) = \emptyset). \end{cases}$$
By Claim 1, $\xi$ is well-defined.

**Claim 2.** $\xi$ is lsc.

*Proof.* Suppose that for some $v \in [0,1)$, $x \in X$ and $t \in \mathbb{I}$ we have $\xi(x,t) > v$. Suppose that there are sequences $(x_n)_n$ in $X$ and $(t_n)_n$ in $\mathbb{I}$ such that

$$x_n \to x, \quad t_n \to t$$

while moreover

$$\xi(x_n, t_n) \leq v$$

for all $n$. We may assume without loss of generality that $\xi(x_n, t_n) \to w \leq v$. Observe that since $v < 1$,

$$\mathrm{diam}\,\bigl(H(\{x_n\} \times [0, \xi(x_n, t_n)])\bigr) \geq t_n$$

for every $n$ so that by continuity of $H$ and diam (Exercise 1.11.12) we get

$$\mathrm{diam}\,\bigl(H(\{x\} \times [0, w])\bigr) \geq t.$$

This shows that $\xi(x,t) \leq w \leq v$, which is a contradiction. $\diamond$

Let $\eta \colon X \times \mathbb{I} \to \mathbb{I}$ be the constant function with values 0. Then $\eta$ is continuous, and hence **usc**. By Corollary A.7.6 there is a continuous function $\delta \colon X \times \mathbb{I} \to \mathbb{I}$ such that for every $(x,t) \in X \times \mathbb{I}$ we have

$$0 \leq \delta(x,t) \leq \xi(x,t)$$

and if $0 < \xi(x,t)$ then

$$0 < \delta(x,t) < \xi(x,t).$$

Now define the homotopy $F \colon X \times \mathbb{I} \to X$ by the formula

$$F(x,t) = H\bigl(x, \delta(x,t)\bigr).$$

Then $F$ is clearly continuous. We claim that it is as required. To this end, pick arbitrary $x \in X$ and $t \in \mathbb{I}$. Then

$$\varrho\bigl(x, F(x,t)\bigr) = \varrho\bigl(x, H(x, \delta(x,t))\bigr)$$

is less than $t$ if $\xi(x,t) > 0$ and $0 \leq t$ otherwise. It therefore suffices to prove that $F$ deforms $X$ through $Y$. It is clear that

$$\varrho\bigl(x, F(x,0)\bigr) = \varrho\bigl(x, H(x, \xi(x,0))\bigr) = \varrho\bigl(x, H(x,0)\bigr) = 0,$$

i.e., $F(x,0) = x$. Finally, take arbitrary $x \in X$ and $t \in (0,1]$. If $\xi(x,t) = 0$ then $0 = \mathrm{diam}(H[\{x\} \times \{0\}]) \geq t > 0$ which is impossible. So $\xi(x,t) > 0$ from which it follows that $\delta(x,t) > 0$ and consequently

$$F(x,t) = H\bigl(x, \delta(x,t)\bigr) \in Y,$$

as required. $\square$

**Corollary 4.1.8.** *Let $Z \subseteq Y \subseteq X$. If $(Y,Z)$ and $(X,Y)$ are A(N)R-pairs then so is $(X,Z)$.*

**Proof.** We prove this for ANR-pairs only, the proof for AR-pairs being entirely similar. Let $F\colon X \times \mathbb{I} \to X$ be a deformation through $Y$ (Theorem 4.1.6). We may assume by Proposition 4.1.7 that $\varrho(H_t(x), t) \leq t$ for all $x \in X$ and $t \in \mathbb{I}$. Similarly, let $G\colon Y \times \mathbb{I} \to Y$ be a deformation through $Z$ having the property that $\varrho(G_t(y), t) \leq t$ for all $y \in Y$ and $t \in \mathbb{I}$. Define the homotopy $H\colon X \times \mathbb{I} \to X$ by

$$H(x,t) = \begin{cases} x & (t=0), \\ G(F(x,t),t) & (t>0). \end{cases}$$

We only need to check that $H$ is continuous at points of the form $(x, 0)$. So let $(x_n)_n$ be a sequence in $X$ converging to $x$ and let $(t_n)_n$ be a sequence in $(0, 1]$ such that $t_n \searrow 0$. Then

$$\begin{aligned}
\varrho\big(H_0(x), H_{t_n}(x_n)\big) &\leq \varrho(x, x_n) + \varrho\big(x_n, H_{t_n}(x_n)\big) \\
&\leq \varrho(x, x_n) + \varrho\big(x_n, F_{t_n}(x_n)\big) + \varrho\big(F_{t_n}(x_n), G_{t_n}(F_{t_n}(x_n))\big) \\
&\leq \varrho(x, x_n) + 2t_n \\
&\overset{n\to\infty}{\Longrightarrow} 0.
\end{aligned}$$

So we are done by another application of Theorem 4.1.6. □

**Partial realization.** The following interesting property of ANR's is not very appealing at first glance. It turns out however that it is of crucial importance in the process of finding a usable topological characterization of ANR's.

We need to introduce some terminology first. Let $X$ be a space and let $\mathcal{U}$ be an open cover of $X$. In addition, let $\mathcal{T}$ be a locally finite simplicial complex and let $\mathcal{S}$ be a subcomplex of $\mathcal{T}$ containing all the vertices of $\mathcal{T}$. A *partial realization* of $\mathcal{T}$ in $X$ relative to $(\mathcal{S}, \mathcal{U})$ is a continuous function

$$f\colon |\mathcal{S}| \to X$$

such that for every $\sigma \in \mathcal{T}$ there exists $U \in \mathcal{U}$ such that

$$f[\sigma \cap |\mathcal{S}|] \subseteq U.$$

In case $\mathcal{S} = \mathcal{T}$ we say that $f$ is a *full realization* of $\mathcal{T}$ in $X$ relative to $\mathcal{U}$.

**Theorem 4.1.9.** *Let $X$ be an ANR. Then for every open cover $\mathcal{U}$ of $X$ there exists an open refinement $\mathcal{V}$ of $\mathcal{U}$ such that for every locally finite simplicial complex $\mathcal{T}$ and every subcomplex $\mathcal{S}$ of $\mathcal{T}$ containing all the vertices of $\mathcal{T}$, every partial realization of $\mathcal{T}$ in $X$ relative to $(\mathcal{S}, \mathcal{V})$ can be extended to a full realization of $\mathcal{T}$ in $X$ relative to $\mathcal{U}$.*

**Proof.** Recall that by Corollary 1.1.8 we may think of $X$ as a closed subspace of a normed linear space $L$. Since $X$ is an ANR, there are a neighborhood $V$ of $X$ in $L$ and a retraction $r\colon V \to X$. Put $\mathcal{E} = \{r^{-1}[U] : U \in \mathcal{U}\}$. Then $\mathcal{E}$ is an open cover of $V$, and since $V$ is open in $L$, there is an open cover $\mathcal{F}$ of $V$ having the following properties:

(1) $\mathcal{F} < \mathcal{E}$,
(2) $\mathcal{F}$ consists of convex sets.

As in the proof of Theorem 4.1.1, put $\mathcal{V} = \mathcal{F} \restriction X$. We claim that $\mathcal{V}$ is the required open cover of $X$.

To this end, let $\mathcal{T}$ be a locally finite simplicial complex, let $\mathcal{S}$ be a subcomplex of $\mathcal{T}$ containing all the vertices of $\mathcal{T}$, and let $f \colon |\mathcal{S}| \to X$ be a partial realization of $\mathcal{T}$ in $X$ relative to $(\mathcal{S}, \mathcal{V})$. For each $\sigma \in \mathcal{T}$ let $\sigma^*$ denote the convex hull of $f[\sigma \cap |\mathcal{S}|]$ in $L$; observe that by (2) $\sigma^*$ is contained in an element of $\mathcal{F}$. For each $n \geq 0$, let

$$\mathcal{K}_n = \mathcal{T}^{(n)} \cup \mathcal{S}$$

(here $\mathcal{T}^{(n)}$ denotes the $n$-skeleton of $\mathcal{T}$ of course, see §2.1). Then $\mathcal{K}_n$ is a subcomplex of $\mathcal{T}$. By induction on $n \geq 0$ we shall construct a continuous function $f_n \colon |\mathcal{K}_n| \to V$ such that the following conditions are satisfied:

(3) $f_0 = f$,
(4) if $n \geq 1$ then $f_n$ extends $f_{n-1}$,
(5) for every $\sigma \in \mathcal{T}$, $f_n[\sigma \cap |\mathcal{K}_n|] \subseteq \sigma^*$.

Observe that $f_0$ is well-defined since $\mathcal{T}^{(0)} \subseteq \mathcal{S}$. Now assume that for certain $n \geq 0$ we constructed the function $f_{n-1}$. We will define $f_n$ on every simplex $\sigma \in \mathcal{K}_n$ separately and will conclude by applying Lemma 2.1.19 that the union of all the constructed functions is continuous. We have to be careful of course since we want the union of all the functions to be well-defined. So take an arbitrary simplex $\sigma \in \mathcal{K}_n = \mathcal{T}^{(n)} \cup \mathcal{S}$. We want to define $f_n$ on $\sigma$. There are two cases to consider. First assume that $\sigma \in \mathcal{K}_{n-1}$. Then put

$$f_n \restriction \sigma = f_{n-1} \restriction \sigma.$$

Suppose therefore that $\sigma \notin \mathcal{K}_{n-1}$, i.e., $\sigma \in \mathcal{T}^{(n)} \setminus \mathcal{K}_{n-1}$. Then $\sigma^\circ = \sigma \setminus \partial \sigma$, the (geometric) interior of $\sigma$, does not intersect $|\mathcal{K}_{n-1}|$. Since $\partial \sigma$ is a subset of $|\mathcal{K}_{n-1}|$, and $f_n$ is not defined yet at any point of $\sigma^\circ$, all we have to do is to extend the restriction $f_{n-1} \restriction \partial \sigma \colon \partial \sigma \to \sigma^*$ over $\sigma$. Since $\sigma^*$ is convex, this can be done by Theorem 1.2.9. This defines $f_n$ on $\sigma$. Observe that the union of all constructed functions is a function since the geometric interiors of all simplexes in $\mathcal{K}_n$ form a pairwise disjoint collection (Exercise 2.1.12) and that it is continuous by Lemma 2.1.19. Since (5) is easily seen to be satisfied, this completes the inductive construction.

Now define $\varphi \colon |\mathcal{T}| \to V$ by

$$\varphi(x) = f_n(x) \qquad (x \in \mathcal{T}^{(n)}).$$

Observe that $\varphi$ is clearly well-defined by (4), that the range of $\varphi$ is contained in $V$ by (5) and that $\varphi$ extends $f$ by (3). In addition, by another appeal to Lemma 2.1.19 it follows that $\varphi$ is continuous.

Finally define $g\colon |\mathcal{T}| \to X$ by
$$g = r \circ \varphi.$$
Since $r$ is a retraction and $\varphi$ extends $f$, $g$ extends $f$ as well.

Now take an arbitrary $\sigma \in \mathcal{T}$ and an element $W \in \mathcal{V}$ such that
$$f[\sigma \cap |\mathcal{S}|] \subseteq W.$$
There exists $n \geq 0$ such that $\sigma \in \mathcal{T}^{(n)}$. By (5) and the definition of $\varphi$ we obtain $\varphi[\sigma] \subseteq \sigma^*$. There exists $F \in \mathcal{F}$ such that $F \cap X = W$; observe that $\sigma^* \subseteq F$. By (1) there exists $U \in \mathcal{U}$ such that $F \subseteq r^{-1}[U]$. We conclude that
$$g[\sigma] = r[\varphi[\sigma]] \subseteq r[\sigma^*] \subseteq r[F] \subseteq U.$$
Consequently, $g$ is the desired extension of $f$. $\square$

From this it is easy to derive a similar result for ANR-pairs. Before we can do this, we need to derive the following lemma.

**Lemma 4.1.10.** *Let $X$ be a space and let $\mathcal{U}$ be an open cover of $X$. Then there exists an open refinement $\mathcal{V}$ of $\mathcal{U}$ such that*

(1) *$\mathcal{V}$ is star-finite,*
(2) *$\mathcal{V}$ is a star-refinement of $\mathcal{U}$.*

*Moreover, if $\mathcal{U}$ is finite then $\mathcal{V}$ can be chosen to be finite as well.*

**Proof.** By Corollary A.7.3 we may assume that $\mathcal{U}$ is locally finite. So in particular, $\mathcal{U}$ is countable. Enumerate $\mathcal{U}$ as $\{U_n : n \in \mathbb{N}\}$. By Proposition A.7.1 there exists a closed shrinking $\mathcal{F} = \{F_n : n \in \mathbb{N}\}$ of $\mathcal{U}$. Fix $x \in X$ for a moment. Since $\mathcal{U}$ is locally finite and $\mathcal{F}$ is a shrinking of $\mathcal{U}$, the set
$$A_x = \{n \in \mathbb{N} : x \in F_n\}$$
is finite. By Exercise A.7.1 it therefore follows that
$$W_x = \bigcap_{n \in A_x} U_n \setminus \Big( \bigcup_{n \notin A_x} F_n \Big)$$
is open.

Put $\mathcal{W} = \{W_x : x \in X\}$. Then $\mathcal{W}$ is an open cover of $X$. We claim that
$$\{\operatorname{St}(\{x\}, \mathcal{W}) : x \in X\} \prec \mathcal{U}.$$
To this end, take $x \in X$ and fix an arbitrary $n \in A_x$. Let $W_y \in \mathcal{W}$ be such that $x \in W_y$. Then clearly $A_x \subseteq A_y$ from which it follows that $W_y \subseteq U_n$. We conclude that $\operatorname{St}(\{x\}, \mathcal{W}) \subseteq U_n$, as required.

Repeating this argument for $\mathcal{W}$ we get an open cover $\mathcal{W}_1$ of $X$ such that for every $x \in X$ there is $W \in \mathcal{W}$ with $\mathrm{St}(\{x\}, \mathcal{W}_1) \subseteq W$. Observe that this implies that $\mathcal{W}_1$ refines $\mathcal{W}$. Now let $W_1 \in \mathcal{W}_1$ be arbitrary. We are interested in computing $\mathrm{St}(W_1, \mathcal{W}_1)$. To this end, pick an arbitrary $x \in W_1$ and let $W_1' \in \mathcal{W}$ be such that $W_1' \cap W_1 \neq \emptyset$, say $y \in W_1' \cap W_1$. There exists by construction an element $W \in \mathcal{W}$ such that $\mathrm{St}(\{y\}, \mathcal{W}_1) \subseteq W$. We conclude that

$$x \in W_1 \subseteq W_1 \cup W_1' \subseteq \mathrm{St}(\{y\}, \mathcal{W}_1) \subseteq W$$

and so $W_1' \subseteq \mathrm{St}(\{x\}, \mathcal{W})$. This shows that $\mathrm{St}(W_1, \mathcal{W}_1) \subseteq \mathrm{St}(\{x\}, \mathcal{W})$ from which it follows that $\mathrm{St}(W_1, \mathcal{W}_1) \subseteq U$ for some $U \in \mathcal{U}$.

By Theorem 2.3.5(1) there is a star-finite refinement $\mathcal{V}$ of $\mathcal{W}_1$. Then $\mathcal{V}$ is clearly as required.

Now assume that $\mathcal{U}$ is finite. Then there is no need to refine $\mathcal{U}$ by a locally finite open cover. Observe that the cover $\mathcal{W}$ is finite in that case and that the same is true for $\mathcal{W}_1$. So $\mathcal{W}_1$ is a star-refinement of $\mathcal{U}$ which is obviously star-finite, being finite. $\square$

**Corollary 4.1.11.** *Let $(X, Y)$ be an ANR-pair. Then for every open cover $\mathcal{U}$ of $X$ there exists an open refinement $\mathcal{V}$ of $\mathcal{U}$ such that for every locally finite simplicial complex $\mathcal{T}$ and every subcomplex $\mathcal{S}$ of $\mathcal{T}$, containing all the vertices of $\mathcal{T}$, and every partial realization $f \colon |\mathcal{S}| \to Y$ of $\mathcal{T}$ in $Y$ relative to $(\mathcal{S}, \mathcal{V} \restriction Y)$ can be extended to a full realization $g \colon |\mathcal{T}| \to Y$ of $\mathcal{T}$ in $Y$ relative to $\mathcal{U} \restriction Y$.*

**Proof.** Let $\mathcal{U}'$ be a star-refinement of $\mathcal{U}$ (Lemma 4.1.10). Since $X$ is an ANR, we let $\mathcal{V}$ be the refinement of $\mathcal{U}'$ we get from Theorem 4.1.9. We claim that this $\mathcal{V}$ also works here.

By Corollary 4.1.5 there exists a deformation $H$ from $X$ through $Y$ which is limited by $\mathcal{U}'$.

Now let a locally finite simplicial complex $\mathcal{T}$ be given, let $\mathcal{S}$ be a subcomplex of $\mathcal{T}$, containing all the vertices of $\mathcal{T}$, and let $f \colon |\mathcal{S}| \to Y$ be a partial realization of $\mathcal{T}$ in $Y$ relative to $(\mathcal{S}, \mathcal{V} \restriction Y)$. Since $f$ is also a partial realization in $X$ relative to $(\mathcal{S}, \mathcal{V})$, $f$ can be extended to a full realization $g$ of $\mathcal{T}$ in $X$ relative to $\mathcal{U}'$. We now use $H$ to push $g$ into $Y$, as follows. First observe that $|\mathcal{T}|$ is a metrizable space, and that $|\mathcal{S}|$ is a closed subspace of $|\mathcal{T}|$ (Lemma 2.1.14 and Proposition 2.1.21). Let $\varrho$ be an admissible metric for $|\mathcal{T}|$ which is bounded by 1 (Exercise A.1.6), and define $\bar{f} \colon |\mathcal{T}| \to X$ by the formula

$$\bar{f}(x) = H\bigl(g(x), \varrho(x, |\mathcal{S}|)\bigr).$$

Then $\bar{f}$ is clearly continuous. If $x \in |\mathcal{S}|$ then $\varrho(x, |\mathcal{S}|) = 0$, and so

$$\bar{f}(x) = g(x) = f(x).$$

Moreover, if $x \notin |\mathcal{S}|$ then $\varrho(x, |\mathcal{S}|) > 0$ and so $\bar{f}(x) \in Y$. We conclude that the range of $\bar{f}$ is contained in $Y$. Moreover, since $H$ is a $\mathcal{U}'$-homotopy, the functions $\bar{f}$ and $g$ are $\mathcal{U}'$-close.

Pick an arbitrary $\sigma \in \mathcal{T}$. There is an element $U_0 \in \mathcal{U}'$ such that $g[\sigma]$ is contained in $U_0$. In addition, $g$ and $\bar{f}$ are $\mathcal{U}'$-close. For every $x \in \sigma$ there consequently exists an element $U'_x \in \mathcal{U}'$ containing $g(x)$ as well as $\bar{f}(x)$. We conclude that $\bar{f}[\sigma]$ is contained in $\mathrm{St}(U_0, \mathcal{U}')$. So we are done since $\mathcal{U}'$ is a star-refinement of $\mathcal{U}$. □

**Dominating polytopes.** Let $X$ and $Y$ be spaces. We say that $X$ *dominates* $Y$ if there are two maps
$$g\colon Y \to X, \quad f\colon X \to Y,$$
such that the composition $f \circ g\colon Y \to Y$ is homotopic to the identity on $Y$. The space $X$ is called a *dominating space* for $Y$.

**Lemma 4.1.12.** *A space $X$ is contractible if and only a one-point space dominates $X$.*

**Proof.** Suppose first that $X$ is contractible and let $H\colon X \times \mathbb{I} \to X$ be a homotopy contracting $X$ to a point, say $p$. Now let $g\colon X \to \{p\}$ be the obvious function and let $f\colon \{p\} \to X$ be the inclusion. Then $H$ connects the identity on $X$ with $f \circ g$, i.e., $\{p\}$ dominates $X$.

Conversely, assume that there exists a point $p$ and functions $g\colon X \to \{p\}$ and $f\colon \{p\} \to X$ such that $f \circ g$ is homotopic to the identity on $X$. Since $f \circ g$ is clearly a constant function we conclude that $X$ is contractible. □

As remarked above, one sometimes needs homotopies 'with control'. For that reason we define a 'controlled version' of the concept of domination. Again let $X$ and $Y$ be spaces and let $\mathcal{U}$ be an open cover of $Y$. We say that $X$ $\mathcal{U}$-*dominates* $Y$ if there are two maps
$$g\colon Y \to X, \quad f\colon X \to Y,$$
such that the composition $f \circ g\colon Y \to Y$ is $\mathcal{U}$-homotopic to the identity on $Y$. The space $X$ is called a $\mathcal{U}$-*dominating space* for $Y$.

**Theorem 4.1.13.** *Let $X$ be a (compact) ANR. Then for every open cover $\mathcal{U}$ of $X$ there exists a (polyhedron) polytope $P$ such that $P$ $\mathcal{U}$-dominates $X$.*

**Proof.** Let $\mathcal{V}$ be an open refinement of $\mathcal{U}$ with the properties stated in Theorem 4.1.1. Let $\mathcal{V}'$ be a star-refinement of $\mathcal{V}$ (Lemma 4.1.10). By Theorem 4.1.9 there exists an open refinement $\mathcal{W}$ of $\mathcal{V}'$ such that for every locally finite simplicial complex $\mathcal{T}$, for every subcomplex $\mathcal{S}$ containing all the vertices of $\mathcal{T}$, and for every partial realization $f\colon |\mathcal{S}| \to X$ of $\mathcal{T}$ in $X$ relative to $(\mathcal{S}, \mathcal{W})$ there exists a full realization $g\colon |\mathcal{T}| \to X$ of $\mathcal{T}$ in $X$ relative to $\mathcal{V}'$ such that $g$ extends $f$.

## 4.1. SOME PROPERTIES OF ANR'S

**Claim 1.** Such a full realization $g$ has the following property:

For every $\sigma \in \mathcal{T}$ and for every $W \in \mathcal{W}$ with $f[\sigma \cap |\mathcal{S}|] \subseteq W$ there exists $V \in \mathcal{V}$ such that $g[\sigma] \cup W \subseteq V$.

*Proof.* Let $\sigma \in \mathcal{T}$ and $W \in \mathcal{W}$ be such that $f[\sigma \cap |\mathcal{S}|] \subseteq W$. There exists an element $V' \in \mathcal{V}'$ such that $g[\sigma] \subseteq V'$. Since $g$ extends $f$, and $f[\sigma \cap |\mathcal{S}|] \subseteq W$ and $f[\sigma \cap |\mathcal{S}|] \neq \emptyset$ since $\mathcal{S}$ contains all the vertices of $\mathcal{T}$, we get $W \cap V' \neq \emptyset$. Since $\mathcal{W}$ refines $\mathcal{V}'$ and $\mathcal{V}'$ is a star-refinement of $\mathcal{V}$, we are done. ◇

By Lemma 4.1.10 there exists a countable open cover $\mathcal{A}$ of $X$ which is both star-finite and a star-refinement of $\mathcal{W}$ (in case that $X$ is compact, $\mathcal{A}$ is finite). Put $P = |N(\mathcal{A})|$ and let $\kappa \colon X \to P$ be the $\kappa$-mapping of $\mathcal{A}$ (see Page 488 for the definition of a $\kappa$-mapping). Now for every vertex $x(A)$ in $N(\mathcal{A})^{(0)}$ choose an arbitrary point $f(x(A)) \in A$. This defines a function

$$f \colon |N(\mathcal{A})^{(0)}| \to X.$$

By Corollary 2.1.15, $|N(\mathcal{A})^{(0)}|$ is a discrete topological space, hence $f$ is continuous.

**Claim 2.** $f$ is a partial realization of $N(\mathcal{A})$ in $X$ relative to $(N(\mathcal{A}^{(0)}), \mathcal{W})$.

*Proof.* This is easy. Let $\sigma = |\{x(A_0), x(A_1), \ldots, x(A_n)\}|$ be a simplex in $N(\mathcal{A})$. Then

$$A_0 \cap A_1 \cap \cdots \cap A_n \neq \emptyset.$$

Since $\mathcal{A}$ is a star-refinement of $\mathcal{W}$ there exists $W \in \mathcal{W}$ such that

$$A_0 \cup A_1 \cup \cdots \cup A_n \subseteq W.$$

Consequently,

$$\begin{aligned} f[\sigma \cap N(\mathcal{A})^{(0)}] &= f[\{x(A_0), x(A_1), \ldots, x(A_n)\}] \\ &\subseteq A_0 \cup A_1 \cup \cdots \cup A_n \\ &\subseteq W. \end{aligned}$$

So we are done. ◇

Now by the special choice of $\mathcal{W}$, the function $f$ can be extended to a full realization $g \colon |N(\mathcal{A})| \to X$ relative to $\mathcal{V}'$.

**Claim 3.** The identity function $1_X$ on $X$ and the function $g \circ \kappa$ are $\mathcal{V}$-close.

*Proof.* Take an arbitrary $x \in X$. Since $\mathcal{A}$ is star-finite, there are only finitely many elements of $\mathcal{A}$ that contain $x$, say $A_0, A_1, \ldots, A_n$. Then

$$\sigma = |\{x(A_0), \ldots, x(A_n)\}|$$

is a simplex in $N(\mathcal{A})$, and by Lemma 2.3.2, $\kappa(x) \in \sigma$. There exists $W \in \mathcal{W}$ such that

$$A_0 \cup A_1 \cup \cdots \cup A_n \subseteq W.$$

Then $x \in W$ and since $f(x(A_i)) \in A_i$ for every $0 \le i \le n$, we have
$$f[\sigma \cap |N(\mathcal{A})^{(0)}|] \subseteq W.$$
By Claim 1 there exists $V \in \mathcal{V}$ such that $g[\sigma] \cup W \subseteq V$. Since $\kappa(x) \in \sigma$ we have $g(\kappa(x)) \in V$ and since $W \subseteq V$ we have $x \in V$. So $\{x, g(\kappa(x))\} \subseteq V$. $\Diamond$

By the special choice of $\mathcal{V}$ we now conclude that the indetity function $1_X$ and $g \circ \kappa$ are $\mathcal{U}$-homotopic, i.e., $P$ $\mathcal{U}$-dominates $X$. $\square$

**Exercises for §4.1.** Let $X$ be a space. An *equiconnecting function* on $X$ is a continuous function $\lambda \colon X \times X \times \mathbb{I} \to X$ such that $\lambda(x, x, t) = x$ for every $t \in \mathbb{I}$.

The spaces $X$ and $Y$ have the same *homotopy type* if there are continuous functions $f \colon X \to Y$ and $g \colon Y \to X$ such that $f \circ g$ is homotopic to the identity function on $Y$ and $g \circ f$ is homotopic to the identity function on $X$.

1. Present a different proof of Lemma 4.1.4 than the one in the text by using Corollary A.7.6.

2. Let $r \colon X \to Y$ be a retraction. Prove that $X$ dominates $Y$.

3. Let $X$ be an ANR with $\dim X \le n$. Prove that for every open cover $\mathcal{U}$ of $X$ there exists a polytope $P$ such that $P$ $\mathcal{U}$-dominates $X$ and $\dim P \le n$.

4. Let $X$ be a compact space with $\dim X = n < \infty$. Prove that there is a finite open cover $\mathcal{U}$ of $X$ such that for every space $Y$ that $\mathcal{U}$-dominates $X$ we have $\dim Y \ge n$.

▶5. Let $X$ be a compact ANR. Prove that for every $\varepsilon > 0$ there exists $\delta > 0$ such that for every space $Y$ and for every closed subspace $A$ of $Y$ and for all maps $f, g \colon A \to X$ with $\hat{\varrho}(f, g) < \delta$, if $f$ has a continuous extension
$$\bar{f} \colon Y \to X$$
then $g$ has a continuous extension
$$\bar{g} \colon Y \to X$$
such that $\hat{\varrho}(\bar{f}, \bar{g}) < \varepsilon$.

6. Let $X$ be an AR. Prove that $X$ has an equiconnecting function.

7. Show that if $(X, Y)$ is an A(N)R-pair then both $X$ and $Y$ are A(N)R's. In addition, prove that $X$ and $Y$ have the same homotopy type.

8. Let $(X, Y)$ be an ANR-pair. Assume that for some closed subspace $A \subseteq X$ there is a retraction $r \colon X \to A$ having the additional property that
$$r[Y] \subseteq Y \cap A.$$
Prove that $(A, Y \cap A)$ is an ANR-pair.

9. Let $X$ be a space with subspace $Y$. Suppose that there is a deformation of $X$ through $Y$. Let $Z$ be a space with closed subspace $K$ and let $\varepsilon \colon Z \to \mathbb{I}$ be continuous such that $\varepsilon^{-1}(0) = K$. Prove that for every continuous function $f \colon Z \to X$ there is a continuous function $g \colon Z \to X$ such that

(1) $\varrho(f(z), g(z)) \leq \varepsilon(z)$ for every $z \in Z$,
(2) $g \restriction K = f \restriction K$,
(3) $g[Z \setminus K] \subseteq Y$.

## 4.2. A characterization of ANR's and AR's

The aim of this section is to present purely topological characterizations of the classes of all ANR's and AR's. We will demonstrate the strength of our characterization by showing that the hyperspace of each *Peano continuum*, i.e., a locally connected continuum, is an AR.

Here is the announced characterization, stated in terms of ANR-pairs.

**Theorem 4.2.1.** *Let $X$ be a space and let $Y \subseteq X$ be dense. The following statements are equivalent:*

(1) $(X, Y)$ *is an* ANR-*pair*,
(2) *for every open cover $\mathcal{U}$ of $X$ there exists an open refinement $\mathcal{V}$ of $\mathcal{U}$ such that for every locally finite simplicial complex $\mathcal{T}$ and every subcomplex $\mathcal{S}$ of $\mathcal{T}$ containing all the vertices of $\mathcal{T}$, every partial realization of $\mathcal{T}$ in $Y$ relative to $(\mathcal{S}, \mathcal{V} \restriction Y)$ can be extended to a full realization of $\mathcal{T}$ in $Y$ relative to $\mathcal{U} \restriction Y$.*

Observe that by Corollary 4.1.11 we only need to verify the implication (2) $\Rightarrow$ (1). This will be done by proving a series of lemmas.

So let $(X, Y)$ be a pair of spaces having the realization property stated in Theorem 4.2.1(2). By Corollary 1.1.8 we may assume that $X$ is a closed subspace of a normed linear space $L$.

We shall prove that there are a neighborhood $U$ of $X$ in $L$ and a retraction $r \colon U \to X$ such that $r[U \setminus X] \subseteq Y$. We first show that this suffices.

Since we will not use anything specific about $L$, we can replace $L$ by the normed space $L' = L \times \mathbb{R}$. We identify $X$ and $X \times \{0\}$ in $L \times \mathbb{I}$. By our claim, there are a neighborhood $U$ of $X$ in $L'$ and a retraction $r \colon U \to X$ such that $r[U \setminus X] \subseteq Y$. This can be used to show that $(X, Y)$ is an ANR-pair, as follows. Let $E$ be a space with closed subspace $F$, and let

$$f \colon F \to X$$

be continuous. Since $X$ is an ANR, being a neighborhood retract of $L$ (Theorem 1.2.9), there are a neighborhood $V$ of $F$ in $E$ such that $f$ can be extended to a continuous function $g \colon V \to X$. Since $U$ is a neighborhood of $X$ in $L'$, the set $U \cap (X \times \mathbb{I})$ is a neighborhood of $X \times \{0\}$ in $X \times \mathbb{I}$. There consequently is a continuous function $\alpha \colon X \to (0, 1]$ such that

$$\{(x, t) : 0 \leq t \leq \alpha(x)\} \subseteq U$$

(Lemma 4.1.4). Let $\varrho$ be an admissible metric on $E$ which is bounded by 1 (Exercise A.1.6), and define $\xi \colon V \to X \times \mathbb{I}$ by
$$\xi(x) = \bigl(g(x), \varrho(x, F) \cdot \alpha\bigl(g(x)\bigr)\bigr).$$
Then $\xi[V] \subseteq U$, $\xi \restriction F = f$ and $\xi[V \setminus F] \subseteq U \setminus (X \times \{0\})$. Now finally define the function $\bar{f} \colon V \to X$ by
$$\bar{f}(x) = (r \circ \xi)(x).$$
Then $\bar{f}$ extends $f$ and by the special property of $r$, $\bar{f}[V \setminus F] \subseteq Y$. So we are done.

For a space $X$ we will let $\tau X$ denote the family of all open subsets of $X$.

**Lemma 4.2.2.** *Let $X$ be a space with subspace $Y$. The function*
$$\sigma \colon \tau Y \to \tau X$$
*defined by*
$$\sigma(A) = \{x \in X : \varrho(x, A) < \varrho(x, Y \setminus A)\}$$
*(where, by convention, $\varrho(x, \emptyset) = \infty$) has the following properties:*
(1) $\sigma(\emptyset) = \emptyset$, $\sigma(Y) = X$,
(2) $\sigma(A) \cap Y = A$ for every $A \in \tau Y$,
(3) if $A, B \in \tau Y$ then $A \subseteq B$ iff $\sigma(A) \subseteq \sigma(B)$,
(4) if $A, B \in \tau Y$ then $\sigma(A \cap B) = \sigma(A) \cap \sigma(B)$.

**Proof.** Let $\kappa \colon \rho Y \to \rho X$ be as in Lemma A.8.1. Then for every $U \in \tau Y$,
$$\sigma(U) = X \setminus \kappa(Y \setminus U).$$
An easy check shows that $\sigma$ is as required. $\square$

It will be convenient to fix some notation. Let $\sigma \colon \tau X \to \tau L$ be the function of Lemma 4.2.2. In addition, by applying our assumptions on $X$ and Lemma 4.1.10, by induction on $n \geq 0$ it is a triviality to construct two sequences of open (= open in $X$) covers $\mathcal{A}_n$ and $\mathcal{B}_n$ of $X$ with the following properties:

(i) $\mathcal{A}_0 = \{X\}$,
(ii) $\mathcal{B}_n < \mathcal{A}_n$ and for every locally finite simplicial complex $\mathcal{T}$ and every subcomplex $\mathcal{S}$ of $\mathcal{T}$ containing all the vertices of $\mathcal{T}$, every partial realization of $\mathcal{T}$ in $Y$ relative to $(\mathcal{S}, \mathcal{B}_n \restriction Y)$ can be extended to a full realization of $\mathcal{T}$ in $Y$ relative to $\mathcal{A}_n \restriction Y$,
(iii) for $n \geq 1$, $\mathcal{A}_n$ is a star-refinement of $\mathcal{B}_{n-1}$,
(iv) $\mathrm{mesh}(\mathcal{A}_n) < 1/n$.

Finally, by Lemma 1.2.1 and the fact that $Y$ is dense in $X$ there exists a countable open cover $\mathcal{U}$ of $L \setminus X$ and a sequence of points $(a_U)_{U \in \mathcal{U}}$ in $Y$ with the following properties:

## 4.2. A CHARACTERIZATION OF ANR'S AND AR'S

(v) if $p \in U \in \mathcal{U}$ then $\varrho(p, a_U) \leq 2\varrho(p, X)$,
(vi) if $U_n \in \mathcal{V}$ for every $n$ and
$$\lim_{n \to \infty} \varrho(U_n, X) = 0$$
then
$$\lim_{n \to \infty} \operatorname{diam}(U_n) = 0.$$

By an appeal to Theorem 2.3.5(1) we find that there exists a countable open cover $\mathcal{V}$ of $L \setminus X$ such that

(vii) $\mathcal{V}$ is star-finite and refines $\mathcal{U}$.

So without loss of generality we may assume that $\mathcal{U}$ is star-finite. We also assume that every $U \in \mathcal{U}$ is nonempty.

**Lemma 4.2.3.** *Let $x \in X$ and let $E$ be a neighborhood of $x$ in $L$. There exists a neighborhood $F$ of $x$ in $L$ such that if $U \in \mathcal{U}$ meets $F$ then $U \subseteq E$.*

**Proof.** Let $\varepsilon > 0$ be such that $B(x, \varepsilon) \subseteq E$. If the lemma is not true then for every $n$ there exists $U_n \in \mathcal{U}$ such that

(1) $U_n \cap B(x, \varepsilon/n) \neq \emptyset$,
(2) $U_n \setminus E \neq \emptyset$.

By (1), $\lim_{n \to \infty} \varrho(U_n, X) = 0$ so by (vi), $\lim_{n \to \infty} \operatorname{diam}(U_n) = 0$. This easily contradicts (1) and (2). □

This simple result is used in the proof of the following lemma.

**Lemma 4.2.4.** *For each $n \geq 0$ there exists an open neighborhood $H_n$ of $X$ in $L$ such that $H_0 = L$ and for $n \geq 1$,*

$(1)_n$ $\overline{H_n} \subseteq B(X, 1/n)$,
$(2)_n$ $\overline{H_n} \subseteq H_{n-1}$,
$(3)_n$ *if $U \in \mathcal{U}$ meets $\overline{H_n}$ then there exists $A \in \mathcal{A}_n$ such that*
$$U \subseteq \sigma(A) \cap H_{n-1}.$$

**Proof.** We shall construct the $H_n$'s by induction on $n \geq 0$. Let $n \geq 1$ and suppose that $H_{n-1}$ has been constructed. By Corollary A.4.3 there exists a neighborhood $V$ of $X$ in $L$ such that
$$X \subseteq V \subseteq \overline{V} \subseteq H_{n-1} \cap B(X, 1/n).$$
For each $x \in X$ there exists $A_x \in \mathcal{A}_n$ such that $x \in A_x$. Then $\sigma(A_x)$ is a neighborhood of $x$ in $L$ and consequently by Lemma 4.2.3 there exists a neighborhood $B_x$ of $x$ in $L$ such that if $U \in \mathcal{U}$ intersects $B_x$ then $U$ is contained in $\sigma(A_x) \cap V$. Put
$$H_n = \bigcup_{x \in X} B_x.$$

We claim that $H_n$ is as required.

Observe that $(1)_n$ and $(2)_n$ are trivially satisfied. For the verification of $(3)_n$, assume that $U \in \mathcal{U}$ meets $\overline{H}_n$. Since $U$ is open, $U \cap H_n \neq \emptyset$ and so there exists $x$ such that $U \cap B_x \neq \emptyset$. From this it follows that $U \subseteq \sigma(A_x) \cap V$. Since $V \subseteq H_{n-1}$, we are done. □

Fix $U \in \mathcal{U}$ for a moment and put

$$E = \{n \geq 0 : U \cap \overline{H}_n \neq \emptyset\}.$$

Since $H_0 \neq \emptyset$, $E \neq \emptyset$. We claim that $E$ is finite. Striving for a contradiction, assume that $E$ is infinite. Then by Lemma 4.2.4(1), $\varrho(U, X) = 0$ from which it follows by (vi) that $\operatorname{diam}(U) = 0$. Since $U$ is nonempty, we find that $U$ consists of a single point which belongs to $X$ since $\varrho(U, X) = 0$; contradiction.

Now for every $U \in \mathcal{U}$ define

$$n(U) = \max\{n \geq 0 : U \cap \overline{H}_n \neq \emptyset\}.$$

Let $\kappa \colon C \setminus X \to |N(\mathcal{U})|$ be the $\kappa$-function of the cover $\mathcal{U}$, cf. §1.1. For every $U \in \mathcal{U}$ select a point $z_U \in U$. By Lemma 4.2.4(3)$_{n(U)}$ we may pick an element $A_U \in \mathcal{A}_{n(U)}$ such that $U \subseteq \sigma(A_U)$.

**Lemma 4.2.5.** *For each $U \in \mathcal{U}$ there exists a point $y_U \in A_U \cap Y$ such that $\varrho(z_U, y_U) \leq 2\varrho(z_U, X)$.*

**Proof.** Let $U \in \mathcal{U}$. If $a_U \in A_U$ then we put $y_U = a_U$ and are done by (v) and the fact that $a_U \in Y$. So assume that $a_U \notin A_U$. Since $z_U \in U \subseteq \sigma(A_U)$ it follows by the definition of $\sigma$ that

$$\varrho(z_U, A_U) < \varrho(z_U, X \setminus A_U).$$

Since $a_U \notin A_U$, (v) consequently implies that

$$\varrho(z_U, A_U) < \varrho(z_U, a_U) \leq 2\varrho(z_U, X).$$

So any point in $A_U \cap Y$ will do. □

Recall that by convention the vertex of $N(\mathcal{U})$ corresponding to $U \in \mathcal{U}$ is denoted by $x(U)$, cf. §2.3. Now we define $\Phi \colon |N(\mathcal{U})^{(0)}| \to Y$ as follows:

$$\Phi\bigl(x(U)\bigr) = y_U.$$

Since $|N(\mathcal{U})^{(0)}|$ is a discrete topological space (Corollary 2.1.15) it follows that $\Phi$ is continuous.

We now aim at extending $\Phi$ over a large part of $|N(\mathcal{U})|$. For each $m \geq 0$ put

$$B_m = \overline{H}_m \setminus H_{m+1},$$

and let
$$\mathcal{P}_m = \{|\{x(U_{i(0)}), \ldots, x(U_{i(k)})\}| \in |N(\mathcal{U})| :$$
$$(\forall\, 0 \leq j \leq k)(U_{i(j)} \cap B_m \neq \emptyset)\}.$$

Clearly $\mathcal{P}_m$ is a subcomplex of $N(\mathcal{U})$.

**Lemma 4.2.6.** *If $|m - n| \geq 2$ then $|\mathcal{P}_n| \cap |\mathcal{P}_m| = \emptyset$.*

**Proof.** This follows directly from Lemma 4.2.4(3). □

**Lemma 4.2.7.** *For each $m > 0$, $\Phi \restriction |(\mathcal{P}_m)^{(0)}|$ is a partial realization of $\mathcal{P}_m$ in $Y$ relative to $((\mathcal{P}_m)^{(0)}, \mathcal{B}_{m-1} \restriction Y)$.*

**Proof.** Let $|\{x(U_{i(0)}), \ldots, x(U_{i(k)})\}|$ be a simplex in $\mathcal{P}_m$. Then by the definition of $\mathcal{P}_m$, for each $0 \leq j \leq k$, $U_{i(j)} \cap \overline{H}_m \neq \emptyset$, so that $m \leq n(U_{i(j)})$. Consequently, (ii) and (iii) imply that for each $0 \leq j \leq k$,
$$\mathcal{A}_{n(U_{i(j)})} < \mathcal{A}_m,$$
so that by Lemma 4.2.4(3) and the fact that $\sigma$ is monotone there exists an element $A_j \in \mathcal{A}_m$ such that
$$U_{i(j)} \subseteq \sigma(A_{U_{i(j)}}) \subseteq \sigma(A_j).$$
Observe that by the definition of $\Phi$ this implies that for every $0 \leq j \leq k$,
$$\Phi(x(U_{i(j)})) \in A_{U_{i(j)}} \subseteq A_j.$$
Since $|\{x(U_{i(0)}), \ldots, x(U_{i(k)})\}|$ is a simplex in $N(\mathcal{U})$,
$$U_{i(0)} \cap \cdots \cap U_{i(k)} \neq \emptyset,$$
so that
$$\sigma(A_0) \cap \cdots \cap \sigma(A_k) \neq \emptyset,$$
from which it follows by Lemma 4.2.2(1),(4) that
$$A_0 \cap \cdots \cap A_k \neq \emptyset.$$
Since by (iii) $\mathcal{A}_m$ is a star-refinement of $\mathcal{B}_{m-1}$, there exists an element of $\mathcal{B}_{m-1}$ which contains all $A_j$ and hence all $\Phi(x(U_{i(j)}))$. □

By (ii) we may extend for every $m \geq 1$ the partial realization $\Phi$ of $\mathcal{P}_{2m}$ in $Y$ relative to $((\mathcal{P}_{2m})^{(0)}, \mathcal{B}_{2m-1} \restriction Y)$ to a full realization
$$\Psi_{2m} \cdot |\mathcal{P}_{2m}| \to Y,$$
relative to $\mathcal{A}_{2m-1} \restriction Y$.

Now for $m \geq 0$ consider the subcomplexes $\mathcal{P}_{2m+1}$ of $N(\mathcal{U})$. Observe that for every $m$ the complex $|\mathcal{P}_{2m+1}|$ meets its neighbors $|\mathcal{P}_{2m}|$ and $|\mathcal{P}_{2m+2}|$ only (Lemma 4.2.6). Define
$$\mathcal{T}_{2m+1} = (\mathcal{P}_{2m+1})^{(0)} \cup (\mathcal{P}_{2m+1} \cap \mathcal{P}_{2m}) \cup (\mathcal{P}_{2m+1} \cap \mathcal{P}_{2m+2}).$$

Clearly, $\mathcal{T}_{2m+1}$ is a subcomplex of $\mathcal{P}_{2m+1}$, containing every vertex of $\mathcal{P}_{2m+1}$. Define a function $\Phi_{2m+1}\colon |\mathcal{T}_{2m+1}| \to Y$ by

$$\Phi_{2m+1}(x) = \begin{cases} \Psi_{2m}(x) & (x \in |\mathcal{P}_{2m}|), \\ \Phi(x) & (x \in |(\mathcal{P}_{2m+1})^{(0)}|), \\ \Psi_{2m+2}(x) & (x \in |\mathcal{P}_{2m+2}|). \end{cases}$$

Since the functions $\Psi_{2m}$ and $\Psi_{2m+2}$ extend $\Phi\!\upharpoonright\!|(\mathcal{P}_{2m})^{(0)}|$ and $\Phi\!\upharpoonright\!|(\mathcal{P}_{2m+2})^{(0)}|$, respectively, and since by Lemma 2.1.14(1) the sets

$$|(\mathcal{P}_{2m+1})^{(0)}|, \quad |\mathcal{P}_{2m+1} \cap \mathcal{P}_{2m}|, \quad |\mathcal{P}_{2m+1} \cap \mathcal{P}_{2m+2}|$$

are closed in $|N(\mathcal{U})|$, the function $\Phi_{2m+1}$ is well-defined and continuous.

**Lemma 4.2.8.** *For each $m \geq 1$, $\Phi_{2m+1}$ is a partial realization of $\mathcal{P}_{2m+1}$ in $Y$ relative to $(\mathcal{T}_{2m+1}, \mathcal{B}_{2m-2}\!\upharpoonright\!Y)$.*

**Proof.** Let $m \geq 1$ and let $\tau \in \mathcal{P}_{2m+1}$. By Lemma 4.2.7 there exists an element $B \in \mathcal{B}_{2m}$ such that

$$\Phi_{2m+1}[\tau \cap |N(\mathcal{U})^{(0)}|] \subseteq B.$$

Since by (ii) $\mathcal{B}_{2m} < \mathcal{A}_{2m}$ there also exists $A \in \mathcal{A}_{2m}$ such that

$$\Phi_{2m+1}[\tau \cap |N(\mathcal{U})^{(0)}|] \subseteq A.$$

Let $\tau'$ be a face of $\tau$ such that $\tau' \in \mathcal{P}_{2m+1} \cap \mathcal{P}_{2m}$. By construction there exists $A_{\tau'} \in \mathcal{A}_{2m-1}$ such that

$$\Psi_{2m}[\tau'] = \Phi_{2m+1}[\tau'] \subseteq A_{\tau'}.$$

Observe that $A \cap A_{\tau'} \neq \emptyset$. Now let $\tau''$ be a face of $\tau$ such that

$$\tau'' \in \mathcal{P}_{2m+1} \cap \mathcal{P}_{2m+2}.$$

By construction there also exists $A_{\tau''} \in \mathcal{A}_{2m+1}$ such that

$$\Psi_{2m}[\tau''] = \Phi_{2m+1}[\tau''] \subseteq A_{\tau''}.$$

As above, $A \cap A_{\tau''} \neq \emptyset$. Now since $\mathcal{A}_{2m+1} < \mathcal{A}_{2m}$, $\mathcal{A}_{2m} < \mathcal{A}_{2m-1}$, and by (iii), $\mathcal{A}_{2m-1}$ is a star-refinement of $\mathcal{B}_{2m-2}$, we are done. $\square$

So by Lemma 4.2.8 and (ii), we may extend for each $m \geq 1$ the function $\Phi_{2m+1}$ to a full realization

$$\Psi_{2m+1}\colon |\mathcal{P}_{2m+1}| \to Y,$$

relative to the covering $\mathcal{A}_{2m-2}\!\upharpoonright\!Y$.

**Lemma 4.2.9.** *For each $n \geq 0$, $\kappa[B_n] \subseteq |\mathcal{P}_n|$.*

**Proof.** Take an arbitrary $x \in B_n = \overline{H}_n \setminus H_{n+1}$. By (vii) there exist finitely many elements of $\mathcal{U}$ that contain $x$ only, say $U_{i(0)}, \ldots, U_{i(k)}$. The simplex
$$\tau = |\{x(U_{i(0)}), \ldots, x(U_{i(k)})\}|$$
is an element of $\mathcal{P}_n$. By Lemma 2.3.2, $\kappa(x)$ belongs to $\tau$, and since $\tau$ is a subset of $|\mathcal{P}_n|$ we are done. $\square$

For each $n \geq 0$ let
$$P_n = \bigcup_{m=n}^{\infty} |\mathcal{P}_m|.$$
Define $\Psi \colon P_2 \to Y$ by
$$\Psi(x) = \Psi_m(x) \qquad (x \in |\mathcal{P}_m|; m \geq 2).$$
Then $\Psi$ is continuous by Lemma 2.1.19.

Notice that for every $m \geq 3$, $\Psi \upharpoonright |\mathcal{P}_m|$ is a full realization of $\mathcal{P}_m$ in $Y$ relative to $\mathcal{A}_{m-3} \upharpoonright Y$.

Finally, define $r \colon \overline{H}_2 \to X$ by
$$r(x) = \begin{cases} x & (x \in X), \\ (\Psi \circ \kappa)(x) & (x \in \overline{H}_2 \setminus X). \end{cases}$$
Then $r$ is well-defined by Lemma 4.2.9. We claim that $r$ is a retraction. It is clear that $r[\overline{H}_2 \setminus X] \subseteq Y$. Since by Theorem 2.3.3, $\kappa$ is continuous and $X$ is closed in $L$, it remains to check the continuity of $r$ at the points of the boundary $\operatorname{Fr} X$ of $X$.

**Lemma 4.2.10.** *$r$ is continuous at every point of $\operatorname{Fr} X$.*

**Proof.** Let $x \in \operatorname{Fr} X$ and let $(t_n)_n$ be a sequence in $\overline{H}_2 \setminus X$ such that $\lim_{n \to \infty} t_n = x$. We shall prove that $\lim_{n \to \infty} r(t_n) = x$. To this end, for every $n$ pick $U_n \in \mathcal{U}$ containing $t_n$. In two steps we shall prove that
$$\lim_{n \to \infty} r(t_n) = \lim_{n \to \infty} \Psi(\kappa(t_n)) = x.$$

**Claim 1.** $\lim_{n \to \infty} \Psi(x(U_n)) = x$.

*Proof.* Since $\lim_{n \to \infty} \varrho(U_n, X) = 0$, by (vi) it follows that
$$\lim_{n \to \infty} \operatorname{diam}(U_n) = 0.$$
From this we conclude that
$$\lim_{n \to \infty} z_{U_n} = x.$$
By Lemma 4.2.5 we therefore obtain
$$\lim_{n \to \infty} \varrho(z_{U_n}, y_{U_n}) \leq 2 \lim_{n \to \infty} \varrho(z_{U_n}, X) = 0,$$

so that by the definition of $\Psi$,
$$\lim_{n\to\infty} \Psi(x(U_n)) = \lim_{n\to\infty} y_{U_n} = \lim_{n\to\infty} z_{U_n} = x,$$
as desired.

As in §2.3 for every $y \in \overline{H}_2 \setminus X$ let $F(y) = \{U \in \mathcal{U} : y \in U\}$ and let $\tau(y)$ be the simplex of $N(\mathcal{U})$ spanned by the (finitely many) vertices
$$\{x(U) : U \in F(y)\}.$$
By Lemma 2.3.2, $\tau(y)$ is the carrier of $\kappa(y)$ in $N(\mathcal{U})$, so in particular,
$$\kappa(y) \in \tau(y).$$

**Claim 2.** $\lim_{n\to\infty} \varrho\left(\Psi(\kappa(t_n)), \Psi(x(U_n))\right) = 0.$

*Proof.* For every $n \in \mathbb{N}$ let $k(n) \in \mathbb{N}$ be such that $t_n \in B_{k(n)}$ (without loss of generality, for every $n$, $k(n) \geq 3$). By Lemma 4.2.4(2) and the fact that the sequence $(t_n)_n$ converges to $x \in X$, it follows easily that $\lim_{n\to\infty} k(n) = \infty$. Observe that for every $n$, $\tau(t_n) \in \mathcal{P}_{k(n)}$. By construction, for every $n$ there exists $A_n \in \mathcal{A}_{k(n)-3}$ such that
$$\Psi[\tau(t_n)] \subseteq A_n,$$
and since $\{x(U_n), \kappa(t_n)\} \subseteq \tau(t_n)$ we therefore obtain
$$\{\Psi(x(U_n)), \Psi(\kappa(t_n))\} \subseteq \Psi[\tau(t_n)] \subseteq A_n.$$
From $\lim_{n\to\infty} k(n) = \infty$ we find by (iv) that
$$\lim_{n\to\infty} \operatorname{diam}(A_n) = 0,$$
so we are done.

This completes the proof. □

**Remark 4.2.11.** Notice that if $\mathcal{U}$ in the above proof has order at most $n+1$, where $n \geq 0$, then the dimension of $N(\mathcal{U})$ is at most $\leq n+1$. This follows from the Countable Closed Sum Theorem 3.2.8 and the obvious fact that $N(\mathcal{U})$ has no simplexes of dimension greater than $n+1$. This will be used in the proof of Theorem 4.2.30.

**Corollary 4.2.12.** *Let $X$ be a space. The following statements are equivalent:*

(1) *$X$ is an ANR,*
(2) *for every open cover $\mathcal{U}$ of $X$ there exists an open refinement $\mathcal{V}$ of $\mathcal{U}$ such that for every locally finite simplicial complex $\mathcal{T}$ and every subcomplex $\mathcal{S}$ of $\mathcal{T}$ containing all the vertices of $\mathcal{T}$, every partial realization of $\mathcal{T}$ in $X$ relative to $(\mathcal{S}, \mathcal{V})$ can be extended to a full realization of $\mathcal{T}$ in $X$ relative to $\mathcal{U}$.*

We shall now derive some applications. Call a space $X$ *homotopically trivial* if for every $n \geq 0$, every continuous function $f \colon \mathbb{S}^n \to X$ can be continuously extended over $B^{n+1}$. So every AR is homopically trivial for obvious reasons.

Notice that a space $X$ is path-connected iff every function $f \colon \mathbb{S}^0 \to X$ can be continuously extended over $B^1$. So a homotopically trivial space is one which is 'path-connected in every dimension'.

**Lemma 4.2.13.** *Every contractible space is homotopically trivial.*

**Proof.** Let $X$ be a contractible space and let $H \colon X \times \mathbb{I} \to X$ be a homotopy contracting $X$ to a single point, say $p$. In addition, let $n \geq 0$ and let $g$ be a continuous function from $\mathbb{S}^n$ to $X$. Define $\bar{g} \colon B^{n+1} \to X$ by

$$\begin{cases} \bar{g}(x) = H\big(g(x/\|x\|), 1 - \|x\|\big) & (x \neq \underline{0}), \\ \bar{g}(\underline{0}) = p. \end{cases}$$

An easy check shows that $\bar{g}$ is the required continuous extension of $g$. □

There exist elementary examples of homotopically trivial spaces that are not contractible. A well-known example of such a space is the so-called *Warsaw circle* in the figure below. See Exercise 4.2.1.

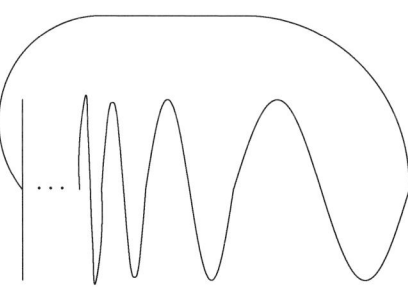

Figure 12.

We now present the following consequence of Theorem 4.2.1.

**Theorem 4.2.14.** *Let $X$ be a locally connected space with dense subspace $Y$. If $X$ has an (open) base $\mathcal{B}$ such that for every finite subfamily $\mathcal{F}$ of $\mathcal{B}$:*

*if $\bigcap \mathcal{F}$ is nonempty then every component $L$ of $\bigcap \mathcal{F}$ has the property that $L \cap Y$ is homotopically trivial.*

*Then $(X, Y)$ is an ANR-pair.*

**Proof.** Let $\mathcal{U}$ be an open cover of $X$. Since $\mathcal{B}$ is an open base for the topology of $X$, there exists a refinement $\mathcal{V}$ of $\mathcal{U}$ consisting entirely of elements of $\mathcal{B}$. By Lemma 4.1.10 there exists a star-refinement $\mathcal{W}$ of $\mathcal{V}$. Since a refinement of $\mathcal{W}$ is also a star-refinement of $\mathcal{V}$, and since by local connectivity of $X$ components of open subsets are open (Exercise A.2.8), we may assume without loss of generality that $\mathcal{W}$ consists of components of elements of $\mathcal{B}$.

We aim at applying Theorem 4.2.1. To this end, let $\mathcal{T}$ be a locally finite simplicial complex, let $\mathcal{S}$ be a subcomplex of $\mathcal{T}$ containing all the vertices of $\mathcal{T}$, and let $f\colon |\mathcal{S}| \to Y$ be a partial realization of $\mathcal{T}$ in $Y$ relative to $(\mathcal{S}, \mathcal{W} \upharpoonright Y)$. For each $\sigma \in \mathcal{T}$ pick $W_\sigma \in \mathcal{W}$ such that $f[\sigma \cap |\mathcal{S}|] \subseteq W_\sigma$. We first want to extend $f$ to a continuous function $g\colon |\mathcal{T}^{(1)} \cup \mathcal{S}| \to Y$ such that $g$ is a partial realization of $\mathcal{T}$ in $Y$ relative to $(\mathcal{T}^{(1)} \cup \mathcal{S}, \mathcal{V} \upharpoonright Y)$. As to be expected, we shall construct the function $g$ 'simplex-wise'. To this end, take an arbitrary

$$\sigma \in \mathcal{T}^{(1)} \cup \mathcal{S}.$$

We distinguish between two subcases. If $\sigma \in \mathcal{S}$ then $f$ is already defined on all of $\sigma$: put $g_\sigma = f \upharpoonright \sigma$. If $\sigma \notin \mathcal{S}$ then $\sigma \in \mathcal{T}^{(1)} \setminus \mathcal{T}^{(0)}$. Since $\mathcal{S}$ contains all the vertices of $\mathcal{T}$, $\sigma \cap |\mathcal{S}| = \partial \sigma$. So by the fact that $W_\sigma \cap Y$ is path-connected, there exists an extension $g_\sigma \colon \sigma \to W_\sigma \cap Y$ of $f \upharpoonright \partial \sigma$. Now define $g\colon |\mathcal{T}^{(1)} \cup \mathcal{S}| \to Y$ as follows:

$$g(x) = g_\sigma(x) \qquad (x \in \sigma \in \mathcal{T}^{(1)} \cup \mathcal{S}).$$

Since $\mathcal{T}^{(1)} \cup \mathcal{S}$ is a simplicial complex, $g$ is well-defined. Also, $g$ is continuous by Lemma 2.1.19. We claim that $g$ is a partial realization of $\mathcal{T}$ in $Y$ relative to $(\mathcal{T}^{(1)} \cup \mathcal{S}, \mathcal{V} \upharpoonright Y)$. To this end, take an arbitrary simplex $\sigma \in \mathcal{T}$. First observe that

$$g[\sigma \cap |\mathcal{T}^{(1)} \cup \mathcal{S}|] = g[\sigma \cap |\mathcal{T}^{(1)}|] \cup f[\sigma \cap |\mathcal{S}|].$$

By construction it therefore follows that

$$g[\sigma \cap |\mathcal{T}^{(1)} \cup \mathcal{S}|] \subseteq \bigcup \{W_\tau : \tau \text{ is a one-dimensional face of } \sigma\} \cup W_\sigma.$$

Again since $\mathcal{S}$ contains all the vertices of $\mathcal{T}$, each of the $W_\tau$'s intersects $W_\sigma$. Since $\mathcal{W}$ is a star-refinement of $\mathcal{V}$, there consequently exists $V_\sigma \in \mathcal{V}$ such that

$$g[\sigma \cap |\mathcal{T}^{(1)} \cup \mathcal{S}|] \subseteq V_\sigma,$$

which is as required.

Now let $\sigma \in \mathcal{T}$ be an arbitrary simplex and put

$$\mathcal{E}(\sigma) = \{\tau \in \mathcal{T} : \sigma \preccurlyeq \tau\}.$$

Since by the local finiteness of $\mathcal{T}$ each vertex of $\sigma$ is contained in at most finitely many simplexes in $\mathcal{T}$, it is clear that $\mathcal{E}(\sigma)$ is finite. Define

$$F(\sigma) = \bigcap_{\tau \in \mathcal{E}(\sigma)} V_\tau.$$

## 4.2. A CHARACTERIZATION OF ANR'S AND AR'S

**Claim 1.** For each $\sigma \in \mathfrak{T}$, every component of $F(\sigma) \cap Y$ is homotopically trivial and if $\sigma \preccurlyeq \tau$ then $F(\sigma) \subseteq F(\tau)$. Moreover, if $\sigma \in \mathfrak{T}^{(1)} \cup \mathcal{S}$ then

$$g[\sigma] \subseteq F(\sigma).$$

*Proof.* That every component of $F(\sigma)$ is homotopically trivial follows by the assumptions on $\mathcal{B}$ and the fact that the $\mathcal{E}(\sigma)$'s are finite. Also, if $\sigma \preccurlyeq \tau$ then

$$\mathcal{E}(\tau) \subseteq \mathcal{E}(\sigma)$$

from which it follows that $F(\sigma) \subseteq F(\tau)$. Now take an arbitrary $\sigma \in \mathfrak{T}^{(1)} \cup \mathcal{S}$ and $\tau \in \mathcal{E}(\sigma)$. Then

$$g[\sigma] \subseteq g[\tau \cap |\mathcal{S} \cup \mathfrak{T}^{(1)}|] \subseteq V_\tau.$$

We conclude that $g[\sigma] \subseteq F(\sigma)$. $\diamondsuit$

Now by induction on $n \geq 1$ we shall construct continuous functions

$$f_n \colon |\mathfrak{T}^{(n)} \cup \mathcal{S}| \to Y$$

such that the following conditions are satisfied:

(1) $f_1 = g$ and $f_{n+1}$ extends $f_n$,
(2) for every $\sigma \in \mathfrak{T}^{(n)} \cup \mathcal{S}$, $f_n[\sigma] \subseteq F(\sigma)$.

Assume that for certain $n \geq 2$, $f_{n-1}$ has been defined. We shall construct $f_n$ 'simplex-wise'. To this end, take an arbitrary $\sigma \in \mathfrak{T}^{(n)} \cup \mathcal{S}$. If $\sigma \in \mathfrak{T}^{(n-1)} \cup \mathcal{S}$ then $f_{n-1}$ is already defined on all of $\sigma$: put $f_\sigma = f_{n-1} \upharpoonright \sigma$. If $\sigma \notin \mathfrak{T}^{(n-1)} \cup \mathcal{S}$ then $\sigma \in \mathfrak{T}^{(n)} \setminus \mathfrak{T}^{(n-1)}$. Observe that $\sigma \cap |\mathfrak{T}^{(n-1)} \cup \mathcal{S}| = \partial\sigma$. Let $\tau$ be a proper face of $\sigma$. Then $\tau \in \mathfrak{T}^{(n-1)}$, which implies, by our inductive assumption and the claim, that

$$f_{n-1}[\tau] \subseteq F(\tau) \subseteq F(\sigma).$$

Consequently,

$$f_{n-1}[\partial\sigma] \subseteq F(\sigma).$$

Since $n \geq 2$ and $\sigma \notin \mathfrak{T}^{(n-1)}$, $\sigma$ is at least two-dimensional. Consequently, $\partial\sigma$ is connected from which it follows that $f_{n-1}[\partial\sigma]$ is connected. By Claim 1 it follows that $f_{n-1}[\partial\sigma]$ is contained in a homotopically trivial subset of the intersection $F(\sigma) \cap Y$. The function $f_{n-1} \upharpoonright \partial\sigma$ can therefore be extended to a continuous function

$$f_\sigma \colon \sigma \to F(\sigma) \cap Y.$$

Now define $f_n \colon |\mathfrak{T}^{(n)} \cup \mathcal{S}| \to Y$ as follows:

$$f_n(x) = f_\sigma(x) \quad (x \in \sigma \in \mathfrak{T}^{(n)} \cup \mathcal{S}).$$

Since $\mathfrak{T}^{(n)} \cup \mathcal{S}$ is a simplicial complex, $f_n$ is well-defined. Also, $f_n$ is continuous by Lemma 2.1.19. We conclude that $f_n$ is as required.

This completes the inductive construction of the functions $f_n$.

As to be expected, define $h: |\mathcal{T}| \to Y$ by
$$h(x) = f_n(x) \quad (x \in |\mathcal{T}^{(n)} \cup \mathcal{S}|).$$
It is clear that by (1), $h$ is well-defined. By another appeal to Lemma 2.1.19 it follows that $h$ is continuous. By construction, $h$ extends $f$ and by (2), it is a full realization of $\mathcal{T}$ in $Y$ relative to $\mathcal{V} \upharpoonright Y$ and therefore to $\mathcal{U} \upharpoonright Y$. □

**Corollary 4.2.15.** *Let $C$ be a convex subspace of a locally convex linear space. If $C \subseteq E \subseteq \overline{C}$ then $(E, C)$ is an AR-pair.*

**Proof.** Since convex sets in a linear space are contractible and hence homotopically trivial, and the linear space under consideration is locally convex, and the intersection of an arbitrary family of convex sets is again convex, this follows immediately from Theorem 4.2.14. □

**Remark 4.2.16.** Since the intersection of an arbitrary family of convex sets in a normed linear space is again convex, it seems that Corollary 4.2.15 gives a new proof of Theorem 1.2.9. This is not true however, since Theorem 1.2.9 was used in the proof of Theorem 4.2.1.

**Corollary 4.2.17.** *Let $0 \leq n < \infty$ and let $A$ be an arbitrary subset of $\mathbb{S}^n$. Then $B^{n+1} \setminus A$ is an AR.*

**Corollary 4.2.18.** *Let $X$ be a locally connected space. If $X$ has an (open) base $\mathcal{B}$ such that for every finite subfamily $\mathcal{F}$ of $\mathcal{B}$:*

*if $\bigcap \mathcal{F}$ is nonempty then every component $C$ of $\bigcap \mathcal{F}$ is homotopically trivial.*

*Then $X$ is an ANR.*

Naturally, Theorem 4.2.1 suggests the question whether there exists a characterization of the class of AR's in the same spirit. In Corollary 1.4.5 we showed that a space $X$ is an AR if and only if $X$ is a contractible ANR. So it seems that our question has already been answered. However, deciding whether a given space is contractible is sometimes quite a complicated task, cf. Corollary 2.4.11. Within the class of ANR's it turns out that contractibility and homotopy triviality are equivalent notions. Since for a given space $X$, verifying that it is homotopically trivial is usually much easier than verifying that it is contractible, we shall present a proof of our assertion in full detail. We first derive the following lemma.

**Lemma 4.2.19.** *Let $X$ be a space. The following statements are equivalent:*

(1) *$X$ is homotopically trivial,*
(2) *for every locally finite simplicial complex $\mathcal{T}$ and every subcomplex $\mathcal{S}$ of $\mathcal{T}$, every continuous function $g: |\mathcal{S}| \to X$ can be extended to a continuous function $\bar{g}: |\mathcal{T}| \to X$.*

**Proof.** The implication (2) ⇒ (1) is a triviality. Simply observe that there is a homeomorphism from $B^n$ onto an $n$-dimensional simplex $\sigma$ taking $\mathbb{S}^{n-1}$ onto the geometric boundary $\partial\sigma$ of $\sigma$ (Exercise 1.1.24).

The implication (1) ⇒ (2) is also very simple to prove. Let us first show that without loss of generality we may assume that $\mathcal{S}$ contains all the vertices of $\mathcal{T}$. Since $|\mathcal{T}^{(0)}|$ is a discrete topological space by Corollary 2.1.15, and since both $|\mathcal{S}|$ and $|\mathcal{T}^{(0)}|$ are closed in $|\mathcal{T}|$ by Lemma 2.1.14(1), we can extend $g$ over $|\mathcal{S} \cup \mathcal{T}^{(0)}|$ in an arbitrary way.

Now, by induction on $n \geq 0$ by a standard procedure we shall construct a continuous function $g_n \colon |\mathcal{S} \cup \mathcal{T}^{(n)}| \to X$ such that

$$g_0 = g, \quad g_{n+1} \text{ extends } g_n.$$

Since $\mathcal{S}$ contains all the vertices of $\mathcal{T}$, we may indeed put $g_0 = g$. Assume that the function $g_{n-1}$ has been constructed for certain $n \geq 1$. As previously in this section, we shall construct $g_n$ 'simplex-wise'. To this end, take an arbitrary

$$\sigma \in \mathcal{S} \cup \mathcal{T}^{(n)}.$$

If $\sigma \in \mathcal{S} \cup \mathcal{T}^{(n-1)}$ put $g_\sigma = g_{n-1} \restriction \sigma = g \restriction \sigma$. So assume that

$$\sigma \in \mathcal{T}^{(n)} \setminus (\mathcal{S} \cup \mathcal{T}^{(n-1)}).$$

Now we use our assumption on $X$ to extend the function $g_{n-1} \restriction \partial\sigma \colon \partial\sigma \to X$ to a continuous function $g_\sigma \colon \sigma \to X$. Define $g_n \colon |\mathcal{S} \cup \mathcal{T}^{(n)}| \to X$ by

$$g_n(x) = g_\sigma(x) \quad (x \in s \in \mathcal{S} \cup \mathcal{T}^{(n)}).$$

As in the previous proofs in this section it follows that $g_n$ is well-defined and continuous. This completes the inductive construction of the functions $g_n$ for $n \in \mathbb{N}$.

Now define $\bar{g} \colon |\mathcal{T}| \to X$ as follows:

$$\bar{g}(x) = g_n(x) \quad (x \in \sigma \in \mathcal{T}^{(n)}).$$

Then $\bar{g}$ is clearly as required. □

We shall now present a proof of the following

**Theorem 4.2.20.** *Let $X$ be a space. The following statements are equivalent:*

(1) *$X$ is an AR,*
(2) *$X$ is a contractible ANR,*
(3) *$X$ is a homotopically trivial ANR.*

**Proof.** Observe that the implications (1) ⇒ (2) ⇒ (3) are trivialities, see Corollary 1.4.5 and Lemma 4.2.13. So let $X$ be a homotopically trivial ANR. We shall prove that $X$ is an AR. Observe that $X$ has the realization property stated in Corollary 4.2.1(2).

It will be convenient to think of $X$ as the space $X$ in the proof of Theorem 4.2.1. We adopt all notation and terminology introduced in that proof. We shall prove that the retraction $r\colon \overline{H}_2 \to X$ can be extended over $L$ so that the desired result follows from Corollary 1.2.9 and the obvious fact that a retract of an AR is an AR.

Recall that
$$P_2 = \bigcup_{m=2}^{\infty} |\mathcal{P}_m|.$$

Define
$$Q = |\mathcal{P}_0| \cup |\mathcal{P}_1| \cup |\mathcal{P}_2|.$$

Observe that by Lemma 4.2.6,
$$P_2 \cap Q = |\mathcal{P}_2|,$$
and also that by definition, $P_2 \cup Q = P_0$. Now since $\mathcal{T} = \mathcal{P}_0 \cup \mathcal{P}_1 \cup \mathcal{P}_2$ is a simplicial complex, and $\mathcal{P}_2$ is a subcomplex of $\mathcal{T}$, by Lemma 4.2.19 we can extend $\Psi \upharpoonright |\mathcal{P}_2|$ to a continuous function
$$\theta \colon Q \to X.$$
By Lemma 4.2.9 it follows that $\kappa[C \setminus H_2] \subseteq Q$. Consequently, the function
$$\bar{r} \colon C \to X$$
defined by
$$\bar{r}(x) = \begin{cases} r(x) & (x \in \overline{H}_2), \\ (\theta \circ \kappa)(x) & (c \in C \setminus H_2), \end{cases}$$
is well-defined if for every $x \in D = \overline{H}_2 \setminus H_2$ we have $r(x) = (\theta \circ \kappa)(x)$. To verify this, take an arbitrary $x \in D$. Then $x \in B_2$. Consequently, Lemma 4.2.9 gives us that $\kappa(x) \in |\mathcal{P}_2|$. By the definition of $\theta$ we therefore have
$$r(x) = (\Psi \circ \kappa)(x) = (\theta \circ \kappa)(x),$$
which is as required. Finally, $\bar{r}$ is continuous since the sets $\overline{H}_2$ and $C \setminus H_2$ are closed in $C$ and the restrictions $\bar{r} \upharpoonright \overline{H}_2$ and $\bar{r} \upharpoonright C \setminus H_2$ are both continuous. □

**Corollary 4.2.21.** *Let $(X, Y)$ be an ANR-pair. If $Y$ is homotopically trivial then $(X, Y)$ is an AR-pair.*

**Proof.** By the previous theorem it suffices to prove that $X$ is homotopically trivial. By Theorem 4.1.6 there is a deformation $H \colon X \times \mathbb{I} \to X$ through $Y$. Let $f \colon \mathbb{S}^n \to X$ be continuous, where $n \geq 0$. For sufficiently small $t > 0$, the functions $f$ and $H_t \circ f$ are as close as we please, so they are homotopic by Theorem 4.1.1 since $X$ is an ANR. Since $Y$ is homotopically trivial, the function $H_t \circ f$ can be extended over $B^{n+1}$. So the functions $H_t \circ f$ and $f$ are homotopic functions from $\mathbb{S}^n$ into $X$, one of which can be extended over $B^{n+1}$. But then the other one can also be extended by the Borsuk Homotopy Extension Theorem 1.4.2. □

**Application to hyperspaces.** As in §1.11, for every compact space $X$ we let $2^X$ denote its hyperspace. As announced above, we shall apply the results in this section to prove that if $X$ is a Peano continuum then $2^X$ is an AR. This demonstrates the power of the techniques derived so far.

We shall conclude from this by using TORUŃCZYK's Theorem from [391] that the hyperspace of each non-degenerate Peano continuum is homeomorphic to the Hilbert cube; this is the Curtis-Schori-West Hyperspace Theorem. (See also [298, Chapter 8] for details.) Since the Hilbert cube is an AR, at first glance there seems no reason for proving the AR-property for hyperspaces. However, that the hyperspace of each Peano continuum is an AR is a fundamental step in the proof of the Curtis-Schori-West Hyperspace Theorem.

**Lemma 4.2.22.** *For each $n \in \mathbb{N}$, there exists a map $r\colon B^{n+1} \to \mathcal{F}_3(\mathbb{S}^n)$ such that for every $x \in \mathbb{S}^n$, $r(x) = \{x\}$.*

**Proof.** First consider the case n = 1. Parametrize $\mathbb{S}^1$ as $[0, 2\pi]$, with 0 and $2\pi$ identified. For $\theta \in [0, 2\pi]$, set $t_\theta = \pi - |\theta - \pi|$. Define a homotopy
$$H\colon \mathbb{S}^1 \times [0, \pi] \to \mathcal{F}_2(\mathbb{S}^1)$$
by
$$H(\theta, t) = \begin{cases} \{\theta - t, \theta + t\} & (0 \leq t \leq t_\theta), \\ \{\theta - t_\theta, \theta + t_\theta\} & (t_\theta \leq t \leq \pi). \end{cases}$$
Observe that $H_0(\theta) = \{\theta\}$ for every $\theta \in \mathbb{S}^1$. In addition, define a homotopy
$$K\colon \mathbb{S}^1 \times [0, \pi] \to \mathcal{F}_3(\mathbb{S}^1)$$
as follows:
$$K(\theta, t) = \{t, 2\pi - t\} \cup \left(\{\theta - t_\theta, \theta + t_\theta\} \cap [t, 2\pi - t]\right).$$
It is easy to see that the homotopy $H$ followed by $K$ provides a homotopy from $\mathbb{S}^1$ into $\mathcal{F}_3(\mathbb{S}^1)$ connecting the inclusion map $\theta \to \{\theta\}$ and the constant map $\theta \to \{\pi\}$.

Now we proceed by induction on $n$. Assume that for certain $n \geq 2$ there exists a map $f\colon B^n \to \mathcal{F}_3(\mathbb{S}^{n-1})$ such that $f(x) = \{x\}$ for every $x \in \mathbb{S}^{n-1}$. We identify $B^{n+1}$ and the two-point compactification of $B^n \times (-1, 1)$. Define a map $g\colon B^{n+1} \to \mathcal{F}_3(\mathbb{S}^n)$ as follows:
$$\begin{cases} g(x, t) = \{(p, t) : p \in f(x)\} & ((x, t) \in B^n \times (-1, 1)), \\ g(+\infty) = \{+\infty\}, \\ g(-\infty) = \{-\infty\}, \end{cases}$$
An easy check shows that $g$ is as required. $\square$

A subcollection $\mathcal{E}$ of $\mathcal{F}_\infty(X)$ is called an *expansion hyperspace* of $X$ if for every $E \in \mathcal{E}$ and $F \in \mathcal{F}_\infty(X)$ such that $E \subseteq F$ we have $F \in \mathcal{E}$. Observe that $\mathcal{F}_\infty(X)$ is an expansion hyperspace.

Dense expansion hyperspaces are AR's, as will be shown now.

**Theorem 4.2.23.** *Let $X$ be a compact space. The following statements are equivalent:*

(1) *$X$ is a Peano continuum,*
(2) *$2^X$ is a Peano continuum,*
(3) *$2^X$ is an AR.*
(4) *If $\mathcal{E} \subseteq 2^X$ is a dense expansion hyperspace then $(2^X, \mathcal{E})$ is an AR-pair.*

**Proof.** We already know that the equivalence (1) $\Leftrightarrow$ (2) is true. See Propositions 1.11.4 and 1.11.13.

Since (4) $\Rightarrow$ (3) $\Rightarrow$ (2) are trivial (Exercise 1.2.6), it remains to verify the implication (1) $\Leftrightarrow$ (2) $\Rightarrow$ (4). Naturally, we aim at applying Theorem 4.2.14 to prove that $(2^X, \mathcal{E})$ is an AR-pair.

**Claim 1.** $\mathcal{E}$ is path-connected and locally path-connected.

*Proof.* Let $F \in \mathcal{E}$, say $F = \{x_1, \ldots, x_n\}$, where $x_i \neq x_j$ if $i \neq j$. If $U$ is a neighborhood of $F$ in $2^X$ then there is a smaller neighborhood of the form $\langle \{V_1, \ldots, V_n\} \rangle$, where $V_i$ is an open neighborhood of $x_i$ for every $i$ (Lemma 1.11.11(2)). We may assume without loss of generality that

$$V_i \cap V_j = \emptyset$$

if $i \neq j$, and that each $V_i$ is connected since $X$ is locally connected. We claim that $\mathcal{A} = \langle \{V_1, \ldots, V_n\} \rangle \cap \mathcal{E}$ is path-connected, which will prove that $\mathcal{E}$ is locally path-connected. This will implicitly also show that $\mathcal{E}$ is path-connected since $\mathcal{E} = \langle \{X\} \rangle \cap \mathcal{E}$ and $X$ is connected. To this end, let $G \in \mathcal{A}$ be arbitrarily chosen. For every $i \leq n$ let $G_i = G \cap V_i$. Then every $G_i \neq \emptyset$ while moreover $G = \bigcup_{i=1}^n G_i$. Let us concentrate on $G_1 = \{a_1, \ldots, a_m\}$ for a moment. By Theorem 1.5.22, $V_1$ is path-connected. So there exists for every $j \leq m$ a path $\lambda_j \colon \mathbb{I} \to V_1$ such that $\lambda_j(0) = a_j$ and $\lambda_j(1) = x_1$. The function $\mu_1 \colon \mathbb{I} \to \langle \{V_1\} \rangle \cap \mathcal{E}$ defined by

$$\mu_1(t) = \{x_1\} \cup \{\lambda_j(t) : j \leq m\}$$

is a path connecting $\{x_1\} \cup G_1$ and $\{x_1\}$ (Corollary 1.11.10). Define the functions $\mu_j$ for $2 \leq j \leq n$ similarly. Then the function $\mu \colon \mathbb{I} \to \mathcal{A}$ defined by

$$\mu(t) = \bigcup_{j=1}^n \mu_j(t)$$

is a path connecting $F$ and $F \cup G$. Using the same technique there exists a path in $\mathcal{A}$ connecting $F \cup G$ and $G$. Indeed, for $j \leq m$ define $\nu_1 \colon \mathbb{I} \to \langle \{V_1\} \rangle \cap \mathcal{E}$ by

$$\nu_1(t) = G_1 \cup \{\lambda_j(1-t) : j \leq m\}.$$

Define the functions $\nu_j$ for $2 \leq j \leq n$ similarly. Then the function $\nu\colon \mathbb{I} \to \mathcal{A}$ defined by
$$\nu(t) = \bigcup_{j=1}^{n} \nu_j(t)$$
is a path connecting $F \cup G$ and $G$.

The join of these paths $\mu$ and $\nu$ is therefore the required path. ◇

So we see in particular that $\mathcal{E}$ is locally connected. Hence components of open sets of $\mathcal{E}$ are open in $\mathcal{E}$ (Exercise A.2.8).

We aim at applying Theorem 4.2.14 to prove that $(2^X, \mathcal{E})$ is an AR-pair. Consider the base $\mathcal{B}(X)$ for $2^X$ identified in Lemma 1.11.11(2). Since $\mathcal{B}(X)$ is closed under finite intersections (Lemma 1.11.11(2)), it suffices to prove that every component of an element of $\mathcal{B}(X) \restriction \mathcal{E}$ is homotopically trivial.

To this end, let $U_1, \ldots, U_n \subseteq X$ be nonempty and open, and consider a component $C$ of $B = \langle \{U_1, \ldots, U_n\} \rangle \cap \mathcal{E}$. Observe that $B \neq \emptyset$ since $\mathcal{E}$ is dense, hence $C \neq \emptyset$.

**Claim 2.** $C$ is homotopically trivial.

*Proof.* We already know by Claim 1 that $C$ is open in $\mathcal{E}$. By connectivity of $C$ and the fact that $\mathcal{E}$ is locally path-connected, it therefore follows by Lemma A.10.8 that $C$ is path-connected. So for the verification that $C$ is homotopically trivial it suffices to consider a continuous function $g\colon \mathbb{S}^m \to C$, where $m \geq 1$. By Lemma 4.2.22 there exists a map $r\colon B^{m+1} \to \mathcal{F}_3(\mathbb{S}^m)$ such that for every $x \in \mathbb{S}^m$, $r(x) = \{x\}$. By Corollary 1.11.8, the function
$$\bar{g}\colon B^{m+1} \to 2^X$$
defined by
$$\bar{g}(p) = \bigcup g[r(p)]$$
is continuous. It is clear that the range of $\bar{g}$ is contained in $\mathcal{E}$ and that
$$\bar{g} \restriction \mathbb{S}^m = g.$$
We claim that $\bar{g}[B^{m+1}] \subseteq B$. To this end, take an arbitrary $p \in B^{m+1}$. Notice that $g[r(p)]$ consists of at most three elements of $B$. So each of these elements meets every $U_i$ and is contained in their union. But then clearly $\bigcup g[r(p)]$ has the same property. By connectivity of $B^{m+1}$ it now follows that
$$\bar{g}[B^{m+1}] \subseteq C,$$
which is as required. ◇

By Theorem 4.2.14 it therefore follows that $(2^X, \mathcal{E})$ is an ANR-pair. However, implicitly, we also proved that $\mathcal{E}$ is homotopically trivial itself since clearly $\mathcal{E} = \langle \{X\} \rangle \cap \mathcal{E}$ and $\mathcal{E}$ is connected by Claim 1. So we conclude that $(2^X, \mathcal{E})$ is an AR-pair by Corollary 4.2.21. □

**Corollary 4.2.24.** *Let $X$ be a Peano continuum, and $\mathcal{E} \subseteq 2^X$ a dense expansion hyperspace. Then there is a deformation of $2^X$ through $\mathcal{E}$.*

**Proof.** Apply the previous theorem and Theorem 4.1.6. □

Observe that since $\mathcal{F}_\infty(X)$ is an expansion hyperspace, this result shows that we can approximate compacta in $X$ by finite sets in a continuous way. This curious result will turn out to be of crucial importance later.

**Toruńczyk's Theorem.** Let $X$ be a space. We say that $X$ has the *disjoint-cells property* provided that for every $n \in \mathbb{N}$, every continuous function $f \colon \mathbb{I}^n \times \{0,1\} \to X$ is approximable (arbitrarily closely) by maps sending $\mathbb{I}^n \times \{0\}$ and $\mathbb{I}^n \times \{1\}$ to disjoint sets. Formally, for every $n$, every continuous function $f \colon \mathbb{I}^n \times \{0,1\} \to X$ and every $\varepsilon > 0$ there is a continuous function $g \colon \mathbb{I}^n \times \{0,1\}$ such that

(1) $\hat{\varrho}(f,g) < \varepsilon$,
(2) $g[\mathbb{I}^n \times \{0\}] \cap g[\mathbb{I}^n \times \{1\}] = \emptyset$.

It is clear that $Q$ has this property. To see this, let $\varepsilon > 0$ and a continuous function $f \colon \mathbb{I}^n \times \{0,1\} \to Q$ be given. There exists $m \in \mathbb{N}$ such that $2^{-(m-1)} < \varepsilon$. Define $g \colon \mathbb{I}^n \times \{0,1\} \to Q$ as follows:

$$g(x,i) = (f(x,i)_1, \cdots, f(x,i)_{m-1}, i, i, \ldots), \quad (i \in \{0,1\}).$$

An easy check shows that $g$ satisfies (1) and (2). Toruńczyk's Theorem states (among other things) that $Q$ is topologically the only AR with the disjoint-cells property.

**Toruńczyk's Theorem 4.2.25.** *Let $X$ be a space. The following statements are equivalent:*

(1) *$X$ is homeomorphic to the Hilbert cube $Q$.*
(2) *$X$ is a compact AR and has the disjoint-cells property.*

For details, see TORUŃCZYK [391] or VAN MILL [298, Chapter 8].

So all there remains to prove for the Curtis-Schori-West Hyperspace Theorem is that $2^X$ has the disjoint-cells property.

**Lemma 4.2.26.** *Let $X$ be a Peano continuum and let $\mathcal{K}$ be a compact subset of $\mathcal{F}_\infty(X)$. Then for every $\varepsilon > 0$ there is a continuous function*

$$f \colon 2^X \to 2^X \setminus \mathcal{K}$$

*such that $\hat{\varrho}_H(f,1) < \varepsilon$.*

**Proof.** Define

$$\mathcal{E} = \{F \in \mathcal{F}_\infty(X) : F \text{ is not contained in an element of } \mathcal{K}\}.$$

Observe that $\mathcal{E}$ is an expansion hyperspace of $X$ and that $\mathcal{E} \cap \mathcal{K} = \emptyset$. We claim that $\mathcal{E}$ is dense in $2^X$. Since $\mathcal{F}_\infty(X)$ is dense in $2^X$ (Lemma 1.11.11(2)), it suffices to prove that for arbitrary $F \in \mathcal{F}_\infty(X)$ and $\varepsilon > 0$ there exists an element $E \in \mathcal{E}$ with $\varrho_H(F, E) < \varepsilon$. So let $F \in \mathcal{F}_\infty(X)$ and $\varepsilon > 0$ be given. Pick a point $x \in F$ and a sequence $(x_n)_n$ in $X \setminus \{x\}$ converging to $x$, with $\varrho(x_n, x) < \varepsilon$ for each $n$. Define

$$F_n = F \cup \{x_1, \ldots, x_n\} \qquad (n \in \mathbb{N}).$$

Observe that for each $n$, $F_n \in \mathcal{F}_\infty(X)$ and $\varrho_H(F, F_n) < \varepsilon$. We claim that for some $n$, $F_n \in \mathcal{E}$. If not, then there exists a sequence $(K_n)_n$ in $\mathcal{K}$ such that $F_n \subseteq K_n$ for each $n$. By compactness of $\mathcal{K}$ we assume without loss of generality that $K_n \to K$, $n \to \infty$.

**Claim 1.** $\{x_1, x_2, x_3, \ldots\} \subseteq K$.

*Proof.* Suppose that for some $n$, $x_n \notin K$. Since $K_n \to K$, $n \to \infty$, there is an $m \geq n$ such that $K_m \subseteq K \setminus \{x_n\}$. But this is impossible since

$$x_n \in F_m \subseteq K_m \subseteq K \setminus \{x_n\}.$$

So we are done. ◇

From the claim we conclude that $K$ is infinite, which contradicts the fact that $K$ is finite.

We see that $\mathcal{E}$ is dense in $2^X$. By Corollary 4.2.24 there are arbitrarily small maps $2^X \to \mathcal{E}$. Since $\mathcal{E} \cap \mathcal{K} = \emptyset$, we are done. □

**The Curtis-Schori-West Hyperspace Theorem 4.2.27.** *Let $X$ be compact space. The following statements are equivalent:*

(1) *$X$ is a Peano continuum,*
(2) *$2^X$ is homeomorphic to the Hilbert cube $Q$.*

**Proof.** As observed above, it suffices to prove that for a Peano continuum $X$, $2^X$ has the disjoint-cells property. Since $2^X$ admits arbitrarily small maps into $\mathcal{F}_\infty(X)$ (Corollary 4.2.24) and since for every compact subset $\mathcal{K}$ of $\mathcal{F}_\infty(X)$ there are arbitrarily small maps from $2^X$ into $2^X \setminus \mathcal{K}$ (Corollary 4.2.26), this is obvious. □

**Characterization of finite dimensional ANR's and AR's.** We shall now characterize the class of all finite dimensional ANR's and AR's in simple topological terms.

Let $X$ be a space and let $0 \leq n < \infty$. We say that $X$ is *connected in dimension $n$*, abbreviated $C^n$, provided that for every $0 \leq m \leq n$, every continuous function $f \colon \mathbb{S}^m \to X$ can be extended to a continuous function $\bar{f} \colon B^{m+1} \to X$. So the homotopically trivial spaces are precisely those spaces that are connected in every dimension. In addition, we say that $X$

is *locally connected in dimension* $n$, abbreviated $LC^n$, provided that for every $x \in X$ and for every neighborhood $U$ of $x$ and for every $0 \le m \le n$ there exists a neighborhood $V$ of $x$ such that every continuous function $f\colon \mathbb{S}^m \to V$ extends to a continuous function $\bar{f}\colon B^{m+1} \to U$. We shall prove that if $\dim X = n < \infty$ then $X$ is an ANR if and only if $X$ is $LC^n$ and also that $X$ is an AR if and only if $X$ is both $LC^n$ and $C^n$.

The proof of the following lemma is precisely the same as the proof of Lemma 4.2.13 and is therefore left as an exercise to the reader.

**Lemma 4.2.28.** *Let $X$ be a space. If $X$ is locally contractible then $X$ is $LC^n$ for every $n$.*

**Proposition 4.2.29.** *Let $X$ be a space and let $0 \le n < \infty$. The following statements are equivalent:*

(1) *$X$ is $LC^n$.*
(2) *for every open cover $\mathcal{U}$ of $X$ there exists an open refinement $\mathcal{V}$ of $\mathcal{U}$ such that for every locally finite simplicial complex $\mathcal{T}$ with*

$$\dim |\mathcal{T}| \le n+1$$

*and every subcomplex $\mathcal{S}$ of $\mathcal{T}$ containing all the vertices of $\mathcal{T}$, every partial realization of $\mathcal{T}$ in $X$ relative to $(\mathcal{S}, \mathcal{V})$ can be extended to a full realization of $\mathcal{T}$ in $X$ relative to $\mathcal{U}$.*

**Proof.** *We prove* (2) $\Rightarrow$ (1).

Take an arbitrary $x \in X$ and a neighborhood $U$ of $x$. There exists an open neighborhood $W$ of $x$ such that $\overline{W} \subseteq U$. Consider the open cover

$$\mathcal{U} = \{U, X \setminus \overline{W}\}$$

of $X$. By assumption, for the open cover $\mathcal{U}$ there exists an open refinement $\mathcal{V}$ such as in (2). Pick $V \in \mathcal{V}$ such that $x \in V$. Fix an integer $m \le n$, and let $\sigma$ be an arbitrary $(m+1)$-dimensional simplex in $\mathbb{R}^{m+1}$. It will be convenient to identify $\mathbb{S}^m$ and $\partial \sigma$ (Exercise 1.1.24). Now let $\mathcal{F}(\sigma)$ be the simplicial complex consisting of all the faces of $\sigma$ and let $\mathcal{S}$ be the subcomplex consisting of all proper faces. Consider a continuous function $f\colon \partial \sigma \to W \cap V$ and observe that $f$ is a partial realization of $\mathcal{F}(\sigma)$ in $X$ relative to $(\mathcal{S}, \mathcal{V})$. By assumption, $f$ can be extended to a full realization $\bar{f}\colon \sigma = |\mathcal{F}(\sigma)| \to X$ relative to $\mathcal{U}$. Consequently, there exists $U' \in \mathcal{U}$ with $\bar{f}[\sigma] \subseteq U'$. Since $\bar{f}$ extends $f$, $U' \cap W \ne \emptyset$, i.e., $U' = U$.

*We prove* (1) $\Rightarrow$ (2).

Let $\mathcal{U}$ be an open cover of $X$. By downward induction, we shall construct open covers $\mathcal{U}_n, \mathcal{V}_n, \mathcal{U}_{n-1}, \mathcal{V}_{n-1}, \ldots, \mathcal{U}_0, \mathcal{V}_0$ of $X$ having the following properties:

(3) $\mathcal{U}_n = \mathcal{U}$,

(4) if $0 \leq i \leq n$ then for every $V \in \mathcal{V}_i$ and for every continuous function $f\colon \mathbb{S}^i \to V$ there exist $U \in \mathcal{U}_i$ with $V \subseteq U$ and a continuous extension $\bar{f}\colon B^{i+1} \to U$ of $f$,
(5) if $0 \leq i \leq n-1$ then $\mathcal{U}_i$ is a star-refinement of $\mathcal{V}_{i+1}$

(Observe that (4) implies $\mathcal{V}_i < \mathcal{U}_i$.) The construction of these covers is a triviality. By Lemma 4.1.10 there is never trouble with the construction of the $\mathcal{U}_n$'s. Let us construct the cover $\mathcal{V}_n$. Take an arbitrary $x \in X$ and pick $U_x \in \mathcal{U}_n$ containing $x$. Since $X$ is $LC^n$ there is an open neighborhood $V_x$ of $x$ such that $V_x \subseteq U_x$ while moreover every continuous function $f\colon \mathbb{S}^n \to V_x$ can be extended to a continuous function $\bar{f}\colon B^{n+1} \to U_x$. The cover

$$\mathcal{V}_n = \{V_x : x \in X\}$$

is clearly as required. The construction of the other $\mathcal{V}_i$'s is similar.

Now put $\mathcal{V} = \mathcal{V}_0$, let $\mathcal{T}$ be a locally finite simplicial complex with $\dim |\mathcal{T}| \leq n+1$, let $\mathcal{S}$ be a subcomplex of $\mathcal{T}$ containing all the vertices of $\mathcal{T}$, and let $f\colon |\mathcal{S}| \to X$ be a partial realization of $\mathcal{T}$ in $X$ relative to $(\mathcal{S}, \mathcal{V})$. By induction, we shall construct for every $0 \leq i \leq n$ a continuous function

$$f_i\colon |\mathcal{S} \cup \mathcal{T}^{(i)}| \to X$$

having the following properties:

(6) $f_0 = f$ and for $1 \leq i \leq n$, $f_i$ extends $f_{i-1}$,
(7) for $0 \leq i \leq n$, $f_i$ is a partial realization of $\mathcal{T}$ in $X$ relative to the pair $(\mathcal{S} \cup \mathcal{T}^{(i)}, \mathcal{V}_i)$.

Since $\mathcal{S}$ contains all the vertices of $\mathcal{T}$, our choice $f_0 = f$ is as required. Now assume that for certain $1 \leq i \leq n$ the function $f_{i-1}$ has been constructed. As in the previous proofs, we shall construct $f_i$ 'simplex-wise'. Let $\sigma$ be an element of $\mathcal{T}^{(i)} \setminus (\mathcal{S} \cup \mathcal{T}^{(i-1)})$. By (7) there exists an element $V \in \mathcal{V}_{i-1}$ such that $f_i[\partial \sigma] \subseteq V$. So by (4) there exist $U \in \mathcal{U}_{i-1}$ with $V \subseteq U$ and a continuous extension $f_\sigma\colon \sigma \to U$ of the function $f_{i-1} \restriction \partial \sigma$. Now define $f_i\colon |\mathcal{S} \cup \mathcal{T}^{(i)}| \to X$ as follows:

$$f_i(x) = \begin{cases} f_{i-1}(x) & (x \in (\mathcal{S} \cup \mathcal{T}^{(i-1)}|), \\ f_\sigma(x) & (x \in \sigma \in \mathcal{T}^{(i)} \setminus \mathcal{T}^{(i-1)})). \end{cases}$$

By Lemma 2.1.19 $f_i$ is continuous. We now claim that $f_i$ is a partial realization of $\mathcal{T}$ in $X$ relative to $(\mathcal{S} \cup \mathcal{T}^{(i)}, \mathcal{V}_i)$. To this end, take an arbitrary simplex $\sigma \in \mathcal{T}$. By (7) there exists $V \in \mathcal{V}_{i-1}$ such that $f_{i-1}[\sigma \cap |\mathcal{S} \cup \mathcal{T}^{(i-1)}|] \subseteq V$. Let $\tau$ be a face of $\sigma$ such that $\tau \in \mathcal{T}^{(i)}$. By construction there exists $U \in \mathcal{U}_{i-1}$ such that $f_i[\tau] \subseteq U$. Observe that $V$ intersects $U$. Consequently,

$$f_i[\sigma \cap |\mathcal{S} \cup \mathcal{T}^{(i)}|] \subseteq \mathrm{St}(V, \mathcal{U}_{i-1}).$$

Now since $\mathcal{V}_{i-1} < \mathcal{U}_{i-1}$ and by (5) $\mathcal{U}_{i-1}$ is a star-refinement of $\mathcal{V}_i$, we are done. This completes the inductive construction.

With the same technique it is clear that we can extend $f_n$ to a continuous function
$$f_{n+1}\colon |\mathcal{S} \cup \mathcal{T}^{(n+1)}| \to X$$
such that for every $\sigma \in \mathcal{T}^{(n+1)}$ there exists $U \in \mathcal{U}_n = \mathcal{U}$ such that
$$f_{n+1}[\sigma] \subseteq U.$$
However, by assumption $\mathcal{T}^{(n+1)} = \mathcal{T}$ so that $f_{n+1}$ is the required full realization of $\mathcal{T}$ in X relative to $\mathcal{U}$. □

We now come to the announced characterization.

**Theorem 4.2.30.** *Let $X$ be a space and let $0 \le n < \infty$. The following statements are equivalent:*

(1) $X$ is $\mathrm{LC}^n$.
(2) *for every space $Y$ and for every closed subspace $A$ of $Y$ with*
$$\dim(Y \setminus A) \le n + 1,$$
*every continuous function $f\colon A \to X$ can be continuously extended over a neighborhood of $A$,*
(3) *for every $x \in X$ and for every neighborhood $U$ of $x$ there exists a neighborhood $V$ of $x$ with $V \subseteq U$ such that for every space $Y$ with $\dim Y \le n$, for every continuous function $f\colon Y \to V$ there exists a homotopy $H\colon Y \times \mathbb{I} \to U$ such that $H_0 = f$ and $H_1$ is a constant function.*

**Proof.** *We prove* (1) $\Rightarrow$ (2).

Let $Y$ be a space, let $A \subseteq Y$ be closed such that $\dim(Y \setminus A) \le n+1$ and let $f\colon A \to X$ be continuous. Without loss of generality we assume that $X$ and $Y$ have empty intersection.

**Claim 1.** *The set $C = (Y \setminus A) \cup X$ can be topologized in such a way that*

(3) *$X$ is a closed subspace of $C$,*
(4) *$Y \setminus A$ is a subspace of $C$,*
(5) *the function $\bar{f}\colon Y \to C$ defined by*
$$\bar{f}(x) = \begin{cases} z & (z \in Y \setminus A), \\ f(z) & (z \in A). \end{cases}$$
*is continuous.*

*Proof.* Let $\sigma\colon \tau A \to \tau Y$ be a function such as in Lemma 4.2.2 and let $\mathcal{B}$ be a countable open base for $X$. In addition, let $\mathcal{F}$ be a countable open base for $Y \setminus A$. For every $B \in \mathcal{B}$ and $n \in \mathbb{N}$ define $B(n) \subseteq C$ by
$$B(n) = \{y \in (Y \setminus A) \cap \sigma(f^{-1}[B]) : \varrho(y, A) < 1/n\} \cup B.$$

## 4.2. A CHARACTERIZATION OF ANR'S AND AR'S

An easy check shows that the collection $\mathcal{F} \cup \{B(n) : B \in \mathcal{B}, n \in \mathbb{N}\}$ is as a base for the required topology on $C$. ◇

We shall now prove that $X$ is a neighborhood retract of $C$. Once this has been established, by part (5) of the claim we are done.

The proof of our assertion is precisely the same as the proof of Theorem 4.2.1. Only two minor changes should be made. We shall adopt the notation and the terminology introduced there, we shall indicate the required changes and we shall leave the precise verification to the reader. By Proposition 4.2.29 we have a realization property available up to dimension $n+1$. So in the construction of the covers $\mathcal{A}_n$ and $\mathcal{B}_n$, one should replace 'locally finite simplicial complex $\mathcal{T}$' by 'locally finite simplicial complex $\mathcal{T}$ with $\dim |\mathcal{T}| \leq n+1$' everywhere. The special open cover $\mathcal{U}$ of $C \setminus X$ used in the proof of Theorem 4.2.1 should have the additional property that its order does not exceed $n+1$. However, by the fact that $\dim(C \setminus X) \leq n+1$, this can be achieved by Theorem 3.2.5(5). This implies that the nerve $N(\mathcal{U})$ has dimension at most $n+1$ (Remark 4.2.11). With these two small changes, the proof of Theorem 4.2.1 can now be followed verbatim.

*We prove* (2) $\Rightarrow$ (3).

Suppose that (3) is not true at the point $p$. Then there exists $\varepsilon > 0$ such that for every $i \in \mathbb{N}$ there exist a space $Y_i$ with $\dim Y_i \leq n$ and a continuous function $f_i \colon Y_i \to B(p, \varepsilon/i)$ that is not homotopic within $B(p, \varepsilon)$ to a constant function. Define $A$ and $Y$ by

$$A = \Big(\sum_{i=1}^{\infty} \big((Y_i \times \{0\}) \cup (Y_i \times \{1\})\big)\Big) \cup \{\infty\},$$

and

$$Y = \Big(\sum_{i=1}^{\infty} (Y_i \times \mathbb{I})\Big) \cup \{\infty\}.$$

Here $\infty \notin \sum_{i=1}^{\infty}(Y_i \times \mathbb{I})$. Topologize $Y$ by requiring that $\sum_{i=1}^{\infty}(Y_i \times \mathbb{I})$ is an open subspace of $Y$ and a basic neighborhood of $\infty$ has the form $\sum_{i=k}^{\infty}(Y_i \times \mathbb{I})$ for $k \in \mathbb{N}$. It is easy to see that $Y$ is a separable metrizable space. Now define the function $f \colon A \to X$ as follows

$$\begin{cases} f(x, 0) = f_i(x) & (x \in Y_i), \\ f(x, 1) = p & (x \in Y_i), \\ f(\infty) = p \end{cases}$$

Then $f$ is clearly continuous. Observe that by Theorems 3.4.12 and 3.2.9 we have $\dim(Y \setminus A) \leq n+1$ so that by assumption there exists a neighborhood $W$ of $A$ in $Y$ such that $f$ can be extended to a continuous function $\bar{f} \colon W \to X$. Then $V = \bar{f}^{-1}[B(p, \varepsilon)]$ is a neighborhood of $\infty$ in $Y$ and therefore contains all but finitely many of the $Y_i \times \mathbb{I}$, $i \in \mathbb{N}$. But this implies that for all but finitely many $i \in \mathbb{N}$ the function $f_i$ is homotopic within $B(p,\varepsilon)$ to a constant function. This is a contradiction.

*We prove* (3) $\Rightarrow$ (1).

Since $\dim \mathbb{S}^m = m$ for all $m$ (Theorem 3.2.12), this is clear. $\square$

**Theorem 4.2.31.** *Let $X$ be a space and let $0 \le n < \infty$. The following statements are equivalent:*

(1) $X$ is $\mathrm{LC}^n$ and $\mathrm{C}^n$.
(2) *For every space $Y$ and for every closed subspace $A$ of $Y$ with*
$$\dim(Y \setminus A) \le n+1,$$
*every continuous function $f\colon A \to X$ can be continuously extended over $Y$,*
(3) $X$ *is* $\mathrm{LC}^n$ *and for every space $Y$ with $\dim Y \le n$, every continuous function $f\colon Y \to X$ is nullhomotopic.*

**Proof.** The proof of (1) $\Rightarrow$ (2) is precisely the same as the proof of the implication (3) $\Rightarrow$ (1) in the proof of Theorem 4.2.20 and is therefore left as an exercise to the reader.

We shall now prove (2) $\Rightarrow$ (3). That $X$ is $\mathrm{LC}^n$ follows from Theorem 4.2.30. Let $Y$ be a space with $\dim Y \le n$ and assume that $f\colon Y \to X$ is continuous. Since by Theorems 3.4.12 and 3.2.8 we have $\dim \triangle(Y) \le n+1$ (here $\triangle(Y)$ denotes the cone over $Y$ of course), by (2) we conclude that $f$ can be extended to a continuous function $\bar{f}\colon \triangle(Y) \to X$. The contractibility of $\triangle(Y)$ (Exercise A.12.5) implies that $\bar{f}$ is nullhomotopic. From this it follows easily that $f$ is nullhomotopic as well.

The proof of (3) $\Rightarrow$ (1) is a triviality of course since $\dim \mathbb{S}^m = m$ for every $m$ (Theorem 3.2.12). $\square$

These results have the following corollaries.

**Corollary 4.2.32.** *Let $X$ be a space. The following statements are equivalent:*

(1) $X$ *is a Peano continuum,*
(2) *there is a continuous surjection $f\colon \mathbb{I} \to X$.*

**Proof.** Assume first that $X$ is a Peano continuum. By Theorem 1.5.10, there exists a continuous surjection $g\colon \mathsf{C} \to X$. In addition, by Theorem 1.5.22, $X$ is $\mathrm{LC}^0$ and $\mathrm{C}^0$. Consequently, an easy application of Theorem 4.2.31 yields that $g$ can be extended to a continuous function $f\colon \mathbb{I} \to X$.

The implication (2) $\Rightarrow$ (1) follows from Exercises A.5.5 and A.2.11. $\square$

We can now state our characterization theorem for finite dimensional ANR's and AR's.

**Theorem 4.2.33.** *Let $X$ be a space, let $0 \leq n < \infty$ and let $\dim X \leq n$. Then*

(1) *$X$ is an ANR iff $X$ is locally contractible iff $X$ is $\mathrm{LC}^n$,*
(2) *$X$ is an AR iff $X$ is $\mathrm{LC}^n$ and $\mathrm{C}^n$.*

**Proof.** We shall first present a proof of (1). To this end, first observe that if $X$ is an ANR then $X$ is locally contractible (Exercise 1.2.1) and consequently $\mathrm{LC}^n$ by Lemma 4.2.28. Conversely, assume that $X$ is $\mathrm{LC}^n$. Then $X$ is locally contractible by Theorem 4.2.30(3). Consequently, $X$ is $\mathrm{LC}^m$ for every $m$ (Lemma 4.2.28). Now let $m = 2n + 1$. By Theorem 3.3.5, we may assume that $X$ is a subspace of $\mathbb{S}^m$. Put $Z = \{x \in B^{m+1} : \|x\| < 1\} \cup X$. Then $Z$ is an AR by Corollary 4.2.17 and $X$ is clearly closed in $Z$. Since

$$\dim\{x \in B^{m+1} : \|x\| < 1\} \leq m + 1,$$

by Theorem 4.2.30 it follows that $X$ is a neighborhood retract of $Z$. Since $Z$ is an AR, Proposition 1.2.10 implies that $X$ is an ANR.

We shall now present a proof of (2). To this end, assume that $X$ is $\mathrm{LC}^n$ and $\mathrm{C}^n$. By Theorem 4.2.31 we conclude that $X$ is contractible and hence $\mathrm{C}^m$ for every $m$ (Lemma 4.2.13). Now define $Z$ as in the proof for (1). By Theorem 4.2.31(2) there is a retraction $r \colon Z \to X$ from which we conclude that X is an AR, as required. $\square$

**Remark 4.2.34.** Unfortunately, a characterization of infinite-dimensional ANR's and AR's as simple as Theorem 4.2.33 does not seem to be possible. BORSUK [70, p. 124] constructed an example of a contractible and locally contractible compactum which is not an ANR.

**Exercise for §4.2.**

1. Give an example of a homotopically trivial compact space which is not contractible

## 4.3. Open subspaces of ANR's

In this section we shall prove that every open subspace of an ANR is again an ANR and that every space that admits an open cover by ANR's is an ANR as well. As an application we conclude that every polytope is an ANR. We also show that every A(N)R has an A(N)R completion.

**Theorem 4.3.1.** *Let $(X, Y)$ be an ANR-pair and let $U$ be an open subspace of $X$. Then $(U, U \cap Y)$ is an ANR-pair.*

**Proof.** Let $Z$ be a space, $A \subseteq Z$ be closed, and let $g\colon A \to U$ be continuous. Since $(X,Y)$ is an ANR-pair, there exists an open neighborhood $V$ of $A$ in $Z$ such that $g$ can be extended to a continuous function $f\colon V \to X$ while moreover $f[V \setminus A] \subseteq Y$. Since $f$ is continuous and extends $g$, and since $U$ is open in $X$, we conclude that $W = f^{-1}[U]$ is an open neighborhood of $A$. Then $\bar{g} = f \upharpoonright W \colon W \to U$ is a continuous extension of $g$. In addition, if $z$ belongs to $W \setminus A$ then $\bar{g}(z) = f(z) \in U \cap Y$. So we are done. □

**Corollary 4.3.2.** *Let $X$ be an ANR and let $U$ be an open subspace of $X$. Then $U$ is an ANR.*

Observe that $[-1,0) \cup (0,1]$ is not an AR but is an open subspace of the AR $\mathbb{J}$. So an open subspace of an AR need not be an AR.

We now aim at proving that a 'local' ANR is an ANR. First we need to derive a few elementary results.

In our proofs in this section we make use of cones. As we saw in Exercise A.12.6, if $A$ is a (closed) subspace of a space $X$ then $\triangle(A)$ can be thought of as a (closed) subspace of $\triangle(X)$. It will be convenient to do that. We will also use the fact that the cone of an an ANR is an AR (Theorem 1.4.6).

**Proposition 4.3.3.** *Let $X$ be a space and assume that $X$ can be written as $U \cup V$, where both $U$ and $V$ are open in $X$. If $U$ and $V$ are ANR's then so is $X$.*

**Proof.** We shall prove that the cone over $X$ is an AR so that $X$ is an ANR by Theorem 1.4.6. By Proposition A.7.1 there exist closed sets $E$ and $F$ in $X$ such that $E \subseteq U, F \subseteq V$ and $E \cup F = X$. Observe that

$$\triangle(E) \subseteq \triangle(U), \quad \triangle(F) \subseteq \triangle(V), \quad \triangle(E) \cup \triangle(F) = \triangle(X).$$

Now let $Y$ be a space, let $A \subseteq Y$ be closed, and let $f\colon A \to \triangle(X)$ be continuous. Put $E' = f^{-1}[\triangle(E)]$ and $F' = f^{-1}[\triangle(F)]$, respectively. By Lemma A.8.1 there exist closed sets $E''$ and $F''$ in $Y$ such that

$$E'' \cap A = E', \quad F'' \cap A = F', \quad E'' \cup F'' = Y.$$

Since $U \cap V$ is an open subspace of $U$, it is an ANR by Corollary 4.3.2. Consequently, Theorem 1.4.6 implies that the function

$$f \upharpoonright E' \cap F' \colon E' \cap F' \to \triangle(E \cap F)$$

can be extended to a continuous function $g\colon E'' \cap F'' \to \triangle(U \cap V)$. Now define $f_E \colon E' \cup (E'' \cap F'') \to \triangle(U)$ by

$$f_E(x) = \begin{cases} f(x) & (x \in E'), \\ g(x) & (x \in E'' \cap F''). \end{cases}$$

Then $f_E$ is obviously continuous. By another application of Theorem 1.4.6, there exists a continuous extension $f_1 \colon E'' \to \triangle(U)$ of $f_E$. Similarly define $f_F$

and a continuous extension $f_2\colon F'' \to \triangle(V)$ of $f_F$. Now define $\bar{f}\colon Y \to \triangle(X)$ in the obvious way, namely,

$$\bar{f}(x) = \begin{cases} f_1(x) & (x \in E''), \\ f_2(x) & (x \in F''). \end{cases}$$

Then $\bar{f}$ is the required continuous extension of $f$. □

**Lemma 4.3.4.** *Let $X$ be a space having an open cover $\mathcal{U}$ by pairwise disjoint ANR's. Then $X$ is an ANR.*

**Proof.** This is a triviality. Assume that $X$ is a closed subspace of a space $Y$. By Lemma 4.2.2, for every $U \in \mathcal{U}$ there exists an open subset $V(U)$ of $Y$ such that

(1) $V(U) \cap X = U$,
(2) the collection $\{V(U) : U \in \mathcal{U}\}$ is pairwise disjoint.

Note that $U$ is closed in $V(U)$ for every $U \in \mathcal{U}$. So by our assumptions on $\mathcal{U}$, for every $U \in \mathcal{U}$ there exist an open subset $W(U)$ of $V(U)$ which contains $U$ and a retraction $r_U\colon W(U) \to U$. Observe that the $W(U)$'s are open in $Y$ since the $V(U)$'s are. Now put $W = \bigcup_{U \in \mathcal{U}} W(U)$. Then $W$ is a neighborhood of $X$ in $Y$ and the function $r\colon W \to X$ defined by

$$r(x) = r_U(x) \qquad (x \in W(U), U \in \mathcal{U})$$

is a retraction. □

We now come to the following

**Theorem 4.3.5.** *Let $X$ be a space and suppose that $X$ admits an open cover $\mathcal{U}$ consisting of ANR's. Then $X$ is an ANR.*

**Proof.** Without loss of generality assume that $\mathcal{U}$ is countable. Enumerate $\mathcal{U}$ as $\{U_n : n \in \mathbb{N}\}$. Since by Proposition 4.3.3 the union of finitely many elements of $\mathcal{U}$ is an ANR, we may assume without loss of generality that for every $n$, $U_n \subseteq U_{n+1}$. For every $n$, put

$$V_n = \{x \in X : \varrho(x, X \setminus U_n) > 1/n\}.$$

It is clear that each $V_n$ is open, that $\overline{V}_n \subseteq V_{n+1}$, and that $\bigcup_{n=1}^{\infty} V_n = X$. Finally define

$$R_n = \begin{cases} V_n & (n = 1, 2), \\ V_n \setminus \overline{V}_{n-2} & (n \geq 3). \end{cases}$$

Then $X = \bigcup_{n=1}^{\infty} R_n$ and for $|m - n| \geq 2$, $R_n \cap R_m = \emptyset$. Also, $R_n$ is an open subset of the ANR $U_n$. So by Corollary 4.3.2, we conclude that $R_n$ is an ANR. Consequently, Lemma 4.3.4 implies that

$$E = \bigcup_{n=1}^{\infty} R_{2n-1}, \quad F = \bigcup_{n=1}^{\infty} R_{2n}$$

are open ANR subspaces of $X$. Since their union equals $X$, we infer by Proposition 4.3.3 that $X$ is an ANR. □

As announced in §2.1, we now get

**Corollary 4.3.6.** *Every polytope is an ANR.*

**Proof.** Let $\mathfrak{T}$ be a locally finite simplicial complex and consider $P = |\mathfrak{T}|$. Take an arbitrary $x \in |\mathfrak{T}^{(0)}|$. Since $\mathfrak{T}$ is locally finite, the collection
$$\mathcal{S} = \{\tau \in \mathfrak{T} : x \in \tau\}$$
is finite. Consequently, $\mathcal{P} = \{\sigma \in \mathfrak{T} : (\exists \tau \in \mathcal{S})(\sigma \preccurlyeq \tau)\}$ is a finite subcomplex of $\mathfrak{T}$. From Theorem 2.1.24 we conclude that $|\mathcal{P}|$ is an ANR. Since $\operatorname{St} x$, the star of $x$, is contained in $|\mathcal{P}|$ and is open in $P$ by Corollary 2.1.17, we conclude from Corollary 4.3.2 that $\operatorname{St} x$ is an ANR. Since by Corollary 2.1.17,
$$\{\operatorname{St} x : x \in |\mathfrak{T}^{(0)}|\}$$
is an open cover of $P$, Theorem 4.3.5 implies that $P$ is an ANR. □

**Enlarging A(N)R's.** We now show that every A(N)R has a completion which is also an A(N)R.

**Theorem 4.3.7.** *Let $Y$ be a space with A(N)R subspace $X$. Then there is a subspace $S$ of $Y$ such that $X \subseteq S \subseteq \overline{X}$ while moreover*

(1) *$S$ is a $G_\delta$-subset of $Y$,*
(2) *$(S, X)$ is an A(N)R-pair.*

**Proof.** It is clear that we may assume without loss of generality that $X$ is dense in $Y$.

We first claim that the theorem for ANR's implies the theorem for AR's. For let $X$ be an AR. By the ANR-case of the theorem, we may assume that for some $S$ with $X \subseteq S \subseteq Y$ we have that $(S, X)$ is an ANR-pair. Since $X$ is homotopically trivial, being an AR, we conclude by Corollary 4.2.21 that $(S, X)$ is an AR-pair.

So now assume that $X$ is an ANR. We begin the proof as the proof of Corollary 1.1.8. Let $aY$ be a compactification of $Y$ (Corollary A.4.8). By Lemma 1.1.6 we may think of $aY$ as a linearly independent subspace of a Banach space of the form $C(A)$ (here $A$ is a compact space).

By linear independence, if $E_0$ is the linear hull of $X$ then $E_0 \cap aY = X$. Hence $X$ is a closed subspace of $E_0$. Since $X$ is an ANR, there are an open neighborhood $U_0$ of $X$ in $E_0$ and a retraction $r: U_0 \to X$. Select an open subset $U$ in $C(A)$ such that $U \cap E_0 = U_0$. Consider the closure $\overline{E_0}$ of $E_0$ in $C(A)$. Since $X$ is dense in $aY$ it follows that $aY \subseteq \overline{E_0}$. Also,
$$U_0 = U \cap E_0 \subseteq U \cap \overline{E_0}.$$

Since $aY$ is topologically complete, being compact, there is a $G_\delta$-subset $T$ of $U \cap \overline{E}_0$ which contains $U_0$ such that $r$ can be extended to a continuous function $\tilde{r} \colon T \to aY$ (Corollary A.8.4). Observe that $T$ is a $G_\delta$-subset of $C(A)$ since $U \cap \overline{E}_0$ obviously is and $T$ is a $G_\delta$-subset of $U \cap \overline{E}_0$. In addition, $X \subseteq T$ since $X \subseteq U_0$.

Put $S = T \cap Y$. Observe that $S$ is obviously a $G_\delta$-subset of $Y$. In addition, $\tilde{r}(y) = y$ for every $y \in S$. This is so since $X$ is dense in $S$ (Exercise A.1.9). Now put
$$S_1 = \tilde{r}^{-1}[S].$$
Then $\tilde{r} \upharpoonright S_1 \colon S_1 \to S$ is a retraction. Observe that
$$U \cap E_0 = U_0 \subseteq S_1 \subseteq U \cap \overline{E}_0.$$
Since $(\overline{E}_0, E_0)$ is an ANR-pair by Corollary 4.2.15 it follows that
$$(U \cap \overline{E}_0, U \cap E_0)$$
is an ANR-pair as well (Theorem 4.3.1). We conclude from this that
$$(S_1, U \cap E_0)$$
is an ANR-pair. Since $\tilde{r}$ is a retraction from $S_1$ onto $S$ such that
$$\tilde{r}[U \cap E_0] = r[U \cap E_0] = X$$
it follows by Exercise 4.1.8 that that $(S, X)$ is an ANR-pair. □

**Corollary 4.3.8.** *Let $X$ be an A(N)R. Then there exists an A(N)R $S$ and an imbedding $i \colon X \hookrightarrow S$ such that*

(1) *$i[X]$ is dense in $S$,*
(2) *$(S, i[X])$ is an A(N)R-pair,*
(3) *$S$ is topologically complete.*

**Proof.** Let $aX$ be a compactification of $X$ (Corollary A.4.8). It follows from Theorem 4.3.7 that there is a $G_\delta$-subset $S$ of $aX$ which contains $X$ while moreover $(S, X)$ is an A(N)R-pair. But then $S$ is an A(N)R and moreover is topologically complete since $aX$ is (Theorem A.6.3). □

**Exercises for §4.3.**

1. Give an example of a space $X$ which can be written as $E \cup F$, where both $E$ and $F$ are closed ANR subspaces of $X$, while $X$ is not an ANR.
2. Let $X$ be the union of two open subspaces $U$ and $V$ such that $U, V$ and $U \cap V$ are AR's. Prove that $X$ is an AR.
▶3. Let $X$ be a space having a point $x$ such that
    (1) $x$ has arbitrarily small homotopically trivial neighborhoods,
    (2) $X \setminus \{x\}$ is an ANR.
   Prove that $X$ is an ANR. (This generalizes Theorem 1.4.6.)

CHAPTER 5

# Basic infinite-dimensional topology

In this chapter we shall prove some elementary results from infinite-dimensional topology. The results are elementary in the sense that no powerful apparatus is needed, but the proofs are not always easy.

In infinite-dimensional topology one studies the topology of objects such as the Hilbert cube $Q$, the Hilbert space $\ell^2$ and its infinite-dimensional linear subspaces, the countably infinite product of lines $\mathbb{R}^\infty$ and its infinite-dimensional linear subspaces, and manifolds modeled on them. See the notes for more information.

In this chapter we develop basic homeomorphism theory in $Q$. Much emphasis is on so-called absorbing systems. They are the basis for the topological classification of function spaces of low Borel complexity in the next chapter.

## 5.1. $Z$-sets

Homeomorphism extension results are useful in infinite-dimensional topology. In this section we shall present a class of subsets of $Q$, the so-called class of $Z$-sets, for which an important homeomorphism extension result shall be derived in §5.3.

The concept of a $Z$-set does not seem to be very appealing, yet it is the most central concept in infinite-dimensional topology. It turns out, although this is not immediately clear from the definition, that $Z$-sets are 'small'. Here smallness has a different meaning than being 'small' with respect to category or measure. It is 'small' in the sense of homotopy. This will be made precise in Exercise 5.1.7 below.

Let $X$ be a space. A closed subset $A \subseteq X$ is called a $Z$ *set* in $X$ provided that for every open cover $\mathcal{U}$ of $X$ and every function $f \in C(Q, X)$ there is a function $g \in C(Q, X)$ such that

(1) $f$ and $g$ are $\mathcal{U}$-close,
(2) $g[Q] \cap A = \emptyset$.

A $\sigma Z$-set is a countable union of $Z$-sets. The collection of $Z$-sets and $\sigma Z$-sets in $X$ are denoted by $\mathcal{Z}(X)$ and $\mathcal{Z}_\sigma(X)$, respectively.

It will sometimes be convenient to have a 'metric' translation of the concept of a $Z$-set.

For compact $X$ the following lemma is a triviality since then every open cover has a Lebesgue number. For general $X$, the proof is only slightly more complicated.

**Lemma 5.1.1.** *Let $X$ be a space and let $A \subseteq X$ be closed. Then the following statements are equivalent:*

(1) $A \in \mathcal{Z}(X)$,
(2) $\forall \varepsilon > 0 \; \forall f \in C(Q, X) \; \exists g \in C(Q, X \setminus A) \colon \hat{\varrho}(f, g) < \varepsilon$.

**Proof.** Since $(1) \Rightarrow (2)$ is trivial, it suffices to establish the reverse implication. Let $\mathcal{U}$ be an open cover of $X$, and let $f \colon Q \to X$ be continuous. By compactness of $f[Q]$, there exists $\delta > 0$ with the property that every $A \subseteq X$ with $\operatorname{diam}(A) < \delta$ and which moreover intersects $f[Q]$, is contained in an element $U \in \mathcal{U}$ (Lemma A.5.3). By (2) there is a function $g \in C(Q, X)$ with $\hat{\varrho}(f, g) < \delta$ and $g[Q] \cap A = \emptyset$. We claim that $f$ and $g$ are $\mathcal{U}$-close. This is a triviality. Pick an arbitrary $x \in Q$. Then $\varrho\bigl(f(x), g(x)\bigr) < \delta$ and since $f(x) \in f[Q]$, $\{f(x), g(x)\}$ is contained in an element of $U$. □

It is also possible to use finite dimensional cubes to characterize $Z$-sets, see Exercise 5.1.3.

We shall now derive some elementary properties of $Z$-sets.

**Lemma 5.1.2.** *Let $X$ be a space. Then*

(1) *If $A \in \mathcal{Z}(X)$ and $B \subseteq A$ is closed in $X$ then $B \in \mathcal{Z}(X)$.*
(2) *If $A \in \mathcal{Z}(X)$ then $A$ has empty interior in $X$.*
(3) *If $(X, \varrho)$ is complete, $A \in \mathcal{Z}_\sigma(X)$ and $A$ is closed then $A \in \mathcal{Z}(X)$; in particular, finite unions of $Z$-sets are again $Z$-sets.*
(4) *If $(X, \varrho)$ is complete and $A \in \mathcal{Z}_\sigma(X)$ then for every $\varepsilon > 0$ and for every $f \in C(Q, X)$ there exists $g \in C(Q, X)$ such that $\hat{\varrho}(f, g) < \varepsilon$ and $g[Q] \cap A = \emptyset$.*
(5) *If $A \in \mathcal{Z}(X)$ and $Y$ is any space then $A \times Y \in \mathcal{Z}(X \times Y)$.*
(6) *If $A \in \mathcal{Z}(X)$ and $h \in \mathcal{H}(X)$ then $h[A] \in \mathcal{Z}(X)$.*
(7) *If $A \subseteq X$ is closed and for every $\varepsilon > 0$ there exists $f \in C(X, X \setminus A)$ such that $\hat{\varrho}(f, 1_X) < \varepsilon$ then $A \in \mathcal{Z}(X)$.*

**Proof.** Statement (1) is trivial.

For (2), (5) and (6), see Exercise 5.1.1.

For (4), write $A = \bigcup_{n=1}^\infty A_n$ with $A_n \in \mathcal{Z}(X)$ for every $n \in \mathbb{N}$. Using Lemma 5.1.1 and Exercise A.5.3, it is clearly possible to construct a sequence of maps $f_n \in C(Q, X \setminus A_n)$, $n \in \mathbb{N}$, such that

(8) $\hat{\varrho}(f_1, f) < \varepsilon/2$,

(9) $\hat{\varrho}(f_{n+1}, f_n) < 3^{-n} \cdot \varepsilon/2$,
(10) $\hat{\varrho}(f_{n+1}, f_n) < 3^{-n} \cdot \min\{\varrho(f_i[Q], A_i) : 1 \leq i \leq n\}$.

By (9) and Proposition 1.3.5, $F = \lim_{n \to \infty} f_n$ exists and is an element of $C(Q, X)$. Take $x \in Q$ arbitrarily. Then by (8) and (9) we find that

$$\varrho(F(x), f(x)) = \lim_{n \to \infty} \varrho(f_n(x), f(x))$$
$$\leq \lim_{n \to \infty} \sum_{m=0}^{n-1} 3^{-m} \cdot \varepsilon/2$$
$$= 3/2 \cdot \varepsilon/2$$
$$= 3/4 \varepsilon.$$

From this we conclude that $\hat{\varrho}(F, f) < \varepsilon$. It now suffices to prove that $F[Q]$ misses $A$. This however is an immediate consequence of (10) and Lemma 1.6.1.

Observe that (3) is a direct consequence of (4).

For (7), let $\varepsilon > 0$ and $g \in C(Q, X)$. Find $f \in C(X, X \setminus A)$ such that $\hat{\varrho}(f, 1_X) < \varepsilon$. Put $h = f \circ g$. Then clearly $h[Q] \cap A = \emptyset$, while moreover,

$$\hat{\varrho}(h, g) = \hat{\varrho}(f \circ g, g) \leq \hat{\varrho}(f, 1_X) < \varepsilon$$

(Exercise 1.3.2). We conclude that $A \in \mathcal{Z}(Q)$ (Lemma 5.1.1). □

We are interested particularly in $Z$-sets in the Hilbert cube $Q$. The following lemma provides us with 'many' $Z$-subsets of $Q$.

**Lemma 5.1.3.** *Let $A \subseteq Q$ be a closed set. Then:*

(1) *$A$ is a $Z$-set if and only if for every $\varepsilon > 0$ there exists $f \in C(Q, Q)$ such that $\hat{\varrho}(f, 1_Q) < \varepsilon$ and $f[Q] \cap A = \emptyset$.*
(2) *If $\pi_n[A] \neq \mathbb{J}_n$ for infinitely many $n \in \mathbb{N}$, then $A \in \mathcal{Z}(Q)$.*
(3) *If $\pi_n[A] \subseteq \{-1, 1\}$ for certain $n \in \mathbb{N}$, then $A \in \mathcal{Z}(Q)$.*

**Proof.** For (1), let $A \in \mathcal{Z}(Q)$. Since $1_Q \in C(Q, Q)$, the definition of a $Z$-set immediately gives us that for every $\varepsilon > 0$ there exists $f \in C(Q, Q)$ with $\hat{\varrho}(f, 1_Q) < \varepsilon$ and $f[Q] \cap A = \emptyset$. The converse is a direct consequence of Lemma 5.1.2(7).

For (2), choose $\varepsilon > 0$ and find $n \in \mathbb{N}$ so large that $2^{-(n-1)} < \varepsilon$ while moreover $\pi_n[A] \neq \mathbb{J}_n$. Take an arbitrary $t \in \mathbb{J}_n \setminus \pi_n[A]$, and define $f : Q \to Q$ by

$$f(x_1, \ldots, x_{n-1}, x_n, x_{n+1}, \ldots) = (x_1, \ldots, x_{n-1}, t, x_{n+1}, \ldots).$$

Then clearly $f[Q] \cap A = \emptyset$ and $\hat{\varrho}(f, 1_Q) < \varepsilon$. So by (1), $A \in \mathcal{Z}(Q)$.

For (3), first observe that $\{-1, 1\} \in \mathcal{Z}(\mathbb{J})$. We consequently get what we want by first applying (5) and then (1) of Lemma 5.1.2. □

**Remark 5.1.4.** Observe that this lemma implies that each singleton subset of $Q$ is a $Z$-set. Similarly, every endface of $Q$ is a $Z$-set. So $Q$ has 'many' $Z$-subsets, There are also spaces with 'few' $Z$-sets. It is easy to see for example that the collection $\mathcal{Z}(\mathbb{J})$ is equal to

$$\{\emptyset, \{-1\}, \{1\}, \{1, -1\}\}.$$

At first glance it seems a little surprising that *all* points in $Q$ are $Z$-sets. However, we already know that $Q$ is homogeneous (Theorem 1.6.6), so once one singleton $Z$-subset has been found it follows automatically that all singleton subsets share this property.

**Corollary 5.1.5.**

(1) $B(Q) \in \mathcal{Z}_\sigma(Q)$,
(2) if $K \subseteq s$ is compact then $K \in \mathcal{Z}(Q)$.

**Exercises for §5.1.** A closed subset $A$ of $Q$ is called an *A-set* if for every open $U \subseteq Q$ such that $U$ is nonempty and homotopically trivial, the set $U \setminus A$ is again nonempty and homotopically trivial.

1. Let $X$ be a space. Prove the following statements:
    (1) If $A \in \mathcal{Z}(X)$ then $A$ has empty interior in $X$.
    (2) If $A \in \mathcal{Z}(X)$ and $Y$ is any space then $A \times Y \in \mathcal{Z}(X \times Y)$.
    (3) If $A \in \mathcal{Z}(X)$ and $h \in \mathcal{H}(X)$ then $h[A] \in \mathcal{Z}(X)$.

2. Give an example of a space $X$ which can be covered by a countable union of $Z$-subsets. (This shows that in Exercise 5.1.1(2) it is not possible to replace $Z$-set by $\sigma Z$-set.)

3. Let $X$ be a space and let $A \subseteq X$ be closed. Prove that the following statements are equivalent:
    (1) $A \in \mathcal{Z}(X)$,
    (2) $\forall n \in \mathbb{N} \; \forall \varepsilon > 0 \; \forall f \in C(\mathbb{I}^n, X) \; \exists g \in C(\mathbb{I}^n, X \setminus A): \; \hat{\varrho}(f, g) < \varepsilon$.

4. Let $n \geq 1$ and
$$A \subseteq B^n = \left\{ x \in \mathbb{R}^n : \sum_{i=1}^n x_i^2 \leq 1 \right\}$$
be closed. Prove that $A \in \mathcal{Z}(B^n)$ iff $A \subseteq \mathbb{S}^{n-1}$.

5. Prove that an $A$-set in $Q$ is nowhere dense.

▶6. Let $A \subseteq Q$ be an $A$-set. Prove that for each polyhedron $P$, for each $\varepsilon > 0$ and for each $f \in C(P, Q)$ there exists $g \in C(P, Q)$ such that $\hat{\varrho}(f, g) < \varepsilon$ and $g[P] \cap A = \emptyset$.

▶7. Let $A \subseteq Q$ be closed. Prove that $A$ is an $A$-set iff $A$ is a $Z$-set.

8. For every $n$ let $X_n$ be a space. Assume that a closed set
$$A \subseteq X = \prod_{n=1}^{\infty} X_n$$
has the property that $\pi_n[A] \neq X_n$ for infinitely many $n$. Prove that $A$ is a $Z$-subset of $X$.

## 5.2. Extending homeomorphisms in $s$

The aim of this section is to prove that if $E, F \subseteq s$ are compact and $f$ is a homeomorphism from $E$ onto $F$ which moves the points less than $\varepsilon$ then it can be extended to a homeomeomorphism of $Q$ that satisfies the same smallness condition.

An element $h \in \mathcal{H}(Q)$ is *boundary preserving* if $h[B(Q)] = B(Q)$, or, equivalently, $h[s] = s$.

**Lemma 5.2.1.** *For every compact subset $K \subseteq s$ such that $K \neq \emptyset$ there is a boundary preserving $h \in \mathcal{H}(Q)$ such that $\pi_1 h[K] = \{0\}$.*

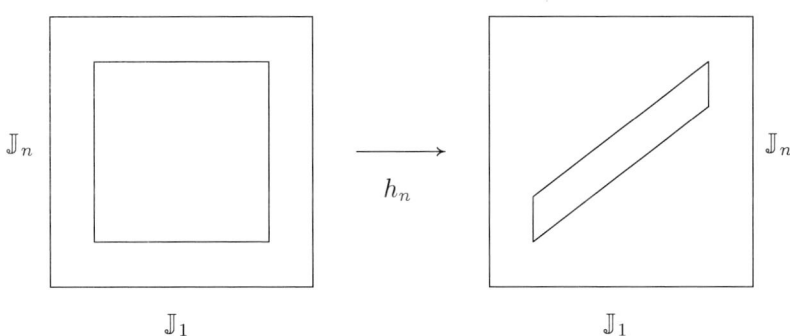

Figure 13.

**Proof.** Without loss of generality we may assume that
$$K = \prod_{n=1}^{\infty} [a_n, b_n],$$
where $-1 < a_n \leq b_n < 1$ for $n \in \mathbb{N}$. For $n > 1$, let $h_n \colon \mathbb{J}_1 \times \mathbb{J}_n \to \mathbb{J}_1 \times \mathbb{J}_n$ be a homeomorphism of $\mathbb{J}_1 \times \mathbb{J}_n$ having the following properties (see Figure 13):

(1) $h_n$ does not change the $\mathbb{J}_1$-coordinate of any point,
(2) every horizontal line intersects $h_n\big[[a_1, b_1] \times [a_n, b_n]\big]$ in a set of diameter at most $1/n$.

It is geometrically obvious how to define $h_n$. Let $U$ be the region in Figure 14 below the dotted line. Similarly, $V$ is the region above the dotted line. We will describe $h_n \restriction U$ only, $h_n \restriction V$ being entirely similar. Let $t \in \mathbb{J}$ be fixed and let $\ell$ be the vertical line through the point $(t, -1)$ in $\mathbb{J}_1 \times \mathbb{J}_n$. In Figure 14 on the line $\ell$ four bullets are drawn, the first one of which is $x_1 = (t, -1)$ and the second one is $x_2 = (t, a_n)$. The third and fourth bullet will be denoted by $x_3 = (t, b)$ and $x_4 = (t, c)$, respectively. Here $a_n < b < c$. Now define $h_n \restriction \{t\} \times \mathbb{J}_n$ by requiring that the interval from $x_1$ to $x_2$ is mapped linearly onto the interval from $x_1$ to $x_3$, and the interval from $x_2$ to $x_4$ linearly onto the interval from $x_3$ to $x_4$. Etc. Then $h_n$ is as required.

Define $f \colon Q \to Q$ by
$$f(x_1, x_2, x_3, \dots) = (x_1, y_2, y_3, \dots),$$
where $(x_1, y_n) = h_n(x_1, x_n)$ for every $n > 1$. The composition $\pi_1 \circ f \colon Q \to \mathbb{J}_1$ is equal to $\pi_1$ and is therefore continuous. For $n > 2$, the composition
$$\pi_n \circ f \colon Q \to \mathbb{J}_n$$
is equal to the composition
$$p \circ h_n \circ \pi_{\{1,n\}},$$
where $p \colon \mathbb{J}_1 \times \mathbb{J}_n \to \mathbb{J}_n$ is the projection.

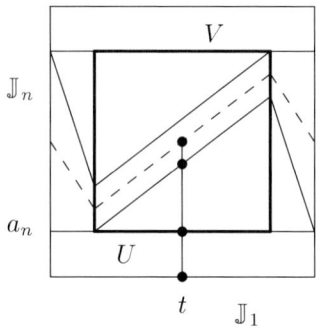

Figure 14.

Hence $\pi_n \circ f$ is continuous, being the composition of three continuous functions. By the definition of the product topology on $Q$ it therefore follows that $f$ is continuous. It is easy to display the continuous inverse of $f$. Alternatively, one can show that $f$ is a bijection. It then suffices to apply Exercise A.5.9 to conclude that $f$ is a homeomorphism. We shall prove here that $f$ is surjective only. To this end, let $y \in Q$ be arbitrary. Put $x_1 = y_1$. Since
$$h_1 \colon \mathbb{J}_1 \times \mathbb{J}_2 \to \mathbb{J}_1 \times \mathbb{J}_2$$

is a homeomorphism, there exists a point $(x_1, a) \in \mathbb{J}_1 \times \mathbb{J}_2$ such that

$$h_1((x_1, a)) = (x_1, y_1).$$

Put $x_2 = a$. Etc. Then clearly $f(x) = y$. That $f$ is boundary preserving follows easily from the fact that all $h_n$ are boundary preserving.

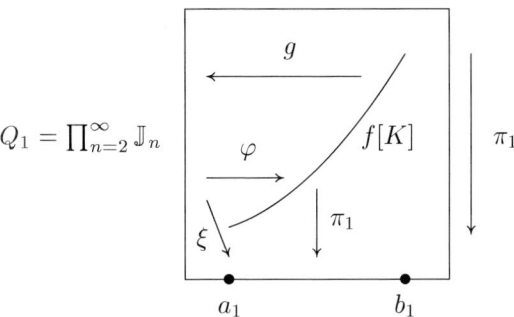

Figure 15.

Now think of $Q$ as $\mathbb{J}_1 \times Q_1$ and take two points in $f[K]$ having the same second coordinates, i.e., points of the form $(x, z)$ and $(y, z)$. If $x \neq y$ then $|x - y| > 1/n$ for certain $n \geq 2$. By the definition of $f$ we have that the points $(x, z_n)$ and $(y, z_n)$ both belong to $h_n[[a_1, b_1] \times [a_n, b_n]]$. But this contradicts (2). So we conclude that the function $g \colon f[K] \to Q_1$ defined by $g(x, y) = y$ is one-to-one and by compactness is therefore an imbedding (Exercise A.5.9). See Figure 15. Let $B = gf[K]$ and let

$$\varphi \colon B \to f[K]$$

be the inverse of $g$. In addition, let $\xi \colon B \to [a_1, b_1]$ be $\varphi$ followed by the projection $\pi_1$ onto $\mathbb{J}_1$. We finish the proof by the same strategy as in the proof of Theorem 1.8.4.

By the Tietze Extension Theorem 1.2.5, we can extend $\xi$ to a continuous function $\lambda \colon Q_1 \to [a_1, b_1]$. For every $t \in [a_1, b_1]$ let $H_t$ be the unique homeomorphism of $\mathbb{J}_1$ taking $[-1, t]$ linearly onto $[-1, 0]$ and $[t, 1]$ linearly onto $[0, 1]$. Then $H \colon \mathbb{J}_1 \times [a_1, b_1] \to \mathbb{J}_1$ defined by $H(x, t) = H_t(x)$ is an isotopy (Exercise 1.8.1). Now put

$$F(x, y) = (H(x, \lambda(y)), y).$$

Then $F$ belongs to $\mathcal{H}(Q)$ by Theorem 1.8.2. In addition, $F$ has the property that for every $(x, y) \in f[K]$,

$$F(x, y) = (H(x, \lambda(y)), y) = (H(x, x), y) = (0, y).$$

Since $F$ is clearly boundary preserving, we find that the function $h = F \circ f$ is as required. □

**Corollary 5.2.2.** *For every compact subset $K \subseteq s$ and $\varepsilon > 0$ there are an infinite $N \subseteq \mathbb{N}$ and a boundary preserving homeomorphism $h: Q \to Q$ such that*

(1) *the complement of $N$ is infinite and $\sum_{n \in N} 2^{-n} < \varepsilon$,*
(2) $\hat{\varrho}(h, 1_Q) < \varepsilon$,
(3) $\pi_n h[K] = \{0\}$ *for every $n \in N$.*

**Proof.** Choose $n \in \mathbb{N}$ so large that

$$\sum_{m=n}^{\infty} 2^{-m} < \varepsilon.$$

Write $\mathbb{N} \setminus \{1, 2, \ldots, n-1\}$ as the disjoint union of infinitely many infinite sets, say $C_1, C_2, \ldots$. Recall that for every $i$

$$\pi_{C_i}: Q \to \prod_{n \in C_i} \mathbb{J}_n$$

denotes the projection. Let

$$h_i: \prod_{n \in C_i} \mathbb{J}_n \to \prod_{n \in C_i} \mathbb{J}_n$$

be a homeomorphism with the following properties:

(1) $h_i$ is boundary preserving (this has its obvious meaning),
(2) $h_i \pi_{C_i}[K]$ projects onto 0 in the first factor of the product $\prod_{n \in C_i} \mathbb{J}_n$

(Lemma 5.2.1). Define $h: Q \to Q$ by

$$h(x) = \begin{cases} x_j & (j \leq n-1), \\ \bigl(h_i(\pi_{C_i}(x))\bigr)_j & (j \in C_i). \end{cases}$$

It is clear that $h$ is as required (observe that $h$ is nothing but the product of the $h_i$'s and the identity on the first $n - 1$ factors).

Finally, observe that the set $N = \{\min(C_i) : i \in \mathbb{N}\}$ has infinite complement in $\mathbb{N}$ and $N \subseteq \{n, n+1, \ldots\}$ so $\sum_{m \in N} 2^{-m} < \varepsilon$. □

The following lemma is the key in deriving our main result. The strategy of the proof is similar to the one in Theorem 1.8.4, but is more complicated.

It will be convenient to introduce some notation that shall be fixed throughout the remaining part of this section. Let $A$ and $B$ be complementary infinite subsets of $\mathbb{N}$. Put

$$Q_A = \{x \in Q : x_n = 0 \text{ if } n \notin A\}, \quad Q_B = \{x \in Q : x_n = 0 \text{ if } n \notin B\},$$

respectively.

It will be convenient to think of $Q$ as $Q_A \times Q_B$. We will specify $A$ and $B$ later. Let $\delta > 0$ be such that

$$\sum_{n \in B} 2^{-n} < \delta/2.$$

The 'origins' of $Q$, $Q_A$ and $Q_B$, i.e., the points having all coordinates 0, shall be denoted by $\mathbf{0}$, or, as we think of $Q$ as $Q_A \times Q_B$, by $(\mathbf{0}, \mathbf{0})$. Let $X, Y$ and $Z$ be compact subsets of $s$ such that $X \cup Y \subseteq Q_A$ and $Z \subseteq Q_B$ and let $p\colon X \to Z$ and $q\colon Y \to Z$ be homeomorphisms such that

(∗) $$\hat{\varrho}(q^{-1}p, 1_X) < \gamma$$

for certain $\gamma > 0$. Again we will specify $p, q, \gamma, X, Y$ and $Z$ later (see Figure 16).

We would like to construct a homeomorphism $h_1$ which takes $X$ onto the 'graph' $G(p)$ of $p$. Similarly, a homeomorphism $h_2$ which takes $Y$ onto the 'graph' $G(q)$ of $q$. And finally, a homeomorphism $h_3$ mapping $G(p)$ onto $G(q)$. Then $h_2^{-1} \circ h_3 \circ h_1$ will a homeomorphism of $Q$ extending the homeomorphism $q^{-1} \circ p\colon X \to Y$.

Before we can make this precise, we define two technical concepts. A homeomorphism $f\colon Q \to Q$ is called an $A$-homeomorphism (respectively, $B$-homeomorphism) if for any $x \in Q$ we have

$$f(x)_n = x_n$$

for all $n \in A$ (respectively, $n \in B$). So an $A$-homeomorphism is a 'vertical action'. Similarly, a $B$-homeomorphism acts 'horizontally'.

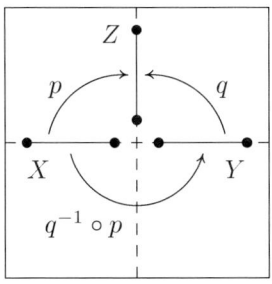

Figure 16.

The desired homeomorphisms are constructed in the following lemma.

**Lemma 5.2.3.**

(1) There is a boundary preserving $A$-homeomorphism $h_1\colon Q \to Q$ such that $h_1(x, \mathbf{0}) = (x, p(x))$ for every $x \in X$.

(2) There is a boundary preserving A-homeomorphism $h_2 \colon Q \to Q$ such that $h_2(y, \mathbf{0}) = (y, q(y))$ for all $y \in Y$.

(3) There is a boundary preserving B-homeomorphism $h_3 \colon Q \to Q$ such that $h_3(x, p(x)) = (q^{-1}p(x), p(x))$ for every $x \in X$ while moreover $\hat{\varrho}(h_3, 1_Q) < \gamma$.

The reader should pause to see what is going on. The homeomorphism $h_1$ indeed takes $X$ onto the 'graph' $G(p)$ of $p$. Similarly, $h_2$ takes $Y$ onto the 'graph' $G(q)$ of $q$. Finally, $h_3$ is a 'small' homeomorphism mapping $G(p)$ onto $G(q)$. So $h_2^{-1} \circ h_3 \circ h_1$ is a 'small' homeomorphism of $Q$ extending the homeomorphism $q^{-1} \circ p \colon X \to Y$ (see Figure 17) if $Q_B$ has 'small' diameter.

**Proof.** Since $X \cup Y \cup Z$ is compact we can find for every $n$ an interval
$$[-r_n, r_n] \subseteq (-1, 1)$$
such that $X \cup Y \cup Z \subseteq \prod_{n=1}^{\infty} [-r_n, r_n]$. Put
$$K_A = \{x \in \prod_{n=1}^{\infty} [-r_n, r_n] : x_n = 0 \text{ if } n \notin A\}$$
and
$$K_B = \{x \in \prod_{n=1}^{\infty} [-r_n, r_n] : x_n = 0 \text{ if } n \notin B\},$$
respectively.

Observe that both $K_A$ and $K_B$ are products of symmetric intervals (the symmetry of the intervals will be important later), that $X \cup Y \subseteq K_A \subseteq Q_A$ and that $Z \subseteq K_B \subseteq Q_B$.

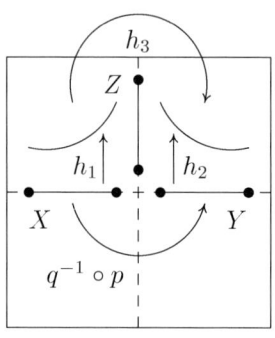

Figure 17.

By applying the Tietze Extension Theorem 1.2.5 to each factor of $K_B$ separately, we can extend $p \colon X \to Z \subseteq K_B$ to a continuous function
$$\bar{p} \colon Q_A \to K_B$$

(alternatively, apply Corollary 1.2.12). For every $t \in (-1,1)$ let $\varphi_t \colon \mathbb{J} \to \mathbb{J}$ be the unique homeomorphism taking the interval $[-1,0]$ linearly onto $[-1,t]$ and $[0,1]$ linearly onto $[t,1]$. For each $n \in B$ let the isotopy

$$H_n \colon \mathbb{J} \times [-r_n, r_n] \to \mathbb{J}$$

be defined by $H_n(x,t) = \varphi_t(x)$. The $H_n$'s define a $K_B$-isotopy

$$H \colon Q_B \times K_B \to Q_B$$

as follows:

$$H(y,x)_n = H^n(y_n, x_n) \qquad (n \in B)$$

(Proposition 1.8.1) (we make an obvious identification here). Now define

$$h_1 \colon Q_A \times Q_B \to Q_A \times Q_B$$

by

$$h_1(x,y) = \bigl(x, H\bigl(y, \bar{p}(x)\bigr)\bigr).$$

Then $h_1$ is an $A$-homeomorphism (Theorem 1.8.2). In addition, if $x \in X$ then

$$h_1(x,\mathbf{0}) = \bigl(x, H(\mathbf{0}, \bar{p}(x))\bigr) = \bigl(x, H(\mathbf{0}, p(x))\bigr) = \bigl(x, p(x)\bigr)$$

(for the last equality, use the definition of the $H_n$'s for $n \in B$). We conclude that $h_1$ is as required. The construction of $h_2$ is precisely the same.

The construction of $h_3$ is similar but slightly more complicated because of the smallness condition involved.

Consider $p^{-1} \colon Z \to X \subseteq K_A$ and $q^{-1} \colon Z \to Y \subseteq K_A$. By Exercise 1.3.8 and $(*)$ on Page 315,

$$\hat{\varrho}(p^{-1}, q^{-1}) = \hat{\varrho}(p^{-1}p, q^{-1}p) = \hat{\varrho}(1_X, q^{-1}p) < \gamma.$$

**Claim 1.** There exist continuous extensions $\xi, \eta \colon Q_B \to K_A$ of $p^{-1}$ and $q^{-1}$, respectively, such that $\hat{\varrho}(\xi, \eta) < \gamma$.

*Proof.* By applying the Tietze Extension Theorem 1.2.5 to each factor of $K_A$ separately (alternatively, apply Corollary 1.2.12), we can extend $p^{-1}$ and $q^{-1}$ to continuous functions $f, g \colon Q_B \to K_A$. Put

$$U = \{x \in Q_B : \varrho\bigl(f(x), g(x)\bigr) < \gamma\}.$$

It is clear that $U$ is an open neighborhood of $Z$ in $Q_B$. Let $\alpha \colon Q_B \to \mathbb{I}$ be a Urysohn function such that

$$\alpha \upharpoonright Z \equiv 1, \quad \alpha \upharpoonright (Q_B \setminus U) \equiv 0$$

(Corollary A.4.1). Now define $\xi, \eta \colon Q_B \to K_A$ by

$$\xi(x) = \alpha(x) \cdot f(x), \quad \eta(x) = \alpha(x) \cdot g(x),$$

respectively. Observe that by the special choice of $K_A$ these functions are well-defined (and obviously continuous). Now if $x \notin U$ then $\alpha(x) = 0$, and so $\xi(x) = \eta(x)$. On the other hand, if $x \in U$ then

$$\varrho\big(\xi(x), \eta(x)\big) = \sum_{n=1}^{\infty} 2^{-n} \alpha(x) \cdot |f(x)_n - g(x)_n|$$
$$= \alpha(x) \cdot \varrho\big(f(x), g(x)\big)$$
$$< 1 \cdot \gamma$$
$$= \gamma.$$

From this we conclude that $\hat{\varrho}(\xi, \eta) < \gamma$ (Exercise A.5.4). If $x \in Z$ then clearly $\alpha(x) = 1$, so $\xi(x) = 1 \cdot f(x) = f(x)$. We conclude that $\xi$ extends $f$, and similarly, that $\eta$ extends $g$. Consequently, $\xi$ and $\eta$ are as required. ◇

For all $(x, y) \in (-1, 1)^2$ let

$$\varphi_{(x,y)} \colon \mathbb{J} \to \mathbb{J}$$

be the unique homeomorphism taking $[-1, x]$ linearly onto $[-1, y]$ and $[x, 1]$ linearly onto $[y, 1]$. Observe the following easy:

**Claim 2.** If $(x, y) \in (-1, 1)^2$ then $\hat{\varrho}(\varphi_{(x,y)}, 1_{\mathbb{J}}) = |x - y|$.

We now define a $(K_A \times K_A)$-isotopy

$$F \colon Q_A \times (K_A \times K_A) \to Q_A$$

coordinatewise as follows:

$$F(q, x, y)_n = \varphi_{(x_n, y_n)}(q_n) \qquad (n \in A),$$

cf. Proposition 1.8.1. Now define

$$h_3 \colon Q_A \times Q_B \to Q_A \times Q_B$$

by

$$h_3(x, y) = \big(F_{(\xi(y), \eta(y))}(x), y\big).$$

Then $h_3$ is a $B$-homeomorphism (Theorem 1.8.2). We shall prove that $h_3$ is as required.

Take an arbitrary $x \in X$. Then

$$h_3\big(x, p(x)\big) = \big(F_{(\xi p(x), \eta p(x))}(x), p(x)\big)$$
$$= \big(F_{(x, q^{-1} p(x))}(x), p(x)\big)$$
$$= \big(q^{-1} p(x), p(x)\big)$$

(for the last equality, use the definition of $F$).

Finally, take an arbitrary $(x,y) \in Q_A \times Q_B$. Then
$$\hat\varrho\big(h_3(x,y),(x,y)\big) = \sum_{n \in A} 2^{-n}|F_{(\xi(y),\eta(y))}(x)_n - x_n|$$
$$= \sum_{n \in A} 2^{-n}|\varphi_{(\xi(y)_n,\eta(y)_n)}(x_n) - x_n|$$
$$\leq \sum_{n \in A} 2^{-n}\hat\varrho(\varphi_{(\xi(y)_n,\eta(y)_n)},1_{\mathbb{J}})$$
$$= \sum_{n \in A} 2^{-n}|\xi(y)_n - \eta(y)_n|$$
$$\leq \hat\varrho(\xi,\eta)$$
$$< \gamma$$
(use Claims 1 and 2). We conclude that $\hat\varrho(h_3, 1_Q) < \gamma$ (Exercise A.5.4). □

We now come to the main result in this section.

**Theorem 5.2.4.** *Let $E, F \subseteq s$ be compact and let $f \colon E \to F$ be a homeomorphism such that $\hat\varrho(f, 1_E) < \varepsilon$. Then $f$ can be extended to a boundary preserving homeomorphism $\bar f \colon Q \to Q$ such that $\hat\varrho(\bar f, 1_Q) < \varepsilon$.*

**Proof.** Let $\varepsilon_1 = \hat\varrho(f, 1_E)$ and put $\delta = \frac{1}{6}(\varepsilon - \varepsilon_1)$ (this specifies $\delta$). By Corollary 5.2.2 we find a boundary preserving homeomorphism $g \colon Q \to Q$ and an infinite subset $B \subseteq \mathbb{N}$ such that

(1) $\hat\varrho(g, 1_Q) < \delta$,
(2) $\pi_n g[E \cup F] = \{0\}$ for every $n \in B$,
(3) $A = \mathbb{N} \setminus B$ is infinite and $\sum_{n \in B} 2^{-n} < \delta/2$.

(this specifies $A$ and $B$). Put $X = g[E]$, $Y = g[F]$ and $h = g \circ f \circ g^{-1}$. Observe that
$$\hat\varrho(h, 1_X) < \varepsilon_1 + 2\delta.$$
Notice that $X \cup Y \subseteq Q_A$. Since $Q_B \approx Q$, we can find a topological copy $Z$ of $X$ in $s$ such that $Z \subseteq Q_B$ (Corollary A.4.4) ($Q_B$ is not a subset of $s$ but a smaller infinite-dimensional 'subcube' is a subset of $s$). Let $p \colon X \to Z$ be any homeomorphism and let $q = p \circ h^{-1}$. Then clearly
$$\hat\varrho(q^{-1} \circ p, 1_X) = \hat\varrho(h, 1_X) < \varepsilon_1 + 2\delta.$$
Put $\gamma = \varepsilon_1 + 2\delta$. This specifies $p, q$ and $\gamma$. Now let $h_1, h_2$ and $h_3$ be as in Lemma 5.2.3. By (3) we have
$$\hat\varrho(h_1, 1_Q) < \delta, \quad \hat\varrho(h_2, 1_Q) < \delta.$$
Consequently, if we put
$$\tau = h_2^{-1} \circ h_3 \circ h_1,$$
then
$$\hat\varrho(\tau, 1_Q) < \gamma + 2\delta = \varepsilon_1 + 4\delta.$$

Since $\tau$ clearly extends $h$, we find that
$$\bar{f} = g^{-1} \circ \tau \circ g$$
extends $f$ and that $\hat{\varrho}(\bar{f}, 1_Q) < \varepsilon_1 + 4\delta + 2\delta = \varepsilon$.

Since it is clear that $\bar{f}$ is boundary-preserving, this finishes the proof. □

**Exercise for §5.2.** Let $\varepsilon > 0$. An isotopy $H\colon Q \times \mathbb{I} \to Q$ is called an $\varepsilon$-*isotopy* provided that for every $x \in Q$ the diameter of the set $H[\{x\} \times \mathbb{I}]$ is less than $\varepsilon$.

▶1. Let $E, F \subseteq s$ be compact and let $f\colon E \to F$ be a homeomorphism such that $\hat{\varrho}(f, 1_E) < \varepsilon$. Prove that there is an $\varepsilon$-isotopy $H\colon Q \times \mathbb{I} \to Q$ such that $H_0 = 1_Q$ and $H_1 \restriction E = f$ (i.e., $H_1$ extends $f$). (This improves Theorem 5.2.4.)

## 5.3. The estimated homeomorphism extension theorem

The aim of this section is among other things to prove that if $E, F \in \mathcal{Z}(Q)$ and if $f\colon E \to F$ is a homeomorphism such that $\hat{\varrho}(f, 1_E) < \varepsilon$ then $f$ can be extended to a homeomorphism $\bar{f}\colon Q \to Q$ such that $\hat{\varrho}(\bar{f}, 1_Q) < \varepsilon$. This result is known as the *(Estimated) Homeomorphism Extension Theorem* and is of fundamental importance in infinite-dimensional topology. The strategy of the proof is that we first push $E$ and $F$ into $s$ by a small motion and then apply Theorem 5.2.4.

**Extending homeomorphisms.** Recall that for $n \in \mathbb{N}$ and $\theta \in \{-1, 1\}$ by $W_n^\theta$ we mean the endface $\pi_n^{-1}(\{\theta\})$. *Throughout, let $K \subseteq s$ be a fixed compact subset.* This set plays no role of importance in this section. But by introducing and dealing with it here we are able to simply our life considerably later.

We will first show that endfaces can be pushed into $s$ by a small movement. This is done by pushing endfaces into endfaces with larger and larger indexes.

**Lemma 5.3.1.** *For each $n \in \mathbb{N}$, $\theta \in \{-1, 1\}$ and $\varepsilon > 0$ there are a homeomorphism $h\colon Q \to Q$ and an $m > n$ such that*

(1) $h[W_n^\theta] \cap \bigcup \{W_i^\mu : i < m, \mu \in \{-1, 1\}\} = \emptyset$,
(2) $h[W_n^\theta] \subseteq W_m^1$,
(3) $\hat{\varrho}(h, 1_Q) < \varepsilon$,
(4) $h \restriction K = 1$.

**Proof.** Let $\varepsilon > 0$ and choose $m > n$ such that
$$2^{-(m-2)} < \min\{\varepsilon/2, \varrho(K, W_n^\theta)\}.$$

## 5.3. THE ESTIMATED HOMEOMORPHISM EXTENSION THEOREM   321

We first push $W_n^\theta$ into $W_m^1$ and then away from the endfaces in the lower coordinate directions. Without loss of generality we assume that $\theta = 1$.

It is geometrically obvious that there is a homeomorphism

$$\varphi \colon \mathbb{J}_n \times \mathbb{J}_m \to \mathbb{J}_n \times \mathbb{J}_m$$

having the following properties:

(5) $\varphi[\{1\} \times \mathbb{J}_m] \subseteq \mathbb{J}_n \times \{1\}$,
(6) if $x \leq 1 - 2^{-(m-1)}$ then $\varphi(x, y) = (x, y)$ for every $y$.

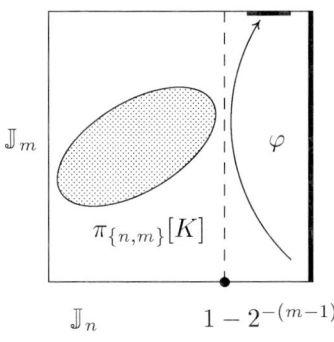

Figure 18.

This is precisely the same homeomorphism as the one in Figure 4 on Page 61.

Let $h_1 \in \mathcal{H}(Q)$ be the homeomorphism of $Q$ that is defined by crossing $\varphi$ with the identity on the other factors of $Q$.

**Claim 1.**

(7) $\hat{\varrho}(h_1, 1_Q) \leq 2^{-(m-2)} < \varepsilon/2$,
(8) $h_1[W_n^\theta] \subseteq W_m^1$,
(9) $h \restriction K = 1_K$.

*Proof.* Observe that (7) follows as in the proof of Lemma 1.6.4. Since (8) and (9) are trivial, we are done.   ◇

By squeezing the $(m-1)$-cell $\mathbb{J}^{m-1}$ into its interior (see Figure 19) we can similarly construct a homeomorphism $\psi \colon \mathbb{J}^m \to \mathbb{J}^m$ such that

(10) $\psi[\mathbb{J}^{m-1} \times \{1\}_m] \subseteq \prod_{j=1}^{m-1}(-1, 1)_j \times \{1\}_m$,
(11) $\hat{\varrho}(\psi, 1) < \varepsilon/2$,
(12) $\psi \restriction \pi_{\{1,2,\ldots,m\}}[K] = 1$.

Now let $h_2 \in \mathcal{H}(Q)$ be the homeomorphism that is defined by crossing $\psi$ with the identity on the other factors of $Q$. It is clear that

(13) $h_2 \restriction K = 1_K$,
(14) $\hat{\varrho}(h_2, 1_Q) < \varepsilon/2$.

From (7), (8), (9), (10), (13) and (14) it follows easily that $h = h_2 \circ h_1$ is as required. □

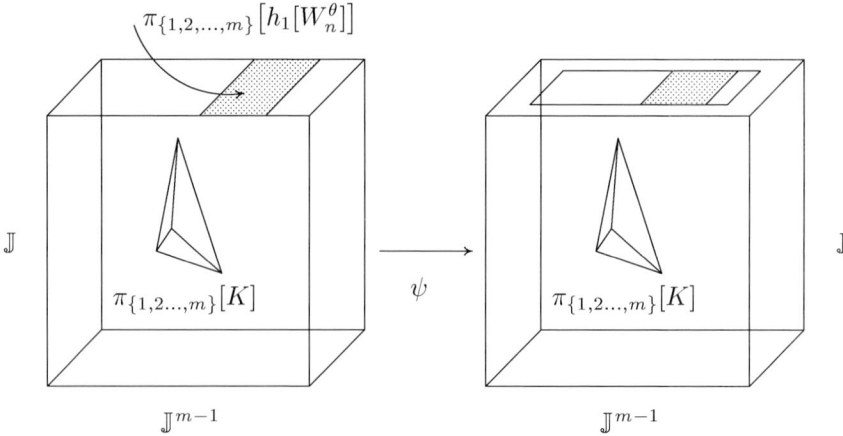

Figure 19.

**Corollary 5.3.2.** *For each $n \in \mathbb{N}$, $\theta \in \{-1, 1\}$ and $\varepsilon > 0$ there is a homeomorphism $h \colon Q \to Q$ with*

(1) $h[W_n^\theta] \subseteq s$,
(2) $\hat{\varrho}(h, 1_Q) < \varepsilon$,
(3) $h \restriction K = 1$.

**Proof.** Put $n_1 = n$. By Lemma 5.3.1 we can push $W_n^\theta$ into some other endface $W_{n_2}^1$ with $n_2 > n_1$, by a homeomorphism $h_1 \in \mathcal{H}(Q)$ which is as close to to the identity as we wish. Repeated applications of this yields a sequence $(h_i)_i \in \mathcal{H}(Q)$, each element of which is close enough to the identity in order to have
$$h = \lim_{i \to \infty} h_i \circ h_{i-1} \circ \cdots \circ h_1 \in \mathcal{H}(Q), \quad \hat{\varrho}(h, 1_Q) < \varepsilon$$
(Theorem 1.6.2). Moreover, each $h_i$ can be taken to be the identity on $K$, so that $h \restriction K = 1_K$.

Finally, each $h_i$ pushes the face $W_{n_i}^1$ off the faces $W_j^\mu$ for $j < n_{i+1}$ and $\mu$ in $\{-1, 1\}$. For $h_i$ close enough to the identity (as prescribed in Lemma 1.6.1), we can keep the limit homeomorphism $h$ 'away from all faces of $Q$', that is, $h[W_n^\theta] \subseteq s$. □

## 5.3. THE ESTIMATED HOMEOMORPHISM EXTENSION THEOREM

We now know how to push an endface into $s$. We are going to use this result to push an arbitrary $Z$-set $A$ away from a fixed endface, and then away from all endfaces, i.e., to push $A$ into $s$.

We need the following result.

**Proposition 5.3.3.** *Let $X$ be a compact space, let $X_0 \subseteq X$ be closed and let $f\colon X \to s$ be continuous such that $f \restriction X_0$ is an imbedding. Then for every $\varepsilon > 0$ there is an imbedding $g\colon X \to s$ such that*

(1) $\hat{\varrho}(f,g) < \varepsilon$,
(2) $g \restriction X_0 = f \restriction X_0$.

**Proof.** Let $\mathcal{B}$ be a countable collection of compacta in $X \setminus X_0$ such that their interiors form a base for $X \setminus X_0$ (observe that $X \setminus X_0$ is locally compact). Assume that $\emptyset \in \mathcal{B}$. In addition, let $\{(F_i, G_i) : i \in \mathbb{N}\}$ enumerate the set

$$\{(F,G) \in \mathcal{B} \times \mathcal{B} : F \cap G = \emptyset\}.$$

Choose $m \in \mathbb{N}$ such that

$$\sum_{n=m+1}^{\infty} 2^{-n} < \tfrac{1}{2}\varepsilon.$$

Since $\pi_{m+i} f[X_0]$ is a compact subset of $(-1,1)_{m+i}$, for every $i \in \mathbb{N}$ we can apply the Tietze Extension Theorem 1.2.5 to construct a continuous function $g_i\colon X \to (-1,1)_{m+i}$ such that

(3) $g_i \restriction X_0 = (\pi_{m+i} \circ f) \restriction X_0$,
(4) $g_i[X_0]$, $g_i[F_i]$ and $g_i[G_i]$ are pairwise disjoint.

(Pick two distinct points $p, q \in (-1,1)_{m+i} \setminus \pi_{m+i} f[X_0]$ and extend the function that is constant $p$ on $F_i$, constant $q$ on $G_i$, and is equal to $\pi_{m+i} \circ f$ on $X_0$.) Define $g\colon X \to s$ by

$$g(x) = (f(x)_1, f(x)_2, \ldots, f(x)_m, g_1(x), g_2(x), \ldots).$$

It is clear that $g$ is continuous, that $g \restriction X_0 = f \restriction X_0$ and that $\hat{\varrho}(f,g) < \varepsilon$. Take $x, y \in X$ such that $x \neq y$. If $x, y \in X_0$ then $g(x) \neq g(y)$ since $f \restriction X_0$ is an imbedding. If $x \in X_0$ and $y \notin X_0$ then we can find an index $i \in \mathbb{N}$ such that $y \in F_i$. By (4) and the definition of $g$ it now follows that

$$g(x)_{m+i} \neq g(y)_{m+i}.$$

Finally, if $x, y \notin X_0$ then we can find an $i \in \mathbb{N}$ such that $x$ belongs to $F_i$ and $y$ belongs to $G_i$. As above it follows that $g(x)_{m+i} \neq g(y)_{m+i}$. We conclude that $g$ is one-to-one and the compactness of $X$ now yields that $g$ is an imbedding (Exercise A.5.9). $\square$

**Lemma 5.3.4.** *Let $A \in \mathcal{Z}(Q)$. Then for each $n \in \mathbb{N}$, $\theta \in \{-1,1\}$ and $\varepsilon > 0$ there is a homeomorphism $h\colon Q \to Q$ with*

(1) $h[A] \cap W_n^\theta = \emptyset$,
(2) $\hat{\varrho}(h, 1_Q) < \varepsilon$,
(3) $h \restriction K = 1$.

**Proof.** By Corollary 5.3.2 we can find a homeomorphism $h_1 \in \mathcal{H}(Q)$ such that $\hat{\varrho}(h_1, 1_Q) < \varepsilon/3$, $h_1 \restriction K = 1_K$ and $h_1[W_n^\theta] \subseteq s$. Since

$$A \cup K \cup B(Q) \in \mathcal{Z}_\sigma(Q)$$

(Corollary 5.1.5), by Lemma 5.1.2(4) there is a continuous function

$$\alpha \colon Q \to Q \setminus (A \cup K \cup B(Q))$$

such that $\hat{\varrho}(\alpha, 1_Q) < \varepsilon/3$. Consequently, $\alpha[Q] \subseteq s$ and hence by compactness, $\varrho(\alpha[Q], A \cup K) > 0$. By Proposition 5.3.3 choosing $X_0 = \emptyset$ there is an imbedding $\beta$ from $Q$ into $s$ with

(4) $\beta[Q] \cap (A \cup K) = \emptyset$,
(5) $\hat{\varrho}(\beta, \alpha) < \varepsilon/3$

(observe that we can achieve (4) since $\varrho(\alpha[Q], A \cup K) > 0$). Notice that

$$\hat{\varrho}(\beta, 1_Q) < 2\varepsilon/3.$$

Put

$$\gamma = \beta \restriction h_1[W_n^\theta] \cup 1_K.$$

Since

$$K \cap (h_1[W_n^\theta] \cup \beta h_1[W_n^\theta]) = \emptyset,$$

the function $\gamma$ is a homeomorphism from $h_1[W_n^\theta] \cup K$ onto $\beta h_1[W_n^\theta] \cup K$ such that $\hat{\varrho}(\gamma, 1) < 2\varepsilon/3$. By Theorem 5.2.4, we can extend $\gamma$ to a homeomorphism $\Gamma \colon Q \to Q$ such that $\hat{\varrho}(\Gamma, 1_Q) < 2\varepsilon/3$. Observe that

(6) $\Gamma \restriction K = 1$,
(7) $\Gamma\big[h_1[W_n^\theta]\big] \cap A = \emptyset$.

Now put $h = (\Gamma \circ h_1)^{-1}$. Then since

$$\hat{\varrho}(h, 1_Q) < 2\varepsilon/3 + \varepsilon/3 = \varepsilon,$$

the function $h$ is as required. □

We can now prove the following:

**Theorem 5.3.5.** *Let $K \subseteq s$ be compact and let $A \in \mathcal{Z}(Q)$. Then for every $\varepsilon > 0$ there is a homeomorphism $h \colon Q \to Q$ with*

(1) $\hat{\varrho}(h, 1_Q) < \varepsilon$,
(2) $h \restriction K = 1_K$,
(3) $h[A] \subseteq s$.

**Proof.** All we have to do is to apply Lemma 5.3.4 inductively to free $A$ from all endfaces $W_n^\theta$ while keeping $K$ pointwise fixed. This has to be done with a little care so that once $A$ is free from an endface, the limit homeomorphism does not carry it back to that endface. But this can be done again with the help of Lemma 1.6.1 (cf. the proof of Corollary 5.3.2). □

This leads us to the following basic result.

**Corollary 5.3.6.** *Let $A \in \mathcal{Z}_\sigma(Q)$. Then for every $\varepsilon > 0$ there exists a homeomorphism $h \colon Q \to Q$ such that $h[A] \subseteq s$ and $\hat{\varrho}(h, 1_Q) < \varepsilon$.*

**Proof.** Write $A = \bigcup_{n=1}^\infty A_n$, where $A_n \in \mathcal{Z}(Q)$ for every $n$. By Theorem 5.3.5 there is a 'small' homeomorphism $f_1 \colon Q \to Q$ such that $f_1[A_1] \subseteq s$. Again by Theorem 5.3.5 there is a 'small' homeomorphism $f_2 \colon Q \to Q$ such that $f_2[f_1[A_2]] \subseteq s$ while moreover $f_2$ restricts to the identity on $f_1[A_1]$. Etc. The infinite left product $h = \lim_{n\to\infty} f_n \circ \cdots \circ f_1$ is a homeomorphism of $Q$ such that $h[A] \subseteq s$ and $\hat{\varrho}(h, 1_Q) < \varepsilon$. □

We can now come to the main result in this section.

**Homeomorphism Extension Theorem 5.3.7.** *Let $E, F \in \mathcal{Z}(Q)$ and let $f$ be a homeomorphism from $E$ onto $F$ such that $\hat{\varrho}(f, 1_E) < \varepsilon$. Then $f$ can be extended to a homeomorphism $\bar{f} \colon Q \to Q$ such that $\hat{\varrho}(\bar{f}, 1_Q) < \varepsilon$.*

**Proof.** Just apply Theorems 5.3.5 and 5.2.4 to $E \cup F$. □

Perhaps the reader feels that it is not worth the trouble in Theorem 5.3.7 to prove that it is possible to extend 'small' homeomorphisms to 'small' homeomorphisms. At first glance it seems that the possibility of extending homeomorphisms is the most important fact and that the extra smallness condition is a technical curiosity. This is not true however. The possibility to extend 'small' homeomorphisms to 'small' homeomorphisms makes it possible to apply the inductive convergence criterion once again to create new homeomorphisms from old ones. This procedure turns out to be extremely powerful.

Observe that the metric $\varrho$ in Theorem 5.3.7 is the standard 'convex' metric on $Q$. It can be shown that the theorem is false if one replaces $\varrho$ by an equivalent 'nonconvex' metric. For details see e.g., ANDERSON, CURTIS and VAN MILL [20]. Of course, homeomorphisms between $Z$-sets can be extended, but problems arise with the smallness condition.

The following result should be compared with Corollary 5.3.6

A *Hilbert cube* is a space homeomorphic to $Q$. We would like to have an estimated homeomorphism extension theorem for Hilbert cubes but run into 'smallness' problems here as well. Fortunately, a weaker version of the estimated homeomorphism extension theorem is true for Hilbert cubes which is powerful enough for the applications that we have in mind.

It will be convenient to let $M^Q$ denote an arbitrary Hilbert cube.

**Corollary 5.3.8.** *Let $X = M^Q$ be a Hilbert cube with admissible metric $\rho$. Then for each $\varepsilon > 0$ there is a $\delta > 0$ with the following property:*

> *if $A, B \in \mathcal{Z}(X)$ are arbitrary and $f: A \to B$ is a homeomorphism with $\hat{\rho}(f, 1_A) < \delta$ then there is a homeomorphism $\bar{f}: X \to X$ such that $\bar{f} \upharpoonright A = f$ and $\hat{\rho}(\bar{f}, 1_X) < \varepsilon$.*

**Proof.** Let $f: Q \to X$ be any homeomorphism. Fix $\varepsilon > 0$ and let $\gamma > 0$ be such that if $A \subseteq Q$ has diameter less than $\gamma$ then $f[A]$ has diameter less than $\varepsilon$. (Here we use that $f$ is uniformly continuous (Exercise A.5.18).) Next, let $\delta > 0$ be a Lebesgue number for the cover

(∗)  $\qquad \{f[U] : U \subseteq Q \text{ is open and } \operatorname{diam} U < \gamma\}$

(Lemma A.5.3). We claim that $\delta$ is as required. To this end, let

$$A, B \in \mathcal{Z}(X)$$

and let $g: A \to B$ be a homeomorphisms such that $\hat{\rho}(g, 1_X) < \delta$. It is clear that $f^{-1}[A]$ and $f^{-1}[B]$ are Z-sets in $Q$ (cf. Lemma 5.1.2(6)) and that the function $\xi: f^{-1}[A] \to f^{-1}[B]$ defined by $\xi(q) = f^{-1}gf(q)$ is a homeomorphism. We claim that $\hat{\varrho}(\xi, 1_Q) < \gamma$. To see this, fix an arbitrary $q \in f^{-1}[A]$. Then $f(q) \in A$ and so $\rho(f(q), gf(q)) < \delta$. There is an open subset $U \subseteq Q$ such that $f[U]$ contains both $f(q)$ and $gf(q)$. This is so because $\delta$ is a Lebesgue number for (∗). But then $q$ and $\xi(q) = f^{-1}gf(q)$ both belong to $U$, i.e., $\varrho(q, \xi(q)) < \gamma$ which is as claimed (cf. Exercise A.5.4). By the Homeomorphism Extension Theorem 5.3.7, we can extend $\xi$ to a homeomorphism

$$\bar{\xi}: Q \to Q$$

such that $\hat{\varrho}(\bar{\xi}, 1_Q) < \gamma$. Now put $\bar{g} = f \circ \bar{\xi} \circ f^{-1}$. It is clear that $\bar{g}$ is a homeomorphism and extends $g$. The only thing that needs to be checked is the smallness condition. To this end, take an arbitrary $x$ in $X$, and let $q$ be $f^{-1}(x)$. Then $\varrho(q, \bar{\xi}(q)) < \gamma$ from which it follows that

$$\rho(x, \bar{g}(x)) = \rho(f(q), \bar{g}(f(q))) = \rho(f(q), f\bar{\xi}(q)) < \varepsilon,$$

as required. $\square$

**Corollaries.** We will now derive some interesting corollaries of the Homeomorphism Extension Theorem.

**Corollary 5.3.9.** *Let $S \in \mathcal{Z}_\sigma(Q)$ and let $\mathcal{A}$ be an infinite collection endfaces of $Q$. Then for every $\varepsilon > 0$ there is a homeomorphism $h: Q \to Q$ such that $\hat{\varrho}(h, 1_Q) < \varepsilon$ while moreover $h[S] \subseteq \bigcup \mathcal{A}$.*

**Proof.** For every $A \in \mathcal{A}$ there are $n(A) \in \mathbb{N}$ and $\theta(A) \in \{-1, 1\}$ such that $A = W_{n(A)}^\theta$. Without loss of generality we assume that there exists an infinite subcollection $\mathcal{A}'$ of $\mathcal{A}$ such that for every $A \in \mathcal{A}'$, $\theta(A) = 1$.

By Corollary 5.3.6 we may assume without loss of generality that $S \subseteq s$. Write $S$ as $\bigcup_{i=1}^\infty S_i$, where $S_i$ is compact for every $i$. Let $\varepsilon > 0$ and pick a natural number $n$ so large that

$$\sum_{m=n}^\infty 2^{-m} < 1/2\varepsilon.$$

Write $\mathbb{N} \setminus \{1, 2, \ldots, n-1\}$ as the disjoint union of infinitely many infinite sets, say $\{E_i : i \in \mathbb{N}\}$, such that for every $i \in \mathbb{N}$, $\min(E_i) = n(A)$ for certain $A \in \mathcal{A}'$. Let

$$h_i \colon \prod_{k \in E_i} \mathbb{J}_k \to \prod_{k \in E_i} \mathbb{J}_k$$

be a homeomorphism such that

$$h_i[\pi_{E_i}[S_i]] \subseteq B\Big(\prod_{k \in E_i} \mathbb{J}_k\Big).$$

Simply observe that

$$\{x \in \mathbb{J}^{E_i} : x_{\min(E_i)} = 1\}$$

is a $Z$-set copy of $Q$ by Lemma 5.1.2(5) and hence contains a topological copy of $\pi_{E_i}[S_i]$ by Corollary A.4.4. So $h_i$ is given by Theorem 5.3.7 since the compact set $\pi_{E_i}[S_i]$ is contained in the pseudo-interior of the Hilbert cube $\prod_{k \in E_i} \mathbb{J}_k$ and hence is a $Z$-set there (Corollary 5.1.5). Define $h \colon Q \to Q$ by

$$h(x) = \begin{cases} x_j & (j \leq n-1), \\ \big(h_i(\pi_{E_i}(x))\big)_j & (j \in E_i). \end{cases}$$

It is clear that $h$ is as required. □

**Corollary 5.3.10.** *If $S \in \mathcal{Z}_\sigma(Q)$ then $Q$ can be deformed through $Q \setminus S$.*

**Proof.** By Corollary 5.3.9 we may assume without loss of generality that $S$ is a subset of $B(Q)$. So the homotopy $H \colon Q \times \mathbb{I} \to Q$ defined by

$$H(x, t) = (1 - t) \cdot x$$

for $t \in \mathbb{I}$ and $x \in Q$ is as required. □

**Mapping Replacement Theorem.** Our final application of the Homeomorphism Extension Theorem is a generalization of Proposition 5.3.3 which will be used several times in the forthcoming.

Let $X$ be a space. A continuous function $f \colon X \to Q$ is called *Z-map* provided that $f[X] \in \mathcal{Z}(Q)$. If $f$ is an imbedding as well as a $Z$-map then it is called *Z-imbedding*.

**Mapping Replacement Theorem 5.3.11.** *Let $X$ be compact with closed subset $A$. Let $f\colon X \to Q$ be a continuous function such that $f \restriction A$ is a $Z$-imbedding. Then for every $\varepsilon > 0$ there is a $Z$-imbedding $g\colon X \to Q$ such that $\hat{\varrho}(f,g) < \varepsilon$ and $g \restriction A = f \restriction A$.*

**Proof.** Since $B(Q) \in \mathcal{Z}_\sigma(Q)$ (Corollary 5.1.5), we can approximate $f$ by a map $\bar{f}\colon X \to s$ (Lemma 5.1.2(4)). By Proposition 5.3.3 we can approximate $\bar{f}$ by an imbedding $\hat{f}\colon X \to s$. Observe that $\hat{f}[X] \in \mathcal{Z}(Q)$ (Corollary 5.1.5). Then $f[A]$ and $\hat{f}[A]$ are homeomorphic $Z$-sets and we can now use Theorem 5.3.7 to find a 'small' homeomorphism $h \in \mathcal{H}(Q)$ extending the obvious homeomorphism between $\hat{f}[A]$ and $f[A]$. Then $g = h \circ \hat{f}$ is as required. □

**Corollary 5.3.12.** *Let $X$ be a compact space with closed subset $A$. In addition, let $f\colon X \to Q$ be a continuous function for which $f \restriction A$ is a $Z$-imbedding. Let $Z \in \mathcal{Z}(Q)$ (possibly empty) be such that $f[A]$ misses $Z$. Then for every $\varepsilon > 0$ there is a $Z$-imbedding $g\colon X \to Q$ such that $\hat{\varrho}(f,g) < \varepsilon$ and $g \restriction A = f \restriction A$ while moreover $g[X] \cap Z = \emptyset$.*

**Proof.** Let $H\colon Q \times \mathbb{I} \to Q$ be a deformation such that $H_0 = 1_Q$ and
$$H_t[Q] \cap Z = \emptyset$$
for every $t \in (0,1]$ (Corollary 5.3.10). Since $H_0 = 1_Q$, by compactness of $Q$ there exists $s \in (0,1]$ such that for every $q \in Q$ we have
$$\operatorname{diam}(\{q\} \times [0,s]) < \tfrac{1}{4}\varepsilon.$$
Let $d$ be an admissible metric for $X$ which is bounded by 1 (Exercise A.1.6). Define $\xi\colon X \to Q$ by
$$\xi(x) = H_{s \cdot d(x, A)}\bigl(f(x)\bigr).$$
Then $\hat{\varrho}(\xi, f) < \tfrac{1}{4}\varepsilon$, $\xi \restriction A = f \restriction A$ and $\xi[X] \cap Z = \emptyset$ since $f[A] \cap Z = \emptyset$. Let $\delta = \varrho(\xi[X], Z)$. By Theorem 5.3.11 there is a $Z$-imbedding $g\colon X \to Q$ such that $g \restriction A = \xi \restriction A = f \restriction A$ while moreover
$$\hat{\varrho}(g, \xi) < \min\{\tfrac{1}{4}\varepsilon, \delta\}.$$
It is clear that $g$ is as required. □

### Exercises for §5.3.

▶1. Prove the following generalization of Theorem 5.3.7: If $E, F \in \mathcal{Z}(Q)$ and $f\colon E \to F$ is a homeomorphism such that $\hat{\varrho}(f, 1_E) < \varepsilon$ then there is an $\varepsilon$-isotopy $H\colon Q \times \mathbb{I} \to Q$ with $H_0 = 1_Q$ and $H_1 \restriction E = f$, i.e., $H_1$ extends $f$.

2. Prove that $Q$ is isotopically homogeneous.

3. Let $X$ be a space. Suppose that $X = Q_0 \cup Q_1$, where
$$Q_0 \approx Q_1 \approx Q_0 \cap Q_1 \approx Q,$$
$Q_0 \cap Q_1 \in \mathcal{Z}(Q_0)$ and $Q_0 \cap Q_1 \in \mathcal{Z}(Q_1)$. Prove that $X \approx Q$.

## 5.4. The compact absorption property

The aim of this section is to characterize all elements $A \in \mathcal{Z}_\sigma(Q)$ for which there is a homeomorphism $h\colon Q \to Q$ such that $h[A] = B(Q)$. As an application, we prove that an infinite product of intervals is homeomorphic to $s$ iff infinitely many of them are not compact and that $s \setminus E \approx s$ for every $\sigma$-compact subset $E \subseteq s$. We also prove the curious result that the subspace
$$\{A \in 2^\mathbb{I} : \dim A = 0\}$$
of $2^\mathbb{I}$ is homeomorphic to $s$. This result will be important later.

**Capsets.** Let $M^Q$ be a Hilbert cube. An element $A \in \mathcal{Z}_\sigma(M^Q)$ is called a *capset*[1] provided that $A$ can be written as the union of an increasing sequence $A_1 \subseteq A_2 \subseteq \cdots \subseteq A_n \subseteq \cdots$ of $Z$-sets in $M^Q$ such that the following absorption property holds:
$$\forall \varepsilon > 0 \ \forall n \in \mathbb{N} \ \forall Z \in \mathcal{Z}(M^Q) \ \exists m > n \ \exists h \in \mathcal{H}(M^Q):$$
(1) $\hat{\varrho}(h, 1) < \varepsilon$,
(2) $h \restriction A_n = 1$,
(3) $h[Z] \subseteq A_m$.

We will prove later that a subset $A \subseteq Q$ is a capset if and only if there exists a homeomorphism $h\colon Q \to Q$ such that $h[A] = B(Q)$. So the property of being a capset characterizes the way $B(Q)$ is topologically placed in $Q$.

This implicitly also characterizes the way $s$ is placed in $Q$, For proving that a subspace $A$ of $Q$ is homeomorphic to $s$ it is sufficient (but not necessary) to prove that the complement $B = Q \setminus A$ is a capset.

**Theorem 5.4.1.** *Let $A$ be a capset in the Hilbert cube $M^Q$. Then for all $K, L \in \mathcal{Z}(M^Q)$ and $\varepsilon > 0$ there is a homeomorphism $h\colon M^Q \to M^Q$ such that*

(1) $\hat{\varrho}(h, 1) < \varepsilon$,
(2) $h \restriction K = 1_K$,
(3) $h[L \setminus K] \subseteq A$.

Roughly speaking, $A$ absorbs $L \setminus K$ by a small motion keeping $K$ pointwise fixed.

---
[1] CAP abbreviates *Compact Absorption Property*.

**Proof.** Let the $Z$-sets $A_n$, $n \in \mathbb{N}$, witness the fact that $A$ is a capset. In addition, let $K$ and $L$ be $Z$-sets and let $\varepsilon > 0$. Finally, let $L_0 \subseteq L_1 \subseteq \ldots$ be a sequence of closed subsets of $L \setminus K$ such that $L_0 = \emptyset$ and $L \setminus K = \bigcup_{i=1}^{\infty} L_i$. We shall inductively construct a sequence $f_0, f_1, \ldots$ in $\mathcal{H}(M^Q)$ and a sequence of natural numbers $n(0) < n(1) < \cdots$ such that $f_0 = 1$ and $n(0) = 1$, while moreover the following statements hold:

(1) $\hat{\varrho}(f_i, 1)$ is so small that the conditions mentioned in the Inductive Convergence Criterion 1.6.2 are satisfied,
(2) $\hat{\varrho}(f_i, 1) < 3^{-i} \cdot \varepsilon$,
(3) $f_i \circ f_{i-1} \circ \cdots \circ f_0[L_i] \subseteq A_{n(i)}$,
(4) $f_i \upharpoonright K \cup A_{n(i-1)} = 1$.

This will establish the theorem since by (1) and (2), $f = \lim_{i \to \infty} f_i \circ \cdots \circ f_1$ exists, belongs to $\mathcal{H}(M^Q)$, and satisfies $\hat{\varrho}(f, 1) < \varepsilon$, while moreover by (3) and (4), $f[L \setminus K] \subseteq A$ and $f \upharpoonright K = 1_K$.

It remains to perform the induction. Since $f_0$ and $n(0)$ are already defined, assume that $f_i$ and $n(i)$ have been chosen up to $i \geq 0$. Find $\delta > 0$ such that if we choose $f_{i+1}$ $\delta$-close to 1 then (1) and (2) are automatically satisfied for $i+1$. Let $B = f_i \circ f_{i-1} \circ \cdots \circ f_0[L_{i+1}]$ and observe that $B \in Z(M^Q)$ (Lemma 5.1.2(6)) and that $B \cap K = \emptyset$. Let $\gamma = \varrho(B, K)$. Let $\xi > 0$ be such that every homeomorphism between $Z$-sets of $M^Q$ that moves the points less than $\xi$ can be extended to a homeomorphism of $M^Q$ moving the points less than $\delta$ (Corollary 5.3.8). Since $A$ is a capset, there are $m > n(i)$ and a homeomorphism $g \colon M^Q \to M^Q$ such that

(5) $\hat{\varrho}(g, 1) < \min\{\xi, \gamma\}$,
(6) $g \upharpoonright A_{n(i)} = 1$,
(7) $g[B] \subseteq A_m$.

Observe that $g[B] \cap K = \emptyset$. It is clear that $g$ is not quite yet as required since $g$ might move $K$. However, clearly

$$(g \upharpoonright B) \cup 1_K \cup 1_{A_{n(i)}}$$

is a homeomorphism of $B \cup K \cup A_{n(i)}$ onto $g[B] \cup K \cup A_{n(i)}$. This homeomorphism can be extended to a homeomorphism $h \colon M^Q \to M^Q$ with $\hat{\varrho}(h, 1) < \delta$. Now if we put $n(i+1) = m$ and $f_{i+1} = h$ then these choices are easily seen to be as required. □

**Corollary 5.4.2.** Let $A$ be a capset in the Hilbert cube $M^Q$. Then for all $K, L \in Z(M^Q)$ such that $K \subseteq A \cap L$ and $\varepsilon > 0$ there is a homeomorphism $h \colon M^Q \to M^Q$ such that

(1) $\hat{\varrho}(h, 1) < \varepsilon$,
(2) $h \upharpoonright K = 1_K$,
(3) $h[L] \subseteq A$.

## 5.4. THE COMPACT ABSORPTION PROPERTY

We now present some fundamental properties of capsets.

**Theorem 5.4.3.** *Let $M^Q$ be a Hilbert cube.*

(1) *If $A \subseteq M^Q$ is a capset and $h\colon M^Q \to M^Q$ is a homeomorphism then $h[A]$ is a capset.*
(2) *If $A \subseteq M^Q$ is a capset and $B \in \mathcal{Z}_\sigma(M^Q)$ then $A \cup B$ is a capset.*
(3) *If $A$ and $B$ are capsets in $M^Q$ and $Z \subseteq M^Q$ is a Z-set (possibly empty) then for each $\varepsilon > 0$ there is a homeomorphism $h\colon M^Q \to M^Q$ such that $h[A \setminus Z] = B \setminus Z$, $\hat\varrho(h,1) < \varepsilon$ and $h \restriction Z = 1_Z$.*

**Remark 5.4.4.** It follows in particular from Theorem 5.4.3(3) that capsets are topologically unique, that is, if $A$ and $B$ are capsets in $Q$ then there is an element $h \in \mathcal{H}(Q)$ such that $h[A] = B$.

**Proof.** For (1), observe that $h$ is uniformly continuous (see e.g., the proof of Corollary 5.3.8).

For (3), write

$$A \setminus Z = \bigcup_{i=1}^\infty A_i, \quad B \setminus Z = \bigcup_{i=1}^\infty B_i,$$

where $A_i, B_i \in \mathcal{Z}(M^Q)$ for every $i \in \mathbb{N}$ and $A_1 = \emptyset = B_1$ (observe that $A$ is $\sigma$-compact, hence so is $A \setminus Z$, apply Lemma 5.1.2, etc.). We shall inductively construct a sequence $f_1, f_2, \dots$ in $\mathcal{H}(M^Q)$ such that $f_1 = 1$ and

(4) $\hat\varrho(f_i, 1)$ is small enough in order to apply Theorem 1.6.2,
(5) $\hat\varrho(f_i, 1) < 3^{-i} \cdot \varepsilon$,
(6) $B_i \subseteq f_i \circ g_{i-1}[A]$,
(7) $f_i \circ g_{i-1}[A_i] \subseteq B$,
(8) $f_i \restriction Z \cup \bigcup_{j=1}^{i-1}(g_{i-1}[A_j] \cup B_j) = 1$,

where $g_{i-1} = f_{i-1} \circ \cdots \circ f_1$.

Assume that $f_1, \dots, f_i$ have been constructed. Find $\delta > 0$ such that if we choose $f_{i+1}$ $\delta$-close to 1 then (4) and (5) are satisfied for $i+1$. Observe that $g_i[A_{i+1}] \in \mathcal{Z}(M^Q)$ (Lemma 5.1.2(6)). Also, by our induction hypothesis,

$$K = \bigcup_{j=1}^i (g_i[A_j] \cup B_j) \subseteq B.$$

Notice that $K \in \mathcal{Z}(M^Q)$ by Lemma 5.1.2(3),(6), that $Z \cap K = \emptyset$ and that $A_{i+1} \cap Z = \emptyset$. We aim at applying Corollary 5.4.2 with the Z-sets $K$ and $K \cup g_i[A_{i+1}]$. Indeed, since $B$ is a capset which contains $K$, there is by Corollary 5.4.2 a homeomorphism $\alpha\colon M^Q \to M^Q$ such that

(9) $\hat\varrho(\alpha, 1) < \delta/2$,
(10) $\alpha[g_i[A_{i+1}]] \subseteq B$,   (i.e., $A_{i+1}$ is absorbed by $B$),

(11) $\alpha \restriction K = 1$.

There is a problem though, since $\alpha$ might move $Z$ and we would like $\alpha$ to have the property that its restriction to $Z$ is the identity. But this can easily be fixed by using the same method as in the proof of Theorem 5.4.1 since by (8), $Z \cap (K \cup g_i[A_{i+1}]) = \emptyset$ (so this again depends on the Homeomorphism Extension Theorem). We may therefore additionally assume that

(11)' $\alpha \restriction (K \cup Z) = 1$.

Since $\alpha \circ g_i$ belongs to $\mathcal{H}(M^Q)$, by (1) it follows that $\alpha \circ g_i[A]$ is a capset. Notice that by (6) and (11)' we have,

$$K' = K \cup \alpha \circ g_i[A_{i+1}] \subseteq \alpha \circ g_i[A]$$

and that $Z \cup K' \cup B_{i+1}$ is a Z-set. By a similar argumentation as the one above, there is a homeomorphism $\beta \colon M^Q \to M^Q$ such that

(12) $\hat{\varrho}(\beta, 1) < \delta/2$,
(13) $\beta[B_{i+1}] \subseteq \alpha \circ g_i[A]$      (i.e., $B_{i+1}$ is absorbed by $\alpha \circ g_i[A]$),
(14) $\beta \restriction Z \cup K' = 1$.

Now define $f_{i+1} = \beta^{-1} \circ \alpha$. Then clearly $\hat{\varrho}(f_{i+1}, 1) < \delta$. Moreover, by (10) and (14),

$$f_{i+1} \circ g_i[A_{i+1}] = \beta^{-1} \circ \alpha \circ g_i[A_{i+1}] = \alpha \circ g_i[A_{i+1}] \subseteq B,$$

and by (13),

$$B_{i+1} \subseteq \beta^{-1} \circ \alpha \circ g_i[A] = f_{i+1} \circ g_i[A].$$

Since clearly $f_{i+1} \restriction Z \cup K = 1$, we see that $f_{i+1}$ is as required.

Now put $f = \lim_{i \to \infty} f_i \circ \cdots \circ f_1 = \lim_{i \to \infty} g_i$. Clearly $\hat{\varrho}(f, 1) < \varepsilon$. By (8) and (7) we have,

(15) $$f[A \setminus Z] = \bigcup_{i=1}^{\infty} f[A_i] = \bigcup_{i=1}^{\infty} g_i[A_i] \subseteq B \setminus Z.$$

In addition, by (8) it follows that for every $i$,

$$f \circ g_i^{-1}[B_i] = \lim_{n > i}(f_n \circ \cdots \circ f_{i+1} \circ g_i)\left[g_i^{-1}[B_i]\right]$$
$$= \lim_{n > i}(f_n \circ \cdots \circ f_{i+1})[B_i]$$
$$= B_i,$$

so that by (6) and (8),

$$B_i = f \circ g_i^{-1}[B_i] \subseteq f[A \setminus Z].$$

We conclude that

(16) $$B \setminus Z = \bigcup_{i=1}^{\infty} B_i \subseteq f[A \setminus Z].$$

By (15) and (16), $f[A \setminus Z] = B \setminus Z$ and so we are done.

Observe that the only thing that was used in the proof of (3) is that every capset satisfies the conclusion of Corollary 5.4.2. It therefore follows that we actually proved a (potentially) stronger result. This will be used in the proof of (2).

For (2), let $A \subseteq M^Q$ be a capset, and let $B \in \mathcal{Z}_\sigma(M^Q)$. Then $A \cup B$ satisfies the conclusion of Corollary 5.4.2. So by the proof of (2), there exists a homeomorphism $h \colon M^Q \to M^Q$ such that $h[A] = A \cup B$. So (1) implies that $A \cup B$ is a capset. □

A combination of (2) and (3) of the above theorem yields the following:

**Corollary 5.4.5.** *Let $M^Q$ be a Hilbert cube, and let $A, B \in \mathcal{Z}_\sigma(M^Q)$ with $A$ a capset. Then there is a homeomorphism $h \colon M^Q \to M^Q$ with $h[A \cup B] = A$.*

**$B(Q)$ is a capset.** The results on capsets proved so far are not very interesting if there were no capsets. Fortunately, $B(Q)$ is a capset which we now aim at proving.

Let $n \in \mathbb{N}$ and define

$$\Xi_n = \prod_{i=1}^\infty [-1 + 2^{-n}, 1 - 2^{-n}]_i,$$

i.e., $\Xi_n = \{x \in Q : (\forall\, i \in \mathbb{N})(|x_i| \leq 1 - 2^{-n})\}$. Put

$$\Xi = \bigcup_{n=1}^\infty \Xi_n$$

and observe that $\Xi$ is a $\sigma$-compact subset of $s$.

**Proposition 5.4.6.** *$\Xi$ is a capset.*

**Proof.** Observe that $\Xi \in \mathcal{Z}_\sigma(Q)$ (Corollary 5.1.5). Choose $K \in \mathcal{Z}(Q)$, $n \in \mathbb{N}$ and $\varepsilon > 0$ arbitrarily. By Theorem 5.3.5 there is a homeomorphism $f \in \mathcal{H}(Q)$ such that $f[K] \subseteq s$, $\hat{\varrho}(f, 1) < \varepsilon/2$ and $f \upharpoonright \Xi_n = 1$. Choose $m \in \mathbb{N}$ so large that

$$\sum_{i=m}^\infty 2^{-i} < \varepsilon/4.$$

There obviously exists $k > n$ such that

$$\pi_i[f[K]] \subseteq [-1 + 2^{-k}, 1 - 2^{-k}]$$

for every $i < m$. For each $j \geq m$ let $h_j \colon \mathbb{J}_j \to \mathbb{J}_j$ be a homeomorphism having the following properties:

(1) $h_j\big[\min \pi_j[f[K]], \max \pi_j[f[K]]\big] \subseteq [-1 + 2^{-k}, 1 - 2^{-k}]$,
(2) $h_j \upharpoonright [-1 + 2^{-n}, 1 - 2^{-n}] = 1$

(It is a triviality that these homeomorphisms exist.) Define $h\colon Q \to Q$ by
$$h(x_1,\cdots,x_{m-1},x_m,\cdots) = (x_1,\ldots,x_{m-1},h_m(x_m),h_{m+1}(x_{m+1}),\ldots)$$
(i.e., $h$ is the product of the $h_m$'s and the identity on the first $m-1$ factors of $Q$). Then $h$ is a homeomorphism, $\hat{\varrho}(h,1) \leq \varepsilon/2 < \varepsilon$, $h \restriction \Xi_n = 1$ and $h[f[K]] \subseteq \Xi_k$. So we are done. □

We are now in a position to prove the converse of Corollary 5.4.2.

**Corollary 5.4.7.** *If an element $A \in \mathcal{Z}_\sigma(Q)$ satisfies the conclusion of Corollary 5.4.2 then it is a capset.*

**Proof.** By Proposition 5.4.6, $\Xi$ is a capset. The proof of Theorem 5.4.3(3) therefore gives us an $h \in \mathcal{H}(Q)$ with $h[A] = \Xi$. So $A$ is a capset by Theorem 5.4.3(1). □

We will now proceed to prove that $B(Q)$ is a capset.

**Corollary 5.4.8.** *Let $\mathcal{A}$ be an infinite collection of endfaces of $Q$. Then $\bigcup \mathcal{A}$ is a capset.*

**Proof.** By Corollary 5.3.9, the capset $\Xi$ can be pushed into $\bigcup \mathcal{A}$ by a homeomorphism of $Q$. Now since $\bigcup \mathcal{A} \in \mathcal{Z}_\sigma(Q)$ (Lemma 5.1.3(3)), the desired result is a direct application of Theorem 5.4.3(2). □

This easily yields the main result in this section.

**Theorem 5.4.9.** *An element $A \in \mathcal{Z}_\sigma(Q)$ is a capset if and only if there is a homeomorphism $h\colon Q \to Q$ such that $h[A] = B(Q)$.*

**Proof.** Apply Corollary 5.4.8 and Theorem 5.4.3(3). □

**A useful characterization of capsets.** We now derive derive an interesting characterization of capsets which can easily be applied in concrete situations.

Let $B = (B_i)_i$ be a *tower* of subsets of $Q$. That is,
$$B_i \subseteq B_{i+1} \subseteq Q$$
for every $i$. We say that $B$ has the *deformation property* if there is a homotopy $H\colon Q \times \mathbb{I} \to Q$ such that $H_0 = 1_Q$ while moreover for each $t \in (0,1]$ there exists an $i \in \mathbb{N}$ such that
$$H[Q \times [t,1]] \subseteq B_i.$$

A tower $(B_i)_i$ of compacta in $Q$ is *expansive* if for each $i$ there exists for some $j$ an imbedding $\varphi_i\colon B_i \times Q \to B_j$ with $\varphi_i(b,\mathbf{0}) = b$ for all $b \in B_i$.

**Theorem 5.4.10.** *Suppose a $\sigma Z$-set $B \subseteq Q$ contains an expansive tower with the deformation property. Then $B$ is a capset.*

## 5.4. THE COMPACT ABSORPTION PROPERTY

**Proof.** By Corollary 5.4.7 and the Homeomorphism Extension Theorem 5.3.7 it suffices to show that for all compacta $L \subseteq Q$ and $K \subseteq L \cap B$, and every $\varepsilon > 0$, there exists an imbedding $e \colon L \to B$ with $\hat{\varrho}(e, 1) < \varepsilon$ and $e \restriction K = 1$.

Assume for convenience that $\operatorname{diam}(Q) \leq \frac{1}{2}$ (simply replace $\varrho$ by the equivalent metric $\varrho/4$). Also assume $\varepsilon < 1$, and set $\delta = \frac{1}{3}\varepsilon$. Define the function $\lambda \colon L \to [0, \delta)$ by $\lambda(y) = \delta \cdot \varrho(y, K)$. By hypothesis, there exists an expansive tower $(B_i)_i$ in $B$ with the deformation property. Let $H \colon Q \times \mathbb{I} \to Q$ be a deformation such that for each $t > 0$, $H[Q \times [t, 1]] \subseteq B_i$ for some $i$, and assume that $\varrho(h(q, t), q) \leq t$ for all $q$ and $t$ (Exercise 5.4.4). Define a map $f \colon L \to B$ by the formula $f(y) = H(y, \lambda(y))$.

**Claim 1.** $f$ is well-defined, $\hat{\varrho}(f, 1) < \delta$ and $f \restriction K = 1$.

*Proof.* If $y \in L \setminus K$ then $\varrho(y, K) > 0$ and so $\lambda(y) > 0$, i.e., $f(y) \in B$. Moreover, if $y \in K$ then $\lambda(y) = 0$ and $f(y) = H(y, 0) = y \in B$. So the range of $f$ is contained in $B$, and $f \restriction K = 1$.

Pick an arbitrary $y \in L$. Then
$$\varrho(y, f(y)) = \varrho(y, H(y, \lambda(y))) \leq \lambda(y) = \delta \cdot \varrho(y, K) \leq \tfrac{1}{2}\delta,$$
hence $\hat{\varrho}(f, 1) \leq \tfrac{1}{2}\delta < \delta$. ◇

For each $i \geq 1$, let
$$E_i = \{y \in L : \tfrac{1}{i+2} \leq \varrho(y, K) \leq \tfrac{1}{i}\}.$$
Thus $L \setminus K = \bigcup_{i=1}^{\infty} E_i$. For every $i$ there exists a map
$$\eta_i \colon E_i \to Q$$
such that $\eta_i^{-1}(\mathbf{0}) = \operatorname{Fr} E_i$ (boundary with respect to $L$) and $\eta_i \restriction E_i \setminus \operatorname{Fr} E_i$ is one-to-one (Exercise 1.6.5). Re-indexing the tower $(B_i)_i$, we may assume for each $i$ that

(a) $\qquad f[E_i] \subseteq H[Q \times [\delta/_{i+2}, 1]] \subseteq B_i$

and that there exists an imbedding
$$\varphi_i \colon B_i \times Q \to B_{i+1}$$
with $\varphi_i(b, \mathbf{0}) = b$ for every $b \in B_i$ and

(b) $\qquad \operatorname{diam} \varphi_i[\{b\} \times Q] < \delta/_{i+2}$

for all $b$. (The last fact follows easily from the compactness of $B_i$.) Now consider the map from $E_1$ to $B_2$ defined by
$$b \mapsto \varphi_1(f(b), \eta_1(b)).$$
Notice that for $b \in E_1$ we have by (a) that $f(b) \in B_1$ and by hence that
$$\varphi_1(f(b), \eta_1(b)) \in B_2.$$
So the map is well-defined, and clearly continuous.

Since $\eta_1[\operatorname{Fr} E_1] \subseteq \{\mathbf{0}\}$, for $b \in \operatorname{Fr} E_1$ we get
$$\varphi_1\bigl(f(b), \eta_1(b)\bigr) = \varphi_1(f(b), \mathbf{0}) = f(b).$$
So the above map may be extended by $f \upharpoonright L \setminus E_1$ to a map $f_1 \colon L \to B$. Note that by (a) and the fact that $B_1 \subseteq B_2$ it follows that $f_1[E_2] \subseteq B_2$. Next, the map from $E_2$ to $B_3$ defined by
$$b \mapsto \varphi_2\bigl(f_1(b), \eta_2(b)\bigr).$$
may be extended by $f_1 \upharpoonright L \setminus E_2$ to a map $f_2 \colon L \to B$ with $f_2[E_3] \subseteq B_3$. Continuing in this fashion, we obtain a sequence of maps $f_i \colon K \to B$.

**Claim 2.** The pointwise limit $e = \lim_{i \to \infty} f_i$ exists and is continuous. It moreover has the following properties:

(1) $e \upharpoonright K = f \upharpoonright K = 1$,
(2) $\hat{\varrho}(e, 1) < \varepsilon$,
(3) $e[L \setminus K] \cap K = \emptyset$.

*Proof.* If $y \in K$ then $\varrho(y, K) = 0$ and $f(y) = y$. From the construction it therefore follows that the sequence $\bigl(f_i(y)\bigr)_i$ is the constant sequence with value $f(y) = y$. This shows that $e$ is well-defined at points of $K$.

Fix $y \in L \setminus K$, with
$$1/_{i+1} \leq \varrho(y, L) < 1/_i.$$
Then
$$f_i(y) = f_{i+1}(y) = \cdots = e(y).$$
This shows that for this point $e$ is also well-defined.

Moreover, observe that
$$\varphi_i(f_{i-1}(y), \mathbf{0}) = f_{i-1}(y)$$
and so
$$f_{i-1}(y) \in \varphi_i[\{f_{i-1}(y)\} \times Q].$$
By the definition of $f_i$ it therefore follows from (b) that
$$\varrho\bigl(f_i(y), f_{i-1}(y)\bigr) < \delta/_{i+2}.$$
Now consider the points $f_{i-1}(y)$ and $f(y)$. By construction we get
$$f(y) = f_{i-2}(y)$$
and so by a similar argumentation as the one above,
$$\varrho\bigl(f_{i-1}(y), f(y)\bigr) < \delta/_{i+1}$$
and so
$$\varrho(e(y), y) \leq \varrho\bigl(f_i(y), f_{i-1}(y)\bigr) + \varrho\bigl(f_{i-1}(y), f(y)\bigr) + \varrho(f(y), y)$$
$$< \delta/_{i+2} + \delta/_{i+1} + \delta \cdot \varrho(y, K).$$

Since $\delta = \frac{1}{3}\varepsilon$ and $\varrho(y, K) < 1$, we have $\varrho(e(y), y) < \varepsilon$. By compactness of $L$ this will imply that $\hat{\varrho}(e, 1) < \varepsilon$ provided we will be able to prove that $e$ is continuous (Exercise A.5.4). Observe that

$$1/_{i+1} \leq \varrho(y, K), \quad \delta < 1/_3\varepsilon, \quad \varepsilon < 1,$$

so that

(c) $\quad \varrho(e(y), y) < \delta/_{i+2} + \delta/_{i+1} + \delta \cdot \varrho(y, K) \leq 3\delta \cdot \varrho(y, K) < \varrho(y, K).$

Continuity of $e$ at points of $L \setminus K$ is clear since each point of $L \setminus K$ has a neighborhood on which $f$ agrees with some $f_i$, which is continuous. So it suffices to prove continuity at the points of $K$. A moments reflection shows that it is enough to show that if a sequence of points $(y_n)_n$ in $L \setminus K$ converges to a point $y \in K$ then $e(y_n) \to e(y)$. From (c) we get $e(y_n) \to y$. Since $e \upharpoonright K = 1$ we have $e(y) = y$ and so we are done.

Observe that (c) also implies that $e[L \setminus K] \cap K = \emptyset$. ◇

We show that $e$ is an imbedding by showing that $e \upharpoonright L \setminus K$ is one-to-one. This clearly suffices since by Claim 2, $e[L \setminus K] \cap K = \emptyset$. Consider distinct points $y, y' \in L \setminus K$. Suppose first that $\varrho(y, K)$ and $\varrho(y', K)$ lie in the same subinterval $[1/_{i+1}, 1/_i)$. Then $e(y) = f_i(y) = \varphi_i(f_{i-1}(y), \eta_i(y))$, and likewise for $e(y')$. Since $\eta_i(y) \neq \eta_i(y')$ it follows that $e(y) \neq e(y')$. On the other hand, if $\varrho(y, K) \in [1/_{i+1}, 1/_i)$ and $\varrho(y', L) \in [1/_{j+1}, 1/_j)$, with $i < j$, then

$$e(y) = f_i(y) \in B_{i+1} \subseteq B_j,$$

and

$$e(y') = f_j(y') \in \varphi_j[B_j \times \{\eta_j(y')\}].$$

Since $\eta_j(y') \neq \mathbf{0}$, $e(y') \notin B_j$. Thus $e(y) \neq e(y')$. □

**Corollary 5.4.11.** *An element $B \in \mathcal{Z}_\sigma(Q)$ is a capset if any only if $B$ can be written as the union of a tower $(B_i)_i$ of $Z$-sets in $Q$ such that:*

(1) *each $B_i$ is homeomorphic to $Q$,*
(2) *each $B_i$ is a $Z$-set in $B_{i+1}$,*
(3) *$(B_i)_i$ has the deformation property.*

**Proof.** First consider the tower $\Xi = (\Xi_n)_{n=1}^\infty$ defined above. It is clear that $\Xi_i \approx Q$ and $\Xi_i$ is a $Z$-set in $\Xi_{i+1}$ for every $i$. For the latter statement, observe that $\Xi_i$ projects onto a proper subset of the projection of $\Xi_{i+1}$ in every coordinate direction (Lemma 5.1.3(2)). In addition, the function

$$H \colon Q \times \mathbb{I} \to Q$$

defined by

$$H(q, t) = (1 - t) \cdot q$$

is a deformation of $Q$ through $\Xi$. If $t \in (0, 1]$ then for $n$ such that $2^{-n} < t$ we have $H\bigl[[Q \times [t, 1]]\bigr] \subseteq \Xi_n$. So the tower $(\Xi_n)_n$ has the deformation property.

Since capsets are topologically unique by Theorem 5.4.3, every capset can be written in the above form.

Conversely, let $(B_i)_i$ be a tower such as the one above. By the $Z$-set Homeomorphism Extension Theorem 5.3.7, the pair $(B_i, B_{i+1})$ is homeomorphic to the pair $(B_i \times \{0\}, B_i \times Q)$. Thus $(B_i)_i$ is an expansive tower. So the desired result follows from Theorem 5.4.10. □

**Some other properties of capsets.** We already know that the union of a capset and a countable union of $Z$-sets is again a capset. We now present a property of capsets in the same spirit.

**Theorem 5.4.12.** *Let $M^Q$ be a Hilbert cube, let $A \subseteq M^Q$ be a capset and let $B \subseteq M^Q$ be a $Z$-set. Then $A \setminus B$ is a capset.*

**Proof.** We may assume that $M^Q = Q$ and that $B \subseteq B(Q)$ (Corollary 5.3.9). So $\Xi \cap B = \emptyset$. Now apply Theorem 5.4.3(3) to get a homeomorphism
$$\alpha \colon Q \to Q$$
such that $\alpha \restriction B$ is the identity, while moreover
$$\alpha[\Xi] = \alpha[\Xi \setminus B] = A \setminus B.$$
We conclude that $A \setminus B$ is a capset by Theorem 5.4.3(1). □

The following corollary, which at first glance seems rather technical, will turn out to be the basis for our considerations in the next section.

**Corollary 5.4.13.** *Let $X$ be compact with closed subspace $K$. In addition, let $A \subseteq X$ be $\sigma$-compact. Then for every map $f \colon X \to Q$ that restricts to a $Z$-imbedding on $K$, there exists for every $\varepsilon > 0$ a $Z$-imbedding $g \colon X \to Q$ such that*

(1) $\hat{\varrho}(g, f) < \varepsilon$,
(2) $g \restriction K = f \restriction K$,
(3) $g^{-1}[B(Q)] \setminus K = A \setminus K$.

**Remark 5.4.14.** This result should be interpreted as follows. The (possibly empty) set $K$ is already in place. So the function $g$ should not interfere with this. The imbedding $g$ pushes $A \setminus K$ in place.

**Proof.** By Theorem 5.3.11 we may assume without loss of generality that $f$ is a $Z$-imbedding. Observe that $f[A] \in \mathcal{Z}_\sigma(Q)$ and that $f[X] \in \mathcal{Z}(Q)$ so that by Theorems 5.4.3(3) and 5.4.12 there exists a homeomorphism $\alpha \colon Q \to Q$ such that $\alpha$ is arbitrarily close to the identity, $\alpha \restriction f[K]$ is the identity and
$$\alpha\bigl[B(Q) \setminus f[K]\bigr] = \bigl((B(Q) \setminus f[X]) \cup f[A]\bigr) \setminus f[K].$$
Now put $g = \alpha^{-1} \circ f$. Observe that
$$g \restriction K = f \restriction K.$$

Also, $f$ and $g$ are arbitrarily close, and

$$x \in g^{-1}[B(Q)] \setminus K \Leftrightarrow g(x) \in B(Q) \setminus g[K]$$
$$\Leftrightarrow f(x) \in \alpha[B(Q)] \setminus \alpha g[K]$$
$$\Leftrightarrow f(x) \in \alpha[B(Q)] \setminus \alpha f[K]$$
$$\Leftrightarrow f(x) \in \alpha[B(Q) \setminus f[K]]$$
$$\Leftrightarrow f(x) \in ((B(Q) \setminus f[X]) \cup f[A]) \setminus f[K]$$
$$\Leftrightarrow f(x) \in f[A] \setminus f[K]$$
$$\Leftrightarrow x \in A \setminus K,$$

which is as required. □

**Application 1.** We shall first present some other simple corollaries to the results obtained so far.

**Corollary 5.4.15.** *Let $A$ be a $\sigma$-compact subset of $s$. Then $s \setminus A$ and $s$ are homeomorphic.*

**Remark 5.4.16.** For a generalization of this result, see Exercise 5.4.6.

**Proof.** By Corollary 5.4.8, $B(Q)$ is a capset. In addition, $A \in \mathcal{Z}_\sigma(Q)$ by Corollary 5.1.5. Consequently,

$$B(Q) \cup A$$

is a capset by Theorem 5.4.3(2). Theorem 5.4.9 therefore implies the existence of an $h \in \mathcal{H}(Q)$ such that

$$h[B(Q)] = B(Q) \cup A,$$

i.e., $h[s] = s \setminus A$. □

**Corollary 5.4.17.** *For every $n$, let $A_n$ be a nondegenerate interval in $\mathbb{R}$. Then*

$$\prod_{n=1}^{\infty} A_n \approx \mathbb{R}^\infty$$

*iff infinitely many of the $A_n$ are not compact.*

**Proof.** If only finitely many of the $A_n$ are not compact then the product of the $A_n$'s is locally compact and hence is certainly not homeomorphic to $\mathbb{R}^\infty$ (Exercise 1.5.16). So assume that infinitely many of the $A_n$'s are not compact. Observe that every $A_n$ is homeomorphic to $(-1,1)$, to $(-1,1]$, or to $[-1,1]$. Also, infinitely often, $A_n$ is not homeomorphic to $[-1,1]$. Consequently, the product of the $A_n$'s is homeomorphic to the complement in $Q$ of a set consisting of infinitely many endfaces. Now apply Corollary 5.4.8 and Theorem 5.4.9. □

**Application 2.** Here we will apply our results to the hyperspace of the unit interval $\mathbb{I}$. The result presented here shall be used in the next section.

If $X$ is a compact space then for $k \in \{0, 1, 2, \ldots, \infty\}$ we let $\dim_{\geq k}(X)$ denote the subspace of $2^X$ consisting of all $\geq k$-dimensional elements. We define $\dim_k(X)$ and $\dim_{\leq k}(X)$ in the same way.

Let $X$ be a compact space. For $k, n \in \mathbb{N}$ let $\mathcal{G}_{n,k}(X)$ be the collection of all nonempty closed subsets $A$ of $X$ having the property that there is a collection open subsets $\mathcal{U}$ of $X$ with $\mathrm{mesh}(\mathcal{U}) \leq 1/n$ and order $\leq k$ such that $A \subseteq \bigcup \mathcal{U}$. Obviously, $\mathcal{G}_{n,k}(X)$ is an open subset of $2^X$ (Exercise 5.4.9).

**Lemma 5.4.18.** *Let $X$ be a compact space. Then*

$$\bigcap_{n=1}^{\infty} \mathcal{G}_{n,k}(X) = \dim_{<k}(X).$$

**Proof.** Let $A \in \dim_{<k}(X)$. Then for every $\varepsilon > 0$ there exists a cover $\mathcal{U}$ of $A$ consisting of relatively open subsets of $A$ such that

$$\mathrm{mesh}(\mathcal{U}) < \varepsilon, \quad \mathrm{ord}(\mathcal{U}) \leq k$$

(Theorem 3.2.5). Let $\kappa \colon \tau A \to \tau X$ be the function in Lemma 4.2.2. For a cover $\mathcal{U}$ as the one above, the collection

$$\{\kappa(U) : U \in \mathcal{U}\}$$

has order at most $k$ by (4) in Lemma 4.2.2. For a sufficiently small open neighborhood $V$ of $A$ we have that $\mathrm{diam}(\kappa(U) \cap V) < \varepsilon$ for every $U \in \mathcal{U}$. The collection

$$\{\kappa(U) \cap V : U \in \mathcal{U}\}$$

therefore shows that $A \in \mathcal{G}_{n,k}(X)$ for every $n$ such that $\varepsilon < 1/n$.

Conversely, if $A \in \bigcap_{n=1}^{\infty} \mathcal{G}_{n,k}(X)$ then since for every collection of subsets $\mathcal{U}$ of $X$ we have that

$$\mathrm{ord}(\mathcal{U} \restriction A) \leq \mathrm{ord}(\mathcal{U}),$$

Theorem 3.2.5 shows that $\dim A < k$. $\square$

**Corollary 5.4.19.** *Let $X$ be a Peano continuum. Then $\dim_{\geq 1}(X)$ is a $\sigma Z$-set in $2^X$.*

**Proof.** By Exercise 5.4.9 and Lemma 5.4.18,

$$\dim_{\geq k}(X) = 2^X \setminus \bigcap_{n=1}^{\infty} \mathcal{G}_{n,k}(X)$$

is an $F_\sigma$-subset of $2^X$. In addition, there is a deformation $H \colon 2^X \times \mathbb{I} \to 2^X$ through $\mathcal{F}_\infty(X)$ by Corollary 4.2.24. Since every element of $\mathcal{F}_\infty(X)$ is zero-dimensional, it follows that $\dim_{\geq 1}(X)$ is a $\sigma Z$-set in $2^X$. $\square$

## 5.4. THE COMPACT ABSORPTION PROPERTY 341

By the Curtis-Schori-West Hyperspace Theorem 4.2.27 we know that $2^\mathbb{I}$ and $Q$ are homeomorphic. So it is natural to ask for 'natural' capset subspaces of $2^\mathbb{I}$. Here is one.

**Theorem 5.4.20.** $\dim_1(\mathbb{I})$ *is a capset in* $2^\mathbb{I}$.

**Proof.** For convenience, let $\mathbb{I}^*$ denote the subset $\{\{t\} : t \in \mathbb{I}\}$ of $2^\mathbb{I}$. It is a $Z$-set in $2^\mathbb{I}$ by Lemma 4.2.26. We will show that $\dim_1(\mathbb{I}) \cup \mathbb{I}^*$ is a capset of $2^\mathbb{I}$. But then
$$\dim_1(\mathbb{I}) = \big(\dim_1(\mathbb{I}) \cup \mathbb{I}^*\big) \setminus \mathbb{I}^*$$
is a capset as well by Theorem 5.4.12.

For $K \subseteq \mathbb{I}$, let $a_K = \inf(K)$ and $b_K = \sup(K)$, respectively. Notice that
$$a_K + (1-t) \cdot (b_K - a_K) \leq b_K$$
for all $K \in 2^\mathbb{I}$ and $t \in \mathbb{I}$. For $t \in \mathbb{I}$ define
$$\mathcal{M}_t = \{K \in 2^\mathbb{I} : [a_K + (1-t) \cdot (b_K - a_K), b_K] \subseteq K\}.$$

**Claim 1.** For $t > 0$, $\mathcal{M}_t$ is a $Z$-set in $2^\mathbb{I}$.

*Proof.* We first claim that $\mathcal{M}_t$ is closed in $2^\mathbb{I}$. Let $(K_n)_n$ be a sequence in $\mathcal{M}_t$ converging to an element $K \in 2^\mathbb{I}$. The points $a_S$ and $b_S$ depend continuously on $S \in 2^\mathbb{I}$ by Exercise 1.11.18. This implies that $a_{K_n} \to a_K$ and $b_{K_n} \to b_K$ and hence
$$[a_{K_n} + (1-t) \cdot (b_{K_n} - a_{K_n}), b_{K_n}] \to [a_K + (1-t) \cdot (b_K - a_K), b_K]$$
in $2^\mathbb{I}$. So
$$[a_K + (1-t) \cdot (b_K - a_K), b_K] \subseteq K$$
by Exercise 1.11.8 and hence $K \in \mathcal{M}_t$.

Every element of $\mathcal{M}_t$ is either a singleton or is one-dimensional. Since $\mathbb{I}^*$ is a $Z$-set, and $\dim_1(\mathbb{I})$ is a $\sigma Z$-set by Corollary 5.4.19, the fact that $\mathcal{M}_t$ is closed implies that it is a $Z$-set (Lemma 5.1.2(3)). ◇

For $t \in \mathbb{I}$ and $K \in 2^\mathbb{I}$ put
$$K_t = \{a_K + (1-t) \cdot (k - a_K) : k \in K\}.$$
Observe that $K_t$ is the image of $K$ under a linear map which sends $a_K$ onto $a_K$ and $b_K$ onto $a_K + (1-t) \cdot (b_K - a_K)$.

Define $H \colon 2^\mathbb{I} \times \mathbb{I} \to 2^\mathbb{I}$ by
$$H(K, t) = K_t \cup [a_K + (1-t) \cdot (b_K - a_K), b_K].$$

**Claim 2.** $H$ is continuous, $H_0$ is the identity, and $H_t$ maps $2^\mathbb{I}$ homeomorphically onto $\mathcal{M}_t$ for every $t > 0$. In addition, $\hat{\varrho}_H(H_t, 1) \leq t$ for $t \in \mathbb{I}$.

*Proof.* That $H$ is continuous follows easily from the fact that $a_K$ and $b_K$ depend continuously on $K$ (Exercise 1.11.18). That $H_0$ is the identity is trivial.

Fix $t \in \mathbb{I}$ and $K \in 2^{\mathbb{I}}$. Observe that
$$\min H_t(K) = a_K, \quad \max H_t(K) = b_K.$$
This implies that $H_t(K) \in \mathcal{M}_t$. We next claim that $H_t$ is one-to-one. To this end, let $K, K' \in 2^{\mathbb{I}}$. If $a_K < a_{K'}$ then obviously $H_t(K) \neq H_t(K')$. So we may assume without loss of generality that $a_K = a_{K'}$ and, similarly, that $b_K = b_{K'}$. The linear maps that send $K$ onto $K_t$ and $K'$ onto $K'_t$ are consequently identical. Since $K_t$ and $K'_t$ are both subsets of
$$[0, a_K + (1-t) \cdot (b_K - a_K)]$$
we get $H_t(K) \neq H_t(K')$.

To show that $H_t$ maps $2^{\mathbb{I}}$ onto $\mathcal{M}_t$, take an arbitrary $K \in \mathcal{M}_t$. Let
$$L = [a_k, a_K + (1-t) \cdot (b_K - a_K)] \cap K$$
and let $S$ be the image of $L$ under the linear map that sends $a_K$ onto $a_K$ and the point $a_K + (1-t) \cdot (b_K - a_K)$ onto $b_K$. Then clearly, $H_t(S) = K$.

That the distance between $H_t$ and the identity is less than or equal $t$ follows easily since
$$b_k - \big(a_K + (1-t) \cdot (b_K - a_K)\big) = t(b_k - a_k) \leq t.$$
So we are done. $\diamondsuit$

Because for $0 \leq s < t \leq 1$ it follows that
$$H_s^{-1}[\mathcal{M}_t] = \mathcal{M}_{(t-s)/(1-s)}$$
is a $Z$-set in $2^{\mathbb{I}}$, this shows that $\mathcal{M}_t$ is a $Z$-set in $\mathcal{M}_s$. So now the desired result follows by Corollary 5.4.11 by using the tower $(\mathcal{M}_{1/i})_i$. $\square$

**Corollary 5.4.21.** *The spaces*
$$\{A \in 2^{\mathbb{I}} : \dim A = 0\}$$
*and $s$ are homeomorphic.*

**Exercises for §5.4.** For each $i \in \mathbb{N}$ let $-1 < a_i < b_i < 1$. The set
$$\{x \in Q : x_i \in [a_i, b_i] \text{ for all but finitely many } i \in \mathbb{N}\}$$
is called the *basic core set* structured on the core $\prod_{i=1}^{\infty} [a_i, b_i]$.

▶1. Prove that $B(Q) \times Q$, $B(Q) \times B(Q)$ and $(Q \times Q) \setminus (s \times s)$ are capsets for $Q \times Q$.

2. Let $A \in \mathcal{Z}_\sigma(Q)$ be a capset and let $B \in \mathcal{Z}(Q)$. Prove that $A \setminus B$ is a capset by following the hint. [*Hint:* Let $A = \bigcup_{n=1}^\infty A_n$, where the $A_n$'s are as in the definition of a capset. For every $n$, put $B_n = \{x \in A_n : \varrho(x, B) \geq 1/n\}$. Prove that the $B_n$'s witness that $A \setminus B$ is a capset.]

3. Prove that each basic core set is a capset in $Q$.

4. Suppose $B = (B_i)_i$ is a tower of subsets of $Q$ with the deformation property. Prove that there is a homotopy $F \colon Q \times \mathbb{I} \to Q$ such that
   (1) $F_0 = 1$ and for every $t \in (0, 1]$ there exists $i \in \mathbb{N}$ such that
   $$F[Q \times [t, 1]] \subseteq B_i,$$
   (2) for all $q \in Q$ and $t \in \mathbb{I}$ we have
   $$\varrho(x, F(x, t)) \leq t.$$

5. Let $A$ be a (relatively) closed subset of $s$ and let $B$ be its closure in $Q$. Prove that $A \in \mathcal{Z}(s)$ iff $B \in \mathcal{Z}(Q)$.

6. Let $A \in \mathcal{Z}_\sigma(s)$. Prove that $\mathbb{R}^\infty \setminus A \approx s$.

7. Let $A \subseteq Q$ be closed and let $f \colon Q \to Q$ be continuous such that $f \upharpoonright A$ is a $Z$-imbedding such that $f^{-1}[s] \cap A = s \cap A$. Prove that $f$ can be approximated arbitrarily closely by a function $g \colon Q \to Q$ having the following properties:
   (1) $g$ is a $Z$-imbedding,
   (2) $g \upharpoonright A = f \upharpoonright A$,
   (3) $g^{-1}[s] = s$.

8. Prove that $C = \{x \in Q : (\exists k \in \mathbb{N})(\forall j \in \mathbb{N})(|x_j| \leq 2^{k-j})\}$ is a capset.

9. Let $X$ be a compact space. Prove that for all $n$ and $k$ the set $\mathcal{G}_{n,k}(X)$ is open in $2^X$.

▶10. Let $f \colon Q \to Q$ be continuous, let $A \subseteq Q$ be $\sigma$-compact and let $K \subseteq Q$ be closed. In addition, let $\mu \colon Q \to [0, \infty)$ be a continuous function such that $\mu^{-1}(0) = K$. Prove that there is a continuous function $g \colon Q \to Q$ such that
   (1) $g \upharpoonright K = f \upharpoonright K$,
   (2) $g \upharpoonright Q \setminus K$ is one-to-one and $g[Q \setminus K] \in \mathcal{Z}_\sigma(Q)$,
   (3) if $x \in Q \setminus K$ then $\varrho(g(x), f(x)) < \mu(x)$,
   (4) $g^{-1}[B(Q)] \setminus K = A \setminus K$.

11. Let $M$ and $N$ be Hilbert cubes containing the capsets $X$ and $Y$, respectively. Prove that if $x \in X$ and $y \in Y$ are arbitrary then there is a homeomorphism $\varphi \colon M \to N$ such that
   (1) $\varphi[X] = Y$,
   (2) $\varphi(x) = y$.

## 5.5. Absorbing systems

In this section we generalize the ideas in the previous section for obtaining more complex structures with 'absorption' properties. As applications we shall prove that the space of all infinite-dimensional compacta and the space $c_0$ endowed with the topology of coordinatewise convergence are homeomorphic to the countably infinite product of copies of $B(Q)$.

Notice that in Theorem 5.4.9 we characterized up to homeomorphism how the pseudo-boundary $B(Q)$ of $Q$ is placed in $Q$. We are interested in the question how the countably infinite product $B(Q)^\infty$ is placed in the Hilbert cube $Q^\infty$. Notice that $B(Q)$ is $\sigma$-compact, but not compact. So $B(Q)^\infty$ is not $\sigma$-compact (Exercise A.5.19) and hence $B(Q)$ and $B(Q)^\infty$ are not homeomorphic.

That this is not so surprising also follows from the following reasoning. Observe that $B(Q)$ is the union of countably many endfaces which are easily seen to be nowhere dense in $B(Q)$. Hence $B(Q)$ does not satisfy the Baire Category Theorem and by Theorem 1.9.12, it consequently contains a closed copy of $\mathbb{Q}$. Hence $B(Q)^\infty$ contains a closed copy of $\mathbb{Q}^\infty$. So by Corollary A.13.4, $B(Q)^\infty$ is an absolute $F_{\sigma\delta}$ but not an absolute $G_{\delta\sigma}$ (Theorem A.13.5).

The conclusion is that we are dealing with something more complex than $B(Q)$ here.

Let $\mathcal{M}$ be a class of spaces. We say that $\mathcal{M}$ is *topological* if for every space $B$ for which there exists an element $A \in \mathcal{M}$ such that $B \approx A$ it follows that $B \in \mathcal{M}$. Similarly, $\mathcal{M}$ is *closed hereditary* if for every $A \in \mathcal{M}$ and closed subset $B \subseteq A$ we have $B \in \mathcal{M}$.

Let $\Gamma$ be an ordered set and let $\mathcal{M}_\gamma$ be a collection of spaces for each $\gamma \in \Gamma$. Each $\mathcal{M}_\gamma$ is assumed to be topological and closed hereditary. Let $\mathcal{M}_\Gamma$ stand for the whole system $(\mathcal{M}_\gamma)_{\gamma\in\Gamma}$. Let $\mathcal{X} = (X_\gamma)_{\gamma\in\Gamma}$ be an *order preserving indexed collection* of subsets of a topological copy $M^Q$ of $Q$. By this we mean that $X_\gamma \subseteq X_{\gamma'}$ if and only if $\gamma \leq \gamma'$.

The system $\mathcal{X}$ is called $\mathcal{M}_\Gamma$-*universal* if for every order preserving system $(A_\gamma)_\gamma$ in $Q$ such that $A_\gamma \in \mathcal{M}_\gamma$ for every $\gamma \in \Gamma$, there is an imbedding $f\colon Q \to M^Q$ with $f^{-1}[X_\gamma] = A_\gamma$ for every $\gamma \in \Gamma$.

So a system is universal if an arbitrary system of appropriate subsets of $Q$ can be pushed in place by means of an imbedding of $Q$ in $M^Q$.

The system $\mathcal{X}$ is called *strongly* $\mathcal{M}_\Gamma$-*universal* if for every order preserving system $(A_\gamma)_\gamma$ in $Q$ such that $A_\gamma \in \mathcal{M}_\gamma$ for every $\gamma \in \Gamma$, and for every map $f\colon Q \to M^Q$ that restricts to a Z-imbedding on some compact set $K$, there exists a Z-imbedding $g\colon Q \to M^Q$ that can be chosen arbitrarily close to $f$ with the properties: $g \restriction K = f \restriction K$ and $g^{-1}[X_\gamma] \setminus K = A_\gamma \setminus K$ for every $\gamma \in \Gamma$.

So roughly speaking, a system is strongly $\mathcal{M}_\Gamma$-universal if an arbitrary system of appropriate subsets of $Q$ can be pushed in place off a Z-set by means of imbeddings approximating a given continuous function.

The system $\mathfrak{X}$ is called *reflexively universal*, if for every continuous function $f: M^Q \to M^Q$ that restricts to a $Z$-imbedding on some compact set $K$, there exists a $Z$-imbedding $g: M^Q \to M^Q$ that can be chosen arbitrarily close to $f$ with the properties: $g \upharpoonright K = f \upharpoonright K$ and $g^{-1}[X_\gamma] \setminus K = X_\gamma \setminus K$ for every $\gamma \in \Gamma$.

So roughly speaking, a system is reflexively universal if it can be pushed in place off a $Z$-set by means of imbeddings approximating a given continuous function.

**Lemma 5.5.1.** *Let $\Gamma$ be an ordered set and for every $\gamma \in \Gamma$ let $\mathfrak{M}_\gamma$ be a topological and closed hereditary class of spaces. Let $\mathfrak{M}_\Gamma$ stand for $(\mathfrak{M}_\gamma)_{\gamma \in \Gamma}$. In addition, let $M^Q$ be a topological copy of $Q$ and let $\mathfrak{X} = (X_\gamma)_{\gamma \in \Gamma}$ be an order preserving indexed collection of subsets of $M^Q$. Then*

(1) *If $\mathfrak{X}$ is both $\mathfrak{M}_\Gamma$-universal and reflexively universal then it is strongly $\mathfrak{M}_\Gamma$-universal.*
(2) *If $\mathfrak{X}$ is strongly $\mathfrak{M}_\Gamma$-universal and $X_\gamma \in \mathfrak{M}_\gamma$ for every $\gamma \in \Gamma$ then $\mathfrak{X}$ is both $\mathfrak{M}_\Gamma$-universal and reflexively universal.*

**Proof.** For (2) use that each $\mathfrak{M}_\gamma$ is topological. For (1), let $(A_\gamma)_\gamma$ be an order preserving system of subsets of $Q$ such that $A_\gamma \in \mathfrak{M}_\gamma$ for every $\gamma \in \Gamma$, and let $f: Q \to M^Q$ be a map that restricts to a $Z$-imbedding on some compact set $K \subseteq Q$. Since $\mathfrak{X}$ is $\mathfrak{M}_\Gamma$-universal there exists an imbedding $\alpha: Q \to M^Q$ such that $\alpha^{-1}[X_\gamma] = A_\gamma$ for every $\gamma \in \Gamma$.

Let $\varepsilon > 0$.

Since $Q$ is an AR (Corollary 1.2.12), we can extend the function

$$\alpha^{-1}: \alpha[Q] \to Q$$

to a continuous function $\tau: M^Q \to Q$. Then obviously $\tau \circ \alpha(q) = q$ for every $q \in Q$. Consider the composition

$$\xi = f \circ \tau: M^Q \to M^Q.$$

Observe that $\xi \upharpoonright \alpha[K]$ is a $Z$-imbedding from $\alpha[K]$ onto $f[K]$. Since $\mathfrak{X}$ is reflexively universal, there consequently is a $Z$-imbedding $g: M^Q \to M^Q$ such that

(3) $\hat{\varrho}(g, f \circ \tau) < \varepsilon$,
(4) $\forall \gamma \in \Gamma : g^{-1}[X_\gamma] \setminus \alpha[K] = X_\gamma \setminus \alpha[K]$,
(5) $g \upharpoonright \alpha[K] = f \circ \tau \upharpoonright \alpha[K]$.

Now put $h = g \circ \alpha: Q \to M^Q$. Observe that by Exercise 1.3.2,

$$\hat{\varrho}(h, f) = \hat{\varrho}(g \circ \alpha, f \circ \tau \circ \alpha) \leq \hat{\varrho}(g, f \circ \tau) < \varepsilon.$$

It is clear that $h$ is a $Z$-imbedding since $h[Q] \subseteq g[M^Q]$, and that $h\restriction K = f\restriction K$. Finally, for every $\gamma \in \Gamma$ we have

$$\begin{aligned} h^{-1}[X_\gamma] \setminus K &= \alpha^{-1}\big[g^{-1}[X_\gamma]\big] \setminus \alpha^{-1}[\alpha[K]] \\ &= \alpha^{-1}\big[g^{-1}[X_\gamma] \setminus \alpha[K]\big] \\ &= \alpha^{-1}\big[X_\gamma \setminus \alpha[K]\big] \\ &= A_\gamma \setminus K. \end{aligned}$$

So we are done. □

The system $\mathcal{X}$ is called $\mathcal{M}_\Gamma$-*absorbing* if

(1) $X_\gamma \in \mathcal{M}_\gamma$ for every $\gamma \in \Gamma$,
(2) $\bigcup_{\gamma \in \Gamma} X_\gamma$ is contained in a $\sigma Z$-set of $M^Q$,
(3) $\mathcal{X}$ is strongly $\mathcal{M}_\Gamma$-universal.

As to be expected, cf. Theorem 5.4.3(3), there is a uniqueness theorem for absorbing systems.

**Theorem 5.5.2.** *Let $\Gamma$ be an ordered set and for every $\gamma \in \Gamma$ let $\mathcal{M}_\gamma$ be a topological and closed hereditary class of spaces. Let $\mathcal{M}_\Gamma$ stand for $(\mathcal{M}_\gamma)_{\gamma \in \Gamma}$. Finally, let $M^Q$ and $M^{Q'}$ be topological copies of $Q$. If*

$$\mathcal{X} = (X_\gamma)_{\gamma \in \Gamma}, \quad \mathcal{Y} = (Y_\gamma)_{\gamma \in \Gamma}$$

*are $\mathcal{M}_\Gamma$-absorbing systems in $M^Q$ respectively $M^{Q'}$ then there is a homeomorphism $h \colon M^Q \to M^{Q'}$ such that $h[X_\gamma] = Y_\gamma$ for all $\gamma \in \Gamma$. If moreover $M^Q = M^{Q'}$ then $h$ can be found arbitrarily close to the identity.*

**Proof.** We use the same back and forth argument as in the proof of Theorem 5.4.3.

Obviously, we may assume that $M^Q = M^{Q'} = Q$. Let $\bigcup_\gamma X_\gamma \subseteq \bigcup_{i=0}^\infty A_i$ and let $\bigcup_\gamma Y_\gamma \subseteq \bigcup_{i=0}^\infty B_i$, where

$$\emptyset = A_0 \subseteq A_1 \subseteq A_2 \subseteq \cdots$$

and

$$\emptyset = B_0 \subseteq B_1 \subseteq B_2 \subseteq \cdots$$

are sequences of $Z$-sets in $Q$.

Let $\varepsilon > 0$. By induction on $i$ we shall construct a sequence of elements $f_i \in \mathcal{H}(Q)$ such that if $g_i = f_i \circ \cdots \circ f_0$ then

(1) $f_0 = 1$,
(2) for $i \geq 1$, $\hat{\varrho}(f_i, 1)$ is small enough in order to apply Theorem 1.6.2 later,
(3) for $i \geq 1$, $\hat{\varrho}(f_i, 1) < 3^{-i} \cdot \varepsilon$,
(4) $A_i \cap X_\gamma = A_i \cap g_i^{-1}[Y_\gamma]$ for all $i \geq 0$ and $\gamma \in \Gamma$,

(5) $B_i \cap g_i[X_\gamma] = B_i \cap Y_\gamma$ for all $i \geq 0$ and $\gamma \in \Gamma$,
(6) for $i \geq 1$, $f_i \restriction (g_{i-1}[A_{i-1}] \cup B_{i-1}) = 1$,

where 1 denotes as usual the identity function.

Assume that $f_i$ has been constructed. Since $X_\gamma \in \mathcal{M}_\gamma$ and $\mathcal{M}_\gamma$ is topological and closed hereditary we have

$$g_i[X_\gamma] \cap (g_i[A_{i+1}] \cup B_i) \in \mathcal{M}_\gamma$$

for every $\gamma \in \Gamma$. Put $K = g_i[A_i] \cup B_i$ and observe that $g_i[X_\gamma] \cap K = Y_\gamma \cap K$. Since $\mathcal{Y}$ is strongly universal we can find an arbitrarily close to the identity $Z$-imbedding $\alpha \colon g_i[A_{i+1}] \cup B_i \to Q$ that fixes $K$ pointwise and that has the additional property that for every $\gamma \in \Gamma$,

$$\alpha^{-1}[Y_\gamma] \cap g_i[A_{i+1}] = g_i[X_\gamma \cap A_{i+1}].$$

Let $\tilde{\alpha}$ be an extension of $\alpha$ to a homeomorphism of $Q$ that is arbitrarily close to $1_Q$ (Theorem 5.3.7). Since $\tilde{\alpha} \circ g_i[\mathcal{X}]$ is just as $\mathcal{X}$ strongly universal, we can find a $Z$-imbedding $\beta \colon \tilde{\alpha} \circ g_i[A_{i+1}] \cup B_{i+1} \to Q$ that fixes $K' = \tilde{\alpha} \circ g_i[A_{i+1}] \cup B_i$ pointwise and that has the property that for every $\gamma \in \Gamma$,

$$\beta^{-1}[\tilde{\alpha} \circ g_i[X_\gamma]] \cap B_{i+1} = Y_\gamma \cap B_{i+1}.$$

Let $\tilde{\beta}$ be an extension of $\beta$ to a homeomorphism of $Q$. Just as in the case of $\tilde{\alpha}$, we may assume that $\tilde{\beta}$ is arbitrarily close to the identity. If we put

$$f_{i+1} = \tilde{\beta}^{-1} \circ \tilde{\alpha}$$

then one can easily verify the induction hypothesis for $i+1$.

So $h = \lim_{i \to \infty} g_i$ is a homeomorphism of $Q$ such that $\hat{\varrho}(h, 1) < \varepsilon$ (this follows from (3)). It is easily verified, cf. the proof of Theorem 5.4.3, that $h$ maps each $X_\gamma$ onto $Y_\gamma$. □

If the $\mathcal{M}_\gamma$ for $\gamma \in \Gamma$ are all equal to a fixed class $\mathcal{M}$ then we use the term $\mathcal{M}$-*absorbing system*. For such systems there are two important special cases. First, $\Gamma = \mathbb{N}$ with an inverted ordering. This means that $\mathcal{X}$ is a decreasing sequence of subsets of $M^Q$. In this case we use the term $\mathcal{M}$-*absorbing sequence*. The second special case is the case where $\Gamma$ is a singleton. We then use the term $\mathcal{M}$-*absorber*.

**An $\mathcal{F}_\sigma$-absorber.** Let $\mathcal{F}_\sigma$ denote the collection of $\sigma$-compact spaces. It follows from Corollary 5.4.13 that every capset in $Q$ is an $\mathcal{F}_\sigma$-absorber. Since by Theorem 5.5.2 $\mathcal{F}_\sigma$-absorbers are topologically unique, it follows that the property mentioned in Corollary 5.4.13 characterizes capsets.

**Theorem 5.5.3.** $B(Q)$ *is an* $\mathcal{F}_\sigma$-*absorber.*

**Absorbing sequences in $Q^\infty$.** We shall now consider the special case of *absorbing sequences*. So our system is a decreasing sequence
$$Q \supseteq X_1 \supseteq X_2 \supseteq \cdots.$$
We put $X_\infty = \bigcap_{n=1}^\infty X_n$. If $\mathcal{M}$ is closed under finite intersections then $\mathcal{M}_\delta$ stands for the collection of countable intersections of elements of $\mathcal{M}$.

**Lemma 5.5.4.** *Let $\mathcal{X}$ be an $\mathcal{M}$-absorbing sequence in $Q$ such that $\mathcal{M}$ is closed under finite intersections. Then $X_\infty$ is an $\mathcal{M}_\delta$-absorber.*

**Proof.** Let $\mathcal{X}$ be the sequence $Q \supseteq X_1 \supseteq X_2 \supseteq \cdots$. Let $M = \bigcap_{n=1}^\infty M_n \subseteq Q$ be such that $M_n \in \mathcal{M}$ for all $n \in \mathbb{N}$. In addition, let $f \colon Q \to Q$ be a map that restricts to a $Z$-imbedding on some closed subset $K \subseteq Q$. There exists by assumption a $Z$-imbedding $\alpha \colon Q \to Q$ which approximates $f$, agrees with $f$ on $K$, and has the property that
$$\alpha^{-1}[X_n] \setminus K = \bigcap_{i=1}^n M_i \setminus K$$
for every $n \in \mathbb{N}$. Then clearly $\alpha^{-1}[X_\infty] \setminus K = \bigcap_{i=1}^\infty M_i \setminus K$, which is as required. □

Let $X$ be a subset of $Q$. We define three decreasing sequences of subsets of $Q^\infty$:
$$S_n(X) = \underbrace{X \times \cdots \times X}_{n \text{ times}} \times Q \times Q \times \cdots,$$
$$S'_n(X) = \{x \in Q^\infty : \text{at least } n \text{ of the } x_i\text{'s are in } X\},$$
$$S''_n(X) = \{x \in Q^\infty : x_i \in X \text{ for some } i \geq n\}.$$
Note that $S_n(X) \subseteq S'_n(X) \subseteq S''_n(X)$, that
$$S_\infty(X) = \bigcap_{n=1}^\infty S_n(X) = X^\infty$$
and that
$$S'_\infty(X) = S''_\infty = \{x \in Q^\infty : \text{infinitely many } x_i \text{ belong to } X\}.$$

**Theorem 5.5.5.** *Let $X \subseteq Q$ be strongly $\mathcal{M}$-universal. Then the sequences*
$$\bigl(S_n(X)\bigr)_n, \quad \bigl(S'_n(X)\bigr)_n, \quad \bigl(S''_n(X)\bigr)_n$$
*are strongly $\mathcal{M}$-universal in $Q^\infty$. If, in addition, $\mathcal{M}$ is closed under finite intersections then $X^\infty$ and $S'_\infty(X)$ are strongly $\mathcal{M}_\delta$-universal in $Q^\infty$.*

**Proof.** Let $(\varrho_n)_n$ be a sequence of admissible metrics on $Q$ such that
$$\varrho(x,y) = \max\{\varrho_n(x_n, y_n) : n \in \mathbb{N}\}$$

is an admissible metric on $Q^\infty$ (Exercise A.5.16). Consider a map
$$f\colon Q \to Q^\infty$$
that restricts to a $Z$-imbedding on some compactum $K$ and a sequence
$$Q \supseteq A_1 \supseteq A_2 \supseteq \cdots$$
of elements of $\mathcal{M}$. We may assume that $f$ is a $Z$-imbedding (Theorem 5.3.11). Write $Q \setminus K$ as a union of a sequence $(F_i)_{i=0}^\infty$ of compacta with $F_i \subseteq \operatorname{Int}(F_{i+1})$ for every $i$ and $F_0 = \emptyset$. Let $\varepsilon > 0$ and put
$$\varepsilon_i = \min\{2^{-i}\varepsilon, \tfrac{1}{2}\varrho(f[K], f[F_i])\}$$
for every $i$. Consider now the $n$-th component $f_n \colon Q \to Q$ of $f$. We shall construct a sequence $\alpha_0, \alpha_1, \ldots$ of continuous functions from $Q$ to $Q$ such that for every and $i$:

(1) $\hat{\varrho}_n(\alpha_i, \alpha_{i-1}) < \varepsilon_i$, $\alpha_i \restriction F_{i-1} = \alpha_{i-1} \restriction F_{i-1}$,
(2) $\alpha_i \restriction Q \setminus F_{i+1} = f_n \restriction Q \setminus F_{i+1}$ and $\alpha_i \restriction F_i$ is a $Z$-imbedding,
(3) $\alpha_i^{-1}[X] \cap F_i = A_n \cap F_i$.

Put $\alpha_0 = f_n$ and assume that $\alpha_i$ has been constructed. Using the strong $\mathcal{M}$-universality of $X$ we find a $Z$-imbedding $\beta \colon F_{i+1} \to Q$, close to $\alpha_i \restriction F_{i+1}$, with $\beta \restriction F_i = \alpha_i \restriction F_i$ and $\beta^{-1}[X] = A_n \cap F_{i+1}$. Using the fact that $Q$ is an AR (Corollary 1.2.12), extend $\beta$ to a map $\alpha_{i+1} \colon Q \to Q$ that restricts to $f$ on $Q \setminus F_{i+2}$ and which is sufficiently close to $\alpha_i$ (Exercise 4.1.5).

The $\alpha_i$'s obviously form a Cauchy sequence and so the function
$$g_n = \lim_{i \to \infty} \alpha_i$$
is continuous (Lemma A.3.1). It is easy to verify that $g_n$ has the following properties:

(4) $\hat{\varrho}_n(g_n, f_n) < \varepsilon$,
(5) if $x \in F_{i+1} \setminus F_i$ then $\varrho_n\bigl(g_n(x), f_n(x)\bigr) < \varrho(f[K], f[F_{i+1}])$,
(6) $g_n \restriction K = f_n \restriction K$,
(6) $g_n \restriction F_i$ is a $Z$-imbedding for every $i$,
(7) $g_n^{-1}[X] \setminus K = A_n \setminus K$.

Define $g = (g_n)_n \colon Q \to Q^\infty$. Note that $g$ is one-to-one and hence an imbedding. The set $g[Q]$ is contained in the $\sigma Z$-set
$$f[K] \cup \bigcup_{i=0}^\omega g_1[F_i] \times Q \times Q \times \cdots$$
and is therefore a $Z$-set (Lemma 5.1.2(3)). The maps $f$ and $g$ are $\varepsilon$-close and $f \restriction K = g \restriction K$. Let $x \in Q \setminus K$. If $x$ is an element of $A_n$ then $x \in \bigcap_{j=1}^n A_j$. Consequently, we have $g_j(x) \in X$ for $j = 1, 2, \ldots, n$. This means that
$$g(x) \in S_n(X) \subseteq S_n'(X) \subseteq S_n''(X).$$

On the other hand, if $g(x)$ is an element of $S_n''(X)$ then $g_j(x) \in X$ for some element $j \geq n$ and hence $x \in A_j \subseteq A_n$. So we are done. □

**An $\mathcal{F}_{\sigma\delta}$-absorber.** Consider the pseudo-boundary $B(Q)$ of $Q$. It is an $\mathcal{F}_\sigma$-absorber in $Q$ by Theorem 5.5.3. Since $\mathcal{F}_\sigma$ is trivially closed under finite intersections we have by Theorems 5.5.2 and 5.5.5:

**Corollary 5.5.6.** *The sequences*
$$\big(S_n(B(Q))\big)_n, \quad \big(S_n'(B(Q))\big)_n, \quad \big(S_n''(B(Q))\big)_n$$
*are $\mathcal{F}_\sigma$-absorbing and hence they are homeomorphic in $Q^\infty$. Moreover, $B(Q)^\infty$ and $S_\infty'\big(B(Q)\big)$ are $\mathcal{F}_{\sigma\delta}$-absorbers.*

So just like $B(Q)$ is the standard model of an $\mathcal{F}_\sigma$-absorber, $B(Q)^\infty$ is the standard model of an $\mathcal{F}_{\sigma\delta}$-absorber.

**Application 1: The space of all infinite-dimensional compacta.** We are interested here in the spaces $\dim_{\geq k}(Q)$, $\dim_{\leq k}(Q)$, etc.

**Theorem 5.5.7.**

(1) *There exists a homeomorphism $\alpha \colon 2^Q \to Q^\infty$ such that for every $k \in \{0, 1, 2, \dots\}$,*
$$\alpha[\dim_{\geq k}(Q)] = \underbrace{B(Q) \times \cdots \times B(Q)}_{k \text{ times}} \times Q \times Q \times \cdots .$$

(2) *There exists a homeomorphism $\beta \colon 2^Q \to Q^\infty$ such that for every $k \in \{0, 1, 2, \dots\}$,*
$$\beta[\dim_{\leq k}(Q)] = \underbrace{Q \times \cdots \times Q}_{k \text{ times}} \times s \times s \times \cdots .$$

Since $\dim_\infty(Q) = \bigcap_{k=0}^\infty \dim_{\geq k}(Q)$, this shows that the space of all infinite-dimensional compacta is homeomorphic to $B(Q)^\infty$.

**Corollary 5.5.8.** *There is a homeomorphism $\alpha \colon 2^Q \to Q^\infty$ such that*
$$\alpha[\dim_\infty(Q)] = B(Q)^\infty.$$

We now present the proofs of these results.

**Proposition 5.5.9.** *The sequence $\big(\dim_{\geq k}(Q)\big)_{k=1}^\infty$ is reflexively universal in $2^Q$.*

**Proof.** Let $F \colon 2^Q \to 2^Q$ be continuous and let $\mathcal{K}$ be a closed subset of $2^Q$ such that $F \upharpoonright \mathcal{K}$ is a $Z$-imbedding. Since $2^Q \approx Q$ (Theorem 4.2.27), we may assume that $F$ is a $Z$-imbedding (Theorem 5.3.11) on all of $2^Q$. Define the function $\varepsilon \colon 2^Q \to \mathbb{I}$ by
$$\varepsilon(A) = \frac{\varrho_H(A, f[\mathcal{K}])}{4}.$$

## 5.5. ABSORBING SYSTEMS

There is by Corollary 4.2.24 a deformation $H\colon 2^Q \times \mathbb{I} \to 2^Q$ such that $H_0$ is the identity and $H_t(A)$ is finite for $t > 0$ and $A \in 2^Q$. By Proposition 4.1.7 we may assume without loss of generality that

$$\hat{\varrho}_H(H_t, 1) \leq t$$

for every $t$ and $A$.

We shall use the vector addition and scalar multiplication operations that $Q$ inherits from $\mathbb{R}^\infty$.

**Claim 1.** We may assume, moreover, that $\hat{\varrho}_H(H_t, 1) \leq 2t$ and that

$$H_t(A) \subseteq [-1+t, 1-t]^\infty$$

for every $t$ and $A$.

*Proof.* We will adjust $H$ a little. Define $S\colon 2^Q \times \mathbb{I} \to 2^Q$ by

$$S(A, t) = (1-t) \cdot H(A, t).$$

Then $S$ is a deformation of $2^Q$ through $\mathcal{F}_\infty(Q)$ such that

$$S_t(A) \subseteq [-1+t, 1-t]^\infty$$

for every $t$ and $A$. Observe that for $t \in \mathbb{I}$ we have

$$\hat{\varrho}_H(S_t, H_t) \leq t$$

so that

$$\hat{\varrho}_H(S_t, 1) \leq \hat{\varrho}_H(S_t, H_t) + \hat{\varrho}_H(H_t, 1) \leq t + t = 2t,$$

as required. ◇

Define the homotopy $\alpha\colon 2^Q \times \mathbb{I} \to 2^Q$ by

$$\alpha_t(A) = \{\mathbf{0}\} \cup \bigcup_{n=1}^{\infty} \{\tfrac{t}{n}\} \times \tfrac{t}{n}\vec{A},$$

where $\vec{A}$ is the subset of $\prod_{i=2}^{\infty} \mathbb{J}_i$ that is obtained from $A$ by a coordinate shift. Note that $\alpha_t(A) \subseteq [-t, t]^\infty$ and that $\alpha_0(A) = \{\mathbf{0}\}$. The map $G\colon 2^Q \to 2^Q$ that approximates $F$ is defined by

$$G(A) = H_{\varepsilon(F(A))}(F(A)) + \alpha_{\varepsilon(F(A))}(A).$$

This function is continuous by Corollary 1.11.8 and the continuity of the homotopies $H$ and $\alpha$. Observe that

$$\varrho_H(G(A), F(A)) \leq 3\varepsilon(F(A))$$

for every $A \in 2^Q$. If $A \in \mathcal{K}$ then $\varepsilon(F(A)) = 0$ and hence $G$ restricts to $F$ on $\mathcal{K}$. Let $A$ be an element of $2^Q \setminus \mathcal{K}$. Then $t = \varepsilon(F(A)) > 0$ and hence $H_t(F(A))$ is finite. So $G(A)$ is a finite union of translates of $\alpha_t(A)$ and consequently is the union of a finite set and a countable collection of copies

of $A$. This means that $G$ preserves dimension by the Countable Closed Sum Theorem 3.2.8 and
$$G^{-1}[\dim_{\geq k}(Q)] \setminus \mathcal{K} = \dim_{\geq k}(Q) \setminus \mathcal{K}.$$
We shall show that $G$ is one-to-one. The restriction of $G$ to $\mathcal{K}$ is obviously one-to-one. If $A \in 2^Q \setminus \mathcal{K}$ then
$$\varrho_H\big(G(A), F(A)\big) \leq 3\varepsilon\big(F(A)\big) < \varrho_H\big(F[\mathcal{K}], F(A)\big)$$
and hence $G(A)$ is not in $G[\mathcal{K}] = f[\mathcal{K}]$. For the remaining case let the elements $A$ and $B$ in $2^Q \setminus \mathcal{K}$ be such that $G(A) = G(B)$. Let $\pi \colon Q \to \mathbb{J}$ be the projection onto the first coordinate. Define $r = \varepsilon\big(F(A)\big)$ and $t = \varepsilon\big(F(B)\big)$. Select a point $y = (a, x) \in G(A) = G(B)$ such that
$$a = \min\big(\pi(G(A))\big) = \min\big(\pi(G(B))\big).$$
Note that $y$ is an element of both $H_r\big(F(A)\big)$ and $H_t\big(F(B)\big)$. Since the latter sets are finite we can define $\lambda > 0$ as one half of the distance of $y$ towards the other points in $H_r\big(F(A)\big) \cup H_t\big(F(B)\big)$.

Let $m$ and $n$ be the first natural numbers that satisfy $r/m \leq \lambda$ and $t/n \leq \lambda$. We now have:
$$(\{y\} + [-\lambda, \lambda]^\infty) \cap G(A) = \{y\} \cup \bigcup_{i=m}^{\infty} \{a + r/i\} \times (x + r/i \vec{A})$$
$$= (\{y\} + [-\lambda, \lambda]^\infty) \cap G(B)$$
$$= \{y\} \cup \bigcup_{i=n}^{\infty} \{a + t/i\} \times (x + t/i \vec{B}).$$
This implies
$$\{a + r/m\} \times (x + r/m \vec{A}) = \{a + t/n\} \times (x + t/n \vec{B})$$
and so $r/m = t/n$ and $r/m \vec{A} = t/n \vec{B}$ and hence that $A = B$. So $G$ is one-to-one and therefore by compactness an imbedding (Exercise A.5.9).

Observe that $\pi[G(A)]$ is countable if $A \in 2^Q \setminus \mathcal{K}$ so $G(A)$ is nowhere dense in $Q$. Since
$$(A, t) \mapsto D_t(A) = \{x \in Q : \varrho(x, A) \leq t\}$$
is a deformation of $Q$ through the complement of $g[2^Q \setminus \mathcal{K}]$ (Exercises A.5.20 and 1.11.19), we have that $G[2^Q \setminus \mathcal{K}]$ is a $\sigma Z$-set. Consequently,
$$G[2^Q] \subseteq F[\mathcal{K}] \cup G[2^Q \setminus \mathcal{K}]$$
is a $Z$-set (Lemma 5.1.2(3)) and so $G$ is a $Z$-imbedding. This completes the proof. $\square$

**Proposition 5.5.10.** *The sequence* $\big(\dim_{\geq k}(Q)\big)_{k=1}^{\infty}$ *is* $\mathcal{F}_\sigma$*-absorbing in* $2^Q$.

**Proof.** By Corollary 5.4.19, $\dim_{\geq 1}(Q)$ is a $\sigma Z$-set in $2^Q$.

In view of the previous proposition it suffices to show that the system $\bigl(\dim_{\geq k}(Q)\bigr)_{k=1}^{\infty}$ is $\mathcal{F}_\sigma$-universal (Lemma 5.5.1).

In Theorem 5.4.20 we showed that $\dim_1(\mathbb{I})$ is a capset for the Hilbert cube $2^{\mathbb{I}}$. So the pairs

$$\bigl(2^{\mathbb{I}}, \dim_1(\mathbb{I})\bigr), \quad \bigl(Q, B(Q)\bigr)$$

are homeomorphic (Theorem 5.4.9). So by Corollary 5.5.6,

$$S'\bigl(\dim_1(\mathbb{I})\bigr)$$

is an $\mathcal{F}_\sigma$-absorbing sequence in $(2^{\mathbb{I}})^\infty$. Define the imbedding $\alpha\colon (2^{\mathbb{I}})^\infty \to 2^Q$ by

$$\alpha\bigl((P_i)_{i=1}^\infty\bigr) = \prod_{i=1}^{\infty} P_i.$$

**Claim 1.** $\prod_{i=1}^{\infty} P_i$ is $k$-dimensional if and only if precisely $k$ of the $P_i$'s are in $\dim_1(\mathbb{I})$.

*Proof.* Consider $\prod_{i=1}^{\infty} P_i$. The dimension of each $P_i$ is either 0 or 1. In the latter case, $P_i$ contains an interval. So the dimension of $\prod_{i=1}^{\infty} P_i$ equals

$$|\{i \in \mathbb{N} : \dim P_i = 1\}|,$$

(Theorem 3.2.12) which proves the claim. ◇

So we have

$$\alpha^{-1}[\dim_{\geq k}(Q)] = S'_k[\dim_1(\mathbb{I})]$$

for every $k$. The sequence $\bigl(\dim_{\geq k}(Q)\bigr)_k$ is therefore $\mathcal{F}_\sigma$-universal because the sequence $\bigl(S'_k\bigl(\dim_1(\mathbb{I})\bigr)\bigr)_k$ is. □

**Corollary 5.5.11.** $\dim_\infty(Q)$ is an $\mathcal{F}_{\sigma\delta}$-absorber.

We find Theorem 5.5.7 by by combining Theorem 5.5.2, Corollary 5.5.6 and Corollary 5.5.11. The fact that $(2^Q, \bigl(\dim_{\geq k}(Q)\bigr)_{k=1}^{\infty})$ is homeomorphic to $(Q^\infty, S(B(Q)))$ means that there is a homeomorphism $\alpha\colon 2^Q \to Q^\infty$ such that

$$\alpha[\dim_{\geq k}(Q)] = \underbrace{B(Q) \times \cdots \times B(Q)}_{k \text{ times}} \times Q \times Q \times \cdots$$

for every $k$. This implies that $\alpha[\dim_\infty(Q)] = B(Q)^\infty$. By comparing the pairs

$$\bigl(2^Q, \dim_{\geq k}(Q)\bigr), \quad \bigl(Q^\infty, S''(B(Q))\bigr)$$

we find part (2) of Theorem 5.5.7.

**Application 2: $c_0$ in the topology of coordinatewise convergence.**
The Banach space $c_0$ was introduced in Example 1.1.9. It is the subset of $s$ consisting of all elements the coordinates of which converge to 0 endowed with a natural norm topology. We will endow $c_0$ here with the subspace topology it inherits from $s$, i.e., the topology of coordinatewise convergence (which incidentally coincides with the *weak topology* on $c_0$, but we will not go into this). We will prove that $c_0$ with this topology is homeomorphic to $B(Q)^\infty$.

It will be convenient here to represent the Hilbert cube $Q$ by $\hat{\mathbb{R}}^\infty$, where $\hat{\mathbb{R}}$ stands for the compactification $[-\infty, \infty]$ of $\mathbb{R}$. Consequently, $\mathbb{R}^\infty$ corresponds to the pseudo-interior of $Q$. In addition, for $n \in \mathbb{N}$ let

$$\Theta_n = \{x \in \mathbb{R}^\infty : |x_i| \leq 2^{-n} \text{ for all but finitely many } i\}.$$

**Lemma 5.5.12.** $\Theta = (\Theta_n)_n$ *is a decreasing sequence of capsets in $Q$ with intersection $c_0$.*

**Proof.** That the sequence $(\Theta_n)_n$ is decreasing and has intersection $c_0$ are trivialities. It is clear that $\Theta_n$ is a basic core set for every $n$. So we are done by Exercise 5.4.3. □

**Proposition 5.5.13.** *The system $\Theta$ is $\mathfrak{F}_\sigma$-universal in $Q$.*

**Proof.** Let $A_1 \supseteq A_2 \supseteq \cdots$ be a sequence of $\sigma$-compacta in $Q$. Let

$$\alpha \colon \mathbb{N} \times \mathbb{N} \to \mathbb{N}$$

be a bijection and put $\mathbb{N}_i = \{\alpha(i,j) : j \in \mathbb{N}\}$ for every $i$. For every $i$ let $Q_i$ denote the Hilbert cube

$$[-2^{-i+1}, 2^{-i+1}]^{\mathbb{N}_i}.$$

**Claim 1.** The set

$$C_i = \{x \in Q_i : (\exists k \in \mathbb{N})(\forall j \in \mathbb{N})(|x_{\alpha(i,j)}| \leq 2^{k-j})\}$$

is an $\mathfrak{F}_\sigma$-absorber in $Q_i$.

*Proof.* This is Exercise 5.4.8.  ◇

Observe that for every $x \in C_i$ we have $\lim_{j \to \infty} x_{\alpha(i,j)} = 0$. Define in $Q_i$ the $\sigma Z$-set

$$D_i = \{x \in Q_i : |x_{\alpha(i,j)}| \leq 2^{-i} \text{ for all but finitely many } i\}.$$

Let $f_i \colon Q \to Q_i$ be an imbedding such that $f_i^{-1}[C_i] = A_i$ and

$$f[Q \setminus A_i] \cap D_i = \emptyset$$

(Exercise 5.5.4). Consider the imbedding $f = (f_i)_i \colon Q \to \prod_{i=1}^\infty Q_i \subseteq Q$. Let $x \in A_n$. If $i > n$ then we have $f_i(x) \in Q_i$ and hence all components of $f_i(x)$ are in $[-2^{-n}, 2^{-n}]$. If $i \leq n$ then we have $x \in A_i$ and hence $f_i(x) \in C_i$. Note that only finitely many components of $f_i(x)$ are outside $[-2^{-n}, 2^{-n}]$ and hence only finitely many components of $f(x)$ are outside this interval. This means that $f(x)$ is an element of $\Theta_n$. If $x \notin A_n$ then we have $f_n(x) \notin D_n$. This means that infinitely many components of $f_n(x)$ have absolute value greater than $2^{-n}$ and hence $f(x) \notin \Theta_n$. So we conclude that $f^{-1}[\Theta_n]$ is equal to $A_n$. □

For a space $X$ and element $* \in X$ we let
$$W(X, *) = \{x \in X^\infty : x_i = * \text{ for all but finitely many } i\}.$$
This set is called the *weak Cartesian product* of $X$.

We will now prove that the system $\Theta$ is reflexively universal. To this end, let $\alpha \colon \mathbb{N}^2 \to \mathbb{N}$ be any bijection, and define $\Phi \colon \hat{\mathbb{R}}^\infty \to (\hat{\mathbb{R}}^\infty)^\infty$ by the formula
$$\Phi(x)_{i,j} = x_{\alpha(i,j)}.$$
So $\Phi$ simply rearranges coordinates. By letting $* = \mathbf{0}$ it is easily seen that with this map the system $\Theta$ satisfies the conditions of Lemma 5.5.14, except maybe for the fact that there is a deformation of $\hat{\mathbb{R}}^\infty$ through $c_0$. To prove this, first observe that there is a deformation of $\hat{\mathbb{R}}^\infty$ through $\mathbb{R}^\infty$ by Corollary 5.3.10. Next, $c_0$ is a dense linear subspace of $\mathbb{R}^\infty$. So there is a deformation of $\mathbb{R}^\infty$ through $c_0$ by Corollary 4.2.15. This is clearly all we need (Corollary 4.1.8).

Again let $\Gamma$ be an ordered set.

**Lemma 5.5.14.** *Let* $\mathfrak{X} = (X_\gamma)_{\gamma \in \Gamma}$ *be an order preserving system in $Q$ such that there is a deformation of $Q$ through $X = \bigcap_{\gamma \in \Gamma} X_\gamma$ and let $* \in \bigcap_{\gamma \in \Gamma} X_\gamma$. Assume that there is a homeomorphism $\Phi \colon Q \to Q^\infty$ satisfying*
$$W(X_\gamma, *) \subseteq \Phi[X_\gamma] \subseteq X_\gamma^\infty$$
*for all $\gamma \in \Gamma$. Then $\mathfrak{X}$ is reflexively universal.*

**Proof.** Let $f \colon Q \to Q$ be a map that restricts to a $Z$-imbedding on some compact set $K$ and let $\varepsilon > 0$. The point $*$ plays a central role in our considerations. If $* \in K$ then there is nothing to worry about. Suppose therefore that $* \notin K$. Since $X$ is dense in $Q$ (otherwise there would not be a deformation of $Q$ through $X$) and $f[K]$ is nowhere dense being a $Z$-set, there is point $p \in X \setminus f[K]$ such that the distance between $p$ and $f(*)$ is arbitrarily small. Consider the function $\hat{f} \colon K \cup \{*\} \to Q$ defined by
$$\hat{f}(x) = \begin{cases} f(x) & (x \in K), \\ p & (x = *). \end{cases}$$
Then $\hat{f}$ is a $Z$-imbedding and is very close to $f \restriction K \cup \{*\}$. We may extend $\hat{f}$ to a function $\tilde{f} \colon Q \to Q$ which agrees with $\hat{f}$ on $K \cup \{*\}$ and which moreover is very close to $f$ (Exercise 4.1.5).

The conclusion of these considerations is that we may assume without loss of generality that

(1) $$* \notin K \implies f(*) \in X \setminus f[K].$$

We put $\hat{K} = K \cup \{*\}$.

Let $d$ be an admissible metric on $Q^\infty$ so that

(2) $$\text{if } x_i = x'_i \text{ for all } i \leq k \text{ then } d(x, x') \leq 2^{-k-2}$$

(cf. Exercise A.1.10).

Let $\mathcal{U}$ be the collection of all open subsets of $Q$ of diameter less than $\frac{1}{2}\varepsilon$, and let
$$\mathcal{V} = \{\Phi[U] : U \in \mathcal{U}\}$$
Then $\mathcal{V}$ is an open cover of the compact space $Q^\infty$ and consequently has a Lebesgue number $\gamma > 0$ (Lemma A.5.3). By Exercise 5.5.1 we may assume without loss of generality that

(3) $$f[Q \setminus \hat{K}] \subseteq X \setminus f[\hat{K}].$$

Define $\delta \colon Q^\infty \to [0, 1)$ by
$$\delta(x) = \min\{\gamma, d(x, \Phi \circ f[\hat{K}])\}.$$

**Claim 1.** If $x \in Q$ then $\delta(\Phi \circ f(x)) = 0$ iff $x \in \hat{K}$.

*Proof.* This is easy. If $x \in \hat{K}$ then clearly $\delta(\Phi \circ f(x)) = 0$. Moreover, if $x \notin \hat{K}$ then $f(x) \notin f[\hat{K}]$ by (3) and so $\Phi \circ f(x) \notin \Phi \circ f[\hat{K}]$ from which we conclude that $\delta(\Phi \circ f(x)) > 0$. ◇

**Claim 2.** There is a homotopy $H \colon Q \times \mathbb{I} \to Q$ with $H[Q \times (0, 1]] \subseteq X$ while moreover $H(x, 0) = x$ and $H(x, 1) = *$ for every $x$.

*Proof.* By our assumptions and Theorem 4.1.6, $(Q, X)$ is an AR-pair, hence $X$ is an AR by Exercise 4.1.7. So $X$ is contractible, and there consequently is a homotopy $S \colon X \times \mathbb{I} \to X$ such that $S_0$ is the identity on $X$ and $S_1$ is the constant function with value $*$ (Exercise A.12.7). There is a homotopy
$$F \colon Q \times \mathbb{I} \to Q$$
such that $F_0$ is the identity on $Q$ and $F[Q \times (0, 1]] \subseteq X$. Define $H \colon Q \times \mathbb{I} \to Q$ by the formula
$$H(x, t) = \begin{cases} F(x, 2t) & (0 \leq t \leq \frac{1}{2}), \\ S(F(x, 1), 2t - 1) & (\frac{1}{2} \leq t \leq 1). \end{cases}$$

An easy check shows that $H$ is as required. ◇

## 5.5. ABSORBING SYSTEMS

Let $\zeta_i$ be the $i$-th component of the map $\Phi \circ f$. For every $k$ define a homotopy $h_k \colon Q \times \mathbb{I} \to Q$ by

$$h_k(x,t) = \begin{cases} H(\zeta_k(x), 2t) & \text{if } 0 \leq t \leq 1/2, \\ H(x, 2-2t) & \text{if } 1/2 \leq t \leq 1. \end{cases}$$

It is easy to see that $h_k$ is well-defined and continuous. Observe that $(h_k)_0$ is $\zeta_k$ and $(h_k)_1$ is the identity function.

For $k \in \mathbb{N}$, put

$$P_k = \{y \in Q : 2^{-k-1} \leq \delta(\Phi \circ f(y)) \leq 2^{-k}\}.$$

Observe that by Claim 1, $\bigcup_{k=1}^{\infty} P_k = Q \setminus \hat{K}$.

For $x \in P_k$ define $f'(x) \in Q^\infty$ by

$$f'(x)_i = \begin{cases} \zeta_i(x) & i \leq k, \\ h_{k+1}(x, 1+k+\log_2 \delta(\Phi \circ f(x))) & i = k+1, \\ x & i \in \{k+2, k+3\}, \\ H(x, 1+k+\log_2 \delta(\Phi \circ f(x))) & i = k+4, \\ * & i > k+4, \end{cases}$$

and extend $f'$ on $\hat{K}$ by $f' \restriction \hat{K} = \Phi \circ f \restriction \hat{K}$.

**Claim 3.** $f'$ is well-defined, continuous and one-to-one. (Hence hence an imbedding by Exercise A.5.9.)

*Proof.* Suppose e.g., that $\delta(\Phi \circ f(x)) = 2^{-k}$ for certain $x \in Q$ and $k \in \mathbb{N}$. Then on the one hand

$$f'(x)_i = \begin{cases} \zeta_i(x) & i \leq k, \\ h_{k+1}(x, 1+k-k) = h_{k+1}(x, 1) = x & i = k+1, \\ x & i \in \{k+2, k+3\}, \\ H(x, 1+k-k) = H(x,1) = * & i = k+4, \\ * & i > k+4, \end{cases}$$

and on the other hand

$$f'(x)_i = \begin{cases} \zeta_i(x) & i \leq k-1, \\ h_k(x, 1+k-1-k) = h_k(x, 0) = \zeta_k(x) & i = k, \\ x & i \in \{k+1, k+2\}, \\ H(x, 1+k-1-k) = H(x,0) = x & i = k+3, \\ * & i > k+3, \end{cases}$$

So both definitions agree.

It is clear that $f' \restriction P_k$ is continuous for every $k$. Hence $f' \restriction Q \setminus \hat{K}$ is continuous. If $(x_n)_n$ is a sequence in $Q \setminus \hat{K}$ converging to an element $x \in \hat{K}$

then $\delta(\Phi \circ f(x_n)) \to \delta(\Phi \circ f(x)) = 0$ (Claim 1) since all functions involved are continuous. We assume without loss of generality that $x_n \in P_n$ for every $n$. Then

$$\begin{aligned}\lim_{n\to\infty} f'(x_n) &= \lim_{n\to\infty} (\zeta_1(x_n), \ldots, \zeta_n(x_n), ?, ?, ?, \ldots) \\ &= (\zeta_1(x), \ldots, \zeta_n(x), \zeta_{n+1}(x), \ldots) \\ &= f'(x).\end{aligned}$$

This proves that $f'$ is continuous.

We next prove that $f'$ is one-to-one. To this end, let $x, x' \in Q$ be distinct points. We distinguish between several subcases.

**Case 1.** There exists $k \in \mathbb{N}$ such that $x, x' \in P_k$.

Then $f'(x)_{k+2} = x \neq x' = f'(x')_{k+2}$.

**Case 2.** There exist $k \in \mathbb{N}$ such that $x \in P_k$ and $x' \in P_{k+1}$.

Then $f'(x)_{k+3} = x$ and $f'(x')_{k+3} = f'(x')_{(k+1)+2} = x'$. So we are again done in this case.

**Case 3.** There exist $k, k' \in \mathbb{N}$ such that $k + 1 < k'$ while moreover $x \in P_k$ and $x' \in P_{k'}$.

Then $k + 4 < k' + 3$ and so $f'(x)_{k'+3} = *$ and $f'(x')_{k'+3} = x'$. But clearly $x' \notin \hat{K}$ and since $* \in \hat{K}$ this proves that $f'(x) \neq f'(x')$.

**Case 4.** There exists $k \in \mathbb{N}$ such that $x \in P_k$ and $x' \in \hat{K}$.

Striving for a contradiction, assume that $f'(x) = f'(x')$. By the definition of $f'$ this implies that $\Phi \circ f(x)$ and $\Phi \circ f(x')$ agree in their first $k$ coordinates. This implies by (2) that

$$d(\Phi \circ f(x), \Phi \circ f[\hat{K}]) \leq 2^{-k-2}$$

and hence

$$\delta(\Phi \circ f(x)) \leq \min\{\gamma, 2^{-k-2}\} \leq 2^{-k-2}.$$

So $x \notin P_k$, which is a contradiction.

Since $f' \restriction \hat{K} = \Phi \circ f \restriction \hat{K}$ is one-to-one, we now conclude that $f'$ is indeed one-to-one. $\diamond$

We next claim that $f'$ is even a $Z$-imbedding.

**Claim 4.** $f'$ is a $Z$-imbedding.

*Proof.* First observe that $f'[\hat{K}] \in \mathcal{Z}(Q^\infty)$ since $f[\hat{K}] \in \mathcal{Z}(Q)$ and $\Phi$ is a homeomorphism (Lemma 5.1.2). Fix $k \in \mathbb{N}$ and consider $f'[P_k]$. Since the coordinates of every point in $f'[P_k]$ are $*$ beyond $k + 4$, it follows that $f'[P_k]$

projects onto a proper subset of $Q$ in infinitely many coordinate directions. So $f'[P_k] \in \mathcal{Z}(Q^\infty)$ by Exercise 5.1.8. This implies that $f'[Q] \in \mathcal{Z}_\sigma(Q^\infty)$ and hence by compactness of $f'[Q]$ it is even a $Z$-set by Lemma 5.1.3(3). ◇

**Claim 5.** $\hat{\varrho}(\Phi^{-1} \circ f', f) < \varepsilon$.

*Proof.* If $x \in Q \setminus \hat{K}$ is arbitrary, say $x \in P_k$, then $f'(x)$ and $\Phi \circ f(x)$ agree in their first $k$ coordinates, so
$$d\big(f'(x), \Phi \circ f(x)\big) \leq 2^{-k-2} < \delta\big(\Phi \circ f(x)\big) \leq \gamma.$$
So there exists $U \in \mathcal{U}$ such that
$$f'(x), \Phi \circ f(x) \in \Phi[U]$$
which implies that
$$\Phi^{-1} \circ f'(x), f(x) \in U,$$
i.e., $\varrho\big(\Phi^{-1} \circ f'(x), f(x)\big) < \frac{1}{2}\varepsilon$. Since the functions $f$ and $\Phi \circ f'$ agree on $\hat{K}$, we are done. ◇

**Claim 6.** If $\gamma \in \Gamma$ then $f'[X_\gamma \setminus \hat{K}] \subseteq W(X_\gamma, *)$.

*Proof.* If $x \in X_\gamma \setminus \hat{K}$ then $f(x) \in X \subseteq X_\gamma$ by (3) and hence
(4)
$$\Phi \circ f(x) \in \Phi[X_\gamma] \subseteq X_\gamma^\infty.$$
Pick $k$ such that $x \in P_k$. From the definition of $f'$ we conclude by (4) that the first $k$ coordinates of $f'(x)$ are in $X_\gamma$. Now observe that $H$ is a deformation through $X$, and hence through $X_\gamma$, and $x, * \in X_\gamma$. So $f'(x) \in W(X_\gamma, *)$, as claimed. ◇

**Claim 7.** For every $\gamma \in \Gamma$ we have $(f')^{-1}[X_\gamma^\infty] \setminus \hat{K} = X_\gamma \setminus \hat{K}$.

*Proof.* Pick an arbitrary point $x \in X_\gamma \setminus \hat{K}$. By Claim 6,
$$f'(x) \in W(X_\gamma, *) \subseteq X_\gamma^\infty.$$
Hence $x \in (f')^{-1}[X_\gamma^\infty] \setminus \hat{K}$.

Conversely, if $x \in (f')^{-1}[X_\gamma^\infty] \setminus \hat{K}$ then $x \notin \hat{K}$ and so $x \in P_k$ for some element $k \in \mathbb{N}$. But then $f'(x)_{k+2} = x$ and so $x \in X_\gamma$. ◇

Now put $g = \Phi^{-1} \circ f'$.

**Claim 8.** $g$ is a $Z$-imbedding which agrees with $f$ on $\hat{K}$, is $\varepsilon$-close to $f$ and moreover satisfies $g^{-1}[X_\gamma] \setminus \hat{K} = X_\gamma \setminus \hat{K}$ for every $\gamma \in \Gamma$.

*Proof.* That $g$ is a $Z$-imbedding follows from Claim 4. In addition, $g$ and $f$ are $\varepsilon$-close by Claim 5. If $\gamma \in \Gamma$ is arbitrary then by Claim 7,

$$g^{-1}[X_\gamma] \setminus \hat{K} = (f')^{-1}[\Phi[X_\gamma]] \setminus \hat{K} \subseteq (f')^{-1}[X_\gamma^\infty] \setminus \hat{K}$$
$$= X_\gamma \setminus \hat{K}.$$

Conversely, if $x \in X_\gamma \setminus \hat{K}$ then by Claim 6,

$$f'(x) \in W(X_\gamma, *) \subseteq \Phi[X_\gamma].$$

From this we conclude that $g(x) \in X_\gamma \setminus \hat{K}$. ◇

So if $* \in K$ then $\hat{K} = K$ and we are done. Assume therefore that we are in the case that $* \notin K$. We ensured that $f(*) \in X \setminus f[K]$ and that $g(*) = f(*)$. Observe that on the one hand for $\gamma \in \Gamma$,

$$g^{-1}[X_\gamma] \setminus (K \cup \{*\}) = (g^{-1}[X_\gamma] \setminus K) \cap (Q \setminus \{*\}),$$

while on the other hand

$$X_\gamma \setminus (K \cup \{*\}) = (X_\gamma \setminus K) \cap (Q \setminus \{*\}).$$

This shows that

$$(g^{-1}[X_\gamma] \setminus K) \setminus \{*\} = (X_\gamma \setminus K) \setminus \{*\}.$$

But $g(*) = f(*) \in X \subseteq X_\gamma$ and so $* \in g^{-1}[X_\gamma] \setminus K$. Also, $* \in X \setminus K \subseteq X_\gamma \setminus K$. The upshot is that

$$g^{-1}[X_\gamma] \setminus K = X_\gamma \setminus K,$$

which is as required. □

So we have:

**Theorem 5.5.15.** *The system $\Theta$ is $\mathcal{F}_\sigma$-absorbing and $c_0$ is an $\mathcal{F}_{\sigma\delta}$-absorber in $Q$.*

**Corollary 5.5.16.** *$c_0$ is homeomorphic to $B(Q)^\infty$.*

**Exercises for §5.5.**

▶1. Suppose that there is a deformation of $Q$ through $A$. In addition, let $X$ be compact, let $K \subseteq X$ be closed, and assume that $f \colon X \to Q$ is continuous while moreover $f \upharpoonright K$ is a $Z$-imbedding. Finally, let $\varepsilon \colon X \setminus K \to (0, \infty)$ be continuous. Prove that there is a continuous function $g \colon X \to Q$ such that
   (1) $g \upharpoonright K = f \upharpoonright K$,
   (2) $g[X \setminus K] \subseteq A \setminus g[K]$,
   (3) for every $x \in X \setminus K$ we have $\varrho(g(x), f(x)) < \varepsilon(x)$.

▶2. Let $A$ be strongly $\mathcal{M}$-universal in $Q$, let $X \subseteq M$ be such that there is a deformation of $M$ through $X$. In addition, assume that $Q \times M \approx Q$. Prove that $A \times X$ is strongly $\mathcal{M}$-universal in $Q \times M$.

▶3. Let $B \subseteq Q$ be an $\mathcal{F}_{\sigma\delta}$-absorber. Prove that there is an $\mathcal{F}_\sigma$-absorber $B' \subseteq Q$ which is contained in $B$.

To prepare ourselves for the applications in Chapter 6, we will identify a few other strongly $\mathcal{F}_{\sigma\delta}$-universal sets in Exercise 5.5.5 below. Some preparatory work has to be done first before we can prove the results we are after.

▶4. Let $A$ be an $\mathcal{F}_\sigma$-absorber and $B$ a $\sigma Z$-set with $A \subseteq B \subseteq Q$. Let $C$ be a $\sigma$-compactum in a compact space $X$. Prove that for each continuous function $f \colon X \to Q$ such that for some compact $K \subseteq X$, $f \restriction K$ is a $Z$-imbedding and for each $\varepsilon > 0$ there is a $Z$-imbedding $g \colon X \to Q$ such that
   (1) $\hat{\varrho}(f, g) < \varepsilon$,
   (2) $g \restriction K = f \restriction K$,
   (3) $g^{-1}[A] \setminus K = C \setminus K$,
   (4) $g[X \setminus (C \cup K)] \cap B = \emptyset$.

▶5. For every $n$ let $E_n \approx Q$. Prove that if $X_n$ is an $\mathcal{F}_\sigma$-absorber which is contained in a $\sigma Z$-set $C_n \subseteq E_n$ for every $n$, then every set $X$ with

$$\prod_{n=1}^\infty X_n \subseteq X \subseteq \prod_{n=1}^\infty C_n$$

is strongly $\mathcal{F}_{\sigma\delta}$-universal in $E = \prod_{n=1}^\infty E_n$.

We will now prove some 'universality properties' of dense linear subspaces of $\mathbb{R}^\infty$. Proposition 5.5.17 below at first glance seems to be very technical. Its strength however is that the metric $\varrho$ under consideration beyond admissibility does not have any relation with the algebraic operations on $L$. For that reason a rather careful analysis is required. Observe that the function $\delta$ in Proposition 5.5.17 is in some vague sense a simultaneous 'parametrization' of many straight-line segments.

**Proposition 5.5.17.** *Let $L$ be a linear space with admissible metric $\varrho$. Then there is a continuous function $\delta \colon L \times L \times \mathbb{I} \to \mathbb{I}$ such that for all $t \in \mathbb{I}$ and $x, p \in L$,*

   (1) $\delta(x, p, t) = 0$ if and only if $t = 0$,
   (2) $\delta(x, p, t) \leq t$
   (3) $\varrho(p, p + \delta(x, p, t)x) \leq t$.

**Proof.** For all $x, p \subset L$ and $t \subset \mathbb{I}$ put

$$G(x, p, t) = \{s \in \mathbb{I} \colon \operatorname{diam}(p + [0, s] \cdot x) \geq t\}.$$

Observe that possibly $G(x, p, t) = \emptyset$ and also that $G(x, p, 0) = \mathbb{I}$.

**Claim 1.** $G(x, p, t)$ *is closed in $\mathbb{I}$ for all $x, p \in L$ and $t \in \mathbb{I}$.*

*Proof.* This follows easily from the continuity of the algebraic operations on $L$ and the function diam (Exercise 1.11.12). ◇

Now define $\xi\colon L \times L \times \mathbb{I} \to \mathbb{I}$ by

$$\xi(x,p,t) = \begin{cases} \min G(x,p,t) & (G(x,p,t) \neq \emptyset), \\ 1 & (G(x,p,t) = \emptyset). \end{cases}$$

By Claim 1, $\xi$ is well-defined. In addition, $\xi(x,p,0) = 0$.

**Claim 2.** $\xi$ is **lsc**.

*Proof.* Suppose that for $v \in [0,1)$, $x, p \in L$ and $t \in \mathbb{I}$ we have that $\xi(x,p,t) > v$. Assume that there are sequences $(x_n)_n, (p_n)_n$ in $L$ and $(t_n)_n$ in $\mathbb{I}$ such that

$$x_n \to x, \quad p_n \to p, \quad t_n \to t$$

while moreover

$$s_n = \xi(x_n, p_n, t_n) \leq v$$

for all $n$. We will derive a contradiction. To this end, without loss of generality assume that $s_n \to w \leq v$. Observe that since $v < 1$,

$$\mathrm{diam}(p_n + [0, s_n] \cdot x_n) \geq t_n$$

for every $n$ so that by continuity of the algebraic operations on $L$ and the function diam (Exercise 1.11.12) we get

$$\mathrm{diam}(p + [0, s] \cdot x) \geq t$$

This shows that $\xi(x,p,t) \leq w \leq v$, which is a contradiction. ◇

Define $\xi'\colon L \times L \times \mathbb{I} \to \mathbb{I}$ by

$$\xi'(x,p,t) = \min\{\xi(x,p,t), t\}.$$

Then $\xi'$ is **lsc** by Exercise A.7.6. Let $\eta\colon L \times L \times \mathbb{I} \to \mathbb{I}$ be the constant function with values 0. Then $\eta$ is continuous, and hence **usc**. By Corollary A.7.6 there is a continuous function $\delta\colon L \times L \times \mathbb{I} \to \mathbb{I}$ such that for every $(x,p,t) \in L \times L \times \mathbb{I}$ with $0 < \xi'(x,p,t)$ we have

$$0 < \delta(x,p,t) < \xi'(x,p,t).$$

Observe that if $t > 0$ and $\xi(x,p,t) = 0$ then

$$0 = \mathrm{diam}(\{p\}) = \mathrm{diam}(p + [0,0] \cdot x) \geq t > 0$$

which is impossible. So we conclude that if $t > 0$ then so are $\xi(x,p,t)$ and $\xi'(x,p,t)$ which implies that $\delta(x,p,t) > 0$. Clearly, $\delta(x,p,0) = 0$. Observe that $\xi'(x,p,t) \leq t$ for all $x, p \in L$. As a consequence, $\delta(x,p,t) \leq t$ for all $x, p \in L$.

To check (3), pick arbitrary $x, p \in L$ and $t \in \mathbb{I}$. If $t = 0$ then $\delta(x,p,t) = 0$ and so there is nothing to prove. So assume that $t > 0$. By the above, $\xi(x,p,t) > 0$ so that by construction,

$$0 < \delta(x,p,t) < \xi(x,p,t).$$

This means that $\delta(x,p,t) \notin G(x,p,t)$, i.e.,

$$\mathrm{diam}(p + [0, \delta(x,p,t)] \cdot x) < t.$$

This means that

$$\varrho(p, p + \delta(x,p,t)x) < t.$$

We conclude that $\delta$ is as required. □

**Corollary 5.5.18.** *Let $Y$ be a dense linear subspace of $\mathbb{R}^\infty$. Then for any admissible metric $\varrho$ on $\mathbb{R}^\infty$ there is a homotopy $H\colon \mathbb{R}^\infty \times \mathbb{R}^\infty \times \mathbb{I} \to \mathbb{R}^\infty$ such that for all $x, y \in \mathbb{R}^\infty$ and $t \in \mathbb{I}$,*

(1) $H_0(x, y) = x$,
(2) *if $t > 0$ then $H_t(x, y) \in Y$ if and only if $y \in Y$,*
(3) $\varrho(H_t(x, y), x) \leq 2t$.

**Proof.** Since $Y$ is a dense linear subspace of $\mathbb{R}^\infty$ it follows that $(\mathbb{R}^\infty, Y)$ is an AR-pair (Corollary 4.2.15). So by Proposition 4.1.7 there is a deformation

$$S\colon \mathbb{R}^\infty \times \mathbb{I} \to \mathbb{R}^\infty$$

such that

(4) $S_0 = 1$,
(5) $S_t[\mathbb{R}^\infty] \subseteq Y$ for every $t > 0$,
(6) $\hat{\varrho}(S_t, 1) \leq t$ for every $t \in \mathbb{I}$.

Let $\delta\colon \mathbb{R}^\infty \times \mathbb{R}^\infty \times \mathbb{I} \to \mathbb{I}$ be continuous such that $\delta(x, p, t) = 0$ if and only if $t = 0$ while moreover $\varrho(p, p + \delta(x, p, t)x) \leq t$ for all $t \in \mathbb{I}$ and $x, p \in \mathbb{R}^\infty$ (Proposition 5.5.17).

Now define $H\colon \mathbb{R}^\infty \times \mathbb{R}^\infty \times \mathbb{I} \to \mathbb{R}^\infty$ by

$$H(x, y, t) = S(x, t) + \delta(y, S(x, t), t)y.$$

We claim that $H$ is as required. If $t = 0$ then $H(x, y, 0) = S(x, 0) = x$ by (4). If $y \in Y$ then clearly $H(x, y, t) \in Y$ by (5) and the fact that $Y$ is a linear subspace of $\mathbb{R}^\infty$. In addition, if $H(x, y, t) \in Y$ and $t > 0$ then $\delta(y, S(x, t), t) > 0$ from which it follows that $y \in Y$ since

$$\delta(y, S(x, t), t)y = H(x, y, t) - S(x, t) \in Y.$$

It therefore suffices to verify (3). But this is trivial, since if $x, y \in \mathbb{R}^\infty$ and $t \in \mathbb{I}$ are arbitrary then

$$\begin{aligned}\varrho\big(x, H_t(x, y)\big) &\leq \varrho\big(x, S(x, t)\big) + \varrho\big(S(x, t), H_t(x, y)\big) \\ &\leq t + \varrho(S(x, t), S(x, t) + \delta(y, S(x, t), t)y) \\ &\leq t + t \\ &= 2t.\end{aligned}$$

So we are done. □

**Corollary 5.5.19.** *Let $Y$ be a dense linear subspace of $\mathbb{R}^\infty$. Then for any admissible metric $\varrho$ on $\mathbb{R}^\infty$ there is a homotopy $F\colon \mathbb{R}^\infty \times \mathbb{I} \to \mathbb{R}^\infty$ such that*

(1) $F_0 = 1$,
(2) *if $x \in \mathbb{R}^\infty$ and $t > 0$ then $F_t(x) \in Y$ if and only if $x \in Y$,*
(3) $\hat{\varrho}(F_t, 1) \leq 2t$ for all $t \in \mathbb{I}$.

**Proof.** Let $H\colon \mathbb{R}^\infty \times \mathbb{R}^\infty \times \mathbb{I} \to \mathbb{R}^\infty$ be the homotopy from Corollary 5.5.18. It suffices to put $F(x, t) = H(x, x, t)$ for all $x \in \mathbb{R}^\infty$ and $t \in \mathbb{I}$. □

We again represent the Hilbert cube $Q$ by $\mathring{\mathbb{R}}^\infty$.

These results enable us to prove the following interesting result:

**Theorem 5.5.20.** *Let $(L_\gamma)_{\gamma \in \Gamma}$ be a system of linear subspaces of $\mathbb{R}^\infty$. Assume that the intersection $\bigcap_{\gamma \in \Gamma} L_\gamma$ is dense in $\mathbb{R}^\infty$ and that there is a continuous function $\psi \colon Q \to \mathbb{R}^\infty$ such that*
$$\psi^{-1}[L_\gamma] \cap \mathbb{R}^\infty = L_\gamma$$
*for each $\gamma \in \Gamma$. Then the system $(L_\gamma \times \mathbb{R}^\infty)_{\gamma \in \Gamma}$ is reflexively universal in $Q \times Q$.*

To start the proof of the theorem, let $f = (f_1, f_2) \colon Q \times Q \to Q \times Q$ be continuous and let $K \subseteq Q \times Q$ be closed such that $f \restriction K$ is a $Z$-imbedding. We let $\varrho$ be an arbitrary admissible metric on $Q$ which is bounded by 1 (Exercise A.1.6). In addition, let $\theta > 0$. We may assume that $f$ is a $Z$-imbedding by Theorem 5.3.11. Let $d$ be the max metric on $Q \times Q$ that corresponds with $\varrho$. Define $\varepsilon \colon Q \times Q \to \mathbb{I}$ by $\varepsilon(x) = \min\{d(f(x), f[K])/4, \theta/4\}$ and observe that $\varepsilon^{-1}(0) = K$ since $f$ is an imbedding. Since $L = \bigcap_\gamma L_\gamma$ is a dense linear subspace of $\mathbb{R}^\infty$, the pair $(\mathbb{R}^\infty, L)$ is an AR-pair by Corollary 4.2.15. In addition, $(Q, \mathbb{R}^\infty)$ is an AR-pair for the same reason. So by Corollary 4.1.8 it follows that $(Q, L)$ is an AR-pair. There consequently is a deformation of $Q$ through $L$. This implies that we can find a continuous function $\tilde{f}_1 \colon Q \times Q \to Q \times Q$ such that
$$\varrho\big(f_1(x), \tilde{f}_1(x)\big) \leq \varepsilon(x)$$
for every $x \in Q \times Q$ and with the property $\tilde{f}_1[(Q \times Q) \setminus K] \subseteq L$ (Exercise 4.1.9).

▶6. Prove that there is a continuous $\xi \colon Q \times Q \to \mathbb{I}$ such that $\xi^{-1}(0) = K$ and
$$\varrho\big(\tilde{f}_1(x,y), \tilde{f}_1(x,y) + \xi(x,y)\psi(x)\big) \leq \varepsilon(x,y)$$
for each $(x,y) \in (Q \times Q) \setminus K$.

Define $g_1 \colon Q \times Q \to Q$ by by
$$g_1(x,y) = \begin{cases} \tilde{f}_1(x,y) & ((x,y) \in K), \\ \tilde{f}_1(x,y) + \xi(x,y)\psi(x) & ((x,y) \notin K). \end{cases}$$
It is clear that $g_1$ is well-defined.

▶7. Prove that $g_1$ is continuous, that $\varrho\big(f_1(x), g_1(x)\big) \leq 2\varepsilon(x)$ for every $x$ in $Q \times Q$ and that
$$\big(g_1^{-1}[L_\gamma] \cap (\mathbb{R}^\infty \times Q)\big) \setminus K = (L_\gamma \times Q) \setminus K$$
for every $\gamma$.

Let $B$ denote the $\sigma$-compact subset $(Q \times Q) \setminus (\mathbb{R}^\infty \times \mathbb{R}^\infty)$ of $Q \times Q$ (it is even a capset of $Q \times Q$ by Exercise 5.4.1, but this is not needed in the proof). By Exercise 5.4.10 there is a continuous function $g_2 \colon Q \times Q \to Q$ such that
$$\varrho\big(g_2(x), f_2(x)\big) \leq \varepsilon(x)$$
for every $x \in Q \times Q$,
$$g_2 \restriction K = f_2 \restriction K,$$

while moreover $g_2 \upharpoonright (Q \times Q) \setminus K$ is one-to-one, $g_2[(Q \times Q) \setminus K] \in \mathcal{Z}_\sigma(Q)$, and
$$g_2^{-1}[B(Q)] \setminus K = B \setminus K.$$

Put $g = (g_1, g_2)$.

▶8. Prove that $g$ is a $Z$-imbedding, that $g \upharpoonright K = f \upharpoonright K$ and that
$$d\big(f(x), g(x)\big) \leq 2\varepsilon(x)$$
for every $x \in Q \times Q$.

The map $g$ satisfies by Exercise 5.5.7 and the properties of the map $g_2$ for every $\gamma \in \Gamma$ the following:
$$\begin{aligned}
g^{-1}[L_\gamma \times \mathbb{R}^\infty] \setminus K &= (g_1^{-1}[L_\gamma] \cap g_2^{-1}[\mathbb{R}^\infty]) \setminus K \\
&= g_1^{-1}[L_\gamma] \cap (g_2^{-1}[\mathbb{R}^\infty] \setminus K) \\
&= g_1^{-1}[L_\gamma] \cap \big((\mathbb{R}^\infty \times \mathbb{R}^\infty) \setminus K\big) \\
&= (L_\gamma \times \mathbb{R}^\infty) \setminus K.
\end{aligned}$$

So this completes the proof of Theorem 5.5.20 since clearly $\hat{d}(f, g) < \theta$.

CHAPTER 6

# Function spaces

In this chapter we will discuss the topology of pointwise convergence on function spaces. During the last decades these function spaces have played a prominent role in both general- and infinite-dimensional topology. The systematic study of them was initiated by ARHANGEL'SKIĬ [24].

Techniques from general topology, infinite-dimensional topology, functional analysis and descriptive set theory are primarily used for the study of function spaces. The mix of methods from several disciplines makes the subject particularly interesting. Several monographs, surveys and research papers were written containing many interesting and diverse results, see e.g., ARHANGEL'SKIĬ [25, 27, 28], LUTZER and MCCOY [264], MCCOY and NTANTU [280], VAN MILL [297], BAARS and DE GROOT [35], DOBROWOLSKI, MARCISZEWSKI and MOGILSKI [123], CAUTY [88], CAUTY, DOBROWOLSKI and MARCISZEWSKI [91] and MARCISZEWSKI [268].

For a Tychonoff space $X$ let $C_p(X)$ be $C(X)$ endowed with the topology of pointwise convergence. The main result in this chapter is Theorem 6.12.15 which implies that if $X$ is nondiscrete and countable and $C_p(X)$ is an $F_{\sigma\delta}$-subset of $\mathbb{R}^X$ then $C_p(X)$ is homeomorphic to $B(Q)^\infty$. Many results of the previous chapters are applied in its proof. To demonstrate its power, observe that it implies that $\mathbb{Q}$ and $\omega + 1$ have homeomorphic function spaces. We will also discuss linear homeomorphisms. It is possible to prove that certain topological properties are shared by spaces having linearly homeomorphic function spaces. We will give three such examples: compactness, dimension and topological completeness. This shows for example that the function spaces of $\mathbb{Q}$ and $\omega+1$ are *not* linearly homeomorphic. An important recent result that we will prove is the statement that for metrizable spaces $X$ and $Y$, if $X$ is countable dimensional and $C_p(X) \approx C_p(Y)$ then $Y$ is countable dimensional. This proves that if $X$ is any finite dimensional metrizable space then the spaces $C_p(X)$ and $C_p(\mathbb{Q})$ are not homeomorphic.

*All spaces under discussion here are Tychonoff, i.e., completely regular and $T_1$.*

## 6.1. Notation

For a Tychonoff space $X$ let $C_p(X)$ be $C(X)$ topologized as a subspace of the full product $\mathbb{R}^X$ endowed with the Tychonoff product topology. So $C_p(X)$ is endowed with the topology of *pointwise convergence*. The spaces $C_p(X)$ are of interest to topologists and functional analysts for various reasons.

It is possible to restricts one's attention to *bounded* functions only. For a space $X$, $C^*(X)$ endowed with the subspace topology of $\mathbb{R}^X$ is denoted by $C_p^*(X)$. Note that $C_p^*(X)$ is a subspace of $C_p(X)$.

If $X$ is compact then $C(X) = C^*(X)$ so then there is no difference.

In this chapter we are primarily interested in the spaces $C_p(X)$. The theory of bounded functions is more complicated and is less well understood. We will demonstrate this with a few examples only, cf. §6.10.

Observe that basic neighborhoods of $f \in \mathbb{R}^X$ have the form

$$N(f, F, \varepsilon) = \{g \in \mathbb{R}^X : (\forall x \in F)(|f(x) - g(x)| < \varepsilon)\},$$

where $F$ is an arbitrary finite subset of $X$, and $\varepsilon$ is an arbitrary positive real number. $N(f, F, \varepsilon)$ is called the *basic neighborhood of $f$ determined by $F$ and $\varepsilon$*.

If $X$ is a countable space which is not necessarily metrizable, then $C_p(X)$ is a linear subspace of the product $\mathbb{R}^X$. So it is a linear subspace of a countable product of real lines, and is therefore a separable metrizable space. The space $X$ can be rather 'bad' from a general topological perspective, its function space is always 'good'. As we will see, certain non-metrizable countable spaces $X$ have very interesting function spaces and this prompts us to study non-metrizable spaces as well.

This means that we sometimes have to be more careful than before. The classical Tietze-Urysohn Extension Theorem (ENGELKING [153, Theorem 2.1.8]) is no longer valid for example since a Tychonoff space need not be normal. But at least a portion of it is true since if $X$ is Tychonoff and $A \subseteq X$ is *compact* then every continuous function $f \colon A \to \mathbb{R}$ can be extended to a (bounded) continuous function $\bar{f} \colon X \to \mathbb{R}$. This can be seen as follows. First think of $X$ as subspace of its *Čech-Stone compactification* $\beta X$. Then $A$, being compact, is closed in $\beta X$. Since $\beta X$ is compact and Hausdorff, it is normal, and so by the Tietze-Urysohn Extension Theorem, $f$ can be extended to a (necessarily bounded) continuous function $\tilde{f} \colon \beta X \to \mathbb{R}$. Hence $\bar{f} = \tilde{f} \restriction X$ is the required (bounded) extension over $X$. We will use this in this chapter in the case that $A$ is a finite set often without explicit reference.

Observe that $\beta X$ is characterized by the fact that it is the unique compactification having the property that every bounded and continuous function $f \colon X \to \mathbb{R}$ can be extended to a continuous function $\bar{f} \colon \beta X \to \mathbb{R}$ (cf. ENGELKING [153, Corollary 3.6.3]).

If $X$ is a space then $X^*$ denotes $\beta X \setminus X$.

We will also frequently use without explicit reference the well-known fact that metrizable spaces are paracompact. This is due to STONE [383]. A simpler proof was found by RUDIN [359]; she in fact proved by a direct construction avoiding theory on paracompact spaces that every open cover of a metrizable space has an open refinement which is both locally finite and $\sigma$-discrete, i.e., the union of countably many discrete subfamilies. See ENGELKING [153, Chapter 4] for additional information.

## 6.2. The spaces $C_p(X)$: Introductory remarks

Since $C_p(X)$ is a dense linear subspace of $\mathbb{R}^X$, it consequently inherits many properties from $\mathbb{R}^X$. For instance, its local convexity, its weight (see Corollary 6.2.2 below) and cellularity, etc.

**Lemma 6.2.1.** *Let $X$ be a Tychonoff space. Then $C_p(X)$ is dense in $\mathbb{R}^X$. Moreover, $C_p^*(X)$ is dense in $C_p(X)$.*

**Proof.** Let $f \in \mathbb{R}^X$, $F \subseteq X$ be finite and $\varepsilon > 0$. The function $g = f \restriction F$ can be extended to a bounded continuous function $\bar{g} \colon X \to \mathbb{R}$ since $F$ is finite. This function is in the basic neighborhood of $f$ determined by $F$ and $\varepsilon$. So this proves both statements simultaneously. □

If $X$ is a space then its *weight* is the smallest cardinality of a base for $X$. Observe that if $X$ is infinite then the weight of $\mathbb{R}^X$ is equal to the cardinality of $X$. In addition, if $Y \subseteq \mathbb{R}^X$ is dense, then $Y$ and $\mathbb{R}^X$ have the same weight.

**Corollary 6.2.2.** *If $X$ is infinite then the weight of $C_p(X)$ is equal to the cardinality of $X$. As a consequence, if $X$ and $Y$ are infinite such that $C_p(X)$ and $C_p(Y)$ are homeomorphic then $|X| = |Y|$.*

**Corollary 6.2.3.** *Let $X$ be a space. The following statements are equivalent:*

(1) $C_p(X)$ *is separable and metrizable.*
(2) $C_p(X)$ *is metrizable.*
(3) $C_p(X)$ *is first countable.*
(4) $X$ *is countable.*

**Proof.** If $\mathbb{R}^X$ contains a dense first countable subspace then it is first countable at some point. This implies that $X$ is countable by the definition of the product topology on $\mathbb{R}^X$. So we are done by Corollary 6.2.2. □

If $L$ and $M$ are linear spaces then $L \sim M$ means that $L$ and $M$ are *linearly* homeomorphic. This means that there is a bijection $\varphi \colon L \to M$ which is simultaneously a homeomorphism and a linear isomorphism.

**ℓ- and t-equivalence.** Topologists are interested in the *classification* of topological spaces. Classifying topological spaces up to homeomorphism or homotopy type etc., is the ultimate goal for a topologist. In this chapter we classify spaces $X$ up to homeomorphism or linear homeomorphism type of their function spaces $C_p(X)$. In its full generality this program would be much too complicated and the complete picture is presently beyond our reach. But for function spaces of low Borel complexity some definitive results are known, and it is our aim to present those in full detail. Many results of the previous chapters will be applied in this chapter.

We say that spaces $X$ and $Y$ are *ℓ-equivalent* provided that $C_p(X)$ and $C_p(Y)$ are *linearly* homeomorphic. Notation: $X\sim_\ell Y$. We define the corresponding notion of *$\ell^*$-equivalence* similarly. Notation: $X\sim_{\ell^*} Y$.

Homeomorphic spaces are obviously $\ell$- and $\ell^*$-equivalent. But the converse need not be true.

**Example 6.2.4.** There are non-homeomorphic $\ell$-equivalent spaces.

**Proof.** Let $X = [0,1] \cup [2,3]$ and $Y = [0,2] \cup \{3\}$. Then evidently, $X$ and $Y$ are not homeomorphic. We will prove that they are $\ell$-equivalent. Indeed, define $\Phi\colon C_p(X) \to C_p(Y)$ by

$$\Phi(f)(y) = \begin{cases} f(y) & (0 \le y \le 1), \\ f(y+1) - \big(f(2) - f(1)\big) & (1 \le y \le 2), \\ f(2) - f(1) & (y = 3). \end{cases}$$

It is left as an exercise to the reader to prove that $\Phi$ is a linear homeomorphism. □

If $X$ is compact then $C(X)$ can also be endowed with its natural Banach space topology (see Page 4). It is clear that the identity $C(X) \to C_p(X)$ is continuous. This has by Corollary 1.1.16 the following interesting consequence.

**Theorem 6.2.5.** *Let $X$ and $Y$ be compact spaces, and let*

$$\varphi\colon C_p(X) \to C_p(Y)$$

*be a linear homeomorphism. Then*

$$\varphi\colon C(X) \to C(Y)$$

*is also a linear homeomorphism.*

The converse of this result is not true. Let us note that MILJUTIN [290] proved that all Banach spaces $C(X)$ with $X$ uncountable and compact metrizable are linearly homeomorphic. Hence $C(\mathbb{I})$ and $C(\mathbb{I}^2)$ are linearly homeomorphic, but $C_p(\mathbb{I})$ and $C_p(\mathbb{I}^2)$ are not, as we will see in §6.9 that $\ell$-equivalent spaces have the same dimension.

## 6.2. THE SPACES $C_p(X)$: INTRODUCTORY REMARKS

**Corollary 6.2.6.** *There are countably infinite compact (necessarily metrizable) spaces $X$ and $Y$ which are not $\ell$-equivalent.*

**Proof.** By BESSAGA and PEŁCZYŃSKI [48] there are such spaces $X$ and $Y$ for which $C(X)$ and $C(Y)$ are not linearly homeomorphic (see also Corollary 6.10.9). So now an appeal to Theorem 6.2.5 finishes the proof. □

We say that $X$ and $Y$ are *t-equivalent* provided that $C_p(X)$ are $C_p(Y)$ are homeomorphic as topological spaces. Notation: $X \sim_t Y$.

**Remark 6.2.7.** Even for simple spaces it is in general difficult to decide whether they are $\ell$- or $t$-equivalent. We will show in §6.12 that all countable non-discrete spaces $X$ such that $C_p(X)$ is an $F_{\sigma\delta}$-subset of $\mathbb{R}^X$ are $t$-equivalent. This result implies that all countable non-discrete metrizable space are $t$-equivalent (Theorem 6.3.10). But they are not all $\ell$-equivalent, as was shown in Corollary 6.2.6.

**Products.** We will prove that the spaces $C_p(X)$ behave well with respect to products.

If $\{X_i : i \in I\}$ are spaces then $\bigoplus_{i \in I} X_i$ denotes their *topological sum*.

**Theorem 6.2.8.** *Let $\{X_i : i \in I\}$ be a family of spaces. Then*
$$C_p\left(\bigoplus_{i \in I} X_i\right) \sim \prod_{i \in I} C_p(X_i).$$
*If $I$ is finite, then $C_p^*(\bigoplus_{i \in I} X_i) \sim \prod_{i \in I} C_p^*(X_i)$.*

**Proof.** It is left as an exercise to the reader to show that the function
$$F \colon C_p(\bigoplus_{i \in I} X_i) \to \prod_{i \in I} C_p(X_i)$$
defined by $F(f)_i = f \restriction X_i$, $i \in I$, is as required. This function also works for bounded functions if $I$ is finite. □

Observe that the same proof does *not* work for bounded functions if $I$ is infinite. The situation for such functions is in fact much more complicated, see the notes for more information.

**Additional structures on $C_p(X)$.** With the operations of pointwise addition and pointwise multiplication, $C_p(X)$ is a commutative topological ring with unit, the unit being the constant function with value 1. It is a famous theorem of GEL'FAND and KOLMOGOROFF [173] that the ring structure by itself determines the topological structure on $X$ provided $X$ is compact. They proved that if $X$ and $Y$ are compact and $C(X)$ and $C(Y)$ are isomorphic as rings then $X$ and $Y$ are homeomorphic. For details, see also

DUGUNDJI [142, Theorem XIII.6.5]. (The proof in [142] makes use of the topology of pointwise convergence.)

For noncompact spaces $X$, the algebraic structure of $C(X)$ is, in general, not strong enough to determine the topology of $X$. For consider the spaces $X = \omega_1$ and $Y = \omega_1 + 1$. Then clearly $C(X)$ and $C(Y)$ are isomorphic as rings, but $X$ and $Y$ are not homeomorphic.

For arbitrary spaces there is a result in the same spirit though. NAGATA [322] proved that $C_p(X)$ and $C_p(Y)$ are *topologically* isomorphic as topological rings if and only if $X$ and $Y$ are homeomorphic. That we deal with *real-valued* functions is essential in this result. It is known that the ring of all continuous functions $X \to \mathbb{R}^\infty$, endowed with the topology of pointwise convergence, does not always determine the topological type of $X$ (ARHANGEL'SKIĬ [27, Page 12]).

In this book we will not discuss these structures and other similar ones. We will study the topological and linear topological properties of $C_p(X)$ exclusively.

### 6.3. The Borel complexity of function spaces

Obviously, $X$ is discrete if and only if $C_p(X) = \mathbb{R}^X$. The question naturally arises how $C_p(X)$ is placed in $\mathbb{R}^X$ if $X$ is not discrete. We show here that it cannot be a $G_{\delta\sigma}$- or $F_\sigma$-subset of $\mathbb{R}^X$ but it can be an $F_{\sigma\delta}$-subset of $\mathbb{R}^X$.

$C_p(X)$ **is not** $G_\delta$. We will first show that $C_p(X)$ cannot be a $G_\delta$-subset of $\mathbb{R}^X$ unless $X$ is discrete. This result will be used in the proof of our main results.

**Lemma 6.3.1.** *Suppose that $C_p(X)$ contains a nonempty $G_\delta$-subset of $\mathbb{R}^X$. Then $X$ is the topological sum of a countable space and a discrete space.*

**Proof.** Let $S$ be a nonempty $G_\delta$-subset of $\mathbb{R}^X$ which is contained in $C_p(X)$. Pick an arbitrary element $f \in S$. Let $\{U_n : n \in \mathbb{N}\}$ be a sequence of open subsets of $\mathbb{R}^X$ such that $\bigcap_{n=1}^\infty U_n = S$. For every $n$ there exists a finite subset $F_n \subseteq X$ such that every $g \in \mathbb{R}^X$ with $g \restriction F_n = f \restriction F_n$ belongs to $U_n$. Put $F = \bigcup_{n=1}^\infty F_n$. Then $F$ is countable, and every $g \in \mathbb{R}^X$ with $g \restriction F = f \restriction F$ belongs to $\bigcap_{n=1}^\infty U_n = S \subseteq C_p(X)$ and hence is continuous.

Let $A$ be any subset of $X \setminus F$ and let $\chi_A : X \to \{0, 1\}$ be the characteristic function. Since $\chi_A \restriction F \equiv 0$ it follows that

$$(f + \chi_A) \restriction F = f \restriction F.$$

So by the above, $f + \chi_A$ is continuous from which it follows that

$$\chi_A = (f + \chi_A) - f$$

is continuous as well. Consequently, every $A \subseteq X \setminus F$ is clopen in $X$. □

**Lemma 6.3.2.** *Let $X$ be countable. If $C_p(X)$ contains a dense $G_\delta$-subset of $\mathbb{R}^X$ then $X$ is discrete.*

**Proof.** Assume that $X$ is not discrete. There is a function $g \in \mathbb{R}^X \setminus C_p(X)$. Define $\Phi \colon \mathbb{R}^X \to \mathbb{R}^X$ by
$$\Phi(f) = f + g.$$
Then $\Phi$ is a homeomorphism of $\mathbb{R}^X$ onto itself and $\Phi[C_p(X)]$ is a dense subspace of $\mathbb{R}^X \setminus C_p(X)$. Thus
$$\mathbb{R}^X \setminus C_p(X)$$
also contains a dense $G_\delta$-subspace of $\mathbb{R}^X$, which violates the Baire Category Theorem. Simply observe that since $X$ is countable, $\mathbb{R}^X$ is topologically complete (Lemma A.6.2 and Exercise A.6.10). □

**Corollary 6.3.3.** *If $X$ is countable and $C_p(X)$ is both a Baire space and a Borel subset of $\mathbb{R}^X$ then $X$ is discrete.*

**Proof.** By Exercise A.13.4, if $C_p(X)$ is both Borel and a Baire space then it contains a dense topologically complete subspace. Since a topologically complete subspace of $\mathbb{R}^X$ is a $G_\delta$-subset of it (Theorem A.6.3), the previous lemma shows that $X$ is discrete. □

These results lead us to our first result on the descriptive complexity of $C_p(X)$ for general $X$.

**Theorem 6.3.4.** *If $C_p(X)$ is a $G_\delta$-subset of $\mathbb{R}^X$ then $X$ is discrete.*

**Proof.** By Lemma 6.3.1 it follows that $X = F \oplus D$, where $F$ is countable and $D$ is discrete. Since $C_p(X)$ is canonically equivalent to the subspace
$$C_p(F) \times \mathbb{R}^D$$
of $\mathbb{R}^F \times \mathbb{R}^D = \mathbb{R}^X$ (Theorem 6.2.8), it follows that $C_p(F)$ is a $G_\delta$-subset of $\mathbb{R}^F$. So $F$ is discrete as well by Lemma 6.3.2. □

**Remark 6.3.5.** The basic ingredient in the proof of Lemma 6.3.2 is that $\mathbb{R}^X$ is a Baire space if $X$ is countable. But $\mathbb{R}^X$ is a Baire space for arbitrary $X$ as well. This follows easily from the fact that $\mathbb{R}^X$ satisfies the countable chain condition from which one gets that every nowhere dense set in $\mathbb{R}^X$ is contained in a nowhere dense $G_\delta$-set. This allows a reduction to the known case of countable products. (Alternatively, one can use the observation in Remark 6.4.4 below or e.g., DE GROOT [178].) So the proof in Lemma 6.3.2 in fact shows that for arbitrary $X$, if $C_p(X)$ contains a dense $G_\delta$-subset of $\mathbb{R}^X$ then $X$ is discrete.

$C_p(X)$ **is not** $F_\sigma$**.** In view of Theorem 6.3.4 the question naturally arises whether $C_p(X)$ can be an $F_\sigma$-subset of $\mathbb{R}^X$.

**Theorem 6.3.6.** *If $C_p(X)$ is an $F_\sigma$-subset of $\mathbb{R}^X$ then $X$ is discrete.*

**Proof.** Assume that $X$ is not discrete, say that $x_0$ is a non-isolated point of $X$, and that $\bigcup_{n=0}^\infty F_n \subseteq C_p(X)$, where each $F_n$ is closed in $\mathbb{R}^X$ and $F_0 = \emptyset$. By induction on $n$ we will construct a function $f_n \colon X \to \mathbb{I}$ and an open neighborhood $U_n$ of $x_0$ such that the following conditions are satisfied:

(1) $f_0 \leq f_1 \leq f_2 \leq \cdots$,
(2) $U_0 \supseteq \overline{U}_1 \supseteq U_1 \supseteq \overline{U}_2 \supseteq U_2 \supseteq \cdots$,
(3) $f_n(x_0) = 1$,
(4) $f_n \restriction U_n \setminus \{x_0\} \equiv 1 - 2^{-n}$,
(5) $f_n \restriction X \setminus \{x_0\}$ is continuous,
(6) $f_{n+1} \restriction X \setminus U_n = f_n$,
(7) if $f \in \mathbb{R}^X$ is such that $f \restriction (X \setminus U_n) \cup \{x_0\} = f_n \restriction (X \setminus U_n) \cup \{x_0\}$ then $f \notin F_n$.

Define $f_0$ by $f_0(x_0) = 1$ and $f_0(x) = 0$ if $x \neq x_0$, and put $U_0 = X$. Suppose that we completed the construction for $0 \leq i \leq n$. Let $g \colon X \to [0, 2^{-n}]$ be continuous such that $g(x_0) = 2^{-n}$ and $g[X \setminus U_n] \subseteq \{0\}$. Define $f_{n+1}$ by

$$f_{n+1}(x) = \begin{cases} 1 & (x = x_0), \\ f_n(x) + \min\{2^{-(n+1)}, g(x)\} & (x \neq x_0). \end{cases}$$

The set $V = g^{-1}\big[(2^{-(n+1)}, 2^{-(n-1)})\big]$ is a neighborhood of $x_0$ such that

$$f_{n+1} \restriction V \setminus \{x_0\} \equiv 1 - 2^{-n} + 2^{-(n+1)} = 1 - 2^{-(n+1)}.$$

Since $x_0$ is not isolated, $f_{n+1}$ is not continuous at $x_0$, hence $f_{n+1} \notin F_{n+1}$. Since $F_{n+1}$ is closed in $\mathbb{R}^X$ there exists a finite subset $A$ of $X$ such that for each $f \in \mathbb{R}^X$, if $f \restriction A = f_{n+1} \restriction A$ then $f$ does not belong to $F_{n+1}$. As a consequence, $U_{n+1} = (V \setminus A) \cup \{x_0\}$ is the required neighborhood of $x_0$. This completes the induction.

Put $f = \lim_{n \to \infty} f_n$. Observe that $f$ exists because of (1). Then for each $n$,

$$f \restriction (X \setminus U_n) \cup \{x_0\} = f_n \restriction (X \setminus U_n) \cup \{x_0\}.$$

By (7) we therefore obtain that $f \notin \bigcup_{n=0}^\infty F_n$.

But $f$ is continuous, as the following argument shows. If $x$ is a point of $X$ which does not belong to $U = \bigcap_{n=0}^\infty U_n$ then by (2), $x \notin \overline{U}_n$ for some $n$. So by (6) and (5), $f$ is continuous at $x$. But if $x \in U$ then $U_n$ is a neighborhood of $x$ such that $f[U_n] \subseteq \{1 - 2^{-n}, 1\}$ by (3) and (4). Hence $f$ is continuous at $x$ by (1), (5), (6) and (3).

So we conclude that $f \in C_p(X) \setminus \bigcup_{n=0}^\infty F_n$. $\square$

$C_p(X)$ is not $G_{\delta\sigma}$. We will now use Lemma 6.3.2 to generalize Theorem 6.3.4.

Let $X$ be a space, and let $Y \subseteq \mathbb{R}$. By $C_p(X,Y)$ we denote the set of continuous functions $f\colon X \to Y$, topologized as a subspace of the full product $Y^X$. Since $Y^X$ is a subspace of $\mathbb{R}^X$, it follows that $C_p(X,Y)$ is simply the subspace $\{f \in C_p(X) : f[X] \subseteq Y\}$ of $C_p(X)$.

Let $X$ be a space and $\varepsilon > 0$. We put
$$U(f,\varepsilon) = \{g \in C_p(X) : (\forall\, x \in X)(|f(x) - g(x)| < \varepsilon)\}.$$

**Lemma 6.3.7.** *Let $X$ be a countable space, let $f \in C_p(X)$ and let $\varepsilon > 0$. Then*

(1) *$U(f,\varepsilon)$ is a $G_\delta$-subset of $C_p(X)$.*
(2) *$U(f,\varepsilon)$ and $C_p(X)$ are homeomorphic.*
(3) *$\bigcup_{0 < \delta < \varepsilon} U(f,\delta)$ is dense in $U(f,\varepsilon)$.*

**Proof.** That $U(f,\varepsilon)$ is a $G_\delta$-subset of $C_p(X)$ is clear since for every $x \in X$ the set
$$\{g \in C_p(X) : |g(x) - f(x)| < \varepsilon\}$$
is open in $C_p(X)$ and $X$ is countable.

To show that $C_p(X)$ and $U(f,\varepsilon)$ are homeomorphic, first consider the translation $\xi\colon \mathbb{R}^X \to \mathbb{R}^X$ defined by
$$\xi(g) = g - f \qquad (g \in \mathbb{R}^X).$$
This function maps $U(f,\varepsilon)$ onto $U(\underline{0},\varepsilon)$, where $\underline{0}$ is the constant function with value 0. So it suffices to show that this set and $C_p(X)$ are homeomorphic. But by observing that $(-\varepsilon,\varepsilon)$ and $\mathbb{R}$ are homeomorphic, this follows by the remark preceding this lemma.

It remains to prove (3). It suffices to prove that $\bigcup_{0 < \delta < \varepsilon} U(\underline{0},\delta)$ is dense in $V = U(\underline{0},\varepsilon)$. So let $g \in V$, $F \subseteq X$ be finite, and $\gamma > 0$. Pick $0 < \lambda < 1$ such that $\varepsilon(1-\lambda) < \gamma$ and put $h = \lambda g$. Then
$$h \in U(\underline{0},\lambda\varepsilon) \cap N\big(g,F,\varepsilon(1-\lambda)\big) \subseteq N(g,F,\gamma) \cap \bigcup_{0 < \delta < \varepsilon} U(\underline{0},\delta).$$
This is clearly as required. □

**Theorem 6.3.8.** *If $C_p(X)$ is a $G_{\delta\sigma}$-subset of $\mathbb{R}^X$ then $X$ is discrete.*

**Proof.** The proof is inspired by the proof of Theorem A.13.5.

As above we may assume without loss of generality that $X$ is countable. So let $X$ be countable and non-discrete such that
$$C_p(X) = \bigcup_{i=0}^{\infty} G_i,$$

where $G_i$ is a $G_\delta$-subset of $\mathbb{R}^X$ for every $i$ and $G_0 = \emptyset$. We will derive a contradiction.

We shall by induction on $i$ define a sequence $\{f_i : i \geq 0\}$ in $C_p(X)$ and a decreasing sequence of positive real numbers $\{\varepsilon_i : i \geq 0\}$ such that

(1) $f_0$ is the constant function with value 0, and $\varepsilon_0 = 2^0$,
(2) $U(f_0, \frac{1}{4}\varepsilon_0) \supseteq U(f_1, \varepsilon_1) \supseteq U(f_1, \frac{1}{4}\varepsilon_1) \supseteq U(f_2, \varepsilon_2) \supseteq \cdots$,
(3) $\varepsilon_{i+1} < 3^{-i-1} \cdot \varepsilon_i$ for every $i$,
(4) $U(f_i, \varepsilon_i) \cap G_i = \emptyset$ for every $i$.

Assume that we have proved the existence of such sequences. Observe that by (1) and (3), $\varepsilon_i \leq 2^{-i}$ for every $i$. So for every $i$ we have by (2) and (3) that
$$\sup_{x \in X} |f_i(x) - f_{i+1}(x)| \leq \tfrac{1}{4}\varepsilon_i \leq 2^{-i-2}.$$

We conclude that the sequence $(f_i)_i$ converges uniformly to a continuous function $f : X \to \mathbb{R}$ (cf. Lemma A.3.1). It follows moreover from (3) that for $x \in X$, $i \geq 0$ and $n \geq 2$ we have
$$|f_i(x) - f_{n+i}(x)| \leq |f_i(x) - f_{i+1}(x)| + |f_{i+1}(x) - f_{i+2}(x)| + \cdots$$
$$\cdots + |f_{n+i-1}(x) - f_{n+i}(x)|$$
$$\leq \tfrac{1}{4}\varepsilon_i + \sum_{j=1}^{n-1} 3^{-j}\varepsilon_i.$$

This implies that
$$\sup_{x \in X} |f_i(x) - f(x)| \leq \tfrac{1}{4}\varepsilon_i + \sum_{j=1}^{\infty} 3^{-j}\varepsilon_i = \tfrac{3}{4}\varepsilon_i < \varepsilon_i$$

for every $i$, i.e., $f \in \bigcap_{i=0}^{\infty} U(f_i, \varepsilon_i) \subseteq C_p(X) \setminus \bigcup_{i=1}^{\infty} G_i = \emptyset$, which is a contradiction.

It remains to perform the induction. Assume that $f_i$ and $\varepsilon_i$ have been determined. By Lemma 6.3.7 it follows that $U(f_i, \frac{1}{4}\varepsilon_i)$ is a $G_\delta$-subset of $C_p(X)$ which in addition is homeomorphic to $C_p(X)$. The latter implies by Lemma 6.3.2 and Theorem A.6.3 that no dense subset of $U(f_i, \frac{1}{4}\varepsilon_i)$ is a $G_\delta$-subset of $\mathbb{R}^X$. The former implies that $G_{i+1} \cap U(f_i, \frac{1}{4}\varepsilon)$ is a $G_\delta$-subset of $G_{i+1}$ and hence is a $G_\delta$-subset of $\mathbb{R}^X$. The conclusion is that
$$G_{i+1} \cap U(f_i, \tfrac{1}{4}\varepsilon)$$
is *not* dense in $U(f_i, \frac{1}{4}\varepsilon)$. So there are by Lemma 6.3.7(3) an element
$$f_{i+1} \in U(f_i, \delta)$$
for some $0 < \delta < \frac{1}{4}\varepsilon_i$, a finite set $F \subseteq X$ and a $\gamma > 0$ such that
$$N(f_{i+1}, F, \gamma) \cap \big(G_{i+1} \cap U(f_i, \tfrac{1}{4}\varepsilon_i)\big) = \emptyset.$$

Now let $\varepsilon_{i+1}$ be a positive real number smaller than
$$\min\{\gamma, \tfrac{1}{4}\varepsilon_i - \delta, 3^{-i-1} \cdot \varepsilon_i\}.$$
It is clear that $f_{i+1}$ and $\varepsilon_{i+1}$ satisfy the inductive hypotheses. □

**Remark 6.3.9.** Observe that for *countable* spaces, Theorem 6.3.6 is a consequence of Theorem 6.3.8.

$C_p(X)$ **can be** $F_{\sigma\delta}$**.** We saw that unless $X$ is discrete, $C_p(X)$ cannot be $F_\sigma$ or $G_{\delta\sigma}$ in $\mathbb{R}^X$. But it can be $F_{\sigma\delta}$ in $\mathbb{R}^X$.

**Theorem 6.3.10.** *Let $X$ be a countable metrizable space. Then $C_p(X)$ is an $F_{\sigma\delta}$-subset of $\mathbb{R}^X$.*

**Proof.** Let $\varrho$ be an admissible metric for $X$, and let $x \in X$. Then the $\varepsilon$-$\delta$ definition of continuity shows that $C_p(X)$ is the set
$$\bigcap_{x \in X} \bigcap_{n=1}^{\infty} \bigcup_{m=1}^{\infty} \{g \in \mathbb{R}^X : g[B(x, 1/m)] \subseteq [g(x) - 1/n, g(x) + 1/n]\}.$$
Since $X$ is countable and each set
$$\{g \in \mathbb{R}^X : g[B(x, 1/m)] \subseteq [g(x) - 1/n, g(x) + 1/n]\}$$
is closed in $\mathbb{R}^X$, $C_p(X)$ is indeed an $F_{\sigma\delta}$-subset of $\mathbb{R}^X$, as claimed. □

So we conclude that all familiar countable metrizable spaces, such as the rational numbers or a convergent sequence with its limit, have function spaces of the same Borel complexity. In fact, as we will see later by using the machinery on $\mathcal{F}_{\sigma\delta}$-absorbers developed in Chapter 5, they are all homeomorphic.

The metrizability is not essential in this result. There are examples of countable, regular, nonmetrizable spaces $X$ for which $C_p(X)$ is an $F_{\sigma\delta}$-subset of $\mathbb{R}^X$. We will come back to this later (see Example **E11** in §6.13).

## 6.4. The Baire property in function spaces

Being a Baire space is an important topological property for a space and it is therefore natural to ask when function spaces are Baire. We already saw in Corollary 6.3.3 that if $X$ is countable and nondiscrete then if $C_p(X)$ is Baire it is not a Borel subset of $\mathbb{R}^X$. This indicates that Baire $C_p(X)$ are rather rare.

A family of subsets $\mathcal{A}$ of a space $X$ is called *strongly discrete* provided that every $A \in \mathcal{A}$ has a neighborhood $U(A)$ such that the family
$$\{U(A) : A \in \mathcal{A}\}$$
is discrete.

**Lemma 6.4.1.** *Let $X$ be a space and suppose that $C_p(X)$ is Baire. Then every pairwise disjoint sequence of finite subsets of $X$ has an infinite strongly discrete subsequence.*

**Proof.** Let $\mathcal{A}$ be an infinite sequence of pairwise disjoint finite subsets of $X$. If $\mathcal{A}$ is finite then there is nothing to prove, so assume it is infinite. For each $n \in \mathbb{N}$ let
$$G_n = \{f \in C_p(X) : (\exists A \in \mathcal{A})(f[A] \subseteq (n, n + \tfrac{1}{2}))\}$$
Then $G_n$ is clearly an open subset of $C_p(X)$ since every $A \in \mathcal{A}$ is finite. We claim that it is also dense in $C_p(X)$. To this end, let $f \in C_p(X)$, $F \subseteq X$ be finite and $\varepsilon > 0$ be arbitrary. Since $\mathcal{A}$ is infinite and pairwise disjoint and $F$ is finite there is an element $A \in \mathcal{A}$ such that $A \cap F = \emptyset$. Define $\hat{f} \colon A \cup F \to \mathbb{R}$ by
$$\hat{f}(x) = \begin{cases} f(x) & (x \in F), \\ n + \tfrac{1}{4} & (x \in A), \end{cases}$$
and let $g \colon X \to \mathbb{R}$ be an arbitrary continuous extension of $\hat{f}$. Then clearly
$$g \in N(f, F, \varepsilon) \cap G_n,$$
and so $G_n$ meets all nonempty basic open sets in $C_p(X)$.

By assumption, $C_p(X)$ is a Baire space and so there is an element
$$f \in \bigcap_{n=1}^{\infty} G_n.$$
For each $n$ pick $A_n \in \mathcal{A}$ such that $f[A_n] \subseteq (n, n + \tfrac{1}{2})$. Since the family
$$\{(n, n + \tfrac{1}{2}) : n \in \mathbb{N}\}$$
is discrete in $\mathbb{R}$ and $f$ is continuous, the same is true for the family
$$\{f^{-1}[(n, n + \tfrac{1}{2})] : n \in \mathbb{N}\}.$$
We conclude that the subsequence $\{A_n : n \in \mathbb{N}\}$ of $\mathcal{A}$ is strongly discrete. $\square$

Interestingly, the converse to Lemma 6.4.1 is also true. So there is a purely topological characterization of the spaces $X$ for which $C_p(X)$ is a Baire space. Before we can prove this, we first formulate and prove the following result.

**Lemma 6.4.2.** *Let $J$ be a closed and bounded interval in $\mathbb{R}$ and let*
$$f \colon X \to J$$
*be continuous. In addition, let $A \subseteq X$ be closed, $F \subseteq X$ finite, $\varepsilon > 0$ and $\tau \colon F \to J$ be such that $|\tau(x) - f(x)| < \varepsilon$ for all $x \in F \cap A$. Then there is a continuous function*
$$g \colon X \to J$$
*such that $g \restriction F = \tau$ and $|g(x) - f(x)| < \varepsilon$ for all $x \in A$.*

## 6.4. THE BAIRE PROPERTY IN FUNCTION SPACES

**Proof.** The function $f\colon X \to J$ can be extended to a continuous function $\bar{f}\colon \beta X \to J$. We let $\bar{A}$ denote the closure of $A$ in $\beta X$. Extend the function
$$\tau - f \restriction F \cap A \colon F \cap A \to (-\varepsilon, \varepsilon)$$
to a continuous function $h\colon \bar{A} \to (-\varepsilon, \varepsilon)$ (use that $F$ is finite). Since $\beta X$ is compact and Hausdorff it is normal from which it follows that the function
$$(\bar{f} + h \restriction \bar{A}) \cup (\tau \restriction F \setminus A)$$
can be extended to a continuous function $\tilde{g}\colon \beta X \to J$. Then $g = \tilde{g} \restriction X$ is clearly as required. □

It will be convenient to introduce the following notation. If $F \neq \emptyset$ and
$$U = N(f, F, \varepsilon)$$
is the basic neighborhood of $f \in \mathbb{R}^X$ determined by $F \subseteq X$ and $\varepsilon > 0$ then we put
$$S(U) = F, \quad m(U) = \sup\{|g(a)| : g \in U, a \in F\}.$$
Observe that $0 \leq m(U) < \infty$, and also that $N(f, F, \varepsilon) = N(g, G, \delta)$ if and only if $F = G$, $\varepsilon = \delta$ and $f \restriction F = g \restriction G$.

**Theorem 6.4.3.** *Let $X$ be a space. The following statements are equivalent:*

(1) *$C_p(X)$ is a Baire space.*
(2) *Every pairwise disjoint sequence of finite subsets of $X$ has an infinite strongly discrete subsequence.*

**Remark 6.4.4.** Since a discrete space obviously has the property that every pairwise disjoint sequence of finite sets has a strongly discrete subsequence, a corollary to this this result is the nontrivial fact that $\mathbb{R}^X$ is a Baire space for arbitrary $X$.

**Proof.** By Lemma 6.4.1, it suffices to prove the implication (2) ⇒ (1). For that it suffices by Exercise 1.6.6 to prove that $C_p(X)$ is not meager in itself.

So let $X$ be a space having the property stated in (2), and let
$$\{F_n : n \in \mathbb{N}\}$$
be a sequence of nowhere dense subsets of $C_p(X)$. Since $C_p(X)$ is dense in $\mathbb{R}^X$, each $F_n$ is nowhere dense in $\mathbb{R}^X$ as well. We may assume without loss of generality that
$$F_1' \subseteq F_2' \subseteq \cdots \subseteq F_n' \subseteq \cdots.$$
We define three sequences by induction as follows: an increasing sequence
$$\{S_n : n \in \mathbb{N}\}$$
of finite subsets of $X$, an increasing sequence $\{m_n : n \in \mathbb{N}\}$ of positive numbers, and a collection $\{\mathcal{U}_n : n \in \mathbb{N}\}$ of finite families of basic open sets in $\mathbb{R}^X$ such that, for each $n$, the following conditions are satisfied:

(i) for every $U \in \mathcal{U}_n$ there exists an element $U' \in \mathcal{U}_{n+1}$ such that $U' \subseteq U$,
(ii) $U \cap F_n = \emptyset$ for all $U \in \mathcal{U}_n$,
(iii) $S(U) \subseteq S_n$ for all $U \in \mathcal{U}_n$,
(iv) $m(U) \leq m_n$ for all $U \in \mathcal{U}_n$.
(v) for all $f \in [-m_n, m_n]^X$ there exists an element $U \in \mathcal{U}_{n+1}$ such that $U \subseteq N(f, S_n, 1/n)$.

Since $F_1$ is nowhere dense, there is a nonempty basic open set $U \in \mathbb{R}^X$ such that $U \cap F_1 = \emptyset$. Put $S_1 = S(U)$, $m_1 = m(U)$ and $\mathcal{U}_1 = \{U\}$. Suppose that $S_n$, $m_n$ and $\mathcal{U}_n$ have been defined satisfying (ii) through (v). Let $Z$ be the compact cube $[-m_n, m_n]^X$. There consequently exists a finite $G \subseteq Z$ such that
$$Z \subseteq \bigcup \{N(g, S_n, 1/2n) : g \in G\}.$$
Consider the finite collection
$$\mathcal{H} = \mathcal{U}_n \cup \{N(g, S_n, 1/2n) : g \in G\}.$$
For every $H \in \mathcal{H}$ let $U_H$ be a nonempty basic open subset of $\mathbb{R}^X$ such that
$$U_H \subseteq H \setminus F_{n+1}$$
(here we use that $F_{n+1}$ is nowhere dense). Now put
$$S_{n+1} = S_n \cup \bigcup_{H \in \mathcal{H}} S(U_H),$$
$$m_{n+1} = m_n + \sum_{H \in \mathcal{H}} m(U_H),$$
$$\mathcal{U}_{n+1} = \{U_H : H \in \mathcal{H}\},$$
respectively.

**Claim 1.** These choices satisfy the inductive requirements.

*Proof.* It is clear that (i) through (iv) are satisfied. To check (v) for $n+1$, let $f \in [-m_n, m_n]^X$. There exists $g \in G$ such that
$$f \in H = N(g, S_n, 1/2n).$$
Pick an arbitrary $h \in U_H$ and let $x \in S_n$. Then
$$|f(x) - h(x)| \leq |f(x) - g(x)| + |g(x) - h(x)|$$
$$< 1/2n + 1/2n$$
$$= 1/n.$$
We conclude that $U_H \subseteq N(f, S_n, 1/n)$ and since $U_H \in \mathcal{U}_{n+1}$, we are done. ◊

The pairwise disjoint sequence
$$\{S_{n+1} \setminus S_n : n \in \mathbb{N}\}$$

has by our assumptions on $X$ a strongly discrete subsequence, say,
$$\{S_{n_k+1} \setminus S_{n_k} : k \in \mathbb{N}\}.$$
We may assume without loss of generality that
$$n_{k+1} \geq \max\{n_k + 2, 2k + 1\}$$
for every $k$. For each $k \in \mathbb{N}$ let
$$T_{2k-1} = S_{n_k},$$
$$T_{2k} = S_{n_k+1},$$
$$M_{2k-1} = m_{n_k},$$
$$M_{2k} = m_{n_k+1},$$
$$\mathcal{V}_{2k-1} = \mathcal{U}_{n_k},$$
$$\mathcal{V}_{2k} = \mathcal{U}_{n_k+1},$$
respectively. We claim that the following statements hold:

(1) $V \cap F_n = \emptyset$ for all $V \in \mathcal{V}_n$,
(2) $S(V) \subseteq T_n$ for all $V \in \mathcal{V}_n$,
(3) $f[S(V)] \subseteq [-M_n, M_n]$ for all $f \in V \in \mathcal{V}_n$,
(4) if $f \in [-M_n, M_n]^X$ then there is an element $V \in \mathcal{V}_{n+1}$ such that $V \subseteq N(f, T_n, 1/n)$.

Statements (1) through (3) are easily verified. Simply use that all sequences considered so far are increasing. For (4), let $f \in [-M_n, M_n]^X$ and assume that $n = 2k$ (the case for odd $n$ follows by a similar argumentation). By (v) there exists an element $U \in \mathcal{U}_{n_k+2}$ such that
$$U \subseteq N(f, S_{n_k+1}, 1/{n_k+1}).$$
Since $n_{k+1} \geq n_k + 2$ it follows that $T_n = S_{2k} \subseteq S_{n_k+1}$. Moreover,
$$1/{n_k+1} \leq 1/{2k} = 1/n,$$
so that
$$U \subseteq N(f, T_n, 1/n),$$
as required.

Let $\{W_{2k} : k \in \mathbb{N}\}$ be a discrete family of open subsets of $X$ such that for each $k$:

(5) $T_{2k} \setminus T_{2k-1} \subseteq W_{2k}$,
(6) $T_{2k-1} \cap W_{2k} = \emptyset$.

For each $k$ let

(7) $D_{2k} = X \setminus \bigcup \{W_{2i} : i > k\}$.

We will now define an increasing sequence $\{j_n : n \in \mathbb{N}\}$ of positive even integers, a sequence $\{V_n : n \in \mathbb{N}\}$ of basic open subsets in $\mathbb{R}^X$, a sequence $\{f_n : n \in \mathbb{N}\}$ of elements of $C_p(X)$ and, finally, a sequence $\{\varepsilon_n : n \in \mathbb{N}\}$ of positive real numbers, having for each $n$ the following properties:

(8) $\varepsilon_{n+1} \leq \varepsilon_n/2$,
(9) $N(f_n, T_{j_n}, 3\varepsilon_n) \subseteq V_n \in \mathcal{V}_{j_n}$,
(10) $f_n \in [-M_{j_n}, M_{j_n}]^X$,
(11) $|f_{n+1}(x) - f_n(x)| < \varepsilon_n$ for all $x \in D_{j_n}$.

To begin the induction, let $j_1 = 2$, let $V_1 \in \mathcal{V}_2$ and $f \in V_1$. By (3),
$$f[S(V_1)] \subseteq [-M_2, M_2].$$
By the Tietze-Urysohn Extension Theorem there is an $f_1 \in C_p(X)$ such that $f_1 \in [-M_2, M_2]^X$ and $f_1 \upharpoonright S(V_1) = f \upharpoonright S(V_1)$. Then $f_1 \in V_1$ and since $S(V_1) \subseteq T_2$ there exists $\varepsilon_1 > 0$ such that
$$N(f_1, T_2, 3\varepsilon_1) \subseteq V_1.$$
Assume that $j_n, V_n, f_n$ and $\varepsilon_n$ have been defined satisfying (8) through (11). Then let $j_{n+1}$ be an even integer greater than $\max\{j_n, 1 + 1/\varepsilon_n\}$. By (10), (4), (i) and the fact that $j_{n+1} - j_n \geq 2$ there exists a $V_{n+1} \in \mathcal{V}_{j_{n+1}}$ such that
$$V_{n+1} \subseteq N\big(f_n, T_{j_{n+1}-1}, 1/(j_{n+1} - 1)\big).$$
Pick an arbitrary element $f \in V_{n+1}$. By (5) and (7),

(12) $\qquad T_{j_{n+1}} \setminus T_{j_{n+1}-1} \subseteq W_{j_{n+1}} \subseteq X \setminus D_{j_n}.$

By (3),
$$f[S(V_{n+1})] \subseteq [-M_{j_{n+1}}, M_{j_{n+1}}]$$
and by (10), $f_n \in [-M_{j_n}, M_{j_n}]^X \subseteq [-M_{j_{n+1}}, M_{j_{n+1}}]^X$. We want to apply Lemma 6.4.2 with the closed set $A = D_{j_n}$ and the functions $f_n$ and
$$\tau = f \upharpoonright T_{j_{n+1}}.$$
Observe that if $x \in D_{j_n} \cap T_{j_{n+1}}$ then by (12), $x \in T_{j_{n+1}-1}$ so that since
$$f \in N\big(f_n, T_{j_{n+1}-1}, 1/(j_{n+1} - 1)\big)$$
we get
$$|f_n(x) - f(x)| = |f_n(x) - \tau(x)| < 1/(j_{n+1} - 1).$$
So by Lemma 6.4.2 there is an element $f_{n+1} \in C_p(X)$ such that
$$f_{n+1} \in [-M_{j_{n+1}}, M_{j_{n+1}}]^X$$
while moreover
$$|f_{n+1}(x) - f_n(x)| < 1/(j_{n+1} - 1) < \varepsilon_n$$
for all $x \in D_{j_n}$ and
$$f_{n+1} \upharpoonright T_{j_{n+1}} = f \upharpoonright T_{j_{n+1}}.$$

Then $f_{n+1} \in V_{n+1}$ and so there exists an $0 < \varepsilon_{n+1} \leq \varepsilon_n/2$ such that
$$N(f_{n+1}, T_{j_{n+1}}, 3\varepsilon_{n+1}) \subseteq V_{n+1}.$$
This completes the inductive construction.

Condition (11) can be modified to say that

(13) $|f_m(x) - f_n(x)| < 2\varepsilon_k$ whenever $m, n \geq k$ and $x \in D_{j_k}$.

This follows from (11) and (8) since
$$\begin{aligned}
|f_m(x) - f_n(x)| &= |f_m(x) - f_{m-1}(x) + \cdots + f_{n+1}(x) - f_n(x)| \\
&\leq |f_m(x) - f_{m-1}(x)| + \cdots + |f_{n+1}(x) - f_n(x)| \\
&< \varepsilon_{m-1} + \cdots + \varepsilon_{n+1} + \varepsilon_n \\
&\leq \frac{\varepsilon_n}{2^{m-n-1}} + \cdots + \frac{\varepsilon_n}{2} + \varepsilon_n \\
&< 2\varepsilon_n \\
&\leq 2\varepsilon_k.
\end{aligned}$$

In particular, (13) says that for each $k$, $\{f_n : n \in \mathbb{N}\}$ is uniformly convergent on $D_{j_k}$.

Therefore, $\{f_n : n \in \mathbb{N}\}$ converges pointwise to an $f \in \mathbb{R}^X$ such that for every $k$ the function $f \upharpoonright D_{j_k}$ is continuous (cf. Lemma A.3.1). Since the collection $\{W_{2k} : k \in \mathbb{N}\}$ is discrete, $f \in C_p(X)$.

Finally, we show that $f \in V_n$ for each $n$. To this end, let $n$ be fixed and let $x \in S(V_n)$. By (2), (6) and (7), $x \in T_{j_n} \subseteq D_{j_n}$. If $m > n$ then by (13),
$$|f_m(x) - f_n(x)| < 2\varepsilon_n.$$
This means that
$$|f(x) - f_n(x)| \leq 2\varepsilon_n < 3\varepsilon_n.$$
So by (9),
$$f \in N(f_n, T_{j_n}, 3\varepsilon_n) \subseteq V_n.$$
So by (1) we get that
$$f \notin \bigcup_{n=1}^{\infty} F_n,$$
as required. □

**Corollary 6.4.5.** Let $X$ be a space for which $C_p(X)$ is Baire. If $Y$ is a subspace of $X$ then $C_p(Y)$ is Baire.

**Proof.** This follows directly from Theorem 6.4.3 since the property that every sequence of pairwise disjoint finite sets has a strongly discrete subsequence is hereditary. □

So a Baire $C_p(X)$ has many Baire subspaces. The question arises whether there is a nondiscrete space $X$ whose function space $C_p(X)$ is hereditary Baire. For countable spaces $X$ this is determined by the question whether $\mathbb{Q}$ admits a closed imbedding in $C_p(X)$ (Corollary 1.9.13). There is a consistent example of such a space, see Page 393 below.

We now turn our attention to the Baire property in products of function spaces.

**Corollary 6.4.6.** *Let $X$ and $Y$ be such that both $C_p(X)$ and $C_p(Y)$ are Baire. Then $C_p(X) \times C_p(Y)$ is a Baire space.*

**Proof.** By Theorem 6.2.8, $C_p(X) \times C_p(Y)$ is homeomorphic to $C_p(X \oplus Y)$. Since $X$ and $Y$ both have the property in Theorem 6.4.3(2), it is clear that $X \oplus Y$ has this property as well. So we are done by the characterization result Theorem 6.4.3. □

Observe that virtually the same proof shows that if $\{X_i : i \in I\}$ is an arbitrary family of spaces such that $C_p(X_i)$ is a Baire space for every $i$ then the space $\prod_{i \in I} C_p(X_i)$ is a Baire space as well. The precise verification of this is left as an exercise to the reader.

In Theorem A.6.10 we proved that the product of two separable metrizable Baire spaces is again a Baire space. It is not possible to generalize the obvious proof of that result to metrizable spaces. The question whether the product of two metrizable Baire spaces is again a Baire space was raised in CHOQUET [99], OXTOBY [331] and SIKORSKI [374]. Using forcing techniques it was answered in the negative by COHEN [101]. It was known earlier from the work of OXTOBY [331] and KROM [232] that the Continuum Hypothesis implies the existence of two metrizable Baire spaces whose product is not Baire. In FLEISSNER and KUNEN [163] such spaces were constructed in ZFC alone. They even constructed in the same paper an example of a single metrizable Baire space whose square is not Baire.

In view of Corollary 6.4.6 the question arises whether the product of *linear* Baire spaces is Baire. If so, then the corollary would be a consequence of the fact that $C_p(X)$ is a linear space. But there are even *normed* Baire spaces whose product is not Baire. This is due independently to VAN MILL and POL [302] and VALDIVIA [92].

The conclusion is that the spaces $C_p(X)$ are linear spaces with a particularly rich structure. (This conclusion is not very surprising.)

We now present some examples of spaces $X$ for which $C_p(X)$ is Baire, or not.

**A countable dense-in-itself space $X$ such that $C_p(X)$ is Baire.** There is according to VAN DOUWEN [130, 134] a countable space $X$ without isolated points having the following curious property: no point of $X$ is simultaneously an accumulation point of two disjoint subsets of $X$ (hence $X$ is perfectly disconnected, see the notes of §A.1 and Exercise A.1.2 for more information). We will show that this space $X$ has the property that its function space $C_p(X)$ is a Baire space. So $C_p(X)$ can be Baire without $X$ being Baire.

We first claim that $X$ is a so-called *nodec space*, that is, a space in which every nowhere dense set is closed. To prove this, let $A \subseteq X$ be nowhere dense. If $A$ is not closed then there exists an $x \in \overline{A} \setminus A$. Then clearly $x \in \overline{A \setminus \{x\}}$ which implies that there is an open neighborhood $U$ of $x$ which is contained in $A \cup \{x\}$ (Exercise A.1.2). Since $X$ has no isolated points this implies that $U \setminus \{x\}$ is a nonempty open subset of $X$ which is contained in $A$, which contradicts the fact that $A$ is nowhere dense.

Observe that in a nodec space every nowhere dense set is not only closed but also discrete.

Let $\mathcal{F}$ be an arbitrary countably infinite collection pairwise disjoint finite subsets of $X$, and let $F_0 \in \mathcal{F}$ be arbitrary.

Suppose first that every neighborhood $U$ of $F_0$ intersects all but finitely many $F \in \mathcal{F} \setminus \{F_0\}$.

**Claim 1.** There are a point $x \in F_0$ and an infinite subcollection $\mathcal{F}'$ of $\mathcal{F} \setminus \{F_0\}$ such that every neighborhood of $x$ intersects all but finitely many members from $\mathcal{F}'$.

*Proof.* We prove this by induction on the cardinality of $F_0$. If $|F_0| = 1$ then $\mathcal{F}' = \mathcal{F} \setminus \{F_0\}$ is as required. Now suppose that we have what we want if $|F_0| = n$, and suppose that $|F_0| = n+1$. Fix an arbitrary $x \in F_0$. If every neighborhood of $F_0 \setminus \{x\}$ intersects all but finitely many members of $\mathcal{F} \setminus \{F_0\}$ then we get what we want by our inductive assumption. If not, then there is a neighborhood $U$ of $F_0 \setminus \{x\}$ such that the collection

$$\{F \in \mathcal{F} \setminus \{F_0\} : F \cap U = \emptyset\}$$

is infinite. But then every neighborhood of $x$ intersects infinitely many members from $\mathcal{F} \setminus \{F_0\}$. ◇

So let $x$ and $\mathcal{F}'$ be as in the claim. Since $x \notin \bigcup \mathcal{F}'$, it follows that $\mathcal{F}'$ is infinite. Split $\mathcal{F}'$ into two disjoint infinite subcollections, say $\mathcal{F}'_0$ and $\mathcal{F}'_1$. Then by Claim 1, $x$ is an accumulation point of $\bigcup \mathcal{F}'_0$ as well as $\bigcup \mathcal{F}'_1$. This is a contradiction since $\bigcup \mathcal{F}'_0$ and $\bigcup \mathcal{F}'_1$ are disjoint and $X$ is perfectly disconnected.

So our assumption that every neighborhood $U$ of $F_0$ intersects all but finitely many $F \in \mathcal{F} \setminus \{F_0\}$ leads to a contradiction. Let $C_0$ be a clopen neighborhood of $F_0$ for which there exists an infinite subcollection $\mathcal{F}_0$ of the collection $\mathcal{F} \setminus \{F_0\}$ such that $\bigcup \mathcal{F}_0 \subseteq X \setminus C_0$. Let $F_1 \in \mathcal{F}_0$ be arbitrary. By the same argument, there is a clopen neighborhood $C_1 \subseteq X \setminus C_0$ of $F_1$ and an infinite subcollection $\mathcal{F}_1$ of $\mathcal{F}_0 \setminus \{F_1\}$ such that $\bigcup \mathcal{F}_1 \subseteq X \setminus C_1$. Observe that in fact $\bigcup \mathcal{F}_1 \subseteq X \setminus (C_0 \cup C_1)$. Continuing in this way inductively, one easily constructs a subsequence $\{F_n : n \in \omega\}$ of $\mathcal{F}$ having pairwise disjoint clopen neighborhoods. But then $S = \bigcup_{n \in \omega} F_n$ is discrete, hence nowhere dense and hence closed since $X$ is nodec. Since $X$ is Lindelöf, being countable, this easily implies that $\{F_n : n \in \omega\}$ is strongly discrete.

For an easier example of a countable nondiscrete space whose function space is Baire, see Corollary 6.5.7.

**A pseudocompact space.** It is clear that a pseudocompact space $X$ has the property that its function space $C_p(X)$ is not Baire. This is so, for example, since in a pseudocompact space any discrete family of open sets is finite. It can also be shown directly, as follows. Simply observe that the subspaces
$$\{f \in C_p(X) : f[X] \subseteq [-n, n]\}$$
are closed and nowhere dense in $C_p(X)$ and cover $C_p(X)$ in case $X$ is pseudocompact.

Our aim now is to construct an example of a particular pseudocompact space which demonstrates that for a space $X$ the property of having a Baire $C_p(X)$ is not determined by its countable subspaces.

A point $p$ of a topological space $X$ is called a *weak P-point* if for every countable $A \subseteq X \setminus \{x\}$ we have $x \notin \overline{A}$.

KUNEN [236] proved that $\mathbb{N}^*$ contains many weak $P$-points. Here $\mathbb{N}$ has the discrete topology. In fact, if $\mathcal{U}$ is a countably infinite collection of pairwise disjoint nonempty clopen subsets of $\mathbb{N}^*$ then
$$\overline{\bigcup \mathcal{U}} \setminus \bigcup \mathcal{U}$$
contains a weak $P$-point. Let $K$ denote the set of all weak $P$-points in $\mathbb{N}^*$ and observe that $K$ is dense in $\mathbb{N}^*$.

**Lemma 6.4.7.** *$K$ is pseudocompact.*

**Proof.** Let $\mathcal{A}$ be a countably infinite discrete collection of nonempty clopen subsets of $K$. We will derive a contradiction. Since $K$ is zero-dimensional, being a subspace of $\mathbb{N}^*$, this guarantees pseudocompactness of $K$. For every $A \in \mathcal{A}$ let $C_A$ be an open subset of $\mathbb{N}^*$ such that $C_A \cap K = A$. Observe that the collection
$$\{C_A : A \in \mathcal{A}\}$$
is pairwise disjoint since $K$ is dense. For every $A \in \mathcal{A}$ let $B_A$ be a nonempty clopen subspace of $\mathbb{N}^*$ such that $B_A \subseteq C_A$. Since the collection
$$\{B_A : A \in \mathcal{A}\}$$

is pairwise disjoint, by the above there is a weak $P$-point in
$$\overline{\bigcup_{A \in \mathcal{A}} B_A} \setminus \bigcup_{A \in \mathcal{A}} B_A,$$
say $x$. So $x \in K$. But this easily contradicts the fact that $\mathcal{A}$ is discrete. □

We conclude that $C_p(K)$ is not Baire since $K$ is pseudocompact. However, every countable subspace of $K$ is discrete. So every countable subspace $A \subseteq K$ has the property that $C_p(A) = \mathbb{R}^A$ is Baire (even topologically complete). This proves indeed that for a space $X$ the property of having a Baire $C_p(X)$ is not determined by its countable subspaces.

## 6.5. Filters and the Baire property in $C_p(\mathbb{N}_{\mathcal{F}})$

For each filter $\mathcal{F}$ on $\mathbb{N}$ we can topologize the set $\hat{\mathbb{N}} = \mathbb{N} \cup \{\infty\}$ as follows: each element of $\mathbb{N}$ is isolated and the collection $\{U \cup \{\infty\} : U \in \mathcal{F}\}$ is a neighborhood base at $\infty$. This topological space will be denoted by $\mathbb{N}_{\mathcal{F}}$. It is an easy exercise to verify that $\mathbb{N}_{\mathcal{F}}$ is Tychonoff if and only if $\mathcal{F}$ is a free filter on $\mathbb{N}$.

The spaces $C_p(\mathbb{N}_{\mathcal{F}})$ are of crucial importance in our analysis of the topological type of function spaces of low Borel complexity in §6.12. Here we are interested in characterizing when such a space has the property that its function space is a Baire space.

**The Cantor set.** We consider the Cantor set $\{0,1\}^\infty$ with its usual Tychonoff product topology. If $\sigma$ and $\tau$ are finite disjoint (possibly empty) subsets of $\mathbb{N}$ then the set
$$\langle \sigma, \tau \rangle = \{x \in \{0,1\}^\infty : (n \in \sigma \Rightarrow x_n = 0) \wedge (n \in \tau \Rightarrow x_n = 1)\}$$
is a basic clopen subset of $\{0,1\}^\infty$. Notice that if $\sigma_1, \tau_1, \sigma_2$ and $\tau_2$ are pairwise disjoint finite subsets of $\mathbb{N}$ then
$$\langle \sigma_1, \tau_1 \rangle \cap \langle \sigma_2, \tau_2 \rangle = \langle \sigma_1 \cup \sigma_2, \tau_1 \cup \tau_2 \rangle.$$

**Lemma 6.5.1.** *Let $F \subseteq \{0,1\}^\infty$ be nowhere dense. Then for every finite subset $A \subseteq \mathbb{N}$ there are disjoint finite subsets $\alpha$ and $\beta$ of $\mathbb{N} \setminus A$ such that if $\gamma$ and $\delta$ are arbitrary complementary subsets of $A$ then*
$$\langle \gamma, \delta \rangle \cap \langle \alpha, \beta \rangle \cap F = \emptyset.$$

**Proof.** The proof is basically a triviality. Let $\gamma_1, \delta_1, \ldots \gamma_n, \delta_n$ list all the complementary pairs of subsets of $A$. Consider the nonempty clopen subset $\langle \gamma_1, \delta_1 \rangle$ of $\{0,1\}^\infty$. Since $F$ is nowhere dense, the set $\langle \gamma_1, \delta_1 \rangle \setminus F$ has nonempty interior. There are consequently disjoint finite subsets $\alpha_1$ and $\beta_1$ of $\mathbb{N} \setminus A$ such that
$$\langle \gamma_1, \delta_1 \rangle \cap \langle \alpha_1, \beta_1 \rangle \cap F = \emptyset.$$

Now consider the nonempty clopen subset
$$\langle \gamma_2, \delta_2 \rangle \cap \langle \alpha_1, \beta_1 \rangle = \langle \gamma_2 \cup \alpha_1, \delta_2 \cup \beta_1 \rangle$$
of $\{0,1\}^\infty$. By a similar argument there are disjoint finite subsets $\alpha_2$ and $\beta_2$ of $\mathbb{N} \setminus (A \cup \alpha_1 \cup \beta_1)$ such that
$$\langle \gamma_2, \delta_2 \rangle \cap \langle \alpha_1, \beta_1 \rangle \cap \langle \alpha_2, \beta_2 \rangle \cap F = \emptyset.$$
If we repeat this process $n$ times, we dealt with all possibilities. It is clear that the finite disjoint sets
$$\alpha = \bigcup_{i=1}^{n} \alpha_i, \quad \beta = \bigcup_{i=1}^{n} \beta_i$$
are as required. □

**Set theoretic characterization of meager filters.** Each subset $A$ of $\mathbb{N}$ can be identified with a point in the Cantor set $\{0,1\}^\infty$, namely with its characteristic function $\chi_A$. So $\mathcal{P}(\mathbb{N})$ can be topologized by using the identification $A \mapsto \chi_A$, as follows: $\mathcal{U} \subseteq \mathcal{P}(\mathbb{N})$ is open if and only if
$$\{\chi_A : A \in \mathcal{U}\}$$
is open in $\{0,1\}^\infty$. The set $\mathcal{P}(\mathbb{N})$ topologized in this way is a Cantor set because $\{0,1\}^\infty$ is.

For $A, S \in \mathcal{P}(\mathbb{N})$ put
$$V(A, S) = \{B \in \mathcal{P}(\mathbb{N}) : B \cap S = A \cap S\}$$
(cf. Page 462). Observe that the topology of $\mathcal{P}(\mathbb{N})$ is generated by all sets of the form $V(A, S)$, where $S$ is finite. Also notice that the finite subsets of $\mathbb{N}$ form a (countable) dense subset of $\mathcal{P}(\mathbb{N})$.

**Lemma 6.5.2.** *Let $\mathcal{F} \subseteq \mathcal{P}(\mathbb{N})$ be nowhere dense. Then for every finite set $A \subseteq \mathbb{N}$ there are a finite set $C \subseteq \mathbb{N} \setminus A$ and a subset $D \subseteq C$ such that for every $B \subseteq \mathbb{N}$ we have*
$$V(B, A) \cap V(D, C) \cap \mathcal{F} = \emptyset.$$

**Proof.** This is exactly the statement in Lemma 6.5.1 translated in terms of the set theoretic language of $\mathcal{P}(\mathbb{N})$. Let $F = \{\chi_H : H \in \mathcal{F}\}$. Then $F$ is nowhere dense in $\{0,1\}^\infty$. Let $\alpha$ and $\beta$ be the sets given by Lemma 6.5.1 and put $D = \alpha$ and $C = \alpha \cup \beta$. Now let $B \subseteq \mathbb{N}$ be arbitrary, and put $\delta = B \cap A$ and $\gamma = A \setminus B$, respectively. Then $\langle \gamma, \delta \rangle \cap \langle \alpha, \beta \rangle \cap F = \emptyset$ implies
$$V(B, A) \cap V(D, C) \cap \mathcal{F} = \emptyset.$$
Simply observe that
$$\langle \alpha, \beta \rangle = \{\chi_H : H \in V(D, C)\},$$
$$\langle \gamma, \delta \rangle = \{\chi_H : H \in V(B, A)\}.$$

So we are done. □

A family $\mathcal{F}$ of subsets of a set $X$ is said to be *closed under supersets* provided that for all $F \in \mathcal{F}$ and $G \in \mathcal{P}(X)$ such that $F \subseteq G$ it follows that $G \in \mathcal{F}$. Note that any filter on $X$ is closed under supersets.

If $\mathcal{F}$ is a collection of subsets of $\mathbb{N}$ then by the phrase '$\mathcal{F}$ is meager' we mean that $\mathcal{F}$ is a meager subset of the space $\mathcal{P}(\mathbb{N})$. Similarly for the phrase '$\mathcal{F}$ is not meager'.

We now come to the following interesting characterization of meager filters on $\mathbb{N}$.

**Theorem 6.5.3.** *Let $\mathcal{F} \subseteq \mathcal{P}(\mathbb{N})$ be closed under supersets. The following statements are equivalent:*

(1) *$\mathcal{F}$ is meager.*
(2) *There is a sequence $(A_n)_n$ of pairwise disjoint finite subsets of $\mathbb{N}$ such that for every $F \in \mathcal{F}$ the set*
$$\{n \in \mathbb{N} : F \cap A_n = \emptyset\}$$
*is finite.*

**Proof.** We first prove that (2) implies (1). This is simple. Indeed, let the sequence $(A_n)_n$ be given. For every $n \in \mathbb{N}$ put
$$\mathcal{D}_n = \{B \in \mathcal{P}(\mathbb{N}) : (\forall m \geq n)(B \cap A_m \neq \emptyset)\}.$$
We will prove that every $\mathcal{D}_n$ is a closed and nowhere dense subset of $\mathcal{P}(\mathbb{N})$, which suffices since by assumption $\mathcal{F}$ is covered by the $\mathcal{D}_n$'s. So fix $n$, and assume that $B \notin \mathcal{D}_n$. Then there exists $m \geq n$ such that $B \cap A_m = \emptyset$. So $V(B, A_m)$ is a clopen neighborhood of $B$ which misses $\mathcal{D}_n$. This proves that $\mathcal{D}_n$ is closed. To prove it is nowhere dense, let $B, A \in \mathcal{P}(\mathbb{N})$ be arbitrary with $A$ finite. We will show that the basic neighborhood $V(B, A)$ of $B$ is not contained in $\mathcal{D}_n$. This clearly suffices. Indeed, since $A$ is finite there is an $m \geq n$ such that $A \cap A_m = \emptyset$. This implies
$$(B \setminus A_m) \cap A = B \cap A,$$
and hence
$$B \setminus A_m \in V(B, A).$$
But clearly, $B \setminus A_m \notin \mathcal{D}_n$.

To prove (1) $\Rightarrow$ (2), let
$$\mathcal{F} \subseteq \bigcup_{n=1}^{\infty} \mathcal{F}_n,$$
where for each $n$ the set $\mathcal{F}_n \subseteq \mathcal{P}(\mathbb{N})$ is nowhere dense. We may assume without loss of generality that

(∗) $\qquad\qquad \mathcal{F}_1 \subseteq \mathcal{F}_2 \subseteq \cdots \subseteq \mathcal{F}_n \subseteq \cdots .$

By Lemma 6.5.2 we may pick by induction on $n$ a finite set $A_n$ with subset $B_n$ such that

(1) $A_n \cap A_m = \emptyset$ if $n \neq m$,
(2) for every $B \in \mathcal{P}(\mathbb{N})$,
$$V\left(B, \bigcup_{k<n} A_k\right) \cap V(B_n, A_n) \cap \mathcal{F}_n = \emptyset.$$

We claim that the sequence $(A_n)_n$ is as required. If not, then for some element $X \in \mathcal{F}$ we have that $E = \{i \in \mathbb{N} : X \cap A_i = \emptyset\}$ is infinite. Since $\mathcal{F}$ is closed under supersets, it follows that $Y = X \cup \bigcup_{i \in E} B_i \in \mathcal{F}$. By (*), there is $i \in E$ such that $Y \in \mathcal{F}_i$. Since $X \cap A_i = \emptyset$ and the sequence $(A_n)_n$ is pairwise disjoint, we get
$$Y \cap A_i = B_i.$$
This shows that $Y \in V(B_i, A_i)$, and consequently that
$$Y \in V\left(Y, \bigcup_{k<i} A_i\right) \cap V(B_i, A_i) \cap \mathcal{F}_i.$$
This contradicts (2). □

This result allows us to 'decompose' meager filters in meager filters.

**Corollary 6.5.4.** *Let $\mathcal{F}$ be a filter on $\mathbb{N}$. In addition, let $\{\mathbb{N}_i : i \in \mathbb{N}\}$ be a partition of $\mathbb{N}$ into infinite pairwise disjoint sets. Put $\mathcal{F}_i = \mathcal{F} \restriction \mathbb{N}_i$ for every $i$. Then*

(1) *if $F_i \in \mathcal{F}_i$ for $i \leq n$ then $\bigcup_{i=1}^n F_i \cup \bigcup_{i>n} \mathbb{N}_i \in \mathcal{F}$.*
(2) *$\mathcal{F}_i$ imbeds as a closed subspace of $\mathcal{F}$ for every $i$,*
(3) *if $\mathcal{F}$ is a free filter on $\mathbb{N}$ and a meager subset of $\mathcal{P}(\mathbb{N})$ then there exists a partition $\{\mathbb{N}_i : i \in \mathbb{N}\}$ of $\mathbb{N}$ such that each $\mathcal{F}_i$ is a free filter on $\mathbb{N}_i$ and a meager subset of $\mathcal{P}(\mathbb{N}_i)$.*

**Proof.** For (1), for every $i \leq n$ let $\hat{F}_i \in \mathcal{F}$ be such that $\hat{F}_i \cap \mathbb{N}_i = F_i$. So
$$\bigcap_{i=1}^n \hat{F}_i \subseteq \bigcup_{i=1}^n F_i \cup \bigcup_{i>n} \mathbb{N}_i$$
and so the desired result follows by observing that $\mathcal{F}$ is closed under supersets.

For (2), simply observe that the function which assigns to each $F \in \mathcal{F}_i$ the set $F \cup (\mathbb{N} \setminus \mathbb{N}_i)$ is a closed imbedding of $\mathcal{F}_i$ into $\mathcal{F}$.

For (3), let $(A_n)_n$ be a pairwise disjoint sequence of finite subsets of $\mathbb{N}$ such as in Theorem 6.5.3. There clearly is a partition $\{\mathbb{N}_i : i \in \mathbb{N}\}$ of $\mathbb{N}$ such that every $\mathbb{N}_i$ contains infinitely many terms of the sequence $(A_n)_n$. Since $\mathcal{F}$ is free, so is every $\mathcal{F}_i$. In addition, by construction and Theorem 6.5.3, every $\mathcal{F}_i$ is a meager subset of $\mathcal{P}(\mathbb{N}_i)$. □

## 6.5. FILTERS AND THE BAIRE PROPERTY IN $C_p(\mathbb{N}_\mathcal{F})$

**Corollary 6.5.5.** *Let $\mathcal{F}$ be an ultrafilter. Then $\mathcal{F}$ is not meager.*

**Proof.** We naturally use the characterization in Theorem 6.5.3. To this end, let $\mathcal{A}$ be a countably infinite pairwise disjoint collection of finite subsets of $\mathbb{N}$. Split $\mathcal{A}$ into two countably infinite subcollections $\mathcal{B}$ and $\mathcal{C}$. Then

$$B = \bigcup \mathcal{B}, \quad C = \bigcup \mathcal{C}$$

are disjoint subsets of $\mathbb{N}$, so at most one of them belongs to $\mathcal{F}$. We may therefore assume without loss of generality that $B \notin \mathcal{F}$. Then $\mathbb{N} \setminus B \in \mathcal{F}$ by Exercise A.1.1 and misses infinitely many elements from $\mathcal{A}$. □

**The Baire property.** We now come to the announced characterization result.

**Theorem 6.5.6.** *Let $\mathcal{F}$ be a free filter on $\mathbb{N}$. The following statements are equivalent:*

(1) $C_p(\mathbb{N}_\mathcal{F})$ *is a Baire space.*
(2) $\mathcal{F}$ *is not meager.*

**Proof.** We first prove that (1) $\Rightarrow$ (2). Assume that $\mathcal{F}$ is meager. We will prove that $C_p(\mathbb{N}_\mathcal{F})$ is meager in itself and so is not a Baire space.

By Theorem 6.5.3, there is a pairwise disjoint sequence $(A_n)_n$ of finite subsets of $\mathbb{N}$ such that for every $F \in \mathcal{F}$ the set

$$\{n \in \mathbb{N} : F \cap A_n = \emptyset\}$$

is finite. If $f \colon \mathbb{N}_\mathcal{F} \to \mathbb{R}$ is continuous then $f^{-1}[(f(\infty) - 1, f(\infty) + 1)]$ is a neighborhood of $\infty$. So there exists an $N \in \mathbb{N}$ such that for every $n \geq N$ there is an element $x_n \in A_n$ such that

$$|f(\infty) - f(x_n)| \leq 1.$$

For every $n$ put

$$B_n = \{f \in \mathbb{R}^{\hat{\mathbb{N}}} : (\forall m \geq n)(\exists x_m \in A_m)(|f(\infty) - f(x_m)| \leq 1)\}.$$

As we just observed,

$$C_p(\mathbb{N}_\mathcal{F}) \subseteq \bigcup_{n=1}^{\infty} B_n.$$

Since $C_p(\mathbb{N}_\mathcal{F})$ is dense in $\mathbb{R}^{\hat{\mathbb{N}}}$ (Lemma 6.2.1), we are done if we can prove that each $B_n$ is meager.

**Claim 1.** *Every $B_n$ is closed and nowhere dense in $\mathbb{R}^{\hat{\mathbb{N}}}$.*

**Proof.** Fix $n$ and assume that $f \in \mathbb{R}^{\hat{\mathbb{N}}} \setminus B_n$. Then there exists $m \geq n$ such that

$$f[A_m] \cap [f(\infty) - 1, f(\infty) + 1] = \emptyset.$$

Since $A_m$ is finite there clearly is an open neighborhood $U$ of $f$ such that for every $g \in U$ we have
$$g[A_m] \cap [g(\infty) - 1, g(\infty) + 1] = \emptyset.$$
So $U$ misses $B_n$ and since $f$ was arbitrary this shows that $B_n$ is closed.

We next prove that $B_n$ is nowhere dense. If not then there are an element $f \in \mathbb{R}^{\hat{\mathbb{N}}}$, a finite subset $F \subseteq \hat{\mathbb{N}}$ and an $\varepsilon > 0$ such that the set

(∗) $$\{g \in \mathbb{R}^{\hat{\mathbb{N}}} : (\forall\, x \in F)(|g(x) - f(x)| < \varepsilon)\}$$

is contained in $B_n$. Without loss of generality, $\infty \in F$. Since $F$ is finite, there is an element $m \geq n$ such that $A_m \cap F = \emptyset$. Define $h\colon F \cup A_m \to \mathbb{R}$ by

$$h(x) = \begin{cases} f(x) & (x \in F). \\ f(\infty) + 2 & (x \in A_m). \end{cases}$$

The function $h$ can be extended to a continuous function $\bar{h}\colon \hat{\mathbb{N}} \to \mathbb{R}$ by the Tietze-Urysohn Extension Theorem. By (∗) it follows that $\bar{h} \in B_n$, but by the definition of $B_n$ it follows that $\bar{h} \notin B_n$. This contradiction establishes the proof of the claim. ◇

We next prove that (2) ⇒ (1). Let $(A_n)_n$ be a pairwise disjoint collection finite subsets of $\hat{\mathbb{N}}$. Our aim is to apply Theorem 6.4.3, i.e., to prove that this sequence has an infinite strongly discrete subsequence, We may assume without loss of generality that $\infty \notin \bigcup_{n=1}^{\infty} A_n$, i.e., that $A_n \subseteq \mathbb{N}$ for every $n$. Since $\mathcal{F}$ is not meager, there exists by Theorem 6.5.3 an element $F \in \mathcal{F}$ such that the set
$$E = \{n \in \mathbb{N} : F \cap A_n = \emptyset\}$$
is infinite. A moments reflection shows that the sequence $(A_n)_{n \in E}$ is strongly discrete in $\mathbb{N}_\mathcal{F}$. □

**Corollary 6.5.7.** *Let $\mathcal{F}$ be an ultrafilter on $\mathbb{N}$. Then $C_p(\mathbb{N}_\mathcal{F})$ is a Baire space.*

**Proof.** This follows directly from Corollary 6.5.5 and Theorem 6.5.6. □

Observe that for an ultrafilter $\mathcal{F}$ the space $C_p(\mathbb{N}_\mathcal{F})$ is not a Borel subset of $\mathbb{R}^{\hat{\mathbb{N}}}$ by Corollary 6.3.3.

**Remark 6.5.8.** For ultrafilters $\mathcal{F}$ there is an alternative way of looking at the spaces $\mathbb{N}_\mathcal{F}$, which we will now explain.

Consider the space $\beta\mathbb{N}$ (here $\mathbb{N}$ is endowed with the discrete topology). If $p \in \mathbb{N}^*$ then the collection
$$\mathcal{F}_p = \{A \subseteq \mathbb{N} : p \in \overline{A}\}$$

is easily seen to be an ultrafilter on $\mathbb{N}$, and the subspace $\mathbb{N} \cup \{p\}$ of $\beta\mathbb{N}$ is canonically homeomorphic to the space $\mathbb{N}_{\mathcal{F}_p}$.

A point $p$ in a topological space $X$ is called a *P-point* if the intersection of countably many neighborhoods of $p$ is again a neighborhood of $p$. It is a well-known result of RUDIN [360] that the space $\mathbb{N}^*$ contains a $P$-point under the Continuum Hypothesis. SHELAH constructed a model of set theory in which there are no $P$-points in $\mathbb{N}^*$ (see e.g., WIMMERS [410]). Notice that every $P$-point is obviously a weak $P$-point and that $\mathbb{N}^*$ contains many weak $P$-points in ZFC alone (see Page 386).

In [184], GUL'KO and SOKOLOV proved that if $X = \mathbb{N} \cup \{p\}$, where $p \in \mathbb{N}^*$, then $C_p(X)$ is hereditary Baire if and only if $p$ is a $P$-point (for a generalization, see MARCISZEWSKI [270]). So if $p$ is not a $P$-point then $C_p(X)$ is Baire, but not hereditary Baire. Observe that there are many non-$P$-points in $\mathbb{N}^*$ so the existence of a Baire but not hereditary Baire $C_p(X)$ does not require any additional set theoretic assumption.

## 6.6. Extenders

Let $A \subseteq X$ be a closed subspace. An *extender* is a function that simultaneously extends real-valued continuous functions on $A$ to real-valued continuous functions on $X$. That is, a function

$$\rho \colon C(A) \to C(X)$$

such that

$$\rho(f) \upharpoonright A = f$$

for every $f \in C(A)$.

**Theorem 6.6.1.** *Let $X$ be a normal space, and let $A \subseteq X$ be closed. Then there is an extender $\rho \colon C(A) \to C(X)$ which is linear.*

**Proof.** This is easy. Simply observe that $C(A)$ is a vector space. So let $\mathcal{F}$ be a Hamel basis for $C(A)$. By the Tietze-Urysohn Theorem, every $f \in \mathcal{F}$ can be extended to a continuous function $\bar{f} \colon X \to \mathbb{R}$. Now define $\rho \colon C(A) \to C(X)$ by the formula

$$\rho\Big(\sum_{i=1}^n \lambda_i f_i\Big) = \sum_{i=1}^n \lambda_i \bar{f}_i,$$

where $f_1, \ldots, f_n \in \mathcal{F}$ and $\lambda_1, \ldots, \lambda_n \in \mathbb{R}$ are arbitrary. It is clear that $\rho$ is as required. $\square$

So linear extenders always exist and hence are not very interesting. For *metrizable* spaces, there always exists an extender which is both linear and continuous. This is much more interesting.

**The Dugundji Theorem (Part 2) 6.6.2.** *Let $X$ be a metrizable space with closed subspace $A$. Then there is a linear extender*
$$\varphi \colon C(A) \to C(X)$$
*having the following properties:*

(1) $\varphi \colon C_p(A) \to C_p(X)$ *is continuous.*
(2) *If $f \in C^*(A)$ then $\|f\| = \|\varphi(f)\|$.*

**Remark 6.6.3.** Observe that by (2),
$$\varphi[C^*(A)] \subseteq C^*(X)$$
and that the function $\varphi$ is also continuous when $C^*(A)$ and $C^*(X)$ are both endowed with their natural Banach space topologies.

**Remark 6.6.4.** By a result of GEBA and SEMADENI [172], the metrizability assumption in Theorem 6.6.2 is essential. Their example is
$$X = \beta\omega, \quad A = \omega^*.$$
(See also ENGELKING [153, Exercise 4.5.20(c)].) In Example **E14** of §6.13 we will show that there is a *countable* space for which the Dugundji Extension Theorem fails. This is of interest since such a space $X$ has the additional property that $C_p(X)$ is separable and metrizable.

**Proof.** We return to the proof of Theorem 1.2.2 for the case $L = \mathbb{R}$. First we note that nowhere in that proof we made an essential use of separability since all metrizable spaces are paracompact (see the introduction to this chapter). Consider the formula (∗) from Page 23:

(∗) $$\bar{f}(x) = \begin{cases} f(x) & (x \in A), \\ \sum_{U \in \mathcal{U}} \kappa_U(x) \cdot f(a_U) & (x \in X \setminus A). \end{cases}$$

It is evident that the assignment $\varphi(f) = \bar{f}$ is linear. We claim that it is also continuous. To prove this, it suffices to prove continuity at $\underline{0}$, the constant function with value 0. To this end, let $P \subseteq X$ be finite and $\varepsilon > 0$. Put
$$Q = (P \cap A) \cup \{a_U : U \cap P \neq \emptyset\}.$$
Then $Q$ is finite since $\mathcal{U}$ is locally finite.

**Claim 1.** $\varphi[N_A(\underline{0}, Q, \varepsilon)] \subseteq N_X(\underline{0}, P, \varepsilon)$.

*Proof.* Let $f \in N_A(\underline{0}, Q, \varepsilon)$. Then $|f(x)| < \varepsilon$ for every $x \in Q$. Consider the function $\varphi(f)$ and let $x \in P$. If $x \in A$ then $x \in Q$ and $\varphi(f)(x) = f(x)$ and so there is nothing to prove. Suppose therefore that $x \notin A$. Let $\mathcal{U}_0$ be the collection of all elements of $\mathcal{U}$ which contain $x$. Observe that $a_U \in Q$

for every $U \in \mathcal{U}_0$ so that $|f(a_U)| < \varepsilon$. Finally, observe that if $U \in \mathcal{U} \setminus \mathcal{U}_0$ then $\kappa_U(x) = 0$. These facts imply that

$$\begin{aligned}|\varphi(f)(x)| &= \left|\sum_{U \in \mathcal{U}} \kappa_U(x) \cdot f(a_U)\right| \\ &= \left|\sum_{U \in \mathcal{U}_0} \kappa_U(x) \cdot f(a_U)\right| \\ &\leq \sum_{U \in \mathcal{U}_0} \kappa_U(x) \cdot |f(a_U)| \\ &< \sum_{U \in \mathcal{U}_0} \kappa_U(x) \cdot \varepsilon \\ &= \varepsilon,\end{aligned}$$

which is as required. $\diamond$

So the claim proves the continuity of $\varphi$.

With a similar calculation we will prove that $\varphi$ is norm preserving. Indeed, if $f \in C^*(A)$, $x \in X \setminus A$ and $t = \|f\|$ then

$$\begin{aligned}|\varphi(f)(x)| &= \left|\sum_{U \in \mathcal{U}} \kappa_U(x) \cdot f(a_U)\right| \\ &\leq \sum_{U \in \mathcal{U}} \kappa_U(x) \cdot |f(a_U)| \\ &\leq \sum_{U \in \mathcal{U}} \kappa_U(x) \cdot t \\ &= t.\end{aligned}$$

This means that $\varphi(f)$ and $f$ have the same norm. $\square$

Using this result in a clever way, it is possible to prove the existence of extenders in various situations.

**Corollary 6.6.5.** *Let $X$ be a space with compact metrizable subspace $K$. Then there is a continuous linear extender $\varphi \colon C_p(K) \to C_p(X)$.*

**Proof.** By Corollary A.4.4 there is an imbedding $i \colon K \to Q$. Since $K$ is compact, the Tietze-Urysohn Theorem shows that the function $i \colon K \to Q$ can be extended to a continuous function $\bar{i} \colon X \to Q$. In addition, Theorem 6.6.2 gives us a continuous linear extender $e \colon C_p(i[K]) \to C_p(Q)$. Now define the function $\varphi \colon C_p(K) \to C_p(X)$ by the formula

$$\varphi(f) = e(f \circ i^{-1}) \circ \bar{i}.$$

It is easy to see that $\varphi$ is as required. $\square$

Various other results in the same spirit can be proved along the same lines. For example, an extender for a pair of spaces $(X,Y)$, where $Y$ is a closed subspace of $X$, exists if $X$ is normal and $Y$ is completely metrizable and separable, or if $X$ is paracompact and $Y$ is completely metrizable. For details, see MICHAEL [285] and ARENS [22].

**Factorizing $C_p(X)$ and $C_p^*(X)$.** Theorem 6.6.2 can be used to prove several interesting factorization properties of function spaces.

Let $X$ be a space and $A \subseteq X$. We say that an element $f \in C_p(X)$ *vanishes* on $A$ provided that $f \restriction A$ is the constant function with value 0.

For a topological space $X$ and subset $A \subseteq X$ we let $C_{p,A}(X)$ denote the subspace of $C_p(X)$ consisting of all functions vanishing on $A$. Similarly for $C_{p,A}^*(X)$. If $A$ is a singleton, say $\{a\}$, then we denote $C_{p,A}(X)$ and $C_{p,A}^*(X)$ simply by $C_{p,a}(X)$ and $C_{p,a}^*(X)$, respectively.

**Proposition 6.6.6.** *Let $X$ be a space and let $A \subseteq X$ be closed. If there is a continuous linear extender $e\colon C_p(A) \to C_p(X)$ then*

$$C_p(X) \sim C_{p,A}(X) \times C_p(A)$$

*In addition, if there is a continuous linear extender $\xi\colon C_p^*(A) \to C_p^*(X)$ then*

$$C_p^*(X) \sim C_{p,A}^*(X) \times C_p^*(A).$$

**Proof.** We will prove this for $C_p^*(X)$ only. The proof for $C_p(X)$ is similar.

Define $\rho\colon C_p^*(X) \to C_p^*(A)$ by $\rho(f) = f \restriction A$. Then $\rho$ is continuous (it is the restriction to $C_p^*(X)$ of a projection) and linear. Define

$$\varphi\colon C_p^*(X) \to C_{p,A}^*(X) \times C_p^*(A)$$

by

$$\varphi(f) = \bigl(f - (\xi \circ \rho)(f), \rho(f)\bigr).$$

We will first prove that $\varphi$ is well-defined. Take an arbitrary $f \in C_p^*(X)$. It is obvious that $\rho(f) \in C_p^*(A)$ and that $f - (\xi \circ \rho)(f) \in C_p^*(X)$. Furthermore,

$$\begin{aligned}
\bigl(f - (\xi \circ \rho)(f)\bigr) \restriction A &= \rho\bigl(f - (\xi \circ \rho)(f)\bigr) \\
&= \rho(f) - \rho \circ \xi \circ \rho(f) \\
&= \rho(f) - \rho(f) \\
&\equiv 0,
\end{aligned}$$

hence $f - (\xi \circ \rho)(f) \in C_{p,A}^*(X)$.

That $\varphi$ is continuous and linear is trivial.

We show that $\varphi$ is a homeomorphism. Define

$$\psi\colon C_{p,A}^*(X) \times C_p^*(A) \to C_p^*(X)$$

by
$$\psi(f,g) = f + \xi(g).$$
It is clear that $\psi$ is well-defined, continuous and linear. Furthermore, $\psi \circ \varphi$ is the identity on $C_p^*(X)$. We show that $\varphi \circ \psi$ is the identity on $C_{p,A}(X) \times C_p^*(A)$. Take arbitrary $f \in C_{p,A}^*(X)$ and $g \in C_p^*(A)$. Note that $\rho(f) = f \upharpoonright A = \underline{0}$, hence by linearity of $\xi$ it follows that $(\xi \circ \rho)(f) = \xi(\underline{0}) = \underline{0}$. So

$$\begin{aligned}(\varphi \circ \psi)(f,g) &= \varphi\big(f + \xi(g)\big) \\ &= \big(f + \xi(g) - (\xi \circ \rho)\big(f + \xi(g)\big), \rho(f + \xi(g))\big) \\ &= \big(f + \xi(g) - (\xi \circ \rho)(f) - (\xi \circ \rho \circ \xi)(g), \rho(f) + (\rho \circ \xi)(g)\big) \\ &= (f + \xi(g) - \underline{0} - \xi(g), \underline{0} + g) \\ &= (f, g),\end{aligned}$$

as required. $\square$

This result will sometimes be used in a very trivial situation, namely that $A$ is a singleton subset of $X$. For such a subset the conditions of Proposition 6.6.6 are trivially satisfied, and so we get:

**Corollary 6.6.7.** *Let $X$ be a space and let $x \in X$. Then*
$$C_p(X) \sim \{f \in C_p(X) : f(x) = 0\} \times \mathbb{R}.$$
*Similarly,*
$$C_p^*(X) \sim \{f \in C_p^*(X) : f(x) = 0\} \times \mathbb{R}.$$

**Proof.** It suffices to observe that $C_p(\{\text{point}\}) = C_p^*(\{\text{point}\}) \sim \mathbb{R}$. $\square$

**Corollary 6.6.8.** *Let $X$ be a metrizable space, and let $A \subseteq X$ be closed. Then*
$$C_p(X) \sim C_{p,A}(X) \times C_p(A), \quad C_p^*(X) \sim C_{p,A}^*(X) \times C_p^*(A).$$

**Proof.** Again, we prove this for $C_p^*(X)$ only. Since $X$ is metrizable, and $A$ in $X$ is closed, there is by Theorem 6.6.2 a continuous linear and norm preserving extender $\xi \colon C_p^*(A) \to C_p^*(X)$. So we are done by an application of Proposition 6.6.6. $\square$

**Corollary 6.6.9.** *Let $X$ be a space containing a nontrivial convergent sequence. Then*
$$C_p(X) \sim C_p(X) \times \mathbb{R}, \quad C_p^*(X) \sim C_p^*(X) \times \mathbb{R}.$$

**Proof.** Let $A \approx \omega+1$ be a nontrivial convergent sequence in $X$. First observe that $(\omega+1) \oplus \{\text{point}\} \approx \omega+1$ so that
$$C_p(\omega+1) \sim C_p((\omega+1) \oplus \{\text{point}\}) \sim C_p(\omega+1) \times \mathbb{R}$$

by Theorem 6.2.8. It now follows from Corollary 6.6.5 and Proposition 6.6.6 that

$$C_p(X) \sim C_{p,A}(X) \times C_p(A)$$
$$\sim C_{p,A}(X) \times (C_p(A) \times \mathbb{R})$$
$$\sim (C_{p,A}(X) \times C_p(A)) \times \mathbb{R}$$
$$\sim C_p(X) \times \mathbb{R}.$$

The proof for bounded functions is similar. □

**Corollary 6.6.10.** *Let $X$ be an infinite first countable space. Then*
$$C_p(X) \sim C_p(X) \times \mathbb{R}, \quad C_p^*(X) \sim C_p^*(X) \times \mathbb{R}.$$

**Proof.** If $X$ is discrete then there is nothing to prove. If $X$ is not discrete then it contains a nontrivial convergent sequence. So then we are done by Corollary 6.6.9. □

**Remark 6.6.11.** This result is not true for all spaces. MARCISZEWSKI [268] proved that there exists a *compact* space $X$ such that $C_p(X)$ is not linearly homeomorphic to $C_p(X) \times \mathbb{R}$. Corollary 6.6.9 indicates that Marciszewski's Example must be rather complex since it does not contain any nontrivial convergent sequence.

If $X$ is a space and $A \subseteq X$ is closed, then $\pi \colon X \to X/A$ denotes the natural decomposition map. The unique point of $X/A$ contained in $\pi[A]$ will be denoted by $\infty$ for convenience. Since $X/A$ is endowed with the quotient topology, it is clear that for every (bounded) continuous function $f \colon X \to \mathbb{R}$ vanishing on $A$ there exists a unique (bounded) continuous function $\tilde{f} \colon X/A \to \mathbb{R}$ vanishing on $\pi[A]$ such that $f = \tilde{f} \circ \pi$. So $C_{p,A}(X)$ and $C_{p,\infty}(X/A)$ can be identified as sets. Fortunately, this identification is a linear homeomorphism. Similarly for bounded functions.

**Proposition 6.6.12.** *Let $X$ be a space, and $A \subseteq X$ be closed. Then*
$$C_{p,A}(X) \sim C_{p,\infty}(X/A), \quad C_{p,A}^*(X) \sim C_{p,\infty}^*(X/A).$$

**Proof.** This will be proved for $C_{p,A}(X)$ only since the proof for bounded functions is entirely similar. Define $\varphi \colon C_{p,A}(X) \to C_{p,\infty}(X/A)$ by $\varphi(f) = \tilde{f}$ (see the remarks above). Then $\varphi$ is a bijection as we just observed, and is clearly linear. It suffices to prove that $\varphi$ is a homeomorphism. Take arbitrary
$$y_1, \ldots, y_n \in X/A$$
and let $x_1, \ldots, x_n \in X$ be such that $\pi(x_i) = y_i$ for every $i$. Then for all $\varepsilon > 0$ and $f \in C_{p,A}(X)$ it is easily seen that
$$\varphi[N(f, \{x_1, \ldots, x_n\}, \varepsilon)] = N(\tilde{f}, \{y_1, \ldots, y_n\}, \varepsilon)].$$
So $\varphi$ is a homeomorphism. □

Again, let $X$ be a space, let $A \subseteq X$ be closed. By Corollary 6.6.7 we get
$$C_p(X/A) \sim C_{p,\infty}(X/A) \times \mathbb{R}$$
and
$$C_p^*(X/A) \sim C_{p,\infty}^*(X/A) \times \mathbb{R}.$$
These trivialities will be used in the proof of the following result.

**Corollary 6.6.13.** *Let $X$ be an infinite metrizable space, and let $A \subseteq X$ be closed. Then*
$$C_p(X) \sim C_p(X/A) \times C_p(A), \quad C_p^*(X) \sim C_p^*(X/A) \times C_p^*(A).$$

**Proof.** We complete the proof in two steps. First, Corollaries 6.6.10 and 6.6.8 give us that
$$C_p(X) \sim C_p(X) \times \mathbb{R} \sim C_{p,A}(X) \times C_p(A) \times \mathbb{R}.$$
But
$$C_{p,A}(X) \sim C_{p,\infty}(X/A)$$
by Proposition 6.6.12. Again by Corollary 6.6.8,
$$C_p(X/A) \sim C_{p,\infty}(X/A) \times C_p(\{\text{point}\}) \sim C_{p,\infty}(X/A) \times \mathbb{R}.$$
It therefore follows by the above remarks that
$$\begin{aligned} C_p(X) &\sim C_{p,A}(X) \times C_p(A) \times \mathbb{R} \\ &\sim C_{p,\infty}(X/A) \times C_p(A) \times \mathbb{R} \\ &\sim (C_{p,\infty}(X/A) \times \mathbb{R}) \times C_p(A) \\ &\sim C_p(X/A) \times C_p(A). \end{aligned}$$
The proof for bounded functions is again similar. $\square$

**Remark 6.6.14.** These simple results are basically the 'only' known general tools for proving that spaces are $\ell$-equivalent. They will be used frequently in the forthcoming sections.

## 6.7. The topological dual of $C_p(X)$

For a space $X$ let $L(X)$ be the *dual* of $C_p(X)$, i.e., the set consisting of all continuous functionals on $C_p(X)$.

For $x \in X$ we define the function $\xi_x \colon C_p(X) \to \mathbb{R}$ by $\xi_x(f) = f(x)$. As usual, $\xi_x$ is called the *evaluation at $x$*.

**Lemma 6.7.1.** *Let $X$ be a space and $x \in X$. Then $\xi_x$ is a continuous functional.*

**Proof.** It is evident that $\xi_x$ is linear. Observe that for $\varepsilon > 0$ and $f \in C_p(X)$ we have
$$\xi_x[N(f, \{x\}, \varepsilon)] \subseteq (f(x) - \varepsilon, f(x) + \varepsilon).$$
From this continuity of $\xi_x$ is obvious. □

If $x, y \in X$ and $x \neq y$ then there is $f \in C(X)$ such that $f(x) \neq f(y)$. As a consequence,
$$\xi_x(f) = f(x) \neq f(y) = \xi_y(f).$$
So we conclude that the assignment $x \mapsto \xi_x$ is one-to-one. This makes it possible to identify $X$ with a subset of $L(X)$. We will show that $X$ is a linearly independent subset of $L(X)$ (hence $X \neq L(X)$) and that it generates $L(X)$. We will topologize $L(X)$ later and prove that $X$ is in fact a subspace of $L(X)$.

**Proposition 6.7.2.** *Let $X$ be a space. Then the collection $\{\xi_x : x \in X\}$ is a Hamel basis for $L(X)$.*

**Proof.** We will first prove linear independence. Take pairwise distinct elements $x_1, \ldots, x_n$ in $X$, and elements $\lambda_1, \ldots, \lambda_n \in \mathbb{R}$ such that $\sum_{i=1}^{n} \lambda_i \xi_{x_i} = 0$. For every $i \leq n$ let $f_i \colon X \to \mathbb{R}$ be continuous such that
$$f_i(x_i) = 1, \qquad f_i(x_j) = 0 \quad (i \neq j).$$
Then
$$0 = \sum_{k=1}^{n} \lambda_k \xi_{x_k}(f_i) = \sum_{k=1}^{n} \lambda_k f_i(x_k) = \lambda_i$$
for every $i \leq n$.

Now let $F \colon C_p(X) \to \mathbb{R}$ be a continuous functional. Then $F(\underline{0}) = 0$, where $\underline{0}$ is the constant function with value $0$. By continuity of $F$ there is a neighborhood of $\underline{0}$ which is mapped by $F$ into the interval $(-1, 1)$. For this neighborhood we may take a basic neighborhood of $\underline{0}$, say $N(\underline{0}, P, \delta)$, where $P \subseteq X$ is finite and $\delta > 0$. Let $P = \{x_1, \ldots, x_n\}$.

**Claim 1.** *If $f, g \in C_p(X)$ and $f \restriction P = g \restriction P$ then $F(f) = F(g)$.*

*Proof.* Let $f, g \in C_p(X)$ be such that $f \restriction P = g \restriction P$. Then $f - g \restriction P \equiv 0$ and hence $k \cdot (f - g) \equiv 0$ for every $k \in \mathbb{N}$. So $k \cdot (f - g) \in N(\underline{0}, P, \delta)$ for all $k$ so that
$$k|F(f) - F(g)| = |F(k \cdot (f - g))| < 1.$$
This shows $|F(f) - F(g)| < 1/k$ for every $k$, hence $F(f) = F(g)$. ◇

Let $\pi \colon C_p(X) \to \mathbb{R}^P$ be the projection
$$f \mapsto f \restriction P.$$

## 6.7. THE TOPOLOGICAL DUAL OF $C_p(X)$

Observe that by the Tietze-Urysohn Theorem it follows that $\pi$ is surjective. Claim 1 consequently shows that there is a unique function $F' \colon \mathbb{R}^P \to \mathbb{R}$ such that
$$F = F' \circ \pi.$$
Then $F'$ is linear since $F$ and $\pi$ are.

Since $P$ is finite, by elementary Linear Algebra there are $\lambda_1, \ldots, \lambda_n \in \mathbb{R}$ such that
$$F'(g) = \sum_{i=1}^{n} \lambda_i g(x_i)$$
for every $g \in \mathbb{R}^P$. From this we conclude that
$$F(f) = F' \circ \pi(f) = \sum_{i=1}^{n} \lambda_i f(x_i) = \Big(\sum_{i=1}^{n} \lambda_i \xi_{x_i}\Big)(f)$$
for every $f \in C_p(X)$. But this is clearly as required. $\square$

We now topologize $L(X)$ in a natural way. For $f \in C(X)$ let
$$L(f) \colon L(X) \to \mathbb{R}$$
be defined by
$$L(f)(F) = F(f).$$
Then $L(f)$ is clearly a functional on $L(X)$. Let $\mathcal{L}(X)$ be the collection of functionals on $L(X)$ obtained in this way, and consider the function
$$e \colon L(X) \to \mathbb{R}^{\mathcal{L}(X)}$$
defined by
$$\big(e(F)\big)_{L(f)} = L(f)(F) = F(f).$$
This function is clearly linear, and we claim that it is one-to-one. Since $e$ is linear, it suffices to prove that $\operatorname{Ker} e = \{\underline{0}\}$. Take an arbitrary $F \in L(X)$ such that $F \neq \underline{0}$. By Proposition 6.7.2 there are elements $x_1, \ldots, x_n \in X$ and $\alpha_1, \ldots, \alpha_n \in \mathbb{R}$ such that
$$F = \sum_{i=1}^{n} \alpha_i \xi_{x_i}.$$
We may assume without loss of generality that the $x$'s are pairwise distinct. Since $F \neq \underline{0}$, we may also assume without loss of generality that $\alpha_1 \neq 0$. Let $f \in C_p(X)$ be such that $f(x_1) = 1$ and $f[\{x_2, \ldots, x_n\}] \subseteq \{0\}$. Then
$$e(F)_{L(f)} = \sum_{i=1}^{n} \alpha_i \xi_{x_i}(f) = \sum_{i=1}^{n} \alpha_i f(x_i) = \alpha_1 \neq 0,$$
from which it follows that $e(F) \neq \underline{0}$.

So we showed that $\mathcal{L}(X)$ separates the points of $L(X)$ which implies that the function $e$ is one-to-one (there are general theorems that allow us to conclude this without any work, but it is a good exercise to perform the above calculations).

The topology on $L(X)$ we are interested in is the weakest topology which makes all functionals $L(f)$, $f \in C(X)$, continuous. That is, we identify $L(X)$ and the linear subspace $e[L(X)]$ of $\mathbb{R}^{\mathcal{L}(X)}$. It is clear that $L(X)$ is a locally convex linear space.

With this topology $L(X)$ is called the *topological dual* of $C_p(X)$. It will be convenient to identify $x \in X$ and $\xi_x \in L(X)$.

**Lemma 6.7.3.** *Let $X$ be a space and $f \in C(X)$. Then $L(f) \colon L(X) \to \mathbb{R}$ is the unique continuous functional that extends $f$.*

**Proof.** That $L(f)$ is a continuous functional is clear. For $x \in X$ we clearly have $L(f)(x) = f(x)$, so $L(f)$ extends $f$. Since $X$ is a Hamel basis for $L(X)$ (Proposition 6.7.2) it follows that $L(f)$ is unique. $\square$

**Proposition 6.7.4.** *The inclusion $X \hookrightarrow L(X)$ is a closed imbedding.*

**Proof.** We show first that $\{\xi_x : x \in X\}$ as subspace of $L(X)$ is homeomorphic to $X$. The homeomorphism that does the job is of course the function

$$x \mapsto \xi_x,$$

which we denote for convenience by $\xi$. Let $U \subseteq X$ be open, and let $x \in U$. We claim that $\xi[U]$ is a neighborhood of $\xi(x) = \xi_x$ in $\xi[X]$. Indeed, let $f \in C(X)$ be such that $f(x) = 1$ and $f[X \setminus U] \subseteq \{0\}$. Put

$$V = L(f)^{-1}[(0, \infty)].$$

Then $V$ is open in $L(X)$ and $\xi_x \in V \cap \xi[X] \subseteq \xi[U]$. This shows that

$$\xi \colon X \to \xi[X] \subseteq L(X)$$

is open. To prove continuity, let $V \subseteq \xi[X]$ be open, and let $\xi(x) = \xi_x \in V$ be arbitrary. By the definition of the topology on $L(X)$, there are functions $f_1, \ldots, f_n \in C(X)$ and open subsets $U_1, \ldots, U_n$ of $\mathbb{R}$ with

$$\xi_x \in \bigcap_{i=1}^{n} L(f_i)^{-1}[U_i] \subseteq V.$$

Then $x \in \bigcap_{i=1}^{n} f_i^{-1}[U_i] \subseteq \xi^{-1}[V]$. This proves continuity, and hence that $\xi$ is a homeomorphism.

This justifies the identification of $X$ and $\xi[X]$.

We show next that $X$ as subspace of $L(X)$ is closed. To this end, take an arbitrary $F \in L(X) \setminus X$. There are $x_1, \ldots, x_n \in X$ and $\lambda_1, \ldots, \lambda_n \in \mathbb{R}$ such that $F = \sum_{i=1}^{n} \lambda_i x_i$ (Proposition 6.7.2). We may assume without loss of generality that $x_i \neq x_j$ if $i \neq j$ and also that $\lambda_i \neq 0$ for every $i$.

We assume first that $n \geq 2$. There are pairwise disjoint open sets $V_i \subseteq X$ with $x_i \in V_i$ for every $i$. Let $f_i \in C(X)$ be such that $f_i(x_i) = 1$ and
$$f_i \restriction X \setminus V_i \equiv 0$$
for every $i$. We claim that $F$ belongs to the open set $\bigcap_{i=1}^{n} L(f_i)^{-1}[\mathbb{R} \setminus \{0\}]$ and that this set is contained in $L(X) \setminus X$. Indeed, for each $i \leq n$,
$$L(f_i)(F) = \sum_{j=1}^{n} \lambda_j L(f_i)(x_j) = \sum_{j=1}^{n} \lambda_j f_i(x_j) = \lambda_i \in \mathbb{R} \setminus \{0\}.$$
Moreover, for an arbitrary $x \in X$ there exists $i \leq n$ with $x \notin V_i$ (since $n \geq 2$) so that $L(f_i)(x) = f_i(x) = 0 \notin \mathbb{R} \setminus \{0\}$ and so
$$X \cap \bigcap_{i=1}^{n} L(f_i)^{-1}[\mathbb{R} \setminus \{0\}] = \emptyset.$$
We conclude that $F$ has a neighborhood which misses $X$.

Assume next that $n = 1$, i.e., $F = \lambda_1 x_1$. Since $F \notin X$, clearly $\lambda_1 \neq 1$. Put $U = \mathbb{R} \setminus \{1\}$ and let $f$ be the constant function with value 1. We claim that $F \in L(f)^{-1}[U] \subseteq L(X) \setminus X$. Indeed,
$$L(f)(F) = \lambda_1 L(f)(x_1) = \lambda_1 f(x_1) = \lambda_1 \in U$$
and for $x \in X$ we have $L(f)(x) = f(x) = 1 \notin U$. So we are done. □

We now come to an interesting 'duality' result.

**Theorem 6.7.5.** *Let $X$ and $Y$ be spaces. Then $C_p(X)$ and $C_p(Y)$ are linearly homeomorphic if and only if $L(X)$ and $L(Y)$ are linearly homeomorphic.*

**Proof.** First suppose that $\varphi \colon C_p(X) \to C_p(Y)$ is a linear homeomorphism. Define $\Psi \colon L(X) \to L(Y)$ by $\Psi(F) = F \circ \varphi^{-1}$. Then $\Psi$ is obviously a well-defined linear function. To see that it is continuous, notice that for $f \in C(Y)$ and open $U \subseteq \mathbb{R}$ we have
$$\Psi^{-1}\big[L(f)^{-1}[U]\big] = (L(f) \circ \Psi)^{-1}[U] = (L(\varphi^{-1}(f)))^{-1}[U],$$
which is open in $L(X)$.

Define $\Phi \colon L(Y) \to L(X)$ by $\Phi(G) = G \circ \varphi$. Then by symmetry it follows that $\Phi$ is continuous. Since trivially $\Phi = \Psi^{-1}$, this shows that $L(X)$ and $L(Y)$ are linearly homeomorphic.

Conversely, suppose that $\psi \colon L(X) \to L(Y)$ is a linear homeomorphism. Define $\varphi \colon C_p(X) \to C_p(Y)$ by $\varphi(f) = (L(f) \circ \psi^{-1}) \restriction Y$. Then $\varphi$ is obviously a well-defined linear function. To prove continuity, take arbitrary finite $P \subseteq Y$ and let $\varepsilon > 0$. For every $y \in Y$ there are
$$x_1^y, \ldots, x_{n_y}^y \in X, \quad \lambda_1^y, \ldots, \lambda_{n_y}^y \in \mathbb{R} \setminus \{0\}$$

such that
$$\psi^{-1}(y) = \sum_{i=1}^{n_y} \lambda_i^y x_i^y$$
(Proposition 6.7.2). Let $N = \max\{\sum_{i=1}^{n_y} |\lambda_i^y| : y \in P\}$, $\delta = \varepsilon/N$ and
$$Q = \{x_i^y : y \in P \text{ and } i \leq n_y\}.$$
We claim that
$$\varphi[N(\underline{0}, Q, \delta)] \subseteq N(\underline{0}, P, \varepsilon).$$
Indeed, if $f \in N(\underline{0}, Q, \delta)$ then for $y \in P$ we have
$$|\varphi(f)(y)| = |(L(f) \circ \psi^{-1})(y)| = \left|L(f)\left(\sum_{i=1}^{n_y} \lambda_i^y x_i^y\right)\right| = \left|\sum_{i=1}^{n_y} \lambda_i^y f(x_i^y)\right|$$
$$\leq \sum_{i=1}^{n_y} |\lambda_i^y| |f(x_i^y)| < \delta \sum_{i=1}^{n_y} |\lambda_i^y| \leq \varepsilon/N \cdot N = \varepsilon.$$
So $\varphi$ is continuous at the origin, and hence is continuous everywhere by linearity.

Define $\vartheta \colon C_p(Y) \to C_p(X)$ by $\vartheta(g) = (L(g) \circ \psi) \restriction X$. By symmetry it follows that $\vartheta$ is continuous and linear. Since clearly $\vartheta = \varphi^{-1}$, we are done. □

**Remark 6.7.6.** It is unknown whether a similar result can be derived for homeomorphisms. Are $C_p(X)$ and $C_p(Y)$ homeomorphic if and only if $L(X)$ and $L(Y)$ are homeomorphic? (Probably not.)

## 6.8. The support function

If $X$ and $Y$ have linearly homeomorphic function spaces $C_p(X)$ and $C_p(Y)$ then $X$ and $Y$ have much in common. A certain multi-valued lower semi-continuous function plays a crucial role in investigating those similarities. In this section we define this function, and derive some of its basic properties. Applications of the results derived here are given in the next sections.

Let $X$ and $Y$ be spaces and let $\varphi \colon C(X) \to C(Y)$ be a linear function. For every $y \in Y$, the *support in $X$ of $y$ with respect to $\varphi$*, abbreviated $\mathrm{supp}_\varphi(y)$, is defined to be set of all $x \in X$ satisfying the condition that for every neighborhood $U$ of $x$, there exists an $f \in C(X)$ such that $f[X \setminus U] \subseteq \{0\}$ and $\varphi(f)(y) \neq 0$.

At first glance this seems to be a very technical concept. But it will turn out to be an important tool in understanding the function spaces $C_p(X)$ a little better. If $\varphi$ is a bijection then $\varphi^{-1} \colon C_p(Y) \to C_p(X)$ is also linear, so we can then consider the support of a point in $Y$ with respect to $\varphi$ and the support of a point in $X$ with respect to $\varphi^{-1}$. When it is clear from the context which support we mean, we will suppress the index.

For $A \subseteq Y$ we define $\operatorname{supp}(A) = \bigcup \{\operatorname{supp}(y) : y \in A\}$.

There is an alternative description of supports in the special case of a *continuous* linear function that is very useful. Let $X$ and $Y$ be spaces and let $\varphi \colon C_p(X) \to C_p(Y)$ be continuous and linear. Fix $y \in Y$. The function $\psi_y \colon C_p(X) \to \mathbb{R}$ defined by $\psi_y = \xi_y \circ \varphi$ is continuous and linear. So $\psi_y \in L(X)$. If $\psi_y \neq 0$ there consequently exist by Proposition 6.7.2 pairwise distinct $x_1, \ldots, x_n \in X$ and $\lambda_1, \ldots, \lambda_n \in \mathbb{R} \setminus \{0\}$ such that

$$\psi_y = \sum_{i=1}^n \lambda_i x_i.$$

(Notice that if $\varphi$ is a bijection then $\psi_y \neq 0$ for every $y \in Y$.) This means that for every $f \in C_p(X)$ we have

(*) $$\varphi(f)(y) = \psi_y(f) = \sum_{i=1}^n \lambda_i f(x_i).$$

We claim that

$$\operatorname{supp}(y) = \{x_1, \ldots, x_n\}.$$

To see this, let $x \in \operatorname{supp}(y)$ and suppose that $x \notin \{x_1, \ldots, x_n\}$. Since

$$X \setminus \{x_1, \ldots, x_n\}$$

is an open neighborhood of $x$, there exists $f \in C_p(X)$ such that $f(x_i) = 0$ for every $i \leq n$ and $\varphi(f)(y) \neq 0$. But by (*),

$$\varphi(f)(y) = \sum_{i=1}^n \lambda_i f(x_i) = 0,$$

which is a contradiction.

Conversely, let $i \leq n$ and let $U \subseteq X \setminus \{x_j : j \leq n \text{ and } j \neq i\}$ be an arbitrary open neighborhood of $x_i$. Let $f \in C(X)$ be such that $f[X \setminus U] \subseteq \{0\}$ and $f(x_i) = 1$. Then again by (*),

$$\varphi(f)(y) = \sum_{i=1}^n \lambda_i f(x_i) = \lambda_i \neq 0,$$

which proves that $x_i \in \operatorname{supp}(y)$.

So we conclude, in particular, that the support of a point is finite.

Let us summarize what we just concluded for continuous linear functions. To this end, let $X$ and $Y$ be spaces and let $\varphi \colon C_p(X) \to C_p(Y)$ be continuous and linear. We proved that for every $y \in Y$ there exists a (possibly empty) finite subset $\operatorname{supp}(y) \subseteq X$ and for every $z \in \operatorname{supp}(y)$ a non-zero real number $\lambda(y, z)$ depending on $y$ and $z$ such that for every $f \in C_p(X)$,

(**) $$\varphi(f)(y) = \sum_{z \in \operatorname{supp}(y)} \lambda(y, z) f(z).$$

Observe that (∗∗) is in fact identical to (∗).

**Lemma 6.8.1.** Let $X$ and $Y$ be spaces, $y \in Y$ and $\varphi \colon C_p(X) \to C_p(Y)$ continuous and linear. If $f, g \in C_p(X)$ coincide on $\mathrm{supp}(y)$ then
$$\varphi(f)(y) = \varphi(g)(y).$$

**Proof.** This follows easily from (∗). Simply observe that if
$$\xi_y \circ \varphi = \sum_{i=1}^{n} \lambda_i x_i$$
then $\mathrm{supp}(y) = \{x_1, \ldots, x_n\}$ and
$$\varphi(f)(y) = \sum_{i=1}^{n} \lambda_i f(x_i) = \sum_{i=1}^{n} \lambda_i g(x_i) = \varphi(g)(y),$$
which is as required. □

We will now present a few other useful properties of the support function.

**Lemma 6.8.2.** Let $X$ and $Y$ be spaces, and let $\varphi \colon C_p(X) \to C_p(Y)$ be continuous and linear. Then

(1) If $\varphi$ is injective then $\mathrm{supp}(Y)$ is dense in $X$.
(2) If $A \subseteq Y$ then $\mathrm{supp}(\overline{A}) \subseteq \overline{\mathrm{supp}(A)}$.
(3) If $\varphi$ is surjective then $\mathrm{supp}(y) \neq \emptyset$ for every $y \in Y$.
(4) If $\mathrm{supp}(y) \neq \emptyset$ for every $y \in Y$ then $\mathrm{supp} \colon Y \Rightarrow X$ is LSC.
(5) If $\varphi$ is a homeomorphism then for every $x \in X$ we have that
$$x \in \mathrm{supp}_\varphi(\mathrm{supp}_{\varphi^{-1}}(x)).$$

**Proof.** For (1), suppose there is an $x \in X$ which is not in the closure of $\mathrm{supp}(Y)$. Let $f \in C(X)$ be such that $f(x) = 1$ and $f[\mathrm{supp}(Y)] \subseteq \{0\}$. Since the function $f$ agrees with the constant function $\underline{0}$ with value 0 on every support set, by Lemma 6.8.1 it follows that $\varphi(f)(y) = \varphi(\underline{0})(y) = 0$ for every $y \in Y$. So $\varphi(f) = \underline{0}$, which contradicts the injectivity of $\varphi$ since $f \neq \underline{0}$.

For (2), assume that there exists an $x \in \mathrm{supp}(\overline{A}) \setminus \overline{\mathrm{supp}(A)}$. Let $y \in \overline{A}$ be such that $x \in \mathrm{supp}(y)$. Let $U$ be an open neighborhood of $x$ such that
$$x \in U \subseteq \overline{U} \subseteq X \setminus \overline{\mathrm{supp}(A)}.$$
Since $x \in \mathrm{supp}(y)$, there exists an element $f \in C(X)$ with $f[X \setminus U] \subseteq \{0\}$ and $\varphi(f)(y) \neq 0$. Since $f \equiv 0$ on $\mathrm{supp}(A)$, it follows by Lemma 6.8.1 that
$$\varphi(f) \upharpoonright A \equiv 0.$$
But then $\varphi(f)(y) = 0$, which is a contradiction.

For (3), take an arbitrary $y \in Y$ such that $\mathrm{supp}(y) = \emptyset$. Let $f \in C(Y)$ be such that $f(y) \neq 0$, and let $g \in C(X)$ be such that $\varphi(g) = f$. Then $g \equiv 0$

on $\emptyset$ which is a neighborhood of $\mathrm{supp}(y)$, and so $\varphi(g)(y) = f(y) = 0$ by Lemma 6.8.1. This is a contradiction.

For (4), let $U \subseteq X$ be open, and put
$$A = \{y \in Y : \mathrm{supp}(y) \cap U \neq \emptyset\}.$$
Let $y \in A$, and take $x \in \mathrm{supp}(y) \cap U$. Let $V \subseteq X$ be open such that
$$x \subseteq V \subseteq \overline{V} \subseteq U.$$
There exists $f \in C(X)$ such that $f[X \setminus V] \subseteq \{0\}$ and $\varphi(f)(y) \neq 0$. Let
$$W = \{z \in Y : \varphi(f)(z) \neq 0\}.$$
Then $W$ is an open neighborhood of $y$. We claim that $W \subseteq A$. Suppose that there is an element $z \in W \setminus A$. Then $\varphi(f)(z) \neq 0$ and $\mathrm{supp}(z) \cap U = \emptyset$. So $X \setminus \overline{V}$ is a neighborhood of $\mathrm{supp}(z)$ and $f[X \setminus \overline{V}] \subseteq \{0\}$, which implies by Lemma 6.8.1 that $\varphi(f)(z) = 0$, contradiction.

For (5), let $x \in X$ and assume that $x \notin A = \mathrm{supp}_\varphi\bigl(\mathrm{supp}_{\varphi^{-1}}(x)\bigr)$. Since $A$ is finite, there exists $f \in C(X)$ such that $f(x) = 1$ and $f[A] \subseteq \{0\}$. By Lemma 6.8.1 it follows that $\varphi(f) \equiv 0$ on $\mathrm{supp}_{\varphi^{-1}}(x)$, and by the same argument that $f(x) = \varphi^{-1}\bigl(\varphi(f)\bigr)(x) = 0$. This is a contradiction. $\square$

Every compact space is obviously pseudocompact, but the converse of this statement is not true as the ordinal space $\omega_1$ shows.

A pseudocompact subspace of a space $X$ is obviously bounded. If $X$ is moreover metrizable then every bounded subset of it is compact (cf. Exercise A.5.14). So for a metrizable space $X$ and subspace $A$, the statements '$A$ is compact', '$A$ is pseudocompact', and '$A$ is bounded' are equivalent.

**Theorem 6.8.3.** *Let $X$ and $Y$ be spaces, and let $\varphi \colon C_p(X) \to C_p(Y)$ be continuous and linear. Then $\mathrm{supp}(A)$ is bounded in $X$ if $A$ is bounded in $Y$.*

**Proof.** Striving for a contradiction, assume that $\mathrm{supp}(A)$ is not bounded. Then there exists an element $f \in C_p(X)$ such that $f[\mathrm{supp}(A)]$ is not bounded in $\mathbb{R}$. We may assume without loss of generality that $\mathbb{N} \subseteq f[\mathrm{supp}(A)]$. For every $n \in \mathbb{N}$, let $V_n = f^{-1}[(n - 1/4, n + 1/4)]$. Then $\mathcal{V} = \{V_n : n \in \mathbb{N}\}$ is a discrete family of open subsets of $X$.

By induction on $k$ we will construct a point $y_k \in A$, an element $n(k) \in \mathbb{N}$, and an element $f_k \in C_p(X)$ such that

(1) $f_k[X \setminus V_{n(k)}] \subseteq \{0\}$,
(2) $n(1) < n(2) < \cdots < n(k) < \cdots$,
(3) if $k > 1$ then $\mathrm{supp}\{y_1, \ldots, y_{k-1}\} \cap \overline{V_i} = \emptyset$ for every $i \geq n(k)$,
(4) $\varphi(f_k)(y_k) = k - h_k$, where

$$h_k = \begin{cases} 0 & (k = 1), \\ \sum_{i=1}^{k-1} \varphi(f_i)(y_k) & (k > 1). \end{cases}$$

For every $n$, let $x_n \in \operatorname{supp}(A)$ be such that $f(x_n) = n$.

There exists $y_1 \in A$ such that $x_1 \in \operatorname{supp}(y_1)$. Put $n(1) = 1$. Since $V_{n(1)}$ is a neighborhood of $x_1$ and $x_1 \in \operatorname{supp}(y_1)$, there exists $h \in C_p(X)$ such that
$$h[X \setminus V_{n(1)}] \subseteq \{0\}, \quad \varphi(h)(y_1) \neq 0.$$
Let
$$\lambda = \frac{1}{\varphi(h)(y_1)},$$
and put $f_1 = \lambda h$. Then $f_1$ is continuous, $f[X \setminus V_{n(1)}] \subseteq \{0\}$, and by linearity of $\varphi$,
$$\varphi(f_1)(y_1) = \lambda \varphi(h)(y_1) = 1 = 1 - h_1.$$
Let $k > 1$ and suppose we completed the construction up to $k - 1$. Let
$$P = \operatorname{supp}\{y_1, \ldots, y_{k-1}\}.$$
Then $P$ is clearly finite, so there exists $n(k) > n(k-1)$ so large that
$$\overline{V_i} \cap P = \emptyset$$
for every $i \geq n(k)$. Since $x_{n(k)} \in \operatorname{supp}(A)$, there is an element $y_k \in A$ such that $x_{n(k)} \in \operatorname{supp}(y_k)$. The set $V_{n(k)}$ is a neighborhood of $x_{n(k)}$ and since $x_{n(k)} \in \operatorname{supp}(y_k)$, there consequently exists an element $h \in C_p(X)$ such that
$$h[X \setminus V_{n(k)}] \subseteq \{0\}, \quad \varphi(h)(y_k) \neq 0.$$
Let
$$\lambda = \frac{k - h_k}{\varphi(h)(y_k)},$$
and $f_k = \lambda h$. Then $f_k[X \setminus V_{n(k)}] \subseteq \{0\}$, and by linearity of $\varphi$,
$$\varphi(f_k)(y_k) = k - h_k.$$

Now let $f = \sum_{k=1}^{\infty} f_k$. To see that $f$ is well-defined and continuous, observe that every $x \in X$ has a neighborhood $U_x$ meeting at most one member of the collection $\{V_{n(k)} : k \in \mathbb{N}\}$. So by (1), $f \upharpoonright U_x$ is a finite sum and hence $f \upharpoonright U_x$ is well-defined and continuous.

For every $k$ let $g_k = \sum_{i=1}^{k} f_i$. Observe that for $j > k$ we have
$$\operatorname{supp}(y_k) \cap \overline{V_{n(j)}} = \emptyset$$
and $f_j \upharpoonright X \setminus \bigcup_{i=k+1}^{\infty} V_{n(i)} \equiv 0$. So $f$ and $g_k$ coincide on $\operatorname{supp}(y_k)$, which implies that $\varphi(f)(y_k) = \varphi(g_k)(y_k)$ (Lemma 6.8.1). But
$$\varphi(g_k)(y_k) = \sum_{i=1}^{k} \varphi(f_i)(y_k) = \sum_{i=1}^{k-1} \varphi(f_i)(y_k) + \varphi(f_k)(y_k)$$
$$= h_k + \varphi(f_k)(y_k)$$
$$= h_k + k - h_k$$
$$= k.$$

So we conclude that $\varphi(f)(y_k) = k$ for every $k$. But this means that $A$ is not bounded in $Y$, which is a contradiction. □

**Corollary 6.8.4.** *Let $X$ and $Y$ be spaces, and let $\varphi\colon C_p(X) \to C_p(Y)$ be continuous, linear, and injective. If $Y$ is pseudocompact then so is $X$.*

**Proof.** Clearly $Y$ is a bounded subset of $Y$. So supp$(Y)$ is a bounded subset of $X$ by Theorem 6.8.3. But supp$(Y)$ is dense in $X$ by Lemma 6.8.2(1). So $X$ has a dense bounded subset, and is therefore bounded itself. We conclude that $X$ is pseudocompact. □

Let $X$ and $Y$ be spaces and let $\varphi\colon C_p(X) \to C_p(Y)$ be continuous and linear. For $U \subseteq X$ let

$$T_U = \{y \in Y : \mathrm{supp}(y) \cap U \neq \emptyset\}.$$

For a family $\mathcal{U}$ of subsets of $X$, let

$$T_\mathcal{U} = \{T_U : U \in \mathcal{U}\}.$$

Observe that if $\mathcal{U}$ covers $X$ and supp$(y) \neq \emptyset$ for every $y \in Y$ then $T_\mathcal{U}$ covers $Y$.

**Corollary 6.8.5.** *Let $X$ and $Y$ be metrizable spaces and let*

$$\varphi\colon C_p(X) \to C_p(Y)$$

*be a continuous linear surjection. If $\mathcal{U}$ is a locally finite open cover of $X$ then every $y \in Y$ has a neighborhood $V_y$ such that the collection*

$$\{U \in \mathcal{U} : V_y \cap T_U \neq \emptyset\}$$

*is finite. As a consequence, $T_\mathcal{U}$ is a locally finite open cover of $Y$.*

**Proof.** By Lemma 6.8.2(3), supp$(y) \neq \emptyset$ for every $y \in Y$. So $T_\mathcal{U}$ covers $Y$. Observe that $T_U$ is open in $Y$ for every $U \in \mathcal{U}$ since supp is LSC (Lemma 6.8.2(4)). If $T_\mathcal{U}$ does not have the property stated in the corollary then there are a point $y \in Y$ and a sequence $(y_n)_n$ in $Y$ converging to $y$ and distinct $U_n$'s in $\mathcal{U}$ such that for every $n$, $y_n \in T_{U_n}$ (here we use that $Y$ is metrizable; in fact, for this argument it suffices to assume that $Y$ is first countable). For every $n$ let $x_n \in \mathrm{supp}(y_n) \cap U_n$ be arbitrary. Since the sequence $(y_n)_n$ is bounded because it converges, supp$\{y_n . n \in \mathbb{N}\}$ is bounded as well by Theorem 6.8.3. Hence $\{x_n : n \in \mathbb{N}\}$ is bounded which implies that

$$Z = \overline{\{x_n : n \in \mathbb{N}\}}$$

is compact since $X$ is metrizable (cf. Exercise A.5.14). Since $\mathcal{U}$ is locally finite, this implies that $Z$ intersects finitely many elements of $\mathcal{U}$ only. This is a contradiction. □

Let $X$ be a space. We say that a subset $A \subseteq X$ is *C-imbedded* in $X$ if every continuous function $f\colon A \to \mathbb{R}$ can be extended to a continuous function $\bar{f}\colon X \to \mathbb{R}$.

Observe that by the Tietze-Urysohn Theorem, every closed subspace of a normal space $X$ is $C$-imbedded in $X$. The situation for nonnormal spaces is more complicated.

Typical examples of closed and discrete $C$-imbedded subsets of an arbitrary space $X$ can be obtained in the following way. Let $f\colon X \to \mathbb{R}$ be continuous and unbounded. It is easy to construct a sequence $(x_n)_n$ in $X$ such that
$$|f(x_{n+1})| \geq |f(x_n)| + 1$$
for every $n$. We claim that $D = \{x_n : n \in \mathbb{N}\}$ is $C$-embedded in $X$. To prove this, first observe that $f[D]$ is a closed discrete subspace of $\mathbb{R}$. Hence $f[D]$ is $C$-imbedded in $\mathbb{R}$ by the Tietze-Urysohn Theorem. Observe that
$$f \restriction D \to f[D]$$
is a bijection. Let $g\colon D \to \mathbb{R}$ be any function. Define the function
$$h\colon f[D] \to \mathbb{R}$$
in the obvious way by
$$h\bigl(f(d)\bigr) = g(d).$$
Then $h$ is continuous, $f[D]$ being discrete. By the above we can therefore extend $h$ to a continuous function $\bar{h}\colon \mathbb{R} \to \mathbb{R}$. Now let
$$\bar{g} = \bar{h} \circ f \colon X \to \mathbb{R}.$$
Then $\bar{g}$ is obviously continuous. In addition, if $d \in D$ then
$$\bar{g}(d) = \bar{h} \circ f(d) = h \circ f(d) = g(d).$$
We conclude that $\bar{g}$ is the required continuous extension of $g$.

**Proposition 6.8.6.** *Let $X$ and $Y$ be spaces, and let*
$$\varphi\colon C_p(X) \to C_p(Y)$$
*be a continuous linear surjection. Then for each closed and bounded subset $K \subseteq X$, the set*
$$L = \{y \in Y : \operatorname{supp}(y) \subseteq K\}$$
*is closed and bounded in $Y$.*

**Proof.** By Lemma 6.8.2(3),(4), supp is LSC and hence $L$ is closed. If $L$ is not bounded, it contains by the above remarks a discrete (faithfully indexed) $C$-imbedded subset $D = \{y_n : n \in \mathbb{N}\}$. For each $n$, let
$$t_n = n \cdot \left( \sum_{z \in \operatorname{supp}(y_n)} |\lambda(y_n, z)| \right)$$

(see (∗∗) on Page 406). Then $t_n > 0$. For if $t_n = 0$ then $\lambda(y_n, z) = 0$ for every $z \in \mathrm{supp}(y_n)$ so that $\varphi(f)(y_n) = 0$ for every $f \in C_p(X)$. Since $\varphi$ is surjective, this means that $g(y_n) = 0$ for every $g \in C_p(Y)$. This is absurd. Let $g \in C(Y)$ be such that $g(y_n) = t_n$. (That $g$ exists follows from the fact that $D$ is discrete and $C$-imbedded.) Since $\varphi$ is a surjection, there exists an element $f \in C(X)$ such that $\varphi(f) = g$. Since $K$ is bounded, there is $c \in \mathbb{R}$ such that $f[K] \subseteq [-c, c]$. Let $n \in \mathbb{N}$ be such that $n > c$. Then

$$\begin{aligned}
t_n &= |g(y_n)| \\
&= |\varphi(f)(y_n)| \\
&= \Big| \sum_{z \in \mathrm{supp}(y_n)} \lambda(y_n, z) f(z) \Big| \\
&\leq \sum_{z \in \mathrm{supp}(y_n)} |\lambda(y_n, z)| \cdot |f(z)| \\
&\leq c \cdot \Big( \sum_{z \in \mathrm{supp}(y_n)} |\lambda(y_n, z)| \Big) \\
&< t_n.
\end{aligned}$$

This contradiction proves the proposition. □

## 6.9. Nonexistence of linear homeomorphisms

It is possible to prove that certain topological properties are shared by spaces having linearly homeomorphic function spaces. We will give three such examples: compactness, dimension and topological completeness. See also the next section for a related result.

We interpret these results as obstacles for obtaining linear homeomorphisms. Since it will follow that if $X$ and $Y$ are $\ell$-equivalent then $X$ is compact if and only $Y$ is compact, $C_p(\mathbb{I})$ is not linearly homeomorphic to $C_p(\mathbb{R})$. Interestingly, these space *are* homeomorphic by a result of GUL'KO and KHMYLEVA [183].

So the results that we prove here do not imply the existence of linear homeomorphisms, but do imply their nonexistence.

There are many other examples and better theorems than the ones presented here. See e.g., the surveys of ARHANGEL'SKIĬ [25, 27, 28] and BAARS and DE GROOT [35] for more information.

**Compactness.**

**Theorem 6.9.1.** *Let $X$ and $Y$ be spaces, and let $\varphi \colon C_p(X) \to C_p(Y)$ be a linear homeomorphism. Then $X$ is $\sigma$-compact if and only if $Y$ is $\sigma$-compact.*

**Proof.** Assume that $X$ is $\sigma$-compact. We claim that $L(X)$ is $\sigma$-compact as well. This is simple. Observe that $X \subseteq L(X)$ is homeomorphic to $X$ (Proposition 6.7.4). Since $L(X)$ is a linear space, for every $n \in \mathbb{N}$ the function
$$\alpha_n \colon X^n \times [-n,n]^n \to L(X)$$
defined by
$$\alpha_n\big((x_1,\ldots,x_n),(t_1,\ldots,t_n)\big) = \sum_{i=1}^n t_i x_i$$
is continuous, and has therefore $\sigma$-compact image. But the images of the $\alpha_n$'s cover $L(X)$ since $X$ generates $L(X)$ (Proposition 6.7.2). As a consequence, the space $L(X)$ is $\sigma$-compact. Now $Y \subseteq L(Y)$ is closed by Proposition 6.7.4, and since $L(X)$ and $L(Y)$ are linearly homeomorphic by Theorem 6.7.5, it follows that $L(X)$ contains a closed homeomorphic copy of $Y$. As a consequence, $Y$ is $\sigma$-compact $\square$

**Corollary 6.9.2.** *Let $X$ and $Y$ be spaces, and let $\varphi \colon C_p(X) \to C_p(Y)$ be a linear homeomorphism. Then $X$ is compact if and only if $Y$ is compact.*

**Proof.** Let $X$ be compact. Then $Y$ is $\sigma$-compact by Theorem 6.9.1. But by an application of Corollary 6.8.4 it also follows that $Y$ is pseudocompact. As a consequence, $Y$ is compact (ENGELKING [153, Theorem 3.11.1]). $\square$

We already know by the result of GUL'KO and KHMYLEVA cited above that $C_p(\mathbb{I})$ and $C_p(\mathbb{R})$ are homeomorphic. So a generalization of Corollary 6.9.2 for (topological) homeomorphisms is not possible. We present other examples of this phenomenon which are based on the results obtained in Chapter 5.

**Example 6.9.3.** *A compact space $X$ and a noncompact space $Y$ such that their function spaces $C_p(X)$ and $C_p(Y)$ are homeomorphic.*

**Proof.** Let $X = \omega+1$. We claim that $C_p(X) \approx B(Q)^\infty$. Indeed, let $A = \{\omega\}$. By Corollary 6.6.8 we have
$$C_p(X) \sim C_{p,A}(X) \times C_p(A) = C_{p,A}(X) \times \mathbb{R}.$$
But $C_{p,A}(X)$ is clearly linearly homeomorphic to
$$c_0 = \{x \in \mathbb{R}^\infty : \lim_{n \to \infty} x_n = 0\}.$$
Since $c_0 \times \mathbb{R} \sim c_0$ it follows by Corollary 5.5.16 that
$$C_p(X) \sim C_{p,A}(X) \times \mathbb{R} \sim c_0 \times \mathbb{R} \sim c_0 \approx B(Q)^\infty.$$

Let $X_n = X$ for every $n$. By Theorem 6.2.8 we consequently obtain that
$$C_p(X) \approx B(Q)^\infty \approx (B(Q)^\infty)^\infty \approx C_p(X)^\infty$$
$$\sim C_p\Big(\bigoplus_{n\in\omega} X_n\Big)$$
$$= C_p(\omega \times (\omega+1)).$$
So the space $Y = \omega \times (\omega + 1)$ is a noncompact space having the property that $C_p(Y)$ is homeomorphic to $C_p(X)$. □

In §6.12 we will obtain a far reaching generalization of this result.

**Dimension.**

**Theorem 6.9.4.** *Let $X$ and $Y$ be separable metrizable spaces with*
$$C_p(X) \sim C_p(Y).$$
*Then $X$ and $Y$ have the same dimension.*

**Proof.** Let $\varphi \colon C_p(X) \to C_p(Y)$ be a linear homeomorphism. For each $n \in \mathbb{N}$, put $Y_n = \{y \in Y : |\operatorname{supp}_\varphi(y)| = n\}$.

**Claim 1.** The set $Z_n = Y \setminus \bigcup_{i=1}^n Y_i$ is open in $Y$ for every $n$.

*Proof.* Let $y \in Z_n$ be arbitrary. Since $\operatorname{supp}_\varphi(y)$ has cardinality greater than $n$, there is a family $\mathcal{V}$ consisting of $n+1$ pairwise disjoint open subsets of $X$ such that $\operatorname{supp}_\varphi(y) \cap V \neq \emptyset$ for every $V \in \mathcal{V}$. Since $\operatorname{supp}_\varphi$ is LSC by Lemma 6.8.2(4), there is a neigborhood $W$ of $y$ such that for every $y' \in W$ and $V \in \mathcal{V}$ we have $\operatorname{supp}_\varphi(y') \cap V \neq \emptyset$. Clearly, $W \subseteq Z_n$. ◇

By Claim 1, $Y_1$ is closed and hence an $F_\sigma$-subset of $Y$. Similarly, $Y_2 \cup Y_1$ is closed in $Y$ and so $Y_2$, being open in $Y_2 \cup Y_1$, is an $F_\sigma$-subset of $Y_2 \cup Y_1$ and hence of $Y$. By the same reasoning it follows that $Y_n$ is an $F_\sigma$-subset of $Y$ for every $n$.

By symmetry, $X_n = \{x \in X : |\operatorname{supp}_{\varphi^{-1}}(x)| = n\}$ is an $F_\sigma$-subset of $X$ for every $n$.

Fix $n \in \mathbb{N}$, and write $Y_n = \bigcup_{i=1}^\infty A_i$, where each $A_i$ is closed in $Y$. Fix $i \in \mathbb{N}$, and pick an arbitrary $y \in A_i$. Let $\operatorname{supp}_\varphi(y) = \{x_1, \dots, x_n\}$. For every $j \leq n$ let $V_j$ be an open neighborhood of $x_j$ such that $V_j \cap V_{j'} = \emptyset$ if $j \neq j'$. By the same reasoning as in the proof of Claim 1, there is a neighborhood $U$ of $y$ such that for every $z \in U$ we have $\operatorname{supp}_\varphi(z) \cap V_j \neq \emptyset$ for every $j \leq n$. As a consequence, $U' = U \cap A_i$ is a neighborhood of $y$ in $A_i$ such that for every $z \in U'$ we have $\operatorname{supp}_\varphi(z) \cap V_j \neq \emptyset$ for every $j \leq n$. We conclude that $\operatorname{supp}_\varphi(z) \cap V_j$ is a single point for every $j \leq n$ since $|\operatorname{supp}_\varphi(z)| = n$. For every $j \leq n$ let $\pi_j \colon U' \to V_j$ be the function that sends $z$ onto the unique point in $\operatorname{supp}_\varphi(z) \cap V_j$. Since supp is LSC, this function is clearly continuous.

Now we do the same thing for every $x_j$. Fix $j \leq n$ and let $n_j \in \mathbb{N}$ be such that $x_j \in X_{n_j}$, and let $\operatorname{supp}_{\varphi^{-1}}(x_j) = \{y_1, \ldots, y_{n_j}\}$. There are a neighborhood $A_j$ of $x_j$ in $X_{n_j}$, and pairwise disjoint neighborhoods $B_1, \ldots, B_{n_j}$ of $y_1, \ldots, y_{n_j}$, respectively, such that for every $k \leq n_j$ the function $\rho_k$ that sends each $t \in A_j$ onto the unique element in $\operatorname{supp}_{\varphi^{-1}}(t) \cap B_k$ is continuous.

Now by Lemma 6.8.2(5) we have $y \in \operatorname{supp}_{\varphi^{-1}}(\operatorname{supp}_{\varphi}(y))$. Assume that $j$ is the index for which $y \in \operatorname{supp}_{\varphi^{-1}}(x_j)$, say $y = y_k$.

Then $\rho_k \circ \pi_j(y) = \rho_k(x_j) = y_k = y$. So the continuous function

$$\rho_k \circ \pi_j \colon U' \to Y$$

has $y$ as fixed-point. The set of all fixed-points $F$ of $\rho_k \circ \pi_j$ is a closed subset of $U'$ (Exercise A.1.9) and is mapped by $\pi_j$ homeomorphically onto a subset of $X$. Observe that $F$ is an $F_\sigma$-subset of $Y$.

Since each neighborhood involved in this process can be picked from a countable base, it follows that we in fact defined countably many functions only, Each of those is defined on an $F_\sigma$-subset of $Y$, and is the composition of a map into $X$ followed by a map into $Y$. Their fixed-points sets are $F_\sigma$-subsets of $Y$, and, as we proved, cover $Y$. In addition, they are mapped homeomorphically onto subsets of $X$.

By the Countable Closed Sum Theorem 3.2.8, one of those fixed-point sets must have the same dimension as $Y$. The Subspace Theorem 3.2.9 therefore implies that $\dim X \geq \dim Y$. By symmetry, $\dim Y \geq \dim X$. So we conclude that $X$ and $Y$ indeed have the same dimension. $\square$

**Remark 6.9.5.** It is an interesting open problem whether $t$-equivalent spaces have the same dimension. CAUTY [90] recently proved that if $X$ and $Y$ are metrizable compacta such that there is an open surjection from $C_p(X)$ onto $C_p(Y)$ then $X^k$ strongly infinite-dimensional for some integer $k$ implies $Y^p$ strongly infinite-dimensional for some integer $p$. As a consequence, the Hilbert cube $Q$ is not $t$-equivalent to any finite dimensional metrizable compactum. MARCISZEWSKI [271] proved, modifying Cauty's technique and using an idea from GUL'KO [182], that if $X$ and $Y$ are metrizable spaces such that $C_p(X) \approx C_p(Y)$ then $X$ is countable dimensional if and only if $Y$ is. See §6.11 for details.

**Completeness.** We will now present our last result in the same spirit. First, let us prove the following preliminary result.

**Lemma 6.9.6.** *Let $(X, \varrho)$ be a metric space which is not completely metrizable, and let $\{\mathcal{U}_n : n \in \mathbb{N}\}$ be a collection of open covers of $X$ such that $\operatorname{mesh}(\mathcal{U}_n) < 1/n$ for every $n$. There are a strictly increasing sequence $(i_n)_n$ of natural numbers and for each $n$ an element $U_n \in \mathcal{U}_{i_n}$ such that $\overline{U}_{n+1} \subseteq U_n$ and $\bigcap_{n=1}^{\infty} U_n = \emptyset$.*

**Proof.** Let $(\tilde{X}, \tilde{\varrho})$ denote the completion of $X$ (ENGELKING [153, p. 273]). For each $U \in \mathcal{U}_n$ there is an open set $V_U \subseteq \tilde{X}$ such that $V_U \cap X = U$. Since $U \in \mathcal{U}_n$ is dense in $V_U$ and $\operatorname{diam}(U) < 1/n$ it follows that $\operatorname{diam} V_U < 1/n$ as well. Let $V_n = \bigcup\{V_U : U \in \mathcal{U}_n\}$. Then $V_n$ is open in $\tilde{X}$, and $X \subseteq V_n$. So

$$V = \bigcap_{n=1}^{\infty} V_n$$

is a $G_\delta$-subset of $\tilde{X}$, and hence is completely metrizable (cf. Theorem A.6.3). So $V \setminus X \neq \emptyset$ and we may pick an arbitrary element $x$ in that set.

**Claim 1.** There are a strictly increasing sequence $(i_n)_n$ of natural numbers and for every $n$ a $U_n \in \mathcal{U}_{i_n}$ such that $x \in V_{U_n}$ and $\overline{V}_{U_{n+1}} \subseteq V_{U_n}$.

*Proof.* Let $i_1 = 1$ and let $U_1 \in \mathcal{U}_1$ be such that $x \in V_{U_1}$. Let $m > 1$ and suppose that $i_1, \ldots, i_{m-1}$ and $U_1, \ldots, U_{m-1}$ have been constructed. Let

$$\delta = \varrho(x, \tilde{X} \setminus V_{U_{m-1}}).$$

Pick $i_m > i_{m-1}$ such that $1/i_m < \delta$. There is an element $U_m \in \mathcal{U}_{i_m}$ such that $x \in V_{U_m}$. Since $\operatorname{diam} \overline{V}_{U_m} \leq 1/i_m < \delta$ and $x \in V_{U_m}$, $\overline{V}_{U_m} \subseteq V_{U_{m-1}}$. So we are done. ◇

Since $\operatorname{diam} V_{U_m} \to 0$, it follows that $\bigcap_{m=1}^{\infty} V_{U_m} \cap X = \{x\} \cap X = \emptyset$. Also, for all $m$ we have

$$\overline{U}_{m+1} = \overline{V}_{U_{m+1}} \cap X \subseteq V_{U_m} \cap X = U_m.$$

(Here the first bar means closure in $X$ and the second bar closure in $\tilde{X}$.) □

**Theorem 6.9.7.** *Let $X$ and $Y$ be metrizable spaces, and let*

$$\varphi \colon C_p(X) \to C_p(Y)$$

*be a continuous linear surjection. If $X$ is completely metrizable then $Y$ is completely metrizable.*

**Proof.** Suppose that $Y$ is not completely metrizable. Let $\varrho$ be an admissible complete metric on $X$. There are by Corollary 6.8.5 and the fact that every metrizable space is paracompact, locally finite open covers $\mathcal{U}_n$ of $X$ and $\mathcal{V}_n$ of $Y$ for $n \subset \mathbb{N}$ such that

(1) $\operatorname{mesh}(\mathcal{U}_n) < 1/n$ and $\operatorname{mesh}(\mathcal{V}_n) < 1/n$,
(2) $\mathcal{U}_{n+1}$ refines $\mathcal{U}_n$ for every $n$, and similarly for the $\mathcal{V}$'s,
(3) for each $V \in \mathcal{V}_n$ the collection

$$\{U \in \mathcal{U}_n : V \cap T_U \neq \emptyset\}$$

is finite.

By Lemma 6.9.6 we may assume without loss of generality that for each $n$ there exists $V_n \in \mathcal{V}_n$ such that $\emptyset \neq \overline{V}_{n+1} \subseteq V_n$ and $\bigcap_{n=1}^{\infty} V_n = \emptyset$.

Let for every $n$,

(4) $$\mathcal{U}'_n = \{U \in \mathcal{U}_n : V_n \cap T_U \neq \emptyset\}.$$

By (3), $\mathcal{U}'_n$ is finite.

**Claim 1.** $\bigcup \mathcal{U}'_{n+1} \subseteq \bigcup \mathcal{U}'_n$.

*Proof.* Pick an arbitrary element $E \in \mathcal{U}'_{n+1}$. Since $\mathcal{U}_{n+1}$ refines $\mathcal{U}_n$ there exists $F \in \mathcal{U}_n$ such that $E \subseteq F$. Since $E \in \mathcal{U}'_{n+1}$ there is $y \in V_{n+1}$ such that $\mathrm{supp}(y) \cap E \neq \emptyset$. But then $\mathrm{supp}(y) \cap F \neq \emptyset$ as well. Since $y \in V_{n+1} \subseteq V_n$, this implies that $F \in \mathcal{U}'_n$ and so $E \subseteq F \subseteq \bigcup \mathcal{U}'_n$, as required. ◇

Notice that by (4), $\mathrm{supp}(V_n) \subseteq \bigcup \mathcal{U}'_n$ for every $n$. For each $n$ pick an arbitrary element $y_n \in V_n$. So $\mathrm{supp}(y_n) \subseteq \bigcup \mathcal{U}'_n$.

Put $K = \overline{\bigcup_{n=1}^{\infty} \mathrm{supp}(y_n)}$.

**Claim 2.** $K$ is compact.

*Proof.* Since $K$ is a closed subspace of the topologically complete space $X$, it suffices to prove that $\varrho \restriction K \times K$ is totally bounded (Theorem A.6.1). Let $\varepsilon > 0$, and let $j \in \mathbb{N}$ be so large that $1/j < \varepsilon/2$. For $U \in \mathcal{U}'_j$, pick an arbitrary element $z(U) \in U$. Since $\mathrm{diam}\, U < 1/j$, clearly $U \subseteq B(z(U), \varepsilon)$. This implies by Claim 1 that

$$\bigcup_{n=j}^{\infty} \mathrm{supp}(y_n) \subseteq \bigcup \{B(z(U), \varepsilon) : U \in \mathcal{U}'_j\}.$$

But $\bigcup_{n=1}^{j-1} \mathrm{supp}(y_n)$ is finite and so we are done. ◇

By Lemma 6.8.6 it follows that $L = \{y \in Y : \mathrm{supp}(y) \subseteq K\}$ is a closed and bounded subset of $Y$. Since $Y$ is metrizable, this means that $L$ is compact (cf. Exercise A.5.14). Since $\bigcap_{n=1}^{\infty} V_n = \emptyset$, $(y_n)_n \subseteq L$ is a sequence without convergent subsequence, which is a contradiction (Theorem A.5.1). □

**Remark 6.9.8.** In §6.12 we will show that $C_p(\mathbb{Q})$ and $C_p(\omega + 1)$ are homeomorphic. So Theorem 6.9.7 is not true for homeomorphisms since $\mathbb{Q}$ is not topologically complete (Corollary A.6.7).

## 6.10. Bounded functions

In this section we derive a result on spaces of bounded functions that is in the same spirit as the results obtained in the previous section. Our results demonstrate a striking difference between $C_p(X)$ and $C_p^*(X)$, and provide for compact spaces new nonexistence results for linear homeomorphisms.

## 6.10. BOUNDED FUNCTIONS

Let $X$ be a space, and let $A \subseteq X$. The *derived set* $A^d$ *of* $A$ *in* $X$ is the set of all accumulation points of $A$ in $X$. That is, the set of all points $x \in X$ such that every neighborhood $U$ of $x$ intersects $A \setminus \{x\}$. Observe that $A^d$ is a closed subset of $X$.

It is not so that $A^d$ is always a subset of $A$. For let $A = \{1/n : n \in \mathbb{N}\}$ and $X = \mathbb{R}$. Then $0 \in A^d$.

For a space $X$ and ordinal number $\alpha$ we define $X^{(\alpha)}$, the $\alpha$-*th derivative of* $X$, by transfinite induction as follows:

(1) $X^{(0)} = X$;
(2) if $\alpha$ is a successor, say $\alpha = \beta+1$, then $X^{(\alpha)}$ is the derived set $(X^{(\beta)})^d$ of $X^{(\beta)}$ in $X$;
(3) if $\alpha$ is a limit ordinal then $X^{(\alpha)} = \bigcap_{\beta<\alpha} X^{(\beta)}$.

**Lemma 6.10.1.** *Let $X$ be a space, and let $\alpha, \beta$ be ordinal numbers. Then*

(1) $X^{(\alpha)}$ *is a closed subset of $X$.*
(2) *If $\beta < \alpha$ then $X^{(\alpha)} \subseteq X^{(\beta)}$.*
(3) $X^{(\alpha+1)}$, *the derived set of $X^{(\alpha)}$ in $X$, coincides with the derived set of $X^{(\alpha)}$ in $X^{(\alpha)}$.*

**Proof.** All three statements follow easily by transfinite induction noting that derived sets are closed. □

Let $A$ be a subspace of $X$. Then clearly $A$ is dense-in-itself iff $A \subseteq A^d$. We say that $A$ is *scattered* if $A$ contains no dense-in-itself subsets, i.e., every subset of $A$ contains a (relative) isolated point.

**Theorem 6.10.2.** *If $X$ is scattered then there exists an ordinal number $\alpha$ such that $X^{(\alpha)} = \emptyset$.*

**Proof.** Since $X$ is scattered, for every ordinal number $\alpha$ such that $X^{(\alpha)} \neq \emptyset$ we have that $X^{(\alpha+1)}$ is a proper subset of $X^{(\alpha)}$. This is so since $X^{(\alpha)}$ has an isolated point. This means that if $X^{(\alpha)}$ is nonempty for every $\alpha$ then

$$X^{(\alpha+1)} \setminus X^{(\alpha)} \neq \emptyset$$

for every $\alpha$. But this is impossible since Lemma 6.10.1(2) shows that the family of subsets

$$\{X^{(\alpha+1)} \setminus X^{(\alpha)} : \alpha \text{ an ordinal number}\}$$

of $X$ is pairwise disjoint. Hence for an ordinal $\alpha > |X|$ we must have

$$X^{(\alpha)} = \emptyset$$

and this is as required. □

The *scattered height* $\kappa(X)$ of a scattered space $X$ is defined to be the smallest ordinal $\alpha$ such that $X^{(\alpha)} = \emptyset$.

It will be convenient to denote $C^*(X)$ with the topology of uniform convergence (see Page 7) by $C_u^*(X)$. Similarly, its subspace consisting of all (bounded) functions vanishing on a closed subset $A$ of $X$ is denoted by $C_{u,A}^*(X)$.

A family $\mathcal{F} \subseteq C(X)$ is *equicontinuous* if for every $x \in X$ and $\varepsilon > 0$ there is a neighborhood $U$ of $x$ in $X$ such that for each $f \in \mathcal{F}$ and $y \in U$,
$$|f(x) - f(y)| < \varepsilon.$$

**Proposition 6.10.3.** *If $\mathcal{F} \subseteq C_u^*(X)$ is compact then it is equicontinuous.*

**Proof.** Let $x \in X$ and $\varepsilon > 0$. The family $\{B(f, \varepsilon/3) : f \in \mathcal{F}\}$ is an open cover of $\mathcal{F}$. Since $\mathcal{F}$ is compact, there are $f_1, \ldots, f_n \in \mathcal{F}$ such that the family $\{B(f_i, \varepsilon/3) : i \leq n\}$ covers $\mathcal{F}$. Since each $f_i$ is continuous, there is a neighborhood $U$ of $x$ such that for all $y \in U$ and for every $i \leq n$,
$$|f_i(x) - f_i(y)| < \varepsilon/3.$$
Now let $f \in \mathcal{F}$ and $y \in U$ be arbitrary. There is $i \leq n$ such that $f \in B(f_i, \varepsilon/3)$. So
$$|f_i(x) - f(x)| < \varepsilon/3, \quad |f_i(y) - f(y)| < \varepsilon/3.$$
Since $y \in U$, we consequently have
$$|f(x) - f(y)| \leq |f(x) - f_i(x)| + |f_i(x) - f_i(y)| + |f_i(y) - f(y)| < \varepsilon,$$
as required. $\square$

We now come to our first main result.

**Theorem 6.10.4.** *Let $X$ and $Y$ be first countable spaces and $\ell^*$-equivalent. Then $X$ is discrete if and only if $Y$ is discrete, or, in terms of $\kappa$, $\kappa(X) < 2$ if and only if $\kappa(Y) < 2$.*

**Proof.** Striving for a contradiction, suppose that $\kappa(X) < 2$ and $\kappa(Y) \geq 2$. Since $X$ is not empty we have $\kappa(X) = 1$ so that $X$ is discrete. Since $\kappa(Y) \geq 2$, some $y \in Y$ is not isolated. Let $\{U_n : n < \omega\}$ be a decreasing open base at $y$. For every $n$ let $f_n \colon Y \to \mathbb{I}$ be a Urysohn function with $f_n(y) = 1$ and $f_n[Y \setminus U_n] \subseteq \{0\}$. Then $f_n \to \chi_{\{y\}}$ pointwise in $\mathbb{R}^Y$, where $\chi_{\{y\}}$ is the characteristic function of $\{y\}$. Since $\chi_{\{y\}} \notin C_p^*(Y)$, $\{f_n : n < \omega\}$ is closed and discrete in $C_p^*(Y)$.

Now let $\varphi \colon C_p^*(X) \to C_p^*(Y)$ be a linear homeomorphism. By Corollary 1.1.16, $\varphi \colon C_u^*(X) \to C_u^*(Y)$ is also a linear homeomorphism. Since $C_u^*(X)$ and $C_u^*(Y)$ are Banach spaces, by Exercise 1.1.31 there is a $k \in \mathbb{N}$ such that for every $f \in C^*(X)$ we have
$$\|f\|/k \leq \|\varphi(f)\| \leq k\|f\|.$$

Let $g_n = \varphi^{-1}(f_n)$ for every $n$. Then $\|g_n\| \leq k\|f_n\| = k$. Hence $(g_n)_n$ is a sequence contained in the compact space $[-k, k]^X$. So the sequence $(g_n)_n$ has an accumulation point $g \in [-k, k]^X$. Since $X$ is discrete, $g$ is continuous (and bounded) on $X$. However, since $\{f_n : n < \omega\}$ is closed and discrete in $C_p^*(Y)$, the same is true for $\{g_n : n < \omega\}$ in $C_p(X)$. This is a contradiction. □

The question naturally arises whether for all scattered metrizable $\ell^*$-equivalent spaces $X$ and $Y$ and for all $n \in \mathbb{N}$ we have $\kappa(X) < n$ iff $\kappa(Y) < n$. For $n = 1$ this is trivially true for then both $X$ and $Y$ are empty, and for $n = 2$ it is true by Theorem 6.10.4. But for larger $n$ it is false.

**Example 6.10.5.** For each $n \geq 2$ there are scattered compact $\ell^*$-equivalent spaces $X$ and $Y$ such that $\kappa(X) = 2$ and $\kappa(Y) = n + 1$.

**Proof.** Indeed, put $X = \omega + 1$ (this $X$ works for all $n$ simultaneously) and $Y_n = (\omega + 1)^n$ for $n \geq 2$. Then $\kappa(X) = 2$ and $\kappa(Y) = n + 1$. We claim that $C_p(Y_n) \sim C_p(X)$ for all $n \geq 2$. We will only prove this for $n = 2$. The remaining part of the proof is left as an exercise to the reader.

Indeed, put $Y = Y_2$ and let
$$A = \{(x, y) \in Y : (x = \omega) \vee (y = \omega)\}.$$
Observe that $A$ is homeomorphic to $\omega + 1$ and that $Y/A \approx \omega + 1$. The last statement follows since $Y/A$ is the one point compactification of a countable discrete space, and one point compactifications are topologically unique. By Corollary 6.6.13 and Theorem 6.2.8 we therefore find that

$$\begin{aligned} C_p(Y) &\sim C_p(Y/A) \times C_p(A) \\ &\sim \big(C_p(\omega + 1)\big)^2 \\ &\sim C_p\big((\omega + 1) \oplus (\omega + 1)\big) \\ &\sim C_p(\omega + 1). \end{aligned}$$

For the last homeomorphism, observe that if $Z = (\omega + 1) \oplus (\omega + 1)$ and $B$ denotes the two non-isolated points in $Z$ then $Z/B \approx \omega + 1$, for which it follows by the same reasoning and Corollary 6.6.10 that
$$C_p(Z) \sim C_p(\omega + 1) \times \mathbb{R}^2 \sim C_p(\omega + 1),$$
as required. □

Theorem 6.10.4 can be generalized though. We will show below that if $X$ and $Y$ are $\ell^*$-equivalent scattered metrizable spaces then $\kappa(X) < \omega$ if and only if $\kappa(Y) < \omega$. Before proving this result, we need to derive two preliminary results first.

**Lemma 6.10.6.** Let $X$ be a metrizable space with $\kappa(X) < \omega$. Then there is a metrizable space $Y$ such that $\kappa(X) = \kappa(Y)$ and $C_p^*(X) \sim C_{p,A}^*(Y)$, where $A = Y^{(1)}$.

**Proof.** We prove the lemma by induction on $\kappa(X)$. If $\kappa(X) = 1$ then $X$ is discrete, so $Y = X$ will do. So suppose the lemma to be true for all metrizable spaces $X$ for which $\kappa(X) < n$, where $n > 1$. Let $X$ be a metrizable space with $\kappa(X) = n$, and let $B = X^{(1)}$. Then by Corollary 6.6.8 we have that

$$C_p^*(X) \sim C_p^*(B) \times C_{p,B}^*(X).$$

Since $\kappa(B) = n - 1$, there is by the induction hypothesis a metrizable space $Z$ such that $\kappa(Z) = \kappa(B)$ and $C_p^*(B) \sim C_{p,C}^*(Z)$, where $C = Z^{(1)}$. Then

$$C_p^*(X) \sim C_{p,C}^*(Z) \times C_{p,B}^*(X) \sim C_{p,B\cup C}^*(Z \oplus X)$$

(cf. Theorem 6.2.8). Let $Y = Z \oplus X$. Then $Y^{(1)} = B \cup C$ and $\kappa(Y) = \kappa(X)$. So we are done. □

If $B$ is a set then we order its finite nonempty subsets by inclusion. Clearly $(\mathcal{F}_\infty(B) \setminus \{\emptyset\}, \subseteq)$ is a directed set.

Let $X$ be a space and $B$ an infinite set. For every $b \in B$ let $f_b \in \mathbb{R}^X$ be such that for every $x \in X$ the set $\{b \in B : f_b(x) \neq 0\}$ is finite. Define the function $\sum_{b \in B} f_b \colon X \to \mathbb{R}$ by the following formula:

$$\left(\sum_{b \in B} f_b\right)(x) = \sum_{\substack{b \in B \\ f_b(x) \neq 0}} f_b(x).$$

Then obviously $\sum_{b \in B} f_b$ is well-defined.

**Lemma 6.10.7.** Let $X$ be a space and $B$ an infinite set. For every $b \in B$ let $f_b \in \mathbb{R}^X$ be such that for every $x \in X$,

$$\{b \in B : f_b(x) \neq 0\}$$

is finite. For $S \in \mathcal{S} = \mathcal{F}_\infty(B) \setminus \{\emptyset\}$ let $f_S = \sum_{b \in S} f_b$. Then $\{f_S : S \in \mathcal{S}\}$ is a net in $\mathbb{R}^X$ and $\lim_{S \in \mathcal{S}} f_S = \sum_{b \in B} f_b$.

**Proof.** It is clear that $\{f_S : S \in \mathcal{S}\}$ is a net in $\mathbb{R}^X$. Now let $\varepsilon > 0$ and $P \subseteq X$ be finite. For every $p \in P$ let $S_p = \{b \in B : f_b(p) \neq 0\}$. Put $S_0 = \bigcup_{p \in P} S_p$. Then $S_0 \in \mathcal{S}$. Let $S \in \mathcal{S}$ be such that $S \supseteq S_0$, $p \in P$, and $f = \sum_{b \in B} f_b$. Then

$$|f(p) - f_S(p)| = \left|\sum_{b \in B} f_b(p) - \sum_{b \in S} f_b(p)\right|$$

$$= \left|\sum_{b \in S_p} f_b(p) - \sum_{b \in S_p} f_b(p)\right|$$

$$= 0$$

$$< \varepsilon.$$

Hence, $\lim_{S \in \mathcal{S}} f_S = f$. □

**Theorem 6.10.8.** *Let $X$ and $Y$ be $\ell^*$-equivalent metrizable spaces. Then*
$$\kappa(X) < \omega \iff \kappa(Y) < \omega.$$

**Proof.** Striving for a contradiction, assume that $\kappa(X) < \omega$ and $\kappa(Y) \geq \omega$ and that $X$ and $Y$ are $\ell^*$-equivalent metrizable spaces. By Lemma 6.10.6 we may replace $C_p^*(X)$ by $C_{p,A}^*(X)$ with $A = X^{(1)}$. Let $\varphi \colon C_{p,A}^*(X) \to C_p^*(Y)$ be a linear homeomorphism. Then by Corollary 1.1.16, $\varphi \colon C_{u,A}^*(X) \to C_u^*(Y)$ is also a linear homeomorphism. So there is by Exercise 1.1.31 a $k \in \mathbb{N}$ such that for every $f \in C_{u,A}^*(X)$ we have
$$\|f\|/k \leq \|\varphi(f)\| \leq k\|f\|.$$

Let $B = X \setminus A$. Since every $x \in B$ is isolated, the characteristic function
$$f_x = \chi_{\{x\}}$$
belongs to $C_{p,A}^*(X)$. Notice that every $f \in C_{p,A}^*(X)$ can be written as
$$f = \sum_{x \in B} \alpha_x f_x,$$
where for $x \in B$, $\alpha_x = f(x)$. Put $g_x = \varphi(f_x)$.

For every $y \in Y$ we put $C_y = \{x \in B : g_x(y) \neq \emptyset\}$.

**Claim 1.** $C_y$ is finite for every $y \in Y$.

*Proof.* Suppose that $C_y$ is infinite for some $y \in Y$. Let $(x_n)_n$ be a sequence in $C_y$ such that $x_n \neq x_m$ if $n \neq m$. Define $h_n \colon X \to \mathbb{R}$ by
$$h_n = \frac{f_{x_n}}{g_{x_n}(y)}.$$
Then $h_n \in C_{p,A}^*(X)$ for every $n$ and the sequence $(h_n)_n$ converges pointwise to $\underline{0}$, the constant function with value 0, in $C_{p,A}^*(X)$. Now
$$\varphi(h_n)(y) = \frac{\varphi(f_{x_n})(y)}{g_{x_n}(y)} = \frac{g_{x_n}(y)}{g_{x_n}(y)} = 1.$$
Hence $\varphi(h_n) \not\to \underline{0}$ in $C_p^*(Y)$. This contradiction proves the claim. $\diamond$

Now define $b \colon Y \to \mathbb{R}$ by
$$b(y) = \sum_{x \in B} |g_x(y)|.$$
Notice that for every $y \in Y$ we have $b(y) = \sum_{x \in C_y} |g_x(y)|$, hence $b$ is well-defined.

**Claim 2.** $\|b\| \leq 2k$.

*Proof.* For $y \in Y$, let
$$C_y^+ = \{x \in B : g_x(y) > 0\},$$
$$C_y^- = \{x \in B : g_x(y) < 0\},$$
respectively. Notice that
$$\Big\|\sum_{x \in C_y^+} g_x\Big\| = \Big\|\varphi\Big(\sum_{x \in C_y^+} f_x\Big)\Big\| \leq k \cdot \Big\|\sum_{x \in C_y^+} f_x\Big\| = k.$$
Similarly we can prove that $\|\sum_{x \in C_y^-} g_x\| \leq k$. So for every $y \in Y$ we have
$$b(y) = \Big|\sum_{x \in C_y^+} g_x(y) - \sum_{x \in C_y^-} g_x(y)\Big| \leq \Big|\sum_{x \in C_y^+} g_x(y)\Big| + \Big|\sum_{x \in C_y^-} g_x(y)\Big| \leq 2k,$$
which proves the claim. $\diamond$

Now for $P \subseteq B$ finite, let
$$\mathcal{M}_P = \Big\{\sum_{x \in P} \tilde{\alpha}_x f_x : |\tilde{\alpha}_x| \leq k \text{ for } x \in P\Big\}.$$
Notice that
$$\mathcal{M}_P = \prod_{x \in P} [-k, k] \times \prod_{x \in X \setminus P} \{0\}$$
is a compact subset of $\mathbb{R}^X$, hence of $C_{p,A}^*(X)$. It is evident that $\mathcal{M}_P$ is also a compact subset of $C_{u,A}^*(X)$.

**Claim 3.** For every $y \in Y$, $P \subseteq B$ finite, and $\varepsilon > 0$, there is a neighborhood $U(y, P, \varepsilon)$ of $y$ in $Y$ such that for each $z \in U(y, P, \varepsilon)$ and $f \in \varphi[\mathcal{M}_P]$ we have $|f(y) - f(z)| < \varepsilon$.

*Proof.* Since $\varphi[\mathcal{M}_P]$ is compact in $C_u^*(Y)$, Proposition 6.10.3 implies that it is equicontinuous. From this, the claim follows immediately. $\diamond$

Now let $N \in \mathbb{N}$ be so large that
$$\frac{3(N+1)}{4k} \geq 2k.$$

**Claim 4.** There are $y_0, \ldots, y_N \in Y$, $P_0, \ldots, P_N \subseteq B$ finite, and $U_0, \ldots, U_N$ neighborhoods of $y_0, \ldots, y_N$ respectively, such that

(1) for every $0 \leq i \leq N$: $C_{y_i} \subseteq P_i$,
(2) $P_0 \subseteq P_1 \subseteq \cdots \subseteq P_N$,
(3) $U_0 \supseteq U_1 \supseteq \cdots \supseteq U_N$,
(4) for every $0 \leq i \leq N$: $U_i \subseteq U(y_i, P_i, 1/4)$,
(5) for every $0 \leq i \leq N$: $y_i \in Y^{(N-i)}$.

## 6.10. BOUNDED FUNCTIONS

*Proof.* We will prove this claim by induction. Since $\kappa(Y) \geq \omega$, $Y^{(N)} \neq \emptyset$ and hence we may pick
$$y_0 \in Y^{(N)}.$$
Let $P_0 = C_{y_0}$ and $U_0 = U(y_0, P_0, 1/4)$. Suppose
$$y_0, \ldots, y_n, \quad P_0, \ldots, P_n, \quad U_0, \ldots, U_n$$
are found for $0 \leq n < N$. Since $y_n \in Y^{(N-n)}$ and $N - n \geq 1$, we can find
$$y_{n+1} \in (U_n \setminus \{y_i : i \leq n\}) \cap Y^{(N-(n+1))}.$$
Let
$$P_{n+1} = P_n \cup C_{y_{n+1}}$$
and
$$U_{n+1} = U_n \cap U(y_{n+1}, P_{n+1}, 1/4).$$
It is easy to see that these choices satisfy our inductive hypotheses. ◇

Now let $g \colon Y \to \mathbb{J}$ be a continuous function such that
$$g(y_i) = (-1)^i \quad (0 \leq i \leq N).$$
Then $\|g\| = 1$, so $\|\varphi^{-1}(g)\| \leq k$. Hence $\varphi^{-1}(g)$ can be written as
$$\sum_{x \in B} \alpha_x f_x,$$
where $|\alpha_x| \leq k$ for every $x \in B$. Notice that
$$\sum_{x \in P_i} \alpha_x f_x \in \mathcal{M}_{P_i}$$
for every $0 \leq i \leq N$.

**Claim 5.** $g = \sum_{x \in B} \alpha_x g_x$.

*Proof.* Let $\mathcal{S} = \mathcal{F}_\infty(B) \setminus \{\emptyset\}$, and for every $S \in \mathcal{S}$ let $f_S = \sum_{x \in S} \alpha_x f_x$. By Lemma 6.10.7, $\varphi^{-1}(g) = \lim_{S \in \mathcal{S}} f_S$ and $\sum_{x \in B} \alpha_x g_x = \lim_{S \in \mathcal{S}} \alpha_x g_x$. So
$$g = \varphi(\varphi^{-1}(y)) = \varphi\left(\lim_{S \in \mathcal{S}} f_S\right)$$
$$= \lim_{S \in \mathcal{S}} \varphi(f_S) = \lim_{S \in \mathcal{S}} \sum_{x \in S} \alpha_x g_x$$
$$= \sum_{x \in B} \alpha_x g_x,$$
which proves the claim. ◇

Let $0 \leq i \leq N$. Since $C_{y_i} \subseteq P_i$ (Claim 4(1)), we have by Claim 5,
$$(-1)^i = g(y_i) = \sum_{x \in B} \alpha_x g_x(y_i) = \sum_{x \in P_i} \alpha_x g_x(y_i).$$
By Claim 4(3) and (4), $y_N \in U(y_i, P_i, 1/4)$. Furthermore,
$$\sum_{x \in P_i} \alpha_x g_x \in \varphi[\mathcal{M}_{P_i}].$$
So by Claim 3,
$$\left| \sum_{x \in P_i} \alpha_x g_x(y_N) - \sum_{x \in P_i} \alpha_x g_x(y_i) \right| < 1/4.$$
If $i > 0$ then by the same argument,
$$\left| \sum_{x \in P_{i-1}} \alpha_x g_x(y_N) - \sum_{x \in P_{i-1}} \alpha_x g_x(y_{i-1}) \right| < 1/4.$$
We conclude that in case $i > 0$, we have by the above and Claim 4(2),
$$\left| \sum_{x \in P_i \setminus P_{i-1}} \alpha_x g_x(y_N) \right| = \left| \sum_{x \in P_i} \alpha_x g_x(y_N) - \sum_{x \in P_{i-1}} \alpha_x g_x(y_N) \right|$$
$$= \left| \sum_{x \in P_i} \alpha_x g_x(y_N) - (-1)^i + (-1)^{i-1} - \sum_{x \in P_{i-1}} \alpha_x g_x(y_N) \pm 2 \right|$$
$$= \left| \sum_{x \in P_i} \alpha_x g_x(y_N) - g(y_i) + g(y_{i-1}) - \sum_{x \in P_{i-1}} \alpha_x g_x(y_N) \pm 2 \right|$$
$$= \left| \sum_{x \in P_i} \alpha_x g_x(y_N) - \sum_{x \in P_i} \alpha_x g_x(y_i) \right.$$
$$\left. + \sum_{x \in P_{i-1}} \alpha_x g_x(y_{i-1}) - \sum_{x \in P_{i-1}} \alpha_x g_x(y_N) \pm 2 \right|$$
$$\geq 2 - \left| \sum_{x \in P_i} \alpha_x g_x(y_N) - \sum_{x \in P_i} \alpha_x g_x(y_i) \right|$$
$$- \left| \sum_{x \in P_{i-1}} \alpha_x g_x(y_N) - \sum_{x \in P_{i-1}} \alpha_x g_x(y_{i-1}) \right|$$
$$> 3/4.$$
If $i = 0$ then with $P_{-1} = \emptyset$ we have
$$\left| \sum_{x \in P_i \setminus P_{i-1}} \alpha_x g_x(y_N) \right| = \left| \sum_{x \in P_0} \alpha_x g_x(y_N) - \sum_{x \in P_0} \alpha_x g_x(y_0) + 1 \right| > 1 - 1/4 = 3/4.$$
So by Claim 4(2),
$$\sum_{x \in P_N} |\alpha_x g_x(y_N)| \geq \sum_{i=0}^{N} \left| \sum_{x \in P_i \setminus P_{i-1}} \alpha_x g_x(y_N) \right| > 3/4(N+1).$$

From this we conclude that
$$b(y_N) = \sum_{x \in P_N} |g_x(y_N)| \geq \sum_{x \in P_N} \left|\frac{\alpha_x}{k} g_x(y_N)\right| > \frac{3}{4k}(N+1) \geq 2k,$$
which contradicts Claim 2. □

**Corollary 6.10.9.** *Let $X$ and $Y$ be $\ell$-equivalent compact metrizable spaces. Then $\kappa(X) < \omega$ if and only if $\kappa(Y) < \omega$.*

So we see for example that if $X = \omega + 1$ and $Y$ is the one point compactification of the topological sum of the spaces $(\omega + 1)^n$, $n \in \mathbb{N}$, then $X \not\sim_\ell Y$ because $\kappa(X) = 2$ and $\kappa(Y) = \omega + 1$.

We are now in a position to present the following:

**Example 6.10.10.** *There are $\ell$-equivalent countable, locally compact (hence metrizable) spaces which are not $\ell^*$-equivalent.*

**Proof.** Let $X$ be the topological sum of countably many copies of $\omega + 1$. Then
$$C_p(X) \sim \prod_{n=1}^{\infty} C_p(\omega + 1)$$
by Theorem 6.2.8.

In addition, let $Y$ be the topological sum of the spaces
$$Y_n = (\omega + 1)^n, \quad (n \in \mathbb{N}).$$
Then
$$C_p(Y) \sim \prod_{n=1}^{\infty} C_p((\omega + 1)^n) \sim \prod_{n=1}^{\infty} C_p(\omega + 1) \sim C_p(X),$$
again by Theorem 6.2.8 and the computations in Example 6.10.5.

Observe that $\kappa(X) = 2$ and $\kappa(Y) = \omega$. So $X$ and $Y$ are not $\ell^*$-equivalent by Theorem 6.10.8. □

**Remark 6.10.11.** It is unknown whether there are $\ell^*$-equivalent (countable) spaces which are not $\ell$-equivalent. BAARS [31] proved the following interesting result:

**Theorem 6.10.12.** *Let $X$ and $Y$ be locally compact zero-dimensional separable metrizable spaces. Then the following statements are equivalent:*

(1) *$X$ and $Y$ are $\ell^*$-equivalent.*
(2) *There is a linear homeomorphism $\varphi \colon C_p(X) \to C_p(Y)$ such that $\varphi[C_p^*(X)] = C_p^*(Y)$.*

So if such $X$ and $Y$ are $\ell^*$-equivalent then they are $\ell$-equivalent. Examples of $\ell^*$-equivalent spaces which are not $\ell$-equivalent must therefore be rather complex. Observe that a linear homeomorphism $\varphi \colon C_p(X) \to C_p(Y)$ does not necessarily map $C_p^*(X)$ onto $C_p^*(Y)$ by Example 6.10.10.

**Remark 6.10.13.** Similar questions about the relation between the function spaces $C_p(X)$ and $C_p^*(X)$ for topological homeomorphisms are also quite interesting. For countable spaces, some definitive answers are known. BANAKH and CAUTY [40] proved the following nice result:

**Theorem 6.10.14.** *If $X$ is countable and nondiscrete then $C_p^*(X)$ is homeomorphic to $C_p(X) \times \sigma$.*

This implies that if $X$ and $Y$ are countable, nondiscrete and $t$-equivalent then they are $t^*$-equivalent. The converse of this statement is false. By a result of MARCISZEWSKI and VAN MILL [272] there are countable nondiscrete $t^*$-equivalent spaces which are not $t$-equivalent.

## 6.11. Nonexistence of homeomorphisms

In view of the results obtained so far the question naturally arises what can be said about homeomorphisms between function spaces which are not necessarily linear. We will approach this as in §6.9, i.e., we try to find topological properties shared by spaces having homeomorphic function spaces. We interpret the results in this section as obstacles for obtaining homeomorphisms. In the next section we will present positive results, i.e., we will prove using the infinite-dimensional apparatus developed in the Chapter 6 that many function spaces are homeomorphic.

Homeomorphisms are much more flexible than linear homeomorphisms. So they are easier to construct but, on the other hand, their nonexistence is harder to prove.

An interesting result is due to OKUNEV: he proved (among other things) that if $X$ is $\sigma$-compact and $C_p(X)$ is homeomorphic to $C_p(Y)$ then $Y$ is $\sigma$-compact (see OKUNEV [328] or ARHANGEL'SKIĬ [27, Corollary III.2.12]). For metrizable spaces $X$ the situation is even better: a metrizable space $X$ is $\sigma$-compact if and only if $C_p(X)$ is analytic (CHRISTENSEN [100]).

Another interesting result in the same spirit was obtained recently by CAUTY [90]. He proved that if $X$ and $Y$ are metrizable compacta such that there is an open surjection from $C_p(X)$ onto $C_p(Y)$ then $X^k$ strongly infinite-dimensional for some integer $k$ implies $Y^p$ strongly infinite-dimensional for some integer $p$. As a consequence, the Hilbert cube $Q$ is not $t$-equivalent to any finite dimensional metrizable compactum. We will present a refinement of the method CAUTY developed in [90] here which is due to MARCISZEWSKI [271] and which will lead us to the following result: if $X$ and $Y$

are metrizable, $X$ is countable dimensional and $C_p(X) \approx C_p(Y)$, then $Y$ is countable dimensional.

We begin by deriving a few auxiliary results. It will be convenient to introduce the following notation. For an integer $k$ let $[X]^k$ be the subspace of $\mathcal{F}_\infty(X)$ consisting of all sets of size precisely $k$.

**Lemma 6.11.1.** *Let $X$ and $Y$ be metrizable spaces and let $\alpha \colon X \to [Y]^n$ be continuous. Then there are a countable open cover $\mathcal{U}$ of $X$ and for every element $U \in \mathcal{U}$ and $k \leq n$ a continuous function $f_U^k \colon U \to Y$ such that for every $x \in U \in \mathcal{U}$ we have*

$$(*) \qquad \alpha(x) = \{f_U^1(x), \ldots, f_U^n(x)\}.$$

**Proof.** Let $x \in X$ be arbitrary and let $\alpha(x) = \{y_x^1, \ldots, y_x^n\}$. There are pairwise disjoint open sets $V_x^1, \ldots, V_x^n$ in $Y$ such that $y_x^k \in V_x^k$ for $k \leq n$. The set

$$W_x = \{x' \in X : (\forall k \leq n)(\alpha(x') \cap V_x^k \neq \emptyset)\}$$

is by continuity of $\alpha$ an open neighborhood of $x$ in $X$ and, for every $k \leq n$, the map

$$x' \mapsto \alpha(x') \cap V_x^k$$

is single-valued and continuous on $W_x$. Since metrizable spaces are paracompact, the cover $\mathcal{W} = \{W_x : x \in X\}$ has a $\sigma$-discrete open refinement, say $\mathcal{V}$. Let $\mathcal{V} = \bigcup_{i=1}^\infty \mathcal{V}_i$, where each $\mathcal{V}_i$ is discrete. For every $i$, put $U_i = \bigcup \mathcal{V}_i$. Then

$$\mathcal{U} = \{U_i : i \in \mathbb{N}\}$$

is a countable open cover of $X$ and we will show that it is as required. To this end, fix $i$ and observe that for every $V \in \mathcal{V}_i$ there exists $x(V) \in X$ such that $V \subseteq W_{x(V)}$. For $k \leq n$ we let $\xi_V^k$ be the restriction of the function

$$x' \mapsto \alpha(x') \cap V_{x(V)}^k$$

to $V$. Then $\xi_V^k$ is evidently continuous. Since $\mathcal{V}_i$ is discrete, the function

$$f_{U_i}^k = \bigcup_{V \in \mathcal{V}_i} \xi_V^k$$

is continuous on $U_i$. It is clear that with these functions $(*)$ is satisfied. $\square$

**Lemma 6.11.2.** *Let $A$ and $B$ be $G_\delta$-subsets of the metrizable spaces $X$ and $Y$, respectively. In addition, let $f \colon A \to Y$ and $g \colon B \to X$ be continuous functions such that $f \circ g$ is the identity on $B$. Then $g \colon B \to g[B]$ is a homeomorphism and $g[B]$ is a $G_\delta$-subset of $X$.*

**Proof.** The map $g \colon B \to g[B]$ is a homeomorphism since by assumption,

$$f \restriction g[B] \colon g[B] \to B$$

is its continuous inverse. It therefore remains to show that $g[B]$ is a $G_\delta$-subset of $X$. Put $S = f^{-1}[B]$. Then $S$ is clearly a $G_\delta$-subset of $X$ since $f$ is continuous. In addition, $g[B]$ is contained in $S$ and $f \restriction g[B] \colon g[B] \to B$ is a homeomorphism. So by Exercise A.8.3 we conclude that $g[B]$ is a $G_\delta$-subset of $S$ and hence of $X$. □

We now formulate the main result in this section.

**Theorem 6.11.3.** *Let $X$ and $Y$ be metrizable spaces. If $C_p(X)$ and $C_p(Y)$ are homeomorphic then $Y$ is a countable union of $G_\delta$-subsets which are homeomorphic to $G_\delta$-subsets of $X$ and vice versa.*

**Corollary 6.11.4.** *Let $X$ and $Y$ be t-equivalent metrizable spaces. Then $X$ is countable dimensional if and only if $Y$ is countable dimensional.*

**Proof.** Suppose that $X$ is countable dimensional. Since countable dimensionality is hereditary, it follows that the $G_\delta$-subsets of $Y$ we get from Theorem 6.11.3 are all countable dimensional. Hence $Y$ is a countable union of countable dimensional subspaces, hence is countable dimensional itself. This completes the proof since the roles of $X$ and $Y$ can be interchanged. □

**Corollary 6.11.5.** *Let $X$ be a finite dimensional metrizable space. Then the function spaces $C_p(Q)$ and $C_p(X)$ are not homeomorphic.*

**Proof.** Simply observe that $Q$ is strongly infinite-dimensional and hence not countable dimensional (Corollary 2.4.13 and Exercise 3.1.6) and that every finite dimensional metrizable space is countable dimensional (cf. Corollary 3.3.9 and ENGELKING [155, Theorem 4.1.17]). □

To begin the proof of Theorem 6.11.3, let us fix some notation. For $A$ and $B$ finite subsets of $X$ and $Y$, respectively, we put for every $n \geq 1$,

$$V(A, n) = \{f \in C_p(X) : |f(x)| < 1/n \text{ for } x \in A\},$$
$$W(B, n) = \{g \in C_p(Y) : |g(y)| < 1/n \text{ for } y \in B\}.$$

Observe that $V(A, n)$ is a basic open neighborhood of the zero $\underline{0}$ of $C_p(X)$, i.e., the constant function with value 0. Similarly for $W(B, n)$. For $x \in X$ and $y \in Y$ we also put for every $n \geq 1$,

$$\overline{V}(x, n) = \{f \in C_p(X) : |f(x)| \leq 1/n\},$$
$$\overline{W}(y, n) = \{g \in C_p(Y) : |g(y)| \leq 1/n\}.$$

Let $\varphi \colon C_p(X) \to C_p(Y)$ be a homeomorphism. Since $C_p(Y)$ is homogeneous, being a linear space, we may assume without loss of generality that $\varphi(\underline{0}) = \underline{0}$. For $n, m \geq 1$ and $y \in Y$ we put

$$\mathcal{C}(y, m, n) = \{A \in \mathcal{F}_\infty(X) \setminus \{\emptyset\} : \varphi[V(A, m)] \subseteq \overline{W}(y, n)\}.$$

In addition, define for $k \geq 1$,

$$C(k, m, n) = \{y \in Y : \bigl(\exists A \in \mathcal{C}(y, m, n)\bigr)\bigl(|A| \leq k\bigr)\}.$$

**Lemma 6.11.6.** $Y = \bigcup_{k=1}^{\infty} \bigcup_{m=1}^{\infty} C(k, m, n)$ for every $n \geq 1$.

**Proof.** Fix $n \geq 1$ and $y \in Y$. Then $\overline{W}(y, n)$ is a neighborhood of $\underline{0}$ in $C_p(Y)$. Hence by continuity of $\varphi$ and the fact that $\varphi(\underline{0}) = \underline{0}$ there are a nonempty finite subset $A \subseteq X$ and an $m \geq 1$ such that
$$\varphi[V(A, m)] \subseteq \overline{W}(y, n).$$
So for $k = |A|$ we have $y \in C(k, m, n)$. □

**Lemma 6.11.7.** *The set $C(k, m, n)$ is closed in $Y$ for every $k, m, n \geq 1$.*

**Proof.** Let $(y_i)_i$ be a sequence in $C(k, m, n)$ converging to a point $y \in Y$. For each $i$ let $A_i \subseteq X$ be a set of size at most $k$ such that
(∗) $$\varphi[V(A_i, m)] \subseteq \overline{W}(y_i, n).$$

**Claim 1.** There are an infinite subset $E \subseteq \mathbb{N}$ and a (possibly empty) subset $A^0$ of $X$ and for every $i \in E$ a partition $A_i^0, A_i^1$ of $A_i$ such that that one of the following statements is true:

(A) Every $x \in X$ has a neighborhood meeting finitely many terms of the sequence $(A_i)_{i \in E}$ only.
(B) $A^0 \neq \emptyset$ and
  (1) the sequence $(A_i^0)_{i \in E}$ converges to $A^0$ in the Vietoris topology. (Observe that $|A^0| \leq k$ because $(A_i^0)_{i \in E} \to A^0$ and $|A_i^0| \leq k$ for every $i \in E$ (Exercise 1.11.3).)
  (2) Every $x \in X$ has a neighborhood meeting finitely many terms of the sequence $(A_i^1)_{i \in E}$ only.

*Proof.* This is precisely the statement of Exercise 1.11.20. ◇

If (A) holds then we redefine $A^0$ to be the empty set. We also put $A_i^0 = \emptyset$ and $A_i^1 = A_i$ for every $i \in E$. So every $x \in X$ has a neighborhood $U_x$ meeting finitely many terms of the sequence $(A_i^1)_{i \in E}$ only. We may assume without loss of generality that $E = \mathbb{N}$.

We claim that
$$\varphi[V(A^0, m)] \subseteq \overline{W}(y, n),$$
which will show that $y \in C(k, m, n)$.

Incidentally, this also shows that $A^0 \neq \emptyset$, and hence that (A) does not hold, for if $A^0 = \emptyset$ then
$$V(A^0, m) = C_p(X)$$
and so $\varphi[V(A^0, m)] = C_p(Y) \not\subseteq \overline{W}(y, n)$. (This observation will be used in the proof of Lemma 6.11.8.)

Striving for a contradiction, suppose there is a function $f \in V(A^0, m)$ such that $\varphi(f) \notin \overline{W}(y, n)$. This means that $|\varphi(f)(y)| > 1/n$. By continuity of $\varphi$ there are $p \geq 1$ and a finite set $B \subseteq X$ such that for every $g \in C_p(X)$ with $f - g \in V(B, p)$ we have $|\varphi(g)(y)| > 1/n$. We may assume that $A^0 \subseteq B$. Since $f$ is continuous there is an open set $U \subseteq X$ containing $A^0$ with

$$f[U] \subseteq (-1/m, 1/m).$$

Since every point in $X$ has a neighborhood meeting finitely many $A_i^1$'s only, there is a neighborhood $V$ of the finite set $B$ and an integer $i_0$ such that

$$\overline{V} \cap A_i^1 = \emptyset$$

for every $i > i_0$. Put $U' = U \cap V$. If (A) holds then $A_i^0 = \emptyset$ for every $i$, and so $A_i^0 \subseteq U'$ for every $i$. If (B) holds, $A_i^0 \to A^0$ which implies that for all but finitely many $i$, $A_i^0 \subseteq U'$. So there is an integer $i_1 > i_0$ such that

$$A_i^1 \cap (B \cup \overline{U'}) = \emptyset$$

and

$$A_i^0 \subseteq U'$$

for every $i > i_1$.

Since $B \cup \overline{U'}$ and $\bigcup_{i > i_1} A_i^1$ are disjoint closed sets in $X$, there is by the Tietze-Urysohn Extension Theorem a function $g \in C_p(X)$ such that $g(x) = f(x)$ for all $x \in B \cup \overline{U'}$ and $g \restriction \bigcup_{i > i_1} A_i^1 \equiv 0$. So $g \restriction A_i^1 \equiv 0$ for $i > i_1$ so that $g \in V(A_i, m)$. But then $|\varphi(g)(y_i)| \leq 1/n$ for $i > i_1$ while on the other hand $f - g \in V(B, p)$ and so $|\varphi(g)(y)| > 1/n$ by the above. This contradicts the continuity of $\varphi(g)$. □

For natural numbers $m, n \geq 1$ and $k \geq 2$ we define

$$E(k, m, n) = C(k, m, n) \setminus C(k - 1, m, n).$$

So $E(k, m, n)$ consists of all points $y \in C(k, m, n)$ such that every element of $\mathcal{C}(y, m, n)$ has size at least $k$. We also put

$$E(1, m, n) = C(1, m, n)$$

for all $m, n \geq 1$. From Lemma 6.11.7 it follows that each $E(k, m, n)$ is a $G_\delta$-subset of $Y$. In addition, by Lemma 6.11.6,

$$Y = \bigcup_{k=1}^{\infty} \bigcup_{m=1}^{\infty} E(k, m, n)$$

for every $n \geq 1$.

For $y \in E(k, m, n)$ we put

$$\mathcal{E}(y, m, n) = \mathcal{C}(y, m, n) \cap [X]^k.$$

**Lemma 6.11.8.** *Let $y_i, y \in E(k, m, n)$ for $i \in \mathbb{N}$ and certain $k, m, n \geq 1$. If $y_i \to y$ and $A_i \in \mathcal{E}(y, m, n)$ for every $i$ then $(A_i)_i$ contains a subsequence which converges to an element of $\mathcal{E}(y, m, n)$.*

**Proof.** By the proof of Lemma 6.11.7 we may assume without loss of generality that there are nonempty finite $A^0 \subseteq X$ and $A_i^0 \subseteq A_i$ such that $A_i^0 \to A^0$ in the Vietoris topology, such that $A^0$ witnesses the fact that $y \in C(k,m,n)$. Since $y \in E(k,m,n)$ this means that $|A^0| \geq k$ and so for all but finitely many $i$ we must have $|A_i^0| \geq k$ by the definition of the Vietoris topology. But then for those infinitely many $i$ we have $|A_i^0| = k$ since $|A_i| = k$ for every $i$. This implies that $|A^0| \leq k$. We obtain $|A^0| = k$ and so $A^0 \in \mathcal{E}(y,m,n)$. □

**Lemma 6.11.9.** *If $k,m,n \geq 1$ and $y \in E(k,m,n)$ then $\mathcal{E}(y,m,n)$ is finite.*

**Proof.** Striving for a contradiction, assume that $\mathcal{E}(y,m,n)$ is infinite.

**Claim 1.** There is a set $A \subseteq X$ with $|A| < k$ and a sequence $(A_i)_i$ of sets from $\mathcal{E}(y,m,n)$ such that $A_i \cap A_j = A$ for all distinct $i,j$.

*Proof.* This follows from Exercise A.1.4. ◇

We will prove that $A \in \mathcal{C}(y,m,n)$ which will contradict $y \in E(k,m,n)$. To this end, take an arbitrary $f \in V(A,m)$. Our task is to prove that
$$\varphi(f) \in \overline{W}(y,n).$$
Striving for a contradiction, assume otherwise, i.e., $|\varphi(f)(y)| > 1/n$. We can find a basic neighborhood $V(B,p)$ of $\underline{0}$, where $B \subseteq X$ is finite and $p \geq 1$, such that for all $g \in C_p(X)$ with $f - g \in V(B,p)$ we have $|\varphi(g)(y)| > 1/n$. Since the sets $A_i \setminus A$ are pairwise disjoint there exists an $i$ such that $(A_i \setminus A) \cap B = \emptyset$. Let $g \in C_p(X)$ be such that
$$g \restriction A \cup B = f \restriction A \cup B, \quad g \restriction A_i \setminus A \equiv 0.$$
Then $g \in V(A_i,m)$ so that $|\varphi(g)(y)| \leq 1/n$. This is a contradiction with the fact that $f - g \in V(B,p)$. □

If $y \in E(k,m,n)$ then $\alpha_{m,n}(y)$ denotes the union of the family $\mathcal{E}(y,m,n)$. So $\alpha_{m,n}(y)$ is a finite set by Lemma 6.11.9. Let $\varrho$ be an arbitrary admissible metric on $X$ and let $p \geq k$ and $q \geq 1$ be natural numbers. Define
$$G(k,m,n,p,q) = \{y \in E(k,m,n) : |\alpha_{m,n}(y)| = p \text{ and }$$
$$\varrho(x,x') \geq 1/q \text{ for all distinct } x,x' \in \alpha_{m,n}(y)\}.$$
Obviously,
$$E(k,m,n) = \bigcup_{p=k}^{\infty} \bigcup_{q=1}^{\infty} G(k,m,n,p,q).$$

**Lemma 6.11.10.** *For all natural numbers $k,m,n,p$ and $q$:*
1. *$G(k,m,n,p,q)$ is a $G_\delta$-subset of $Y$.*
2. *$\alpha_{m,n} \restriction G(k,m,n,p,q) \colon G(k,m,n,p,q) \to [X]^p$ is continuous.*

**Proof.** Fix $k, m, n \geq 1$. For $p \geq k$ and $q \geq 1$ let $K(p,q)$ be the set of all points $y \in E(k, m, n)$ such that $\alpha_{m,n}(y)$ has a subset $C$ of cardinality $p$ such that $\varrho(x, x') \geq 1/q$ for all distinct $x, x' \in C$.

**Claim 1.** $K(p,q)$ is closed in $E(k, m, n)$.

*Proof.* Indeed, let $(y_i)_i$ be a sequence of points in $K(p,q)$ converging to an element $y \in E(k, m, n)$. For every $i$ let
$$C_i = \{x_i^1, \ldots, x_i^p\} \subseteq \alpha_{m,n}(y_i)$$
witness the fact that $y_i \in K(p,q)$. Consider the sequence $(x_i^1)_i$. For every $i$ there is an element $A_i \in \mathcal{E}(y_i, m, n)$ containing $x_i^1$. By Lemma 6.11.8 we may assume without loss of generality that $A_i \to A$, where $A \in \mathcal{E}(y, m, n)$. By the definition of the Vietoris topology this implies that the sequence $(x_i^1)_i$ has a convergent subsequence (Lemma 1.11.2(3)), say converging to an element $x^1 \in \alpha_{n,m}(y)$. Let $E_1$ be the infinite set of indices that correspond to the members of that convergent subsequence. Consider the sequence $(x_i^2)_{i \in E_1}$. By the same reasoning, it contains a subsequence converging to an element $x^2 \in \alpha_{m,n}(y)$. Etc. Since $p$ is finite, we can therefore assume without loss of generality that
$$x_i^j \to x^j \in \alpha_{m,n}(y)$$
for every $j \leq p$. Put $C = \{x^1, \ldots, x^p\}$. Then $|C| \leq p$ by construction. But the elements of every $C_i$ are at least $1/q$ apart. This shows that $C$ also has $p$ elements which are all at least $1/q$ apart, i.e., $C$ is a witness of the fact that $y \in K(p,q)$. ◇

Since from Lemma 6.11.7 it follows that each $E(k, m, n)$ is a $G_\delta$-subset of $Y$, (1) follows from Claim 1 and the equality
$$G(k, m, n, p, q) = K(p, q) \setminus \bigcup_{r=1}^{\infty} K(p+1, r).$$

For (2), in the above argument, if we take $y_i$ and $y$ from $G(k, m, n, p, q)$ then
$$C_i = \alpha_{m,n}(y_i), \quad C = \alpha_{m,n}(y).$$
So by Lemma 6.11.8 we conclude that $\alpha_{m,n}(y_i) \to \alpha_{m,n}(y)$. □

Consider the inverse $\varphi^{-1}: C_p(Y) \to C_p(X)$. We define for $\varphi^{-1}$ the sets
$$D(k, n, m), \quad F(k, m, n), \quad H(k, m, n, p, q)$$
and the functions $\beta_{n,m}$ as the sets
$$C(k, n, m), \quad E(k, m, n), \quad G(k, m, n, p, q)$$
and the functions $\alpha_{n,m}$ were defined for $\varphi$. In fact, only the value $n = 1$ will be important in the remaining part of this section.

## 6.11. NONEXISTENCE OF HOMEOMORPHISMS

**Lemma 6.11.11.** Let $y \in E(k,m,1)$ for certain $k,m \geq 1$. For $x \in \alpha_{m,1}(y)$ let $k_x, \ell_x \in \mathbb{N}$ be such that $x \in F(k_x, \ell_x, m+1)$. There exists an $x \in \alpha_{m,1}(y)$ such that $y \in \beta_{\ell_x, m+1}(x)$.

**Proof.** Suppose that
$$y \notin \bigcup\{\beta_{\ell_x, m+1}(x) : x \in \alpha_{m,1}(y)\}.$$
By the definition of $\alpha_{m,1}$ there is a subset $A \subseteq \alpha_{m,1}(y)$ such that
$$\varphi[V(A,m)] \subseteq \overline{W}(y,1).$$
Similarly, for every $x \in A$ we can pick a set $B_x \subseteq \beta_{\ell_x, m+1}(x)$ for which
$$\varphi^{-1}[W(B_x, \ell_x)] \subseteq \overline{V}(x, m+1).$$
By our assumption, $y \notin \bigcup\{B_x : x \in A\}$. We can therefore find $g \in C_p(X)$ such that $g(y) = 2$ and $g \restriction \bigcup\{B_x : x \in A\} \equiv 0$. For every $x \in A$ we have $g \in W(B_x, \ell_x)$, hence
$$|\varphi^{-1}(g)(x)| \leq 1/m+1 < 1/m.$$
This means that $\varphi^{-1}(g) \in V(A,m)$ and consequently
$$g = \varphi(\varphi^{-1}(g)) \in \overline{W}(y,1),$$
which is a contradiction. $\square$

We are now finally in the position to present the proof of the main result in this section.

**Proof of Theorem 6.11.3.** Since $Y$ is the union of the countably many $G_\delta$-sets $G(k,m,1,p,q)$ it suffices to prove that each of them itself is a countable union of $G_\delta$-sets homeomorphic to $G_\delta$-subsets of $X$. Fix $k,m,p$ and $q$. By Lemmas 6.11.10 and 6.11.1 it follows that
$$G(k,m,1,p,q) = \bigcup_{r=1}^{\infty} G_r,$$
where for each $r \in \mathbb{N}$, $G_r$ is open in $G(k,m,1,p,q)$ (hence is a $G_\delta$-subset of $Y$) while moreover there exist continuous functions $f_i \colon G_r \to X$ for $i = 1, \ldots, p$ such that for every $y \in G_r$,
$$\alpha_{m,1}(y) = \{f_1(y), \ldots, f_p(y)\}.$$
Fix $r \in \mathbb{N}$. Consider the family of $G_\delta$-sets
$$\{H(k', m', m+1, p', q') : k', m', q' \geq 1, p' \geq k'\}$$
which covers $X$, and the corresponding continuous functions $\beta_{m', m+1}$. By an application of Lemma 6.11.1 for these families of sets and functions, we can represent $X$ as a countable union of $G_\delta$-sets $H_s$, $s \in \mathbb{N}$, such that for every $s$ there are continuous functions $g_j^s \colon H_s \to Y$ for $j = 1, \ldots, p_s'$ having the following property: for every $y \in G_r$ and $s_1, \ldots, s_p \in \omega$ with $f_i(y) \in H_{s_i}$,

there exist $i \leq p$ and $j \leq p'_{s_i}$ such that $g_j^{s_i}(f_i(y)) = y$. For this last property use Lemma 6.11.11. The set $G_r$ is the union of the $G_\delta$-sets

$$G_{\bar{s}} = \{y \in G_r : f_i(y) \in H_{s_i}, i = 1, \ldots, p\},$$

where $\bar{s} = (s_1, \ldots, s_p) \in \mathbb{N}^p$. For a fixed $\bar{s}$ we have

$$G_{\bar{s}} = \bigcup_{i=1}^{p} \bigcup_{j=1}^{p'_{s_i}} \{y \in G_{\bar{s}} : g_j^{s_i}(f_i(y)) = y\}.$$

These sets are closed in $G_{\bar{s}}$ and by Lemma 6.11.2 are homeomorphic to $G_\delta$-subsets of $X$. □

**Remark 6.11.12.** The converse of Theorem 6.11.3 does not hold. To see this, let $X$ and $Y$ be countable spaces. Then $X$ is the union of the countably many $G_\delta$-sets $\{\{x\} : x \in X\}$. Similarly, $Y$ is the union of the countably many $G_\delta$-sets $\{\{y\} : y \in Y\}$. Since points are topologically unique, $X$ and $Y$ satisfy the conclusion of Theorem 6.11.3. But it is not so that $C_p(X)$ and $C_p(Y)$ are necessarily homeomorphic. Simply observe that if $X = \mathbb{N}$ then $C_p(X)$ is topologically complete and if $Y = \mathbb{Q}$ then $C_p(Y)$ is not a Baire space.

## 6.12. Topological equivalence of certain function spaces

In this section we prove that if $X$ is nondiscrete and countable and $C_p(X)$ is an $F_{\sigma\delta}$-subset of $\mathbb{R}^X$ then $C_p(X)$ is homeomorphic to $B(Q)^\infty$. This implies among other things that all function spaces of countable nondiscrete metrizable spaces (e.g., $\mathbb{Q}$, $\omega+1$, etc.) are homeomorphic (Theorem 6.3.10). This interesting result is due to DOBROWOLSKI, MARCISZEWSKI and MOGILSKI [123] who proved it via the BESTVINA-MOGILSKI [53] technique. Our method of proof is via the absorber method developed in Chapter 5. This yields a slightly stronger result which is due independently to BAARS, GLADDINES and VAN MILL [32] and DIJKSTRA and MOGILSKI [118].

We have to make several detours in order to arrive at the desired conclusion.

**Topological characterization of meager filters.** We characterized meager filters in purely set theoretic terms in §6.5. We will now do the same in purely topological terms.

**Theorem 6.12.1.** *Let $\mathcal{F}$ be a free filter on $\mathbb{N}$. Then the following conditions are equivalent:*

(1) $\mathcal{F} \in \mathbf{C}(\mathcal{P}(\mathbb{N}))$,
(2) $\mathcal{F}$ *is a meager subset of* $\mathcal{P}(\mathbb{N})$.

**Proof.** It suffices to prove the implication (1) $\Rightarrow$ (2). Let $\mathcal{U}, \mathcal{X}, \mathcal{Y}$ be subsets of $\mathcal{P}(\mathbb{N})$ such that $\mathcal{U}$ is open, $\mathcal{X}$ and $\mathcal{Y}$ are meager, and $\mathcal{F} = (\mathcal{U} \setminus \mathcal{X}) \cup \mathcal{Y}$. Striving for a contradiction, assume that $\mathcal{F}$ is not a meager subset of $\mathcal{P}(\mathbb{N})$. Then $\mathcal{U} \neq \emptyset$, and hence there exist an element $B \in \mathcal{U}$ and an element $\ell \in \mathbb{N}$ such that $V(B, S) \subseteq \mathcal{U}$, where $S = \{1, 2, \ldots, \ell\}$. Write $\mathbb{N}_0 = \mathbb{N} \setminus S$ and let $\mathcal{F}_0 = \mathcal{F} \restriction \mathbb{N}_0$. Observe that the function $\pi \colon \mathcal{P}(\mathbb{N}) \to \mathcal{P}(\mathbb{N}_0)$ defined by

$$A \mapsto A \cap \mathbb{N}_0$$

corresponds to the projection from $\{0, 1\}^{\mathbb{N}}$ onto $\{0, 1\}^{\mathbb{N}_0}$. Hence it is an open continuous surjection (Exercise A.5.5(3)). In addition, every fiber of $\pi$ has evidently cardinality $2^\ell$. We claim that

$$\pi[V(B, S)] = \mathcal{P}(\mathbb{N}_0).$$

Indeed, if $A \subseteq \mathbb{N}_0$ is arbitrary then $A \cup (B \cap S) \in V(B, S)$ since $A \cap S = \emptyset$ and $\pi(A \cup (B \cap S)) = A$. Since $\mathcal{F}$ is a free filter, so is $\mathcal{F}_0$ and hence it is dense in $\mathcal{P}(\mathbb{N}_0)$. Moreover, since $\mathcal{F}$ contains a dense $G_\delta$-subset of $V(B, S)$ and $\pi[V(B, S)] = \mathcal{P}(\mathbb{N}_0)$ and $\pi$ is a continuous open surjection with finite fibers, $\mathcal{F}_0$ contains a dense $G_\delta$-subset of $\mathcal{P}(\mathbb{N}_0)$ by Exercise A.1.17. (This can also be proved directly.) Let $\xi \colon \mathcal{P}(\mathbb{N}_0) \to \mathcal{P}(\mathbb{N}_0)$ be the homeomorphism assigning to each $B \in \mathcal{P}(\mathbb{N}_0)$ the set $\mathbb{N}_0 \setminus B$. Since $\mathcal{F}_0$ is a filter,

$$\xi[F_0] \cap F_0 = \emptyset.$$

Consequently, the compact space $\mathcal{P}(\mathbb{N}_0)$ contains two disjoint dense $G_\delta$-subsets which by Exercise A.6.10 is impossible. $\square$

**Sequence spaces from filters.** Let $\mathcal{F}$ be a free filter on a set $E \subseteq \mathbb{N}$. Define

$$c_\mathcal{F} = \{x \in \mathbb{R}^E : (\forall \varepsilon > 0)(\{i \in E : |x_i| < \varepsilon\} \in \mathcal{F}\}.$$

Observe that $c_\mathcal{F}$ is a dense linear subspace of $\mathbb{R}^E$ (use that $\mathcal{F}$ is free).

It is clear that if $\mathcal{F}$ is the so-called *Fréchet filter*

$$\mathfrak{F} = \{A \subseteq \mathbb{N} : \mathbb{N} \setminus A \text{ is finite}\}$$

then $c_\mathcal{F}$ is $c_0$.

Finally, observe that if $\mathcal{F}$ is a free filter on $\mathbb{N}$ then $c_\mathcal{F}$ is a closed linear subspace of the space $C_p(\mathbb{N}_\mathcal{F})$ considered in §6.5. Indeed, $c_\mathcal{F}$ is canonically homeomorphic to the subspace

$$\{f \in C_p(\mathbb{N}_\mathcal{F}) : f(\infty) = 0\}$$

of $C_p(\mathbb{N}_\mathcal{F})$. This implies by Corollary 6.6.7 that

$$C_p(\mathbb{N}_\mathcal{F}) \sim c_\mathcal{F} \times \mathbb{R}.$$

**Lemma 6.12.2.** *Let $\mathcal{F}$ be a free filter on $\mathbb{N}$. If $c_\mathcal{F}$ is analytic then so is $\mathcal{F}$ (topologized as subspace of $\mathcal{P}(\mathbb{N})$).*

**Proof.** Define $\xi\colon \mathcal{F} \to c_\mathcal{F}$ as follows:

$$\xi(F)_n = \begin{cases} 0 & (n \in F), \\ 1 & (n \notin F). \end{cases}$$

It is easy to see that $\xi$ is an imbedding of $\mathcal{F}$ into $c_\mathcal{F}$. To prove that it is a closed imbedding, simply observe that $\xi[\mathcal{F}] = c_\mathcal{F} \cap \{0,1\}^\mathbb{N}$. So $\mathcal{F}$ is homeomorphic to a closed subspace of an analytic space, and hence is analytic itself (Lemma A.13.11). □

**Corollary 6.12.3.** *Let $\mathcal{F}$ be a free filter on $\mathbb{N}$. If $c_\mathcal{F}$ is analytic then $\mathcal{F}$ is a meager subset of $\mathcal{P}(\mathbb{N})$.*

**Proof.** If $c_\mathcal{F}$ is analytic then so is $\mathcal{F}$ by Lemma 6.12.2. This implies by Theorem A.13.13 that $\mathcal{F} \in \mathbf{C}\big(\mathcal{P}(\mathbb{N})\big)$. So $\mathcal{F}$ is meager by Theorem 6.12.1. □

Note that for any free filter $\mathcal{F}$ we have that $c_0 \subseteq c_\mathcal{F}$.

We now aim at proving 'universality' properties for spaces of the form $c_\mathcal{F}$. The following simple corollary to Lemma 6.5.2 will be useful.

**Corollary 6.12.4.** *Let $\mathcal{F}$ be a closed and nowhere dense subset of $\mathcal{P}(\mathbb{N})$. Then for every $n \in \mathbb{N}$, there is a set $X \subseteq \{n+1, n+2, \dots\}$ such that the collection*

$$\{E \cup X : E \subseteq \{1, 2, \dots, n\}\}$$

*misses $\mathcal{F}$.*

**Proof.** Let $A = \{1, 2 \dots, n\}$. By Lemma 6.5.2 there are a finite set $C \subseteq \mathbb{N}\setminus A$ and a subset $D \subseteq C$ such that for every $B \subseteq \mathbb{N}$ we have

(∗) $\qquad\qquad V(B, A) \cap V(D, C) \cap \mathcal{F} = \emptyset.$

Let $X = D$. We claim that $X$ is as required. To this end, let $E \subseteq \{1, 2, \dots, n\}$ be arbitrary, and put $Y = E \cup X$. Since $Y \cap C = B$ we have

$$Y \in V(Y, A) \cap V(B, C).$$

As a consequence, $Y \notin \mathcal{F}$ by (∗). □

As in §5.5 we let $\hat{\mathbb{R}}$ be the two-point compactification $[-\infty, \infty]$ of $\mathbb{R}$. We represent $Q$ by $\hat{\mathbb{R}}^\mathbb{N}$. We use the following arithmetic on $Q$:

$$1/0 = \infty, \quad \pm\infty + a = \pm\infty$$

($a$ finite), and $\infty - \infty = 0$, etc.

The following result is needed in several forthcoming results.

**Proposition 6.12.5.** *Let $\mathcal{F}$ be a free filter on $\mathbb{N}$ which is a meager subset of $\mathcal{P}(\mathbb{N})$. Then $c_\mathcal{F}$ is contained in a $\sigma Z$-set of $Q$.*

**Remark 6.12.6.** We already know from Theorem 6.5.6 and the fact that
$$C_p(\mathbb{N}_{\mathcal{F}}) \sim c_{\mathcal{F}} \times \mathbb{R}$$
that $c_{\mathcal{F}}$ is not a Baire space. For if it would be a Baire space then $C_p(\mathbb{N}_{\mathcal{F}})$ would be a Baire space as well by Theorem A.6.10. So $c_{\mathcal{F}}$ is a meager subset of $Q$ (Exercise 1.6.6) and hence is meager in itself. That it is contained in a $\sigma Z$-set of $Q$ therefore comes as no surprise.

One could guess from these observations that no extra work is required since it is tempting to think that a linear space which is meager in itself is in fact a countable union of $Z$-subsets. If this would be true then some proofs in this section would have been simpler. However, it is *not* true, as was shown by BANAKH [39]. His example is the linear span in $\ell^2$ of Erdős' space (Example 1.5.18). It is even absolutely Borel. See also BANAKH, RADUL and ZARICHNYI [41, Theorem 5.5.19] for details.

**Proof.** Put $\hat{T} = \hat{\mathbb{R}} \setminus ((-3,-1) \cup (1,3))$. Define $g \colon \hat{T} \to \hat{\mathbb{R}}$ by
$$g(x) = \begin{cases} x & (x \in \mathbb{J}), \\ x - 2 & (x \in [3, \infty]), \\ x + 2 & (x \in [-\infty, -3]). \end{cases}$$

Let $g^\infty \colon \hat{T}^\infty \to \hat{\mathbb{R}}^\infty$ be defined by $g^\infty(x)_i = g(x_i)$ for $i \in \mathbb{N}$. Then $g^\infty$ is obviously continuous.

There is for every $n$ a nowhere dense closed subset $\mathcal{A}_n$ of $\mathcal{P}(\mathbb{N})$ such that $\mathcal{F} \subseteq \bigcup_{n=1}^\infty \mathcal{A}_n$. Put
$$B_n = \{x \in \hat{T}^\infty : \{i \in \mathbb{N} : |x_i| < 2\} \in \mathcal{A}_n\}$$
for $n \in \mathbb{N}$.

**Claim 1.** $B_n$ is closed in $\hat{T}^\infty$.

*Proof.* Let $x \in \hat{T}^\infty \setminus B_n$. Then
$$K = \{i \in \mathbb{N} : |x_i| < 2\} \notin \mathcal{A}_n$$
and so there is an element $m \in \mathbb{N}$ such that if $S = \{1, 2, \ldots, m\}$ then
$$V(K, S) \cap \mathcal{A}_n = \emptyset.$$
For $i \leq m$ such that $|x_i| \leq 1$ let $V_i = \mathbb{J}$. In addition, for $i \leq m$ with $|x_i| \geq 3$ let $V_i = [3, \infty]$ if $x_i > 0$ and $V_i = [-\infty, -3]$ if $x_i < 0$. Then
$$V = \prod_{i=1}^m V_i \times \hat{T}^{\mathbb{N}\setminus S}.$$
is an open neighborhood of $x$ in $\hat{T}^\mathbb{N}$ which misses $B_n$. For if $y \in V$ and
$$L = \{i \in \mathbb{N} : |y_i| < 2\}$$

then $L \cap S = K \cap S$. This implies that $L \in V(K, S)$ and hence $L \notin \mathcal{A}_n$. ◇

**Claim 2.** $g^\infty[B_n] \in \mathcal{Z}(Q)$.

*Proof.* For each $m$ put Put $S_m = \{1, 2, \ldots, m\}$. By Corollary 6.12.4 there exists $A_m \subseteq \{m+1, m+2, \ldots\}$ such that the collection
$$\{T \cup A_m : T \subseteq S_m\}$$
misses $\mathcal{A}_n$. Define $f_m \colon Q \to Q$ by
$$f_m(x)_j = \begin{cases} x_j & (j \leq m), \\ 0 & (j \in A_m), \\ 1081 & (j \notin A_m). \end{cases}$$

Then $f_m$ is clearly continuous and the sequence $(f_m)_m$ approximates the identity on $\hat{Q}$. We claim that $f_m[Q] \cap g^\infty[B_n] = \emptyset$ for every $m$, which is sufficient by Lemma 5.1.3(1). Striving for a contradiction, suppose that there are $y \in Q$ and $x \in B_n$ such that
$$f_m(y) = g^\infty(x).$$
Put
$$C = \{k \in \mathbb{N} : |x_k| < 2\}.$$
Then $C \in \mathcal{A}_n$ and so
$$C \notin \{T \cup A_m : T \subseteq S_m\}.$$
There consequently exists $j > m$ such that $j \in C \triangle A_m$. If $j \in C \setminus A_m$ then $f_m(y)_j = 1081$. In addition, $|x_j| < 2$ since $j \in C$ and so $|g^\infty(x)_j| \leq 1$. This is a contradiction. If $j \in A_m \setminus C$ then $f_m(y)_j = 0$. Moreover, since $j \notin C$ we have $|x_j| \geq 3$ and so $|g^\infty(x)_j| \geq 1$. This also leads to a contradiction. ◇

**Claim 3.** $c_\mathcal{F} \subseteq \bigcup_{n=1}^\infty g^\infty[B_n]$.

*Proof.* Pick an arbitrary $x \in c_\mathcal{F} \subseteq \mathbb{R}^\infty$. Define the point $y \in \hat{T}^\mathbb{N}$ by
$$y_i = \begin{cases} x_i & (x_i \in \mathbb{J}), \\ x_i + 2 & (x_i \in (1, \infty)), \\ x_i - 2 & (x_i \in (-\infty, -1)). \end{cases}$$

Then $g^\infty(y) = x$. Since $x \in c_\mathcal{F}$,
$$K = \{i \in \mathbb{N} : |x_i| < 1\} \in \mathcal{F}.$$

Since $K$ is clearly equal to the set
$$\{i \in \mathbb{N} : |y_i| < 2\}$$
we conclude that $y \in \bigcup_{n=1}^\infty B_n$. ◇

This completes the proof of the lemma. □

**Lemma 6.12.7.** *Let $\mathcal{F}$ be a free filter on $\mathbb{N}$ such that $c_\mathcal{F}$ is analytic. Then $c_\mathcal{F}$ is both reflexively universal and $\mathcal{F}_{\sigma\delta}$-universal in $Q$.*

**Proof.** We first prove that $c_\mathcal{F}$ is reflexively universal. To this end, first observe that if $\mathcal{F} = \mathfrak{F}$ then $c_\mathcal{F} = c_0$ and so we are done by Theorem 5.5.15. We may therefore assume that $\mathcal{F} \neq \mathfrak{F}$, i.e., that there is an element $F \in \mathcal{F}$ such that $\mathbb{N} \setminus F$ is infinite. Let $\mathcal{G} = \mathcal{F} \restriction F$. Observe that $\mathcal{G} \subseteq \mathcal{F}$ and that $c_\mathcal{F}$ is canonically homeomorphic to $c_\mathcal{G} \times s$. We aim at applying Theorem 5.5.20. For that we need to show that there is a function $\psi \colon \hat{\mathbb{R}}^F \to \mathbb{R}^F$ such that
$$\psi^{-1}[c_\mathcal{G}] \cap \mathbb{R}^F = c_\mathcal{G}.$$
But this is easily done. If $\alpha \colon \hat{\mathbb{R}} \to [-1081, 1081]$ is a homeomorphism then the function $\psi \colon \hat{\mathbb{R}}^F \to [-1081, 1081]^F \subseteq \mathbb{R}^F$ defined by
$$\bigl(\psi(x)\bigr)_n = \alpha(x_n), \quad n \in F,$$
has the desired property.

We next prove that $c_\mathcal{F}$ is $\mathcal{F}_{\sigma\delta}$-universal. First observe that $\mathcal{F}$ is a meager subset of $\mathcal{P}(\mathbb{N})$ by Lemma 6.12.2 and Corollary 6.12.3. Find a partition
$$\{\mathbb{N}_i : i \in \mathbb{N}\}$$
of $\mathbb{N}$ such as in Corollary 6.5.4(3). As on Page 354, for every $n \in \mathbb{N}$, put
$$\Theta_n = \{x \in \mathbb{R}^{\mathbb{N}_i} : |x_i| \leq 2^{-n} \text{ for all but finitely many } i\}.$$
Let $\mathcal{F}_i = \mathcal{F} \restriction \mathbb{N}_i$ for every $i$. Then every $\mathcal{F}_i$ is meager. It is easily seen that
$$c_0 \subseteq c_\mathcal{F} \subseteq \prod_{i=1}^\infty c_{\mathcal{F}_i}.$$
For each $i \in \mathbb{N}$, let $Q_i = [-2^{-i+1}, 2^{-i+1}]^{\mathbb{N}_i}$. Then $\prod_{i=1}^\infty Q_i \subseteq Q$. As in the proof of Proposition 6.12.5 it follows that $c_{\mathcal{F}_i} \cap Q_i$ is contained in a $\sigma Z$-set $D_i$ of $Q_i$.

Let $A = \bigcap_{n=1}^\infty A_n$ where each $A_n$ is a $\sigma$-compact subset of $Q$ while moreover $A_{n+1} \subseteq A_n$ for every $n \in \mathbb{N}$. Define for every $i$,
$$C_i = \{x \in Q_i : (\forall j \in \mathbb{N}_i)(|x_j| \leq 2^{k-j} \text{ for some } k)\}.$$
As in the proof of Lemma 5.5.12, $C_i$ is an $\mathcal{F}_\sigma$-absorber in $Q_i$. By Exercise 5.5.4, there is an imbedding $f_i \colon Q \to Q_i$ such that $f_i^{-1}[C_i] = A_i$ and
$$f_i[Q \setminus A_i] \cap D_i = \emptyset.$$
Define $f = (f_i)_i \colon Q \to \prod_{i=1}^\infty Q_i \subseteq Q$. Then $f$ is an imbedding and we claim that $f^{-1}[c_\mathcal{F}] = A$. First, let $x \in A$ and $n \in \mathbb{N}$. If $i > n$, then $f_i(x) \in Q_i$, so all components of $f_i(x)$ are in $[-2^{-n}, 2^{-n}]$. If $i \leq n$, then since $x \in A_i$ we have $f_i(x) \in C_i$. In this case only finitely many components of $f_i(x)$ are outside the interval $[-2^{-n}, 2^{-n}]$. We conclude that only finitely many components of $f(x)$ are outside $[-2^{-n}, 2^{-n}]$. This implies that $f(x) \in \Theta_n$.

Since $n$ was arbitrary we get from Lemma 5.5.12 that $f(x) \in c_0 \subseteq c_{\mathcal{F}}$. Second, let $f(x) \in c_{\mathcal{F}}$ and fix $n \in \mathbb{N}$. Then $f_n(x) \in c_{\mathcal{F}_n} \cap Q_n \subseteq D_n$, hence $x \in A_n$. We conclude that $x \in A$. □

**Function spaces of low Borel complexity: the metrizable case.**
We will now make our first steps towards proving that all $F_{\sigma\delta}$-function spaces of countable spaces are homeomorphic.

Some of the methods used here were used before, for example in §6.6. Since we are dealing with $Q$ instead of $s$, things are slightly more complicated. For completeness sake we will include the routine verifications.

We define the continuous function $\Psi \colon \hat{\mathbb{R}} \to \hat{\mathbb{R}}^{\mathbb{N}}$ by

$$\Psi(r)(n) = \operatorname{sign}(r) \cdot \min\{|r|, n\}.$$

Note that $\Psi(r)(n)$ is finite for all $n \neq \infty$ and

(∗) $$r = \lim_{n \to \infty} \Psi(r)(n) = \Psi(r)(\infty).$$

Hence $\Psi[\mathbb{R}] \subseteq \mathbb{R}^{\hat{\mathbb{N}}}$ and $\Psi[\mathbb{R}] \subseteq C_p(\mathbb{N}_{\mathcal{F}})$.

For $f \in Q = \hat{\mathbb{R}}^{\mathbb{N}}$ let $\hat{f}$ be the extension of $f$ over $\hat{\mathbb{N}}$ that assigns 0 to $\infty$. Define a function $\varphi \colon \hat{\mathbb{R}}^{\mathbb{N}} \times \hat{\mathbb{R}} \to \hat{\mathbb{R}}^{\hat{\mathbb{N}}}$ by

$$\varphi(f, r) = \hat{f} + \Psi(r).$$

**Lemma 6.12.8.** $\varphi$ *is a homeomorphism and* $\varphi[c_{\mathcal{F}} \times \mathbb{R}] = C_p(\mathbb{N}_{\mathcal{F}})$.

**Proof.** It is clear that $\varphi$ is well-defined. To prove that it is surjective, take an arbitrary $g \in \hat{\mathbb{R}}^{\hat{\mathbb{N}}}$. Put $r = g(\infty)$ and $f = g - \Psi(r)$. Observe that by (∗),

$$f(\infty) = g(\infty) - \Psi(g(\infty))(\infty) = 0.$$

This means that if $h = f \restriction \mathbb{N}$ then $\hat{h} = f$ and so

$$\varphi(h, r) = f,$$

as required.

It is clear that $\varphi$ is continuous. The above considerations easily allow to display the inverse of $\varphi$, as follows:

(∗∗) $$\varphi^{-1}(h) = \left(\left[h - \Psi(h(\infty))\right] \restriction \mathbb{N}, h(\infty)\right).$$

This means that $\varphi$ is a homeomorphism.

Take arbitrary $x \in c_{\mathcal{F}}$ and $r \in \mathbb{R}$. Then $\varphi(x, r) = \hat{x} + \Psi(r)$. By the above, both $\hat{x}$ and $\Psi(r)$ are continuous functions on $\mathbb{N}_{\mathcal{F}}$. So the same is true for their sum.

Conversely, let $f \in C_p(\mathbb{N}_{\mathcal{F}})$ and let

$$r = f(\infty).$$

Then $r \in \mathbb{R}$ and by (**),
$$\varphi^{-1}(f) = ([f - \Psi(r)] \restriction \mathbb{N}, r).$$
Since
$$f - \Psi(r)(\infty) = 0,$$
we have by (*) that $f - \Psi(r) \in c_{\mathcal{F}}$, which is as required. □

We will now investigate which $C_p(X)$ are contained in a $\sigma Z$-set of $Q$. Before we can say more about this, we need to derive a simple result first.

If $\mathcal{F}$ is closed under supersets then a subcollection $\mathcal{G}$ of $\mathcal{F}$ is said to be a *base* for $\mathcal{F}$ provided that for every $F \in \mathcal{F}$ there exists $G \in \mathcal{G}$ such that $G \subseteq F$.

**Lemma 6.12.9.** *Let $\mathcal{F}$ be a free filter on $\mathbb{N}$ with base $\mathcal{B}$. If $\mathcal{B}$ is an analytic subset of $\mathcal{P}(\mathbb{N})$ then so is $\mathcal{F}$.*

**Proof.** Define a function $\varphi \colon \mathcal{P}(\mathbb{N}) \times \mathcal{P}(\mathbb{N}) \to \mathcal{P}(\mathbb{N})$ by
$$\varphi(A, B) = A \cup B.$$
We first claim that $\varphi$ is continuous. Assume that
$$(A, B) \in \varphi^{-1}[V(E, F)]$$
for certain $E, F \in \mathcal{P}(\mathbb{N})$ such that $F$ is finite. Put
$$\mathcal{K} = V(A, F), \quad \mathcal{L} = V(B, F),$$
respectively. We claim that
$$\mathcal{K} \times \mathcal{L} \subseteq \varphi^{-1}[V(E, F)],$$
which is obviously as required. To this end, pick arbitrary $A' \in \mathcal{K}$ and $B' \in \mathcal{L}$. Then
$$A' \cap F = A \cap F, \qquad B' \cap F = B \cap F$$
and so
$$(A' \cup B') \cap F = (A \cup B) \cap F = E \cap F.$$
This implies that $\varphi\big((A', B')\big) \in V(E, F)$, as claimed.

It is clear that $\mathcal{B} \times \mathcal{P}(\mathbb{N})$ is analytic, being the product of two analytic spaces. Since
$$\varphi[\mathcal{B} \times \mathcal{P}(\mathbb{N})] = \mathcal{F}$$
the continuity of $\varphi$ shows that $\mathcal{F}$ is analytic as well. □

This yields the following interesting result.

**Proposition 6.12.10.** *Let $X$ be a countable nondiscrete space such that its function space $C_p(X)$ is analytic. Then $C_p(X)$ is contained in a $\sigma Z$-set of $Q$.*

**Proof.** Let $x \in X$ be a nonisolated point of $X$. Identify $X \setminus \{x\}$ with $\mathbb{N}$ and $x$ with $\infty$; put
$$\mathcal{F} = \{U \subseteq \mathbb{N} : U \cup \{x\} \text{ is a neighborhood of } x\}.$$
Then $\mathcal{F}$ is a free filter on $\mathbb{N}$ and $C_p(X) \subseteq C_p(\mathbb{N}_{\mathcal{F}})$.

Let
$$\mathcal{B} = \{B \subseteq \mathbb{N} : B \cup \{x\} \text{ is a } \textit{clopen} \text{ neighborhood of } x\}.$$
Clearly, $\mathcal{B}$ is a base for $\mathcal{F}$ since $X$ is zero-dimensional being countable (cf. Exercise 1.5.1). The subspace $\mathcal{B}$ of $\mathcal{P}(\mathbb{N})$ and the subspace
$$\mathcal{E} = \{f \in C_p(X) : (f[X] \subseteq \{0,1\}) \wedge (f(x) = 1)\}$$
of $C_p(X)$ are canonically homeomorphic (identify each $B \in \mathcal{B}$ with its characteristic function $\chi_B$), and moreover, $\mathcal{E}$ is closed in $C_p(X)$ (cf. the proof of Lemma 6.12.2). So $\mathcal{B}$ is analytic since $\mathcal{E}$ is. But then $\mathcal{F}$ is analytic as well by Lemma 6.12.9. By Proposition 6.12.5, $c_{\mathcal{F}}$ is contained in a $\sigma Z$-set of $Q$. So by Lemma 6.12.8 the same is true for $C_p(\mathbb{N}_{\mathcal{F}})$ and hence in turn for its subspace $C_p(X)$ (Lemma 5.1.2). □

**Lemma 6.12.11.** *Let $X$ be a countable space. If $X$ contains a clopen subset $Y$ containing exactly one accumulation point and $C_p(X)$ is an $\mathcal{F}_{\sigma\delta}$-subset of $\hat{\mathbb{R}}^X$ then $C_p(X)$ is an $\mathcal{F}_{\sigma\delta}$-absorber in $\hat{\mathbb{R}}^X$.*

**Proof.** By Proposition 6.12.10, $C_p(X)$ is contained in a $\sigma Z$-set of $Q$. There is a free filter $\mathcal{F}$ on $\mathbb{N}$ such that $\mathbb{N}_{\mathcal{F}}$ is homeomorphic to $Y$. It easily follows that $c_{\mathcal{F}}$ is an $\mathcal{F}_{\sigma\delta}$-subset of $Q$, hence by Lemmas 6.12.7 and 5.5.1, $c_{\mathcal{F}}$ is strongly $\mathcal{F}_{\sigma\delta}$-universal. Since there clearly is a deformation of $\hat{\mathbb{R}}$ through $\mathbb{R}$, by Exercise 5.5.2 and Lemma 6.12.8 we obtain that $C_p(\mathbb{N}_{\mathcal{F}})$ is strongly $\mathcal{F}_{\sigma\delta}$-universal. Since $Y$ is clopen in $X$, $C_p(X)$ is canonically homeomorphic to the product
$$C_p(Y) \times C_p(X \setminus Y)$$
(Theorem 6.2.8). Since $C_p(X \setminus Y)$ is a dense linear subspace of $\mathbb{R}^{X \setminus Y}$, there is a deformation of $\hat{\mathbb{R}}^{X \setminus Y}$ through $C_p(X \setminus Y)$ by Corollaries 4.2.15 and 4.1.8. By an application of Exercise 5.5.2 we consequently obtain that $C_p(X)$ is strongly $\mathcal{F}_{\sigma\delta}$-universal in $Q$. □

**Proposition 6.12.12.** *Let $X$ be a countable nondiscrete metrizable space. Then $C_p(X)$ is an $\mathcal{F}_{\sigma\delta}$-absorber in $\hat{\mathbb{R}}^X$.*

**Proof.** By Theorem 6.3.10 it follows that $C_p(X)$ is an $\mathcal{F}_{\sigma\delta}$-subset of $\hat{\mathbb{R}}^X$. By Corollary A.13.12 and Proposition 6.12.10, $C_p(X)$ is contained in a $\sigma Z$-set of $\hat{\mathbb{R}}^X$ (there is a simple direct proof of this fact, but Proposition 6.12.10 is needed in full generality later so we might as well use it here). It remains to show that $C_p(X)$ is strongly $\mathcal{F}_{\sigma\delta}$-universal.

## 6.12. TOPOLOGICAL EQUIVALENCE OF CERTAIN FUNCTION SPACES

Consider the space $\mathbb{N}_{\mathfrak{F}}$. Topologically this is a simple convergent sequence with its limit and hence it is a countable nondiscrete metrizable space. So $C_p(\mathbb{N}_{\mathfrak{F}})$ is an $\mathcal{F}_{\sigma\delta}$-absorber by Lemma 6.12.11.

We now reduce the problem for $C_p(X)$ to $C_p(\mathbb{N}_{\mathfrak{F}})$, as follows. Let $\varrho$ be an admissible metric on $X$ and let $A$ be a nontrivial convergent sequence in $X$. Observe that such a sequence exists since $X$ is nondiscrete. By the above it follows that $C_p(A)$ is strongly $\mathcal{F}_{\sigma\delta}$-universal in $\hat{\mathbb{R}}^A$.

**Claim 1.** There is a retraction $r \colon X \to A$.

*Proof.* Let $a$ be the unique nonisolated point of $A$. Since $X$ is metrizable (hence first countable) and zero-dimensional (cf. Exercise 1.5.1) there is a decreasing clopen neighborhood base $(U_n)_n$ of $a$ in $X$ such that $U_n \setminus U_{n+1}$ contains exactly one point from $A \setminus \{a\}$ for every $n$, say $a(n)$. Now define the retraction $r \colon X \to A$ by the following formula:
$$r(x) = \begin{cases} a(n) & (x \in U_n \setminus U_{n+1}), \\ a & (x = a). \end{cases}$$
An easy check shows that $r$ is as required. ◇

The formula
$$\Psi(g)(x) = \operatorname{sign}\bigl(g(r(x))\bigr) \min\left\{|g(r(x))|, \frac{1}{\varrho(x, r(x))}\right\}$$
defines a continuous extender that extends every $g \in \hat{\mathbb{R}}^A$ to an element of $\hat{\mathbb{R}}^X$. The map $\Psi$ has the following properties: $\Psi(g) \restriction A = g$, $\Psi(g) \restriction X \setminus A$ has its values in $\mathbb{R}$ and $\psi[C_p(A)] \subseteq C_p(X)$. If $f \in \hat{\mathbb{R}}^{X \setminus A}$ then $\hat{f}$ is the extension of $f$ over $X$ with zeros. As in the proof of Lemma 6.12.8 it is easily seen that
$$\Phi(f, g) = \hat{f} + \Psi(g)$$
is a homeomorphism from $\hat{\mathbb{R}}^{X \setminus A} \times \hat{\mathbb{R}}^A$ onto $\hat{\mathbb{R}}^X$. Let
$$C_p(X, A) = \{f \restriction X \setminus A : f \in C_p(X) \text{ and } f \restriction A \equiv 0\}$$
and note that
$$C_p(X) = \Phi[C_p(X, A) \times C_p(A)].$$
Since $C_p(X, A)$ is a dense linear subspace of $\hat{\mathbb{R}}^{X \setminus A}$, there is as in the proof of Lemma 6.12.11 a deformation of $\hat{\mathbb{R}}^{X \setminus A}$ through $C_p(X, A)$. So Exercise 5.5.2 implies that $C_p(X)$ is strongly $\mathcal{F}_{\sigma\delta}$-universal in $\hat{\mathbb{R}}^X$. ⊓

**Function spaces of low Borel complexity: the general case.** We now come to the main result in this section.

**Proposition 6.12.13.** *Let $X$ be a countable nondiscrete space. Then $X$ has a clopen subspace $Y$ with exactly one accumulation point or there is a partition of $X$ into infinitely many pairwise disjoint nondiscrete clopen sets.*

**Proof.** Suppose that $X$ does not have a clopen subspace with exactly one accumulation point. Enumerate $X$ as $\{x_n : n \in \mathbb{N}\}$. Put $X_0 = \emptyset$. Assume that we constructed pairwise disjoint nondiscrete clopen sets $X_1, \ldots, X_n$ of $X$ such that $Y_n = X \setminus \bigcup_{i=0}^{n} X_i$ is also nondiscrete and

$$\{x_1, \ldots, x_n\} \subseteq \bigcup_{i=0}^{n} X_i$$

if $n \geq 1$. By assumption, $Y_n$ contains at least two accumulation points. By the zero-dimensionality of $X$ (cf. Exercise 1.5.1), we can split $Y_n$ into two nondiscrete clopen sets, say $A$ and $B$. If

$$x_{n+1} \in A \cup \bigcup_{i=0}^{n} X_i$$

then put $X_{n+1} = A$. If not then $x_{n+1} \in B$ and we put $X_{n+1} = B$. □

**Lemma 6.12.14.** *Let $X$ be a countable nondiscrete space. Then $C_p(X)$ contains an $\mathcal{F}_{\sigma\delta}$-absorber for $\hat{\mathbb{R}}^X$.*

**Proof.** Since $X$ is zero-dimensional (cf. Exercise 1.5.1), for every pair of distinct points $x, y \in X$ there is a continuous function $f \colon X \to \{0,1\}$ such that $f(x) = 0$ and $f(y) = 1$. Picking such a function for each pair of distinct points of $X$ yields a continuous injective function $f \colon X \to \{0,1\}^\infty$. Let $Y$ be the image of $X$ under $f$; clearly, $Y$ is a countable metrizable space. In addition, $Y$ is not discrete since $f$ is injective. By Proposition 6.12.12 we have that $C_p(Y)$ is an $\mathcal{F}_{\sigma\delta}$-absorber in $\hat{\mathbb{R}}^Y$.

Let $\tau$ be the metrizable topology on $X$ that it inherits from $Y$. So a set $U$ in $X$ is $\tau$-open if and only if $f[U]$ is open in $Y$. The set $X$ endowed with the topology $\tau$ will be denoted by $X^\tau$. Then clearly $X^\tau \approx Y$ and hence $C_p(X^\tau)$ is an $\mathcal{F}_{\sigma\delta}$-absorber in $\hat{\mathbb{R}}^X$. Observe that $\tau$ is weaker than the original topology on $X$ since $f$ is continuous. This implies that every function $g \in C_p(X^\tau)$ is also continuous with respect to the original topology on $X$. From this it follows that $C_p(X^\tau) \subseteq C_p(X)$. Since $C_p(X^\tau)$ is an $\mathcal{F}_{\sigma\delta}$-absorber in $\hat{\mathbb{R}}^X$, we are done. □

We now come to the main result in this section.

**Theorem 6.12.15.** *Let $X$ be a countable nondiscrete space such that $C_p(X)$ is an $F_{\sigma\delta}$-subset of $Q$. Then $C_p(X)$ is an $\mathcal{F}_{\sigma\delta}$-absorber in $Q$.*

**Proof.** By Proposition 6.12.13 and Lemma 6.12.11 we may assume without loss of generality that $X$ admits a clopen partition

$$\{X_n : n \in \mathbb{N}\}$$

such that every $X_n$ is not discrete. By Lemma 6.12.14 there is for every $n$ an $\mathcal{F}_{\sigma\delta}$-absorber $B_n$ in $\hat{\mathbb{R}}^{X_n}$ such that $B_n \subseteq C_p(X_n)$. So by Exercise 5.5.3

we may pick an $\mathcal{F}_\sigma$-absorber $B'_n$ in $\hat{\mathbb{R}}^{X_n}$ such that $B'_n \subseteq B_n$. Finally, Proposition 6.12.10 gives us that $C_p(X_n)$ is contained in a $\sigma Z$-set $C_n \subseteq \hat{\mathbb{R}}^{X_n}$. We now have by Theorem 6.2.8 the following inclusions:

$$\prod_{n=1}^{\infty} B'_n \subseteq \prod_{n=1}^{\infty} C_p(X_n) \equiv C_p(X) \subseteq \prod_{n=1}^{\infty} C_n.$$

By Exercise 5.5.5 we consequently get that $C_p(X)$ is strongly $F_{\sigma\delta}$-universal, which finishes the proof. □

**Remark 6.12.16.** In view of Theorem 6.12.15 the question naturally arises whether for a countable space $X$ the Borel type of $C_p(X)$ determines the topological type of $C_p(X)$. But this is not true, as was shown by CAUTY [88]. As we saw in §6.3, nontrivial function spaces are at least of '$F_{\sigma\delta}$ Borel complexity'. Theorem 6.12.15 proves that at the first nontrivial stage all function spaces are homeomorphic. What happens at more complex situations is a mystery, although Cauty's result just quoted shows that a result as elegant as Theorem 6.12.15 is not possible.

## 6.13. Examples

In this section we collect several examples dealing with function spaces, some with and some without details. Several of the examples are already dealt with earlier in this chapter.

**E1.** *Non-homeomorphic $\ell$-equivalent spaces.*

See Example 6.2.4. This shows that $\ell$-equivalence is not the same as topological equivalence.

**E2.** *Countable compact (necessarily metrizable) spaces $X$ and $Y$ which are not $\ell$-equivalent.*

See Corollary 6.2.6.

**E3.** *Countable nondiscrete spaces $X_n$ for every $n$ such that*

$$C_p^*\Big(\bigoplus_{n\in\mathbb{N}} X_i\Big) \not\approx \prod_{n\in\mathbb{N}} C_p^*(X_n).$$

This shows that Theorem 6.2.8 is in general not true for spaces of bounded functions, even if one considers homeomorphisms instead of linear homeomorphisms. See MARCISZEWSKI and VAN MILL [272] for details. There are some special cases for which the version of Theorem 6.2.8 for homeomorphisms does hold for spaces of bounded functions. See e.g., BANAKH and CAUTY [40].

**E4.** $C(\omega_1)$ *and* $C(\omega_1+1)$ *are isomorphic as rings, yet* $\omega_1$ *and* $\omega_1+1$ *are not homeomorphic.*

See Page 371 to see why this is of interest.

**E5.** *Countable spaces $X_\alpha$ for which $C_p(X_\alpha)$ is of arbitrary multiplicative Borel complexity.*

See CALBRIX [83, 84] and LUTZER, VAN MILL and POL [265] for details. This is of interest since on Page 375 it was shown that $C_p(X)$ cannot be a $G_{\delta\sigma}$-subset of $\mathbb{R}^X$, unless $X$ is discrete. In addition, CAUTY, DOBROWOLSKI and MARCISZEWSKI [91] used this to prove the interesting result that if $X$ is countable and $C_p(X)$ is a Borel subset of $\mathbb{R}^X$ then it is of multiplicative Borel class. So the results of Calbrix and Lutzer et all show that any possible Borel complexity for $C_p(X)$ can be attained.

**E6.** *A countable regular, nonmetrizable space $X$ for which $C_p(X)$ is an $F_{\sigma\delta}$-subset of $\mathbb{R}^X$.*

Let $X_k = \{(1/n, 1/nk) : 1 \leq n < \infty\} \cup \{(0,0)\}$ for $k \geq 1$ and let
$$X = \bigcup\{X_k : 1 \leq k < \infty\}.$$
Each $X_k$ is topologized in such a way that $X_k$ becomes a convergent sequence with limit $(0,0)$. A subset $S \subseteq X$ is closed in $X$ if and only if $S \cap X_k$ is closed in $X_k$ for each $k$. Then $X$ is nonmetrizable (it is not first countable at $(0,0)$) (cf. Exercise A.11.3)). A function $f \colon X \to \mathbb{R}$ is continuous iff $f \restriction X_k \colon X_k \to \mathbb{R}$ is continuous for each $k$. For $k \geq 1$ let
$$C_k = \{g \in \mathbb{R}^X : g \restriction X_k \text{ is continuous on } X_k\}.$$
The proof of Theorem 6.3.10 can easily be modified to show that each $C_k$ is an $F_{\sigma\delta}$-subset of $\mathbb{R}^X$. Hence so is $C_p(X) = \bigcap_{k=1}^\infty C_k$. (Observe that we can think of $X$ as $\mathbb{N}_{\mathcal{F}}$ for some free filter $\mathcal{F}$ on $\mathbb{N}$.)

**E7.** *A countable nondiscrete space $X$ such that $C_p(X)$ is a Baire space.*

See Page 392 for details.

**E8.** *A countable nondiscrete space $X$ such that $C_p(X)$ is a hereditary Baire space.*

See Page 393 for details. This is an example under the assumption of the existence of a $P$-point in $\omega^*$. As was remarked on Page 393, this assumption is consistent with and independent from the usual axioms for set theory. It is unknown whether there is a hereditary Baire $C_p(X)$ in ZFC alone.

**E9.** *A space $X$ such that $C_p(X)$ is not Baire but $C_p(A)$ is Baire for every countable subspace $A \subseteq X$.*

See Page 387 for details.

**E10.** *A countable dense-in-itself space $X$ such that $C_p(X)$ is Baire.*

See Page 384 for details.

**E11.** *An infinite compact space $X$ such that $C_p(X)$ is not linearly homeomorphic to $C_p(X) \times \mathbb{R}$.*

Such an example was constructed by MARCISZEWSKI [268]. It is known that if $X$ contains a nontrivial convergent sequence then $C_p(X) \sim C_p(X) \times \mathbb{R}$ (Corollary 6.6.9). So Marciszewski's example does not contain nontrivial convergent sequences. The first example that comes to mind as a possible candidate is $\beta\omega$. But this space is homeomorphic to the topological sum of itself and a one-point space and so $C_p(\beta\omega) \sim C_p(\beta\omega) \times \mathbb{R}$. The second space that comes to mind as a possible example is $\omega^* = \beta\omega \setminus \omega$. But $\omega^*$ contains a copy of $\beta\omega$ which is a retract. So $C_p(\omega^*) \sim C_p(\omega^*) \times \mathbb{R}$.

**E12.** *A nonmetrizable compact space $K$ such that $C_p(K)$ is not homeomorphic to $C_p(K) \times C_p(K)$.*

See MARCISZEWSKI [266] and GUL'KO [181].

In view of Corollary 6.6.10 and Examples **E11** and **E13**, the question naturally arises whether there is an infinite *metrizable* space $X$ for which the spaces $C_p(X)$ and $C_p(X) \times C_p(X)$ are not (linearly) homeomorphic.

Observe that this question would be trivial if every infinite-dimensional linear space $L$ would have the property that $L$ is (linearly) homeomorphic to its own square $L \times L$. But this is not true as is clear from Example **E13**. It is not even true if $L$ is metrizable, as was first shown by POL [342]. By a refinement of Pol's method, VAN MILL [296] proved that there is an infinite-dimensional normed linear space $L$ such that $L$ is not homeomorphic to $L \times \mathbb{R}$.

The question was answered by POL [344] who presented three examples, one of which is COOK's continuum from [103] and another one is a very rigid subspace of $\mathbb{R}$, basically due to KURATOWSKI [237]. We will present the second example below.

**E13.** *An infinite subspace $X$ of $\mathbb{R}$ such that $C_p(X)$ is not linearly homeomorphic to $C_p(X) \times C_p(X)$.*

We will first show that there is a 'very rigid' subspace of $\mathbb{R}$.

**Example 6.13.1.** There exists a subset $B \subseteq \mathbb{R}$ having the following properties:

(1) No nonempty open subset of $B$ is meager in $\mathbb{R}$.
(2) If $V \subseteq B$ is open and nonempty then each continuous function
$$f \colon V \to B$$
is either the identity or constant on a nonempty open subset $W \subseteq V$.

**Proof.** Let $\mathcal{K}$ be the family of Cantor subsets of $\mathbb{R}$. Put
$$\mathcal{C} = \{(K, f) : K \in \mathcal{K}, f \in C(K), f[K] \text{ is uncountable}, f[K] \cap K = \emptyset\}.$$

Since $|\mathcal{K}| = \mathfrak{c}$ (Exercise A.2.16) and $|C(K)| \le \mathfrak{c}$ for every $K \in \mathcal{K}$ (Corollary 1.3.4), we can enumerate $\mathcal{C}$ as
$$\{(K_\alpha, f_\alpha) : \alpha < \mathfrak{c}\}.$$
By transfinite induction on $\alpha < \mathfrak{c}$, we will pick points $x_\alpha \in K_\alpha$ such that
(∗)  $\quad\quad\quad \{x_\beta : \beta \le \alpha\} \cap \{f_\beta(x_\beta) : \beta \le \alpha\} = \emptyset.$

Let $x_0$ be an arbitrary element of $K_0$, and assume that for some $\alpha < \mathfrak{c}$ we picked $x_\beta$ for every $\beta < \alpha$. Now observe that $f_\alpha[K_\alpha]$ is uncountable, and consequently has cardinality $\mathfrak{c}$ (Corollary 1.5.13). So there is a set $B \subseteq f_\alpha[K_\alpha]$ such that $|B| = \mathfrak{c}$ while moreover
$$B \cap \{x_\beta, f_\beta(x_\beta) : \beta < \alpha\} = \emptyset.$$
There exists a set $A \subseteq K_\alpha$ such that $f_\alpha \restriction A$ is one-to-one and $f_\alpha[A] = B$. Some $A' \subseteq A$ has $|A'| = \mathfrak{c}$ while moreover
$$A' \cap \{x_\beta, f_\beta(x_\beta) : \beta < \alpha\} = \emptyset.$$
Let $x_\alpha$ be any point of $A'$. It is easily seen that $x_\alpha$ is as required.

Put $B = \{x_\alpha : \alpha < \mathfrak{c}\}$. We claim that $B$ is the 'very rigid' subspace of $\mathbb{R}$ we are after.

Observe that $B$ has the property that for every $\alpha < \mathfrak{c}$ the point $f_\alpha(x_\alpha)$ does not belong to $B$.

**Claim 1.** $B$ and $\mathbb{R} \setminus B$ intersect every Cantor set in $\mathbb{R}$.

*Proof.* For a Cantor set $K \subseteq \mathbb{R}$, let $f : K \to \mathbb{R}$ be an imbedding such that
$$K \cap f[K] = \emptyset.$$
(For this, all one needs to observe is that $K$ is bounded.) The pair $(K, f)$ belongs to $\mathcal{C}$ and hence there exists $\alpha < \mathfrak{c}$ such that
$$(K, f) = (K_\alpha, f_\alpha).$$
By construction, $x_\alpha \in K_\alpha$. The pair $(f[K], f^{-1})$ also belongs to $\mathcal{C}$ and so the same argument shows that $K \cap (\mathbb{R} \setminus B) \ne \emptyset$. ◇

**Claim 2.** If $U \subseteq B$ is nonempty and open (in $B$) then it is not meager in $\mathbb{R}$.

*Proof.* Let $U' \subseteq \mathbb{R}$ be open such that $U' \cap B = U$. Suppose that $U$ is meager in $\mathbb{R}$. Then there is a countable collection $\mathcal{E}$ of nowhere dense closed subsets of $\mathbb{R}$ such that $U \subseteq \bigcup \mathcal{E}$. By the Baire Category Theorem,
$$G = U' \setminus \bigcup \mathcal{E}$$
is nonempty. In fact, the Baire Category Theorem implies that $G$ is uncountable. So $G$ is an uncountable $G_\delta$-subset of $\mathbb{R}$ and consequently contains a Cantor set $K$ by Theorems A.6.3 and 1.5.12. So $G$ intersects $B$ by Claim 1. But this is obviously a contradiction since $G \cap B = \emptyset$. ◇

**Claim 3.** If $U \subseteq B$ is nonempty and open (in $B$) and $f: U \to B$ is continuous, then either $f$ is the identity on $U$, or is constant on some nonempty open subset $V \subseteq U$.

*Proof.* Suppose that $f$ is not the identity. Then there exists a nonempty open subset $W \subseteq U$ such that the closure of $W$ in $B$ is contained in $U$ while moreover $\overline{W} \cap \overline{f[W]} = \emptyset$ (the closure is taken in $\mathbb{R}$). We claim that $f[W]$ is countable. Striving for a contradiction, suppose that it is uncountable. By Corollary A.8.4 there is a $G_\delta$-subset $S \subseteq \overline{W}$ such that $W \subseteq S$ while moreover $f \restriction W$ can be extended to a continuous function $\bar{f}: S \to \overline{f[W]}$. Since $\bar{f}[S]$ contains $f[W]$, it is uncountable. By Theorem 1.5.12 there consequently exists a Cantor set $K \subseteq S$ such that $\bar{f} \restriction K$ is one-to-one. There exists $\alpha < \mathfrak{c}$ such that
$$(K, \bar{f} \restriction K) = (K_\alpha, f_\alpha).$$
As a consequence, $x_\alpha \in K \cap B$, which is contained in the closure of $W$ in $B$, which is contained in $U$. But by construction, $f_\alpha(x_\alpha) \notin B$, and so, since
$$\bar{f} \restriction U = f,$$
it also follows that
$$f_\alpha(x_\alpha) = (\bar{f} \restriction K)(x_\alpha) = f(x_\alpha) \in B,$$
which is a contradiction.
Now for every $q$ in the countable set $f[W]$ let $A_q = \{x \in W : f(x) = q\}$. Then $\mathcal{A} = \{A_q : q \in f[W]\}$ is a countable cover of $W$ by (relative) closed sets. By Claim 2, one of those sets must have nonempty interior in $B$, so we are done. $\diamondsuit$

This completes the construction of $B$. $\square$

We now proceed with the proof that $C_p(B)$ is not linearly homeomorphic to its own square $C_p(B) \times C_p(B)$. For this, we need to do some preparatory work first.

**Lemma 6.13.2.** *Let $K, L$ be separable metrizable spaces. Assume that no nonempty open subset of $K$ is meager in $K$. Let $S: K \to \mathcal{F}_\infty(L) \setminus \{\emptyset\}$ be* LSC. *Then there exist a nonempty open set $G \subseteq K$ and finitely many pairwise disjoint open sets in $L$, say $H_1, \ldots, H_n$, and for every $i \le n$ continuous functions $s_i: G \to H_i$ such that*
$$S(t) = \{s_1(t), \ldots, s_n(t)\}$$
*for each $t \in G$.*

*Proof.* For every $n \in \mathbb{N}$, the set $F_n = \{t \in G : |S(t)| = n\}$ is an $F_\sigma$-subset of $G$. So one of them has nonempty interior, say $F_n$. So we can find a nonempty open set $G \subseteq F_n$, small enough that there are $n$ pairwise disjoint open sets $H_i \subseteq L$ such that each $S(t)$ with $t \in G$ intersects every $H_i$ in a singleton. The functions $s_i$ are defined by letting $s_i(t)$ be the unique point in $H_i \cap S(t)$ for $t \in G$. □

**Theorem 6.13.3.** *Let $B$ be the space constructed in Example 6.13.1. Then its function space $C_p(B)$ is not linearly homeomorphic to $C_p(B) \times C_p(B)$.*

**Proof.** First observe that
$$C_p(B \oplus B) \sim C_p(B) \times C_p(B)$$
(Theorem 6.2.8). So it suffices to show that $C_p(B)$ is not linearly homeomorphic to $C_p(B \oplus B)$.

Let $B \oplus B = B' \cup B''$, where $B'$ and $B''$ are disjoint open copies of $B$. For each $x \in B$ and $A \subseteq B$, we denote by $x'$, $x''$, $A'$ and $A''$ the corresponding copies of the point $x$ and the set $A$ in $B'$ and $B''$, respectively.

Striving for a contradiction, assume that there is a linear homeomorphism
$$\varphi \colon C_p(B) \to C_p(B' \cup B'').$$
We will apply Lemma 6.13.2 to $K = B'$ and the LSC function
$$F \colon B' \to \mathfrak{F}_\infty(B) \setminus \{\emptyset\}$$
defined by
$$F(y) = \mathrm{supp}_\varphi(y).$$
Observe that $F$ is well-defined by Lemma 6.8.2(3). So we get a nonempty open set $U' \subseteq B'$, and for some $n \in \mathbb{N}$, continuous mappings
$$s_i \colon U' \to B,$$
such that the family $\{s_i[U'] : i \leq n\}$ is pairwise disjoint and
$$F(y) = \{s_1(y), \ldots, s_n(y)\}$$
for every $y \in U'$. Observe that $U'$ contains a nonempty open set $W$ such that $s_1(y') = y$ for every $y \in W$, or $s_1 \upharpoonright W$ is constant. So by repeating the same observation $n$ times, we get a nonempty open set $V' \subseteq U'$ such that for each $i \leq n$, $s_i$ is constant on $V'$ or $s_i(y') = y$ for every $y' \in V'$.

Since $V'$ is infinite and supports of points are finite, $\bigcup\{\mathrm{supp}_\varphi(y') : y' \in V'\}$ is infinite as well (Lemma 6.8.2(5)). So the functions $s_i$ are not all constant. So there is an index $i_0 \leq n$ such that $s_{i_0}(y') = y$ for every $y' \in V'$. Since the images of the functions $s_i$ are pairwise disjoint, there is exactly one $s_i$ with this property, say $s_1$. Let $s_i$ be the constant function with value $a_i$ for $2 \leq i \leq n$. So

(1)  $\qquad \mathrm{supp}_\varphi(y') = \{y, a_2, \ldots, a_n\}$ for $y \in V$.

We now perform a similar construction with $K = V''$ – the copy of the set $V$ in $B''$. We end up with a nonempty open set $W'' \subseteq V''$ and points $b_i \in B$ for $2 \leq i \leq m$ such that

(2)  $\qquad \mathrm{supp}_\varphi(y'') = \{y, b_2, \ldots, b_m\}$ for $y \in W \subseteq V$.

Let $J = \{a_2, \ldots, a_n, b_2, \ldots, b_m\}$, and take an arbitrary element $c \in W \setminus J$ having the property that

(3) $\qquad c', c'' \notin \{\mathrm{supp}_{\varphi^{-1}}(x) : x \in J\}$.

(Again we use here that supports of points are finite.) We define a linear function $T \colon \mathbb{R} \to \mathbb{R}^2$ in the following way. Given $r \in \mathbb{R}$, choose any $f \in C_p(B)$ such that $f \restriction J \equiv 0$ and $f(c) = r$. Let $T(r) = \big(\varphi(f)(c'), \varphi(f)(c'')\big)$. We claim that $T(r)$ does not depend on the choice of $f$, and that the correspondence

$$r \mapsto T(r)$$

is linear. To see that this is true, let $g \in C(X)$ also be such that $g \restriction J \equiv 0$ and $g(c) = r$. Then

$$\varphi(f)(c') = \sum_{z \in \mathrm{supp}_\varphi(c')} \lambda(c', z) f(z) = \lambda(c', c) f(c) = \lambda(c', c) \cdot r$$

(see formula (∗∗) on Page 406) and, similarly,

$$\varphi(g)(c') = \lambda(c', c) \cdot r.$$

By making a similar calculation for $c''$, we conclude that $T$ is well-defined. Now that we know this, linearity is clear.

We will now show that $T$ is surjective, which by linearity obviously yields a contradiction. Pick arbitrary $r', r'' \in \mathbb{R}$. There exists by (3) a function

$$g \in C_p(B' \cup B'')$$

such that

$$g \restriction \bigcup \{\mathrm{supp}_{\varphi^{-1}}(x) : x \in J\} \equiv 0,$$

while moreover

$$g(c') = r', \quad g(c'') = r''.$$

Let $f = \varphi^{-1}(g)$ and $r = f(c)$. Then again by using formula (∗∗) on Page 406, we get that if $x \in J$ then

$$f(x) = \varphi^{-1}(g)(x) = \sum_{z \in \mathrm{supp}_{\varphi^{-1}}(x)} \lambda(x, z) g(z) = 0.$$

From this we conclude that

$$T(r) = \big(\varphi(f)(c'), \varphi(f)(c'')\big) = \big(g(c'), g(c'')\big) = (r', r''),$$

which is as required. $\qquad \square$

**E14.** *A countable space for which the Dugundji Extension Theorem fails.*

Countable spaces have separable metrizable function spaces and so in view of the result of GEBA and SEMADENI [172] mentioned in Example 6.6.4, the question naturally arises (see ARHANGEL'SKIĬ [26, Problem 58]) whether some form of the Dugundji Extension Theorem holds for that class of spaces. Observe that there are always linear extenders by Lemma 6.6.1. But, as we will show, there are not always extenders which are continuous, even for countable spaces.

One can interpret this from the point of view of selection theory, as follows. If $X$ is countable then $C_p(X)$ is a separable metrizable locally convex vector space. In addition, if $A \subseteq X$ is closed then the restriction operator

$$\rho \colon C_p(X) \to C_p(A)$$

defined by $\rho(f) = f \restriction A$ is bounded, linear, onto (by the Tietze Extension Theorem) and open. So by our claim there does not always exist a continuous selection for the continuous set-valued function

$$f \mapsto \rho^{-1}(f), \quad f \in C_p(A).$$

Interestingly, a result of MICHAEL [288] implies that for every *countable* subspace $C \subseteq C_p(A)$ there is a continuous selection for the set-valued function

$$f \mapsto \rho^{-1}(f), \quad f \in C.$$

We will now present the construction of the desired example.

Let $\mathcal{D}$ be an almost disjoint family of infinite subsets of $\omega$, i.e., $|D| = \omega$ for all $D \in \mathcal{D}$ and $|D \cap D'| < \omega$ for all distinct $D, D' \in \mathcal{D}$.

There are 'large' almost disjoint families of subsets of $\omega$. For example, if for every $t \in \mathbb{P}$, $S(t)$ is a sequence of rational numbers converging to $t$, then the family

$$\{S(t) : t \in \mathbb{P}\}$$

is an almost disjoint family of $\mathfrak{c}$ subsets of the countable set $\mathbb{Q}$. Let $\mathcal{F}$ be a collection of functions from $\omega$ to $\omega$ and let $\varphi \colon \mathcal{D} \to \mathcal{F}$ be a surjection. Finally, let $q$ be any point not in $\omega \times (\omega + 1)$. Topologize

$$\Delta(\varphi) = \{q\} \cup \big(\omega \times (\omega + 1)\big)$$

as follows. The set $\omega \times (\omega + 1)$ is an open subspace of $\Delta(\varphi)$, carrying the usual product topology. For finite $\mathcal{E} \subseteq \mathcal{D}$ and $n \in \omega$ define

$$V(\mathcal{E}) = \{q\} \cup \Big[\big(\omega \setminus \bigcup \mathcal{E}\big) \times (\omega + 1)\Big] \cup \bigcup_{E \in \mathcal{E}} \varphi(E) \restriction E$$

and

$$U(\mathcal{E}, n) = V(\mathcal{E}) \setminus \big(\{0, 1, \ldots, n-1\} \times (\omega + 1)\big),$$

respectively. (We identify here the functions $\varphi(E) \restriction E$ with their respective graphs.) The $U(\mathcal{E}, n)$'s form a neighborhood base for $q$. It is easy to see that $\Delta(\varphi)$ is regular and $T_1$.

**Lemma 6.13.4.** *Let $D \in \mathcal{D}$. Then $\varphi(D) \restriction D \subseteq \omega \times \omega \subseteq \omega \times (\omega+1)$ converges to $q$ in $\Delta(\varphi)$.*

**Proof.** Let $U(\mathcal{E}, n)$ be a basic open neighborhood of $q$. If $D \in \mathcal{E}$ then

$$\varphi(D) \restriction D \subseteq V(\mathcal{E})$$

and so all but finitely many elements of $\varphi(D) \restriction D$ are contained in $U(\mathcal{E}, n)$. Suppose therefore that $D \notin \mathcal{E}$. Then $D \cap \bigcup \mathcal{E}$ is finite since $\mathcal{D}$ is pairwise disjoint and $\mathcal{E}$ is finite. So all but finitely elements of $\varphi(D) \restriction D$ are contained in $V(\mathcal{E})$ and the same is true for $U(\mathcal{E}, n)$. □

We will always let $A$ denote the subspace $(\omega \times \{\omega\}) \cup \{q\}$ of $\triangle(\varphi)$. Observe that the subspace topology that $A$ inherits from $\triangle(\varphi)$ is independent of $\mathcal{F}$. Indeed, $A \setminus \{q\}$ is discrete and a basic neighborhood of $q$ has the form

$$\{q\} \cup (E(\mathcal{E}, n) \times \{\omega\}),$$

where

$$E(\mathcal{E}, n) = \omega \setminus \left(\bigcup \mathcal{E} \cup \{0, 1, \ldots, n-1\}\right),$$

for certain finite $\mathcal{E} \subseteq \mathcal{D}$ and $n < \omega$.

We will now describe our example. As on Page 46, let $\Sigma$ denote $\bigcup_{n<\omega} 2^n$, the set of finite sequences of 0's and 1's. For each $f \in 2^\omega$ we define

$$I(f) = \{f \restriction \{0, 1, \ldots, n-1\} : n < \omega\},$$

the set of initial sequences of $f$; $I(f)$ can be seen as the set of finite approximations to $f$. It is clear that

if $f, g \in 2^\omega$ are distinct, then $I(f) \cap I(g)$ is finite.

It will be convenient to identify $\omega$ and $\Sigma$. So we think of

$$\mathcal{D} = \{I(f) : f \in 2^\omega\}$$

as an almost disjoint collection of infinite subsets of $\omega$.
We are interested in $\triangle = \triangle(\varphi)$, where $\mathcal{D}$ is as above, $\mathcal{F} = \omega^\omega$; the function

$$\varphi \colon \mathcal{D} \to \mathcal{F}$$

will be determined below. Since the subspace topology that $A$ inherits from $\triangle(\varphi)$ is independent of $\mathcal{F}$ we already know $A$. For every $D \in \mathcal{D}$ we let $D_\omega$ denote the subset $D \times \{\omega\}$ of $A$. Observe that $D_\omega$ is clopen in $A$; we let $D_\omega^* \in C_p(A)$ denote its characteristic function. Our choice of $\mathcal{D}$ easily implies that the set $\mathcal{D}^* = \{D_\omega^* : D \in \mathcal{D}\} \subseteq C_p(A)$ is homeomorphic to the Cantor set $\{0, 1\}^\infty$. We now let

$$\varphi \colon \mathcal{D} \to \omega^\omega$$

be a function such that for each $f \in \omega^\omega$ the set $\{D_\omega^* : \varphi(D) = f\}$ intersects every Cantor set in $\mathcal{D}^*$. That such a function exists is not obvious. We will construct it by applying the following result which is known as the *Disjoint Refinement Lemma*.

**Lemma 6.13.5.** *Let $X$ be a set, and let $\mathcal{A}$ be a family of at most $\mathfrak{c}$ subsets of $X$ each of size $\mathfrak{c}$. Then for every $A \in \mathcal{A}$ there is a subset $B(A) \subseteq A$ of size $\mathfrak{c}$ such that the collection*

$$\{B(A) : A \in \mathcal{A}\}$$

is pairwise disjoint.

**Proof.** Enumerate $\mathcal{A}$ as $\{A_\alpha : \alpha < \mathfrak{c}\}$ such that each $A \in \mathcal{A}$ is listed $\mathfrak{c}$ times. By transfinite induction on $\alpha < \mathfrak{c}$ we will pick
$$x_\alpha \in A_\alpha \setminus \{x_\beta : \beta < \alpha\}.$$
This is a triviality. Let $x_0$ be any point of $A_0$. If $x_\beta$ has been defined for every $\beta < \alpha$, where $\alpha < \mathfrak{c}$, then proceed as follows. The set
$$A'_\alpha = A_\alpha \setminus \{x_\beta : \beta < \alpha\}$$
is nonempty since $|A_\alpha| = \mathfrak{c}$ and $|\alpha| < \mathfrak{c}$. So we can let $x_\alpha$ be any point of $A'_\alpha$. For every $A \in \mathcal{A}$ let
$$B(A) = \{x_\gamma : A_\gamma = A\}.$$
Then the collection $\{B(A) : A \in \mathcal{A}\}$ is clearly as required. □

Observe that the cardinality of $\mathcal{D}^*$ is equal to $\mathfrak{c}$, that the family of all Cantor subsets of $\mathcal{D}^*$ has size $\mathfrak{c}$ (Exercise A.2.16) and that every Cantor set has size $\mathfrak{c}$. So by Lemma 6.13.5 there is a disjoint family $\mathcal{H}$ of subsets of $\mathcal{D}^*$ such that every Cantor subset of $\mathcal{D}^*$ contains an element of $\mathcal{H}$; moreover, each $H \in \mathcal{H}$ has size $\mathfrak{c}$. For every $H \in \mathcal{H}$ let $\varphi_H \colon H \to \omega^\omega$ be a surjection. Now define
$$\varphi \colon \bigcup \mathcal{H} \to \omega^\omega$$
by the rule $\varphi(x) = \varphi_H(x)$ iff $x \in H$ and extend $\varphi$ over the rest of $\mathcal{D}^*$ in an arbitrary way. Then this function is clearly as desired.

**Theorem 6.13.6.** *If $e \colon \mathcal{D}^* \to C_p(\Delta)$ is an extender and $\mathcal{S} \subseteq \mathcal{D}^*$ is a dense $G_\delta$-subset then $e \restriction \mathcal{S}$ is not continuous.*

**Proof.** Striving for a contradiction, assume that there exist a subcollection $\mathcal{G}$ of $\mathcal{D}$ and an extender $e \colon \mathcal{D}^* \to C_p(\Delta)$ such that

(1) $\mathcal{G}^* = \{G_\omega^* : G \in \mathcal{G}\}$ is a dense $G_\delta$-subset of $\mathcal{D}^*$;
(2) $e \restriction \mathcal{G}^*$ is continuous.

For every $\sigma \in \Sigma$, the set
$$\mathcal{D}^*(\sigma) = \{D_\omega^* \in \mathcal{D}^* : \sigma \subseteq D\}$$
is a basic clopen subset of $\mathcal{D}^*$ and consequently intersects $\mathcal{G}^*$. For convenience, put
$$\mathcal{G}^*(\sigma) = \mathcal{D}^*(\sigma) \cap \mathcal{G}^* \qquad (\sigma \in \Sigma).$$
We will now construct a function $f \colon \Sigma \to \omega$ having some special properties. To this end, take an arbitrary $\sigma \in \Sigma$. The value $f(\sigma)$ is determined as follows. Take an arbitrary $G_\omega^* \in \mathcal{G}^*(\sigma)$. Since $e(G_\omega^*)\big((\sigma,\omega)\big) = 1$, there exists $n \in \omega$ such that $e(G_\omega^*)\big((\sigma,n)\big) > \frac{1}{2}$. Put $f(\sigma) = n$.
For $\sigma \in \Sigma$ put

(∗) $\quad \mathcal{U}(\sigma) = \{u \in C_p(\Delta) : u\big((\sigma,\omega)\big) > \frac{1}{2}, \ u\big((\sigma,f(\sigma))\big) > \frac{1}{2}\}.$

Then $\mathcal{U}(\sigma)$ is clearly open and contains $e(G_\omega^*)$. In particular,
$$e[\mathcal{G}^*(\sigma)] \cap \mathcal{U}(\sigma) \neq \emptyset,$$
or, equivalently,
$$\mathcal{G}^*(\sigma) \cap e^{-1}[\mathcal{U}(\sigma)] \neq \emptyset.$$

For every $k \in \omega$ define
$$(**) \qquad \mathcal{U}_k = \bigcup \{\mathcal{U}(\sigma) : \text{length } \sigma \geq k\}.$$
By the above and the continuity of $e$ on $\mathcal{G}^*$, for every $k$ we have that
$$e^{-1}[\mathcal{U}_k] \cap \mathcal{G}^*$$
is dense and open in $\mathcal{G}^*$. We conclude that
$$\bigcap_{k \in \omega} e^{-1}[\mathcal{U}_k] \cap \mathcal{G}^*$$
is a dense $G_\delta$-subset of $\mathcal{G}^*$, and hence of $\mathcal{D}^*$, and it therefore contains a Cantor set (Theorem 1.5.12). So by construction, there exists
$$D_\omega^* \in \bigcap_{k \in \omega} e^{-1}[\mathcal{U}_k] \cap \mathcal{G}^*, \quad \varphi(D) = f.$$
It follows that $e(D_\omega^*) \in \mathcal{U}_k$ for all $k$, and hence, by $(*)$ and $(**)$,
$$e(D_\omega^*)\big((\sigma, \omega)\big) > 1/2, \quad e(D_\omega^*)\big((\sigma, f(\sigma))\big) > 1/2$$
for infinitely many $\sigma$. However,
$$e(D_\omega^*)\big((\sigma, \omega)\big) = D_\omega^*\big((\sigma, \omega)\big) > 1/2$$
implies that $\sigma \in D$, so in fact,
$$(\dagger) \qquad e(D_\omega^*)\big((\sigma, f(\sigma))\big) > 1/2$$
for infinitely many $\sigma \in D$. But $e(D_\omega^*)(q) = 0$ and $f = \varphi(D)$ and so $f \upharpoonright D$ converges to $q$. By continuity of $e(D_\omega^*)$ we must consequently have that
$$(\ddagger) \qquad e(D_\omega^*)\big((\sigma, f(\sigma))\big) < 1/2$$
for all but finitely many $\sigma \in D$. Conditions $(\dagger)$ and $(\ddagger)$ provide a contradiction, which establishes the proof. $\square$

**Remark 6.13.7.** Suppose that $A \subseteq X$ is a closed subset of a countable space $X$ such that the function space $C_p(X)$ is Borel in $\mathbb{R}^X$. Then by the Yankov–von Neumann Theorem, the restriction map $f \mapsto f \upharpoonright A$ from $C_p(X)$ to $C_p(A)$ has a selection $e : C_p(A) \to C_p(X)$ (an extender) measurable with respect to the $\sigma$-algebra generated by the analytic sets in $C_p(A)$, cf. ROGERS ET AL [352, p. 212] or ARVESON [29, Theorem 3.4.3]. Therefore, for each subset $\mathcal{E} \subseteq C_p(A)$ which is a $G_\delta$-set in $\mathbb{R}^A$, the restriction $e : \mathcal{E} \to C_p(X)$ is measurable with respect to the $\sigma$-algebra of sets in $\mathcal{E}$ that are open modulo a first category set (in $\mathcal{E}$). As a consequence, $e$ is continuous on a dense $G_\delta$-subset of $\mathcal{E}$.

Let us now return to the space $\triangle$ and its closed subset $A$. Since $\mathcal{D}^*$ is a Cantor set, by the above remarks and Theorem 6.13.6 it follows that $C_p(\triangle)$ is not Borel. Also, there does not even exist a measurable extender from $C_p(A)$ to $C_p(\triangle)$. Interestingly, $C_p(A)$ *is* Borel since the space $A$ is similar to one of the examples considered in LUTZER, VAN MILL and POL [265].

> **E15.** *A compact space $X$ and a noncompact space $Y$ such that $C_p(X)$ and $C_p(Y)$ are homeomorphic.*

Let $X = \mathbb{I}$ and $Y = \mathbb{R}$. Then $C_p(X) \approx C_p(Y)$, as was shown by GUL'KO and KHMYLEVA [183]. See also Example 6.9.3.

> **E16.** *An infinite-dimensional compact space $X$ such that $C_p(X)$ is not homeomorphic to $C_p(Y)$ for any finite dimensional compact metrizable space $Y$.*

By CAUTY [90], $X = Q$ is as required. See also Corollary 6.11.4.

> **E17.** *Scattered compact $\ell^*$-equivalent spaces $X$ and $Y$ such that $\kappa(X) = 2$ and $\kappa(Y) = n + 1$.*

See Example 6.10.5.

> **E18.** *$\ell$-equivalent countable, locally compact (hence metrizable) spaces which are not $\ell^*$-equivalent.*

See Example 6.10.10.

> **E19.** *Countable $t^*$-equivalent spaces which are not $t$-equivalent.*

See MARCISZEWSKI and VAN MILL [272].

> **E20.** *Countable spaces $X$ and $Y$ such that $C_p(X)$ and $C_p(Y)$ are nonhomeomorphic absolutely Borel sets of the same Borel type.*

See CAUTY [88].

APPENDIX A

# Preliminaries

In this appendix we will collect for easy reference and for completeness sake some facts that are important throughout this book. It is certainly not our intention that the results presented here will be used as a text for an introductory course in topology. For the more experienced reader there is no need to read this chapter in full detail. She or he should read §A.1 only to get our notation straight. *All spaces under discussion in Chapters 1 through 5 of the present book are separable and metrizable exclusively.*

## A.1. Prerequisites and notation

We assume that the reader understands the material of basic courses in general topology, calculus and linear algebra. Concepts such as topological space, metric space, metrizable space, complete metric space, completely metrizable space, compact space, product space, continuous function, etc., should be familiar. For all undefined notions see any textbook on General Topology (e.g., DUGUNDJI [142], ENGELKING [153] and NAGATA [324]).

**Notation.** We let $\mathbb{I}$, $\mathbb{J}$, $\mathbb{R}$, $\mathbb{N}$ and $\mathbb{Z}$ denote the interval $[0,1]$, the interval $[-1,1]$ and the sets of real and natural numbers and integers, respectively. In addition, $\mathbb{Q}$ denotes the subset of $\mathbb{R}$ consisting of all rational numbers, and $\mathbb{P}$ denotes $\mathbb{R} \setminus \mathbb{Q}$. That is, $\mathbb{P}$ is the subset of $\mathbb{R}$ consisting of all irrational numbers.

**Set theoretic notions.** If $A$ and $B$ are sets then $A \triangle B$ denotes their *symmetric difference*, i.e.,

$$A \triangle B = (A \setminus B) \cup (B \setminus A).$$

If $\{X_i : i \in I\}$ is a family of sets then

$$\prod_{i \in I} X_i$$

is their *(Cartesian) product*. For every $j \in I$, the projection $\prod_{i \in I} X_i \to X_j$ will be denoted by $\pi_j$. If $E$ is an arbitrary subset of $I$ then $\pi_E$ denotes the projection from $\prod_{i \in I} X_i$ onto its subproduct $\prod_{i \in E} X_i$.

Let $\{x_i : i \in I\}$ be an indexing of a set $X$. We say that $X$ is *faithfully indexed* by $I$ provided that the function $I \to X$ defined by

$$i \mapsto x_i$$

is a bijection.

If $\mathcal{U}$ is a family of sets and $X$ is a set then $\mathcal{U} \upharpoonright X$ denotes the collection

$$\{U \cap X : U \in \mathcal{U}\}.$$

If $f \colon X \to Y$ is a function and $A \subseteq X$ then $f \upharpoonright A$ denotes the restriction of $f$ to $A$. For $t \in \mathbb{R}$, the symbol $f \equiv t$ means that $f$ is the constant function with value $t$. The *identity function* from $X$ to $X$ shall be denoted by $1$ or by $1_X$.

A function $f \colon X \to Y$ is sometimes identified with its graph

$$\Gamma = \{(x, f(x)) : x \in X\} \subseteq X \times Y.$$

By this identification we can use the standard set theoretic terminology for functions. For example, if $A, B$ are subsets of $X$ and $f \colon A \to Y$, $g \colon B \to Y$ are functions such that $f(x) = g(x)$ for every $x \in A \cap B$ then

$$f \cup g$$

is the function from $A \cup B$ to $Y$ defined as follows:

$$\begin{cases} f(x) & (x \in A), \\ g(x) & (x \in B). \end{cases}$$

Etc.

If $X$ and $Y$ are sets and $f \colon X \to Y$ then a *fiber* of $f$ is a set of the form $f^{-1}[\{y\}]$, where $y \in f[X]$. If $y \in f[X]$ then its fiber $f^{-1}[\{y\}]$ will for simplicity be denoted by $f^{-1}(y)$.

If $X$ is a set then $\mathcal{P}(X)$ denotes the *power set* of $X$.

Each subset $A$ of a set $X$ can be identified with its so-called *characteristic function* $\chi_A \colon X \to \{0,1\}$ defined by $\chi_A(x) = 1$ if and only if $x \in A$. This means that we can identify $A \subseteq X$ with $\chi_A \in \{0,1\}^X$. The function $A \mapsto \chi_A$ is clearly a bijection from $\mathcal{P}(X)$ onto $\{0,1\}^X$.

It will be convenient to introduce the following notation. If $\mathcal{U}$ and $\mathcal{V}$ are covers of a set $X$ then $\mathcal{U} < \mathcal{V}$ means that $\mathcal{U}$ is a *refinement* of $\mathcal{V}$ (i.e., for every $U \in \mathcal{U}$ there exists $V \in \mathcal{V}$ such that $U \subseteq V$).

A nonempty subcollection $\mathcal{A}$ of $\mathcal{P}(X)$ is a *$\sigma$-algebra* provided that

(1) if $A \in \mathcal{A}$ then $X \setminus A \in \mathcal{A}$,
(2) if $A_n \in \mathcal{A}$ for every $n$ then $\bigcup_{n=1}^\infty A_n \in \mathcal{A}$.

In addition, a subcollection $\mathcal{F} \subseteq \mathcal{P}(X)$ is called a *filter* on $X$ provided that it has the following properties:

(1) $\emptyset \notin \mathcal{F}$,

(2) if $F, G \in \mathcal{F}$ then $F \cap G \in \mathcal{F}$,
(3) if $F \in \mathcal{F}$ and $F \subseteq G \subseteq X$ then $G \in \mathcal{F}$.

A filter $\mathcal{F}$ is called *free* if $\bigcap \mathcal{F} = \emptyset$. If a filter is not free then it is called *fixed*.

An *ultrafilter* is a filter not properly contained in another filter. The Kuratowski-Zorn Lemma easily implies that every filter is contained in some ultrafilter. It is easy to see that the intersection of a fixed ultrafilter is a singleton. So a fixed ultrafilter on a set $X$ has the form

$$\{A \subseteq X : x \in A\}$$

for some $x \in X$.

Notice that each filter has the finite intersection property. If $\mathcal{F} \subseteq \mathcal{P}(X)$ has the finite intersection property then the collection

$$\mathcal{G} = \{A \subseteq X : (\exists F \in \mathcal{F})(F \subseteq A)\}$$

is easily seen to be a filter. It is called the filter *generated* by $\mathcal{F}$.

The first infinite cardinal is $\omega$ and $\mathfrak{c}$ denotes $2^\omega$, i.e., the cardinality of the continuum. We usually identify $\omega$ and the set of positive integers. By $\omega + 1$ we mean the sequence of positive integers converging to the point $\omega$. So $\omega + 1$ is homeomorphic to the subspace

$$\{1/n : n \in \mathbb{N}\} \cup \{0\}$$

of $\mathbb{R}$. That is, $\omega + 1$ is nothing but a prototype of a convergent sequence with its limit.

If $X$ is a set then $|X|$ denotes its cardinality.

**Metric spaces.** Let $(X, \varrho)$ be a metric space. If $A \subseteq X$ then $\operatorname{diam}(A)$ denotes the *diameter* of $A$. If $\mathcal{U}$ is a collection of subsets of $X$ then the *mesh* of $\mathcal{U}$ is the number

$$\operatorname{mesh}(\mathcal{U}) = \sup\{\operatorname{diam}(U) : U \in \mathcal{U}\}.$$

We allow $\operatorname{mesh}(\mathcal{U})$ to be equal to $\infty$. In addition, $\varrho$ is called *bounded* if the diameter of $X$ is finite.

For $A \subseteq X$ and $\varepsilon > 0$ we let

$$B(A, \varepsilon) = \{y \in X : \varrho(A, y) < \varepsilon\}$$

and

$$D(A, \varepsilon) = \{y \in X : \varrho(A, y) \leq \varepsilon\}$$

denote the *open* and *closed* balls about $A$ with radius $\varepsilon$, respectively. If $A$ is a singleton, say $\{x\}$, then we write $B(x, \varepsilon)$ instead of $B(\{x\}, \varepsilon)$ and $D(x, \varepsilon)$ instead of $D(\{x\}, \varepsilon)$. If it is unclear to which metric $\varrho$ we are referring, we occasionally use the notation $B_\varrho(A, \varepsilon)$ and $B_\varrho(x, \varepsilon)$. Similarly for $D$.

We sometimes denote $B(A, \varepsilon)$ by $B_\varepsilon(A)$ and $D(A, \varepsilon)$ by $D_\varepsilon(A)$.

Let $(X, \varrho)$ and $(Y, d)$ be metric spaces. A surjective function $f\colon X \to Y$ is an *isometry* provided that
$$d\bigl(f(x), f(y)\bigr) = \varrho(x, y)$$
for all $x, y \in X$.

**Metrizable spaces.** Let $X$ be a metrizable space. We say that a metric $\varrho$ on $X$ is *admissible* if it generates the topology on $X$. Each admissible metric can be replaced by one which in addition is bounded, see Exercise A.1.6.

**Topological notions.** Let $X$ be a topological space with subset $A$. Then $\overline{A}$ and $\mathrm{Int}(A)$ denote its *closure* and *interior*, respectively. In addition, $\mathrm{Fr}(A)$ is the *boundary* of $A$.

A subset of a space $X$ is called *clopen* if it is both closed and open.

If $\mathcal{U}$ is a collection of subsets of $X$ then $\overline{\mathcal{U}}$ denotes the collection
$$\{\overline{U} : U \in \mathcal{U}\}.$$

Let $A \subseteq X$ and let $\mathcal{U}$ be an open cover of $X$. The *star* of $A$ with respect to $\mathcal{U}$ is the set
$$\mathrm{St}(A, \mathcal{U}) = \bigcup \{U \in \mathcal{U} : U \cap A \neq \emptyset\}.$$
The cover $\{\mathrm{St}(U, \mathcal{U}) : U \in \mathcal{U}\}$ is denoted by $\mathrm{St}(\mathcal{U})$. We say that an open cover $\mathcal{V}$ of $X$ is a *star-refinement* of $\mathcal{U}$ if $\mathrm{St}(\mathcal{V}) < \mathcal{U}$, i.e., if for every $V \in \mathcal{V}$ there exists $U \in \mathcal{U}$ such that
$$\mathrm{St}(V, \mathcal{V}) \subseteq U.$$

An important characterization of compactness is the well-known fact that a space $X$ is compact if and only if every collection of closed subsets of $X$ with the finite intersection property has nonempty intersection. This will be used without explicit reference in the remaining part of this book.

Let $f\colon X \to Y$. We say that $f$ is *closed* provided that $f[A]$ is closed for every closed set $A \subseteq X$. Similarly for *open*.

If $X$ and $Y$ are spaces then $f\colon X \to Y$ is an *imbedding* if the function
$$f\colon X \to f[X]$$
is a homeomorphism. If $f\colon X \to Y$ is an imbedding and $f[X]$ is a closed subset of $Y$ then $f$ is called a *closed imbedding*. Similarly for *open*. Moreover, if $f\colon X \to Y$ is an imbedding and $f[X]$ is dense in $Y$ then $f$ is called a *dense imbedding*.

The symbol '$X \approx Y$' means that $X$ and $Y$ are homeomorphic spaces.

If $X$ and $Y$ are spaces then $C(X, Y)$ denotes the collection of all *continuous* functions from $X$ to $Y$.

We assume that all topological spaces under discussion are nonempty.

**Topological products.** Let $X = \prod_{i=1}^{\infty} X_i$, where $X_i$ is a nonempty topological space for every $i$, and $X$ is endowed with the Tychonoff product topology. It is possible to identify each $X_i$ in a natural way with a subspace of $X$, as follows. Pick an arbitrary element $x_i \in X_i$ for every $i$. The subspace

$$\{x_1\} \times \cdots \times \{x_{i-1}\} \times X_i \times \{x_{i+1}\} \times \cdots$$

is clearly homeomorphic to $X_i$. Observe that the composition

$$X_i \xhookrightarrow{i} X \xrightarrow{\pi_i} X_i,$$

where $i$ denotes the 'inclusion', is the identity on $X_i$.

If $\{A, B\}$ is a partition of $I$ then $X$ can be naturally be identified with the product

$$\prod_{i \in A} X_i \times \prod_{i \in B} X_i.$$

We freely will make such identifications without warning. So we will permute or reorganize the factors of a product in any way that is convenient to us.

A countably infinite product of copies of the same space $X$ shall sometimes for convenience be denoted by $X^{\infty}$.

The $i$-th coordinate of a point $x$ in a product space shall be denoted by $x_i$.

**The Euclidean spaces $\mathbb{R}^n$.** The standard metric on $\mathbb{R}^n$ is the Euclidean metric which is defined by

$$\varrho(x,y) = \sqrt{\sum_{i=1}^{n} |x_i - y_i|^2}.$$

Let $B^n$ and $\mathbb{S}^{n-1}$ abbreviate the unit ball and unit sphere in $\mathbb{R}^n$, respectively. Two points $x, y \in \mathbb{S}^n$ will be called *antipodal* provided that $x = -y$. The function $\mathbb{S}^n \to \mathbb{S}^n$ defined by $x \mapsto -x$ is called the *antipodal mapping* on $\mathbb{S}^n$.

**The Hilbert cube.** The *Hilbert cube*, abbreviated $Q$, is the product $\mathbb{J}^{\infty}$ endowed with the Tychonoff product topology. Observe that $Q$ is compact, being a product of compact spaces. A natural metric for $Q$ that generates its topology is

$$\varrho(x,y) = \sum_{i=1}^{\infty} 2^{-i}|x_i - y_i|$$

(Exercise A.1.10). Unless otherwise stated, $Q$ will always be endowed with this metric. The diameter of $Q$ is 2.

The point in $Q$ all coordinates of which are 0 will be denoted by **0**.

Geometrically one should think of $Q$ as an infinite-dimensional brick the sides of which get shorter and shorter. This can be demonstrated in the following way. Let $x(n) \in Q$ be the point having all coordinates 0 except for the $n$-th coordinate which equals 1. So $x(n)$ is the 'endpoint' of the $n$-th axis in $Q$. Intuitively, each $x(n)$ has distance 1 from $\mathbf{0}$ and hence $x(n)$ and $\mathbf{0}$ are far apart. However, the appearance of the factor $2^{-i}$ in the definition of $\varrho$ implies that
$$\varrho(x(n), \mathbf{0}) = 2^{-n},$$
whence the sequence $(x(n))_n$ converges to $\mathbf{0}$ in $Q$.

If $n \in \mathbb{N}$ then by the above remarks, $Q$ is naturally homeomorphic to
$$\mathbb{J}^n \times \prod_{m=n+1}^{\infty} \mathbb{J}_m.$$

It will be convenient to let $Q_n$ denote the product $\prod_{m=n+1}^{\infty} \mathbb{J}_m$.

For $n \in \mathbb{N}$ and $\theta \in \{-1, 1\}$ by $W_n^\theta$ we denote the *endface*
$$\pi_n^{-1}(\{\theta\}) = \{x \in Q : x_n = \theta\}$$
of the Hilbert cube $Q$.

**Groups.** A group $G$ which is also a topological space, is called a *topological group* if the function $G \times G \to G$ defined by
$$(x, y) \mapsto x \cdot y^{-1}$$
is continuous. This easily implies that a *translation* on $G$, i.e., a function from $G$ into $G$ of the form
$$x \mapsto x \cdot a$$
or of the form
$$x \mapsto a \cdot x$$
for some fixed $a \in G$, is a homeomorphism.

**Exercises for §A.1.** Let $X$ be a space with subset $A$. We say that $p \in X$ is an *accumulation point* of $A$ provided that $p \in \overline{A \setminus \{p\}}$.

A topological space $X$ is called *homogeneous* provided that for all $x, y \in X$ there exists a homeomorphism $h \colon X \to X$ such that $h(x) = y$. So, loosely speaking, a space $X$ is homogeneous if from the topological standpoint all points in $X$ behave in the same way.

For $A, S \in \mathcal{P}(\mathbb{N})$, put $V(A, S) = \{B \in \mathcal{P}(\mathbb{N}) : B \cap S = A \cap S\}$.

1. Let $X$ be a set and let $\mathcal{F} \subseteq \mathcal{P}(X)$. Prove that the following statements are equivalent:
    (1) $\mathcal{F}$ is an ultrafilter.
    (2) $\mathcal{F}$ has the finite intersection property, and if $A \subseteq X$ then either $A \in \mathcal{F}$ or $X \setminus A \in \mathcal{F}$.

▶2. Let $X$ be a space without isolated points. Prove that the following statements are equivalent:
   (1) No point of $X$ is simultaneously an accumulation point of two disjoint subsets of $X$.
   (2) If $p$ is an accumulation point of $A \subseteq X$ then $p \in \text{Int}(A \cup \{p\})$.
   (3) If $x \in X$ then the collection
   $$\{A \subseteq X : x \text{ is an accumulation point of } A\}$$
   is an ultrafilter on $X$.

3. Let $X$ be a space and let $\mathcal{V}$ be a countable (finite) collection of clopen subsets $X$. Prove that there is a countable (finite) collection clopen subsets $\mathcal{W}$ of $X$ such that
   (1) every $W \in \mathcal{W}$ is contained in an element $V \in \mathcal{V}$,
   (2) $\mathcal{W}$ is pairwise disjoint,
   (3) $\bigcup \mathcal{V} = \bigcup \mathcal{W}$.

▶4. Let $X$ be a set, let $k \in \mathbb{N}$ and let $\mathcal{A}$ be an infinite collection of subsets of $X$ each of cardinality $k$. Prove that there are a set $A \subseteq X$ with $|A| < k$ and a sequence $(A_i)_i$ of elements of $\mathcal{A}$ such that $A_i \cap A_j = A$ for all distinct $i$ and $j$.

5. Let $A, B, E, F \in \mathcal{P}(\mathbb{N})$. Prove that the following statements are equivalent:
   (1) $V(A, E) \subseteq V(B, F)$,
   (2) $B \cap F \subseteq A \cap E$ and $F \setminus B \subseteq E \setminus A$
   (hence $F \subseteq E$).

6. Let $X$ be a space and let $\varrho$ be an admissible metric for $X$ and let $a \in (0, \infty)$. Prove that the function $\rho \colon X \to X \times [0, \infty)$ defined by
$$\rho(x, y) = \min\{a, \varrho(x, y)\}$$
is an admissible bounded metric for $X$. Prove that if $\varrho$ is complete, then so is $\rho$.

7. Let $A$ and $B$ be subsets of a space $X$. Prove that for every $x \in X$ we have
$$\varrho(x, A \cup B) \in \{\varrho(x, A), \varrho(x, B)\}.$$

8. Let $X$ and $Y$ be spaces with $Y$ Hausdorff. Prove that if $f \colon X \to Y$ is continuous then its graph
$$\Gamma = \{(x, f(x)) : x \in X\}$$
is a closed subspace of the product $X \times Y$. (For a related result, see Theorem 1.1.15.)

9. Let $X$ and $Y$ be spaces with $Y$ Hausdorff and let $f, g \colon X \to Y$ be continuous. Prove that the set
$$\{x \in X : f(x) = g(x)\}$$
is closed in $X$.

▶10. For every $n$ let $\varrho_n$ be an admissible metric for $X_n$. Assume that for some constant $c \in (0, \infty)$ all the metrics $\varrho_n$ are bounded by $c$. Define
$$\varrho \colon \prod_{n=1}^{\infty} X_n \times \prod_{n=1}^{\infty} X_n \to \mathbb{R}$$
by the formula
$$\varrho(x, y) = \sum_{n=1}^{\infty} 2^{-n} \varrho_n(x_n, y_n).$$
Prove that $\varrho$ is an admissible metric for the Tychonoff product topology on $\prod_{n=1}^{\infty} X_n$.

11. Let $X$ be a metrizable space and let $(x_n)_n$ be a Cauchy sequence in $X$. Assume that $(x_n)_n$ has a convergent subsequence, say with limit $x$. Prove that $\lim_{n \to \infty} x_n = x$.

12. Let $G$ be a topological group. Prove that $G$ is homogeneous.

13. For every $n$ let $X_n$ and $Y_n$ be spaces and let $f_n \colon X_n \to Y_n$ be continuous. Prove that the function
$$f \colon \prod_{n=1}^{\infty} X_n \to \prod_{n=1}^{\infty} Y_n$$
defined by
$$f(x_1, x_2, \dots) = (f_1(x_1), f_2(x_2), \dots)$$
is continuous. Also prove that if every $f_n$ is an imbedding (homeomorphism) then $f$ is an imbedding (homeomorphism). Conclude that if $A_n$ is contained in $X_n$ for every $n$ then $\prod_{n=1}^{\infty} A_n$ is a subspace of $\prod_{n=1}^{\infty} X_n$.

14. Let $A_1, \dots, A_n$ be a collection of subsets of a metric space $X$ such that
$$A_i \cap \bigcup_{j=1}^{i-1} A_j \neq \emptyset$$
for every $i \leq n$ with $i \geq 2$. Prove that
$$\operatorname{diam}\left(\bigcup_{i=1}^{n} A_i\right) \leq \sum_{i=1}^{n} \operatorname{diam}(A_i).$$

15. Let $f \colon X \to Y$ be a closed map, let $A \subseteq Y$ and let $U$ be an open neighborhood of $f^{-1}[A]$ in $X$. Prove that there is an open neighborhood $V$ of $A$ in $Y$ such that $f^{-1}[V] \subseteq U$.

16. Let $X$ be a space with subspace $Y$. Prove that if $U \subseteq X$ is open then the boundary of $U \cap Y$ in $Y$ is a subset of the boundary of $U$ in $X$.

17. Let $f \colon X \to Y$ be a continuous surjection. Assume moreover that $f$ is open and that $f^{-1}(y)$ is finite for every $y$. Prove that if $S$ is a dense $G_\delta$-subset of $X$ then $f[S]$ is a dense $G_\delta$-subset of $Y$. (See Exercise A.6.12 for an example showing that the finiteness of the fibers of the map $f$ is essential.)

## A.2. Separable metrizable topological spaces

Our primary interest in this book is in infinite-dimensional and related topology and hence in separable metrizable spaces. This is why we do not bother to state our theorems in the first five chapters in their most general forms for this would only distract the reader from the essential parts of our considerations. For simplicity we therefore call a separable metrizable space simply a *space*.

Non-separable metrizable spaces and non-metrizable spaces are considered in Chapter 6, where we study the topology of pointwise convergence on function spaces. If $X$ is a countable space which is not necessarily metrizable, then its function space of real-valued continuous functions endowed with the topology of pointwise convergence is a linear subspace of the product $\mathbb{R}^X$. So it is a linear subspace of a countable product of real lines, and is therefore a separable metrizable space. The space $X$ can be rather 'bad' from a general topological perspective, its function space is always 'good'. As we will see, certain non-metrizable countable spaces $X$ have very interesting function spaces and this prompts us to study non-metrizable spaces as well.

It is not so that we consider the theory of separable metrizable spaces to be more important than the theory of general topological spaces. Topological spaces with applications in computer science have very bad separation properties and consequently are very far from being separable metrizable, yet they are important. Similar remarks can be made for spaces endowed with a weak topology of some sort in functional analysis, or CW-complexes in algebraic topology. However, we believe that for a good understanding of topology it is essential to make a thorough study of separable metrizable spaces first. In our belief generalizations should be postponed until having mastered the separable metrizable spaces.

The separable metrizable spaces can be characterized in general topological terms, as follows. A topological space $X$ is separable and metrizable if and only if $X$ is a regular $T_1$-space with a countable base. This is called *Urysohn's Metrization Theorem*. If $X$ is separable and metrizable then $X$ is regular and $T_1$ for obvious reasons and has a countable base by Theorem A.2.1 below. Conversely, if $X$ is regular and $T_1$ and has a countable base then it is normal, being Lindelöf, so by Urysohn's Lemma for normal spaces there are enough real-valued continuous functions to separate closed sets. The proof of Corollary A.4.4 then shows that $X$ can be imbedded in $Q$ and hence is separable and metrizable. (See ENGELKING [153, Theorem 4.2.9] for more details.)

For any space $X$ we let $\varrho$ denote an admissible metric on $X$.

**Theorem A.2.1.** *Let $X$ be a space. There is a countable collection of open subsets of $X$ which is a base for its topology.*

**Remark A.2.2.** This result will be used without explicit reference throughout the remaining part of this book.

**Proof.** Let $\varrho$ be an admissible metric on $X$. There is by assumption a countable dense subset of $X$, say $D$. Put
$$\mathcal{B} = \{B(x,q) : x \in D, q \in \mathbb{Q}\}.$$
Then $\mathcal{B}$ is a countable collection open subsets of $X$ and we claim that it is a base. To this end, let $x \in X$ and $U \subseteq X$ be open such that $x \in U$. There is $\varepsilon > 0$ such that $B(x,\varepsilon) \subseteq U$. Let $d \in D$ be such that $\varrho(x,d) < \frac{1}{2}\varepsilon$ and let $q$ be rational such that $\varrho(x,d) < q < \frac{1}{2}\varepsilon$. If $z \in B(d,q)$ then
$$\varrho(x,z) \leq \varrho(x,d) + \varrho(d,z)$$
$$< \frac{1}{2}\varepsilon + \frac{1}{2}\varepsilon$$
$$= \varepsilon.$$
So we conclude that
$$B(d,q) \subseteq B(x,\varepsilon) \subseteq U.$$
In addition, $x \in B(d,q)$ since $\varrho(x,d) < q$. □

The following corollary to this result will be used quite frequently.

**Corollary A.2.3.** *Let $X$ be a space and let $\mathcal{U}$ be a collection of open subsets of $X$. Then there is a countable subcollection $\mathcal{V}$ of $\mathcal{U}$ such that $\bigcup \mathcal{V} = \bigcup \mathcal{U}$.*

**Proof.** Let $\mathcal{B}$ be a countable base for $X$ and put
$$\mathcal{B}' = \{B \in \mathcal{B} : (\exists U \in \mathcal{U})(B \subseteq U)\}.$$
Then $\mathcal{B}'$ is countable since $\mathcal{B}$ is. For every $B \in \mathcal{B}'$ pick $U(B) \in \mathcal{U}$ such that $B \subseteq U(B)$. We claim that the countable subcollection
$$\mathcal{V} = \{U(B) : B \in \mathcal{B}'\}$$
of $\mathcal{U}$ is as required. To this end, let $x \in \bigcup \mathcal{U}$ be arbitrary. Pick $U \in \mathcal{U}$ such that $x \in U$. Since $\mathcal{B}$ is a base, there exists $B \in \mathcal{B}$ such that $x \in B \subseteq U$. So $B \in \mathcal{B}'$ from which it follows that
$$x \in B \subseteq U(B).$$
We conclude that $x$ is contained in an element of $\mathcal{V}$. □

**Exercises for §A.2.** A subset $D$ of a space $X$ is called *discrete* if every $x \in D$ has a neighborhood $U_x$ such that $U_x \cap D = \{x\}$.

If $X$ is a space then $A \subseteq X$ is called a $G_\delta$-*subset* of $X$ if there are open subsets $U_n \subseteq X$, $n \in \mathbb{N}$, such that $A = \bigcap_{n=1}^\infty U_n$. The complement of a $G_\delta$-set is called an $F_\sigma$-*subset* and is characterized by being the union of a countable collection of closed subsets of $X$.

Let $X$ be a space and let $x \in X$. A collection open subsets $\mathcal{U}$ of $X$ is called a *local base at $x$* if for open set $V$ in $X$ with $x \in V$ there is an element $U \in \mathcal{U}$ such that $x \in U \subseteq V$.

1. Let $X$ be a space, and suppose that $A \subseteq X$ is not closed. Prove that there is a sequence $(a_n)_n$ of points in $A$ such that $x = \lim_{n \to \infty} a_n$ exists and does not belong to $A$.

2. For each $i \in \mathbb{N}$ let $X_i$ be a space. Show that a sequence $\bigl(x(n)\bigr)_n$ in the product $\prod_{k=1}^{\infty} X_k$ converges to a point $x \in \prod_{k=1}^{\infty} X_k$ if and only if the sequence $(x(n)_i)_n$ converges to $x_i$ for every $i \in \mathbb{N}$. (We say that $\prod_{k=1}^{\infty} X_k$ is endowed with the topology of *coordinatewise convergence*.)

3. Let $X$ be a space. Prove that every closed subspace of $X$ is a $G_\delta$-subset of $X$. Conclude that every open subspace of $X$ is an $F_\sigma$-subset of $X$.

4. Let $X$ be a space. Prove that the intersection of a countable family of $G_\delta$-subsets of $X$ is again a $G_\delta$-subset of $X$. Conclude that the union of a countable family of $F_\sigma$-subsets of $X$ is an $F_\sigma$-subset of $X$.

5. Let $X$ be a space with $G_\delta$-subset $A$. Prove that if $B$ is a $G_\delta$-subset of $A$ then $B$ is a $G_\delta$-subset of $X$. Conclude that a $G_\delta$-subset of a closed subset of $X$ is a $G_\delta$-subset of $X$.

6. Let $X$ be a space. Prove that if $D \subseteq X$ is discrete then $D$ is a $G_\delta$-subset of $X$.

7. Let $X$ be a space and let $\mathcal{K}$ be a chain of closed subsets of $X$. Prove that there is a countable subfamily $\mathcal{K}'$ of $\mathcal{K}$ such that $\bigcup \mathcal{K}' = \bigcup \mathcal{K}$.

8. Prove that a space $X$ is locally connected if and only if for every non-empty open subset $U \subseteq X$ every component $C$ of $U$ is open.

9. Let $A$ be a closed subset of a locally connected space $X$. Prove that if $U$ is a component of $X \setminus A$ then $U \cup A$ is closed.

10. Let $X$ be locally compact and locally connected. Prove that $X$ has a base consisting of connected open sets with compact closures.

11. Let $f \colon X \to Y$ be a continuous closed surjection. Prove that if $X$ is locally connected then so is $Y$.

12. Let $\mathcal{U}$ be a collection of open subsets of a space $X$ which is a local base at every point of a subset $A \subseteq X$. Prove that there is a countable subcollection of $\mathcal{U}$ having the same property. Conclude that if $\mathcal{U}$ is a base for a space $X$ then there is a countable subcollection of $\mathcal{U}$ which is also a base for $X$.

▶13. Prove that each uncountable space can be written as the union of a countable subspace and a dense-in-itself subspace (this is the so-called *Cantor-Bendixson Theorem*).

14. Let $X$ be a space, and let $A \subseteq X$ be uncountable. Prove that if $<$ is any linear order on $A$ then there is an element $x \in A$ such that
$$x \in \overline{\{a \in A : a < x\}} \cap \overline{\{a \in A : x < a\}}.$$

15. Let $X$ be a space and let $E \subseteq X$. For every $n$ let $\mathcal{U}_n$ be a collection of open subsets of $X$ such that $E \subseteq \bigcup \mathcal{U}_n$ and $\operatorname{mesh}(\mathcal{U}_n) < 1/n$. Prove that the collection $\bigcup_{n=1}^{\infty} \mathcal{U}_n$ is a local base at every point of $E$.

16. Let $X$ be a space. Prove that $X$ has at most $\mathfrak{c}$ closed subsets. (Hence $X$ has cardinality at most $\mathfrak{c}$ and $X$ has at most $\mathfrak{c}$ open subsets.)

17. Let $X$ and $Y$ be spaces. Prove that the collection of all continuous functions $X \to Y$ has cardinality at most $\mathfrak{c}$.

18. Let $X$ be a space with closed subspace $K$. Let $\varepsilon \colon X \setminus K \to (0, \infty)$ be continuous. Prove that there is a continuous function $\delta \colon X \to [0, \infty)$ such that
   (1) $\delta \restriction K \equiv 0$,
   (2) if $x \in X \setminus K$ then $0 < \delta(x) \leq \varepsilon(x)$.

## A.3. Limits of continuous functions

Let $X$ and $(Y, \varrho)$ be spaces. For all $f, g \in C(X, Y)$ put
$$\hat\varrho(f, g) = \sup\{\varrho(f(x), g(x)) : x \in X\}.$$
Observe that $\hat\varrho(f, g) \in [0, \infty]$.

If $X = Y = \mathbb{N}$ with the Euclidean metric $\varrho$, $f$ is the identity and $g$ is the function $n \mapsto 2n$, then clearly $\hat\varrho(f, g) = \infty$.

As a consequence, $\hat\varrho$ need not be a metric on $C(X, Y)$, it is an 'extended' metric. Yet it is convenient to adopt some of the terminology of metrics and to treat $\hat\varrho$ as some sort of generalized metric. For example, we call a sequence $(f_n)_n$ in $C(X, Y)$ *Cauchy* if for each $\varepsilon > 0$ there exists an $N \in \mathbb{N}$ such that $\hat\varrho(f_n, f_m) < \varepsilon$ for all $n, m \geq N$. This is of course nothing but the ordinary definition of a Cauchy sequence in a metric space.

**Lemma A.3.1.** *Let $X$ and $(Y, \varrho)$ be spaces. Let $(f_n)_n$ be a $\hat\varrho$-Cauchy sequence in $C(X, Y)$ such that for every $x \in X$, $\lim_{n \to \infty} f_n(x)$ exists. Then the function $f \colon X \to Y$ defined by $f(x) = \lim_{n \to \infty} f_n(x)$ is continuous.*

**Proof.** Let $x \in X$. We shall prove that $f$ is continuous at $x$. To this end, let $\varepsilon > 0$ be arbitrary. There exists by assumption an $N \in \mathbb{N}$ such that $\hat\varrho(f_n, f_m) < \varepsilon/3$ for all $n, m \geq N$.

**Claim 1.** For every $y \in X$ and $m \geq N$, $\varrho(f(y), f_m(y)) \leq \varepsilon/3$.

*Proof.* Let $y \in X$. Since $\varrho(f_n(y), f_m(y)) < \varepsilon/3$ for all $n, m \geq N$ and since $f(y)$ is equal to $\lim_{n \to \infty} f_n(y)$, we obtain that indeed $\varrho(f(y), f_m(y)) \leq \varepsilon/3$ for every $m \geq N$. ◇

So the sequence $(f_n)_n$ converges uniformly to $f$ on $X$. Fix an admissible metric $d$ on $X$. Since $f_N$ is continuous at $x$, there exists $\delta > 0$ such that if $d(x,z) < \delta$ then $\varrho(f_N(x), f_N(z)) < \varepsilon/3$. Now for an arbitrary $z \in X$ with $d(x,z) < \delta$ we have

$$\varrho(f(x), f(z)) \leq \varrho(f(x), f_N(x)) + \varrho(f_N(x), f_N(z)) + \varrho(f_N(z), f(z))$$
$$< \varepsilon/3 + \varepsilon/3 + \varepsilon/3$$
$$= \varepsilon.$$

Since $\varepsilon$ was arbitrary, we conclude that $f$ is continuous at $x$. □

**Exercises for §A.3.**

1. Let $f_n \colon X \to \mathbb{R}$ be continuous for every $n$. Let $(M_n)_n$ be a sequence of real numbers such that for every $x \in X$ and $n \in \mathbb{N}$, $|f_n(x)| < M_n$. Prove that if $\sum_{n=1}^{\infty} M_n$ is convergent then $\sum_{n=1}^{\infty} f_n$ is uniformly convergent (and hence continuous by Lemma A.3.1).

2. Give an example of a sequence $(f_n)_n \in C(\mathbb{I}, \mathbb{I})$ such that for every $x \in \mathbb{I}$ the limit $f(x) = \lim_{n \to \infty} f_n(x)$ exists while the function $f$ is not continuous. (This shows that the condition that the sequence $(f_n)_n$ in Lemma A.3.1 is Cauchy cannot be omitted.)

## A.4. Normality type properties

Real-valued continuous functions are very important in topology. The following lemma shows that disjoint closed subsets can always be separated by such a function.

**Lemma A.4.1.** *Let $X$ be a space. If $A$ and $B$ are pairwise disjoint nonempty closed subsets of $X$, then there exists a continuous function $u \colon X \to \mathbb{I}$ such that $u^{-1}(0) = A$ and $u^{-1}(1) = B$.*

**Proof.** Let $\varrho$ be an admissible metric for $X$. It is easily seen that the function $\alpha \colon X \to \mathbb{I}$ defined by

$$u(x) = \frac{\varrho(x, A)}{\varrho(x, A) + \varrho(x, B)}$$

is as required. □

Let $X$ be a topological space, and let $A$ and $B$ be disjoint closed subsets of $X$. A continuous function $f \colon X \to \mathbb{R}$ for which there are distinct elements $a, b \in \mathbb{R}$ such that $f[A] \subseteq \{a\}$ and $f[B] \subseteq \{b\}$, is called a *Urysohn function*. So the function $\alpha$ in Lemma A.4.1 is a 'special' Urysohn function since it has the extra property that it 'separates' $A$ from $B$ in a precise way.

We will need a second type of Urysohn functions in the following situation later. Suppose that $A$ and $B$ are closed subsets of a space $X$. In addition, let $A' \subseteq A \setminus B$ and $B' \subseteq B \setminus A$ be closed sets which are nonempty. Let $\varrho$ be an admissible metric for $X$. Define the sets $A''$ and $B''$ in $X$ by

$$A'' = \{x \in X : \varrho(x, A) \leq \varrho(x, B)\},$$
$$B'' = \{x \in X : \varrho(x, A) \geq \varrho(x, B)\}.$$

Observe that these sets are closed in $X$, and that

$$A'' \cap (A \cup B) = A, \quad B'' \cap (A \cup B) = B, \quad A'' \cup B'' = X.$$

By the above there are Urysohn functions $\alpha, \beta \colon X \to \mathbb{I}$ such that

$$\alpha^{-1}(0) = A', \quad \alpha^{-1}(1) = B'', \quad \beta^{-1}(0) = B', \quad \beta^{-1}(1) = A''.$$

Define $u \colon X \to \mathbb{I}$ by the following formula:

$$u(x) = \begin{cases} \tfrac{1}{2}\alpha(x) & (x \in A''), \\ 1 - \tfrac{1}{2}\beta(x) & (x \in B''). \end{cases}$$

If $x \in A'' \cap B''$ then $\tfrac{1}{2}\alpha(x) = \tfrac{1}{2}$ and $1 - \tfrac{1}{2}\beta(x) = \tfrac{1}{2}$. Since $A''$ and $B''$ are closed in $X$, this easily implies that $u$ is well defined and continuous. So it is a Urysohn function. It is special since

$$u^{-1}(0) = A', \quad u^{-1}(1) = B',$$

and

$$u[A \setminus B] \subseteq [0, \tfrac{1}{2}), \quad u[B \setminus A] \subseteq (\tfrac{1}{2}, 1].$$

This can be verified quite easily. So we completed the proof of the following result:

**Lemma A.4.2.** *Let $X$ be a space containing the closed sets $A$ and $B$. If*

$$A' \subseteq A \setminus B, \quad B' \subseteq B \setminus A$$

*are nonempty closed sets then there is a Urysohn function $u \colon X \to \mathbb{I}$ such that*

(1) $u^{-1}(0) = A'$ and $u^{-1}(1) = B'$,
(2) $u[A \setminus B] \subseteq [0, \tfrac{1}{2})$ and $u[B \setminus A] \subseteq (\tfrac{1}{2}, 1]$.

We will now present some simple consequences.

**Corollary A.4.3.** *Let $X$ be a space and let $A$ and $B$ be disjoint closed subsets of $X$. Then there are disjoint open subsets $U$ and $V$ in $X$ such that*

$$A \subseteq U, \quad B \subseteq V.$$

**Proof.** By Lemma A.4.1 there exists a continuous function $u \colon X \to \mathbb{I}$ such that $u \restriction A \equiv 0$, and $u \restriction B \equiv 1$. Put $U = u^{-1}\bigl[[0, \tfrac{1}{2})\bigr]$ and $V = u^{-1}\bigl[(\tfrac{1}{2}, 1]\bigr]$, respectively. Then $U$ and $V$ are clearly as required. □

**Corollary A.4.4.** *Let $X$ be a space. Then $X$ can be imbedded in $Q$ and in $\mathbb{R}^\infty$.*

**Proof.** Let $\mathcal{B}$ be a countable open base for $X$. For every pair $(E, F)$ of elements of $\mathcal{B}$ with $\overline{E} \subseteq F$ pick a Urysohn function $u \colon X \to \mathbb{I}$ such that
$$u[\overline{E}] \subseteq \{0\}, \quad u[X \setminus F] \subseteq \{1\}$$
(Lemma A.4.1). Let $\mathcal{U}$ denote the countable collection of functions obtained in this way. The function $e \colon X \to \mathbb{I}^{\mathcal{U}}$ defined by
$$e(x) = (u(x))_{u \in \mathcal{U}}$$
is easily seen to be an imbedding (Exercise A.4.1).

Since $Q$ clearly imbeds in $\mathbb{R}^\infty$, we are done. $\square$

We will now generalize Lemma A.4.1 in an interesting way.

**Lemma A.4.5.** *Let $X$ be a space with closed subspace $A$, and let $g \colon A \to \mathbb{R}$ be continuous such that $|g(a)| \leq K < \infty$ for every $a \in A$. Then there is a continuous function $h \colon X \to \mathbb{R}$ such that*

$$\begin{aligned}
|h(x)| &\leq \tfrac{1}{3}K && \text{for all } x \in X, \\
|g(a) - h(a)| &\leq \tfrac{2}{3}K && \text{for all } a \in A.
\end{aligned}$$

**Proof.** Let $B_1 = g^{-1}[-K, -\tfrac{1}{3}K]$, $B_2 = g^{-1}[\tfrac{1}{3}K, K]$. Then $B_1$ and $B_2$ are closed and disjoint in $A$ and so in $X$. By Lemma A.4.1 there exists a continuous function $\bar{h} \colon X \to \mathbb{I}$ such that $\bar{h}[B_1] \subseteq \{0\}$ and $\bar{h}[B_2] \subseteq \{1\}$. Now let $h = \tfrac{2}{3}K(\bar{h} - \tfrac{1}{2})$. We claim that $h$ is as required. Indeed, if $x \in X$ is arbitrary then
$$|h(x)| = \tfrac{2}{3}K \left| \bar{h}(x) - \tfrac{1}{2} \right| \leq \tfrac{2}{3}K \tfrac{1}{2} = \tfrac{1}{3}K.$$
Moreover, if $a \in A$ then there are three cases to consider. First assume that $a \in B_1$. Then $\bar{h}(a) = 0$ from which it follows that $h(a) = -\tfrac{1}{3}K$. Since $g(a) \in [-K, -\tfrac{1}{3}K]$ this shows that $|g(a) - h(a)| \leq \tfrac{2}{3}K$. A similar calculation can be made if $a \in B_2$. Finally, if $a \notin B_1 \cup B_2$ then $|g(a)| \leq \tfrac{1}{3}K$, and since by the above $|h(a)| \leq \tfrac{1}{3}K$, we get $|g(a) - h(a)| \leq \tfrac{2}{3}K$ as well. $\square$

This result enables us to prove that certain continuous functions can be extended.

**Theorem A.4.6.** *Let $X$ be a space with closed subspace $A$. If $f \colon A \to \mathbb{J}$ is continuous then there is a continuous function $F \colon X \to \mathbb{J}$ such that $F \restriction A = f$.*

**Remark A.4.7.** For nontrivial generalizations of this result, see §1.2.

**Proof.** By induction on $n \geq 0$, we define functions $h_n \colon X \to \mathbb{R}$ such that

$$\begin{aligned}
|h_n(x)| &\leq \tfrac{1}{3} \cdot (\tfrac{2}{3})^n && \text{for all } x \in X, \\
\left| f(a) - \sum_{i=0}^{n} h_i(a) \right| &\leq (\tfrac{2}{3})^{n+1} && \text{for all } a \in A.
\end{aligned}$$

We get the function $h_0$ by applying Lemma A.4.5 with $g = f$ and $K = 1$. If $h_i$ has been properly defined for $i \leq n$, then $h_{n+1}$ is obtained by applying Lemma A.4.5 with $g = f - \sum_{i=0}^{n} h_i$ and $K = (2/3)^{n+1}$. We claim that

$$F = \sum_{i=0}^{\infty} h_i$$

is as required. Indeed, by Exercise A.3.1, the series $\sum_{i=0}^{\infty} h_i$ is uniformly convergent, and so its sum $F \colon X \to \mathbb{R}$ is continuous. Also, observe that

$$|F(x)| \leq 1/3 \sum_{n=0}^{\infty} (2/3)^n = 1$$

for every $x \in X$. Finally, by construction it follows that $F \upharpoonright A = f$. □

Let $X$ be a space. A *compactification* of $X$ is a pair $(K, i)$, where $K$ is a compact space and $i \colon X \to K$ is a dense imbedding. It will be convenient to identify $X$ and $i[X]$. Compactifications are important because they allow us to study spaces as dense subspaces of compact spaces. This sometimes simplifies life considerably.

If $aX$ and $bX$ are compactifications of $X$ then we say that $aX \leq bX$ provided that there is a continuous function $f \colon bX \to aX$ which extends the identity on $X$. If $aX \leq bX$ and $bX \leq aX$ then we call $aX$ and $bX$ *equivalent* compactifications. This terminology is justified by the following argumentation. Let $f \colon bX \to aX$ witness the fact that $aX \leq bX$. Similarly for $g \colon aX \to bX$. Then

$$f \circ g \colon aX \to aX$$

extends the identity on $X$ and hence has to be the identity on all of $aX$ since by Exercise A.1.9 continuous functions agree on a closed set (here we use that $X$ is dense in $aX$). It follows similarly that $g \circ f \colon bX \to bX$ is the identity. We conclude that $f$ is a homeomorphism that leaves $X$ pointwise fixed; its inverse is $g$.

**Corollary A.4.8.** *Every space has a compactification.*

**Proof.** Let $X$ be a space. By Corollary A.4.4 we may assume that $X$ is a subspace of the compact space $Q$. So the closure of $X$ in $Q$ is a compactification of $X$. □

If $X$ is a noncompact locally compact space then it can be compactified through the addition of a single point. This is the so-called *Alexandroff (one-point) compactification* $\alpha X$ of $X$. The unique point in $\alpha X \setminus X$ is usually denoted by $\infty$. Observe that $X$ is necessarily open in $\alpha X$ (Exercise A.4.3) so that a basic neighborhood of a point $x$ in $\alpha X$ that belongs to $X$ has the form $U$ with $U$ an arbitrary open neighborhood of $x$ in $X$. In addition, a basic neighborhood of $\infty$ in $\alpha X$ has the form $\alpha X \setminus K$, where $K \subseteq X$ is an arbitrary compact set.

**Exercises for §A.4.**

1. Prove that the function $e \colon X \to \mathbb{I}^{\mathcal{U}}$ in the proof of Corollary A.4.4 is an imbedding.

2. Let $aX$ and $bX$ be compactifications of $X$. Assume that there is a continuous function $f \colon aX \to bX$ such that $f$ restricts to the identity on $X$. Prove that $f[aX \setminus X] = bX \setminus X$.

3. Let $X$ be locally compact, and let $aX$ be a compactification of $X$. Prove that $X$ is open in $aX$.

▶4. Let $X$ be a space with subset $U$, and $f \colon X \to \mathbb{I}$ continuous. Prove that the set
$$B = \left\{\delta \in (0,1) : \overline{f^{-1}[[0,\delta))} \cap \overline{U} \neq \overline{f^{-1}[[0,\delta)) \cap U}\right\}$$
is countable.

▶5. Let $X$ be a space and let $\mathcal{U}$ be a countable collection subsets of $X$. Prove that for every closed set $A \subseteq X$ and neighborhood $V$ of $A$ there is an open neighborhood $W$ of $A$ such that
   (1) $\overline{W} \subseteq V$,
   (2) $\overline{W \cap U} = \overline{U} \cap \overline{W}$ for every $U \in \mathcal{U}$.

6. Prove that every (separable metrizable) connected space contains one point only or has cardinality precisely $\mathfrak{c}$.

7. Observe that any endface $W$ of $Q$ is homeomorphic to $Q$. As a consequence, every space can also be imbedded in any endface of $Q$.

8. Prove that for every $n$ the one-point compactification $\mathbb{R}^n \cup \{\infty\}$ of $\mathbb{R}^n$ is homeomorphic to $\mathbb{S}^n$.

9. Let $X$ be a space containing the closed sets $A$ and $B$ such that $A \cup B = X$. Prove that if $A' \subseteq A \setminus B$ and $B' \subseteq B \setminus A$ are nonempty closed sets then there is a Urysohn function $u \colon X \to \mathbb{I}$ such that
$$u^{-1}(0) = A', \quad u^{-1}(1) = B', \quad u^{-1}(\tfrac{1}{2}) = A \cap B.$$

## A.5. Compactness type properties

A space $X$ is called *countably compact* if every countable open cover of $X$ has a finite subcover. In addition, $X$ is called *sequentially compact* if every sequence in $X$ has a convergent subsequence.

**Theorem A.5.1.** *Let $X$ be a space. The following statements are equivalent:*

(1) *$X$ is countably compact.*

(2) $X$ is sequentially compact.
(3) $X$ is compact.

**Proof.** *We prove* (3) $\Rightarrow$ (2).

Let $(x_n)_n$ be a sequence in $X$ having no convergent subsequence. Then by Exercise A.2.1, every subset of $A = \{x_n : n \in \mathbb{N}\}$ is closed. As a consequence, for every $n$ there exists $\varepsilon_n > 0$ such that
$$B(x_n, \varepsilon_n) \cap (A \setminus \{x_n\}) = \emptyset.$$
Then $\mathcal{U} = \{(X \setminus A) \cup B(x_n, \varepsilon_n) : n \in \mathbb{N}\}$ is a countable open cover of $X$ having no finite subcover.

*We prove* (2) $\Rightarrow$ (1).

Let $\mathcal{U} = \{U_n : n \in \mathbb{N}\}$ be a countable open cover of $X$. If $\mathcal{U}$ has no finite subcover, then for every $n$ there exists

(∗) $$x_n \in X \setminus \bigcup_{i=1}^{n} U_i.$$

By assumption, there exists a convergent subsequence $(x_{n_k})_k$ of $(x_n)_n$, say
$$x_{n_k} \to x.$$
There exists $m$ with $x \in U_m$ and so there exists $K \in \mathbb{N}$ with $x_{n_k} \in U_m$ for $k \geq K$. Now pick $k \geq K$ such that $n_k \geq m$. Then on the one hand
$$x_{n_k} \in U_m,$$
but on the other hand
$$x_{n_k} \notin U_m$$
by (∗). So we arrive at a contradiction.

Finally (1) $\Rightarrow$ (3) is trivial since $X$ is Lindelöf. □

So in order to prove that a given space is compact, all one needs to do is to show that every sequence has a convergent subsequence.

**Corollary A.5.2.** *Let $X$ be a space with complete admissible metric $\varrho$. Suppose that for every $n$, $\mathcal{U}_n$ is a finite collection of open subsets of $X$ such that*

(1) $\mathrm{mesh}(\mathcal{U}_n) < 1/n$,
(2) *for every $V \in \mathcal{U}_{n+1}$ there exists $U \in \mathcal{U}_n$ such that $\overline{V} \subseteq U$.*

*Then*
$$K = \bigcap_{n=1}^{\infty} \bigcup \mathcal{U}_n$$
*is a compact subset of $X$.*

**Proof.** By (2) and the finiteness of the collections $\mathcal{U}_n$ it follows that $K$ is closed in $X$, whence $\varrho \restriction K \times K$ is complete. Let $(x_n)_n$ be any sequence in $K$. We shall prove that it has a convergent subsequence. Since $\mathcal{U}_1$ is finite, there are an infinite set $E_1 \subseteq \mathbb{N}$ and an element $U_1 \in \mathcal{U}$ such that
$$\{x_n : n \in E_1\} \subseteq U_1.$$
In addition, since $E_1$ is infinite, and $\mathcal{U}_2$ is finite, there are an infinite subset $E_2 \subseteq E_1$ and an element $U_2 \in \mathcal{U}_2$ such that
$$\{x_n : n \in E_2\} \subseteq U_2.$$
Proceeding in this way recursively, we get a sequence
$$E_1 \supseteq E_2 \supseteq \cdots \supseteq E_n \supseteq \cdots$$
of infinite subsets of $\mathbb{N}$ and elements $U_n \in \mathcal{U}_n$ such that for every $n$,

(∗) $$\{x_i : i \in E_n\} \subseteq U_n.$$

Pick $n_1 < n_2 < \cdots$ such that $n_i \in E_i$ for every $i$. By (1) and (∗) it clearly follows that $(x_{n_i})_i$ is Cauchy, and hence converges to a point $p$ of $X$. Since $K$ is closed, $p$ belongs to $K$. So we are done by Theorem A.5.1. □

The following simple result is quite important.

**Lemma A.5.3.** *Let $X$ be a compact subspace of a space $Y$ and let $\mathcal{U}$ be a collection of open subsets of $Y$ which covers $X$. Then there exists $\delta > 0$ with the property that every $A \subseteq Y$ with $\mathrm{diam}(A) < \delta$ and which moreover intersects $X$ is contained in an element $U \in \mathcal{U}$.*

**Proof.** Suppose, to the contrary, that such $\delta$ does not exist. Then for every $n \in \mathbb{N}$ we can find a subset $A_n$ of $Y$ such that

(1) $\mathrm{diam}(A_n) < 1/n$,
(2) $A_n$ intersects $X$, say $x_n \in A_n \cap X$,
(3) $A_n$ is not contained in any element of $\mathcal{U}$.

Since $X$ is compact, every sequence in $X$ has a convergent subsequence (Theorem A.5.1), so without loss of generality we may assume that
$$x = \lim_{n \to \infty} x_n$$
exists and belongs to $X$. There exists $U \in \mathcal{U}$ such that $x \in U$. Since $U$ is open, there exists $\varepsilon > 0$ such that $B(x, \varepsilon) \subseteq U$. In addition, there exists $N \in \mathbb{N}$ such that $x_m \in B(x, \varepsilon/2)$ for every $m \geq N$. Now choose $m \geq N$ so large that $1/m \leq \varepsilon/2$. Since the diameter of $A_m$ is less than $1/m \leq \varepsilon/2$, it follows easily that $A_m \subseteq B(x, \varepsilon) \subseteq U$, which is a contradiction. □

A number $\delta > 0$ such as the one in the above lemma is called a *Lebesgue number* for $\mathcal{U}$.

**Corollary A.5.4.** *Let $X$ be a compact subspace of a space $Y$ and let $U$ be an open neighborhood of $X$ in $Y$. Then there exists $\delta > 0$ such that $B(X, \delta) \subseteq U$.*

**Proof.** Let $\delta > 0$ be a Lebesgue number for $\{U\}$ (Lemma A.5.3). We claim that $\delta$ is as required. To this end, take an arbitrary $y \in B(X, \delta)$. There exists $x \in X$ such that $\varrho(x, y) < \delta$. As a consequence, $\{x, y\} \subseteq U$ since $\operatorname{diam}(\{x, y\}) < \delta$ and $\{x, y\} \cap X \neq \emptyset$. □

**Corollary A.5.5.** *Let $\mathcal{M}$ be a finite closed cover of a compact space $X$. Then there exists $\delta > 0$ such that if $\mathcal{N} \subseteq \mathcal{M}$ and $A \subseteq X$ satisfy*

(1) $\operatorname{diam}(A) < \delta$,
(2) $\forall N \in \mathcal{N}$: $N \cap A \neq \emptyset$,

*then $\bigcap \mathcal{N} \neq \emptyset$.*

**Proof.** Fix a subcollection $\mathcal{A} \subseteq \mathcal{M}$ with $\bigcap \mathcal{A} = \emptyset$ for a moment. Then

$$\{X \setminus A : A \in \mathcal{A}\}$$

covers $X$. Hence there there exists by Lemma A.5.3 a Lebesgue number

$$\varepsilon(\mathcal{A}) > 0$$

for that cover. This implies that if $B \subseteq X$ has diameter less than $\varepsilon(\mathcal{A})$ then it is contained in an element of the cover and hence misses the complement of that element. We conclude that if $B \subseteq X$ has diameter less than $\varepsilon(\mathcal{A})$ then for some $A \in \mathcal{A}$ we have $B \cap A = \emptyset$.

The same argument can be repeated for every subcollection of $\mathcal{M}$ with empty intersection. Put

$$\delta = \min\left\{\varepsilon(\mathcal{A}) : \mathcal{A} \subseteq \mathcal{M} \text{ and } \bigcap \mathcal{A} = \emptyset\right\}.$$

Since $\mathcal{M}$ is finite, $\delta > 0$. It is clear that $\delta$ is as required. □

We now state a result that at first glance seems rather curious. It will be used several times in the remaining part of this book.

**Lemma A.5.6.** *Let $X$ be a compact space. There is a (countable) collection*

$$\{(A_n, B_n) : n \in \mathbb{N}\}$$

*of pairs of disjoint closed subsets of $X$ such that for any pair $(E, F)$ of disjoint closed subsets in $X$ there is $n \in \mathbb{N}$ such that $E \subseteq A_n$ and $F \subseteq B_n$.*

**Proof.** Let $\mathcal{U}$ be a countable open base for $X$. The collection $\mathcal{V}$ of all finite unions of the collection

$$\overline{\mathcal{U}} = \{\overline{U} : U \in \mathcal{U}\}$$

is clearly countable. Let $E$ and $F$ be arbitrary disjoint closed subsets of $X$. Since $\mathcal{U}$ is a base and $X$ is compact, $E$ is the intersection of all elements of $\mathcal{V}$ which contain $E$. Similarly for $F$. Now if the collection
$$\mathcal{W} = \{V \in \mathcal{V} : (E \subseteq V) \vee (F \subseteq V)\}$$
has the finite intersection property then by compactness
$$\emptyset \neq \bigcap \mathcal{W} = E \cap F = \emptyset,$$
which is a contradiction. Since $\mathcal{V}$ is closed under finite intersections it therefore follows that $E$ and $F$ are contained in disjoint members from $\mathcal{V}$. □

**Exercises for §A.5.** Let $(X, \varrho)$ and $(Y, \rho)$ be spaces. A function $f \colon X \to Y$ is called *uniformly continuous* provided that for every $\varepsilon > 0$ there exists $\delta > 0$ such that for *all* $a, b \in X$ such that $\varrho(a, b) < \delta$ we have $\rho(f(x), f(y)) < \varepsilon$.

A metric $\varrho$ for a space $X$ is called *convex* provided that
$$D_\delta(D_\varepsilon(A)) = D_{\delta+\varepsilon}(A)$$
for all closed $A \subseteq X$ and $\delta, \varepsilon \geq 0$.

Let $f \colon X \to Y$ be a continuous surjection. We say that $f$ is *irreducible* if
$$f[A] \neq Y$$
for all proper closed subsets $A \subseteq X$.

If $X$ is a space then $A \subseteq X$ is called *bounded* in $X$ provided that $f[A]$ is bounded in $\mathbb{R}$ for every $f \in C(X)$.

1. Let $X$ be a compact space with admissible metric $\varrho$. Prove that $\varrho$ is complete.

2. Let $X$ be a space and let $K$ be a compact subset of $X$. Prove that there exist $x, y \in K$ such that $\varrho(x, y) = \text{diam}(K)$.

3. Let $X$ be a space and let $A$ and $B$ be disjoint closed subsets of $X$ such that $A$ is compact. Prove that $\varrho(A, B) > 0$.

4. Let $X$ and $(Y, \varrho)$ be spaces, where $X$ is compact. Suppose that for $f$ and $g$ in $C(X, Y)$ and $t \in \mathbb{R}$ we have
$$\varrho(f(x), g(x)) < t$$
for every $x \in X$. Prove that
$$\hat{\varrho}(f, g) < t.$$

5. (1) Let $X$ and $Y$ be spaces where $X$ is compact and let $f \colon X \to Y$ be continuous. Prove that $f$ is closed.
   (2) Give an example of a continuous surjection between compact spaces which is not open.
   (3) Let $X$ and $Y$ be spaces. Prove that the projection $\pi \colon X \times Y \to X$ is open.

6. Let $X, Y$ and $Z$ be compact spaces. Assume that $f\colon X \to Y$ and $g\colon Y \to Z$ are surjections such that $f$ and $g \circ f$ are continuous. Prove that $g$ is continuous as well.

7. Let $X$ and $Y$ be spaces and let $A \subseteq X$ and $B \subseteq Y$ be compact. Prove that for every neighborhood $E$ of $A \times B$ in $X \times Y$ there are open sets $U \subseteq X$ and $V \subseteq Y$ such that $A \times B \subseteq U \times V \subseteq E$.

8. Let $X$ and $K$ be spaces with $K$ compact. Prove that the projection
$$\pi\colon X \times K \to X$$
is closed.

9. Let $X$ be compact and let $f\colon X \to Y$ be one-to-one. Prove that if $f[X]$ is dense in $Y$ then $f$ is a homeomorphism from $X$ onto $Y$.

10. Prove that the Alexandroff compactification is topologically unique, i.e., prove that if $aX$ and $bX$ are compactifications of the locally compact space $X$ such that
$$|aX \setminus X| = 1 = |bX \setminus X|$$
then there is a homeomorphism $f\colon aX \to bX$ which restricts to the identity on $X$.

11. Give an example of a locally compact space $X$ and compactifications $aX$ and $bX$ of $X$, respectively, such that
$$|aX \setminus X| = 2 = |bX \setminus X|$$
while $aX \not\approx bX$.

12. Let $X$ be a space and assume that $X$ is not compact. Prove that $X$ contains an infinite closed discrete subspace.

13. Let $X$ be a non-compact space. Prove that there exists a continuous function $\lambda\colon X \to (0,1]$ such that $\inf \lambda[X] = 0$.

14. Let $X$ be a space. Prove that a closed subset of $X$ is bounded in $X$ iff it is compact.

▶15. Let $X$ be a compact space, and let $\varrho$ be a metric on $X$. Assume that $\varrho\colon X \times X \to [0, \infty)$ is continuous. Prove that $\varrho$ is an admissible metric.

▶16. For every $n$, let $X_n$ be a space. Prove that there are admissible metrics $\varrho_n$ for $X_n$ for all $n$ such that the formula
$$\varrho(x,y) = \max\{\varrho_n(x_n, y_n) : n \in \mathbb{N}\}$$
defines an admissible metric for $\prod_{n=1}^{\infty} X_n$.

17. Let $(C_n)_n$ be a decreasing sequence of compacta in a space $X$. Prove that for every neighborhood $U$ of $C = \bigcap_{n=1}^{\infty} C_n$ there exists an $n \in \mathbb{N}$ such that $C_n \subseteq U$.

18. Let $X$ and $Y$ be spaces, where $X$ is compact, and let $f\colon X \to Y$ be continuous. Prove that $f$ is uniformly continuous, no matter which admissible metrics on $X$ and $Y$ are chosen.

19. For every $n$ let $X_n$ be a space which is not compact. Prove that $\prod_{n=1}^{\infty} X_n$ is not $\sigma$-compact.

20. Prove that the standard metric $\varrho$ of $Q$ is convex.

21. Let $X$ be a compact space and let $f\colon X \to Y$ be a continuous surjection. Prove that there is a closed subspace $A \subseteq X$ such that $f \restriction A \to Y$ is irreducible.

22. Let $X$ be compact and let $f\colon X \to Y$ be irreducible. Prove that for every nonempty open subset $U \subseteq X$ there is a nonempty open subset $U^{\#} \subseteq Y$ such that $f^{-1}[U^{\#}]$ is a dense subset of $U$.

## A.6. Completeness type properties

Let $X$ be a space with admissible metric $\varrho$. As usual, we say that $\varrho$ is *totally bounded* if for every $\varepsilon > 0$ the open cover

$$\{B(x,\varepsilon) : x \in X\}$$

of $X$ has a finite subcover.

The following result improves Exercise A.5.1.

**Theorem A.6.1.** *Let $(X, \varrho)$ be a space. The following statements are equivalent:*

(1) *$\varrho$ is totally bounded and complete.*
(2) *$X$ is compact.*

**Proof.** *We prove* (1) $\Rightarrow$ (2).

Let $(x_n)_n$ be a sequence in $X$. We shall prove that it has a convergent subsequence. An appeal to Theorem A.5.1 then does the job for us. Since $\varrho$ is complete, it suffices to prove that $(x_n)_n$ has a Cauchy subsequence.

Since $\varrho$ is totally bounded, the open cover $\{B(x, 2^{-2}) : x \in X\}$ has a finite subcover. The Pigeon Hole Principle gives us an infinite subset $A_1$ of $\mathbb{N}$ such that for all $n, m \in A_1$ we have

$$\varrho(x_n, x_m) < 2^{-1}.$$

Continuing in this way recursively, we can construct a sequence of infinite subsets $A_1 \supseteq A_2 \supseteq \cdots \supseteq A_k \supseteq \cdots$ of $\mathbb{N}$ such that for all $n, m \in A_k$ we have

$$\varrho(x_n, x_m) < 2^{-k}.$$

The subsequence $(x_{n_k})_k$ of $(x_n)_n$, where $n_k = \min A_k$, is clearly Cauchy.

*We prove* (2) $\Rightarrow$ (1).

This is Exercise A.5.1. $\square$

So if one wants to verify that a given space is compact, all one needs to do is to show that it has an admissible metric $\varrho$ which is both totally bounded and complete. Sometimes this is more convenient than proving that every open cover has a finite subcover, or showing that every sequence has a convergent subsequence (Theorem A.5.1).

A space $X$ is called *topologically complete* if it admits an admissible complete metric. Observe that a closed subspace of a topologically complete space is topologically complete as well. But even more is true, we shall prove below that every $G_\delta$-subset of a topologically complete space is topologically complete.

Basic examples of topologically complete spaces are the compact spaces (Exercise A.5.1). They even have the curious property that no matter which admissible metric is chosen, it is always complete. We will generalize this in Corollary A.6.4 below by showing that *locally compact* spaces are topologically complete as well.

**Lemma A.6.2.** *If $X_n$ is topologically complete for every $n \in \mathbb{N}$, then so is the product $\prod_{n=1}^{\infty} X_n$.*

**Proof.** For every $n \in \mathbb{N}$ let $\varrho_n$ be an abmissible complete metric on $X_n$. By Exercise A.1.6 we may assume that every $\varrho_n$ is bounded by 1. Now define

$$\varrho \colon \prod_{n=1}^{\infty} X_n \times \prod_{n=1}^{\infty} X_n \to \mathbb{R}$$

as in Exercise A.1.10 by

$$\varrho(x, y) = \sum_{n=1}^{\infty} 2^{-n} \varrho_n(x_n, y_n).$$

We already know by that exercise that $\varrho$ is admissible on $\prod_{n=1}^{\infty} X_n$. That it is complete is left to the reader in Exercise A.6.8. □

**Theorem A.6.3.** *Let $X$ be a space. The following statements are equivalent:*

(1) *$X$ is topologically complete.*
(2) *$X$ can be imbedded in $\mathbb{R}^\infty$ as a closed subspace.*
(3) *$X$ is a $G_\delta$-subset of any space $Y$ containing $X$.*
(4) *$X$ is homeomorphic to a $G_\delta$-subset of a topologically complete space $Y$.*

**Proof.** To begin with, let us establish the following

**Claim 1.** *Let $S$ be a $G_\delta$-subset of a space $X$. Then $S$ can be imbedded in $X \times \mathbb{R}^\infty$ as a closed subset.*

*Proof.* There exist closed sets $G_i$, $i \in \mathbb{N}$, in $X$ such that
$$X \setminus \bigcup_{i=1}^{\infty} G_i = S.$$
For $i \in \mathbb{N}$, let $f_i \colon S \to \mathbb{R}$ be defined by
$$f_i(x) = \frac{1}{\varrho(x, G_i)}.$$
In addition, define $f \colon S \to X \times \mathbb{R}^\infty$ by
$$f(x) = (x, f_1(x), f_2(x), \dots).$$
Since the inclusion $S \hookrightarrow X$ is an imbedding, so is $f$. We shall prove that $f[S]$ is closed in $X \times \mathbb{R}^\infty$. To this end, let $(x_n)_n$ be a sequence in $S$ such that
$$f(x_n) \to (a, b) \in X \times \mathbb{R}^\infty.$$
Then $(x_n)_n$ converges to $a$. Suppose that $a \notin S$, say $a \in G_i$. Since $(x_n)_n$ converges to $a$ we have $\lim_{n \to \infty} \varrho(x_n, G_i) = 0$ from which it follows that
$$\lim_{n \to \infty} f_i(x_n) = \infty.$$
So by continuity of $f$, the $i$-th coordinate of $b$ must be equal to $\infty$. This is a contradiction and we conclude that $a \in S$. By continuity of $f$ we have
$$f(a) = \lim_{n \to \infty} f(x_n) = (a, b),$$
so $(a, b) \in f[S]$. ◇

We prove (1) ⇒ (3).

Let $Y$ be a space containing $X$. It suffices to prove that $X$ is a $G_\delta$-subset of its closure (Exercise A.2.5). So assume, for convenience, that $X$ is dense in $Y$. Let $\varrho$ be an admissible complete metric on $X$. For each $n \in \mathbb{N}$ let $\mathcal{B}_n$ be a cover of $X$ by open subsets of $X$ each of $\varrho$-diameter less than $1/n$. There exists for each $n \in \mathbb{N}$ a collection $\mathcal{E}_n$ of open subsets of $Y$ such that
$$\mathcal{E}_n \restriction X = \mathcal{B}_n.$$
Let $E_n$ be the union of the family $\mathcal{E}_n$. We claim that $X$ is equal to the intersection of the $E_n$'s and hence is indeed a $G_\delta$-subset of $Y$. To see that this is true, take an arbitrary point $y \in \bigcap_{n=1}^{\infty} E_n$. Since $X$ is dense in $Y$, there exists a sequence $(x_i)_i$ in $X$ whose limit is $y$. This sequence is $\varrho$-Cauchy. For take $\varepsilon > 0$ and let $n \in \mathbb{N}$ be so large that $1/n < \varepsilon$. There exists $E \in \mathcal{E}_n$ such that $y \in E$. Since $(x_i)_i$ converges to $y$, all but finitely many of the $x_i$ belong to $E \cap X$, which has $\varrho$-diameter less than $1/n < \varepsilon$ by construction. So the sequence $(x_i)_i$ is $\varrho$-Cauchy which implies that $y \in X$ because $\varrho$ is complete.

We prove (3) ⇒ (2).

By Corollary A.4.4 we may assume that $X$ is a subspace of $\mathbb{R}^\infty$. So (3) implies that $X$ is a $G_\delta$-subset of $\mathbb{R}^\infty$. It consequently follows from the claim that $X$ can be imbedded as a closed subspace of $\mathbb{R}^\infty \times \mathbb{R}^\infty$ which is obviously homeomorphic to $\mathbb{R}^\infty$.

*We prove* (2) $\Rightarrow$ (4).

This is trivial.

*We prove* (4) $\Rightarrow$ (1).

By the claim, $X$ is homeomorphic to a closed subspace of $Y \times \mathbb{R}^\infty$. Since the product of two topologically complete spaces is topologically complete (Lemma A.6.2), this establishes (1). □

**Corollary A.6.4.** *Every locally compact space is topologically complete.*

**Proof.** Let $X$ be locally compact. Its Alexandroff compactification $\alpha X$ is topologically complete by Exercise A.5.1. Since $\alpha X \setminus X$ consists of one point only, $X$ is open is $\alpha X$. So $X$ is topologically complete by Theorem A.6.3. □

**Remark A.6.5.** Since $\mathbb{R}$ is topologically complete, and $\mathbb{P}$ is a $G_\delta$-subset of $\mathbb{R}$, being the complement of the countable set $\mathbb{Q}$, it follows that $\mathbb{P}$ is topologically complete by Theorem A.6.3(2). So it follows, in particular, that $\mathbb{P} \times \mathbb{P}$ is topologically complete by Lemma A.6.2. This will be used in Exercise A.6.7[1].

A space $X$ is a *Baire space* if the intersection of countably many dense open subsets of $X$ is again dense. Or, equivalently, if the union of any countable family of nowhere dense subsets of $X$ has empty interior.

**Theorem A.6.6.** *Every topologically complete space is a Baire space.*

**Proof.** Let $\varrho$ be an admissible complete metric on $X$, and let $\{U_n : n \in \mathbb{N}\}$ be a family of dense open subsets of $X$, where $U_1 = X$. Let $x \in X$ and $\varepsilon > 0$. We will prove that $B(x,\varepsilon) \cap \bigcap_{n=1}^\infty U_n \neq \emptyset$. By induction on $n$ we will construct a point $y_n \in X$ and a real number $\delta_n > 0$ such that

(1) $y_1 = x$ and $\delta_1 = \varepsilon$,
(2) $B(y_{n+1}, \delta_{n+1}) \subseteq B(y_n, 1/2 \delta_n) \cap U_{n+1}$ for all $n$,
(3) $\delta_{n+1} \leq \min\{2^{-n}, 1/2 \delta_n\}$ for all $n$.

(We will not loose generality by assuming that $\varepsilon < 1$. If we do so then we do not need to require in (3) that $\delta_{n+1} \leq 2^{-n}$.) Suppose that $y_1, \ldots, y_n$ and $\delta_1, \ldots, \delta_n$ have been constructed satisfying (1) through (3). Notice that $B(y_n, 1/2 \delta_n)$ is a non-empty open set and consequently intersects the

---

[1] We will show later that $\mathbb{P}$ is homeomorphic to $\mathbb{N}^\infty$ (Corollary 1.9.9) from which these statements are trivial corollaries.

dense set $U_{n+1}$. Let $y_{n+1}$ in $U_{n+1} \cap B(y_n, 1/2\delta_n)$ be arbitrary. Since the intersection $U_{n+1} \cap B(y_n, 1/2\delta_n)$ is open, there exists $\delta > 0$ such that
$$B(y_{n+1}, \delta) \subseteq U_{n+1} \cap B(y_n, 1/2\delta_n).$$
Put $\delta_{n+1} = \min\{\delta, 2^{-n}, 1/2\delta_n\}$. Then $\delta_{n+1}$ is as required.

The sequence $(y_n)_n$ is clearly Cauchy since $\varrho(y_{n+1}, y_n) < 2^{-n}$ for every $n$. Hence it converges, say with limit $y$. If $n \in \mathbb{N}$ then $y_m \in B(y_n, 1/2\delta_n)$ for every $m \geq n$ and so
$$\varrho(y, y_n) = \lim_{m \geq n} \varrho(y_m, y_n) \leq 1/2\delta_n < \delta_n.$$
Hence $\varrho(y, x) < \varepsilon$ and $y \in \bigcap_{n=1}^{\infty} U_n$. This is clearly as required. □

**Corollary A.6.7.** $\mathbb{Q}$ *is not topologically complete and so it is not a $G_\delta$-subset of $\mathbb{R}$.*

**Proof.** It is clear that $\mathbb{Q}$ is not a Baire space. The collection
$$\{\mathbb{Q} \setminus \{q\} : q \in \mathbb{Q}\}$$
consists of countably many dense open subsets of $\mathbb{Q}$ and has empty intersection. So we are done by Theorem A.6.6. □

**Remark A.6.8.** So in some vague sense $\mathbb{Q}$ and $\mathbb{P}$ are of different 'complexity'. The field of mathematics dealing with such phenomena is called *Descriptive Set Theory*. See §A.13 for more information.

It is not true that every Baire space is topologically compete. See Exercise A.6.7 for a counterexample. Since the product of topologically complete spaces is topologically complete (Lemma A.6.2), the question therefore naturally arises whether the product of Baire spaces is a Baire space. We will answer this question in Corollary A.6.10 below.

A subset $A$ of a space $X$ is called *meager* if it is a countable union of nowhere dense sets. Clearly, a countable union of meager sets is meager and so is a subset of a meager set. Also observe that a set $A \subseteq X$ is meager if and only if it is a subset of a countable union of *closed* and nowhere dense subsets of $X$.

**Lemma A.6.9.** *Let $X$ and $Y$ be spaces and let $\mathcal{U}$ be a countable collection dense open subsets of $X \times Y$. Then there is a meager subset $E \subseteq X$ such that for every $x \in X \setminus E$ and $U \in \mathcal{U}$ we have that $U \cap (\{x\} \times Y)$ is dense in $\{x\} \times Y$.*

**Proof.** Let $\mathcal{F}$ be a countable base for $Y$ consisting of nonempty sets. For every $F \in \mathcal{F}$ and $U \in \mathcal{U}$ let
$$A(F, U) = \{x \in X : U \cap (\{x\} \times F) = \emptyset\}.$$

Observe that $U \cap (X \times F)$ is dense in $X \times F$. Since the projection
$$\pi \colon X \times F \to X$$
is open and continuous (Exercise A.5.5), it follows that $\pi[U \cap (X \times F)]$ is a dense and open subset of $X$. Since
$$A(F, U) = X \setminus \pi[U \cap (X \times F)]$$
it follows that $A(F, U)$ is a closed and nowhere dense subset of $X$.

Pick an arbitrary
$$x \in X \setminus \bigcup \{A(F, U) : F \in \mathcal{F}, U \in \mathcal{U}\}.$$
We will show that $U \cap (\{x\} \times Y)$ is dense in $\{x\} \times Y$ for every $U \in \mathcal{U}$, which is as required. Indeed, if this is not true for certain $U \in \mathcal{U}$ then there is an $F \in \mathcal{F}$ such that
$$(\{x\} \times F) \cap U = \emptyset.$$
But then $x \in A(F, U)$, which is a contradiction. $\square$

This lemma easily implies the following interesting result.

**Theorem A.6.10.** *Let $X$ and $Y$ be Baire spaces. Then $X \times Y$ is a Baire space as well.*

**Exercises for §A.6.** Let $(X, \varrho)$ be a space and let $\varepsilon > 0$. A subset $F \subseteq X$ is called an *$\varepsilon$-net* if for every $x \in X$ there exists $y \in F$ such that $\varrho(x, y) < \varepsilon$.

1. Let $X$ be a space with subset $A$. Prove that if for some $\varepsilon > 0$ the subset $F$ of $A$ is an $\varepsilon$-net for $A$ then $F$ is an $2\varepsilon$-net for $\overline{A}$.

2. Let $(X, \varrho)$ be a space. Prove that $\varrho$ is totally bounded if and only if for every $\varepsilon > 0$, $X$ admits a finite $\varepsilon$-net.

▶3. Let $X$ be a space with admissible metric $\varrho$ and subspace $A$. Prove that if $\varrho$ is totally bounded then so is its restriction to $A$.

4. Prove that every space has an admissible totally bounded metric.

5. Let $X$ be a space. Prove that $X$ has an admissible metric $\varrho$ for which there is a sequence $\{\mathcal{U}_n : n \in \mathbb{N}\}$ of finite open covers of $X$ such that $\text{mesh}(\mathcal{U}_n) < 1/n$ for every $n$.

6. For every $n$ let $X_n$ be topologically complete. Prove that the topological sum $\bigoplus_{n=1}^{\infty} X_n$ is topologically complete as well.

7. Prove that a space $X$ is a Baire space if it contains a dense topologically complete subspace. Prove that the subspace
$$Z = \{(x, y) \in \mathbb{R}^2 : (x, y \in \mathbb{Q}) \text{ or } (x, y \in \mathbb{P})\}$$
of $\mathbb{R}^2$ is an example of a Baire space which is not topologically complete.

8. Prove that the admissible metric $\varrho$ defined on $\prod_{n=1}^{\infty} X_n$ in the proof of Lemma A.6.2 is complete.

▶9. Let $(X, \varrho)$ and $Y$ be spaces with $\varrho$ complete and let $f\colon X \to Y$ be continuous. Prove that the following statements are equivalent:
   (1) $f$ is not both one-to-one and closed.
   (2) For some $\varepsilon > 0$ there exist sequences $(x_k)_k$ and $(y_k)_k$ in $X$ such that $\varrho(x_k, y_k) > \varepsilon$ for each $k$ and $\lim_{k\to\infty} f(x_k) = \lim_{k\to\infty} f(y_k)$.

10. Let $X$ be a Baire space. Prove that if $\mathcal{S}$ is a countable family of dense $G_\delta$-subsets of $X$ then $\bigcap \mathcal{S}$ is dense in $X$. (So in a Baire space any two dense $G_\delta$-subsets intersect.)

11. Let $X_n$ be a Baire space for every $n$. Prove that $\prod_{n=1}^{\infty} X_n$ is a Baire space.

▶12. Give an example of a compact space $X$ and an open continuous surjection $f\colon X \to Y$ and a dense $G_\delta$-subset $S \subseteq X$ such that $f[S]$ is not a $G_\delta$-subset of $Y$.

## A.7. A covering type property

Let $X$ be a space and let $\mathcal{A}$ and $\mathcal{B} = \{B(A) : A \in \mathcal{A}\}$ be covers of $X$ (not necessarily by open or closed sets). We say that $\mathcal{B}$ is a *shrinking* of $\mathcal{A}$ if for every $A \in \mathcal{A}$, $B(A) \subseteq A$. Observe that if $B(A_0), B(A_1) \in \mathcal{B}$ are distinct then so are $A_0$ and $A_1$ and that the converse need not hold. We call $\mathcal{B}$ an *open shrinking* if $\mathcal{B}$ consists of open subsets of $X$. A *closed shrinking* is a shrinking consisting of closed sets.

**Proposition A.7.1.** *Let $X$ be a space and let $\mathcal{U}$ be an open cover of $X$. Then $\mathcal{U}$ admits a closed shrinking.*

**Proof.** First assume that $\mathcal{U}$ is countable. Enumerate it as $\{U_n : n \in \mathbb{N}\}$. By Lemma A.4.1, for each $n$ there exists a continuous function $f_n\colon X \to \mathbb{I}$ such that $f_n^{-1}[(0,1]] = U_n$. Define $f\colon X \to \mathbb{I}$ by

$$f(x) = \sum_{n=1}^{\infty} 2^{-n} f_n(x).$$

Since

$$\sum_{n=1}^{\infty} 2^{-n} = 1,$$

we get by Exercise A.3.1 that $f$ is continuous. As $\mathcal{U}$ covers $X$, $f(x) > 0$ for every $x \in X$. Put

$$A_n = \left\{ x \in X : f_n(x) \geq \frac{f(x)}{2} \right\} \qquad (n \in \mathbb{N}).$$

By continuity of the functions $f_n$ and $f$, it follows easily that each $A_n$ is closed. Also, since $f(x) > 0$ for all $x$, $A_n \subseteq U_n$.

We claim that the collection $\{A_n : n \in \mathbb{N}\}$ covers $X$. To this end, assume that there exists $x \in X$ such that for every $n$, $x \notin A_n$. Then

$$f_n(x) < \frac{f(x)}{2},$$

for every $n$, so that

$$f(x) = \sum_{n=1}^{\infty} 2^{-n} f_n(x) \leq \frac{1}{2} \cdot \sum_{n=1}^{\infty} 2^{-n} f(x) = \frac{f(x)}{2},$$

which is a contradiction since $f(x) > 0$.

Now let $\mathcal{U}$ be an arbitrary open cover of $X$. Let $\mathcal{V}$ be a countable subcover of $\mathcal{U}$ (Corollary A.2.3). By the above, there exists a closed shrinking

$$\mathcal{W} = \{W(V) : V \in \mathcal{V}\}$$

of $\mathcal{V}$. For each $U \in \mathcal{U}$ define the subset $E(U) \subseteq X$ by

$$E(U) = \begin{cases} W(U) & (U \in \mathcal{V}), \\ \emptyset & (U \notin \mathcal{V}). \end{cases}$$

Since $\mathcal{W}$ covers $X$, the collection $\mathcal{E} = \{E(U) : U \in \mathcal{U}\}$ is clearly a closed shrinking of $\mathcal{U}$. □

Let $\mathcal{A}$ be a collection of subsets of a space $X$. We say that $\mathcal{A}$ is *locally finite* provided that for every $x \in X$ there is a neighborhood $U_x$ of $x$ such that the collection

$$\{A \in \mathcal{A} : A \cap U_x \neq \emptyset\}$$

is finite. Observe that $\mathcal{A}$ is countable. This can be seen as follows: every point in $X$ has a neighborhood meeting only finitely many elements of $\mathcal{A}$ and countably many of these neighborhoods cover $X$ by Corollary A.2.3.

**Proposition A.7.2.** *Let $X$ be a space. Then for every open cover $\mathcal{U}$ of $X$ there is an open shrinking $\mathcal{V} = \{V(U) : U \in \mathcal{U}\}$ of $\mathcal{U}$ which is locally finite.*

**Proof.** Let $X$ be a space and let $\mathcal{U}$ be an open cover of $X$. If $\mathcal{U}$ has a finite subcover, say $\mathcal{U}'$ then there is nothing to prove. Simply put $V(U) = U$ if $U \in \mathcal{U}'$ and $V(U) = \emptyset$ if $U \notin \mathcal{U}'$. So assume without loss of generality that $\mathcal{U}$ does not have a finite subcover.

By Proposition A.7.1 there is a closed shrinking $\{W(U) : U \in \mathcal{U}\}$ of $\mathcal{U}$. For each $U \in \mathcal{U}$ let $E(U)$ be an open subset of $X$ such that

$$W(U) \subseteq E(U) \subseteq \overline{E(U)} \subseteq U$$

(Corollary A.4.3). The cover $\{E(U) : U \in \mathcal{U}\}$ of $X$ has a countable subcover by Corollary A.2.3, say $\mathcal{E} = \{E(U) : U \in \mathcal{U}'\}$, where $\mathcal{U}' \subseteq \mathcal{U}$ is countable.

Observe that $\mathcal{U}'$ covers $X$ as well. Let $\{U_n : n \in \mathbb{N}\}$ be a faithful indexing of $\mathcal{U}'$ (observe that $\mathcal{U}'$ is infinite). For every $n$ put
$$V_n = U_n \setminus \bigcup_{m<n} \overline{E(U_m)}.$$
We claim that
$$\mathcal{V} = \{V_n : n \in \mathbb{N}\}$$
is a locally finite open cover of $X$. Clearly $\mathcal{V}$ consists of open sets. For each $x \in X$ let $n(x)$ be the smallest integer with $x \in U_{n(x)}$. For this number $n(x)$ we clearly have $x \in V_{n(x)}$. Consequently, $\mathcal{V} = \{V_n : n \in \mathbb{N}\}$ is a cover of $X$. We shall prove that $\mathcal{V}$ is locally finite as well. Take an arbitrary $x \in X$. Since $\mathcal{E}$ covers $X$, there is an $n$ with $x \in E(U_n)$. Clearly $E(U_n) \cap V_m = \emptyset$ for all $m > n$. Consequently, $E(U_n)$ is a neighborhood of $x$ which intersects at most $n$ members from $\mathcal{V}$.

Now for each $U \in \mathcal{U}$ define $V(U) \subseteq X$ by
$$V(U) = \begin{cases} V_n & (U = U_n), \\ \emptyset & (U \notin \mathcal{U}'). \end{cases}$$
Since $\mathcal{V} = \{V_n : n \in \mathbb{N}\}$ is a shrinking of $\mathcal{U}' = \{U_n : n \in \mathbb{N}\}$, this assignment is clearly as required. □

A space $X$ is called *paracompact* if for every open cover $\mathcal{U}$ of $X$ there exists an open refinement $\mathcal{V}$ of $\mathcal{U}$ such that $\mathcal{V}$ is locally finite.

**Corollary A.7.3.** *Every space is paracompact.*

Let $X$ be a space. A family $\mathcal{F}$ of continuous functions from $X$ to $\mathbb{I}$ is called a *partition of unity on* $X$ if for each $x \in X$ there exist a neighborhood $U_x$ of $x$ and a *finite* subset $\mathcal{F}(x)$ of $\mathcal{F}$ such that

(1) $\sum_{f \in \mathcal{F}(x)} f(y) = 1$ for each $y \in U_x$,
(2) if $f \in \mathcal{F} \setminus \mathcal{F}(x)$ and $y \in U_x$ then $f(y) = 0$.

Each partition of unity on $X$ is countable. This is so because countably many $U_x$'s cover $X$ (Corollary A.2.3) and the corresponding $\mathcal{F}(x)$'s cover $\mathcal{F}$ with the possible exception of the constant function with value 0.

If $\mathcal{F}$ is a partition of unity on $X$ then we define
$$\mathcal{U}(\mathcal{F}) = \{f^{-1}[(0,1]] : f \in \mathcal{F}\}.$$

**Lemma A.7.4.** *Let $X$ be a space and let $\mathcal{F}$ be a partition of unity on $X$. Then $\mathcal{U}(\mathcal{F})$ is a locally finite open cover of $X$.*

**Proof.** That $\mathcal{U}(\mathcal{F})$ consists of open sets is clear. Take an arbitrary $x \in X$ and let $U_x$ and $\mathcal{F}(x)$ be as in the above definition. By (1) there exists an $f \in \mathcal{F}(x)$ such that $f(x) \neq 0$. We conclude that $\mathcal{U}(\mathcal{F})$ covers $X$. That $\mathcal{U}(\mathcal{F})$ is locally finite follows immediately from (2). □

Let $\mathcal{U}$ be a locally finite open cover of a space $(X, \varrho)$. We shall associate with $\mathcal{U}$ a certain family of continuous functions which will be useful here as well as in Chapter 3, as follows. For every $U \in \mathcal{U}$ define $\kappa_U \colon X \to \mathbb{R}$ by

$$(*) \qquad \kappa_U(x) = \frac{\varrho(x, X \setminus U)}{\sum_{V \in \mathcal{U}} \varrho(x, X \setminus V)}.$$

These functions are called the $\kappa$-*functions with respect to the cover* $\mathcal{U}$. Observe that the sum in the denominator of $(*)$ contains at least one but at most finitely many non-zero terms, so that $\kappa_U$ is well-defined. Also observe that $\kappa_U(x) \geq 0$.

We next claim that each $\kappa_U$ is continuous. This is easy. Take an arbitrary $x \in X$. There is an open neighborhood $W$ of $x$ such that the set

$$\mathcal{F} = \{U \in \mathcal{U} : U \cap W \neq \emptyset\}$$

is finite. Let $\rho_U$ denote the restriction $\kappa_U \upharpoonright W$. Then for every $y \in W$ we have

$$\rho_U(y) = \frac{\varrho(y, X \setminus U)}{\sum_{V \in \mathcal{F}} \varrho(y, X \setminus V)}.$$

Since $\mathcal{F}$ is finite, $\rho_U$ is a continuous function on $W$. Since $W$ is a neighborhood of $x$ this implies that $\kappa_U$ is continuous at $x$.

Finally, observe that $\sum_{U \in \mathcal{U}} \kappa_U(x) = 1$ for every $x \in X$ and that for every $U \in \mathcal{U}$,

$$\kappa_U^{-1}\big[(0,1]\big] \subseteq U.$$

We say that a partition of unity $\mathcal{F}$ on a space $X$ is *subordinated to a cover* $\mathcal{V}$ of $X$ if the cover $\mathcal{U}(\mathcal{F})$ is a shrinking of $\mathcal{V}$, in the following sense. We demand that $\mathcal{F}$ can be listed as $\{f_V : V \in \mathcal{V}\}$ such that

$$f_V^{-1}\big[(0,1]\big] \subseteq V$$

for every $V \in \mathcal{V}$.

**Theorem A.7.5.** *Let $X$ be a space and let $\mathcal{U}$ be an open cover of $X$. Then there exists a partition of unity on $X$ which is subordinated to $\mathcal{U}$.*

**Proof.** By Proposition A.7.2, we may assume without loss of generality that $\mathcal{U}$ is locally finite. Let $\mathcal{K} = \{\kappa_U : U \in \mathcal{U}\}$ be the set of $\kappa$-functions with respect to $\mathcal{U}$. Since $\mathcal{U}$ is locally finite, it is clear that $\mathcal{K}$ is a partition of unity on $X$ which is subordinated to $\mathcal{U}$. $\square$

Partitions of unity are very useful. We will demonstrate this in §1.1 by using them for obtaining a classical result in selection theory. Here we will present another useful application.

Let $X$ be a space. A function (not necessarily continuous) $f \colon X \to \mathbb{R}$ is called *lower semi-continuous* (abbreviated **lsc**) if $f^{-1}\big[(t, \infty)\big]$ is open in $X$

for every $t \in \mathbb{R}$. Similarly, $f$ is called *upper semi-continuous* (abbreviated **usc**) if $f^{-1}\big[(-\infty, t)\big]$ is open in $X$ for every $t \in \mathbb{R}$.

**Corollary A.7.6.** *Let $X$ be a space and let $f, g\colon X \to \mathbb{R}$ be functions such that $g \leq f$ while moreover $f$ is lower semi-continuous and $g$ is upper semi-continuous. Then there exists a continuous $h\colon X \to \mathbb{R}$ such that*
$$g \leq h \leq f.$$
*In addition, if for some $x \in X$ we have $g(x) < f(x)$ then $g(x) < h(x) < f(x)$.*

**Proof.** For each $q \in \mathbb{Q}$, put
$$U_q = \{x \in X : g(x) < q\} \cap \{x \in X : f(x) > q\}.$$
Due to the lower and upper semi-continuity of $f$ and $g$, respectively, this is an open set. Let $\mathcal{U} = \{U_q : q \in \mathbb{Q}\}$ and $U = \bigcup \mathcal{U}$. It is possible of course that the function $\mathbb{Q} \to \mathcal{U}$ defined by $q \mapsto U_q$ is not one-to-one. But there is a subset $E \subseteq \mathbb{Q}$ on which it is one-to-one and surjective.

If $x \in U$ then there is an element $q \in \mathbb{Q}$ such that $x \in U_q$ and hence
$$g(x) < f(x).$$
So $g < f$ on $U$. Suppose that there exists $x \in X \setminus U$ such that $g(x) < f(x)$. Then for some $q \in \mathbb{Q}$ we have $g(x) < q < f(x)$, showing that
$$X \setminus U \ni x \in U_q \subseteq U.$$
This is a contradiction. So $f = g$ on the set $X \setminus U$.

Let $\mathcal{F} = \{\kappa_q : q \in E\}$ be a partition of unity for $U$ which is subordinated to $\{U_q : q \in E\}$ (Theorem A.7.5). Here $\kappa_q$ corresponds to $U_q$ for $q \in E$. Define $\varphi \colon U \to \mathbb{R}$ by
$$\varphi(x) = \sum_{q \in E} q \cdot \kappa_q(x).$$
Then $\varphi$ is clearly continuous.

Let $x \in U$ be given and let $\mathcal{F}(x) = \{\kappa_{q_1}, \ldots, \kappa_{q_n}\}$ be the subset of $\mathcal{F}$ consisting of those functions $\kappa_q$ for which $\kappa_q(x) \neq 0$. Then $x \in U_{q_1} \cap \cdots \cap U_{q_n}$ so that $g(x) < q_i < f(x)$ for each $i = 1, \ldots, n$. Hence
$$g(x) = g(x) \cdot \sum_{i=1}^n \kappa_{q_i}(x) < \sum_{i=1}^n q_i \kappa_{q_i}(x) = \varphi(x) < f(x) \cdot \sum_{i=1}^n \kappa_{q_i}(x) = f(x).$$
We conclude that $g < \varphi < f$ on $U$.

Define the function $h\colon X \to \mathbb{R}$ by
$$h(x) = \begin{cases} \varphi(x) & (x \in U), \\ g(x) = f(x) & (x \notin U). \end{cases}$$
We shall prove that $h$ is continuous. Observe that $h \!\restriction\! U$ and $h \!\restriction\! X \setminus U$ are continuous. Let $(x_n)_n$ be an arbitrary sequence in $U$ converging to an arbitrary

element $x \notin U$. It suffices to prove that $h(x_n) \to h(x)$. There are two cases to consider. Suppose first that for infinitely many $n$ we have $h(x_n) \leq h(x)$. Then $g(x_n) \leq g(x)$ for those $n$. Since $g$ is **usc**, $g(x_n) \to g(x)$ and so

$$h(x_n) \to g(x) = h(x).$$

The case that for infinitely many $n$ we have $h(x) \leq h(x_n)$ can be completed similarly by using the fact that $f$ is **lsc**.

So $h$ has all desired properties. □

**Exercises for §A.7.** Let $(X, \varrho)$ and $(Y, d)$ be metric spaces. Then $f \colon X \to Y$ is called *Lipschitz* provided that

$$d(f(x), f(y)) \leq \varrho(x, y)$$

for all $x, y \in X$. Observe that a Lipschitz function is continuous.

1. Let $\mathcal{F}$ be a locally finite collection of closed subsets of a space $X$. Prove that $\bigcup \mathcal{F}$ is closed in $X$.

2. Let $X$ be a space. Prove that the following statements are equivalent:
   (1) $X$ is compact.
   (2) Every open cover of $X$ has a locally finite subcover.

3. Let $\mathcal{F}$ be a discrete collection of closed subsets of a space $X$. Prove that every $F \in \mathcal{F}$ has an open neighborhood $U(F)$ such that the collection
   
   $$\{U(F) : F \in \mathcal{F}\}$$
   
   is discrete as well.

4. Let $X$ be a space. Prove that the following statements are equivalent:
   (1) $X$ is compact.
   (2) Every locally finite open cover of $X$ is finite.

▶5. Let $(X, \varrho)$ be a space and $\mathcal{U}$ a collection of open subsets of $X$. Prove that there is a Lipschitz function $\varepsilon \colon X \to \mathbb{I}$ such that $\varepsilon^{-1}[(0, 1]] = \bigcup \mathcal{U}$ while moreover for every $x \in X$ the open ball about $x$ with radius $\varepsilon(x)$ is contained in an element of $\mathcal{U}$.

6. Let $X$ be a space and let $f, g \colon X \to \mathbb{R}$ be **lsc**. For every $x \in X$ put
   
   $$h(x) = \min\{f(x), g(x)\}.$$
   
   Prove that $h \colon X \to \mathbb{R}$ is **lsc**.

## A.8. Extension type properties

Let $X$ be a space. It will be convenient to let $\rho X$ denote the family of all closed subsets of $X$. In addition, for every $x \in X$, we put $\varrho(x, \emptyset) = \infty$.

**Lemma A.8.1.** *Let $X$ be a space with subspace $Y$. The function*
$$\kappa\colon \rho Y \to \rho X$$
*defined by*
$$\kappa(A) = \{x \in X : \varrho(x, A) \le \varrho(x, Y \setminus A)\}$$
*has the following properties:*

(1) $\kappa(\emptyset) = \emptyset$, $\kappa(Y) = X$,
(2) $\kappa(A) \cap Y = A$ for every $A \in \rho Y$,
(3) if $A, B \in \rho Y$ and $A \subseteq B$ then $\kappa(A) \subseteq \kappa(B)$,
(4) if $A, B \in \rho Y$ then $\kappa(A \cup B) = \kappa(A) \cup \kappa(B)$.

**Proof.** The straightforward verification that $\kappa$ is well-defined is left to the reader. Clearly, (1), (2), and (3) hold. For (4), take $A, B \in \rho Y$. Observe that by (3), $\kappa(A) \cup \kappa(B) \subseteq \kappa(A \cup B)$. Let $x \in \kappa(A \cup B)$. Since
$$\varrho(x, A \cup B) \in \{\varrho(x, A), \varrho(x, B)\}$$
(Exercise A.1.7), without loss of generality $\varrho(x, A \cup B) = \varrho(x, A)$. We shall prove that $x \in \kappa(A)$. Since $x \in \kappa(A \cup B)$,

(5) $\qquad \varrho(x, A) = \varrho(x, A \cup B) \le \varrho\big(x, Y \setminus (A \cup B)\big),$

and trivially,

(6) $\qquad \varrho(x, A) = \varrho(x, A \cup B) \le \varrho(x, B).$

Since $Y \setminus A \subseteq \big(Y \setminus (A \cup B)\big) \cup B$, (5) and (6) and Exercise A.1.7 imply that
$$\varrho(x, A) \le \varrho\big(x, \big(Y \setminus (A \cup B)\big) \cup B\big) \le \varrho(x, Y \setminus A),$$
so $x \in \kappa(A)$. $\qquad\square$

Let $X$ be a space. Subsets $A$ and $B$ of $X$ are called *separated* if
$$\overline{A} \cap B = \emptyset = A \cap \overline{B}.$$
It is clear that if $A$ and $B$ are disjoint and are both closed, or both open, then $A$ and $B$ are separated. More interesting examples of separated sets are obtained in the following way. Let $Y$ be a subspace of $X$ and let $U$ and $V$ be disjoint subsets of $Y$ that are open *in* $Y$. Then $U$ and $V$ are separated *in* $X$. For let $U'$ be an open subset of $X$ such that $U' \cap Y = U$. Then $U' \cap V = \emptyset$, i.e., $\overline{V} \cap U = \emptyset$. It follows similarly that $\overline{U} \cap V = \emptyset$.

**Corollary A.8.2.** *Let $A$ and $B$ be separated subsets of a space $X$. Then $A$ and $B$ can be separated by disjoint open subsets of $X$.*

**Proof.** It is clear that $A$ and $B$ are closed in their union $A \cup B$. By Lemma A.8.1, there exist closed subsets $A'$ and $B'$ of $X$ such that
$$A \subseteq A', \quad B \subseteq B', \quad B \cap A' = \emptyset, \quad B' \cap A = \emptyset, \quad A' \cup B' = X.$$

So $U = X \setminus B'$ and $V = X \setminus A'$ are disjoint open neighborhoods of $A$ and $B$, respectively. □

We now turn to extendable continuous functions.

**Lemma A.8.3.** *Let $Y$ be a dense subspace of $X$, let $Z$ be compact, and let $f\colon Y \to Z$ be continuous. The following statements are equivalent:*
  (1) *$f$ can be extended to a continuous function $\bar{f}\colon X \to Z$.*
  (2) *For all closed sets $A, B \subseteq Z$ we have*
$$A \cap B = \emptyset \implies \overline{f^{-1}[A]} \cap \overline{f^{-1}[B]} = \emptyset.$$
*(Here closure means closure in $X$.)*

**Proof.** For (1) ⇒ (2), pick disjoint closed subsets $A, B \subseteq Z$. Since $\bar{f}$ extends $f$ we have
$$(*) \qquad f^{-1}[A] \subseteq \bar{f}^{-1}[A] \text{ and } f^{-1}[B] \subseteq \bar{f}^{-1}[B].$$
Observe that by continuity of $\bar{f}$, $\bar{f}^{-1}[A]$ and $\bar{f}^{-1}[B]$ are disjoint closed sets of $X$. So we get what we want from $(*)$.

The proof of (2) ⇒ (1) is more complicated. Pick an arbitrary $x \in X$ and let $\mathcal{B}(x)$ be the collection of all its open neighborhoods in $X$. Put
$$\mathcal{F}(x) = \{\overline{f[U \cap Y]} : U \in \mathcal{B}(x)\}.$$
(Here closure means closure in $Z$.) Since for a finite subcollection $\mathcal{U} \subseteq \mathcal{B}(x)$ we have $\bigcap \mathcal{U} \in \mathcal{B}(x)$ and
$$\overline{f\left[\bigcap \mathcal{U} \cap Y\right]} \subseteq \bigcap_{U \in \mathcal{U}} \overline{f[U \cap Y]},$$
the family $\mathcal{F}(x)$ has the finite intersection property (here we use that $Y$ is dense in $X$). By compactness of $Z$, $\mathcal{F}(x)$ consequently has non-empty intersection.

We shall prove that $\bigcap \mathcal{F}(x)$ consists of a single point. Let $y_1$ and $y_2$ be distinct elements of $Z$. There exist closed neighborhoods $V_1$ and $V_2$ of $y_1$ and $y_2$ respectively, such that $V_1 \cap V_2 = \emptyset$. From (2) it follows that
$$\overline{f^{-1}[V_1]} \cap \overline{f^{-1}[V_2]} = \emptyset.$$
We may assume without loss of generality that $x \notin \overline{f^{-1}[V_1]}$. Thus
$$X \setminus \overline{f^{-1}[V_1]} \in \mathcal{B}(x).$$
Since $f\big[Y \setminus f^{-1}[V_1]\big]$ misses $V_1$, its closure misses the interior of $V_1$. This means that $y_1 \notin \bigcap \mathcal{F}(x)$.

Observe that if $x \in Y$ then $f(x) \in \bigcap \mathcal{F}(x)$ so that, by what we just proved, the unique point in $\bigcap \mathcal{F}(x)$ is $f(x)$.

Assigning to $x \in X$ the unique point in $\bigcap \mathcal{F}(x)$, we define a function $\bar{f}\colon X \to Z$ which extends $f$. It remains to check that $\bar{f}$ is continuous.

Let $V$ be an open neighborhood of $\bar{f}(x)$ in the space $Z$. Then
$$\{\bar{f}(x)\} = \bigcap \mathcal{F}(x) \subseteq V,$$
and so by compactness of $Z$ there is a finite subfamily $\mathcal{U} \subseteq \mathcal{B}(x)$ such that
$$\bigcap_{U \in \mathcal{U}} \overline{f[Y \cap U]} \subseteq V.$$
It is clear that $\bigcap \mathcal{U}$ is a neighborhood of $x$ for which $\bar{f}[\bigcap \mathcal{U}] \subseteq V$. □

This result has some very interesting consequences.

**Corollary A.8.4.** *Let $Y$ be topologically complete. If $X$ is a space, $A \subseteq X$ and $f\colon A \to Y$ is continuous, then there exists a $G_\delta$-subset $S$ of $X$ containing $A$ so that $f$ can be extended to a continuous function $\bar{f}\colon S \to Y$.*

**Proof.** We may assume without loss of generality that $A$ is dense in $X$ (Exercise A.2.5).

By Corollary A.4.8 we may assume that $Y$ is a subspace of a compact space $Z$. Let $\{(A_n, B_n) : n \in \mathbb{N}\}$ be a collection of pairs of disjoint closed subsets of $Z$ such as in Lemma A.5.6. For every $n \in \mathbb{N}$ put
$$D_n = \overline{f^{-1}[A_n]} \cap \overline{f^{-1}[B_n]}.$$
Then $D_n$ is closed in $X$ and is disjoint from $A$. Now put $T = X \setminus \bigcup_{n=1}^\infty D_n$. Then $T$ is a $G_\delta$-subset of $X$ which contains $A$ as a dense subset and $f$ can be extended to a continuous function $g\colon T \to Z$ by Lemma A.8.3. Since $Y$ is complete it is a $G_\delta$-subset of $Z$ by Theorem A.6.3. So $S = T \cap g^{-1}[Y]$ is a $G_\delta$-subset of $X$ containing $A$ and $\bar{f} = g \restriction S$ is as required. □

These elementary results imply an important result on extending homeomorphisms.

**The Lavrentieff Theorem A.8.5.** *Let $X$ and $Y$ be topologically complete spaces, let $A \subseteq X$ and $B \subseteq Y$. For every homeomorphism $f\colon A \to B$ there exist $G_\delta$-subsets $S \subseteq X$ and $T \subseteq Y$, respectively, such that $A \subseteq S$ and $B \subseteq T$ while moreover $f$ can be extended to a homeomorphism $\bar{f}\colon S \to T$.*

**Proof.** We may assume without loss of generality that $\overline{A} = X$ and $\overline{B} = Y$ (Exercise A.2.5). Let $g\colon B \to A$ be the inverse of $f$. By Lemma A.8.4 there exist $G_\delta$-subsets $S_0$ of $X$ and $T_0$ of $Y$ such that $A \subseteq S_0$ and $B \subseteq T_0$ while moreover $f$ can be extended to a continuous function $F\colon S_0 \to Y$ and $g$ can be extended to a continuous function $G\colon T_0 \to X$. Put
$$S = S_0 \cap F^{-1}[T_0], \quad T = T_0 \cap G^{-1}[S_0],$$

respectively. Then $S$ and $T$ are $G_\delta$-subsets of $X$ and $Y$, respectively, such that $A \subseteq S$ and $B \subseteq T$. Observe that $G \circ F \upharpoonright S \colon S \to X$ is well-defined and is the identity on $A$. Since $A$ is dense in $S$, Exercise A.1.9 gives us that $G \circ F \upharpoonright S$ is the identity on $S$. Likewise, $F \circ G \upharpoonright T$ is the identity on $T$. □

**Exercises for §A.8.**

▶1. Construct a continuous function $f \colon \mathbb{P} \to \mathbb{I}$ such that for no $q \in \mathbb{Q}$ there exists a continuous extension $\bar{f} \colon \mathbb{P} \cup \{q\} \to \mathbb{I}$.

2. Let $f \colon \mathbb{Q} \to \mathbb{I}$ be continuous. Prove that there exists an element $p \in \mathbb{P}$ such that $f$ can be extended to a continuous function $\bar{f} \colon \mathbb{Q} \cup \{p\} \to \mathbb{I}$.

▶3. Let $X$ and $Y$ be spaces and let $f \colon X \to Y$ be a continuous surjection. In addition, let $S$ be a subset of $X$ such that $f \upharpoonright S \colon S \to Y$ is a homeomorphism. Prove that $S$ is a $G_\delta$-subset of $X$.

## A.9. Wallman compactifications

Compactifications are very important in topology. They allow to study spaces as subspaces of compact spaces in a convenient way. There is a simple method of compactifying an arbitrary space. This is the so-called *Wallman compactification method*, which we will describe here.

Let $X$ be a space. A collection $\mathcal{F}$ of closed subsets of $X$ is a *base for the closed subsets of* $X$ (abbreviated: closed base) if the collection $\{X \setminus F : F \in \mathcal{F}\}$ is a base for the open subsets of $X$. It is clear that $\mathcal{F}$ is a closed base if and only if for every $x \in X$ and closed subset $A \subseteq X$ such that $x \notin A$ there is an element $F \in \mathcal{F}$ such that $A \subseteq F$ and $x \notin F$.

A collection $\mathcal{F}$ of subsets of $X$ is called a *Wallman base* of $X$ if

(1) $\mathcal{F}$ is countable and a closed base of $X$,
(2) $\mathcal{F}$ is closed under finite intersections and finite unions,
(3) if $F_0, F_1 \in \mathcal{F}$ are disjoint then there exist $F_0', F_1' \in \mathcal{F}$ such that
$$F_0 \subseteq F_0', \quad F_1 \subseteq F_1', \quad F_0' \cup F_1' = X$$
and
$$F_0 \cap F_1' = \emptyset = F_0' \cap F_1,$$
(4) if $F \in \mathcal{F}$ and $x \in X \setminus F$ then there is an element $F' \in \mathcal{F}$ with $x \in F'$ and $F' \cap F = \emptyset$.

**Lemma A.9.1.** *Let $\mathcal{S}$ be an arbitrary countable family closed subsets of $X$. Then there is a Wallman base $\mathcal{F}$ of $X$ such that $\mathcal{S} \subseteq \mathcal{F}$.*

**Proof.** Let $\mathcal{B}$ be a countable open base of $X$, and put
$$\mathcal{S}_1 = \{X \setminus B : B \in \mathcal{B}\} \cup \mathcal{S} \cup \{\overline{B} : B \in \mathcal{B}\}.$$

Then $\mathcal{S}_1$ is a closed base of $X$ and contains $\mathcal{S}$. Now let $\mathcal{S}_2$ be the collection of all finite intersections of finite unions of elements of $\mathcal{S}_1$. Observe that $\mathcal{S}_2$ is closed under finite intersections and finite unions. Let $S_0, S_1 \in \mathcal{S}$ be disjoint, and let $f \colon X \to \mathbb{I}$ be a Urysohn function such that

$$f[S_0] \subseteq \{0\}, \quad f[S_1] \subseteq \{1\}$$

(Lemma A.4.1). Put $A = f^{-1}\big[[0, \frac{1}{2}]\big]$ and $B = f^{-1}\big[[\frac{1}{2}, 1]\big]$, respectively. Then $A$ and $B$ are obviously closed, while moreover

$$S_0 \subseteq A, \quad S_1 \subseteq B, \quad A \cup B = X,$$

and

$$S_0 \cap B = \emptyset = A \cap S_1.$$

So if we replace $\mathcal{S}_2$ by $\mathcal{S}_2 \cup \{A, B\}$ then (3) in the definition of a Wallman base is satisfied for the pair $S_0, S_1 \in \mathcal{S}_2$. So by adding countably many closed sets to $\mathcal{S}_2$, condition (3) in the definition of a Wallman base is satisfied for *all* pairs of disjoint elements of $\mathcal{S}_2$. Let $\mathcal{S}_3$ be the collection of all finite intersections of finite unions of elements of the new collection. Now continue as above with $\mathcal{S}_3$, add new sets, and take finite unions of intersections, to create $\mathcal{S}_4$. Etc. At the end of this process, the collection

$$\mathcal{F} = \bigcup_{n=1}^{\infty} \mathcal{S}_n$$

is the desired Wallman base of $X$. (Observe that (4) is trivially satisfied since $\mathcal{F}$ contains the collection $\{\overline{B} : B \in \mathcal{B}\}$.) □

Let $X$ be a space with Wallman base $\mathcal{F}$. A subcollection $\mathcal{S} \subseteq \mathcal{F}$ is called *centered* if it has the finite intersection property. In addition, $\mathcal{S} \subseteq \mathcal{F}$ is called *maximally centered* provided that $\mathcal{S}$ is centered but not properly contained in another centered subcollection of $\mathcal{F}$. A maximally centered subcollection of $\mathcal{F}$ is also sometimes referred to as an $\mathcal{F}$-*ultrafilter*. It is easy to prove the existence of $\mathcal{F}$-ultrafilters. Indeed, if $x \in X$ then put

$$i(x) = \{F \in \mathcal{F} : x \in F\}.$$

This is an $\mathcal{F}$-ultrafilter. To show this, first observe that $i(x)$ is clearly centered. To check maximality, let $F \in \mathcal{F}$ be such that $i(x) \cup \{F\}$ is centered. Our task is to prove that $F \in i(x)$. This is equivalent to proving that $x \in F$. But this is trivial. For if $x \notin F$ then there is by (4) in the definition of a Wallman base an element $S \in \mathcal{F}$ with $x \in S \subseteq X \setminus F$. But then $S \in i(x)$, which violates the fact that $i(x) \cup \{F\}$ is centered.

The following lemma shows that there are in fact 'many' $\mathcal{F}$-ultrafilters.

**Lemma A.9.2.** *Let $\mathcal{F}$ be a Wallman base of a space $X$. Then every centered subfamily $\mathcal{S} \subseteq \mathcal{F}$ is contained in an $\mathcal{F}$-ultrafilter.*

**Proof.** Enumerate $\mathcal{F}$ as $\{F_n : n \in \mathbb{N}\}$, and, recursively, define subcollections $\mathcal{S}_n$ of $\mathcal{F}$, as follows. Put $\mathcal{S}_1 = \mathcal{S}$. Now assume that $\mathcal{S}_n$ has been defined. If $\mathcal{S}_n \cup \{F_n\}$ is centered then $\mathcal{S}_{n+1} = \mathcal{S}_n \cup \{F_n\}$, otherwise $\mathcal{S}_{n+1} = \mathcal{S}_n$. So we try to 'catch' as many elements of $\mathcal{F}$ as possible. It is clear that

$$\bigcup_{n=1}^{\infty} \mathcal{S}_n$$

is a maximally centered subcollection of $\mathcal{F}$ which contains $\mathcal{S}$. □

We now study the $\mathcal{F}$-ultrafilters a little closer.

**Lemma A.9.3.** *Let $\mathcal{F}$ be a Wallman base of $X$. If $\mathcal{S}$ is an arbitrary $\mathcal{F}$-ultrafilter then*

(1) $\emptyset \notin \mathcal{F}$,
(2) $\mathcal{F}$ *is closed under finite unions and finite intersections,*
(3) *if $S \in \mathcal{S}$ and $F \in \mathcal{F}$ contains $S$ then $F \in \mathcal{S}$,*
(4) *if $F \in \mathcal{F} \setminus \mathcal{S}$ then there is an element $S \in \mathcal{S}$ with $F \cap S = \emptyset$,*
(5) *if $A, B \in \mathcal{F}$ and $A \cup B = X$ then $A \in \mathcal{F}$ of $B \in \mathcal{F}$.*

*Moreover, if $\mathcal{S}'$ is an $\mathcal{F}$-ultrafilter such that $\mathcal{S} \neq \mathcal{S}'$ then there exist $A \in \mathcal{S}$ and $B \in \mathcal{S}'$ such that $A \cap B = \emptyset$.*

**Proof.** Since $\mathcal{S}$ is centered, (1) is trivial.

If $\mathcal{G} \subseteq \mathcal{S}$ is an arbitrary finite subcollection, then

$$\mathcal{S} \cup \{\bigcap \mathcal{G}\}$$

is centered since $\mathcal{S}$ is. Since $\bigcap \mathcal{G} \in \mathcal{F}$ this shows that $\bigcap \mathcal{G} \in \mathcal{S}$. So $\mathcal{S}$ is closed under finite intersections. That $\mathcal{S}$ is also closed under finite unions, follows by a similar argument. This proves (2). Statement (3) is trivial.

For (4), observe that

$$\mathcal{S} \cup \{F\}$$

is not centered since $\mathcal{S}$ is maximal. So there is a finite subcollection $\mathcal{G} \subseteq \mathcal{S}$ such that

$$\bigcap \mathcal{G} \cap F = \emptyset.$$

But $S = \bigcap \mathcal{G} \in \mathcal{S}$ by (2). Hence $S$ is as required.

For (5), assume that $A \notin \mathcal{S}$. By (4) for some $S \in \mathcal{S}$ we have $S \cap A = \emptyset$. So $S \subseteq B$ which implies by (3) that $B \in \mathcal{S}$.

Since $\mathcal{S} \neq \mathcal{S}'$ we may without loss of generality assume that there is an element $A \in \mathcal{S} \setminus \mathcal{S}'$. There exists by (5) an element $B \in \mathcal{S}'$ such that $A \cap B = \emptyset$. So we are done. □

Let $X$ be a space with Wallman base $\mathcal{F}$. Put
$$\omega(X, \mathcal{F}) = \{\mathcal{S} \subseteq \mathcal{F} : \mathcal{S} \text{ is an } \mathcal{F}\text{-ultrafilter}\}.$$
To every $x \in X$ we associated an element $i(x) \in \omega(X, \mathcal{F})$ above. If $x, y \in X$ are distinct then there is an element $F \in \mathcal{F}$ which contains $y$ but not $x$. So $F$ witnesses the fact that $i(y) \neq i(x)$. We therefore conclude that the function
$$i \colon X \to \omega(X, \mathcal{F})$$
is injective. It will be convenient to identify $x \in X$ with $i(x) \in \omega(X, \mathcal{F})$. So we regard $X$ to be a subset of $\omega(X, \mathcal{F})$.

For every $F \in \mathcal{F}$ put
$$F^* = \{\mathcal{S} \in \omega(X, \mathcal{F}) : F \in \mathcal{S}\}.$$

**Lemma A.9.4.** *Let $X$ be a space with Wallman base $\mathcal{F}$. If $F, G \in \mathcal{F}$ then*
  (1) $F^* \cap X = F$,
  (2) *if $F \cap G = \emptyset$ then $F^* \cap G^* = \emptyset$,*
  (3) $F^* \cap G^* = (F \cap G)^*$,
  (4) $F^* \cup G^* = (F \cup G)^*$,
  (5) *if $F \cup G = X$ then $F^* \cup G^* = \omega(X, \mathcal{F})$.*

**Proof.** This is basically a reformulation of Lemma A.9.3.

The proof of (1) is trivial. Similarly for (2). In addition, (3) and (4) follow from Lemma A.9.3(2) and (3). Finally, (5) follows from (5) of that same lemma. □

So the collection
$$(*) \qquad \mathcal{F}^* = \{F^* : F \in \mathcal{F}\}$$
is closed under finite unions and finite intersections. So the collection of its complements is closed under finite intersections and finite unions as well. As is thought in every first year topology course, such a collection $\mathcal{U}$ of subsets of a set $Y$ can be used to define a topology on $Y$, as follows: a set $O \subseteq Y$ is open if and only if for every $y \in O$ there is an element $U \in \mathcal{U}$ such that
$$y \in U \subseteq O.$$
We can therefore topologize $\omega(X, \mathcal{F})$ by using the collection
$$\{\omega(X, \mathcal{F}) \setminus F^* : F \in \mathcal{F}\}$$
as a base for its topology.

So the collection $\mathcal{F}^*$ is a base for the closed subsets of $\omega(X, \mathcal{F})$. By Lemma A.9.4(1) this implies that the injective function
$$i \colon X \to \omega(X, \mathcal{F})$$
is an imbedding. We conclude that $X$ is a subspace of $\omega(X, \mathcal{F})$.

**Lemma A.9.5.** *Let $X$ be a space with Wallman base $\mathcal{F}$. Then*

(1) *$X$ is dense in $\omega(X, \mathcal{F})$,*
(2) *$\omega(X, \mathcal{F})$ is a separable metrizable space,*
(3) *$\omega(X, \mathcal{F})$ is compact.*

**Proof.** For (1), let $F \in \mathcal{F}$ be such that
$$U = \omega(X, \mathcal{F}) \setminus F^* \neq \emptyset.$$
Our task is to prove that $U \cap X \neq \emptyset$. This is trivial. For take an arbitrary element $\mathcal{S} \in U$. Then $\mathcal{S} \notin F^*$ or, equivalently, $F \notin \mathcal{S}$. So there is an element $S \in \mathcal{S}$ with $S \cap F = \emptyset$ by Lemma A.9.3(4). Then $S$ is not empty and is contained in $U \cap X$.

We claim that $X$ is Hausdorff. To this end, let $\mathcal{S}_0, \mathcal{S}_1$ be distinct elements in $\omega(X, \mathcal{F})$. There are elements $S_0 \in \mathcal{S}_0$ and $S_1 \in \mathcal{S}_1$ such that $S_0 \cap S_1 = \emptyset$. By (3) in the definition of a Wallman base there are elements $A, B \in \mathcal{F}$ such that
$$S_0 \subseteq A, \quad S_1 \subseteq B, \quad A \cup B = X,$$
and
$$B \cap S_0 = \emptyset = S_1 \cap A.$$
But then by Lemma A.9.4,
$$\omega(X, \mathcal{F}) \setminus B^*, \quad \omega(X, \mathcal{F}) \setminus A^*$$
are disjoint open neighborhoods of $\mathcal{S}_0$ and $\mathcal{S}_1$, respectively.

We next claim that $\omega(X, \mathcal{F})$ is compact. This is simple. If $\mathcal{U}$ is an open cover of $\omega(X, \mathcal{F})$ by basic open sets having no finite subcover then the collection of its complements is centered but has empty intersection. So there is a subfamily $\mathcal{H}$ of $\mathcal{F}$ such that the collection
$$\{H^* : H \in \mathcal{H}\}$$
is centered, but has empty intersection. This leads to a contradiction, as follows. Notice that the collection $\mathcal{H}$ is centered by Lemma A.9.4. So by Lemma A.9.2 it is contained in an $\mathcal{F}$-ultrafilter $\mathcal{S}$. But then
$$\mathcal{S} \in \bigcap \{H^* : H \in \mathcal{H}\},$$
which is a contradiction.

So $\omega(X, \mathcal{F})$ is a compact Hausdorff space with a countable base. Since a compact Hausdorff space is regular, it must be separable and metrizable by Urysohn's Metrization Theorem (see Page 465). $\square$

So we conclude that $\omega(X, \mathcal{F})$ is a compactification of $X$. It is called the *Wallman compactification of $X$ with respect to $X$*.

Perhaps the reader feels that Wallman compactifications are esoteric objects. But this is not true: we will show that *every* compactification of an arbitrary space is a Wallman compactification. Once one is used to the language of Wallman compactifications, they are a very convenient tool.

**Proposition A.9.6.** *Let $aX$ be a compactification of $X$. Suppose that $aX$ has Wallman base $\mathcal{F}$ such that for every nonempty $F \in \mathcal{F}$ we have $F \cap X \neq \emptyset$. Then the compactifications $aX$ and $\omega(X, \mathcal{F})$ of $X$ are equivalent.*

**Proof.** Define $\varphi \colon aX \to \omega(X, \mathcal{F})$ by
$$\varphi(p) = \{F \cap X : p \in F\}.$$
A moments reflection shows that $\varphi$ is a homeomorphism which extends the identity on $X$. □

So now that we know how Wallman compactifications are characterized, we aim at proving that all compactifications are Wallman.

**Theorem A.9.7.** *Every space $X$ has a Wallman base every nonempty element of which has nonempty interior in $X$.*

**Proof.** This will follow by using the same technique as in the proof of Lemma A.9.1 and by applying Exercise A.4.5. Indeed, let $\mathcal{B}$ be a countable base for $X$. Let $\{(B_0^n, B_1^n) : n \in \mathbb{N}\}$ list all pairs of elements $B_0, B_1 \in \mathcal{B}$ such that $\overline{B_0} \subseteq B_1$. By Exercise A.4.5 it is easy to construct for every $n$ an open set $U_n \subseteq X$ such that
$$\overline{B_0^n} \subseteq U_n \subseteq \overline{U_n} \subseteq B_1^n$$
while moreover for any finite subset $F \subseteq \mathbb{N}$ the equality $\overline{\bigcap_{i \in F} U_i} = \bigcap_{i \in F} \overline{U_i}$ holds. So if $\mathcal{V}$ is the collection of all finite unions of finite intersections of elements of $\{U_n : n \in \mathbb{N}\}$ then $\mathcal{V}$ has the following property: if $\mathcal{E} \subseteq \mathcal{V}$ is an arbitrary finite subcollection then $\overline{\bigcap \mathcal{E}} = \bigcap_{E \in \mathcal{E}} \overline{E}$. So if a finite intersection of closures of elements of $\mathcal{V}$ is nonempty then this intersection has nonempty interior.

Precisely such as in the proof of Lemma A.9.1 we can continue this process, using Exercise A.4.5 at each stage of the construction to ensure that we do not create sets with empty interior. □

**Corollary A.9.8.** *Every compactification is a Wallman compactification.*

**Proof.** Let $aX$ be a compactification of a space $X$. By Theorem A.9.7 the space $aX$ has a Walman base $\mathcal{F}$ each nonempty element has nonempty interior in $aX$. Hence each nonempty element of $\mathcal{F}$ intersects the dense set $X$. So Proposition A.9.6 now gives us that $aX$ and $\omega(X, \mathcal{F} \upharpoonright X)$ are equivalent compactifications. □

**Exercises for §A.9.**

1. Let $X$ be a space with Wallman base $\mathcal{F}$. Prove that for every $F \in \mathcal{F}$ the set $F^*$ coincides with the closure of $F$ in $\omega(X, \mathcal{F})$.

2. Let $X$ be a space with Wallman base $\mathcal{F}$. Prove that for all disjoint nonempty closed sets $A$ and $B$ in $\omega(X, \mathcal{F})$ there are disjoint $F_0, F_1 \in \mathcal{F}$ such that $A \subseteq F_0^*$ and $B \subseteq F_1^*$.

3. Let $X$ be a space and let $\mathcal{S}$ be a countable collection of closed subsets of $X$. Prove that $X$ has a compactification $\gamma X$ such that if $S_0, S_1 \in \mathcal{S}$ are disjoint then their closures in $\gamma X$ are disjoint as well.

4. Let $\mathcal{F}$ and $\mathcal{G}$ be Wallman bases of $X$. Prove that the following statements are equivalent:
   (1) $\omega(X, \mathcal{F}) \leq \omega(X, \mathcal{G})$,
   (2) for all disjoint $F_0, F_1 \in \mathcal{F}$ there exist disjoint $G_0, G_1 \in \mathcal{G}$ such that $F_0 \subseteq G_0$ and $F_1 \subseteq G_1$.

## A.10. Connectivity

If $x \in X$ then it is easy to see that the *component* of $x$ is equal to

$$\bigcup \{C \subseteq X : C \text{ connected and } x \in C\}.$$

Since if $A \subseteq X$ is connected then so is $\overline{A}$, it follows that the component of $x$ is a closed subset of $X$.

**Lemma A.10.1.** *Let $C$ be a component of the compact space $X$. Then for every neighborhood $U$ of $C$ there exists a clopen subset $E$ of $X$ such that $C \subseteq E \subseteq U$.*

**Proof.** Let $C(x)$ denote the component of the point $x \in X$. In addition, let $\mathcal{G}$ be the family of all clopen neighborhoods of $x$, and put

$$Q(x) = \bigcap \mathcal{G}.$$

**Claim 1.** *For every neighborhood $U$ of $Q(x)$ there exists a clopen subset $E$ of $X$ such that $Q(x) \subseteq E \subseteq U$.*

*Proof.* Observe that the intersection of finitely many elements of $\mathcal{G}$ again belongs to $\mathcal{G}$. Without loss of generality we assume that $U$ is open. Since

$$(X \setminus U) \cap \bigcap \mathcal{G} = \emptyset$$

it follows by compactness that $U$ contains an element of $\mathcal{G}$. ◇

It is clear that $C(x)$ is a subset of $Q(x)$, since if $A$ is any clopen subset of $X$ then either $C(x) \cap A = \emptyset$ or $C(x) \subseteq A$. We will prove that $Q(x)$ is connected. If this is true then $Q(x) \subseteq C(x)$, i.e., $C(x) = Q(x)$. So then we get what we want by Claim 1.

Striving for a contradiction, assume that $Q(x)$ is not connected. We shall derive a contradiction. We can write $Q(x)$ as $E \cup F$, where both $E$ and $F$ are nonempty closed subsets of $X$ and $E \cap F = \emptyset$. Without loss of generality, we may assume that $x \in E$. There exist disjoint open neighborhoods $U$ and $V$ of $E$ and $F$, respectively (Corollary A.4.3). Since $U \cup V$ is open and contains $Q(x)$, by Claim 1 there exists a clopen subset $G$ of $X$ such that $Q(x) \subseteq G \subseteq U \cup V$. Observe that

$$G \cap U = G \setminus V$$

is both open and closed. But this implies that $Q(x) \subseteq G \cap U$ and so $F = \emptyset$, which is a contradiction. □

**Corollary A.10.2.** *Let $X$ be a compact space every component of which has diemater less than $\varepsilon$. Then $X$ can be split into finitely many clopen sets of diameter less than $\varepsilon$.*

**Proof.** Let $C$ be an arbitrary component of $X$. By Lemma A.10.1 there is a clopen neighborhood $U_C$ of $C$ with diameter less than $\varepsilon$. Finitely many $U_C$'s cover $X$ by compactness of $X$. So we are done by Exercise A.1.3. □

Let $\mathcal{P}$ be a decomposition of $X$ into pairwise disjoint nonempty closed subsets. We say that $\mathcal{P}$ is *upper semi-continuous* provided that for every closed subset $A$ of $X$, $\bigcup \{P \in \mathcal{P} : P \cap A \neq \emptyset\}$ is closed in $X$.

**Proposition A.10.3.** *Let $X$ be compact. Then the collection of all components of $X$ is an upper semi-continuous decomposition of $X$.*

**Remark A.10.4.** The compactness of $X$ is an essential in this result. See Exercises A.10.9 and A.10.11.

**Proof.** Let $\mathcal{C}$ denote the collection of all components of $X$ and let $A$ be closed. We shall prove that if $\mathcal{P} = \{C \in \mathcal{C} : C \cap A \neq \emptyset\}$ then $\bigcup \mathcal{P}$ is closed in $X$. Take $x \notin \bigcup \mathcal{P}$. Then the component $C(x)$ of $X$ which contains $x$ misses $A$, hence by Lemma A.10.1, for some clopen $E$ in $X$,

$$C(x) \subseteq E \subseteq X \setminus A.$$

But then $E \cap \bigcup \mathcal{P} = \emptyset$ and so $x \notin \overline{\bigcup \mathcal{P}}$. □

A *continuum* is a space which is both compact and connected.

**Corollary A.10.5.** *Let $A$ be a closed subspace of a continuum $X$ such that $\emptyset \neq A \neq X$. Then every component $C$ of $A$ meets $\operatorname{Fr} A$.*

**Proof.** First observe that if $\operatorname{Fr} A = \emptyset$ then $A = X$ by connectivity of $X$. So $\operatorname{Fr} A \neq \emptyset$. Striving for a contradiction, suppose that $C$ is a component of $A$ such that $C \cap \operatorname{Fr} A = \emptyset$. Then $X \setminus \operatorname{Fr} A$ is a neighborhood of $C$. By compactness of $A$ there exists by Proposition A.10.3 and Lemma A.10.1 a relatively clopen subset $K$ of $A$ such that $C \subseteq K \subseteq X \setminus \operatorname{Fr} A$. Then $K$ is closed in $X$ since it is closed in $A$ and $A$ is closed in $X$. On the other hand, $K$ misses $\operatorname{Fr} A$ and hence is contained in the interior of $A$ in $X$ which is open in $X$. So $K$ is a relatively open subset of the interior of $A$ in $X$ and hence is open in $X$ as well. We conclude that $K$ is a proper nonempty clopen subspace of $X$ since $\operatorname{Fr} A \neq \emptyset$, which contradicts connectivity. $\square$

This result enables us to formulate and prove the following interesting result. To put it into perspective, first observe that no connected space admits a partition into finitely many but at least two nonempty closed sets. With the extra condition of compactness added to connectivity we can do a little better.

**The Sierpiński Theorem A.10.6.** *No continuum can be partitioned into countably many pairwise disjoint closed and nonempty sets.*

**Proof.** To begin the proof, we first establish the following.

**Claim 1.** *Let $X$ be a continuum which admits a partition $\mathcal{E}$ consisting of countably many closed sets of which at least two are nonempty. Then for every $E \in \mathcal{E}$ there is a continuum $C \subseteq X \setminus E$ meeting at least two distinct elements of $\mathcal{E}$.*

*Proof.* If $E = \emptyset$ then we let $C = X$. So we may assume that $E \neq \emptyset$. Pick a nonempty element $E' \in \mathcal{E} \setminus \{E\}$. There are by Lemma A.4.1 disjoint open neighborhoods $U$ and $V$ of $E$ and $E'$, respectively. Let $x$ be an arbitrary point in $E'$ and let $C$ be the component of $x$ in $\overline{V}$. Then clearly $C \cap E = \emptyset$ and $E' \cap C \neq \emptyset$. Also, $C \cap \operatorname{Fr} \overline{V} \neq \emptyset$ by Corollary A.10.5. Since $E'$ is contained in the interior of $\overline{V}$ it follows that $\operatorname{Fr} \overline{V} \cap E' = \emptyset$. As a consequence, there exists $y \in C \setminus E'$. There consequently has to be an element $E'' \in \mathcal{E} \setminus \{E'\}$ such that $C \cap E'' \neq \emptyset$. $\diamond$

The proof can now easily be completed by recursion, as follows. Assume that $X$ is a continuum admitting a partition $\{X_i : i \in \mathbb{N}\}$ consisting of closed sets of which at least two are nonempty. By the Claim there exists a decreasing sequence of continua $C_1 \supseteq C_2 \supseteq \cdots$ in $X$ such that $C_i \cap X_i = \emptyset$. To keep the recursion going, we assume that each $C_i$ meets at least two distinct $X_j$'s. Then by compactness, $\emptyset \neq \bigcap_{i=1}^{\infty} C_i \subseteq X \setminus \bigcup_{i=1}^{\infty} X_i = \emptyset$. This is a contradiction. $\square$

The following simple result is basic.

**Proposition A.10.7.** *Let $X$ be a space and let $(C_n)_n$ be a decreasing sequence of subcontinua of $X$. Then $C = \bigcap_{n=1}^{\infty} C_n$ is a continuum.*

**Proof.** To the contrary, assume that $C$ is not connected. Let $E$ and $F$ be disjoint relatively clopen nonempty subsets of $C$ such that $E \cup F = C$. By Lemma A.8.1 there exist disjoint open subsets $E'$ and $F'$ of $X$ such that
$$E' \cap C = E, \quad F' \cap C = F.$$
Then $U = E' \cup F'$ is a neighborhood of $C$ and hence by Exercise A.5.17 there exists $n \in \mathbb{N}$ such that $C_n \subseteq E' \cup F'$. But this contradicts the connectivity of $C_n$ since $C \subseteq C_n$ and $C$ meets both $E'$ and $F'$. □

Let $X$ be a space, and let $x$ and $y$ be elements of $X$. We say that $x$ and $y$ can be *joined by a path* provided that there is a continuous function $\alpha \colon \mathbb{I} \to X$ such that $\alpha(0) = x$ and $\alpha(1) = y$. A function such as $\alpha$ is called a *path* from $x$ to $y$. Observe that if $x$ and $y$ can be joined by a path then so are $y$ and $x$.

Exercise A.10.1 below implies that for a space $X$ the relation 'can be joined by a path' is an equivalence relation. An equivalence class of this relation is called a *path-component* of $X$. Observe that path-components are connected.

A space $X$ is called *path-connected* if it has one path-component only. A path-connected space is clearly connected. But the converse need not be true. Consider e.g., the well-known $\sin(1/x)$-continuum
$$\{(x, \sin(1/x)) : 0 < x \leq 1/\pi\} \cup \{(0, y) : -1 \leq y \leq 1\}$$
in the plane (Exercise A.10.10).

A space $X$ is called *locally path-connected* if every point has arbitrarily small path-connected neighborhoods.

**Lemma A.10.8.** *Let $X$ be a connected space. Suppose that every $x \in X$ has a path-connected neighborhood. Then $X$ is path-connected.*

**Proof.** Pick an arbitrary $x \in X$ and put
$$U = \{y \in X : x \text{ and } y \text{ can be joined by a path}\}.$$
Choose an arbitrary $y \in \overline{U}$ and let $V$ be a path-connected neighborhood of $y$. We claim that $V \subseteq U$. To prove this, let $w \in V$ be arbitrary. Since $y \in \overline{U}$ and $V$ is a neighborhood of $y$, there is a point $v \in V \cap U$. Since $v, w \in V$ and $V$ is path-connected, $w$ and $v$ can be joined by a path. Moreover, since $v \in U$ it follows that $v$ and $x$ can be joined by a path. By Exercise A.10.1 we get that $w$ and $x$ can be joined by a path, i.e., $w \in U$. We conclude that indeed, $W \subseteq U$.

Since $y \in \overline{U}$ was chosen arbitrarily, it follows that $\overline{U}$ is both open and closed. Since clearly $x \in U$, the connectivity of $X$ implies that $U = X$.

Now choose arbitrary $a, b \in X$. Then $a$ and $x$ can be joined by a path. Similarly, $b$ and $x$ can be joined by a path. By Exercise A.10.1 we consequently get that $a$ and $b$ can be joined by a path. □

**Corollary A.10.9.** *Every connected, locally path-connected space is path-connected.*

Let $X$ be a space and let $A$ and $B$ be two disjoint closed subsets of $X$. A *partition* between $A$ and $B$ is a closed subset $S \subseteq X$ such that $X \setminus S$ can be written as the disjoint union of open sets $U$ and $V$ with $A \subseteq U$ and $B \subseteq V$.

A typical partition between $A$ and $B$ has the form $f^{-1}(1/2)$, where $f$ is a Urysohn function with $f[A] \subseteq \{0\}$ and $f[B] \subseteq \{1\}$. So by Lemma A.4.1 any pair of disjoint closed sets has a partition.

Let $X$ be a space and let $A$ and $B$ be disjoint closed subsets of $X$. By a *continuum from $A$ to $B$* we mean a continuum $C$ in $X$ meeting $A$ as well as $B$.

Observe that if $S$ is a partition between the closed sets $A$ and $B$ then every continuum from $A$ to $B$ meets $S$.

Let $A$ and $B$ be disjoint subsets of a space $X$. A closed set $K \subseteq X$ is called a *cut* in $X$ between $A$ and $B$ if $K \cap (A \cup B) = \emptyset$ and $K \cap Y \neq \emptyset$ for every continuum $Y \subseteq X$ from $A$ to $B$. In particular, if there is no continuum in $X$ which meets both $A$ and $B$ then $\emptyset$ is a cut in $X$ between $A$ and $B$. So a partition is a cut; the converse need not be true as we will see in Exercise A.10.7.

A set $T$ in a space $X$ *separates* $X$ if $X \setminus T$ is disconnected.

A point $x$ in a space $X$ is said to be a *cut point* of $X$ if $\{x\}$ separates $X$. In Exercise A.10.4 we will show that if $x$ is a cut point of $X$ then $X \setminus \{x\}$ can be written as $U \cup V$, where $U$ and $V$ are disjoint, open and nonempty subsets of $X$.

**Exercises for §A.10.** A continuous surjection $f \colon X \to Y$ is called *monotone* provided that for every $y \in Y$ the fiber $f^{-1}(y)$ is connected.

1. Let $X$ be a space. If $x, y \in X$ and $y, z \in X$ are both joined by a path then so are $x$ and $z$.

2. Let $X$ be connected and let $\mathcal{F}$ be a finite cover of $X$ consisting of nonempty closed sets. Let $n = |\mathcal{F}|$. Prove that we can list $\mathcal{F}$ as
$$F_1, \ldots, F_n$$
such that for any $i \leq n$ with $i \geq 2$ we have
$$F_i \cap \bigcup_{j=1}^{i-1} F_j \neq \emptyset.$$

3. Let $X$ be a space with connected subset $A$. Prove that if $\{F, \ldots, F_n\}$ is a collection of closed subsets of $X$ which covers $A$ then
$$\operatorname{diam}(A) \leq \sum_{i=1}^{n} \operatorname{diam}(F_i).$$

4. Prove that if $x$ is a cut point of $X$ then $X \setminus \{x\}$ can be written as $U \cup V$, where $U$ and $V$ are disjoint, open and nonempty subsets of $X$.

5. Let $X$ be a continuum and let $U$ and $V$ be disjoint nonempty open subsets of $X$ such that $X \setminus (U \cup V)$ is a single point, say $p$. Prove that both $U \cup \{p\}$ and $V \cup \{p\}$ are continua.

▶6. Let $X$ be a nontrivial continuum. Prove that $X$ contains at least two noncut points.

7. Give an example of a continuum $X$ containing three distinct points $p, q$ and $r$ such that $r$ is a cut between $p$ and $q$ but not a partition.

8. Let $X$ be a continuum and let $\mathcal{F}$ be a countable family pairwise disjoint closed subsets of $X$. Prove that $X \setminus \bigcup \mathcal{F}$ is uncountable.

9. Construct an example of a space $X$ having a point $x$ for which the component $C(x)$ of $x$ in $X$ does not satisfy the conclusion of Lemma A.10.1.

10. Prove that the $\sin(1/x)$-continuum in the plane is not path-connected.

11. Construct an example of a space $X$ the components of which do not form an upper semi-continuous decomposition of $X$.

12. Give an example of a path-connected compact space which is not locally connected.

13. Let $f \colon X \to Y$ be a continuous surjection which is both closed and monotone. Prove that for any connected $E \subseteq Y$ we have that $f^{-1}[E]$ is connected.

14. Let $n > 1$ and let $K \subseteq \mathbb{R}^n$ be compact. Prove that $\mathbb{R}^n \setminus K$ has a component $U$ such that $\mathbb{R}^n \setminus U$ is bounded. Conclude that $U$ is the only unbounded component of $\mathbb{R}^n \setminus K$.

## A.11. The quotient topology

Let $\mathcal{P}$ be a decomposition of a space $X$ into pairwise disjoint nonempty closed sets. In addition, let $q \colon X \to \mathcal{P}$ be the function sending $x \subset X$ to the unique element of $\mathcal{P}$ containing $x$. Call a subset $\mathcal{U}$ of $\mathcal{P}$ *open* if and only if $q^{-1}[\mathcal{U}]$ is open in $X$. In this way we endow $\mathcal{P}$ with the *quotient topology* derived from $X$ and $q$. With this topology we denote $\mathcal{P}$ by $X/\mathcal{P}$ and call it $X$ *modulo* $\mathcal{P}$.

A function $f \colon X \to Y$ is called *quotient* provided that $U \subseteq X$ is open if and only if $f^{-1}[U]$ is open. Observe that a quotient map is continuous. It

is clear that if $f\colon X \to Y$ is quotient and surjective then $Y$ is canonically homeomorphic to the space $X/\mathcal{P}$, where $\mathcal{P}$ is the decomposition
$$\{f^{-1}(y) : y \in Y\}$$
of $X$.

If $X$ is a space, and $A \subseteq X$ is closed, then $\mathcal{P} = \{A\} \cup \{\{x\} : x \in X \setminus A\}$ is a partition. It will be convenient to denote $X/\mathcal{P}$ by $X/A$ in this case. So $X/A$ is the quotient space obtained from $X$ by collapsing $A$ to a single point.

**Lemma A.11.1.** *Let $\mathcal{P}$ be a decomposition of a space $X$ into pairwise disjoint closed sets. The following statements are equivalent.*

(1) *The natural quotient mapping $q\colon X \to X/\mathcal{P}$ is closed.*
(2) *$\mathcal{P}$ is upper semi-continuous.*

**Proof.** For every closed set $A \subseteq X$ we have
$$q^{-1}[q[A]] = \bigcup \{P \in \mathcal{P} : P \cap A \neq \emptyset\}.$$
From this the implication (1) $\Rightarrow$ (2) is obvious. On the other hand, if (2) holds then for every closed set $A \subseteq X$ we have that $q^{-1}[q[A]]$ is closed from which it follows that $q[A]$ is closed since we are dealing with the quotient topology. □

Due to our self-chosen convention to deal with separable metrizable spaces exclusively, we are in a very unpleasant situation now. It is rather trivial to prove that the quotient topology just defined is indeed a topology, but it is not necessarily separable and metrizable (see Exercises A.11.1 and A.11.3). We are primarily interested in decompositions of which the individual elements are compact. Fortunately, in that particular situation the decomposition spaces involved are separable and metrizable, as the following result shows.

**Theorem A.11.2.** *Let $X$ be a space and let $\mathcal{P}$ be an upper semi-continuous decomposition of $X$ consisting of compact sets. Then $X/\mathcal{P}$ is separable and metrizable.*

**Proof.** First observe that $Y = X/\mathcal{P}$ is $T_1$ since the elements of $\mathcal{P}$ are closed sets. We next claim that $Y$ is regular. Consider the natural quotient map
$$q\colon X \to Y.$$
Then $q$ is closed by Lemma A.11.1. Let $y \in Y$ and let $U$ be an open neighborhood of it. Since $q^{-1}(y)$ is closed, and $q^{-1}[U]$ is open, there is a closed neighborhood $V$ of $q^{-1}(y)$ such that $V \subseteq q^{-1}[U]$ (Corollary A.4.3). Since $q$ is closed, by Exercise A.1.15 there is a neighborhood $W$ of $y$ such that $q^{-1}[W] \subseteq V$. But then $y \in W \subseteq \overline{W} \subseteq U$ since $W \subseteq q[V] \subseteq U$ and $q[V]$ is closed.

So $Y$ is regular and $T_1$ and it consequently suffices to prove by Urysohn's Metrization Theorem mentioned on Page 465 that it has a countable base. To this end, let $\mathcal{B}$ be a countable open base for $X$ which is closed under finite unions. For every $B \in \mathcal{B}$ put

(*) $$U(B) = Y \setminus q[X \setminus B].$$

Since $q$ is closed, $U(B)$ is open.

Now if $U$ is an open neighborhood of $y$ in $Y$ then $q^{-1}[U]$ is an open neighborhood of the compact set $q^{-1}(y)$ in $X$. So there is a finite subcollection $\mathcal{E}$ of $\mathcal{B}$ such that $q^{-1}(y) \subseteq \bigcup \mathcal{E} \subseteq q^{-1}[U]$. Since $\mathcal{B}$ is closed under finite unions, clearly $\bigcup \mathcal{E} \in \mathcal{B}$. So there exists an element $B \in \mathcal{B}$ such that

$$q^{-1}(y) \subseteq B \subseteq q^{-1}[U].$$

Since $q^{-1}(y) \subseteq B$ it follows that $y \in U(B) \subseteq U$. We conclude that the collection $\{U(B) : B \in \mathcal{B}\}$ is a countable open base for $Y$. □

We shall now present a few important examples of upper semi-continuous decompositions. Our first example is the collection of all components of a compact space.

**Corollary A.11.3.** *Let $X$ be a compact space and let $\mathcal{C}$ be its collection of components. Then $X/\mathcal{C}$ is a compact metrizable space with a base consisting of open and closed sets[2].*

**Proof.** That $X/\mathcal{C}$ is compact and metrizable follows from Proposition A.10.3 and Theorem A.11.2.

Let $q \colon X \to Y = X/\mathcal{C}$ be the natural quotient map. If $y \in Y$ and $U$ is a neighborhood of $y$ in $Y$ then $q^{-1}[U]$ is a neighborhood of the component $q^{-1}(y)$ of $X$. By Proposition A.10.3 and Lemma A.10.1 there is a clopen set $W \subseteq X$ such that $q^{-1}(y) \subseteq W \subseteq q^{-1}[U]$. Since $\mathcal{C}$ consists of connected sets, it easily follows that $q^{-1}[q[W]] = W$. This shows that $q[W]$ is a clopen neighborhood of $y$ which is contained in $U$. □

Our second example are the so-called *adjunction spaces*. Let $X$ be a space with compact subspace $A$, and let $f$ be a continuous surjection from $A$ to a space $B$. We shall prove below that the decomposition

$$\mathcal{P} = \{f^{-1}(b) : b \subset B\} \cup \{\{x\} : x \subset X \setminus A\}$$

is upper semi-continuous. The space $X/\mathcal{P}$ is denoted by $X \cup_f B$ and is called '$X$ attached to $B$ by $f$', or the *adjunction space* obtained from $X$ by attaching $B$ to $X$ by $f$. Observe that $X \setminus A$ is naturally imbedded in $X \cup_f B$. So in a sense $X/\mathcal{P}$ can be seen as $X$ where $A$ is replaced by $B$ via the map $f$.

---

[2] Spaces with such a base are called *zero-dimensional*.

**Corollary A.11.4.** *Let $X$ be a space, $A \subseteq X$ be compact and $f: A \to B$ be a continuous surjection. The decomposition*
$$\mathcal{P} = \{f^{-1}(b) : b \in B\} \cup \{\{x\} : x \in X \setminus A\}$$
*of $X$ is upper semi-continuous. Hence $X \cup_f B$ is a separable metrizable space. In addition, if $\pi: X \to X \cup_f B$ is the natural quotient map then $\pi \restriction X \setminus A$ is a homeomorphism and $\pi[A]$ is homeomorphic to $B$.*

**Remark A.11.5.** This result shows that we may identify $X \setminus A$ and $\pi[X \setminus A]$ as well as $B$ and $\pi[A]$. It will sometimes be convenient to do that.

**Proof.** Let $E \subseteq X$ be closed. Then
$$\bigcup \{P \in \mathcal{P} : P \cap E \neq \emptyset\} = E \cup f^{-1}\big[f[E \cap A]\big]$$
is closed since $f[E \cap A]$ is closed in $B$ by Exercise A.5.5(1). Hence $X \cup_f B$ is separable and metrizable by Theorem A.11.2.

The second part of the corollary is a triviality. Observe that
$$\pi \restriction X \setminus A : X \setminus A \to \pi[X \setminus A]$$
is a bijective quotient map, and hence a homeomorphism (Exercise A.11.2). A moments reflection shows that there is a canonical continuous bijection from $B$ onto $\pi[B]$. So $B$ and $\pi[B]$ are homeomorphic by compactness of $B$ (Exercise A.5.9). $\square$

We shall now discuss our last important example of an upper semi-continuous decomposition.

A *null-sequence* in a space $X$ is a sequence $(A_n)_n$ of closed subsets of $X$ such that $\lim_{n \to \infty} \operatorname{diam}(A_n) = 0$.

**Corollary A.11.6.** *Let $X$ be a space, and let $(A_n)_n$ be a null-sequence consisting of pairwise disjoint nonempty compact subsets of $X$. Then the decomposition*
$$\mathcal{P} = \{A_n : n \in \mathbb{N}\} \cup \left\{\{x\} : x \notin \bigcup_{n=1}^{\infty} A_n\right\}$$
*is upper semi-continuous (and so $X/\mathcal{P}$ is separable and metrizable).*

**Proof.** Let $B \subseteq X$ be closed. We will prove that
$$C = B \cup \bigcup \{A_n : A_n \cap B \neq \emptyset, n \in \mathbb{N}\}$$
is closed. If not, then there is a sequence $(x_n)_n$ in $C$ whose limit $x$ does not belong to $C$ (Exercise A.2.1). If infinitely many terms of the sequence belong to $B$ then $x \in B$ since $B$ is closed. If some $A_n$ that meets $B$ contains

infinitely many terms of the sequence then $x \in A_n$ since $A_n$ is closed. So we may assume without loss of generality that there is a sequence of integers
$$m(1) < m(2) < \cdots$$
such that $x_n \in A_{m(n)}$ for every $n$. But since
$$\lim_{n \to \infty} \operatorname{diam}(A_{m(n)}) = 0$$
and $A_{m(n)} \cap B \neq \emptyset$ for every $n$, this implies that $x \in B$. □

**Exercises for §A.11.**

1. Let $\mathcal{P}$ be a partition of a space into nonempty closed sets. Prove that the quotient topology on $X/\mathcal{P}$ is indeed a topology.

2. Let $f \colon X \to Y$ be a bijective quotient map. Prove that $f$ is a homeomorphism.

▶3. Prove that the quotient topology is not necessarily metrizable, even if $X$ is a separable metrizable topological space. Demonstrate this by proving that $\mathbb{R}/\mathbb{N}$ is not first countable at the point $\{\mathbb{N}\}$.

4. Let $X$ be a space, and let $(A_n)_n$ be a null-sequence consisting of pairwise disjoint nonempty compact subsets of $X$. In addition, for every $n$ let $f_n \colon A_n \to B_n$ be a continuous surjection. Prove that the decomposition
$$\mathcal{P} = \{f_n^{-1}(y) : y \in A_n, n \in \mathbb{N}\} \cup \left\{\{x\} : x \notin \bigcup_{n=1}^{\infty} A_n\right\}$$
is upper semi-continuous (and so $X/\mathcal{P}$ is separable and metrizable).

▶5. Let $X$ be a compact space with disjoint closed subsets $A$ and $B$. In addition, let $S$ be a closed subset of $X$ such that every component of $S$ misses $A$ or $B$. Prove that $S$ can be covered by a finite disjoint collection $\mathcal{N}$ of closed sets such that every $N \in \mathcal{N}$ misses $A$ or $B$.

6. Let $X$ and $Y$ be compact spaces, and let $f \colon X \to Y$ be a continuous surjection. Prove that $f$ is quotient.

7. Let $X$ and $Y$ be compact spaces, and let $f \colon X \to Y$ be a continuous surjection. In addition, let $A \subseteq X$ and $C \subseteq Y$ be closed sets such that $f[A] = C$. Finally, let $B$ be a compact space for which there exist continuous surjections
$$\xi \colon A \to B, \quad \eta \colon B \to C$$
such that
$$f \restriction A = \eta \circ \xi.$$
Prove that if $\pi \colon X \to X \cup_\xi C$ is the natural quotient map, then there is a unique continuous function $\gamma \colon X \cup_\xi B \to Y$ such that
$$f = \gamma \circ \pi.$$
Observe that $\gamma \restriction B = \eta$.

▶8. Let $X$ be a space with closed subspace $F$, and let $\gamma X$ be a compactification of $X$. Prove that if $aF$ is a compactification of $F$ such that $aF \geq \overline{F}$ (here $\overline{F}$ is the closure of $F$ in $\gamma X$) then there is a compactification $bX$ of $X$ with $bX \geq \gamma X$ while moreover the closure of $F$ in $bX$ is $aF$.

## A.12. Homotopies

Let $X$ and $Y$ be spaces. A *homotopy* from $X$ to $Y$ is a continuous function $H\colon X \times \mathbb{I} \to Y$. If $t \in \mathbb{I}$ then the function $H_t\colon X \to Y$ defined by

$$H_t(x) = H(x,t)$$

is called the *t-level* of $H$. One should think of a homotopy $H$ as a continuous family of functions connecting $H_0$ and $H_1$.

A homotopy from $X$ to $X$ is also called a *deformation* of $X$.

Two continuous functions $f, g\colon X \to Y$ are called *homotopic*, in symbols

$$f \simeq g,$$

if there is a homotopy $H\colon X \times \mathbb{I} \to Y$ such that $H_0 = f$ and $H_1 = g$. We also say that $f$ is *homotopic* to $g$. A continuous function that is homotopic to a constant function is said to be *nullhomotopic*

**Lemma A.12.1.** *Let $X$ and $Y$ be spaces and let $f, g, h\colon X \to Y$ be continuous. Then*

(1) $f \simeq f$,
(2) *if $f \simeq g$ then $g \simeq f$,*
(3) *if $f \simeq g$ and $g \simeq h$ then $f \simeq h$.*

**Proof.** For (1), define $H\colon X \times \mathbb{I} \to X$ by $H(x,t) = f(x)$.

For (2), let $H\colon X \times \mathbb{I} \to Y$ be a homotopy such that $H_0 = f$ and $H_1 = g$. Define $F\colon X \times \mathbb{I} \to Y$ by

$$F(x,t) = H(x, 1-t).$$

Then $F$ is a homotopy with $F_0 = g$ and $F_1 = f$.

For (3), let $H, F\colon X \times \mathbb{I} \to Y$ be homotopies with

$$H_0 = f, \quad H_1 = g, \quad F_0 = g, \quad F_1 = h.$$

Define $S\colon X \times \mathbb{I} \to Y$ by

$$S(x,t) = \begin{cases} H(x, 2t) & (0 \leq t \leq 1/2), \\ F(x, 2t-1) & (1/2 \leq t \leq 1). \end{cases}$$

Then $S$ is a homotopy with $S_0 = f$ and $S_1 = h$. □

From the above lemma we conclude that the homotopy relation '$\simeq$' is an equivalence relation in $C(X,Y)$. If $X$ and $Y$ are spaces and if $f\colon X \to Y$ is continuous then $[f]$ denotes its $\simeq$-equivalence class and $[X,Y]$ denotes
$$\{[f] : f \in C(X,Y)\}.$$
We say that $[X,Y]$ is *trivial* if it contains one point only (i.e., all continuous functions from $X$ to $Y$ are homotopic and hence nullhomotopic).

As we will see, homotopic functions have much in common.

A space $X$ is called *contractible* provided that there exists a homotopy $H\colon X \times \mathbb{I} \to X$ such that $H_0$ is the identity and $H_1$ is a constant function; the homotopy $H$ is called a *contraction* of $X$. So a space $X$ is contractible if and only if the identity function on $X$ is nullhomotopic. It is clear that every contractible space is path-connected from which it follows that $\mathbb{S}^0 = \{-1, 1\}$ is not contractible. Path-connected spaces need not be contractible since no $\mathbb{S}^n$ is contractible (Corollary 2.4.11).

Observe that $\mathbb{I}$ is contractible. The homotopy
$$(x,t) \mapsto t \cdot x$$
contracts $\mathbb{I}$ to 0. Since products of contractible spaces are contractible (Exercise A.12.2), it follows that $\mathbb{I}^n$ is contractible for every $n \leq \infty$.

We shall now present a simple but useful characterization of contractible spaces.

**Proposition A.12.2.** *Let $X$ be a path-connected space. The following statements are equivalent:*

(1) *$X$ is contractible,*
(2) *if $Y$ is any path-connected space then $[X,Y]$ is trivial,*
(3) *if $Y$ is any space then $[Y,X]$ is trivial.*

**Proof.** For (1) $\Rightarrow$ (2), let $F\colon X \times \mathbb{I} \to X$ be a contraction. If $f\colon X \to Y$ is continuous then $f \circ F$ is a homotopy from $X$ to $Y$ that connects $f$ with a constant function. It therefore suffices to prove that any two constant functions from $X$ to $Y$ are homotopic. Let $c_p, c_q \colon X \to Y$ be the constant functions with values $p$ and $q$, respectively. Since $Y$ is path-connected, there is a path $\alpha\colon \mathbb{I} \to Y$ with $\alpha(0) = p$ and $\alpha(1) = q$. Now define the required homotopy $H\colon X \times \mathbb{I} \to Y$ by $H(x,t) = \alpha(t)$.

For (2) $\Rightarrow$ (3), first observe that $X$ is contractible since $[X,X]$ is trivial. Let $F\colon X \times \mathbb{I} \to X$ be a contraction such that $F_1$ is the constant function with value $c$. If $Y$ is any space and $f\colon Y \to X$ is continuous then
$$H = F \circ (f \times 1_\mathbb{I}) \colon Y \times \mathbb{I} \to X$$
is a homotopy that connects $f$ with the constant function with value $c$.

Since (3) $\Rightarrow$ (1) is a triviality, we are done. $\square$

Let $X$ be a space with subspace $A$. We say that $A$ is a *retract* of $X$ provided that there is a continuous function $r\colon X \to A$ such that $r$ restricted to $A$ is the identity on $A$. Such a function $r$ is called a *retraction*.

**Lemma A.12.3.** *Let $X$ be a space with subspace $A$. If $A$ is a retract of $X$ then $A$ is a closed subset of $X$.*

**Proof.** Let $r\colon X \to A$ be a retraction. Consider the composition
$$X \xrightarrow{r} A \xhookrightarrow{i} X.$$
This function is continuous and agrees with the identity function on $X$ precisely on the set $A$. That $A$ is closed is therefore a consequence of Exercise A.1.9. □

We say that $A$ is a *neighborhood retract* of $X$ provided that there exists a neighborhood $U$ of $A$ in $X$ such that $A$ is a retract of $U$.

Retractions are very interesting functions since they preserve many topological properties. For example, if $r\colon X \to A$ is a retraction and $X$ has the fixed-point property, then $A$ has the fixed-point property as well, etc. See Exercise A.12.10(2).

**Theorem A.12.4.** *Every retract of a contractible space is contractible.*

**Proof.** Let $Y$ be a contractible space, let $X$ be a subspace of $Y$, and let
$$r\colon Y \to X$$
be a retraction. By definition, there exists a contraction
$$H\colon Y \times \mathbb{I} \to Y.$$
Now define $F\colon X \times \mathbb{I} \to X$ by the formula
$$F(x,t) = r\bigl(H(x,t)\bigr).$$
Obviously, $F$ is a contraction of $X$. □

We shall sometimes identify $\mathbb{R}^2$ and the complex plane $\mathbb{C}$. We will make use of the exponential function
$$z \mapsto e^z \quad (z \in \mathbb{C}).$$
As usual, if $z = a + bi$ is a complex number then
$$|z| = \sqrt{a^2 + b^2}, \quad \bar{z} = a - bi$$
denote its *absolute value* and its *complex conjugate*, respectively.

Let $X$ be a space. A continuous function $f\colon X \to \mathbb{S}^1$ is called *inessential* if $f$ is nullhomotopic. A continuous function that is not inessential is called *essential*. This divides the class of continuous functions from $X$ to $\mathbb{S}^1$ into two classes. That for some $X$ there are essential maps $X \to \mathbb{S}^1$ will be shown in Exercise A.12.8.

A second, more technical concept is needed. We say that a function
$$f \in C(X, \mathbb{S}^1)$$
has a *continuous logarithm* provided that there is a continuous function
$$\varphi \colon X \to \mathbb{R}$$
such that $f = e^{i\varphi}$, i.e., $f(x) = e^{i\varphi(x)}$ for every $x \in X$. We shall prove that for compact $X$ a function $f \in C(X, \mathbb{S}^1)$ is inessential if and only if it has a continuous logarithm.

If $X$ is a space and $f, g \colon X \to \mathbb{C}$ are continuous then $|f - g|$ denotes $\sup\{|f(x) - g(x)| : x \in X\}$ whenever this supremum exists.

**Lemma A.12.5.** *Let $X$ be a space and let $f \colon X \to \mathbb{S}^1$ be continuous such that $f[X] \neq \mathbb{S}^1$. Then $f$ has a continuous logarithm.*

**Proof.** Choose $q \in \mathbb{R}$ such that $e^{iq} \notin f[X]$. The function $t \mapsto e^{it}$ is a homeomorphism from the interval $(q, q + 2\pi)$ onto $\mathbb{S}^1 \setminus \{e^{iq}\}$. Let
$$L \colon \mathbb{S}^1 \setminus \{e^{iq}\} \to (q, q + 2\pi)$$
be the inverse of this mapping. Define $\varphi \colon X \to \mathbb{R}$ by $\varphi(x) = L(f(x))$. Then $\varphi$ is clearly continuous and
$$e^{i\varphi(x)} = e^{iL(f(x))} = f(x),$$
for every $x \in X$. For this simply observe that if $z \in \mathbb{S}^1 \setminus \{e^{iq}\}$ then $z$ coincides with $e^{iL(z)}$. □

**Lemma A.12.6.** *Let $X$ be a space and let $f_1, f_2 \colon X \to \mathbb{S}^1$ be continuous functions such that*
$$|f_1 - f_2| < 1.$$
*Then $f_1$ has a continuous logarithm if and only if $f_2$ has a continuous logarithm.*

**Proof.** Define $h \colon X \to \mathbb{S}^1$ by $h = f_1/f_2$. Observe that $h$ is well-defined. If $x \in X$ then
$$|h(x) - 1| = \left| \frac{(f_1 - f_2)(x)}{f_2(x)} \right| = \frac{|(f_1 - f_2)(x)|}{|f_2(x)|} = |(f_1 - f_2)(x)| < 1,$$
by assumption. It follows that the image of $h$ is contained in the half plane
$$\{z = a + bi : a > 0\},$$
and consequently $h$ is not surjective. According to Lemma A.12.5 there is a continuous function $\varphi \colon X \to \mathbb{R}$ such that $h = e^{i\varphi}$. Since $f_1 = f_2 \cdot h$, the proof is complete. □

**Lemma A.12.7.** *Let $X$ be a compact space and let $H\colon X \times \mathbb{I} \to \mathbb{S}^1$ be a homotopy. Then $H_0$ has a continuous logarithm if and only if $H_1$ has one.*

**Proof.** Since $H$ is uniformly continuous by Exercise A.5.18, there is an $n \in \mathbb{N}$ such that if $s, t \in \mathbb{I}$ and $|s - t| \leq 1/n$ then
$$|H(x,t) - H(x,s)| < 1$$
for every $x \in X$ (in fact, much more is true). Put $f_j = H_{j/n}$ for $0 \leq j \leq n$. Then
$$|f_{j+1}(x) - f_j(x)| = |H_{(j+1)/n}(x) - H_{j/n}(x)| < 1,$$
for every $0 \leq j \leq n-1$ and $x \in X$. By applying Lemma A.12.6 successively to the level functions $f_j$ for $0 \leq j \leq n$, we arrive at the desired result. $\square$

We now come to the following result.

**Theorem A.12.8.** *Let $X$ be compact and let $f\colon X \to \mathbb{S}^1$ be continuous. The following statements are equivalent:*

(1) *$f$ is inessential,*
(2) *$f$ has a continuous logarithm.*

**Proof.** For (1) $\Rightarrow$ (2), let $H\colon X \times \mathbb{I} \to \mathbb{S}^1$ be a homotopy such that $H_0 = f$ and $H_1$ is constant. Since $H_1$ has clearly a continuous logarithm we find that $f = H_0$ has one by Lemma A.12.7.

For (2) $\Rightarrow$ (1), suppose that $f = e^{i\varphi}$ for a certain continuous function $\varphi\colon X \to \mathbb{R}$. It is easy to see that the function $H\colon X \times \mathbb{I} \to \mathbb{S}^1$ defined by
$$H(x,t) = e^{ti\varphi(x)}$$
is a homotopy such that $H_0$ is constant with value 1 and $H_1 = f$. $\square$

A continuum $X$ is *unicoherent* provided that whenever $A$ and $B$ are subcontinua of $X$ such that $A \cup B = X$ then $A \cap B$ is connected.

**Theorem A.12.9.** *Let $X$ be a compact space. Assume that every continuous function $f\colon \mathbb{S}^1 \to X$ is nullhomotopic. Then $X$ is unicoherent.*

**Proof.** Put
$$\mathbb{S}^1_+ = \{(x,y) \in \mathbb{S}^1 : y \geq 0\}, \quad \mathbb{S}^1_- = \{(x,y) \in \mathbb{S}^1 : y \leq 0\}.$$
Observe that $\mathbb{S}^1_+$ and $\mathbb{S}^1_-$ are arcs and that
$$\mathbb{S}^1_+ \cup \mathbb{S}^1_- = \mathbb{S}^1, \quad \mathbb{S}^1_+ \cap \mathbb{S}^1_- = \{(1,0), (-1,0)\} \approx \mathbb{S}^0.$$
Assume that $X$ is a continuum that is not unicoherent. Let $A$ and $B$ be subcontinua of $X$ such that $A \cup B = X$ while $A \cap B$ is not connected. Observe that since $X$ is a continuum we have $A \cap B \neq \emptyset$. Write $A \cap B$ as the disjoint

union of two nonempty compacta, say $E$ and $F$. Define $f\colon A\cap B \to \mathbb{S}^1_+\cap\mathbb{S}^1_-$ by
$$f[E] = (1,0). \quad f[F] = (-1,0).$$
Since $\mathbb{S}^1_+$ and $\mathbb{S}^1_-$ are arcs we can extend $f$ to continuous functions
$$f_1\colon A \to \mathbb{S}^1_+, \quad f_2\colon B \to \mathbb{S}^1_-$$
(Theorem A.4.6). Let $g\colon X \to \mathbb{S}^1$ be the function $f_1 \cup f_2$. Then $g$ is clearly continuous and we claim that it is essential. Observe that

(∗)  $\quad g[A\cap B] = \{(1,0),(-1,0)\} = g[A]\cap g[B].$

Striving for a contradiction, assume that $g$ is inessential. By Theorem A.12.8 there consequently is a continuous function $\varphi\colon X \to \mathbb{R}$ such that $g = e^{i\varphi}$. Since $A$ is connected, so is $\varphi[A]$. Similarly for $\varphi[B]$. Since $A\cap B \neq \emptyset$, $\varphi[A]$ and $\varphi[B]$ are intersecting intervals in $\mathbb{R}$. As a consequence, $S = \varphi[A]\cap\varphi[B]$ is connected. By continuity of the exponential function, this implies that the set $e^{iS}$ is connected. We will show that this leads to a contradiction. Indeed, by (∗) and the fact that $g = e^{i\varphi}$ we obtain

$$\begin{aligned}\{(1,0),(-1,0)\} = g[A\cap B] &= e^{i\varphi[A\cap B]} \\ &\subseteq e^{i(\varphi[A]\cap\varphi[B])} = e^{iS} \\ &\subseteq e^{i\varphi[A]}\cap e^{i\varphi[B]} \\ &= g[A]\cap g[B] = \{(1,0),(-1,0)\}.\end{aligned}$$

This shows that the connected set $e^{iS}$ is also disconnected, being equal to the doubleton $\{(1,0),(-1,0)\}$. This is absurd.  □

**Corollary A.12.10.** *Every contractible continuum is unicoherent.*

**Proof.** This is a trivial consequence of Theorem A.12.9 for if $X$ is a contractible continuum then $[X,\mathbb{S}^1]$ is trivial by Proposition A.12.2.  □

**Remark A.12.11.** We will prove later that $\mathbb{S}^n$ is unicoherent for every $n \geq 2$ (Exercise 3.6.7) and that no $\mathbb{S}^n$ is contractible (Corollary 2.4.11). This proves that unicoherent continua need not be contractible.

**Cones.** An important construction in topology is that of a *cone* over a topological space, which we will now describe.

Let $X$ be a space and let $\infty$ be a point not in $X \times [0,1)$. Topologize
$$\triangle(X) = \bigl(X \times [0,1)\bigr) \cup \{\infty\}$$
as follows: points of the form $(x,t)$ have their usual product neighborhoods and a basic neighborhood of $\infty$ has the form
$$\bigl(X \times (s,1)\bigr) \cup \{\infty\},$$
where $0 < s < 1$. We call $\triangle(X)$ the *cone* over $X$.

Observe that $X \times \{0\}$ is a closed subset of $\triangle(X)$, hence $X$ can be thought of as a closed subspace of its own cone. It is easy to see that $\triangle(X)$ is contractible, see Exercise A.12.5 below.

If $X$ is compact then $\triangle(X)$ is the one-point compactification of $X \times [0,1)$ and $\triangle(X)$ is homeomorphic to the quotient space obtained from $X \times \mathbb{I}$ by identifying the subset $X \times \{1\}$ to a single point. Observe that $\triangle(X)$ has a countable base and therefore is a separable metrizable space (for compact spaces, this also follows from Theorem A.11.2.)

**Exercises for §A.12.** A space $X$ is called *locally contractible* (abbreviated LC) *at* $x \in X$ if for every neighborhood $U$ of $x$ in $X$ there is a neighborhood $V$ of $x$ and a homotopy $H \colon V \times \mathbb{I} \to U$ such that $H_1$ is the identity and $H_0$ is constant. In addition, $X$ is called *locally contractible* if $X$ is locally contractible at every point.

A space $X$ has the *fixed-point property* if every continuous function $f \colon X \to X$ has a fixed-point, i.e., a point $x \in X$ with $f(x) = x$.

1. Prove Lemma A.12.1.
2. For each $n \in \mathbb{N}$, let $X_n$ be contractible. Prove that the product $\prod_{n=1}^{\infty} X_n$ is contractible.
3. For each $n \in \mathbb{N}$, let $X_n$ be a space. Prove that the following statements are equivalent:
    (1) $\prod_{n=1}^{\infty} X_n$ is LC.
    (2) Each $X_n$ is LC and there is $N \in \mathbb{N}$ such that $X_m$ is contractible for every $m \geq N$.
4. Let $X$ be locally connected. Prove that if $Y$ is a subspace of $X$ which is a retract of $X$ then $Y$ is locally connected as well.
5. Let $X$ be a space. Prove that $\triangle(X)$ is contractible.
6. Let $X$ be a space with subspace $A$. Prove that subspace
$$\nabla(A) = (A \times [0,1)) \cup \{\infty\}$$
of $\triangle(X)$ is homeomorphic to the cone $\triangle(A)$ over $A$. Moreover, if $A$ is closed in $X$ then $\nabla(A)$ is closed in $\triangle(X)$.
7. Let $X$ be a contractible space, and let $x \in X$ be an arbitrary point. Prove that there is a homotopy $H \colon X \times \mathbb{I} \to X$ such that $H_0$ is the identity and $H_1$ is the constant function with value $x$.
▶8. Let $n \in \mathbb{Z} \setminus \{0\}$. Prove that the mapping $\psi_n \colon \mathbb{S}^1 \to \mathbb{S}^1$ defined by
$$\psi_n(z) = z^n$$
is essential.
9. Prove that $\mathbb{S}^1$ is not contractible.
▶10. (1) Prove that $\mathbb{I}$ has the fixed-point property.

(2) Let $Y$ be a retract of $X$. Prove that if $X$ has the fixed-point property then $Y$ has the fixed-point property.
(3) Prove that the $\sin(1/x)$-continuum in the plane has the fixed-point property.

▶11. Let $D = \{z \in \mathbb{C} : |z| \leq 1\}$. Prove for every continuous function $f \colon D \to \mathbb{C}$ with $f[\mathbb{S}^1] \subseteq D$ there is a point $x \in D$ with $f(x) = x$.

12. Prove that $\mathbb{I}^2$ has the fixed-point property.

## A.13. Borel and similar sets

We will now briefly discuss Borel sets, analytic sets and sets with the property of Baire.

**Borel sets.** Let $X$ be a space. A *Borel set* in $X$ is an element of the $\sigma$-algebra of subsets of $X$ generated by the open subsets of $X$. The family of all Borel subsets of $X$ is denoted by $\mathcal{B}(X)$.

It is clear that every closed subset of $X$ is Borel, being the complement of an open set. So the smallest $\sigma$-algebra of subsets of $X$ generated by the closed subsets of $X$ is contained in $\mathcal{B}(X)$. It moreover contains all open sets, so it must be equal to $\mathcal{B}(X)$. The conclusion is that we could also have started with the closed sets instead of the open sets in our definition of Borel set.

Countable unions of closed sets are Borel, as well as countable intersections of open sets. So all $F_\sigma$-subsets and all $G_\delta$-subsets of $X$ are Borel. Consider the subspace $\mathbb{Q}$ of $\mathbb{R}$. It is an $F_\sigma$-set and is therefore Borel. But it is not a $G_\delta$-subset of $\mathbb{R}$ by Corollary A.6.7. The field in mathematics that studies among other things the fine distinctions between various Borel sets is called *Descriptive Set Theory*. See e.g., KURATOWSKI and MOSTOWSKI [244], MOSCHOVAKIS [315], KECHRIS [218] and MILLER [310] for more information. We will only discuss some very basic material here.

A family $\mathcal{R}$ of sets is $\sigma$-*additive* ($\delta$-*multiplicative*) if for every *countable* subcollection $\mathcal{S} \subseteq \mathcal{R}$ we have that $\bigcup \mathcal{S} \in \mathcal{R}$ (respectively, $\bigcap \mathcal{S} \in \mathcal{R}$). It is clear that the union and the intersection of an arbitrary family of $\sigma$-additive collections of subsets of a set is again $\sigma$-additive. Similarly for $\delta$-multiplicative. Since $\mathcal{P}(X)$ is both $\sigma$-additive and $\delta$-multiplicative, it follows that every collection of sets $\mathcal{S} \subseteq \mathcal{P}(X)$ is contained in a smallest (with respect to inclusion) collection $\mathcal{T} \subseteq \mathcal{P}(X)$ which is $\sigma$ additive as well as $\delta$ multiplicative.

**Theorem A.13.1.** *Let $X$ be a space. Then $\mathcal{B}(X)$ is the smallest $\sigma$-additive and $\delta$-multiplicative collection of subsets of $X$ which contains all open subsets of $X$ (respectively, all closed subsets of $X$).*

**Proof.** Let $\mathcal{E}$ (respectively, $\mathcal{F}$) be the smallest $\sigma$-additive and $\delta$-multiplicative collection of subsets of $X$ which contains all open subsets of $X$ (respectively, all closed subsets of $X$). Observe that

$$\mathcal{G} = \{X \setminus F : F \in \mathcal{F}\}$$

is also $\sigma$-additive and $\delta$-multiplicative. Since $\mathcal{G}$ contains all open sets, this implies that $\mathcal{E} \subseteq \mathcal{G}$.

Since every open set is a countable union of closed sets (Exercise A.2.3), the collection of all open subsets of $X$ is contained in $\mathcal{F}$. So $\mathcal{E} \subseteq \mathcal{F}$. Since every closed set is a countable intersection of open sets, it follows similarly that $\mathcal{F} \subseteq \mathcal{E}$. We conclude that $\mathcal{E} = \mathcal{F}$.

So if $E \in \mathcal{E}$ then $E \in \mathcal{G}$ and hence there exists $F \in \mathcal{F}$ such that $E = X \setminus F$. Since $F \in \mathcal{E}$ this shows that $X \setminus E \in \mathcal{E}$.

From this we conclude that $\mathcal{E}$ is a $\sigma$-algebra containing all the open subsets of $X$, hence $\mathcal{B}(X)$. On the other hand, $\mathcal{B}(X)$ is $\sigma$-additive and $\delta$-multiplicative and contains all open subsets of $X$, hence $\mathcal{E} \subseteq \mathcal{B}(X)$. Since we already showed that $\mathcal{E} = \mathcal{F}$, this finishes the proof. □

A countable intersection of $F_\sigma$-subsets of $X$ is called an $F_{\sigma\delta}$-set. Observe that a countable intersection of $F_{\sigma\delta}$-sets is again an $F_{\sigma\delta}$-set. The complement of an $F_{\sigma\delta}$-set is called a $G_{\delta\sigma}$-set. It is clear that a $G_{\delta\sigma}$-set is the union of countably many $G_\delta$-subsets of $X$ which implies that the union of countably many $G_{\delta\sigma}$-sets is again a $G_{\delta\sigma}$-set.

Since closed subsets of $X$ are $G_\delta$-subsets (Exercise A.2.3), it follows that every $F_\sigma$-subset of $X$ is a $G_{\delta\sigma}$-set. It follows similarly that every $G_\delta$-subset is an $F_{\sigma\delta}$-set.

Finite intersections of $G_{\delta\sigma}$-sets are again $G_{\delta\sigma}$, and similarly for finite unions of $F_{\sigma\delta}$-sets.

A space $X$ is called an *absolute $G_\delta$* if it is a $G_\delta$-subset of every space it is imbedded in. We define the notions *absolute $F_\sigma$*, *absolute $F_{\sigma\delta}$* and *absolute $G_{\delta\sigma}$* similarly.

**Theorem A.13.2.** *Let $X$ be a space.*

(1) *$X$ is an absolute $G_\delta$ if and only if $X$ is topologically complete.*
(2) *$X$ is an absolute $F_\sigma$ if and only if $X$ is $\sigma$-compact.*
(3) *$X$ is an absolute $G_{\delta\sigma}$ if and only if $X$ is the union of countably many topologically complete subspaces.*

**Proof.** Observe that (1) follows directly from Theorem A.6.3. For (2), first observe that if $X$ is $\sigma$-compact then it is an absolute $F_\sigma$. Conversely, use that $X$ can be imbedded in $Q$ (Corollary A.4.4), and observe that $Q$ is compact and hence every $F_\sigma$-subset of it is $\sigma$-compact. Statement (3) can be proved by similar considerations. □

It is also possible to characterize the class of all absolute $F_{\sigma\delta}$'s 'internally', but this would lead us to far. For our purposes the following result suffices.

**Theorem A.13.3.** *Let $X$ be a space. The following statements are equivalent:*

(1) *$X$ is an absolute $F_{\sigma\delta}$.*
(2) *$X$ can be imbedded in some topologically complete space as an $F_{\sigma\delta}$-subset.*

**Proof.** Since $X$ can be imbedded in the topologically complete space $Q$ (Corollary A.4.4), the implication (1) $\Rightarrow$ (2) is obvious. For (2) $\Rightarrow$ (1), assume that $X$ is contained as an $F_{\sigma\delta}$-subset of some topologically complete space $Y$, as well as in a space $Z$. We may assume that $Z$ is a subspace of $Q$. Since being an $F_{\sigma\delta}$-subset is inherited by subspaces, it clearly suffices to prove that $X$ is an $F_{\sigma\delta}$-subset of $Q$. By Theorem A.8.5, there are $G_\delta$-subsets $S$ of $Y$ and $T$ of $Q$, respectively, such that $X \subseteq S$ and $X \subseteq T$ while moreover the identity on $X$ can be extended to a homeomorphism $f \colon S \to T$. It is clear that $S \setminus X$ is a $G_{\delta\sigma}$-subset of $S$, hence $f[S \setminus X]$ is a $G_{\delta\sigma}$-subset of $T$. Since a $G_\delta$-subset of a $G_\delta$-subset is a $G_\delta$-subset, it follows that $f[S \setminus X]$ is a $G_{\delta\sigma}$ subset of $Q$. Since $Q \setminus T$ is an $F_\sigma$-subset of $Q$, it is also a $G_{\delta\sigma}$ subset of $Q$. So $X$ is indeed an $F_{\sigma\delta}$-subset of $Q$. $\square$

**Corollary A.13.4.** *If $X_n$ is $\sigma$-compact for every $n$ then $\prod_{n=1}^{\infty} X_n$ is an absolute $F_{\sigma\delta}$.*

**Proof.** Assume that $X_n \subseteq Q$ for every $n$ (Corollary A.4.4). Then

$$\Pi = \prod_{n=1}^{\infty} X_n$$

is a subspace of $Q^\infty$ by Exercise A.1.13. Observe that the complement of $\Pi$ in $Q^\infty$ is equal to

$$\bigcup_{n=1}^{\infty} \underbrace{Q \times \cdots \times Q}_{n-1 \text{ times}} \times (Q \setminus X_n) \times Q \times Q \times \cdots$$

and hence is $G_{\delta\sigma}$ by Lemma A.6.2. So $\Pi$ is an $F_{\sigma\delta}$-subset of $Q^\infty$ and hence is an absolute $F_{\sigma\delta}$ by Theorem A.13.3. $\square$

So $\mathbb{Q}^\infty$ is an absolute $F_{\sigma\delta}$. The question naturally arises whether it is an absolute $G_{\delta\sigma}$ as well.

**Theorem A.13.5.** *$\mathbb{Q}^\infty$ is not an absolute $G_{\delta\sigma}$.*

**Proof.** Let $\{A_i : i \in \mathbb{N}\}$ be a collection of topologically complete subspaces of $\mathbb{Q}^\infty$. We claim that they do not cover $\mathbb{Q}^\infty$, which suffices by Theorem A.13.2(3). Indeed, since $\mathbb{Q}^\infty$ is not Baire, $A_1$ is not dense in it by

Exercise A.6.7. So there is a non-empty basic open subset $U$ of $\mathbb{Q}^\infty$ such that $U \cap A_1 = \emptyset$. Let $n_1 \in \mathbb{N}$ and $(q_1, \ldots, q_{n_1}) \in \mathbb{Q}^{n_1}$ be such that
$$X_1 = (q_1, \ldots, q_{n_1}) \times \mathbb{Q} \times \mathbb{Q} \times \cdots \subseteq U.$$
Since $A_2 \cap X_1$ is closed in $A_2$ it is topologically complete. Since $X_1 \approx \mathbb{Q}^\infty$ is not Baire, as above we can find $n_1 < n_2$, $(q_{n_1+1}, \ldots, q_{n_2}) \in \mathbb{Q}^{n_2-n_1}$ such that
$$X_2 = (q_1, \ldots, q_{n_2}) \times \mathbb{Q} \times \mathbb{Q} \times \cdots$$
misses $A_2$. Continuing in this way we find a point $q = (q_i)_i \in \mathbb{Q}^\infty$ which does not belong to $\bigcup_{i=1}^\infty A_i$. □

**Corollary A.13.6.** $\mathbb{R}^\infty \setminus \mathbb{Q}^\infty$ is an absolute $G_{\delta\sigma}$ which is not an absolute $F_{\sigma\delta}$.

Summarizing:

- $\mathbb{Q}$ is absolute $F_\sigma$ but not $G_\delta$,
- $\mathbb{P}$ is absolute $G_\delta$ but not $F_\sigma$,
- $\mathbb{Q}^\infty$ is absolute $F_{\sigma\delta}$ but not $G_{\delta\sigma}$,
- $\mathbb{R}^\infty \setminus \mathbb{Q}^\infty$ is absolute $G_{\delta\sigma}$ but not $F_{\sigma\delta}$.

**The property of Baire.** If $X$ is a space and $A, B \subseteq X$ then we say that
$$A =^* B$$
if $A \triangle B$ is meager.

Observe that if $A =^* B$ and $B =^* C$ then $A =^* C$ (Exercise A.13.11). This will be used without explicit reference in the remaining part of this section.

**Lemma A.13.7.** Let $X$ be a space with subsets $A$, $B$ and $M$.

(1) If $M = A \triangle B$ then $A = M \triangle B$.
(2) If $M \subseteq X$ is meager then $A =^* A \cup M =^* A \setminus M$.

**Proof.** For (1), observe that
$$M \triangle B = (A \triangle B) \triangle B = A \triangle (B \triangle B) = A \triangle \emptyset = A.$$

For (2), simply observe that $(A \cup M) \triangle A \subseteq M$ and $(A \setminus M) \triangle A \subseteq M$, and that $M$ is meager. □

Let $\mathbf{C}(X)$ be the collection of all subsets $V$ of $X$ for which there exist an *open* set $U \subseteq X$ such that $V =^* U$. We say that the elements of $\mathbf{C}(X)$ have the *property of Baire*[3].

---

[3]This has nothing to do with Baire spaces. Being a Baire space is a topological notion, while having the property of Baire says something about how a subset of a topological space is placed therein.

## A.13. BOREL AND SIMILAR SETS

**Proposition A.13.8.** *If $X$ is a space then $\mathbf{C}(X)$ is a $\sigma$-algebra which contains all open subsets as well as all closed subsets of $X$.*

**Proof.** That $\mathbf{C}(X)$ contains all open sets is clear. So if we prove that it is a $\sigma$-algebra then it also contains all closed subsets of $X$.

For any $A \subseteq X$, Fr $A$ is nowhere dense, so
$$\text{Int } A =^* \overline{A}.$$
This implies that if $U \subseteq X$ is open and $F \subseteq X$ is closed then
$$U =^* \overline{U}, \quad F =^* \text{Int } F.$$

**Claim 1.** *If $A \in \mathbf{C}(X)$ then $X \setminus A \in \mathbf{C}(X)$.*

*Proof.* Let $A =^* U$ for some open $U \subseteq X$. Then $M = A \triangle U$ is meager. In addition,
$$A \triangle U = (X \setminus A) \triangle (X \setminus U)$$
and
$$X \setminus U =^* \text{Int}(X \setminus U).$$
So we conclude from Lemma A.13.7 that
$$\text{Int}(X \setminus U) =^* X \setminus U = M \triangle (X \setminus A) =^* X \setminus A,$$
as required. ◇

**Claim 2.** *If $A_n \in \mathbf{C}(X)$ for every $n$ then $\bigcup_{n=1}^{\infty} A_n \in \mathbf{C}(X)$.*

*Proof.* For every $n$ let $A_n = U_n \triangle M_n$ for $U_n, M_n \subseteq X$ with $U_n$ open and $M_n$ meager (Lemma A.13.7(1)). We will prove that
$$\bigcup_{n=1}^{\infty} A_n =^* \bigcup_{n=1}^{\infty} U_n.$$
First observe that
$$\bigcup_{n=1}^{\infty} A_n = \bigcup_{n=1}^{\infty} (U_n \setminus M_n) \cup \bigcup_{n=1}^{\infty} (M_n \setminus U_n).$$
The set $\bigcup_{n=1}^{\infty} (M_n \setminus U_n)$ is meager so that by Lemma A.13.7(2),
$$\bigcup_{n=1}^{\infty} A_n =^* \bigcup_{n=1}^{\infty} (U_n \setminus M_n).$$
In addition, by the same lemma,
$$\bigcup_{n=1}^{\infty} U_n =^* \bigcup_{n=1}^{\infty} U_n \setminus \bigcup_{n=1}^{\infty} M_n.$$

Moreover,
$$\Big(\bigcup_{n=1}^{\infty}(U_n \setminus M_n)\Big) \triangle \Big(\bigcup_{n=1}^{\infty} U_n \setminus \bigcup_{n=1}^{\infty} M_n\Big) \subseteq \bigcup_{n=1}^{\infty} M_n$$
is meager, which gives us what we want. ◇

This completes the proof. □

**Corollary A.13.9.** *If $X$ is a space then $\mathcal{B}(X) \subseteq \mathbf{C}(X)$.*

**Proof.** This is clear by Proposition A.13.8 since $\mathcal{B}(X)$ is the $\sigma$-algebra of subsets of $X$ generated by its open subsets. □

**Proposition A.13.10.** *Let $X$ be a space and $A \subseteq X$. The following statements are equivalent:*

(1) $A \in \mathbf{C}(X)$,
(2) $A$ can be written as $G \cup M$ where $G$ is a $G_\delta$-subset of $X$ and $M$ is meager.
(3) $A$ can be written as $F \setminus M$, where $F$ is an $F_\sigma$-subset of $X$ and $M$ is meager.

**Proof.** By Proposition A.13.8, (2) ⇒ (1) and (3) ⇒ (1). For (1) ⇒ (2), let $U$ be open such that $A \triangle U$ is meager. Hence $A \triangle U$ is contained in a meager $F_\sigma$-subset, say $F$. Then $G = U \setminus F$ is $G_\delta$ and $G \subseteq A$ while moreover $M = A \setminus G$ is meager. For (1) ⇒ (3), observe that $X \setminus A \in \mathbf{C}(X)$ (Proposition A.13.8). Hence we can use (2) for $X \setminus A$ to get what we want. □

**Analytic sets.** Let $X$ be a space. We say that it is *analytic* provided that it is a continuous image of a topologically complete space. We will show later that every topologically complete space is a continuous image of $\mathbb{P}$. So an analytic space can also be defined as one which is a continuous image of $\mathbb{P}$ (Corollary 1.9.11).

**Lemma A.13.11.** *Let $X$ be topologically complete. Then the collection of all analytic subspaces of $X$ is $\sigma$-additive and $\delta$-multiplicative and moreover contains all closed subsets of $X$.*

**Proof.** Every closed subspace of a topologically complete space is topologically complete, hence analytic for trivial reasons.

Suppose that $\mathcal{A}$ is a countable collection of analytic subspaces of $X$. For every $A$ there exists a topologically complete space $Y_A$ and a continuous surjection $f_A \colon Y_A \to A$.

Then $\bigcup \mathcal{A}$ is a continuous image of the topological sum $\bigoplus_{A \in \mathcal{A}} Y_A$ which is topologically complete by Exercise A.6.6. So $\bigcup \mathcal{A}$ is analytic.

The proof that $\bigcap \mathcal{A}$ is analytic as well, is slightly more complicated. Consider the subspace
$$Z = \left\{ y \in \prod_{A \in \mathcal{A}} Y_A : (\forall A, A' \in \mathcal{A})(f_A(y_A) = f_{A'}(y_{A'})) \right\}$$
of $\prod = \prod_{A \in \mathcal{A}} Y_A$. By continuity of the functions $f_A$, $A \in \mathcal{A}$, it easily follows that $Z$ is a closed subspace of $\prod$ (cf. Exercise A.1.9). So we conclude that $Z$ is topologically complete since $\prod$ is (Lemma A.6.2).

Fix $B \in \mathcal{A}$. The function $\varphi \colon Z \to X$ defined by
$$\varphi(y) = f_B(y_B)$$
is well-defined and continuous since it is equal to the composition of the projection $\prod \to Y_B$ and $f_B$. We claim that $\varphi$ maps $Z$ onto $\bigcap \mathcal{A}$. If we can prove this, then we are done since we already observed that $Z$ is topologically complete. So let $x \in \bigcap \mathcal{A}$ be arbitrary. For every $A \in \mathcal{A}$ pick a point $y_A \in Y_A$ such that $f_A(y_A) = x$. Then $y = (y_A)_{A \in \mathcal{A}} \in Z$ and $\varphi(y) = x$. Conversely, if
$$y = (y_A)_{A \in \mathcal{A}} \in Z$$
then for every $A \in \mathcal{A}$ we have
$$\varphi(y) = f_B(y_B) = f_A(x_A) \in A.$$
So we conclude that $\varphi(y) \in \bigcap \mathcal{A}$. □

**Corollary A.13.12.** *Let $X$ be a topologically complete space. Then every Borel subset of $X$ is analytic.*

**Proof.** This is follows from Theorem A.13.1 and Lemma A.13.11. □

We finish this section with an important result.

**Theorem A.13.13.** *Let $X$ be a space with analytic subspace $A$. Then $A$ belongs to $\mathbf{C}(X)$.*

**Proof.** We first claim that we may assume without loss of generality that $X$ is compact. For let $aX$ be a compactification of $X$ (Corollary A.4.8), and suppose that we are able to prove that $A \in \mathbf{C}(aX)$. Then there are subsets $U$ and $M$ of $aX$ with $U$ open and $M$ meager such that $A = U \triangle M$. Since $X$ is dense in $aX$, it follows that $M \cap X$ is meager in $X$ (Exercise A.13.7) and since $A \subseteq X$ we obtain
$$A = (U \cap X) \triangle (M \cap X) \in \mathbf{C}(X),$$
as required (Lemma A.13.7).

We next claim that we may assume without loss of generality that $A$ is dense in $X$. For if we can show that $A \in \mathbf{C}(\overline{A})$ then there are $U, M \subseteq \overline{A}$ such that $U$ is relatively open and $M$ is relatively meager such that $M = A \triangle U$.

But this clearly implies that $M$ is meager in $X$. Let $U'$ be an open subset of $X$ such that $U' \cap \overline{A} = U$. Then
$$A = (U' \triangle M) \cap \overline{A} \in \mathbf{C}(X)$$
by Proposition A.13.8.

So from now on we assume that $X$ is compact and that $A$ is dense in $X$.

Let $Z$ be a topologically complete space which admits a continuous surjection $f \colon Z \twoheadrightarrow A$.

We will discuss one more reduction. Let $\mathcal{U}$ be the collection of all open subsets $U$ of $Z$ such that $f[U]$ is meager in $X$. Then $\mathcal{U}$ has a countable subcollection $\mathcal{U}'$ such that $\bigcup \mathcal{U}' = \bigcup \mathcal{U}$ (Corollary A.2.3). Hence $S = f[\bigcup \mathcal{U}]$ is a countable union of meager sets, and hence is meager itself. Since
$$A \setminus S \subseteq A' = f[Z \setminus \bigcup \mathcal{U}] \subseteq A$$
and $S$ is meager, it follows that $A =^* A'$. In addition, $Z \setminus \bigcup \mathcal{U}$ is closed in $Z$ and hence is topologically complete. If $V \subseteq Z' = Z \setminus \bigcup \mathcal{U}$ is relatively open and nonempty then $f[V]$ is not meager. For assume otherwise, and let $V' \subseteq Z$ be open such that $V' \cap Z' = V$. Then clearly $f[V']$ is meager, being the union of the countably many meager sets
$$\{f[U \cap V'] : U \in \mathcal{U}'\} \cup \{f[V]\}.$$
So $V' \in \mathcal{U}$ which implies that $V' \cap Z' = \emptyset$. But this contradicts the fact that $V$ is nonempty.

These considerations show that we may assume without loss of generality that for every nonempty open subset $U \subseteq Z$ we have that $f[U]$ is not a meager subset of $X$. So in particular, if $U \subseteq Z$ is nonempty and open then the interior of $\overline{f[U]}$ in $X$ is nonempty.

Let $\varrho$ be an admissible complete metric on $Z$ which is bounded by 1 and let $d$ be an admissible metric on $X$ which is also bounded by 1 (Exercise A.1.6).

We claim that there is a meager $F_\sigma$-subset $F \subseteq X$ such that the dense $G_\delta$-subset $T = X \setminus F$ of $X$ is contained in $A$. (Observe that $T$ is dense by the Baire Category Theorem A.6.6.)

To prove this, let $\mathcal{B}$ be a countable open base for $Z$ such that $Z \in \mathcal{B}$, and for every $n$ let $\mathcal{B}_n$ be an open cover of $Z$ having the following properties:

(1) $\mathcal{B}_n \subseteq \mathcal{B}$,
(2) $\operatorname{mesh}(\mathcal{B}_n) < 2^{-n}$,
(3) $\operatorname{mesh}(f[\mathcal{B}_n]) < 2^{-n}$.

If $B \in \mathcal{B}_n$ then put
$$\mathcal{E}(B) = \{B' \in \mathcal{B}_{n+1} : B' \cap B \neq \emptyset\}$$

Observe that $B \subseteq \bigcup \mathcal{E}(B)$.

**Claim 1.** For every $n$ and $B \in \mathcal{B}_n$ the intersection of the open set
$$E(B) = \bigcup \{\operatorname{Int} \overline{f[B']} : B' \in \mathcal{E}(B)\}.$$
with $\overline{f[B]}$ is dense in $\overline{f[B]}$.

*Proof.* Let $V \subseteq X$ be open such that $V \cap \overline{f[B]} \neq \emptyset$. In addition, let $W$ in $X$ be open such that $\overline{W} \subseteq V$ and $W \cap f[B] \neq \emptyset$. So $f^{-1}[W] \cap B$ is a nonempty open subset of $Z$. There is an element $B' \in \mathcal{B}_{n+1}$ such that
$$E = B' \cap (f^{-1}[W] \cap B) \neq \emptyset.$$
So $B' \in \mathcal{E}(B)$ and since
$$\emptyset \neq \operatorname{Int} \overline{f[E]} \subseteq \operatorname{Int} \overline{f[B']} \cap \operatorname{Int} \overline{f[B]} \cap V \subseteq (E(B) \cap \overline{f[B]}) \cap V,$$
we are done. ◇

For every $n$ and $B \in \mathcal{B}_n$, put
$$D(B) = \overline{f[B]} \setminus E(B).$$
Then $D(B)$ is a nowhere dense closed subspace of $X$ by Claim 1. So
$$F = \bigcup_{n=1}^{\infty} \bigcup_{B \in \mathcal{B}_n} D(B)$$
is a meager $F_\sigma$-subset of $X$.

**Claim 2.** $T = X \setminus F \subseteq A$.

*Proof.* Pick an arbitrary element $p \in T$. Put $U_1 = Z$. Then $p \in \overline{f[U_1]}$ since $A$ is dense in $X$. By induction on $n \geq 2$ we will construct an element $U_n \in \mathcal{B}_n$ such that $U_{n-1} \cap U_n \neq \emptyset$ and $p \in \overline{f[U_n]}$. Indeed, suppose that for some $n \geq 1$ we constructed $U_n$. Since $p \notin D(U_n)$, it follows that $p \in E(U_n)$. So there exists $U_{n+1} \in \mathcal{E}(U_n)$ such that $p \in \overline{f[U_{n+1}]}$. But then $U_{n+1} \cap U_n \neq \emptyset$, which shows that $U_{n+1}$ is as required.

For every $n \geq 2$ pick a point $z_n \in U_n \cap U_{n-1}$. By (2), the sequence $(z_n)_{n \geq 2}$ is Cauchy and hence converges to an element $z \in Z$. We claim that $f(z) = p$. Let $V$ be an arbitrary neighborhood of $p$. Since $\lim_{n \to \infty} \operatorname{diam} \overline{f[U_n]} = 0$ by (3), there is an $N \in \mathbb{N}$ such that $\overline{f[U_n]} \subseteq V$ for all $n \geq N$. But this implies that all but finitely many terms of the sequence $\big(f(z_n)\big)_{n \geq 2}$ belong to $V$. So
$$f(z_n) \to p$$
and since $z_n \to z$, by continuity of $f$ we have $f(z) = p$. ◇

We conclude that $A$ contains a dense $G_\delta$-subset of $X$, from which it follows that $A =^* X \in \mathbf{C}(X)$. □

**Exercises for §A.13.** A space $X$ is called *absolutely Borel* if it is a Borel subset of every space it is imbedded in.

Let $A \subseteq X$. Then $A$ is said to be *meager at the point $p \in X$* if there is a neighborhood $U$ of $p$ such that $U \cap A$ is meager. The set of points of $X$ at which $A$ is not meager is denoted by $D(A)$.

1. Let $X$ be a space with subspace $Y$. Prove that
$$\mathcal{B}(Y) = \mathcal{B}(X) \restriction Y.$$

2. Let $Z$ be a space with subspaces $X$ and $Y$ such that $X \subseteq Y$.
   (1) Prove that if $X \in \mathcal{B}(Z)$ then $X \in \mathcal{B}(Y)$.
   (2) Prove that if $Y \in \mathcal{B}(Z)$ and $X \in \mathcal{B}(Y)$ then $X \in \mathcal{B}(Z)$.

3. Let $X$ be a space. Prove that the following statements are equivalent:
   (1) $X$ is an absolute Borel set.
   (2) $X$ can be imbedded as a Borel set in a topologically complete space $Y$.

▶4. Let $X$ be an absolute Borel set. Prove that if $X$ is a Baire space if and only if it contains a dense topologically complete subspace.

5. (1) Prove that the number of $G_\delta$-subsets of $Q$ is $\mathfrak{c}$.
   (2) Prove that the number of analytic subsets of $Q$ is $\mathfrak{c}$.
   (3) Let $X$ be a space. Prove that $|\mathcal{B}(X)| = \mathfrak{c}$ if and only if $X$ is infinite.

6. Prove that there is a subset of $\mathbb{R}$ which is not Borel.

7. Let $X$ be a dense subspace of $Y$. Prove that if $M$ is meager in $Y$ then $M \cap X$ is meager in $X$.

8. Let $X$ be a space with subspace $A$. Then
   (1) $D(A) = \emptyset \Leftrightarrow A$ is meager.
   (2) $D(A \setminus D(A)) = \emptyset$.
       (Hence $A \setminus D(A)$ is meager.)
   (3) $D(A) \subseteq \overline{A}$.
   (4) $D(A)$ is closed.
   (5) If $E \subseteq X$ is meager then $D(A) = D(A \setminus E) = D(A \cup E)$.

9. Let $X$ be compact and let $f \colon X \to Y$ be irreducible. Prove that for $A \subseteq X$ the following statements are equivalent:
   (1) $A$ is meager,
   (2) $f[A]$ is meager.
   Conclude that for $B \subseteq Y$ the following statements are equivalent:
   (1) $B$ is meager,
   (2) $f^{-1}[B]$ is meager.

10. Prove that if $X$ is absolutely Borel then it is analytic.

11. Prove that if $A, B$ and $C$ are subsets of a space $X$ such that
$$A =^* B, \quad B =^* C$$
then $A =^* C$.

# APPENDIX B

# Answers to selected exercises

It is of course best to try to figure out each exercise before looking for the answer here. But these answers are intended to be read, since some of them are used in the text and some of them occasionally provide additional information not covered by the text.

▶**Exercise 1.1.10:**
Define $i\colon X \to C(X)$ by $i(x)(y) = \varrho(x,y)$. We will show that $i$ is an isometry. For take arbitrary $x_1, x_2 \in X$. Then
$$\hat{\varrho}(i(x_1), i(x_2)) \geq |i(x_1)(x_2) - i(x_2)(x_2)| = |\varrho(x_1, x_2)| = \varrho(x_1, x_2).$$
In addition, for every $y \in X$ we have
$$|i(x_1)(y) - i(x_2)(y)| = |\varrho(x_1, y) - \varrho(x_2, y)| \leq \varrho(x_1, x_2),$$
from which it follows that $\hat{\varrho}(i(x_1), i(x_2)) \leq \varrho(x_1, x_2)$. Consequently,
$$\hat{\varrho}(i(x_1), i(x_2)) = \varrho(x_1, x_2),$$
which is as required.
Observe that if $x \in X$ then $i(x)$ is a function having nonnegative values only. As a consequence, a limit of functions of the form $i(x)$ has nonnegative values only.
To prove (1), let $Y \subseteq X$ be an arbitrary subset. Assume that $f \in \operatorname{conv}(i[Y])$ and that for some sequence $(y_n)_n$ in $Y$, $f = \lim_{n\to\infty} i(y_n)$. We shall prove that $f \in i[Y]$. Since $f \in \operatorname{conv}(i[Y])$ there exist distinct $a_1, \ldots, a_m \in Y$ and elements $t_1, \ldots, t_m \in \mathbb{I}$ such that
$$f = \sum_{j=1}^m t_j i(a_j), \quad \sum_{j=1}^m t_j = 1$$
(Lemma 1.1.1). Without loss of generality we may assume that $t_1 \neq 0$. Then for every $n \in \mathbb{N}$,
$$\hat{\varrho}(f, i(y_n)) \geq |f(y_n) - i(y_n)(y_n)|$$
$$= |f(y_n)|$$
$$= f(y_n)$$
$$\geq t_1 i(a_1)(y_n)$$
$$= t_1 \varrho(a_1, y_n).$$
Since the sequence $(i(y_n))_n$ converges to $f$, it therefore follows that
$$\lim_{n\to\infty} y_n = a_1.$$

This obviously implies that $f = i(a_1)$.
(2) is a triviality.

▶**Exercise 1.1.13:**
We first prove that $\ell^p$ is a vector subspace of $s$. To this end, simply observe that since $0 < p < 1$ for all $x, y \in \mathbb{R}$ we have
$$|x+y|^p \leq |x|^p + |y|^p.$$
This implies that if $x, y \in \ell^p$ then
$$\sum_{n=1}^{\infty} |x_n + y_n|^p \leq \sum_{n=1}^{\infty} (|x_n|^p + |y_n|^p) = \sum_{n=1}^{\infty} |x_n|^p + \sum_{n=1}^{\infty} |y_n|^p < \infty,$$
and so $x + y \in \ell^p$. Similarly, $tx \in \ell^p$ for all $x \in \ell^p$ and $t \in \mathbb{R}$.
That $\varrho$ is a metric and that the topology derived from this metric is compatible with the linear structure on $\ell^p$ follows from straightforward calculations which are left to the reader.
We proceed to prove that $\ell^p$ is not locally convex. To this end, let $U = B(\underline{0}, 1)$, the open ball of radius 1 about the origin of $\ell^p$. We claim that $\underline{0}$ does not have a convex neighborhood which is contained in $U$. Striving for a contradiction, assume that $V$ is such a neighborhood. There exists $\varepsilon > 0$ such that $D(\underline{0}, \varepsilon) \subseteq V$. Let $n \in \mathbb{N}$ be so large that
$$n^{1-p}\varepsilon > 1,$$
and consider for every $i \leq n$ the point $x_i \in s$ having all coordinates 0 except for the $i$-th coordinate which equals $\varepsilon^{1/p}$. Observe that $\varrho(x_i, \underline{0}) = (\varepsilon^{1/p})^p = \varepsilon$ and so $x_i \in D(\underline{0}, \varepsilon) \subseteq V$. Consider the point
$$x = \sum_{i=1}^{n} 1/n \, x_i.$$
Then $x$ is a convex combination of $x_1, \ldots, x_n$ and by convexity of $V$ it follows that $x \in V$ (Lemma 1.1.1). But
$$\varrho(\underline{0}, x) = \sum_{i=1}^{n} (1/n)^p \varepsilon = n^{1-p}\varepsilon > 1.$$
Since $V \subseteq B(\underline{0}, 1)$, this is a contradiction.

▶**Exercise 1.1.18:**
We prove by induction on $n$ that $\|\cdot\|$ is continuous on $\mathbb{R}^n$. We first claim that it suffices to prove continuity at the origin of $\mathbb{R}^n$. To see this, let $(x_n)_n$ be a sequence in $\mathbb{R}^n$ converging to an element $p \in \mathbb{R}^n$. Then $p - x_n \to \underline{0}$, from which it follows by assumption that $\|p - x_n\| \to \|\underline{0}\| = 0$. Since $|\|p\| - \|x_n\|| \leq \|p - x_n\|$ for every $n$, this shows that $\|x_n\| \to \|p\|$, which is as required.
Now let $n = 1$, and $f(x) = \|x\|$ for every $x \in \mathbb{R}$. Put $p = \|1\|$. Then for every $x \in \mathbb{R}$ we have
$$f(x) = f(x \cdot 1) = |x| \cdot f(1) = p \cdot |x|.$$
Since $p$ is fixed, this shows that $f$ is continuous.
Now assume the statement is true for $n$ and consider a norm $\|\cdot\|$ on $\mathbb{R}^{n+1}$. We think of $\mathbb{R}^{n+1}$ as $\mathbb{R}^n \times \mathbb{R}$. Let $(x_i)_i$ be a sequence in $\mathbb{R}^{n+1}$ converging to the origin of $\mathbb{R}^{n+1}$. Write every $x_i$ as $(a_i, b_i)$, where $a_i \in \mathbb{R}^n$ and $b_i \in \mathbb{R}$. Clearly $a_i \to \underline{0}$

and $b_i \to \underline{0}$. Since both $\|\cdot\| \restriction \mathbb{R}^n \times \{\underline{0}\}$ and $\|\cdot\| \restriction \{\underline{0}\} \times \mathbb{R}$ are continuous by our inductive assumption,

$$\|x_i\| \leq \|(a_i, \underline{0})\| + \|(\underline{0}, b_i)\| \to \|\underline{0}\| + \|\underline{0}\| = 0 + 0 = 0,$$

as required. This completes the inductive proof.

We claim that the metric $d\colon \mathbb{R}^n \times \mathbb{R}^n \to [0, \infty)$ defined by $d(x,y) = \|x - y\|$ is continuous. This follows easily because $d$ is the composition of the function

$$\varphi\colon \mathbb{R}^n \times \mathbb{R}^n \to \mathbb{R}^n$$

defined by $\varphi(x,y) = x - y$ and the norm $\|\cdot\|$. So by Exercise A.5.15 the topology generated by $d$ on every compact ball of $\mathbb{R}^n$ is the Euclidean topology. But this implies that the topology on $\mathbb{R}^n$ generated by $d$ is the Euclidean topology since $\mathbb{R}^n$ is locally compact.

▶**Exercise 1.1.20:**
By an argument similar to the one in the first part of the solution of Exercise 1.1.18 it follows that $f$ is continuous. In addition, $D(\underline{0}, 1)$ is compact by the same exercise, hence so is $f[D(\underline{0}, 1)]$. Let $t$ be the maximum of $f[D(\underline{0}, 1)]$. Since $f$ is not identically 0, there is an element $y \in \mathbb{R}^n$ such that $f(y) \neq 0$. We may clearly assume that $f(y) > 0$ (replace if necessary $y$ by $-y$). By linearity, $y \neq \underline{0}$ and so

$$f\left(y/\|y\|\right) = 1/\|y\| f(y) > 0.$$

We conclude that $t > 0$ and claim that it is as required. For this it suffices to prove that $f^{-1}(t) \cap B(\underline{0}, 1) = \emptyset$. Striving for a contradiction, assume that there is a vector $x \in f^{-1}(t) \cap B(\underline{0}, 1)$. Since $t > 0$, $x \neq \underline{0}$. Observe that

$$f\left(x/\|x\|\right) = t/\|x\|.$$

Since $\|x\| < 1$ we get

$$t/\|x\| > t,$$

which contradicts the fact that $t$ is the maximum of $f[D(\underline{0}, 1)]$.

▶**Exercise 1.1.21:**
We construct the $e_n$ by induction on $n$. Let $e_1$ be any unit vector in $V$. Suppose that the vectors $e_1, \ldots, e_{n-1}$ have been constructed for some $n \geq 2$. The linear subspace $W$ generated by $e_1, \ldots, e_{n-1}$ is not equal to $V$ by infinite-dimensionality. Take an arbitrary vector $p \in V \setminus W$, and let $R$ be the linear subspace of $V$ generated by $W$ and $p$. Then $R$ is both topologically and algebraically isomorphic to $\mathbb{R}^n$ (Exercise 1.1.18). Let $f\colon R \to \mathbb{R}$ be a linear function with kernel $W$. By Exercise 1.1.20 there is $t > 0$ such that $f^{-1}(t) \cap B(\underline{0}, 1) = \emptyset$ and $f^{-1}(t) \cap D(\underline{0}, 1) \neq \emptyset$. Let $x$ be an arbitrary vector in $f^{-1}(t) \cap D(\underline{0}, 1)$. So $\|x\| = 1$ and $x \notin W$, i.e., $\{e_1, \ldots, e_{n-1}, x\}$ is linearly independent. Suppose that for some $i \leq n-1$ we have $\|x - e_i\| < 1$. Since $f(x - e_i) = f(x) - f(e_i) = t - 0 = t$ we get $x - e_i \in f^{-1}(t) \cap B(\underline{0}, 1)$ which is a contradiction. So by putting $e_n = x$ we satisfy our inductive demands.

▶**Exercise 1.1.24:**
We may assume without loss of generality that the closed ball $D(\underline{0}, 2)$ is contained in $A$. Suppose that $x_0 \in \mathbb{R}^n$ is such that for some $b \in [0, \infty)$ we have $bx_0 \in A$.

Then by convexity of $A$ and the fact that $\underline{0} \in A$ it clearly follows that $ax_0 \in A$ for all $a \in [0, b]$. By convexity of $A$ we also have that

$$\bigcup \{I(x, bx_0) : x \in \mathbb{S}^{n-1}\} \subseteq A.$$

By Exercise 1.1.23 it consequently follows that every point of the form $\alpha x_0$ such that $\alpha \in [0, b)$, belongs to the interior of $A$. Hence if $bx_0 \in \operatorname{Fr} A$ then

(1) $0 \leq a < b$ implies that $ax_0$ belongs to the interior of $A$,
(2) $b < a$ implies that $ax_0$ does not belong to $A$.

Now for an arbitrary $x_0 \in \mathbb{R}^n$ put

$$g(x_0) = \sup\{a \in [0, \infty) : ax_0 \in A\}.$$

Observe that $0x_0 \in A$ and that $A$ is bounded, so $g(x_0)$ exists. Also,

$$g(x_0) \cdot x_0 \in \overline{A} = A$$

which implies that $g(x_0) \cdot x_0 \in A$. Similarly, $g(x_0) \cdot x_0 \in \overline{\mathbb{R}^n \setminus A}$. We conclude that $g(x_0) \cdot x_0 \in \operatorname{Fr} A$. So for every $x \in \mathbb{S}^{n-1}$ the element $g(x) \in [0, \infty)$ is the only element $r \in [0, \infty)$ such that $rx \in \operatorname{Fr} A$. If $g(x) > a \geq 0$ then $ax \in \operatorname{Int} A$ and so there exists $\varepsilon > 0$ such that $B(ax, \varepsilon) \subseteq \operatorname{Int} A$. The set

$$\{y \in \mathbb{R}^n : y \in B(ax, \varepsilon) \text{ and } \|y\| = \|ax\| = a\}$$

is open in

$$\{y \in \mathbb{R}^n : \|y\| = \|ax\| = a\}$$

so

$$O = \{\tfrac{1}{a} y : y \in B(ax, \varepsilon) \text{ and } \|y\| = a\}$$

is open in $\mathbb{S}^{n-1}$. And if $z \in O$ then $az \in B(ax, \varepsilon) \subseteq \operatorname{Int} A$ so $g(z) > a$. We conclude that $g^{-1}[(a, \infty)]$ is open. It follows similarly that $g^{-1}[(-\infty, a)]$ is open. We conclude that $g \colon \mathbb{S}^{n-1} \to \mathbb{R}$ is continuous. Define $f \colon D(\underline{0}, 1) \to A$ by

$$f(x) = \begin{cases} g\left(\frac{x}{\|x\|}\right) \cdot x & (x \neq \underline{0}), \\ \underline{0} & (x = \underline{0}), \end{cases}$$

and $h \colon A \to D(\underline{0}, 1)$ by

$$h(x) = \begin{cases} \frac{1}{g\left(\frac{x}{\|x\|}\right)} \cdot x & (x \neq \underline{0}), \\ \underline{0} & (x = \underline{0}), \end{cases}$$

respectively. There exists $M \geq 0$ such that $1 \leq g(x) \leq M$ for every $x \in \mathbb{S}^{n-1}$. So

$$\|f(x)\| \leq M \cdot \|x\|$$

for every $x \in D(\underline{0}, 1)$ and

$$\|h(x)\| \leq \|x\|$$

for every $x \in A$. Hence $f$ and $h$ are continuous at $\underline{0}$. Notice that if $x \neq \underline{0}$ then

$$x \mapsto \frac{1}{\|x\|} \mapsto \frac{x}{\|x\|} \mapsto g\left(\frac{x}{\|x\|}\right) \mapsto g\left(\frac{x}{\|x\|}\right) \cdot x$$

is a composition of continuous functions, whence $f$ is continuous at all points. By the same argument it follows that $h$ is continuous. Finally notice that $h(f(x)) = x$

and $f(h(y)) = y$ for every $x \in D(\mathbf{0}, 1)$ and $y \in A$. We conclude that $f$ and $h$ are homeomorphisms, and $f[\mathbb{S}^{n-1}] = \operatorname{Fr} A$ since $g(x_0) \cdot x_0 \in \operatorname{Fr} A$ for every $x_0 \in \mathbb{R}^n$.

▶ **Exercise 1.1.26:**
We prove $(1) \Rightarrow (2)$. The triangle inequality for $\|\cdot\|$ directly implies that

$$\forall x, y \in \ell^2 : \Big|\|x\| - \|y\|\Big| \leq \|x - y\|.$$

Therefore, for every $n \in \mathbb{N}$,

$$0 \leq \Big|\|x\| - \|x(n)\|\Big| \leq \|x - x(n)\| = \varrho\big(x, x(n)\big).$$

Since $\lim_{n \to \infty} \varrho(x, x(n)) = 0$ this implies that $\lim_{n \to \infty} \|x(n)\| = \|x\|$.
That $\lim_{n \to \infty} x(n)_i = x_i$ for every $i \in \mathbb{N}$ is a triviality.
We prove $(2) \Rightarrow (1)$. Let $\varepsilon > 0$. As $x \in \ell^2$ there is a $p \in \mathbb{N}$ such that

$$\sum_{i=p}^{\infty} x_i^2 < \varepsilon. \tag{3}$$

Now $\|x(n)\| \to \|x\|$ as $n \to \infty$ and hence

$$\exists m_0 > 0 : n \geq m_0 \Rightarrow \Big|\|x(n)\|^2 - \|x\|^2\Big| < \varepsilon. \tag{4}$$

Finally, $x(n)_i \to x_i$ for $i = 1, \ldots, p-1$ as $n \to \infty$, and hence

$$\exists m_1 : n \geq m_1 \Rightarrow \sum_{i=1}^{p-1} (x(n)_i - x_i)^2 < \varepsilon, \tag{5}$$

and

$$\exists m_2 : n \geq m_2 \Rightarrow \sum_{i=1}^{p-1} \big|x(n)_i^2 - x_i^2\big| < \varepsilon. \tag{6}$$

Then for $n \geq \max(m_0, m_1, m_2)$ we have by (4) and (6),

$$\left|\sum_{i=p}^{\infty} x(n)_i^2 - \sum_{i=p}^{\infty} x_i^2\right| = \left|\|x(n)\|^2 - \sum_{i=1}^{p-1} x(n)_i^2 + \sum_{i=1}^{p-1} x_i^2 - \|x\|^2\right|$$

$$\leq \Big|\|x(n)\|^2 - \|x\|^2\Big| + \left|\sum_{i=1}^{p-1}(x(n)_i^2 - x_i^2)\right|$$

$$< \varepsilon + \varepsilon$$

$$= 2\varepsilon,$$

from which it follows by (3) that

$$\sum_{i=p}^{\infty} x(n)_i^2 < 3\varepsilon. \tag{7}$$

Since $(a-b)^2 \leq 2a^2 + 2b^2$ for all $a, b \in \mathbb{R}$,

$$\sum_{i=1}^{\infty}(x(n)_i - x_i)^2 = \sum_{i=1}^{p-1}(x(n)_i - x_i)^2 + \sum_{i=p}^{\infty}(x(n)_i - x_i)^2$$

$$\leq \sum_{i=1}^{p-1}(x(n)_i - x_i)^2 + 2\sum_{i=p}^{\infty} x(n)_i^2 + 2\sum_{i=p}^{\infty} x_i^2.$$

Consequently, for $n \geq \max(m_0, m_1, m_2)$ we obtain by (5), (7) and (3) that

$$\sum_{i=1}^{\infty}(x(n)_i - x_i)^2 < \varepsilon + 6\varepsilon + 2\varepsilon = 9\varepsilon.$$

We conclude that $\lim_{n\to\infty} x(n) = x$ (in $\ell^2$).

▶ **Exercise 1.1.29:**
Recall that
$$\Delta(X) = (X \times [0,1)) \cup \{\infty\},$$
where $\infty \notin X \times [0,1)$, topologized as explained on Page 515. The function
$$\varphi \colon \Delta(X) \to L \times \mathbb{I}$$
defined by
$$\varphi(x, t) = ((1-t)x, t) \qquad (t \in [0,1)),$$
$$\varphi(\infty) = (\underline{0}, 1),$$
is clearly a bijection. To prove that it is a homeomorphism, we only need to worry about the point $\infty$. Let $((x_n, t_n))_n$ be a sequence in $X \times [0,1)$ converging to the point $\infty$. It is clear that the sequence $(t_n)_n$ converges to 1. Since $X$ is bounded, there is an element $\varepsilon \in [0, \infty)$ such that $\|x\| \leq \varepsilon$ for every $x \in X$. As a consequence,
$$0 \leq \|(1-t_n)x_n\| \leq (1-t_n)\varepsilon \to 0,$$
so that the sequence $((1-t_n)x_n)_n$ converges to $\underline{0}$ in $L$. This shows that
$$\varphi(x_n, t_n) \to \varphi(\infty) = (\underline{0}, 1),$$
which proves that $\varphi$ is continuous. To prove that $\varphi^{-1}$ is continuous at $(\underline{0}, 1)$, let $t_n$ in $[0,1)$ and $x_n \in X$ be arbitrary such that $((1-t_n)x_n, t_n) \to (\underline{0}, 1)$. Then $t_n \to 1$ and hence regardless of the behaviour of the sequence $(x_n)_n$, we have $(x_n, t_n) \to \infty$ in $\Delta(X)$.

▶ **Exercise 1.2.9:**
By Theorem 1.2.2 there exist continuous functions $f_1 \colon L_1 \to L_2$ and $f_2 \colon L_2 \to L_1$ such that $f_1 \restriction A_1 = h$ and $f_2 \restriction A_2 = h^{-1}$. Define $H_1, H_2 \colon L_1 \times L_2 \to L_1 \times L_2$ by
$$H_1(x_1, x_2) = (x_1, x_2 + f_1(x_1)), \quad H_2(x_1, x_2) = (x_1 - f_2(x_2), x_2).$$
That $H_1$ and $H_2$ are homeomorphisms of $L_1 \times L_2$ follows by elementary considerations (continuity is clear, and it is easy to display their inverses). Then $H = H_2 \circ H_1$ is as required.

▶ **Exercise 1.4.2:**
Let $t_n \to t$ in $\mathbb{I}$ and $x_n \to x$ in $L$. We have to prove that $\beta(t_n, x_n) \to \beta(x, t)$ in $L$.

There are several cases to consider. Suppose first that $t = 0$. We may assume without loss of generality that $t_n \neq 0$ for every $n$. Observe that
$$t_n \alpha\left(\frac{x_n}{t_n}\right) \longrightarrow 0$$
since $\alpha[L]$ is a bounded subset of $\mathbb{R}$. But then since $X$ is a bounded subset of $L$, it follows by Exercise 1.1.12 that
$$\beta(t_n, x_n) \longrightarrow (0, \underline{0}) = \beta(t, x).$$
So from now on assume that $t > 0$. If $x/t \in X \setminus U \subseteq V$ then since
$$\frac{x_n}{t_n} \longrightarrow \frac{x}{t}$$
and $V$ is open, there is an $N_1 \in \mathbb{N}$ such that
$$\frac{x_n}{t_n} \in V$$
for all $n \geq N_1$. This show that
$$\alpha\left(\frac{x_n}{t_n}\right) \longrightarrow 0$$
and hence
$$\beta(t_n, x_n) \longrightarrow (0, \underline{0}) = \beta(t, x).$$
Now assume that $x/t \in U$. There is an $N_2 \in \mathbb{N}$ such that
$$\frac{x_n}{t_n} \in U$$
for all $n \geq N_2$. But now the continuity of all functions involved clearly give us what we want.

▶**Exercise 1.6.1:**
Conditions (2) and (4) and Lemma A.3.1 show that $h$ is continuous with dense image. We show that $h$ is one-to-one and closed. Suppose that, for some $\varepsilon > 0$ there exist sequences $(x_k)_k$ and $(y_k)_k$ in $X$ such that $\varrho(x_k, y_k) > \varepsilon$ for each $k$ and
$$\lim_{k \to \infty} h(x_k) = z = \lim_{k \to \infty} h(y_k)$$
(Exercise A.6.9). Choose $n$ such that $2^{-n} < \varepsilon/5$, and pick $U \in \mathcal{U}_n$ such that $z \in U$. Then for some $k > n$,
$$\{h_k \circ \cdots \circ h_1(x_k), h_k \circ \cdots \circ h_1(y_k)\} \subseteq U.$$
By conditions (1) and (4) there exist $V, W \in \mathcal{U}_n$ such that
$$\{h_n \circ \cdots \circ h_1(x_k), h_k \circ \cdots \circ h_1(x_k)\} \subseteq V$$
and
$$\{h_k \circ \cdots \circ h_1(y_k), h_k \circ \cdots \circ h_1(y_k)\} \subseteq W.$$
There also exist $V', W' \in \mathcal{U}_n$ such that
$$\{h_{n-1} \circ \cdots \circ h_1(x_k), h_n \circ \cdots \circ h_1(x_k)\} \subseteq V'$$
and
$$\{h_{n-1} \circ \cdots \circ h_1(y_k), h_n \circ \cdots \circ h_1(y_k)\} \subseteq W'.$$
Thus $\{V', V, U, W, W'\}$ is a chain in $\mathcal{U}_n$ with $h_{n-1} \circ \cdots \circ h_1(x_k) \in V'$ and
$$h_{n-1} \circ \cdots \circ h_1(y_k) \in W'.$$

Applying $(h_{n-1} \circ \cdots \circ h_1)^{-1}$, we obtain a 5-chain in
$$(h_{n-1} \circ \cdots \circ h_1)^{-1}[\mathcal{U}_n]$$
between the points $x_k$ and $y_k$. By condition (3), $\varrho(x_k, y_k) < 5 \cdot 2^{-n} < \varepsilon$, a contradiction. It follows that $h$ is one-to-one and closed, and therefore a homeomorphism of $X$.

▶**Exercise 1.8.2:**
Let $q \in Q$ and let $V$ be an open neighborhood of $q$ in $Q$. There exists by Theorem 1.6.6 an element $\psi \in \mathcal{H}(Q)$ such that $\psi(q) = \mathbf{0}$. So $\psi[V]$ is a neighborhood of $\mathbf{0}$ and so there consequently exist $0 < \delta < 1$ and $N \in \mathbb{N}$ such that
$$U = \{x \in Q : x_i \in (-\delta, \delta) \text{ for } i = 1, \ldots, N\} \subseteq V.$$
It is clear that
$$\overline{U} = \{x \in Q : x_i \in [-\delta, \delta] \text{ for } i = 1, \ldots, N\}$$
is a Hilbert cube. The pseudo-interior of $\overline{U}$ shall be denoted by $s(\overline{U})$ and the boundary of $\overline{U}$ by $\partial U$, i.e.,
$$\partial U = \{x \in Q : (\exists i \leq N)(|x_i| = \delta)\}.$$

**Claim 1.** For arbitrary $x, y \in s(\overline{U})$ there is an element $f \in \mathcal{H}(\overline{U})$ such that

- $f(x) = y$,
- $f \upharpoonright \partial U = 1$.

Observe such that such $f$ can be extended to an element $\bar{f} \in \mathcal{H}(Q)$ such that
$$\bar{f} \upharpoonright Q \setminus U = 1_Q.$$

*Proof.* Let $i \in \{1, \ldots, N\}$ and factorize $\overline{U}$ as $[-\delta, \delta]_i \times \overline{U}_i$, where $\overline{U}_i$ is the product of the remaining factors of $\overline{U}$. Let $p_i$ be the projection from $\overline{U}_i$ onto the first $N-1$ factors, and let $\rho_i$ be the projection from $\overline{U}$ onto $\overline{U}_i$. For every $z \in \overline{U}_i$ put
$$\lambda_i(z) = \frac{\varrho(p_i(z), p_i \circ \rho_i[B])}{\varrho(p_i \circ \rho_i(x), p_i \circ \rho[B])}$$
and define the continuous function $\alpha \colon \overline{U}_i \to \mathbb{I}$ by
$$\alpha_i(z) = \min\{\lambda_i(z), 1\}.$$
Define the isotopy $g_i \colon [-\delta, \delta] \times \mathbb{I} \to [-\delta, \delta]$ by requiring that for every $t \in \mathbb{I}$ the function $g_{it}$ maps the intervals $[-\delta, x_i]$ and $[x_i, \delta]$ linearly onto the intervals
$$[-\delta, (1-t)x_i + ty_i], \quad [(1-t)x_i + ty_i, \delta].$$

Define the function $f_i \colon \overline{U} \to \overline{U}$ by
$$f_i(u, v) = \bigl(g_i(u, \alpha(v)), v\bigr).$$
Notice that $f_i \in \mathcal{H}(\overline{U})$ by Theorem 1.8.2, that $f_i \upharpoonright \partial U = 1$ and that $f_i(x_i) = y_i$. Identify $\overline{U}$ and $\prod_{i=1}^{N}[-\delta, \delta]_i \times \prod_{i>N} \mathbb{J}_i$. By a similar construction as the one above, it is clearly possible to construct an element $F \in \mathcal{H}(\overline{U})$ such that $F \upharpoonright B = 1$ and $F(x)_i = y_i$ for all $i > N$.
Now put
$$f = F \circ f_N \circ f_{N-1} \circ \cdots \circ f_1.$$

Then $f$ is as required. ◇

So just as in the case of the homogeneity of $Q$ we are done if we can prove that for every $x \in U$ there exists $g \in \mathcal{H}(\overline{U})$ such that $g(x) \in s(\overline{U})$ and $g \upharpoonright \partial U = 1$. (Again we extend such homeomorphism to a homeomorphism of $Q$ by letting it restrict to the identity on $Q \setminus U$.)
Take an arbitrary $x \in U$ and let $\omega_i \colon [-\delta, \delta] \to [-1, 1]$ be the homeomorphism taking the intervals $[-\delta, x_i]$ and $[x_i, \delta]$ linearly onto the intervals $[-1, 0]$ and $[0, 1]$. Define the homeomorphism $\Omega \colon \overline{U} \to Q$ by

$$\Omega(z)_i = \begin{cases} \omega_i(z_i) & (i = 1, \ldots, N), \\ z_i & (i > N). \end{cases}$$

Let $x_0 = \Omega(x)$ and

$$W = \Omega[B] = \bigcup_{i=1}^{N} \left( W_i^{-1} \cup W_i^i \right).$$

For every $0 \leq t \leq 1$ put

$$K_t = \{x \in Q : x_i \in [-1+t, 1-t] \text{ for } i = 1, \ldots, N\}.$$

Notice that $x_0 \in K_1$. The function $\beta \colon Q \to \mathbb{I}$ defined by

$$\beta(z) = \sup\{t \in \mathbb{I} : z \in K_t\}$$

is obviously continuous. It therefore suffices to prove the following:

**Claim 2.** Let $z \in K_1$. Then there exists $f \in \mathcal{H}(Q)$ such that

- $f(z) \in s$,
- $f \upharpoonright W = 1$.

*Proof.* The proof is similar to the proof of Lemma 1.6.5. Lemma 1.6.4 is not directly applicable, but instead we use the following analogous statement:

> Let $m > N$, $z \in K_1$ and $\varepsilon > 0$. Then there exists $h \in \mathcal{H}(Q)$ such that
> (1) $\hat{\varrho}(h, 1) < \varepsilon$,
> (2) $h(z)_m \in (-1, 1)$,
> (3) $h(u)_i = u_i$ for all $u \in Q$ and $i = 1, \ldots, m-1$,
> (4) $h \upharpoonright W = 1$ and $h[K_1] = K_1$.

An inspection of the proof of Lemma 1.6.4 will show that this can be proved along the same lines. ◇

The completes the solution to the first part of the exercise. The second part clearly follows from the first part and from Theorem 1.6.9.

▶**Exercise 1.10.8:**
Let $f_i \colon X \to X$ ($i \in \{1, 2\}$) be continuous functions such that the function

$$x \mapsto (f_1(x), f_2(x))$$

maps $X$ onto $X^2$. Define the functions $g_n\colon X \to X$ ($n \in \mathbb{N}$) as follows:
$$g_1 = f_1$$
$$g_2 = f_1 \circ f_2$$
$$g_3 = f_1 \circ f_2 \circ f_2$$
$$g_n = f_1 \circ \underbrace{f_2 \circ \cdots \circ f_2}_{n-1 \text{ times}}, \quad \text{for } n \in \mathbb{N}.$$

We claim that the function $f\colon X \to X^\infty$ defined by
$$x \mapsto (g_1(x), g_2(x), \ldots, g_n(x), \ldots)$$
maps $X$ onto $X^\infty$. Since $X$ is compact, it suffices to show that the image of $X$ under this map is dense in $X^\infty$ (Exercise A.5.5(1)), and it turn it suffices to show that for every sequence $y \in X^\infty$ and every $n \in \mathbb{N}$ there is an $x \in X$ such that $f(x)_i = y_i$ for all $i \leq n$. Equivalently, we need to prove that for every $n$-tuple $(y_1, \ldots, y_n) \in X^n$ there is an $x \in X$ such that $g_i(x) = y_i$ for $i \leq n$. We prove this by induction on $n$. For $n = 1$ this is just the fact that $f_1$ is onto. Now we assume the claim is true for $n$ and prove it for $n+1$. By the induction hypothesis, find $x' \in X$ such that $g_i(x') = y_{i+1}$, for $1 \leq i \leq n$. Now pick $x$ such that $f_1(x) = y_1$ and $f_2(x) = x'$. Then for $2 \leq i \leq n+1$ we have
$$g_i(x) = g_{i-1}(f_2(x)) = g_{i-1}(x') = y_i,$$
and this completes the proof.

▶**Exercise 1.11.19:**
Let $U \subseteq X$ be open. We claim that $e^{-1}[\langle\{U\}\rangle]$ and $e^{-1}[\langle\{U, X\}\rangle]$ are both open. This establishes continuity of $e$ by Lemma 1.11.11.
Let $(A, t) \in e^{-1}[\langle\{U, X\}\rangle]$. Pick a point
$$x \in e(A, t) \cap U = D(A, t) \cap U.$$
There is $\varepsilon > 0$ such that $D(x, \varepsilon) \subseteq U$. Pick $a \in A$ such that $\varrho(x, a) \leq t$. We claim that
$$e[\langle\{D(a, \varepsilon/2), X\}\rangle \times (t - \varepsilon/2, t + \varepsilon/2)] \subseteq \langle\{U, X\}\rangle.$$
Observe that this is as required since $\langle\{D(a, \varepsilon/2), X\}\rangle$ is a neighborhood of $A$ and
$$(t - \varepsilon/2, t + \varepsilon/2)$$
is a neighborhood of $t$. Indeed, pick an arbitrary
$$(A', t') \in \langle\{D(a, \varepsilon/2), X\}\rangle \times (t - \varepsilon/2, t + \varepsilon/2).$$
We have to prove that $D(A', t') \cap U \neq \emptyset$. There exists $a' \in D(a, \varepsilon/2) \cap A'$. Then
$$\varrho(a', x) \leq \varrho(a', a) + \varrho(a, x) \leq \varrho(a, x) + \varepsilon/2 \tag{1}$$
and
$$\varrho(a, x) \leq \varrho(a', a) + \varrho(a', x)$$
and therefore
$$\varrho(a', x) \geq \varrho(a, x) - \varrho(a', a) \geq \varrho(a, x) - \varepsilon/2.$$
We conclude that
$$\varrho(a', x) \in [\varrho(a, x) - \varepsilon/2, \varrho(a, x) + \varepsilon/2]. \tag{2}$$
**Case 1.** $\varrho(a, x) - \varepsilon/2 \leq 0$.

Then by (1),
$$\varrho(x, a') \leq \varrho(x, a) + \varepsilon/2 = \varepsilon.$$
Hence since $D(x, \varepsilon) \subseteq U$ we have
$$a' \in A' \cap U \subseteq D(A', t') \cap U,$$
and so $e(A', t') \cap U \neq \emptyset$.

**Case 2.** $\varrho(a, x) - \varepsilon/2 > 0$.

Observe that by (2) we have
$$\varrho(a', x) \leq (\varrho(a, x) - \varepsilon/2) + \varepsilon$$
and hence, by convexity of $\varrho$,
$$a' \in D\big(D(x, \varepsilon), \varrho(a, x) - \varepsilon/2\big).$$
So there is an element $x' \in D(x, \varepsilon)$ such that
$$\varrho(a', x') \leq \varrho(a, x) - \varepsilon/2.$$
Observe that
$$\varrho(a, x) - \varepsilon/2 \leq t - \varepsilon/2 < t'$$
from which it follows that
$$x' \in D(A', t') \cap D(x, \varepsilon) \subseteq D(A', t') \cap U.$$
We conclude that in this case $e(A', t')$ intersects $U$ as well.

To prove that $e^{-1}[\langle\{U\}\rangle]$ is open, let $(A, t) \in e^{-1}[\langle\{U\}\rangle]$. Then $D(A, t) \subseteq U$. Since $X$ is compact, there exists $\varepsilon > 0$ such that $D\big(D(A, t), \varepsilon\big) \subseteq U$ (Corollary A.5.4). This shows that
$$e\big[\langle\{D(A, \varepsilon/2)\}\rangle\big] \times [0, t + \varepsilon/2]\big] \subseteq \langle\{U\}\rangle,$$
which completes the solution to this exercise.

▶**Exercise 1.11.20:**
If $k = 1$ then there is nothing to prove. Simply observe that in that case we are dealing with a sequence with a convergent subsequence, in which case we are done, or with a sequence having no convergent subsequence. In the latter case, our sequence is discrete by Exercise A.2.1, in which case we are done as well.

Assume that the statement of the exercise is true for $k-1$, where $k \geq 2$. For every $i$ pick an arbitrary element $x_i \in A_i$ and let $B_i = A_i \setminus \{x_i\}$. Let the infinite set $E \subseteq \mathbb{N}$ and the (possibly empty) subset $B^0 \subseteq X$ be given by our induction hypothesis for the sequence $(B_i)_i$. Assume first that every point in $X$ has a neighborhood meeting finitely many terms of the sequence $(B_i)_{i \in E}$ only. We distinguish between two subcases. If there is an infinite subset $E_0 \subseteq E$ such that the sequence $(x_i)_{i \in E_0}$ is convergent, say to $x \in X$, then $E_0$, $\{x\}$ and for every $i \in E_0$ the partition $B_i, \{x_i\}$ of $A_i$ satisfy our inductive requirements. If there is no such $E_0$ then the sequence $(x_i)_{i \in E}$ is discrete (Exercise A.2.1) from which it follows that the sequence $(A_i)_{i \in E}$ has the property that every $x \in X$ has a neighborhood meeting finitely terms of the sequence $(A_i)_{i \in E}$ only. Simply observe that by assumption every $x \in X$ has a neighborhood $U_x$ meeting finitely many terms of the sequence $(B_i)_{i \in E}$ only. In addition, each $x \in X$ has a neighborhood $V_x$ meeting at most one element of the set $\{x_i : i \in E\}$. So $U_x \cap V_x$ is a neighborhood of $x$ meeting finitely many terms of

the sequence $(A_i)_{i \in E}$ only.
Suppose next that there are a nonempty subset $B^0$ of $X$ and for every $i \in E$ a partition $B_i^0, B_i^1$ of $B_i$ such that that

(1) $(B_i^0)_{i \in E}$ converges to $B^0$ in the Vietoris topology.
(2) Every $x \in X$ has a neighborhood meeting finitely many terms of the sequence $(B_i^1)_{i \in E}$ only.

We argue as before.

**Case 3.** There is an infinite subset $E_0 \subseteq E$ such that $(x_i)_{i \in E_0}$ converges to a point $x \in X$.

We then put $A^0 = \{x\} \cup B^0$ and $A_i^0 = \{x_i\} \cup B_i^0, A_i^1 = B_i^1$ for every $i \in E_0$. An easy check shows that our choices are as required (cf. Proposition 1.11.7).

**Case 4.** $(x_i)_{i \in E}$ does not have a convergent subsequence.

So the set $\{x_i : i \in E\}$ is closed and discrete in $X$ by Exercise A.2.1. Now put $A_i^0 = B_i^0$, $A^0 = B^0$ and $A_i^1 = B_i^1 \cup \{x_i\}$ for every $i \in E$. As in the case $k = 1$ it follows that the sequence $(A_i^1)_{i \in E}$ has the property that every $x \in X$ has a neighborhood meeting finitely many of its terms only.

▶**Exercise 1.11.23:**
Observe that $\pi$ is upper semi-continuous by Corollary 1.11.17. Let $Y = X/\mathcal{E}$ and define $f \colon Y \to 2^X$ by
$$f(y) = \pi^{-1}(y).$$
We claim that $f$ is continuous. An application of Exercise 1.11.16 then does the job.

To this end, let $(y_n)_n$ be a sequence in $Y$ converging to $y \in Y$. Put $E_n = \pi^{-1}(y_n)$ for every $n$, and $E = \pi^{-1}(y)$, respectively. Let $S$ be an arbitrary limit point of the sequence $(E_n)_n$ in $2^X$. That is, $S$ is the limit of a subsequence of $(E_n)_n$. Observe that by compactness of $2^X$ (Corollary 1.11.4) such limits exist.

**Claim 1.** $S \subseteq E$.

*Proof.* Striving for a contradiction, assume that there is an element $x \in S \setminus E$. Then $z = \pi(x) \neq y$, hence there is a closed neighborhood $V$ of $y$ which does not contain $z$. Since $(y_n)_n$ converges to $y$, all but finitely many $y_n$ belong to $V$. So $\pi^{-1}[V]$ is a closed subset of $X$ which contains all but finitely many $E_n$ (here we use that $\mathcal{E}$ is upper semi-continuous, or, equivalently, that $\pi$ is continuous). But then each limit point of the sequence $(E_n)_n$ must be contained in $\pi^{-1}[V]$ by Exercise 1.11.2. Since $x$ is a witness of the fact that $S$ is not contained in $\pi^{-1}[V]$, we reached the desired contradiction. ◇

Since $\mathcal{C}(X)$ is closed in $2^X$ (Lemma 1.11.14), it follows that $S$ is a subcontinuum of $E$. There is a subsequence of $(E_n)_n$ which converges to $S$. Since $\omega$ is continuous, this shows that $\omega(S) = t$. Hence $S$ cannot be a proper subcontinuum of $E$ for otherwise
$$t = \omega(S) < \omega(E) = t.$$

We conclude that $S = E$.
This proves that the sequence $(E_n)_n$ has only one limit point in $2^X$. Hence $(E_n)_n$ converges to $E$ (in $2^X$) and this is what we had to prove.

▶**Exercise 2.1.9:**
Let $\sigma = \{v_0, \ldots, v_n\}$, where the $v$'s are geometrically independent. By Theorem 2.1.3 it follows that the vectors $v_1 - v_0, \ldots, v_n - v_0$ are linearly independent in $\mathbb{R}^n$. By a change of base we may assume that $v_i - v_0$ is equal to the $i$-th unit vector

$$e_i = (\underbrace{0, \ldots, 0}_{i-1 \text{ times}}, 1, 0, \ldots, 0)$$

of $\mathbb{R}^n$. Observe that a change of base is done via a linear isomorphism which consequently is a homeomorphism since linear maps are continuous (see Page 113). So the composition of the translation $x \mapsto x - v_0$ and this linear isomorphism is a homeomorphism which maps $|\sigma|$ onto the standard simplex

$$|\tau| = \left\{ x \in \mathbb{R}^n : \forall i \leq n, x_i \geq 0 \text{ and } \sum_{i=1}^n x_i \leq 1 \right\}.$$

In addition, this homeomorphism sends $|\sigma|^\circ$ onto $|\tau|^\circ$. But for $|\tau|$ it is clear that its geometric interior coincides with its topological interior. So the same is true for $|\sigma|$.

▶**Exercise 2.1.16:**
Let $U = \{x \in |\mathcal{S}| : x \in \sigma \in \mathcal{T} \Rightarrow \sigma \in \mathcal{S}\}$. We claim that $U$ is open in $|\mathcal{T}|$. To prove this, let $\tau$ be an arbitrary element of $\mathcal{T}$. If $U \cap \tau = \emptyset$ then $U \cap \tau$ is certainly open in $\tau$. Suppose therefore that $U \cap \tau \neq \emptyset$. Then $\tau \in \mathcal{S}$ by the definition of $U$. Suppose that $y \in \tau \setminus U$. Then there has to be a simplex $\sigma \in \mathcal{T}$ such that $y \in \sigma$ and $\sigma \notin \mathcal{S}$. If $\tau \subseteq \sigma$ then $U \cap \sigma \neq \emptyset$ from which it follows that $\sigma \in \mathcal{S}$ which is a contradiction. So $\sigma \cap \tau$ is a proper face of $\tau$. In addition, every point of $\sigma \cap \tau$ does not belong to $U$ as $\sigma$ is a witness of that. So we conclude that $\tau \setminus U$ is the union of finitely many proper faces of $\tau$ and hence is closed, i.e., $\tau \cap U$ is open.
Let $y$ be an arbitrary element of $|\mathcal{S}| \setminus U$. There is a simplex $\sigma \in \mathcal{T}$ such that $y \in \sigma \notin \mathcal{S}$ and so $y \in \bigcup(\mathcal{T} \setminus \mathcal{S}) \cap |\mathcal{S}|$. In addition, every neighborhood of $y$ intersects the geometrical interior $\sigma^\circ$ of $\sigma$. We claim that $\sigma^\circ$ misses $|\mathcal{S}|$ from which it will follow that $y$ belongs to the topological boundary of $|\mathcal{S}|$. To the contrary, assume that there is an element $x \in \sigma^\circ \cap |\mathcal{S}|$. Let $\tau \in \mathcal{S}$ be such that $x \in \tau$. Then $x \in \sigma \cap \tau$ and hence $\sigma \cap \tau$ is a face of $\sigma$. But it cannot be a proper face since $x \in \sigma^\circ$. But this implies that $\sigma \subseteq \tau$ and in turn that $\sigma \in \mathcal{S}$ since $\mathcal{S}$ is a subcomplex (this is the only place in the solution where we use that $\mathcal{S}$ is a subcomplex). This contradiction establishes the solution.

▶**Exercise 2.2.1:**
This problem is equivalent to the following set-theoretical statement: *Let $V$ be a finite set and let $\mathcal{K}$ be a chain of subsets of $V$. In addition, let $T$ be an arbitrary nonempty subset of $V$ not belonging to $\mathcal{K} \cup \{V\}$. Then there exists a linear order $\leq$ on $V$ such that every element of $\mathcal{K}$ is an initial segment, while $T$ is not an initial segment.* We will solve this problem instead of the one in the text.
There is a linear order $\leq$ on $V$ such that every element of $\mathcal{K}$ is an initial $\leq$-segment.

If $T$ is not an $\leq$-initial segment then we are done. If not then let $t$ be the last element of $T$. Observe that $t$ is not the last element of $V$ for otherwise $T = V$. So let $s$ be the successor of $t$, interchange $s$ and $t$, and observe that the linear order we obtain in this way does the job for us.

▶**Exercise 2.4.1:**
Our notation is the same as in the proof of Lemma 2.4.1. Take an arbitrary $x \in U$ and observe that
$$\sum_{i=1}^{n} |x_i| < \sum_{i=1}^{n-1} \left(\frac{1}{n} + \frac{1}{n^2}\right) + \frac{1}{n^2} = 1. \tag{$*$}$$
Assume without loss of generality that $x_n \leq 0$ (the proof for $x_n \geq 0$ is simpler). Then the equality
$$x = (1 + x_n - \sum_{i=1}^{n-1} x_i) \cdot \underline{0} + \sum_{i=1}^{n-1} x_i \cdot e_i + (-x_n) \cdot (-e_n)$$
shows that $x$ is an affine combination of the $\{\underline{0}, e_1, \ldots, e_{n-1}, -e_n\}$ with non-negative coefficients by $(*)$. As a consequence,
$$f(x) = f^2(x) = (1 + x_n - \sum_{i=1}^{n-1} x_i) \cdot a_0 + \sum_{i=1}^{n-1} x_i \cdot a_i + (-x_n) \cdot a_n^2$$
belongs to $\sigma_2$.

▶**Exercise 2.4.3:**
By compactness there exists a finite open cover $\mathcal{E}$ of $C$ by nonempty open sets such that $\text{mesh}(\mathcal{E}) < \varepsilon/2$. Let $\mathcal{E} = \{E_1, \ldots, E_m\}$. For each $i \leq m$, pick an arbitrary point $y_i \in E_i$. Let $\kappa_1, \ldots, \kappa_m \colon C \to \mathbb{I}$ denote the $\kappa$-functions with respect to the cover $\mathcal{E} = \{E_1, \ldots, E_m\}$. Define $f_\varepsilon \colon C \to L$ by
$$f_\varepsilon(x) = \sum_{i=1}^{m} \kappa_i(x) \cdot y_i.$$
Then $f_\varepsilon$ is continuous and we claim that it is as required. It is clear that the range of $f_\varepsilon$ is contained in $C$ and is contained in a finite dimensional linear subspace of $L$ because each $f_\varepsilon(x)$ is a convex combination of the points $y_1, \ldots, y_m$. So it suffices to verify that $f_\epsilon$ does not move the points too far. This will be done in Claim 1 below.

For every $x \in C$, let $F_x = \{i \leq m : x \in E_i\}$. Observe that
$$\kappa_i(x) > 0 \quad \Longleftrightarrow \quad i \in F_x.$$
This shows that $\sum_{i \in F_x} \kappa_i(x) = 1$.

**Claim 1.** $\hat{\varrho}(f_\varepsilon, 1) < \varepsilon$.

*Proof.* Fix an arbitrary $x \in C$. Since $\text{mesh}(\mathcal{E}) < \varepsilon/2$ it follows that that
$$\|x - y_i\| < \varepsilon/2$$

for every $i \in F_x$. Consequently,

$$\|x - f_\varepsilon(x)\| = \left\|\sum_{i \in F_x} \kappa_i(x) \cdot x - \sum_{i \in F_x} \kappa_i(x) \cdot y_i\right\|$$
$$\leq \sum_{i \in F_x} \kappa_i(x)\|x - y_i\|$$
$$< \sum_{i \in F_x} \kappa_i(x) \cdot \varepsilon/2$$
$$= \varepsilon/2.$$

We conclude that $\hat{\varrho}(f_\varepsilon, 1) \leq \varepsilon/2 < \varepsilon$, as required. ◇

▶**Exercise 2.4.8:**
By Exercise A.10.14, $\mathbb{R}^n \setminus A$ has a unique unbounded component. Since $A$ separates $\mathbb{R}^n$, there consequently is a bounded component of $\mathbb{R}^n \setminus A$, say $U$. Pick an arbitrary element $p \in U$. Define $\beta \colon \mathbb{R}^n \setminus \{p\} \to \mathbb{S}^{n-1}$ by

$$\beta(x) = \frac{x-p}{\|x-p\|}.$$

It is clear that $\beta$ is well-defined and continuous. We claim that $\beta \restriction A$ it cannot be extended over $U \cup A$, and hence not over $\mathbb{R}^n$. Striving for a contradiction, assume that $f \colon U \cup A \to \mathbb{S}^{n-1}$ is a continuous extension of $\beta \restriction A$. Since both $A$ and $U$ are bounded, we can find $r > 0$ such that

$$B = B(p,r) \supseteq A \cup U.$$

Let
$$S = \{x \in \mathbb{R}^n : \|x-p\| = r\},$$

and define $g \colon D(p,r) \to S$ by the following formula:

$$g(x) = \begin{cases} p + r \cdot \beta(x) & (x \in D(p,r) \setminus U), \\ p + r \cdot f(x) & (x \in \overline{U}). \end{cases}$$

Then $g$ is well-defined and continuous since the intersection of the two closed sets $\overline{D(p,r) \setminus U}$ and $\overline{U}$ is contained in $\overline{U} \setminus U$ which is in turn contained in $A$ by Exercise A.2.9.
Let $x \in S$ be arbitrary. Then

$$g(x) = p + r \cdot \frac{x-p}{r} = x.$$

So $g$ is a retraction from $D(p,r)$ onto $S$, which contradicts the No-Retraction Theorem 2.4.10.
We conclude that indeed $\beta \restriction A \colon A \to \mathbb{S}^{n-1}$ cannot be extended over $\mathbb{R}^n$. So it cannot be nullhomotopic by Corollaries 1.2.13 and 1.4.3.

▶**Exercise 2.5.2:**
For (1) ⇒ (2), assume that there exists an antipodal-preserving map $f \colon \mathbb{S}^n \to \mathbb{S}^{n-1}$. By Exercise 2.5.1 we can cover $\mathbb{S}^{n-1}$ by a family $\mathcal{M}$ of $n+2$ closed sets such that no $M \in \mathcal{M}$ contains a pair of antipodal points. Since $f$ is antipodal preserving, the same is true for the family $f^{-1}[\mathcal{M}]$. But this contradicts (1). For (2) ⇒ (3), assume that there is an antipodal-preserving map $f \colon \mathbb{S}^{n-1} \to \mathbb{S}^{n-1}$ which is nullhomotopic.

Identify $\mathbb{S}^{n-1}$ and the subspace $\mathbb{S}^{n-1} \times \{0\}$ of $\mathbb{S}^n$. By Theorem 1.4.2 we can extend $f$ over the northern hemisphere of $\mathbb{S}^n$ to a continuous function $\bar{f}$. The function $\bar{f}$ also gives us a function on the southern hemisphere of $\mathbb{S}^n$ via the formula $x \mapsto -\bar{f}(-x)$. Since $f$ is antipodal preserving, these two functions agree on $\mathbb{S}^{n-1}$ and so their union is continuous and is obviously antipodal-preserving. But this contradicts (2). For (3) $\Rightarrow$ (4), assume that $f \colon \mathbb{S}^n \to \mathbb{R}^n$ is such that $f(x) \neq f(-x)$ for every $x \in \mathbb{S}^n$. Define $\hat{f} \colon \mathbb{S}^n \to \mathbb{S}^{n-1}$ by

$$\hat{f}(x) = \frac{f(x) - f(-x)}{\|f(x) - f(-x)\|}.$$

This function is antipodal-preserving and hence so is its restriction to $\mathbb{S}^{n-1}$. (We make the same identification as the one above.) But this restriction is nullhomotopic as it can be extended over the northern hemisphere of $\mathbb{S}^n$. This contradicts (3). Finally, for (4) $\Rightarrow$ (1), let $\mathcal{M}$ be a closed covering of $\mathbb{S}^n$ such that $|\mathcal{M}| = n+1$ and no $M \in \mathcal{M}$ contains an antipodal pair. Fix $M_0 \in \mathcal{M}$ and let $f_M \colon \mathbb{S}^n \to \mathbb{I}$ for $M \in \mathcal{M} \setminus \{M_0\}$ be a Urysohn function such that $f_M \restriction M \equiv 0$ and $f_M \restriction -M \equiv 1$. Define $f \colon \mathbb{S}^n \to \mathbb{R}^n$ by $f(x) = \bigl(f_M(x)\bigr)_{M \in \mathcal{M}}$. It suffices to observe that this function does not identify any antipodal pair of points.

▶**Exercise 3.1.9:**
Suppose that such component does not exist. Fix a component $C$ of $X$ for the time being. There exist partitions $D_i$ in $C$ between $A_i \cap C$ and $B_i \cap C$ for every $i$ such that $\bigcap_{i \in \Gamma} D_i = \emptyset$. There consequently are by Corollary 3.1.5 for all $i$ partitions $E_i$ in $X$ between $A_i$ and $B_i$ such that $\bigcap_{i \in \Gamma} E_i \cap C = \emptyset$. By Lemma A.10.1, there is a clopen neighborhood $U_C$ of $C$ such that $\bigcap_{i \in \Gamma} E_i \cap U_C = \emptyset$. This shows that the collection

$$\{(A_i \cap U_C, B_i \cap U_C) : i \in \Gamma\}$$

is inessential in $U_C$.
By compactness, finitely many of the clopen sets $U_C$ cover $X$. So there is a finite clopen cover $\mathcal{V}$ of $X$ such that

$$\{(A_i \cap V, B_i \cap V) : i \in \Gamma\}$$

is inessential in $V$ for every $V \in \mathcal{V}$. By Exercises A.1.3 and 3.1.8 we may assume without loss of generality that $\mathcal{V}$ is pairwise disjoint. For every $i \in \Gamma$ and $V \in \mathcal{V}$ let $D_i(V)$ be a partition in $V$ between $A_i \cap V$ and $B_i \cap V$ such that for every $V \in \mathcal{V}$,

$$\bigcap_{i \in \Gamma} D_i(V) = \emptyset.$$

Since $\mathcal{V}$ is a finite disjoint clopen cover of $X$, it is easy to show that for every $i \in \Gamma$ the set

$$E_i = \bigcup_{V \in \mathcal{V}} D_i(V)$$

is a partition in $X$ between $A_i$ and $B_i$. Since obviously $\bigcap_{i \in \Gamma} E_i = \emptyset$, we arrived at the desired contradiction.
The conclusion that every strongly infinite-dimensional compact space has a strongly infinite-dimensional component is trivial.

▶**Exercise 3.1.15:**
Let $E$ be Erdős' space. Then $E$ is totally disconnected and not zero-dimensional

since if $U \subseteq E$ is open such that
$$\underline{0} \in U \subseteq \{x \in E : \|x\| < 1\}$$
then $U$ is not clopen (see Example 1.5.18). This means that if $S$ is any partition between the closed sets $\{\underline{0}\}$ and $A = \{x \in E : \|x\| \geq 1\}$ of $E$ then $S \neq \emptyset$.
Let $S$ be any partition between $\{\underline{0}\}$ and $A$. We claim that $S$ is reducible. As we observed, $S \neq \emptyset$, so let $x \in S$ be arbitrary. Since $E$ is totally disconnected, there is a clopen set $C \subseteq E$ such that $x \in C$ but $\underline{0} \notin C$. We claim that $T = S \setminus C$ is a partition between $\{\underline{0}\}$ and $A$. Since $T$ is by construction a proper subset of $S$, this will do. Write $E \setminus S$ as the disjoint union of open sets $U$ and $V$ such that $\underline{0} \in U$ and $A \subseteq V$. Put $U' = U \setminus C$ and $V' = V \cup C$, respectively. Then $\underline{0} \in U'$ since $\underline{0} \notin C$ and $A \subseteq V'$. Also, clearly, $U' \cap V' = \emptyset$. So we done since $E \setminus (U' \cup V') = T$.

▶**Exercise 3.1.16:**
By Theorem 3.1.3, $\dim \mathbb{I}^n = n$ for every $n$. So the topological sum
$$X = \bigoplus_{n=1}^{\infty} \mathbb{I}^n$$
contains for every $n$ an essential family of $n$ pairs of disjoint closed subsets. Hence $\dim X = \infty$.
Conversely, if $X$ is infinite-dimensional then $X$ contains for every $n$ an essential family $\tau_n = \{(A_i, B_i) : i \leq n\}$ of pairs of disjoint closed subsets. Fix $n$. There clearly is a continuous function $\alpha \colon X \to \mathbb{I}^n$ sending for every $i \leq n$ the pair $(A_i, B_i)$ into the $i$-th pair of opposite faces of $\mathbb{I}^n$. So the collection of pairs of opposite faces of $\mathbb{I}^n$ must be essential, for otherwise $\tau_n$ is inessential (Exercise 3.1.11). But this clearly implies that $\partial \mathbb{I}^n$ is not a retract of $\mathbb{I}^n$ and so the Fixed-Point Theorem in dimension $n$ (Exercise 2.4.2).

▶**Exercise 3.2.3:**
The assumptions imply that we may pick a point $x_i \in U_i$ for every $i$ such that
$$i \neq j \implies x_i \neq x_j.$$
Let $V_i$ be an open neighborhood of $x_i$ such that $\overline{V}_i \subseteq U_i$ while moreover the collection
$$\{\overline{V}_1, \ldots, \overline{V}_m\}$$
is pairwise disjoint. There are two cases. First assume that $m \leq n + 1$. Then condition (3) is empty and so we need only to worry about (1) and (2). But it is clear how to do it. By Proposition A.7.1 there exist closed sets $G_i \subseteq U_i$ for every $i$ such that $\bigcup_{i=1}^m G_i = X$. The problem of course is that the proposition does not guarantee that the $G_i$'s are nonempty. But that is easily fixed by putting $F_i = G_i \cup \overline{V}_i$ for every $i$. Assume next that $m \geq n + 2$. We will deal with the indices $1, \ldots, n+2$ only. From the construction it will be clear how to proceed and to deal with all collections of $n + 2$ indices.
By Proposition A.7.1 there exist closed sets $G_i \subseteq U_i$ for every $i$ such that
$$\bigcup_{i=1}^m G_i = X.$$

For every $i$ put
$$B_i = (G_i \cup \overline{V_i}) \setminus \bigcup_{j \neq i} V_j.$$
Then $B_i$ is closed and nonempty because $\overline{V_i} \subseteq B_i$, and is contained in $U_i$. In addition, the collection $\{B_i : i \leq m\}$ covers $X$. For take an arbitrary $x \in X$, and let $i \leq m$ be such that $x \in G_i \cup \overline{V_i}$. If $x \notin \bigcup_{j \neq i} V_j$ then $x \in B_i$. If there exists $j \neq i$ such that $x \in V_j$ then $x \in B_j$.

For every $i \leq m$ let $W_i$ be an open neighborhood of $x_i$ such that $\overline{W_i} \subseteq V_i$ and put
$$A_i = (X \setminus U_i) \cup \bigcup_{j \neq i} \overline{W_j}.$$
Then $\{(A_i, B_i) : i \leq n+1\}$ is a family of $n+1$ pairs of disjoint closed subsets of $X$. Since $\dim X \leq n$, there exist open sets $S_i \subseteq X$ for $i \leq n+1$ such that

(1) $B_i \subseteq S_i \subseteq \overline{S_i} \subseteq X \setminus A_i$,
(2) $\bigcap_{i=1}^{n+1} \operatorname{Fr} S_i = \emptyset$.

Now consider the closed sets $B_1, \ldots, B_{n+2}$. For $i \leq n+1$ put $\hat{B}_i = \overline{S_i}$ and let
$$\hat{B}_{n+2} = B_{n+2} \setminus \bigcup_{i=1}^{n+1} S_i.$$

Observe that the $\hat{B}$'s are the 'improved' sets we get from Lemma 3.2.3 by considering the collections of closed sets $B_{n+2}, B_1, \ldots, B_{n+1}$ and open sets $S_1, \ldots, S_{n+1}$. So we get by (2) that
$$\bigcap_{i=1}^{n+2} \hat{B}_i \subseteq \bigcap_{i=1}^{n+1} \operatorname{Fr} S_i = \emptyset. \tag{3}$$

Observe that
$$\hat{\mathcal{B}} = \{\hat{B}_1, \hat{B}_2, \ldots, \hat{B}_{n+2}, B_{n+3}, B_{n+4}, \ldots, B_m\} \tag{4}$$
covers $X$ since $\bigcup_{i=1}^{n+2} B_i \subseteq \bigcup_{i=1}^{n+2} \hat{B}_i$. In addition, the first $n+2$ members of $\hat{\mathcal{B}}$ have empty intersection by (3).

It is clear that $\hat{B}_i \neq \emptyset$ for every $i \leq n+1$. But $\hat{B}_{n+2}$ is nonempty as well since $B_{n+2}$ contains $W_{n+2}$ and $W_{n+2} \cap \bigcup_{i=1}^{n+1} S_i = \emptyset$.

It is clear how to proceed. We first use Corollary 3.2.2 to swell the collection (4) to appropriate open sets. The first $n+2$ members of the new open cover have empty intersection. Then we pick another collection of $n+2$ indices and proceed as above. After finitely many steps we are done.

▶ **Exercise 3.2.4:**
Let $\tau = \{(A_i, B_i) : i \leq n\}$ be an essential family of pairs of disjoint closed subsets of $X$. By Exercise 3.1.9 there is a component $C$ of $X$ such that the collection
$$\{(A_i \cap C, B_i \cap C) : i \leq n\}$$
is essential in $C$. This prove that $\dim C \geq n$. That $\dim C \leq n$ is a consequence of the Subspace Theorem 3.2.9.

▶ **Exercise 3.2.7:**
The identity from $E \subseteq \ell^2$ to $E \subseteq \mathbb{Q}^\infty \subseteq \mathbb{R}^\infty$ is continuous by Lemma 1.1.12. Let

us denote this function by $f$. For every $\varepsilon > 0$, consider the open ball
$$B(\underline{0}, \varepsilon) = \{x \in E : \|x\| < \varepsilon\}.$$
Its closure in $E$ is the closed ball
$$D(\underline{0}, \varepsilon) = \{x \in E : \|x\| \leq \varepsilon\}.$$
We claim that $f[D]$ is a nowhere dense closed subset of $\mathbb{Q}^\infty$, where $D = D(\underline{0}, \varepsilon)$. To this end, let $q \in \mathbb{Q}^\infty \setminus f[D]$. Then $\|q\| > \varepsilon$, which implies that there exists $N \in \mathbb{N}$ such that
$$\sqrt{\sum_{i=1}^{N} q_i^2} > \varepsilon.$$
There clearly exists $\delta > 0$ such that if $|y_i - q_i| < \delta$ for every $i \leq N$ then
$$\sqrt{\sum_{i=1}^{N} y_i^2} > \varepsilon.$$
So
$$\prod_{i=1}^{N}(q_i - \delta, q_i + \delta) \times \prod_{i>N} \mathbb{R}_i$$
is a neighborhood of $q$ in $\mathbb{R}^\infty$ which misses $f[D]$. So we conclude that $f[D]$ is closed in $\mathbb{Q}^\infty$. It is clearly nowhere dense since every basic open subset of $\mathbb{Q}^\infty$ depends on finitely many coordinates only and hence contains elements of arbitrarily large norm.

Since $\mathbb{Q}^\infty$ is zero-dimensional, the set $\mathbb{Q}^\infty \setminus f[D]$ can be covered by (countably many) clopen sets. So by continuity of $f$ we get that in $E \subseteq \ell^2$ the complement of the closed ball $D(\underline{0}, \varepsilon)$ is the union of clopen subsets of $E$. So $\underline{0}$ has a neighborhood base of the required type. But since $E$ is a topological group it is homogeneous, so the same phenomenon is true at any point.

▶**Exercise 3.3.4:**
Observe that $C(\mathsf{C}, X)$ is topologically complete by Corollary 1.3.7. Since $\mathcal{S}(\mathsf{C}, X)$ is closed in $C(\mathsf{C}, X)$ by Proposition 1.3.8 and nonempty by Theorem 1.5.10, it is a Baire space by Theorem A.6.6.
For every $k \in \mathbb{N}$ put
$$\mathcal{C}_k = \{f \in \mathcal{S}(\mathsf{C}, X) : (\exists\, x \in X)(\exists\, x_1, \ldots, x_{n+2} \in f^{-1}(x)$$
$$\text{such that } i \neq j \text{ implies } \varrho(x_i, x_j) \geq 1/k)\}.$$

We first claim that $\mathcal{C}_k$ is closed. Suppose that $(f_m)_m$ is a sequence in $\mathcal{C}_k$ with limit $f \in \mathcal{S}(\mathsf{C}, X)$. For every $m$ there exists $x_m \in X$ and a subset $F_m \subseteq f_m^{-1}(x_m)$ of cardinality $n+2$, each two distinct points of which have distance at least $1/k$. By compactness of $X$ we may assume without loss of generality that there is an $x \in X$ such that $x_m \to x$ (Theorem A.5.1), and by compactness of $2^X$ (Corollary 1.11.4) we may assume for the same reason that without loss of generality there exists an $F$ such that $F_m \to F$. Since $\mathcal{F}_{n+2}(X)$ is closed in $2^X$ (Exercise 1.11.3), it follows that $|F| \leq n+2$. Let $\varepsilon > 0$ be so small that the collection $\mathcal{B} = \{B(x, \varepsilon) : x \in F\}$ is pairwise disjoint. We may assume without loss of generality that $\varepsilon < 1/{2k}$. By the definition of the topology of $2^X$ there exists $m \in \mathbb{N}$ such that $F_m \subseteq \bigcup \mathcal{B}$

and $F_m \cap B \neq \emptyset$ for every $B \in \mathcal{B}$. But every $B \in \mathcal{B}$ has diameter less than $1/k$ and hence contains at most one point of $F_m$. So $\mathcal{B}$ must have cardinality at least $n+2$ which shows that $|F| = n+2$. This argument also gives us that distinct points of $F$ are at least $1/k$ apart. Since $x_m \to x$, $f_m \to f$, $F_m \to F$ and $F_m \subseteq f_m^{-1}(x_m)$ for every $m$, we get $F \subseteq f^{-1}(x)$, i.e., $f \in \mathcal{C}_k$.

We next claim that $\mathcal{C}_k$ is nowhere dense in $\mathcal{S}(\mathsf{C}, X)$. To this end, let $f \in \mathcal{S}(\mathsf{C}, X)$ be arbitrary and let $\varepsilon > 0$. Let $\mathcal{E} = \{E_1, \ldots, E_m\}$ be a clopen partition of $\mathsf{C}$ of mesh less than $1/k$ such that $\mathrm{diam}(f[E_i]) < \varepsilon$ for every $i \leq m$. For every $i \leq m$ let $V_i$ be an open neighborhood of $f[E_i]$ of diameter less than $\varepsilon$. By Exercise 3.2.3 there exist for $i \leq m$ nonempty closed sets $B_i \subseteq V_i$ such that $\bigcup_{i=1}^m B_i = X$ and for any choice of indices $i_1 < \cdots < i_{n+2} \leq m$ we have $\bigcap_{j=1}^{n+2} B_{i_j} = \emptyset$. For every $i \leq m$ let $g_i \colon E_i \to B_i$ be a continuous surjection (Theorem 1.5.10). The union $g$ of the functions $g_i$ is a continuous surjection from $\mathsf{C}$ onto $X$. If $x \in \mathsf{C}$, say $x \in E_i$, then both $f(x)$ and $g(x)$ belong to $V_i$. So $\varrho(f(x), g(x)) < \varepsilon$ from which it follows that $\hat{\varrho}(f, g) < \varepsilon$ (cf. Exercise A.5.4). It remains to verify that $g \notin \mathcal{C}_k$. Striving for a contradiction, assume to the contrary that $g \in \mathcal{C}_k$. Then there exist $x \in X$ and a subset $F$ of $g^{-1}(x)$ of size $n+2$ that witness this. Since the mesh of $\mathcal{E}$ is less than $1/k$, every $E_i$ can contain at most one element of $F$. So $F$ is covered by precisely $n+2$ elements of $\mathcal{E}$, say $E_{i_1}, \ldots, E_{i_{n+2}}$. But then

$$g(x) \in \bigcap_{j=1}^{n+2} B_{i_j},$$

which is a contradiction since $B$'s with $n+2$ different indices have empty intersection.

By the Baire Category Theorem there consequently exists an element

$$\xi \in \mathcal{S}(\mathsf{C}, X) \setminus \bigcup_{k=1}^\infty \mathcal{C}_k.$$

If some fiber $\xi^{-1}(x)$ contains the pairwise distinct points $p_1, \ldots, p_{n+2}$ then for $k \in \mathbb{N}$ so large that

$$1/k < \min\{\varrho(p_i, p_j) : i \neq j \in \{1, 2, \ldots, n+2\}\}$$

we have $\xi \in \mathcal{C}_k$. So this is impossible and we conclude that $\xi$ is at most $(n+1)$-to-one.

▶**Exercise 3.3.7:**
Let $X$ be an $(n+1)$-dimensional compactum with essential family

$$\{(A_i, B_i) : i \leq n+1\}.$$

By Corollary 3.3.9 there are zero-dimensional subspaces

$$N_1, \ldots, N_{n+2}$$

of $X$ with $\bigcup_{k=1}^{n+2} N_i = X$. There is a partition $T$ between $A_1$ and $B_1$ such that $T$ misses $N_{n+2}$ (Corollary 3.1.6). Hence $T$ is contained in the $\bigcup_{i=1}^{n+1} N_i$ from which it follows that $\dim T \leq n$ by another application of Corollary 3.3.9. On the other hand, $\dim T \geq n$ by Theorem 3.1.9.

▶**Exercise 3.4.4:**
For (1), Let $T = \mathbb{R} \times X$. Then $T$ is topologically complete by Lemma A.6.2. The

Lavrentieff Theorem A.8.5 implies that the identity $i\colon Y \subseteq S \to Y \subseteq T$ can be extended to a homeomorphism over $G_\delta$-subsets $T'$ of $T$ and $S'$ of $S$, respectively. This shows that we are done, once we establish the following result.

**Claim 1.** *If $T' \subseteq T$ be a $G_\delta$-subset containing $Y$. Then there exists $t \in \mathbb{P}$ such that $\{y\} \times X \subseteq T'$.*

*Proof.* For every $n$ let $U_n \subseteq T$ be open such that $T' = \bigcap_{n=1}^\infty U_n$. Fix $n \in \mathbb{N}$ and let $q \in \mathbb{Q}$ be arbitrary. By Exercise A.5.8 the projection $\pi\colon T \to \mathbb{R}$ is closed. There consequently exists by Exercise A.5.5 an open neighborhood $V_q$ of $q$ such that $\pi^{-1}[V_q] = V_q \times X \subseteq U_n$. The union $V_n$ of the collection $\{V_q : q \in \mathbb{Q}\}$ is a open neighborhood of $\mathbb{Q}$ such that $V_n \times X \subseteq U_n$.
The intersection $V = \bigcap_{n=1}^\infty V_n$ is a $G_\delta$-subset of $\mathbb{R}$ such that $V \times X \subseteq T'$. There exists $t \in V \setminus \mathbb{Q}$ by (Corollary A.6.7). Such a $t$ is clearly as required. ◇

For (2), observe that $\dim Y \leq n$ (Theorem 3.4.12). As a consequence, $Y$ has a compactification $\gamma Y$ such that $\dim \gamma Y \leq n$ (Corollary 3.3.8). So $\dim(\gamma Y \setminus Y) \leq n$ by the Subspace Theorem 3.2.9. This implies that $\operatorname{def} Y \leq n$. However, $\operatorname{def} Y \geq n$ as well since if $\alpha Y$ is an arbitrary compactification of $Y$ then $\alpha Y \setminus Y$ contains a copy of $X$ by (1) and hence has to be at least $n$-dimensional again by the Subspace Theorem 3.2.9.

▶**Exercise 3.5.2:**
Let
$$X_1 \xleftarrow{f_1} X_2 \xleftarrow{f_2} X_3 \xleftarrow{f_3} \cdots,$$
be an inverse system consisting of at most $m$-dimensional spaces. By Corollary 3.3.8 the space $X_1$ has an at most $m$-dimensional compactification $a_1 X_1$. By Exercise 3.5.1 there exists an $m$-dimensional compactification $a_2 X_2$ such that $f_1$ can be extended to a continuous function $\bar{f}_1\colon a_2 X_2 \to a_1 X_1$. Etc. This yields an inverse sequence
$$a_1 X_1 \xleftarrow{\bar{f}_1} a_2 X_2 \xleftarrow{\bar{f}_2} a_3 X_3 \xleftarrow{\bar{f}_3} \cdots,$$
of at most $m$-dimensional compact spaces the inverse limit of which is at most $m$-dimensional by Exercise 3.2.2. But the inverse limit of the original sequence is a subspace of that inverse limit. So we are done by the Subspace Theorem 3.2.9.

▶**Exercise 3.5.4:**
As in the solution to Exercise 3.1.15, we start with Erdős' space $E$. Since $E$ admits an injective function into the zero-dimensional space $\mathbb{Q}^\infty$ (see Example 1.5.18), there is a countable family $\mathcal{C}$ of clopen subsets of $E$ such that if $x$ and $y$ are any two distinct points of $E$ then for some $C \in \mathcal{C}$ we have $x \in C \subseteq E \setminus \{y\}$.
If $U \subseteq E$ is open such that
$$\underline{0} \subseteq U \subseteq \{x \in E : \|x\| < 1\}$$
then $U$ is not clopen (see Example 1.5.18). This means that no partition between the closed sets $\{\underline{0}\}$ and $A = \{x \in E : \|x\| \geq 1\}$ of $E$ is empty.
Now let $X = \gamma E$ be a compactification of $E$ such that for every $C \in \mathcal{C}$ the sets $C$ and $E \setminus C$ have disjoint closures in $\gamma E$ (Theorem 3.5.1). This means that $\overline{C}$ is clopen in $\gamma E$ for every $C \in \mathcal{C}$. We claim that $X$ is as required.
To this end, let let $S$ be any partition between $\{\underline{0}\}$ and $\overline{A}$. Then $S$ is reducible.

Since $S \cap E$ is a partition between $\{\underline{0}\}$ and $A$ in $E$ we have $S \cap E \neq \emptyset$. Pick an arbitrary $x \in S \cap E$. There is an element $C \in \mathcal{C}$ such that $x \in C \setminus \{\underline{0}\}$. Now as in the solution to Exercise 3.1.15 it follows that $T = S \setminus \overline{C}$ is a partition between $\{\underline{0}\}$ and $\overline{A}$.

▶**Exercise 3.5.6:**
This is basically a triviality. We use Exercise 3.5.5. Let $\varrho$ and $\rho$ be admissible metrics for $X$ and $Y$, respectively. We endow $X \times Y$ with the admissible metric

$$d\big((x,y),(a,b)\big) = \max\{\varrho(x,a), \rho(y,b)\}$$

(cf. Exercise A.5.16). Let $\varepsilon > 0$. Let $U$ be an open neighborhood of $A$ in $X$ such that every continuum in $U$ has $\varrho$-diameter less than $\varepsilon$. Pick $V$ similarly for $B$. Now if $C$ is a continuum in $U \times V$ then $\pi_1[C]$, where $\pi_1 \colon X \times Y \to X$ is the projection, is a continuum in $U$, hence has $\varrho$-diameter at most $\varepsilon$. It follows similarly that the projection of $C$ into $Y$ has $\rho$-diameter less than $\varepsilon$. So $C$ has obviously $d$-diameter less than $\varepsilon$.

▶**Exercise 3.6.1:**
By induction on $n \geq 0$ we shall prove that every continuous function from $F_1$ into $\mathbb{S}^n \times M$, where $M$ is any AR, is continuously extendable over $X$. By taking

$$M = \{\text{pt}\}$$

we get the statement in the exercise.
Consider the case $n = 0$. So the collection $\{F_i : i \in \mathbb{N}\}$ is pairwise disjoint. Let

$$f \colon F_1 \to \mathbb{S}^0 \times M$$

be continuous, where $M$ is an AR. Assume that the closed set

$$A = f^{-1}(-1) \times M \subseteq F_1$$

is nonempty. Let $\tilde{X}$ be the space we obtain from $X$ by identifying $A$ to a single point $a$, i.e., $\tilde{X}$ is the space $X/A$, let $q \colon X \to \tilde{X}$ be the decomposition map and let $C$ be the component of $a$ in $\tilde{X}$. Then $C$ is a continuum which admits the pairwise disjoint, closed covering

$$\{\{a\}, q[B] \cap C\} \cup \{q[F_i] \cap C : i \geq 2\},$$

where $B = f^{-1}[\{1\} \times M]$. By Theorem A.10.6 we have $C = \{a\}$. So there is by Proposition A.10.3 a clopen subset $O$ of $\tilde{X}$ with $a \in O$ and $O \cap B = \emptyset$. Because $M$ is an AR we can find continuous functions

$$g_1 \colon q^{-1}[O] \to \{-1\} \times M$$

and

$$g_2 \colon X \setminus q^{-1}[O] \to \{1\} \times M$$

such that $g_1 \restriction A = f \restriction A$ and $g_2 \restriction B = f \restriction B$. Then $g_1 \cup g_2 \colon X \to \mathbb{S}^0 \times M$ is as required.
Assume now that we have what we want for $n$. So let $X$ be a compact space, let

$$\{F_i : i \in \mathbb{N}\}$$

be a closed cover of $X$ such that $\dim(F_i \cap F_j) \leq n$ for all distinct indices $i$ and $j$, let $M$ be an AR and let $f\colon F_1 \to \mathbb{S}^{n+1} \times M$ be continuous. By the Countable Closed Sum Theorem 3.2.8 we have that
$$B = \bigcup \{F_i \cap F_j : i, j \in \mathbb{N} \text{ and } i \neq j\}$$
has dimension $\leq n$. Let $x_1$ and $x_2$ be distinct points in $\mathbb{S}^{n+1}$ and note that
$$\mathbb{S}^{n+1} \setminus \{x_1, x_2\} \approx \mathbb{S}^n \times \mathbb{R}.$$
By Corollary 3.3.10 there is a closed covering $\{H_1, H_2\}$ of $X$ such that
$$H_j \cap f^{-1}[\{x_j\} \times M] = \emptyset$$
for $j = 1, 2$ and $\dim(H_1 \cap H_2 \cap B) < n$. Consider the compact space $X' = H_1 \cap H_2$ and its closed covering $\{F_i \cap X' : i \in \mathbb{N}\}$. Then
$$\dim(F_i \cap F_j \cap X') \leq \dim(B \cap X') < n$$
for $i \neq j$. Observe that $f \restriction F_1 \cap X'$ is a continuous function into
$$(\mathbb{S}^{n+1} \setminus \{x_1, x_2\}) \times M,$$
which is homeomorphic to $\mathbb{S}^n \times \mathbb{R} \times M$. Since $\mathbb{R} \times M$ is an AR, being a product of AR's, we may apply our induction hypothesis to find a continuous function
$$g\colon X' \to (\mathbb{S}^{n+1} \setminus \{x_1, x_2\}) \times M$$
with $g \restriction F_1 \cap X' = f \restriction F_1 \cap X'$. Observing that $\mathbb{S}^{n+1} \setminus \{x_j\}$ is homeomorphic to $\mathbb{R}^{n+1}$, we may select continuous functions
$$h_j\colon H_j \to (\mathbb{S}^{n+1} \setminus \{x_j\}) \times M, \quad j = 1, 2,$$
with $h_j \restriction X' = g$ and $h_j \restriction H_j \cap F_1 = f \restriction H_j \cap F_1$ (Corollary 1.2.12). Then $h = h_1 \cup h_2$ is a continuous mapping from $X$ into $\mathbb{S}^{n+1} \times M$ which extends $f$. So we are done.

▶**Exercise 3.6.4:**
We prove this by induction on $k + n$. Assume first that $k + n = 0$, i.e., $k = n = 0$. Then $f$ can be extended to a continuous function $\bar{f}\colon X \to \mathbb{S}^0$ by Theorem 3.6.3. So we put $B = \emptyset$ in this case.
Assume that we are done for all $0 \leq k \leq n$, where $k + n$ is less than or equal to $m \geq 0$.
Take $0 \leq k \leq n$ such that $k + n \leq m + 1$, and let $A, X$ and $f$ be as in the formulation of the exercise. Suppose first that $k = 0$. Put $E = f^{-1}(-1)$ and $F = f^{-1}(1)$, respectively. By Lemma 3.6.1 there is a partition $B$ between $E$ and $F$ such that $\dim B \leq n - 1$. Since $X \setminus B$ is the disjoint union of two open sets one of which contains $E$ and the other one contains $F$, it is clear that $f$ can be extended to a continuous function $\bar{f}\colon X \setminus B \to \mathbb{S}^0$. This completes the case that $k = 0$
Suppose next that $k > 0$.
We argue as in the proof of Theorem 3.6.3. Define
$$\mathbb{S}^k_1 = \{x \in \mathbb{S}^k : x_1 \geq 0\}, \quad \mathbb{S}^k_2 = \{x \in \mathbb{S}^k : x_1 \leq 0\},$$
respectively. Observe that $\mathbb{S}^k_1$ and $\mathbb{S}^k_2$ are both homeomorphic to $\mathbb{I}^k$ and that $\mathbb{S}^k_1 \cap \mathbb{S}^k_2$ is homeomorphic to $\mathbb{S}^{k-1}$. Now put $A_1 = g^{-1}[\mathbb{S}^k_1]$ and $A_2 = g^{-1}[\mathbb{S}^k_2]$. Let $X_1$ and $X_2$ be such as in Corollary 3.6.2. By our inductive assumption, there is a closed subset $B$ of $X_1 \cap X_2$ of dimension at most
$$(n - 1) - (k - 1) - 1 = n - k - 1$$

and which moreover misses $A_1 \cap A_2$ so that we can extend $g \restriction A_1 \cap A_2$ to a continuous function $h \colon (X_1 \cap X_2) \setminus B \to \mathbb{S}_1^n \cap \mathbb{S}_2^n$. Now for $i = 1, 2$ define
$$g_i \colon A_i \cup [(X_1 \cap X_2) \setminus B] \to \mathbb{S}_i^n$$
by $g_i = (g \restriction A_i) \cup h$. Since $\mathbb{S}_i^k$ is homeomorphic to $\mathbb{I}^k$, we can extend $g_i$, $i = 1, 2$, to a continuous function $\bar{g}_i \colon X_i \setminus B \to \mathbb{S}_i^k$ (Corollary 1.2.12). Clearly,
$$\bar{g} = \bar{g}_1 \cup \bar{g}_2 \colon X \setminus B \to \mathbb{S}^k$$
is a continuous extension of $g$. Since $B$ is closed and has the correct dimension, we are done.

▶**Exercise 3.7.4:**
Striving for a contradiction, suppose that every continuum in $L \setminus \bigcup \mathcal{N}$ misses one of the sets $A_{n_2}$ and $B_{n_2}$. Since $\mathcal{N}$ is finite and disjoint, there is a finite disjoint collection $\mathcal{N}'$ of closed subsets of $X$ of such that $\bigcup \mathcal{N}$ is contained in the interior $W$ of $\bigcup \mathcal{N}'$ while moreover every continuum in $\bigcup \mathcal{N}'$ misses $A_{n_1}$ or $B_{n_1}$ (Exercise 3.7.3). Each of the sets in the collection $\mathcal{N}'$ can be split into finitely many disjoint closed subsets each of which misses misses $A_{n_2}$ or $B_{n_2}$ (Exercise A.11.5). All these sets together form the collection $\mathcal{N}_1$. Since by assumption every continuum in $L \setminus \bigcup \mathcal{N}$ misses misses $A_{n_2}$ or $B_{n_2}$, the same is true for $L \setminus W$. Hence by the same argument as the one above, $L \setminus W$ can be covered by a finite disjoint collection $\mathcal{N}_2$ of closed sets such that every $N \in \mathcal{N}_2$ misses $A_{n_2}$ or $B_{n_2}$ (Exercise A.11.5). For $i = 1, 2$ let $S_i$ be a closed set in $X$ separating the disjoint closed sets
$$C_i = A_{n_i} \cup \bigcup \{N \in \mathcal{N}_i : N \cap B_{n_i} = \emptyset\},$$
$$D_i = B_{n_i} \cup \bigcup \{N \in \mathcal{N}_i : N \cap B_{n_i} \neq \emptyset\}.$$
Then clearly $S_1 \cap S_2 \cap L = \emptyset$. But this is a contradiction since the collection
$$\{(A_i, B_i) : i \in \Gamma(0) \cup \{n_1, n_2\}\}$$
is essential by Exercise 3.1.1.

▶**Exercise 3.7.6:**
Since $\dim X \leq 1$ there is a finite closed cover $\mathcal{F}$ of $X$ such that

(1) $\mathrm{mesh}(\mathcal{F}) < \varepsilon$.
(2) $\mathrm{ord}(\mathcal{F}) \leq 1$.

Let $\mathcal{N}$ be the collection of all intersections of precisely two distinct elements from $\mathcal{F}$. By (2), $\mathcal{N}$ is pairwise disjoint, and by (1), $\mathrm{mesh}(\mathcal{N}) < \varepsilon$. For every $F \in \mathcal{F}$ let
$$U(F) = F \setminus \bigcup \{F' \in \mathcal{F} : F' \neq F\}.$$
Then $U(F)$ is open since $\mathcal{F}$ is finite and the collection
$$\mathcal{U} = \{U(F) : F \in \mathcal{F}\}$$
is pairwise disjoint. Let $x \in X \setminus \bigcup \mathcal{N}$ be arbitrary. Then there exists $F \in \mathcal{F}$ containing $F$. But then for no $F' \in \mathcal{F} \setminus \{F\}$ we have $x \in F'$ for otherwise $x \in \bigcup \mathcal{N}$. We conclude that $x \in U(F)$.
Now let $C \subseteq X \setminus \bigcup \mathcal{N}$ be any continuum. Then $C$ is covered by the disjoint collection $\mathcal{U}$ and therefore has to be contained in an element of $\mathcal{U}$ and hence in an element of $\mathcal{F}$. We conclude that $\mathrm{diam}\, U < \varepsilon$, which is as required.

## ▶Exercise 3.10.2:

If the empty set is a partition between $A$ and $B$ then we are clearly done. So assume that this is not the case. Let $\{B_n : n \in \mathbb{N}\}$ be a countable base for $X$. Recursively, we will construct closed subsets $T_n \subseteq S$, as follows. Put $T_0 = S$. If $T_n$ has been defined, we perform the following test. If $T_n \setminus B_n$ cuts between $A$ and $B$ then we put $T_{n+1} = T_n \setminus B_n$, otherwise we put $T_{n+1} = T_n$. Then $(T_n)_n$ is a decreasing intersection of nonempty cuts between $A$ and $B$. Put $T = \bigcap_{n=1}^{\infty} T_n$. If $C$ is a continuum in $X$ meeting both $A$ and $B$ but missing $T$ then the sequence $(T_n \cap C)_n$ is a decreasing sequence of closed subsets of the compact space $C$ with empty intersection. There consequently is an $N \in \mathbb{N}$ such that $T_N \cap C = \emptyset$ (Exercise A.5.17). But this contradicts the fact that $T_N$ cuts between $A$ and $B$.
By Lemma 3.10.2 it follows that $T$ is a partition between $A$ and $B$. If $T' \subseteq T$ is a proper closed set which is also a partition between $A$ and $B$ then there exists an $n \in \mathbb{N}$ such that $B_n \cap T \neq \emptyset$ and $B_n \cap T' = \emptyset$. At stage $n$ of our construction we looked at the set $E = T_n \setminus B_n$. If $E$ is a cut between $A$ and $B$ then $T_{n+1}$ misses $B_n$ and hence $T$ misses $B_n$ which is a contradiction. On the other hand, if $E$ is not a cut between $A$ and $B$ then there is a continuum $C$ between $A$ and $B$ such that $C \cap E = \emptyset$. But then
$$C \cap T' \subseteq C \cap T_n \subseteq C \cap B_n \subseteq X \setminus T'.$$
So $C \cap T' = \emptyset$ which contradicts the fact that $T'$ is a cut between $A$ and $B$.

## ▶Exercise 3.10.3:

Let us begin by making a few general remarks that will be used without explicit reference throughout the remaining part of the solution to this exercise. First, $X$ is locally connected and topologically complete (Exercise A.5.1), so the notions of partition and cut agree in $X$ (Lemma 3.10.2). Second, the local connectivity of $X$ implies that components of open sets are open (Exercise A.2.8). This implies that if $E \subseteq X$ is closed and $C$ is an arbitrary component of $X \setminus E$ then $C \cup E$ is closed (Exercise A.2.9).

Striving for a contradiction, assume that $S$ is not connected. Write $S$ as the disjoint union of the nonempty closed sets $E$ and $F$. Since $F$ is a proper closed subset of $S$, it is not a partition between $A$ and $B$. As a consequence, there is a continuum $C$ in $X \setminus F$ intersecting both $A$ and $B$. Let $U$ be the component of $X \setminus F$ which contains $C$. Then $U$ is open and contains $A \cup B$ since $A \cup C \cup B$ is a continuum contained in $X \setminus F$. Observe that the sets $U \cup F$ and $E$ are closed. So if $E' = U \cap E$ then
$$\overline{E'} \subseteq (U \cup F) \cap E = U \cap E = E' \subseteq \overline{E'}.$$
We conclude that $E'$ is a clopen subset of $E$ and hence of $S$. So $E'$ is a proper closed subset of $S$. By irreducibility of $S$, it is not a partition between $A$ and $B$ and hence not a cut. Let $\mathcal{V}$ be the collection of all components of $X \setminus E'$. As above, we can find an element $V \in \mathcal{V}$ containing $A \cup B$. Let $V'$ be a component of $X \setminus E'$ different from $V$. Then $V' \cup E'$ is closed and hence $\operatorname{Fr}(V') \cap E' \neq \emptyset$ since $X$ is connected. Hence $U \cup V'$ is connected since $E' \subseteq U$. We conclude that
$$K = U \cup \bigcup \{V' \in \mathcal{V} : V' \neq V\}$$
is connected. In addition, $K \cup V = X$ since $E' \subseteq U \subseteq K$.

**Claim 1.** $K \cap V \cap S = \emptyset$.

*Proof.* If $x \in K \cap V \cap S$ then since components are disjoint we have $x \in U \cap V$. Since $E' \subseteq U$ and $V \cap E' = \emptyset$ it follows that $x \in F \cup (E \setminus E')$. But
$$U \cap (F \cup (E \setminus E')) = \emptyset.$$
This is a contradiction. ◇

Since $A \cup B \subseteq K$ and $A \cup B \subseteq V$ there are by connectivity of $K$ and $V$ and Exercise 1.5.9 subcontinua $A'$ and $B'$ of $X$ such that
$$A \cup B \subseteq A' \subseteq K, \quad A \cup B \subseteq B' \subseteq V, \quad A' \cup B' = X.$$
By Claim 1, $A' \cap B' \cap S = \emptyset$. Since $A \cup B \subseteq A' \cap B'$ and $S$ is a partition between $A$ and $B$ this proves that the set $A' \cap B'$ is disconnected, contradicting the fact that $X$ is unicoherent.

▶**Exercise 3.10.5:**
We may assume without loss of generality that the spaces $X_i$ are compact (Corollary 3.3.8).
Assume that
$$h = (h_1, h_2) \colon \mathbb{S}^n \to X \times Y$$
is an imbedding, where $X$ and $Y$ are compact and $\dim X \leq 1$. We shall show that $\dim Y \geq n$. This clearly suffices since the product of at most $n-1$ at most one-dimensional spaces is at most $(n-1)$-dimensional (Theorem 3.4.12).
Consider the compactum $h_1[\mathbb{S}^n]$. If it is a singleton then $\mathbb{S}^n$ clearly imbeds in $Y$. So then $\dim Y \geq n$ by the Subspace Theorem 3.2.9 since $\dim \mathbb{S}^n = n$ (Theorem 3.2.12). So we assume without loss of generality that $h_1[\mathbb{S}^n]$ is not a singleton. Since $X_1$ is at most one-dimensional, there is an at most zero-dimensional closed subset $A \subseteq X_1$ which separates $h_1[\mathbb{S}^n]$. Then $h_1^{-1}[A]$ separates $\mathbb{S}^n$. Since by Exercise 3.6.7 $\mathbb{S}^n$ is unicoherent, there is a continuum $C \subseteq h_1^{-1}[A]$ which also separates $\mathbb{S}^n$ (Exercises 3.10.2 and 3.10.3). By Exercise 2.4.9 there is a continuous function $f \colon C \to \mathbb{S}^{n-1}$ which cannot be extended over $\mathbb{S}^n$. Since $h_1[C]$ is a connected subset of the at most zero-dimensional set $A$, it is a singleton and hence $h_2 \restriction C \colon C \to Y$ is an imbedding. Put $D = h_2[C]$. Let $u \colon D \to C$ be the inverse of the imbedding $h_2 \restriction C$. Because the function
$$f = (f \circ u) \circ (h_2 \restriction C) \colon C \to \mathbb{S}^{n-1}$$
cannot be extended over $\mathbb{S}^n$, the map $f \circ u \colon D \to \mathbb{S}^{n-1}$ cannot be extended over $Y$. But this implies that $\dim Y \geq n$ (Theorem 3.6.5).

▶**Exercise 3.13.3:**
Write $G = \bigcap_{i=1}^{\infty} U_i$, where $U_i$ is open for every $i$. For every $n$ let $K_n = [-1/n, 1/n]$. Since
$$\bigcap_{n=1}^{\infty} K_n^{\infty} = \{\mathbf{0}\},$$
by compactness of the set $K_1^{\infty}$ it follows that for some index $n_1 \in \mathbb{N}$ we have that $K_{n_1}^{\infty} \subseteq U_1$. Observe that
$$K_{n_1} \times \{\mathbf{0}\} \subseteq U_2.$$
So again by compactness, there is an integer $n_2 > n_1$ such that
$$K_{n_1} \times K_{n_2} \times K_{n_2} \times \cdots \subseteq U_2.$$

Observe that
$$K_{n_1} \times K_{n_2} \times \{\mathbf{0}\} \subseteq U_3.$$
So by the same argumentation there is an integer $n_3 > n_2$ such that ..., etc. At the end of our process we obtain that
$$\prod_{i=1}^{\infty} K_{n_i} \subseteq \bigcap_{i=1}^{\infty} U_i = G,$$
which proves our claim.

▶**Exercise 3.13.5:**
Let $f\colon \mathsf{C} \to Q$ be a continuous surjection (Theorem 1.5.10) and define
$$u\colon Q \to \mathsf{C} \subseteq \mathbb{R}$$
by $u(q) = \max f^{-1}(q)$.

**Claim 1.** $u$ is upper semi-continuous.

*Proof.* Let $r \in \mathbb{R}$ and $q \in u^{-1}[(-\infty, r)]$. Then $u(q) < r$ and so $f^{-1}(q) \subseteq (-\infty, r)$. By compactness of $\mathsf{C}$ the function $f$ is closed (Exercise A.5.5). There consequently exists a neighborhood $V$ of $q$ in $Q$ such that $f^{-1}[V] \subseteq (-\infty, r)$ (Exercise A.1.15). Now for every $p \in V$ we have $u(p) < r$. We conclude that $u^{-1}[(-\infty, r)]$ is open. ◇

**Claim 2.** $u$ is not countably continuous.

*Proof.* Let $Q = E_1 \cup E_2 \cup \cdots$. Since $Q$ is not the union of countably many zero-dimensional subspaces (Corollary 3.13.6), for some $i$ we have $\dim E_i > 0$. We claim that $u \upharpoonright E_i$ is not continuous. Observe that the composition
$$f \circ (u \upharpoonright E_i)$$
is the identity on $E_i$ and that $f$ is continuous. But then if $u \upharpoonright E_i$ were continuous this would imply that $u \upharpoonright E_i \colon E_i \to u[E_i] \subseteq \mathsf{C}$ is an imbedding. But this is a contradiction since $\dim E_i > 0$ but $\dim \mathsf{C} = 0$. ◇

So we are done.

▶**Exercise 4.1.5:**
Let $\mathcal{U}$ be the collection of all open subsets of $X$ of diameter less than $\varepsilon$. By Theorem 4.1.1 there is an open refinement $\mathcal{V}$ of $\mathcal{U}$ such that for every space $Y$, any two $\mathcal{V}$-close mappings $f, g\colon Y \to X$ are $\mathcal{U}$-homotopic. Let $\delta > 0$ be a Lebesgue number for $\mathcal{V}$ (Lemma A.5.3). We claim that this $\delta$ works. To this end, let $Y$ be a space, $A \subseteq Y$ be closed and $f, g\colon A \to X$ continuous functions with $\hat{\varrho}(f, g) < \delta$. Assume moreover that $f$ can be extended to a continuous function $\bar{f}\colon Y \to X$. First observe that $f$ and $g$ are $\mathcal{V}$-close, so they are $\mathcal{U}$-homotopic, say by the $\mathcal{U}$-homotopy $H$. By the 'controlled' Borsuk Homotopy Extension Theorem 4.1.3 there is a homotopy $\bar{H}\colon Y \times \mathbb{I} \to X$ which is limited by $\mathcal{U}$ while moreover $\bar{H}_0 = \bar{f}$ and $\bar{H} \upharpoonright (A \times \mathbb{I}) = H$. Clearly $\bar{H}_0$ and $\bar{H}_1$ are $\mathcal{U}$-close, so
$$\hat{\varrho}(\bar{H}_0, \bar{H}_1) < \varepsilon.$$
Since $\bar{H}_0 = f$ and $\bar{H}_1 \upharpoonright A = g$, we conclude that $\bar{g} = \bar{H}_1$ is as required.

▶**Exercise 4.3.3:**
Assume that $X \setminus \{x\}$ is an ANR. We aim at applying the characterization Theorem 4.2.1. To that end, let $\mathcal{U}$ be an open cover of $X$. Fix $U \in \mathcal{U}$ such that $x \in U$. There exists a homotopically trivial neighborhood $A$ of $x$ such that $A \subseteq U$. Now let $B$ be an open neighborhood of $x$ with $\overline{B} \subseteq A$. There exists an open refinement $\mathcal{U}_0$ of $\mathcal{U}$ such that for every $U_0 \in \mathcal{U}_0$ with $U_0 \cap B \neq \emptyset$, we have $U_0 \subseteq A$. Since $X \setminus \{x\}$ is an ANR, there exists an open cover $\mathcal{V}$ of $X \setminus \{x\}$ such that for every locally finite simplicial complex $\mathcal{T}$ and every subcomplex $\mathcal{S}$ of $\mathcal{T}$ containing all the vertices of $\mathcal{T}$, every partial realization of $\mathcal{T}$ in $X \setminus \{x\}$ relative to $(\mathcal{S}, \mathcal{V})$ can be extended to a full realization of $\mathcal{T}$ in $X \setminus \{x\}$ relative to $\mathcal{U}_0 \cap (X \setminus \{x\})$. Put $\mathcal{W} = \mathcal{V} \cup \{B\}$. Then $\mathcal{W}$ is an open cover of $X$. Now let $\mathcal{T}$ be a locally finite simplicial complex, let $\mathcal{S}$ be a subcomplex of $\mathcal{T}$ containing all the vertices of $\mathcal{T}$, and let $f\colon |\mathcal{S}| \to X$ be a partial realization of $\mathcal{T}$ in $X$ relative to $(\mathcal{S}, \mathcal{W})$. Define

$$\mathcal{T}_0 = \{\sigma \in \mathcal{T} : (\exists V \in \mathcal{V})(f[\sigma \cap |\mathcal{S}|] \subseteq V)\}, \quad \mathcal{T}_1 = \mathcal{T} \setminus \mathcal{T}_0.$$

In addition, put $\mathcal{S}_0 = \mathcal{S} \cap \mathcal{T}_0$. It is clear that $\mathcal{T}_0$ is a subcomplex of $\mathcal{T}$ and that $\mathcal{S}_0$ contains all the vertices of $\mathcal{T}_0$. Also, $g = f \restriction |\mathcal{S}_0|$ is a partial realization of $\mathcal{T}_0$ relative to $(\mathcal{S}_0, \mathcal{V})$. Consequently, $g$ can be extended to a full realization $\bar{g}$ of $\mathcal{T}_0$ in $X \setminus \{x\}$ relative to $\mathcal{U}_0 \restriction (X \times [0,1))$.
Now put $\mathcal{S}_1 = \mathcal{S} \cup \mathcal{T}_0$ and define $f_1 \colon |\mathcal{S}_1| \to X$ by

$$f_1(x) = \begin{cases} f(x) & (x \in |\mathcal{S}|), \\ \bar{g}(x) & (x \in |\mathcal{T}_0|). \end{cases}$$

It is clear that $f_1$ is well-defined and continuous. By a standard argument, by induction on $n \geq 0$, we shall construct a continuous function

$$\bar{f}_{n+1} \colon |\mathcal{S}_1 \cup \mathcal{T}^{(n)}| \to X$$

having the following properties:

(1) $\bar{f}_1 = f_1$ and for $n \geq 1$, $\bar{f}_{n+1}$ extends $\bar{f}_n$,
(2) for every $\sigma \in \mathcal{T}^{(n)} \cap \mathcal{T}_1$, $\bar{f}_{n+1}[\sigma] \subseteq A$.

Observe that since $\mathcal{S}$ contains all the vertices of $\mathcal{T}$, our choice $\bar{f}_1 = f_1$ is possible. Now assume that for certain $n \geq 0$, the function $\bar{f}_{n+1}$ has been constructed. Naturally, we shall construct $\bar{f}_{n+2}$ 'simplex-wise'. To this end, take an arbitrary simplex

$$\sigma \in \mathcal{S}_1 \cup \mathcal{T}^{(n+1)}.$$

If $\sigma \in \mathcal{S}_1 \cup \mathcal{T}^{(n)}$ then put $f_\sigma = \bar{f}_{n+1} \restriction \sigma$. If $\sigma \in \mathcal{T}^{(n+1)} \setminus (\mathcal{S}_1 \cup \mathcal{T}^{(n)})$, then $\sigma$ belongs to $\mathcal{T}_1$, from which it follows that $f[\sigma \cap |\mathcal{S}|] \subseteq B$.

**Claim 1.** $\bar{f}_{n+1}[\partial \sigma] \subseteq A$.

*Proof.* Let $\tau$ be a proper face of $\sigma$. There are two cases to consider.
First assume that $\tau \in \mathcal{T}_0$. Then there exists an element

$$U \in \mathcal{U}_0 \cap (X \times [0,1))$$

such that $\bar{f}_{n+1}[\tau] = f_1[\tau] \subseteq U$. Since, as was just observed, $f_1$ maps every vertex of $\tau$ in $B$, we conclude that $U \cap B \neq \emptyset$. Consequently, by the special choice of $\mathcal{U}_0$

we obtain, $\bar{f}_{n+1}[\tau] \subseteq A$.
Second, assume that $\tau \notin \mathcal{T}_0$. Then by our inductive hypothesis (2) we get
$$\bar{f}_{n+1}[\tau] \subseteq A,$$
which is again as required.

Now since $A$ is homotopically trivial, there exists a continuous extension $f_\sigma \colon \sigma \to A$ of $\bar{f}_{n+1} \upharpoonright \partial \sigma$ (Lemma 4.2.13). Define $\bar{f}_{n+2} \colon |\mathcal{S}_1 \cup \mathcal{T}^{(n+1)}| \to X$ by
$$\bar{f}_{n+2}(x) = f_\sigma(x) \qquad (x \in \sigma \in \mathcal{S}_1 \cup \mathcal{T}^{(n+1)}).$$
An application of Lemma 2.1.19 establishes the required continuity of $\bar{f}_{n+2}$. This completes the inductive construction.
Now define $\bar{f} \colon |\mathcal{T}| \to X$ by
$$\bar{f}(x) = \bar{f}_{n+1}(x) \qquad (x \in |\mathcal{T}^{(n)}|).$$
Applying Lemma 2.1.19 again gives us that $\bar{f}$ is a full realization of $\mathcal{T}$ in $\triangle(X)$ relative to $\mathcal{U}$. Since $\bar{f}$ clearly extends $f$, we are done.

▶**Exercise 5.1.6:**
Let $P$ be a polyhedron, $\varepsilon > 0$ and $f \in C(P, Q)$ be arbitrary. Let $\mathcal{U}$ be a finite open cover of $Q$ consisting of convex sets of diameter less than $\varepsilon$. Let $\varepsilon_0 > 0$ be a Lebesgue number for $\mathcal{U}$ (Lemma A.5.3). In addition, let $\delta > 0$ be such that if $A \subseteq P$ and $\operatorname{diam} A < \delta$ then $\operatorname{diam} f[A] < \varepsilon_0$. Here we use that $f$ is uniformly continuous (Exercise A.5.18). Let $\mathcal{T}$ be a triangulation of $P$ such that every $\sigma \in \mathcal{T}$ has diameter less than $\delta$ (Theorem 2.2.10).
We will now proceed by induction on the dimension of $P$. If $\dim P = 0$ then $P$ is finite and we obviously get what we want since $A$ is nowhere dense (Exercise 5.1.5). Now assume that we have what we want for all polyhedra which are $n$-dimensional, where $n \geq 0$, and assume that $\dim P = n+1$. Let $R$ be the union of all at most $n$-dimensional simplices of $\mathcal{T}$. Then $R$ is an at most $n$-dimensional subpolyhedron of $P$. For every $\sigma \in \mathcal{T}$ there exists by construction an element $U_\sigma \in \mathcal{U}$ such that $f[\sigma] \subseteq U_\sigma$. Since $U_\sigma$ is open and $f[\sigma]$ is compact,
$$\gamma_\sigma = \varrho(f[\sigma], Q \setminus U_\sigma) > 0$$
(Exercise A.5.3). Since $\mathcal{T}$ is finite,
$$\gamma = \min\{\gamma_\sigma : \sigma \in \mathcal{T}\} > 0.$$
We can apply our inductive hypothesis to $R$ and the function $f \upharpoonright R$ to get a continuous function $\alpha \colon R \to Q \setminus A$ such that
$$\hat{\varrho}(\alpha, f \upharpoonright R) < \gamma. \tag{1}$$
Now pick an arbitrary $(n+1)$-dimensional simplex in $\mathcal{T}$. Since $f[\sigma] \subseteq U_\sigma$ it follows from (1) that
$$\alpha[\partial \sigma] \subseteq U_\sigma \setminus A.$$
Since $U_\sigma$ is convex it is homotopically trivial, and by assumption, so is $U_\sigma \setminus A$. So $U_\sigma \setminus A$ is a homotopically trivial **ANR** by Corollary 4.3.2 and hence is an **AR** by Theorem 4.2.20. We can therefore extend the function $\alpha \upharpoonright \partial \sigma \colon \partial \sigma \to U_\sigma \setminus A$ to a continuous function $\beta_\sigma \colon \sigma \to U_\sigma \setminus A$. Now let $g$ be the union of the functions
$$\beta_\sigma, \quad \sigma \in \mathcal{T}.$$

Then $g$ is continuous by Lemma 2.1.19. It clearly suffices to prove that $\hat{\varrho}(f, g) < \varepsilon$. To this end, let $x \in P$ be arbitrary and let $\sigma \in \mathcal{T}$ be a simplex with maximal dimension which contains $x$. Then both $f[\sigma]$ and $g[\sigma]$ are contained in $U_\sigma$ and hence $\varrho(f(x), g(x)) < \varepsilon$ since $\operatorname{diam} U_\sigma < \varepsilon$.

▶ **Exercise 5.1.7:**
Let $A \subseteq Q$ be an $A$-set. In addition, let $\varepsilon > 0$. For every $n$ we identify $\mathbb{J}^n$ and the subspace
$$\mathbb{J}^n \times \{(0, 0, \dots)\}$$
of $Q$. For a sufficiently large $n$ the projection $\pi \colon Q \to \mathbb{J}^n$ moves the points less than $\frac{1}{2}\varepsilon$. Since $\mathbb{J}^n$ is a polyhedron, there exists by Exercise 5.1.6 a continuous function
$$\beta \colon \mathbb{J}^n \to Q \setminus A$$
which moves the points less than $\frac{1}{2}\varepsilon$. So the function $\beta \circ \pi \colon Q \to Q \setminus A$ moves the points less than $\varepsilon$. Now apply Lemma 5.1.2(7).

Conversely, let $A \in \mathcal{Z}(X)$, and let $U \subseteq Q$ be nonempty and homotopically trivial. Fix $n \geq 1$ and let $f \colon \mathbb{S}^{n-1} \to U \setminus A$ be continuous. As in the solution to Exercise 5.1.6 it follows that $U$ is an AR. So we can extend $f$ to a continuous function $\bar{f} \colon B^n \to U$. Let $\delta = \varrho(f[B^n], Q \setminus U)$. Since $A$ is a $Z$-set, we may by Exercise 5.1.3 approximate $\bar{f}$ by a continuous function $g \colon B^n \to Q \setminus A$ such that $\hat{\varrho}(g, \bar{f}) < \delta$. Observe that this implies that $g[B^n] \subseteq U \setminus A$. There are two function from $\mathbb{S}^{n-1}$ to $U \setminus A$, namely, $f$ and $g \restriction \mathbb{S}^{n-1}$. Since $U \setminus A$ is an ANR, being an open subspace of an AR (Corollary 4.3.2), it follows that we can choose the approximation $g$ so close to $\bar{f}$ that the functions $f$ and $g \restriction \mathbb{S}^{n-1}$ are homotopic (Theorem 4.1.1). By the Borsuk Homotopy Extension Theorem 1.4.2, it therefore follows that we may extend $f$ to a continuous function $F \colon B^n \to U \setminus A$.

▶ **Exercise 5.2.1:**
Let $\varepsilon_1$ be such that $\hat{\varrho}(f, 1_E) < \varepsilon_1 < \varepsilon$ and let $n \in \mathbb{N}$ be so large that
$$2^{-n} < \frac{\varepsilon - \varepsilon_1}{\varepsilon_1 + 5}.$$
Let $K = E \cup F$. Then $K$ is a compact subset of $s$. Define an imbedding $\phi \colon K \to s$ by the formula
$$\phi(x) = (x_1, \dots, x_n, 0, x_1, x_2, \dots).$$
It is clear that
$$\hat{\varrho}(\phi, 1_K) \leq \sum_{i=n+1}^{\infty} 2^{-i} \cdot 2 = 2^{-n+1}.$$
So by Theorem 5.3.3 there is an element $\bar{\phi} \in \mathcal{H}(Q)$ such that $\bar{\phi} = \phi$ and
$$\hat{\varrho}(\bar{\phi}, 1_Q) < 2^{-n+1}.$$
For $0 \leq t \leq 1$ let $\theta_t \colon \mathbb{J} \to \mathbb{J}$ be the unique homeomorphism that takes the intervals $[-1, 0]$ and $[0, 1]$ linearly onto the intervals $[-1, \frac{1}{2}t]$ and $[\frac{1}{2}t, 1]$, respectively. Define the isotopy $G \colon Q \times \mathbb{I} \to Q$ by
$$\begin{cases} G(x, t)_i = \theta_t(x_i) & (i = n + 1), \\ G(x, t)_i = x_i & (i \neq n + 1). \end{cases}$$

It is clear that
$$\hat{\varrho}(G_t, 1_Q) = \tfrac{1}{2} t \cdot 2^{-n-1} \leq 2^{-n-2}$$
for all $0 \leq t \leq 1$. Finally, let $G' \colon Q \times \mathbb{I} \to Q$ be the isotopy $G'_t = G_t^{-1}$ for $0 \leq t \leq 1$. Put $A = \phi[E]$ and $B = G_1 \circ \phi[F]$. Then $A$ and $B$ are compact disjoint subsets of $s$. This is clear since $\pi_{n+1}[A] \subseteq \{0\}$ and $\pi_{n+1}[B] \subseteq \{\tfrac{1}{2}\}$. The functions $\alpha \colon E \to Q$ and $\beta \colon F \to Q$ defined by
$$\alpha(x) = (\tfrac{1}{2} x_1 + \tfrac{1}{2} f(x)_1, \ldots, \tfrac{1}{2} x_n + \tfrac{1}{2} f(x)_n, 0, x_1, x_2, \ldots),$$
$$\beta(x) = G_1 \circ \alpha \circ f^{-1}(x),$$
are well-defined imbeddings. Observe that $\alpha[E]$ and $\beta[F]$ are disjoint compact subsets of $s$. Now define a function $h \colon A \cup B \to s$ by
$$h(x) = \begin{cases} \alpha \circ \phi^{-1}(x) & (x \in A), \\ \beta \circ \phi^{-1} \circ G_1^{-1}(x) & (x \in B). \end{cases}$$
It is easy to see that $h \restriction A$ and $h \restriction B$ are imbeddings. Moreover,
$$h[A] = \alpha[E], \quad h[B] = \beta[B]$$
from which it follows that $h$ is an imbedding as well. We claim that since
$$A = \{(x_1, \ldots, x_n, 0, x_1, x_2, \ldots) : x \in E\},$$
$$B = \{(x_1, \ldots, x_n, \tfrac{1}{2}, x_1, x_2, \ldots) : x \in F\},$$
it follows from the definitions of $\alpha$ and $\beta$ and
$$\hat{\varrho}(f, 1_E) = \hat{\varrho}(f^{-1}, 1_F) < \varepsilon_1$$
that
$$\hat{\varrho}(h \restriction A, 1_A) < \tfrac{1}{2}\varepsilon_1, \quad \hat{\varrho}(h \restriction B, 1_B) < \tfrac{1}{2}\varepsilon_1 + 2^{-n-1}\varepsilon_1,$$
so $\hat{\varrho}(h, 1_{A \cup B}) < \tfrac{1}{2}\varepsilon_1 + 2^{-n-1}\varepsilon_1 = \varepsilon_2$. For let
$$y = (x_1, \ldots, x_n, 0, x_1, x_2, \ldots) \in A$$
for certain $x \in E$. Then $\phi^{-1}(y) = x$ and
$$\alpha(x) = (\tfrac{1}{2} x_1 + \tfrac{1}{2} f(x)_1, \ldots, \tfrac{1}{2} x_n + \tfrac{1}{2} f(x)_n, 0, x_1, x_2, \ldots)$$
so that
$$\varrho(y, \alpha(x)) = \sum_{i=1}^{\infty} 2^{-i} |y_i - \alpha(x)_i|$$
$$= \sum_{i=1}^{n} 2^{-i} |x_i - \tfrac{1}{2} x_i - \tfrac{1}{2} f(x)_i| + 0$$
$$= \sum_{i=1}^{n} 2^{-i} \tfrac{1}{2} |x_i - f(x)_i|$$
$$\leq \tfrac{1}{2} \varrho(x, f(x))$$
$$< \tfrac{1}{2} \varepsilon_1.$$
Moreover, if
$$z = (x_1, \ldots, x_n, \tfrac{1}{2}, x_1, x_2, \ldots)$$
for certain $x \in F$ then
$$G_1^{-1}(z) = (x_1, \ldots, x_n, 0, x_1, x_2, \ldots)$$

and $\phi^{-1}(G_1^{-1}(z)) = x$ so that for $s = f^{-1}(x)$ we have
$$\alpha(s) = (\tfrac{1}{2}s_1 + \tfrac{1}{2}x_1, \ldots, \tfrac{1}{2}s_n + \tfrac{1}{2}x_n, 0, s_1, s_2, \ldots),$$
and so
$$h(z) = (\tfrac{1}{2}s_1 + \tfrac{1}{2}x_1, \ldots, \tfrac{1}{2}s_n + \tfrac{1}{2}x_n, \tfrac{1}{2}, s_1, s_2, \ldots).$$
This shows that
$$\varrho(z, h(z)) = \sum_{i=1}^{n} 2^{-i}|\tfrac{1}{2}x_i - \tfrac{1}{2}s_i| + 0 + 2^{-n-1}\varrho(x, s)$$
$$< \tfrac{1}{2}\varepsilon_1 + 2^{-n-1}\varepsilon_1.$$

So by Theorem 5.2.4 there exists an element $\bar{h} \in \mathcal{H}(Q)$ such that $\bar{h} \upharpoonright A \cup B = h$ and $\hat{\varrho}(\bar{h}, 1_Q) < \varepsilon_2$.

Define the function $H \colon Q \times \mathbb{I} \to Q$ by
$$H_t = \bar{\phi}^{-1} \circ G_t^{-1} \circ \bar{h}^{-1} \circ G_t \circ \bar{h} \circ \bar{\phi}$$
for all $0 \leq t \leq 1$. We claim that $H$ is as required. Since
$$\bar{\phi}, \quad \bar{\phi}^{-1}, \quad G, \quad G', \quad \bar{h}, \quad \bar{h}^{-1},$$
are uniformly continuous, we see that $H$ is continuous, i.e., $H$ is an isotopy. It remains to verify the following:

(1) $H_0 = 1_Q$,
(2) $H_1 \upharpoonright E = f$,
(3) $\hat{\varrho}(H_u, H_v) < \varepsilon$ for all $u, v \in \mathbb{I}$.

Observe that (1) is trivial since $G_0 = 1_Q$.
For (2), let $x \in E$. Then $\bar{\phi}(x) = \phi(x) \in A$, hence $\bar{h}(\bar{\phi}(x)) = \alpha(x)$. Also,
$$G_1 \circ \phi(f(x)) \in B$$
and so $h(G_1 \circ \phi(f(x))) = \beta(f(x)) = G_1(\alpha(x))$. So $G_1(\alpha(x)) \in h[B]$ from which it follows that
$$\bar{h}^{-1}(G_1(\alpha(x))) = G_1 \circ \phi(\beta^{-1}(G_1 \circ \alpha(x))) = G_1 \circ \phi(f(x)).$$
Hence
$$H_1(x) = \bar{\phi}^{-1} \circ G_1^{-1}(G_1 \circ \phi(f(x))) = f(x),$$
which is as required.

For (3), take arbitrary $u, v \in \mathbb{I}$. For every $0 \leq t \leq 1$ we have
$$\hat{\varrho}(\bar{\phi}^{-1} G_t^{-1} \bar{h} G_t, 1_Q) \leq \hat{\varrho}(\bar{\phi}, 1_Q) + 2\hat{\varrho}(G_t, 1_Q) + \hat{\varrho}(\bar{h}, 1_Q)$$
$$< 2^{-n+1} + 2^{-n-1} + \tfrac{1}{2}\varepsilon_1 + 2^{-n-1}\varepsilon_1$$
$$= 2^{-n-1}(\varepsilon_1 + 5) + \tfrac{1}{2}\varepsilon_1$$
$$< \tfrac{1}{2}\varepsilon$$

which implies that

$$\hat{\varrho}(H_u, H_v) = \hat{\varrho}(\bar{\phi}^{-1}G_u^{-1}\bar{h}^{-1}G_u\bar{h}\bar{\phi}, \bar{\phi}^{-1}G_v^{-1}\bar{h}^{-1}G_v\bar{h}\bar{\phi})$$
$$= \hat{\varrho}(\bar{\phi}^{-1}G_u^{-1}\bar{h}^{-1}G_u, \bar{\phi}^{-1}G_v^{-1}\bar{h}^{-1}G_v)$$
$$\leq \hat{\varrho}(\bar{\phi}^{-1}G_u^{-1}\bar{h}^{-1}G_u, 1_Q) + \hat{\varrho}(1_Q, \bar{\phi}^{-1}G_v^{-1}\bar{h}^{-1}G_v)$$
$$< \varepsilon.$$

So $H$ is an $\varepsilon$-isotopy.

▶**Exercise 5.3.1:**
Pick $\varepsilon_1 > 0$ such that $\hat{\varrho}(f, 1_E) < \varepsilon_1 < \varepsilon$ and put $\delta = \frac{1}{4}(\varepsilon - \varepsilon_1)$. Let $K = E \cup F$. Then $K \in \mathcal{Z}(Q)$ by Lemma 5.1.2(3). By Theorem 5.3.5 there is an element

$$\varphi \in \mathcal{H}(Q)$$

such that $\varphi[K] \subseteq s$ and $\hat{\varrho}(\varphi, 1_Q) < \delta$. Let $A = \varphi[E] \subseteq s$, $B = \varphi[F] \subseteq s$ and define the homeomorphism $g\colon A \to B$ by

$$g = \varphi \circ f \circ (\varphi^{-1} \restriction A).$$

Notice that

$$\hat{\varrho}(g, 1_A) < 2\hat{\varrho}(\varphi, 1_Q) + \hat{\varrho}(f, 1_E) < 2\delta + \varepsilon_1 = \varepsilon_2.$$

By Exercise 5.2.1 there is an $\varepsilon_2$-isotopy $G\colon Q \times \mathbb{I} \to Q$ such that

$$G_0 = 1_Q, \quad G_1 \restriction A = g.$$

Now define the isotopy $H\colon Q \times \mathbb{I} \to Q$ in the obvious way by

$$H_t = \varphi^{-1} \circ G_t \circ \varphi \qquad (0 \leq t \leq 1).$$

Then $H_0 = 1_Q$, $H_1(x) = \varphi^{-1}(g(\varphi(x))) = f(x)$ for $x \in E$, and for all $u, v \in \mathbb{I}$ we have

$$\hat{\varrho}(H_u, H_v) = \hat{\varrho}(\varphi^{-1}G_u\varphi, \varphi^{-1}G_v\varphi)$$
$$= \hat{\varrho}(\varphi^{-1}G_u, \varphi^{-1}G_v)$$
$$\leq \hat{\varrho}(\varphi^{-1}G_u, G_u) + \hat{\varrho}(G_u, G_v) + \hat{\varrho}(G_v, \varphi^{-1}G_v)$$
$$< 2\delta + \varepsilon_2$$
$$= \varepsilon.$$

So we are done.

▶**Exercise 5.4.1:**
We will make use of the capset $\Xi$ defined on Page 333.
Define the homeomorphism $\varphi\colon Q \times Q \to Q$ by

$$\varphi(x, y) = (x_1, y_1, x_2, y_2, \dots).$$

An easy observation shows that $\varphi[B(Q) \times Q]$ is the union of an infinite collection endfaces of $Q$, hence is a capset by Corollary 5.4.8.
This also shows that $\varphi[(B(Q) \times Q) \cup (Q \times B(Q))]$ is a capset since it is the union of a capset and a $\sigma Z$-set and hence is a capset (Theorem 5.4.3(2) and Lemma 5.1.2(5)).
It remains to show that its subset $\varphi[B(Q) \times B(Q)]$ is a capset as well. To this end,

let $\{E_n : n \in \mathbb{N}\}$ be a partition of $\mathbb{N}$ such that $E_n$ is infinite and $\min(E_n) + 1 \in E_n$ for every $n$. Factorize $Q$ as

$$\mathbb{J}^{E_1} \times \mathbb{J}^{E_2} \times \cdots \times \mathbb{J}^{E_n} \times \cdots .$$

Define

$$\mathcal{A} = \{W^1_{\min(E_n)} \cap W^1_{\min(E_n)+1} : n \in \mathbb{N}\}$$

and notice that

$$\bigcup \mathcal{A} \subseteq \varphi[B(Q) \times B(Q)].$$

For every $n \in \mathbb{N}$ we have that

$$\prod_{i \in E_n} [-1 + 2^{-n}, 1 - 2^{-n}]_i$$

and

$$\{x \in \mathbb{J}^{E_n} : x_{\min(E_n)} = x_{\min(E_n)+1} = 1\}$$

are homeomorphic to $Q$ and are elements of $\mathcal{Z}(\mathbb{J}^{E_n})$. There consequently exists by the Homeomorphism Extension Theorem 5.3.7 a homeomorphism $h \colon \mathbb{J}^{E_n} \to \mathbb{J}^{E_n}$ such that

$$h_n\Big[\prod_{i \in E_n}[-1 + 2^{-n}, 1 - 2^{-n}]_i\Big] = \{x \in \mathbb{J}^{E_n} : x_{\min(E_n)} = x_{\min(E_n)+1} = 1\}.$$

It follows that the homeomorphism

$$h = h_1 \times h_2 \times \cdots \times h_n \times \cdots \colon Q \to Q$$

has the property that $h[\Xi] \subseteq \bigcup \mathcal{A} \subseteq \varphi[B(Q) \times B(Q)]$. So $B(Q) \times B(Q)$ is a capset by Proposition 5.4.6 and Theorem 5.4.3(2). [Remark: This exercise can also be solved by using the characterization of capsets in Corollary 5.4.11.]

▶ **Exercise 5.4.10:**
If $K = Q$ then there is nothing to prove. So assume without loss of generality that $K \neq Q$.
Write $Q \setminus K$ as $\bigcup_{i=0}^{\infty} C_i$, where each $C_i$ is compact, $C_0 = \emptyset$ and each $C_i$ is contained in the interior of $C_{i+1}$. For every $i$ let

$$\delta_i = \min(\mu[C_i]).$$

Observe that $\delta_1 \geq \delta_2 \geq \cdots \geq \delta_i \geq \cdots > 0$, and that $\lim_{i \to \infty} \delta_i = 0$.
Put $g_0 = f$. By induction on $i \geq 1$ we will construct continuous functions

$$g_i \colon Q \to Q$$

having the following properties:

(1) $g_i \restriction K = f \restriction K$ and $g_i \restriction C_i$ is a $Z$-imbedding,
(2) $g_i \restriction C_{i-1} = g_{i-1} \restriction C_{i-1}$,
(3) $\hat{\varrho}(g_i, g_{i-1}) < 2^{-i}$,
(4) if $x \in Q \setminus K$ then $\varrho\big(g_i(x), g_{i-1}(x)\big) < 2^{-i}\mu(x)$,
(5) $(C_i \cap g_i^{-1}[B(Q)]) \setminus K = (A \cap C_i) \setminus K$.

Suppose that for some $i \geq 0$ we constructed $g_i$ satisfying our inductive requirements. We will construct $g_{i+1}$. Let $\varepsilon = \min(2^{-(i+1)}, 2^{-(i+1)}\delta_{i+2})$.
Let $\delta > 0$ be such that for every space $Y$ and every closed subspace $A \subseteq Y$ and all continuous functions $\xi, \eta \colon A \to Q$ with $\hat{\varrho}(\xi, \eta) < \delta$, if $\xi$ has a continuous extension $\bar{\xi} \colon Y \to Q$ then $\eta$ has a continuous extension $\bar{\eta} \colon Y \to Q$ with $\hat{\varrho}(\bar{\xi}, \bar{\eta}) < \varepsilon$ (Exercise 4.1.5).
The function $g_i \colon C_{i+1} \to Q$ can be approximated by a $Z$-imbedding $h \colon C_{i+1} \to Q$ such that

(6) $h \upharpoonright C_i = g_i \upharpoonright C_i$,
(7) $(C_{i+1} \cap h^{-1}[B(Q)]) \setminus K = (A \cap C_{i+1}) \setminus K$

(Corollary 5.4.13). So we may assume that $\hat{\varrho}(h, g_i \upharpoonright C_{i+1}) < \delta$. Let $V$ be the interior of $C_{i+2}$. By construction, $C_{i+1} \subseteq V$. Put $A = (Q \setminus V) \cup C_{i+1}$. Define a function $\xi \colon A \to Q$ by

$$\xi(q) = \begin{cases} g_i(q) & (q \in Q \setminus V), \\ h(q) & (q \in C_{i+1}). \end{cases}$$

The distance between $\xi$ and $g_i \upharpoonright A$ is obviously smaller than $\delta$. Since $g_i \upharpoonright A$ can be extended to $g_i \colon Q \to Q$ we find that $\xi \colon A \to Q$ can be extended to a continuous function $\bar{\xi} \colon Q \to Q$ such that $\hat{\varrho}(\bar{\xi}, g_i) < \varepsilon$. We claim that by putting $g_{i+1} = \bar{\xi}$ we satisfy our inductive demands. Observe that $(1)_{i+1}$, $(2)_{i+1}$, $(3)_{i+1}$ and $(5)_{i+1}$ are trivially satisfied. We will check $(4)_{i+1}$. Since $g_{i+1}$ and $g_i$ agree on $Q \setminus V$, it suffices to consider a point $x \in C_{i+2}$. Observe that for such $x$,

$$\varrho(g_i(x), g_{i+1}(x)) < \varepsilon \leq 2^{-(i+1)}\delta_{i+2}.$$

Since $\delta_{i+2} \leq \mu(x)$ for every $x \in C_{i+2}$, we are done.
Put $g = \lim_{i \to \infty} g_i$. Then $g$ is continuous by (3) and Lemma A.3.1. Clearly,

$$g \upharpoonright K = f \upharpoonright K.$$

By (1) and (2) it follows that $g \upharpoonright Q \setminus K$ is one-to-one. In addition, for $x \in C_i$ we get by (4) and (2) that

$$\varrho(f(x), g(x)) \leq \sum_{j=1}^{i} \varrho(g_{j-1}(x), g_j(x))$$
$$< \sum_{j=1}^{i} 2^{-j} \mu(x)$$
$$\leq \mu(x).$$

It therefore suffices to prove that $g^{-1}[B(Q)] \setminus K = A \setminus K$. That this is true follows easily from (5).

▶ **Exercise 5.5.1:**
By Exercise A.2.18 we may assume that $\varepsilon$ is defined on all of $X$ and that $\varepsilon \upharpoonright K \equiv 0$. By Corollary 5.3.10 there is a deformation $S \colon Q \times \mathbb{I} \to Q$ through $Q \setminus f[K]$. We may assume without loss of generality that

$$\varrho(S_t(x), x) \leq t$$

for every $x \in Q$ and $t \in \mathbb{I}$ (Proposition 4.1.7). Define $f'\colon X \to Q$ by the formula
$$f'(x) = S_{\varepsilon(x)/4}(f(x)).$$
Then $f'$ is continuous, $f' \restriction K = f$ and
$$\varrho(f(x), f'(x)) \leq \tfrac{1}{4}\varepsilon(x) \tag{1}$$
for every $x \in X$. In addition, if $x \in X \setminus K$ then $\varepsilon(x) > 0$ and so $f'(x) \notin f[K]$. Let $H\colon Q \times \mathbb{I} \to Q$ be a deformation through $A$. Again by Proposition 4.1.7 we may assume that
$$\varrho(H_t(x), x) \leq \tfrac{1}{2}t$$
for every $x \in Q$ and $t \in \mathbb{I}$. Define $\xi\colon X \to [0, \infty)$ by
$$\xi(x) = \min\{\tfrac{1}{4}\varepsilon(x),\, \tfrac{1}{2}\varrho(f'(x), f'[K])\}.$$
Then $\xi$ is clearly continuous. Finally, define $g\colon X \to Q$ by
$$g(x) = H_{\xi(x)}(f'(x)).$$
Then $g$ is continuous, $g \restriction K = f' \restriction K = f$. Also,
$$\varrho(f'(x), g(x)) \leq \min\{\tfrac{1}{4}\varepsilon(x),\, \tfrac{1}{2}\varrho(f'(x), f'[K])\} \tag{2}$$
for every $x \in X$. So (1) and (2) show that for arbitrary $x$ we have
$$\varrho(f(x), g(x)) \leq \varrho(f(x), f'(x)) + \varrho(f'(x), g(x)) \leq \tfrac{1}{4}\varepsilon + \tfrac{1}{4}\varepsilon < \varepsilon.$$
In addition, if $x \in X \setminus K$ then $\varepsilon(x) > 0$ and so $f'(x) \notin f'[K]$. As a consequence, by (2) we get $g(x) \notin f'[K] = g[K]$. So we are done.

▶**Exercise 5.5.2:**
The solution to this exercise is similar to the proof of Theorem 5.5.5. Let
$$f = (f_1, f_2)\colon Q \to Q \times M$$
be a $Z$-imbedding. In addition, let $K$ and $C$ be subsets of $Q$ such that $K$ is closed and $C \in \mathcal{M}$. Just as in the proof of Theorem 5.5.5 we can find a continuous function $g_1\colon Q \to Q$ closely approximating $f_1$ such that
$$g_1^{-1}[A] \setminus K = C \setminus K, \qquad g_1 \restriction K = f_1 \restriction K$$
and $g_1 \restriction Q \setminus K$ is a one-to-one map whose range is a $\sigma Z$-set. Since there is a deformation of $M$ through $X$, we can find a continuous function $g_2\colon Q \to M$ which closely approximates $f_2$ such that $g_2 \restriction K = f_2 \restriction K$ while moreover $g_2[Q \setminus K] \subseteq X$ (Exercise 4.1.9). The map $g = (g_1, g_2)$ is a $Z$-imbedding of $Q$ into $Q \times M$ which closely approximates $f$ while moreover $g \restriction K = f \restriction K$ and $g^{-1}[A \times X] \setminus K = C \setminus K$.

▶**Exercise 5.5.3:**
We adopt the notation we introduced on Page 333. We proved in Proposition 5.4.6 that the set
$$\Xi = \bigcup_{n=1}^{\infty} \Xi_n,$$
where
$$\Xi_n = \prod_{i=1}^{\infty} [-1 + 2^{-n},\, 1 - 2^{-n}]_i,$$

is a capset for $Q$. By the same argument as in the proof of that proposition, it follows that
$$\Xi(\infty) = \bigcup_{n=1}^{\infty} \Xi_n^\infty$$
is a capset for $Q^\infty$. But $\Xi(\infty)$ is contained in $\Xi^\infty$, which is an $\mathcal{F}_{\sigma\delta}$-absorber in $Q^\infty$ by Theorem 5.5.5. So the $\mathcal{F}_{\sigma\delta}$-absorber $\Xi^\infty$ contains a capset. By the topological uniqueness of $\mathcal{F}_{\sigma\delta}$-absorbers (Theorem 5.5.2), this consequently holds for all $\mathcal{F}_{\sigma\delta}$-absorbers.

▶**Exercise 5.5.4:**
Let $A_1 \supseteq A_2$ be an $(\mathcal{F}_\sigma, \mathcal{F}_\sigma)$-absorbing pair in $Q$ (Corollary 5.5.6). By the proof of Theorem 5.5.2 there is a homeomorphism $h \colon Q \to Q$ such that
$$h[A] = A_2, \quad h[B] \subseteq A_1, \quad \hat{\varrho}(h,1) < \varepsilon/3.$$
Since $A_1 \supseteq A_2$ is an $(\mathcal{F}_\sigma, \mathcal{F}_\sigma)$-absorbing pair in $Q$ and $C \supseteq C$ is a decreasing pair of $F_\sigma$-subsets of $X$, there is a $Z$-imbedding $p \colon X \to Q$ such that

(5) $p^{-1}[A_1] \setminus K = C \setminus K = p^{-1}[A_2] \setminus K$,
(6) $p \upharpoonright K = (h \circ f) \upharpoonright K$,
(7) $\hat{\varrho}(p, h \circ f) < \varepsilon/3$.

Let $g = h^{-1} \circ p$. Then $g \upharpoonright K = f \upharpoonright K$. Furthermore
$$\hat{\varrho}(g, f) = \hat{\varrho}(h^{-1} \circ p, f) \leq \hat{\varrho}(h^{-1} \circ p, p) + \hat{\varrho}(p, h \circ f) + \hat{\varrho}(h \circ f, f)$$
$$\leq \hat{\varrho}(h^{-1}, 1) + \hat{\varrho}(p, h \circ f) + \hat{\varrho}(h, 1)$$
$$< \varepsilon.$$

Finally,
$$g^{-1}[A] \setminus K = p^{-1}\big[h[A]\big] \setminus K = p^{-1}[A_2] \setminus K = C \setminus K$$
and
$$g[X \setminus (C \cup K)] \cap B \subseteq h^{-1}\big[p[X \setminus (C \cup K)]\big] \cap h^{-1}[A_1]$$
$$= h^{-1}\big[p[X \setminus C] \cap p[X \setminus K] \cap A_1\big]$$
$$= \emptyset.$$

So we are done.

▶**Exercise 5.5.5:**
Let $\varrho_n$ be an admissible metric on $E_n$ such that
$$\varrho(x,y) = \max\{\varrho_n(x_n, y_n) : n \in \mathbb{N}\}$$
is an admissible metric on $E$ (Exercise A.5.16). Pick an arbitrary subset $X \subseteq E$ such that
$$\prod_{n=1}^{\infty} X_n \subseteq X \subseteq \prod_{n=1}^{\omega} C_n.$$
We will modify the proof of Theorem 5.5.5 to obtain the desired result. Consider a map $f \colon Q \to E$ that restricts to a $Z$-imbedding on some compactum $K$ and a sequence $Q \supseteq A_1 \supseteq A_2 \supseteq \cdots$ of $F_\sigma$-subsets. Put $A_\infty = \bigcap_{n=1}^\infty A_n$. We may assume that $f$ is a $Z$-imbedding (Theorem 5.3.11). Write $Q \setminus K$ as a union of a

sequence $(F_i)_{i=0}^{\infty}$ of compacta with $F_i \subseteq \text{Int}(F_{i+1})$ for every $i$ and $F_0 = \emptyset$. Let $\varepsilon > 0$ and put
$$\varepsilon_i = \min\{2^{-i}\varepsilon, \tfrac{1}{2}\varrho(f[K], f[F_i])\}$$
for every $i$. Consider now the $n$-the component $f_n \colon Q \to Q$ of $f$. We can construct just as in the proof of Theorem 5.5.5 a sequence $\alpha_0, \alpha_1, \ldots$ of continuous functions from $Q$ to $E_n$ such that for every and $i$:

(1) $\hat{\varrho}_n(\alpha_i, \alpha_{i-1}) < \varepsilon_i$, $\alpha_i \restriction F_{i-1} = \alpha_{i-1} \restriction F_{i-1}$,
(2) $\alpha_i \restriction Q \setminus F_{i+1} = f_n \restriction Q \setminus F_{i+1}$ and $\alpha_i \restriction F_i$ is a $Z$-imbedding,
(3) $\alpha_i^{-1}[X_n] \cap F_i = A_n \cap F_i$.

By Exercise 5.5.4, we may assume that the function $\alpha_i \colon Q \to E_n$ additionally satisfies the following property:

(4) $\alpha_i[F_i \setminus A_n] \cap C_n = \emptyset$.

This has the effect that for the function $g_n = \lim_{i \to \infty} \alpha_i$ we have

(5) $g_n[Q \setminus (A_n \cup K)] \cap C_n = \emptyset$.

We claim that the function $g = (g_n)_n$ has the property that $g^{-1}[X] \setminus K = A_\infty \setminus K$. Let $x \in g^{-1}[X] \setminus K$. Then $g(x) \in X \subseteq \prod_{i=1}^{\infty} C_i$. So for every $n \in \mathbb{N}$, $g_n(x) \in C_n$, hence $g_n(x) \notin g_n[Q \setminus (A_n \cup K)]$. This gives $x \notin Q \setminus (A_n \cup K)$ for all $n$. Since $x \notin K$ we consequently obtain $x \in A_n$ for all $n$, hence $x \in A_\infty$. Conversely, if $x \in A_\infty \setminus K$ then for each $n$, $x \in A_n \setminus K$. Consequently, $g_n(x) \in B_n$ for every $n$ which implies that
$$g(x) \in \prod_{i=1}^{\infty} B_i \subseteq X,$$
i.e., $x \in g^{-1}[X]$.

▶**Exercise 5.5.6:**
By Proposition 5.5.17 there is a continuous function $\delta \colon s \times s \times \mathbb{I} \to \mathbb{I}$ such that

(1) $\delta(x, p, t) = 0$ if and only if $t = 0$,
(2) $\delta(x, p, t) \leq t$,
(3) $\varrho(p, p + \delta(x, p, t)x) \leq t$ for every $t \in \mathbb{I}$ and $x, p \in s$.

Define $\xi \colon Q \times Q \to \mathbb{I}$ by
$$\xi(x, y) = \begin{cases} 0 & ((x,y) \in K), \\ \delta(\psi(x), \tilde{f}_1(x,y), \varepsilon(x,y)) & ((x,y) \notin K). \end{cases}$$

It is clear that $\xi$ is well-defined. In addition, if $(x, y) \notin K$ then
$$\varrho(\tilde{f}_1(x,y), \tilde{f}_1(x,y) + \xi(x,y)\psi(x)) \leq \varepsilon(x,y)$$
by the definition of $\xi$ and property (3) of $\delta$.
If $(x, y) \in K$ then $\xi(x, y) = 0$. Moreover, if $(x, y) \notin K$ then $\varepsilon(x, y) > 0$ and so
$$\xi(x, y) = \delta(\psi(x), \tilde{f}_1(x,y), \varepsilon(x,y)) > 0,$$
by property (1) of $\delta$.
It suffices to prove that $\xi$ is continuous and for this it suffices to prove its continuity at points of $K$. To this end, let $((x_n, y_n))_n$ be a sequence in $(Q \times Q) \setminus K$ converging to

an element $(x,y) \in K$. Then $\varepsilon(x_n, y_n) \to 0$ from which it follows that $\xi(x_n, y_n) \to 0$ by property (2) of $\delta$.

▶**Exercise 5.5.7:**
The continuity of $g_1$ need only be checked at points of $K$. To this end, let $((x_n, y_n))_n$ be a sequence in $(Q \times Q) \setminus K$ converging to an element $(x,y) \in K$. Then by Exercise 5.5.6,
$$\varrho\big(\tilde{f}_1(x_n, y_n), g_1(x_n, y_n)\big) \leq \varepsilon(x_n, y_n) \to 0$$
from which it follows by continuity of $\tilde{f}_1$ that $g_1(x_n, y_n) \to \tilde{f}_1(x,y) = g_1(x,y)$, which is as required.
It is trivial that $\varrho(f_1(x), g_1(x)) \leq 2\varepsilon(x)$ for every $x \in Q \times Q$.
Fix $\gamma \in \Gamma$. If
$$(x,y) \in \big(g_1^{-1}[L_\gamma] \cap (\mathbb{R}^\infty \times Q)\big) \setminus K$$
then $x \in \mathbb{R}^\infty$, $(x,y) \notin K$, hence $\xi(x,y) > 0$, and $g_1(x,y) \in L_\gamma$. This implies that
$$L_\gamma \ni g_1(x,y) = \tilde{f}_1(x,y) + \xi(x,y)\psi(x).$$
Since $\tilde{f}_1(x,y) \in L \subseteq L_\gamma$ and $L_\gamma$ is a linear subspace of $\mathbb{R}^\infty$ we conclude that
$$\xi(x,y)\psi(x) \in L_\gamma$$
and since $\xi(x,y) > 0$, we get $\psi(x) \in L_\gamma$. So $x \in \psi^{-1}[L_\gamma] \cap \mathbb{R}^\infty = L_\gamma$, i.e., $x \in L_\gamma$. Consequently, $(x,y) \in (L_\gamma \times Q) \setminus K$. Since $(x,y)$ was arbitrary, this shows that
$$\big(g_1^{-1}[L_\gamma] \cap (\mathbb{R}^\infty \times Q)\big) \setminus K \subseteq (L_\gamma \times Q) \setminus K.$$
Conversely, if $(x,y) \in (L_\gamma \times Q) \setminus K$ then $x \in L_\gamma = \psi^{-1}[L_\gamma] \cap \mathbb{R}^\infty$ and $(x,y) \notin K$. So $x \in \mathbb{R}^\infty$ and $\psi(x) \in L_\gamma$, hence
$$g_1(x,y) = \tilde{f}_1(x,y) + \xi(x,y)\psi(x) \in L_\gamma.$$
We conclude that
$$(x,y) \in \big(g_1^{-1}[L_\gamma] \cap (\mathbb{R}^\infty \times Q)\big) \setminus K,$$
which is as required.

▶**Exercise 5.5.8:**
Since $f_i \upharpoonright K = g_i \upharpoonright K$ for $i = 0, 1$, it clearly follows that $f \upharpoonright K = g \upharpoonright K$. It is also clear that $d(f(x), g(x)) \leq 2\varepsilon(x)$ for every $x \in Q \times Q$.
We shall prove that $g$ is one-to-one. To this end, take distinct points
$$(x,y), (x',y') \in Q \times Q.$$
There are several cases to consider. If both $(x,y)$ and $(x',y')$ are in $K$ then there is nothing to prove since $f \upharpoonright K = g \upharpoonright K$ is an imbedding. Next, assume that e.g.,
$$(x,y) \in K, \quad (x',y') \notin K.$$
Then $\varepsilon(x',y') = \min\{d(f(x',y'), f[K])/4, \theta/4\} > 0$ and so
$$d(g(x',y'), g[K]) \geq d(f(x',y'), f[K])/2 > 0.$$
Since $g(x,y) \in g[K]$ this proves that $g(x',y') \neq g(x,y)$. Assume finally that
$$(x,y), (x',y') \notin K.$$
But then $g_2(x,y) \neq g_2(x',y')$ by construction and so $g(x,y) \neq g(x',y')$ as well. So $g$ is an imbedding by Exercise A.5.9.
It remains to prove that $g[Q \times Q]$ is a $Z$-set in $Q \times Q$. Since $f[K] = g[K]$ is a $Z$-set, it

suffices to prove that $L = g[(Q \times Q)] \setminus K$ is a $\sigma Z$-set (Lemma 5.1.2(3)). Since $L$ is $\sigma$-compact, it consequently suffices to prove by Lemma 5.1.2(1),(3) that it is contained in a $\sigma Z$-set. But this is clear from Lemma 5.1.2(5) since $g_2[(Q \times Q) \setminus K] \in \mathcal{Z}_\sigma(Q)$.

▶**Exercise A.1.2:**
For (1) ⇒ (2), observe that $A$ misses $X \setminus A$. Since $p$ is an accumulation point of $A$, by (1) it follows that $p$ is not an accumulation point of $X \setminus A$. So there is a neighborhood $U$ of $p$ which misses

$$(X \setminus A) \setminus \{p\} = X \setminus (A \cup \{p\})$$

and hence is contained in $A \cup \{p\}$.
For (2) ⇒ (3), for some $x \in X$ let

$$\mathcal{F} = \{A \subseteq X : x \in \overline{A \setminus \{x\}}\}.$$

We will first show that $\mathcal{F}$ has the finite intersection property. To this end, let $\mathcal{G}$ be an arbitrary finite subcollection of $\mathcal{F}$. Then by (2), for every $G \in \mathcal{G}$ there is a neighborhood $U_G$ of $x$ such that

$$U_G \subseteq G \cup \{x\}.$$

Put $U = \bigcap_{G \in \mathcal{G}} U_G$. Then $U$ is a neighborhood of $x$ and since $X$ has no isolated points, there consequently is a point $p \in U \setminus \{x\}$. So clearly $\bigcap \mathcal{G} \neq \emptyset$ since this intersection contains $p$. We next prove that $\mathcal{F}$ is an ultrafilter by showing that if $A$ and $B$ are arbitrary complementary subsets of $X$ then either $A \in \mathcal{F}$ or $B \in \mathcal{F}$ (Exercise A.1.1). Assume the contrary, that is, there are complementary sets $A$ and $B$ in $X$ such that $p$ is neither an accumulation point of $A$ nor $B$. Then there is a neighborhood $U$ of $p$ which misses

$$(A \setminus \{p\}) \cup (B \setminus \{p\}) = X \setminus \{p\},$$

hence is contained in $\{p\}$. But this contradicts the fact that $X$ has no isolated points.
Since (3) ⇒ (1) is trivial, this completes the solution.

▶**Exercise A.1.4:**
We prove this by induction on $k$. If $k = 1$ then $\mathcal{A}$ contains infinitely many singletons and so there is nothing to prove (let $A = \emptyset$). Suppose that we are done for $k - 1$, where $k \geq 2$. For $i \in \mathbb{N}$ let $A_i \in \mathcal{A}$ be such that $A_i \neq A_j$ if $i \neq j$. If $\{A_i : i \in \mathbb{N}\}$ contains an infinite pairwise disjoint subfamily, then we are clearly done. So assume that this is not true. Then there has to be an $i \in \mathbb{N}$ such that the set

$$E = \{j \in \mathbb{N} \setminus \{i\} : A_j \cap A_i \neq \emptyset\}$$

is infinite. We may assume without loss of generality that $i = 1$. Since $A_1$ is finite there consequently has to an infinite subset $F \subseteq E$ and an element $x \in A_1$ such that $x \in A_j$ for every $j \in F$. For every $i$ put $B_i = A_i \setminus \{x\}$. By our induction hypothesis, there are an infinite subset $E \subseteq F$ and a set $B' \subseteq X$ of size at most $k-2$ such that if $i, j \in E$ are distinct then

$$B_i \cap B_j = B'.$$

Now put $B = B' \cup \{x\}$. Then if $i, j \in E$ are distinct, we have
$$A_i \cap A_j = (B_i \cup \{x\}) \cap (B_j \cup \{x\})$$
$$= (B_i \cap B_j) \cup \{x\}$$
$$= B' \cup \{x\}$$
$$= B.$$
So we are done since clearly $|B| \leq k - 1$.

▶**Exercise A.1.10:**
That $\varrho$ is a metric is easily seen. To show it is admissible, consider a basic open subset $U$ of $\prod = \prod_{n=1}^{\infty} X_n$ first. It has the form
$$U = \prod_{n=1}^{N} U_i \times \prod_{n=N+1}^{\infty} X_n,$$
where $N \in \mathbb{N}$ and $\emptyset \neq U_i \subseteq X_i$ is open for every $i \leq N$. Pick an arbitrary $x \in U$, and let $\varepsilon > 0$ be such that $B_{\varrho_i}(x_i, \varepsilon) \subseteq U_i$ for every $i \leq N$. Pick an arbitrary element $y \in \prod$ such that $\varrho(x,y) < 2^{-N}\varepsilon$. Then for $i \leq N$ we have
$$2^{-i} \varrho_i(x_i, y_i) \leq \varrho(x,y) < 2^{-N}\varepsilon$$
and so
$$\varrho_i(x_i, y_i) < 2^{-N+i}\varepsilon \leq \varepsilon.$$
We conclude that $y \in U$, i.e., $B_{\varrho}(x, 2^{-N}\varepsilon) \subseteq U$.
Conversely, take an arbitrary $x \in \prod$, and let $\varepsilon > 0$. There exists $N \in \mathbb{N}$ such that
$$\sum_{i=N+1}^{\infty} 2^{-i}c < 1/2\varepsilon.$$
For $i \leq N$ consider the ball $U_i = B_{\varrho_i}(x_i, 1/2N\varepsilon)$. We claim that the basic open neighborhood
$$U = \prod_{n=1}^{N} U_i \times \prod_{n=N+1}^{\infty} X_n,$$
of $x$ is contained in $B_{\varrho}(x, \varepsilon)$. Indeed, take an arbitrary $y \in U$. Then
$$\varrho(x,y) = \sum_{i=1}^{\infty} 2^{-i}\varrho_i(x_i,y_i) \leq \sum_{i=1}^{N} 2^{-i}\varrho_i(x_i,y_i) + \sum_{i=N+1}^{\infty} 2^{-i}c$$
$$< \sum_{i=1}^{N} 2^{-i} \cdot 1/2N\varepsilon + 1/2\varepsilon \leq 1/2\varepsilon + 1/2\varepsilon$$
$$= \varepsilon.$$
So we are done.

▶**Exercise A.2.13:**
Let $X$ be uncountable, and put
$$\mathcal{U} = \{U \subseteq X : U \text{ is open and countable}\}.$$
The collection $\mathcal{U}$ of has a countable subcollection with the same union by Corollary A.2.3. So $\bigcup \mathcal{U}$ is a countable union of countable sets and is therefore countable

itself. We claim that $Y = X \setminus \bigcup \mathcal{U}$ is dense-in-itself. In fact we will show that it is locally uncountable, that is: every non-empty (relatively) open set is uncountable. This is clearly as required. Striving for a contradiction, assume that $V \subseteq Y$ is a non-empty relatively open set which is countable. There is an open set $V' \subseteq X$ such that $V' \cap Y = V$. But then $V'$, being the union of $V$ and the countable set

$$V' \cap \bigcup \mathcal{U},$$

is countable. This shows that $V' \in \mathcal{U}$ and hence that $\emptyset \neq V = V' \cap Y = \emptyset$, which is a contradiction.

▶**Exercise A.4.4:**
Observe that if $E$ and $F$ are subsets of $X$ then $\overline{E \cap F} \subseteq \overline{E} \cap \overline{F}$. So if $\overline{E \cap F} \neq \overline{E} \cap \overline{F}$ then

$$(\overline{E} \cap \overline{F}) \setminus \overline{E \cap F} \neq \emptyset.$$

Striving for a contradiction, assume that $B$ is uncountable. By the above, for every $\delta \in B$ there exists an element

$$a(\delta) \in \left(\overline{f^{-1}\big[[0,\delta)\big]} \cap \overline{U}\right) \setminus \overline{f^{-1}\big[[0,\delta)\big] \cap U}. \tag{1}$$

If $f\big(a(\delta)\big) < \delta$ then

$$a(\delta) \in f^{-1}\big[[0,\delta)\big] \cap \overline{U} \subseteq \overline{f^{-1}\big[[0,\delta)\big] \cap U}$$

which contradicts (1). So $f\big(a(\delta)\big) = \delta$ for every $\delta \in B$ which means that we can order the set

$$\{a(\delta) : \delta \in B\}$$

by the following rule:

$$a(\delta_0) < a(\delta_1) \quad \Leftrightarrow \quad \delta_0 < \delta_1.$$

By Exercise A.2.14 there is an element $\delta_1 \in B$ such that

$$a(\delta_1) \in \overline{\{a(\delta) : \delta \in B, \delta < \delta_1\}}.$$

Let $V$ be an arbitrary open neighborhood of $a(\delta_1)$. There exists $\delta_0 \in B$ such that $\delta_0 < \delta_1$ while moreover $a(\delta_0) \in V$. But then

$$a(\delta_0) \in V \cap f^{-1}\big[[0,\delta_1)\big] \cap \overline{U}.$$

This proves that

$$V \cap f^{-1}\big[[0,\delta_1)\big] \cap U \neq \emptyset.$$

Since $V$ was an arbitrary neighborhood of $a(\delta_1)$, we get

$$a(\delta_1) \in \overline{f^{-1}\big[[0,\delta_1)\big] \cap U}.$$

But this contradicts (1).

▶**Exercise A.4.5:**
We may assume without loss of generality that $V$ is open. Let

$$f : X \to \mathbb{I}$$

by a Urysohn function with $f(A) \subseteq \{0\}$ and $f[X \setminus V] \subseteq \{1\}$ (Lemma A.4.1). Since $\mathcal{U}$ is countable and $(0,1)$ is uncountable, there is an element $\delta \in (0,1)$ such that

$$\overline{f^{-1}\big[[0,\delta_1)\big] \cap U} = \overline{f^{-1}\big[[0,\delta_1)\big]} \cap \overline{U}$$

for every $U \in \mathcal{U}$ (Exercise A.4.4). Hence
$$W = f^{-1}\big[[0, \delta)\big]$$
is as required.

▶**Exercise A.5.15:**
Let $Y$ be $X$ with the topology generated by the metric $\varrho$. We claim that the identity $X \to Y$ is continuous. This suffices since by compactness of $X$ it will then follow that the identity is a homeomorphism (Exercise A.5.9). It is clearly enough to prove that if $x \in X$ and $\delta > 0$ then $B(x, \delta)$ is open in $X$. Take an arbitrary $y \in B(x, \delta)$. Then $\varrho(x, y) < \delta$. Hence by continuity of $\varrho$ there are neighborhoods $U$ and $V$ of $x$ and $y$, respectively, such that
$$\varrho[U \times V] \subseteq [0, \delta).$$
So $\varrho[\{x\} \times V] \subseteq [0, \delta)$ showing that $V \subseteq B(x, \delta)$. We conclude that $B(x, \delta)$ is a neighborhood of $y$.

▶**Exercise A.5.16:**
By Corollary A.4.8 we may assume that for every $n$, $X_n$ is a subspace of a compact space $Z_n$. (This also follows from Corollary A.4.4.) By Exercise A.1.6 there is an admissible metric $\varrho_n$ for $Z_n$ which are bounded by $2^{-n}$. We claim that
$$\varrho(x, y) = \max\{\varrho_n(x_n, y_n) : n \in \mathbb{N}\}$$
is as required. That it defines a metric on $Z = \prod_{n=1}^\infty Z_n$ is left to the reader. To show that it is admissible on $Z$, simply observe that
$$\varrho \colon Z \times Z \to [0, \infty)$$
is continuous since $Z$ is endowed with the topology of coordinatewise convergence (Exercise A.2.2) and the functions $\varrho_n$ are continuous on $X_n \times X_n$. So $\varrho$ is admissible on $Z$ by Exercise A.5.15. Hence the restriction of $\varrho$ to $X = \prod_{n=1}^\infty X_n$ is admissible on $X$ since $X$ is a subspace of $Z$ (Exercise A.1.13).

▶**Exercise A.6.3:**
Let $\varepsilon > 0$. Since $\varrho$ is totally bounded, we may choose by Exercise A.6.2 a finite $\tfrac{1}{2}\varepsilon$-net $B$ for $X$. Put
$$B' = \{x \in B : \varrho(x, A) < \tfrac{1}{2}\varepsilon\}.$$
For every $x \in B'$ pick a point $a_x \in A$ such that $\varrho(x, a_x) < \tfrac{1}{2}\varepsilon$. We claim that the finite set
$$\{a_x : x \in B'\}$$
is an $\varepsilon$-net for $A$. If so then we are done by another application of Exercise A.6.2. Let $a \in A$ be arbitrary. There exists $x \in B$ such that $\varrho(x, a) < \tfrac{1}{2}\varepsilon$. Hence $x \in B'$ and
$$\varrho(a, a_x) < \varrho(a, x) + \varrho(x, a_x) < \tfrac{1}{2}\varepsilon + \tfrac{1}{2}\varepsilon = \varepsilon.$$
So we are done.

▶**Exercise A.6.9:**
For (1) $\Rightarrow$ (2), assume first that $f$ is not one-to-one. There are distinct $x, y \in X$ such that $f(x) = f(y)$. Let $x_k = x$ and $y_k = y$ for every $k$. The sequences
$$(x_k)_k, \quad (y_k)_k,$$

and the number
$$\varepsilon = 2\varrho(x,y)$$
demonstrate that (2) holds. Assume therefore that $f$ is not closed. There is a closed set $A \subseteq X$ for which $f[A]$ is not closed. By Exercise A.2.1 there is a sequence $(a_n)_n$ in $A$ such that
$$\lim_{n \to \infty} f(a_n) = z \notin f[A].$$
If $(a_n)_n$ converges say to $p$, then $p \in A$ since $A$ is closed and so $f[A] \ni f(p) = z$ which is a contradiction. The completeness of $\varrho$ implies that $(a_n)_n$ is not Cauchy. There consequently exists $\varepsilon > 0$ such that for every $N \in \mathbb{N}$ there exist $n, m \geq N$ with $\varrho(a_n, a_m) > \varepsilon$. A simple inductive argument proves the existence of subsequences
$$(a_{n_k})_k, \quad (a_{m_k})_k$$
of $(a_n)_n$ such that for every $k$ we have $\varrho(a_{n_k}, a_{m_k}) > \varepsilon$. The existence of the sequences $(a_{n_k})_k$ and $(a_{m_k})_k$ and the point $z$ proves (2).

Conversely, assume that (2) holds. We shall prove (1). We assume that $f$ is closed as well as one-to-one and will derive a contradiction.

**Claim 1.** *The sequence $(x_k)_k$ has a convergent subsequence.*

*Proof.* If not then for every $E \subseteq \mathbb{N}$ we have that the set then the set $\{x_n : n \in E\}$ is closed (Exercise A.2.1). Consider the set $\{f(x_n) : n \in \mathbb{N}\}$. Since $f$ is closed, it follows that for every $E \subseteq \mathbb{N}$ the set $\{f(x_n) : n \in E\}$ is closed as well. But then the sequence $(f(x_n))_n$ is eventually constant since it is convergent. But this contradicts $f$ being one-to-one. ◇

So we may assume without loss of generality that $(x_k)_k$ converges, say to $x$. We may assume similarly that $(y_k)_k$ converges, say to $y$. But then $\varrho(x,y) \geq \varepsilon > 0$ and $f(x) = f(y)$. This is a contradiction.

▶**Exercise A.6.12:**
Consider the projection $\pi \colon \mathbb{I}^2 \to \mathbb{I}$. We first claim that the set $\mathbb{Q}_0 = \mathbb{Q} \cap \mathbb{I}$ is the $\pi$-image of a $G_\delta$-subset of $\mathbb{I}^2$. This is easy. Enumerate $\mathbb{Q}_0$ as $\{q_n : n \in \mathbb{N}\}$. For every $n$ let
$$z_n = (q_n, 1 - 1/n) \in \mathbb{I}^2.$$
The set $Z = \{z_n : n \in \mathbb{N}\}$ is discrete and is therefore a $G_\delta$-subset of $\mathbb{I}^2$ (Exercise A.2.6). Now let $\Pi \colon \mathbb{I}^3 \to \mathbb{I}^2$ be the projection onto the first two coordinates, i.e.,
$$\Pi(x,y,z) = (x,y).$$
Then $\Pi$ is open by Exercise A.5.5(3). Put
$$S = \big((0,1] \times \mathbb{I}^2\big) \cup \{(0, q_n, 1 - 1/n) : n \in \mathbb{N}\}.$$
Then $S$ is a dense $G_\delta$-subset of $\mathbb{I}^3$ (since its complement is evidently a countable union of compact nowhere dense subsets of $\mathbb{I}^2$) and
$$T = \Pi[S] = \{(x,y) \in \mathbb{I}^2 : (x = 0 \Rightarrow y \in \mathbb{Q}_0)\}.$$
But $T$ is not a $G_\delta$-subset of $\mathbb{I}^2$. Observe that $\{0\} \times \mathbb{I}$ is a $G_\delta$-subset of $\mathbb{I}^2$. So if $T$ would be a $G_\delta$-subset of $\mathbb{I}^2$ then $\mathbb{Q}_0$ would be a $G_\delta$-subset of $\mathbb{I}$. This leads to a contradiction in the same way as in the proof of Corollary A.6.7.

# B. ANSWERS TO SELECTED EXERCISES

▶**Exercise A.7.5:**
We may assume without loss of generality that $\mathcal{U}$ is locally finite in $\bigcup \mathcal{U}$ (Corollary A.7.3). For every $U \in \mathcal{U}$ define $f_U \colon X \to \mathbb{I}$ by
$$f_U(x) = \min\{1, \varrho(x, X \setminus U)\}.$$
The function $\varepsilon \colon X \to \mathbb{I}$ defined by
$$\varepsilon(x) = \max\{f_U(x) : U \in \mathcal{U}\}$$
is well-defined since $\mathcal{U}$ is locally finite. Clearly, $\varepsilon^{-1}\big[(0,1]\big] = \bigcup \mathcal{U}$. In addition, $\varepsilon$ has the property that for every $x \in X$,
$$\{y \in X : \varrho(y, x) < \varepsilon(x)\} \qquad (*)$$
is contained in an element of $\mathcal{U}$. To prove this, first observe that if $x \in A = X \setminus \bigcup \mathcal{U}$ then $\varepsilon(x) = 0$. So if $x \in A$ then the set in $(*)$ is empty, and is therefore contained in every $U \in \mathcal{U}$. Moreover, if $x \in \bigcup \mathcal{U}$ then let $U \in \mathcal{U}$ be such that
$$\varepsilon(x) = f_U(x).$$
Then clearly
$$\{y \in X : \varrho(y, x) < \varepsilon(x) = \min\{1, \varrho(x, X \setminus U)\}\} \subseteq U.$$
It remains to verify that $\varepsilon$ is Lipschitz. To this end, take arbitrary $x, y \in X$. We have to distinguish several subcases. If $x, y \in A$ then $\varepsilon(x) = \varepsilon(y) = 0$ and so there is nothing to prove. Assume next that $x \notin A$, say $U \in \mathcal{U}$ is such that
$$\varepsilon(x) = f_U(x).$$
If $y \notin A$ then $y \notin U$ and so
$$|\varepsilon(x) - \varepsilon(y)| = \varepsilon(x) = f_U(x) \leq \varrho(x, X \setminus U) \leq \varrho(x, y),$$
as desired. So we can assume that $y \notin A$ as well, say $U' \in \mathcal{U}$ is such that
$$\varepsilon(y) = f_{U'}(y).$$
We may assume without loss of generality that $\varepsilon(x) \leq \varepsilon(y)$. If $x \notin U'$ then
$$|\varepsilon(x) - \varepsilon(y)| \leq \varepsilon(y) = f_{U'}(y) \leq \varrho(y, X \setminus U') \leq \varrho(y, x) = \varrho(x, y).$$
So assume that $x \in U'$.

**Claim 1.** $\varepsilon(y) \leq \varrho(x, y) + \varepsilon(x)$.

*Proof.* If $\varepsilon(x) = 1$ then there is nothing to prove. So we may assume that
$$1 > \varepsilon(x) = \varrho(x, X \setminus U).$$
Since
$$\min\{1, \varrho(x, X \setminus U')\} \leq \varepsilon(x) < 1$$
we get
$$\varrho(x, X \setminus U') \leq \varepsilon(x). \qquad (1)$$
Observe that
$$\varrho(y, X \setminus U') \leq \varrho(x, y) + \varrho(x, X \setminus U') \qquad (2)$$
and hence
$$\varepsilon(y) = \min\{1, \varrho(y, X \setminus U')\} \leq \min\{1, \varrho(x, y) + \varrho(x, X \setminus U')\}. \qquad (3)$$

If $\varepsilon(y) = 1$ then by (3) and (1),
$$\varepsilon(y) = 1 \leq \varrho(x,y) + \varrho(x, X \setminus U') \leq \varrho(x,y) + \varepsilon(x).$$
On the other hand, if $\varepsilon(y) < 1$ then $\varepsilon(y) = \varrho(y, X \setminus U')$ and we are done as well by (2) and (1). ◇

So by the Claim and $\varepsilon(x) \leq \varepsilon(y)$, we get
$$|\varepsilon(x) - \varepsilon(y)| = \varepsilon(y) - \varepsilon(x) \leq \varrho(x,y),$$
as required.

▶**Exercise A.8.1:**
Let $\{q_n : n \in \mathbb{N}\}$ be a faithful indexing of $\mathbb{Q}$. For every $n \in \mathbb{N}$ let $(p_i^n)_i$ be a sequence of pairwise distinct irrational numbers converging to $q_n$. By Lemma A.4.1 there is for every $n \in \mathbb{N}$ a continuous function
$$f_n \colon \mathbb{R} \setminus \{q_n\} \to \mathbb{I}$$
such that
$$\begin{cases} f_n(p_i^n) = 0 & (i \text{ even}), \\ f_n(p_i^n) = 1 & (i \text{ odd}). \end{cases}$$
Define $f \colon \mathbb{P} \to \mathbb{I}$ by
$$f = \sum_{n=1}^{\infty} 5^{-n} (f_n \restriction \mathbb{P}).$$
Then $f$ is continuous by Exercise A.3.1. We claim that $f$ is as required. To this end, fix $m \in \mathbb{N}$. By continuity of the functions $(f_n)_{n<m}$ we can find a neighborhood $U$ of $x = q_m$ in $\mathbb{R}$ such that
$$\left| \sum_{n<m} 5^{-n} (f_n(x) - f_n(y)) \right| < 1/4 \cdot 5^{-m} \qquad (*)$$
for every $y \in U \setminus \{q_1, \ldots, q_{m-1}\}$. We pick arbitrary
$$y_0 \in U \cap \{p_i^m : i \text{ even}\}, \quad y_1 \in U \cap \{p_i^m : i \text{ odd}\}.$$
It follows that
$$f(y_0) = \sum_{n<m} 5^{-n} f_n(y_0) + 0 + \sum_{n>m} 5^{-n} f_n(y_0)$$
$$\leq \sum_{n<m} 5^{-n} f_n(y_0) + 1/4 \cdot 5^{-m},$$
$$f(y_1) = \sum_{n<m} 5^{-n} f_n(y_1) + 5^{-m} + \sum_{n>m} 5^{-n} f_n(y_1)$$
$$\geq \sum_{n<m} 5^{-n} f_n(y_1) + 5^{-m}.$$
Since $(*)$ implies
$$\left| \sum_{n<m} 5^{-n} (f_n(y_1) - f_n(y_0)) \right| < 1/2 \cdot 5^{-m}$$
it follows that
$$|f(y_1) - f(y_0)| \geq 5^{-m} - 1/2 \cdot 5^{-m} - 1/4 \cdot 5^{-m} = 1/4 \cdot 5^{-m}.$$

Since $y_1$ and $y_0$ were chosen arbitrarily, this shows that $f$ cannot be extended to a continuous function $\bar{f} \colon \mathbb{P} \cup \{q_m\} \to \mathbb{I}$.

▶**Exercise A.8.3:**
*First Solution:* We may assume without loss of generality that $S$ is dense in $X$ (Exercise A.2.5). By Corollary A.4.8 we may assume that $X$ is a subspace of some compact space $\tilde{X}$, and $Y$ is a subspace of some compact subspace $\tilde{Y}$ (the proof in fact already 'works' if $\tilde{X}$ and $\tilde{Y}$ are just topologically complete). By the Lavrentieff Theorem A.8.5, there are $G_\delta$-subsets $S_0$ of $\tilde{X}$ and $Y_0$ of $\tilde{Y}$ containing $S$ and $Y$, respectively, such that $g = f \restriction S \colon S \to Y$ can be extended to a homeomorphism $\bar{g} \colon S_0 \to Y_0$. The continuous functions

$$f \restriction S_0 \cap X \colon S_0 \cap X \to \tilde{Y}$$

and

$$\bar{g} \restriction S_0 \cap X \colon S_0 \cap X \to \tilde{Y}$$

agree on a closed subspace of $S_0 \cap X$ (Exercise A.1.9). Since they agree on $S$ and $S$ is dense in $X$, it follows that they agree on all of $S_0 \cap X$. Assume that there exists an element $x \in (S_0 \cap X) \setminus S$. Then $f(x) \in Y$ and hence there is an element $s \in S$ such that $f(s) = f(x)$ (here we use that $f$ is surjective). But then $s \ne x$ and

$$\bar{g}(s) = f(s) = f(x) = \bar{g}(x)$$

yields a contradiction since $\bar{g}$ is injective. So $S_0 \cap X = S$, proving that $S$ is a $G_\delta$-subset of $X$.

*Second Solution:* We may assume without loss of generality that $S$ is dense in $X$ (Exercise A.2.5). Put $g = f \restriction S$ and let $\mathcal{B}$ be a countable open basis for $Y$. For every pair $E, F$ of elements of $\mathcal{B}$ such that $\overline{E} \subseteq F$, consider the disjoint closed sets

$$g^{-1}[\overline{E}], \quad g^{-1}[Y \setminus F]$$

of $S$. Observe that the closed set

$$V(E, F) = \overline{g^{-1}[\overline{E}]} \cap \overline{g^{-1}[Y \setminus F]}$$

misses $S$. Since $\mathcal{B}$ is countable, the collection $\mathcal{G}$ of all $V(E, F)$'s obtained in this way is countable as well. It suffices to prove that $X \setminus S = \bigcup \mathcal{G}$. To this end, assume that there is an element

$$x \in X \setminus (S \cup \bigcup \mathcal{G}).$$

Since $S$ is dense in $X$ there is a sequence $(x_n)_n$ in $S$ with limit $x$. The set

$$\{x_n : n \in \mathbb{N}\}$$

is an infinite, closed and discrete subset of $S$, hence

$$B = \{f(x_n) : n \in \mathbb{N}\}$$

is an infinite, closed and discrete subset of $Y$. Simply observe that $f \restriction Y$ is a homeomorphism. There is an element $F \in \mathcal{B}$ such that $f(x) \in F$ while moreover $F$ misses a neighborhood of infinitely many elements of $B$ (in fact, all but one). Let $E \in \mathcal{B}$ be such that $f(x) \in E \subseteq \overline{E} \subseteq F$. Then $f^{-1}[E]$ is an open subset of $X$ which contains $x$. Since $S$ is dense in $X$, this implies that

$$x \in \overline{f^{-1}[E] \cap S} = \overline{g^{-1}[E]} \subseteq \overline{g^{-1}[\overline{E}]}.$$

Since $F$ misses a neighborhood of infinitely many elements of $B$, by similar considerations we obtain that for infinitely many $n$ we have
$$x_n \in \overline{g^{-1}[Y \setminus F]}.$$
This gives us that $x \in \overline{g^{-1}[Y \setminus F]}$, hence $x \in V(E,F) \subseteq \bigcup \mathcal{G}$. This is a contradiction.

▶**Exercise A.10.6:**
Let $x \in X$ be arbitrary. We claim that there is a noncut point $y \in X \setminus \{x\}$. This clearly suffices.
Striving for a contradiction, assume that every point $y \in X \setminus \{x\}$ is a cut point. Let $\mathcal{B} = \{B_n : n \in \mathbb{N}\}$ be a countable base for $X$ such that $B_n \neq \emptyset$ for every $n$. Put $C_0 = \{x\}$. Recursively, we will construct for every $n$ a subcontinuum $C_n$ of $X$ having the following properties:

(1) $C_{n-1}$ is contained in the interior of $C_n$,
(2) $C_n \cap B_n \neq \emptyset$,
(3) $C_n \neq X$.

Suppose $C_n$ has been constructed. Pick a point $p \in X \setminus C_n$ in the following way. If $B_{n+1} \setminus C_n \neq \emptyset$ then pick $p$ from that set. If $B_{n+1} \subseteq C_n$ then let $p \in X \setminus C_n$ be arbitrarily chosen (here we use (3)). By assumption, $p$ is a cut point of $X$. So we can write $X \setminus \{p\}$ as the union of two disjoint nonempty open sets, say $U$ and $V$. We assume without loss of generality that $x \in U$. Since $p \notin C_n$ and $C_n$ is a continuum, it consequently follows that $C_n \subseteq U$. Put $C_{n+1} = U \cup \{p\}$. Then $C_{n+1}$ is a continuum by Exercise A.10.5. Moreover, $C_{n+1} \neq X$ since it misses $V$. The interior of $C_{n+1}$ contains $U$ and hence $C_n$. If $B_{n+1} \subseteq C_n$ then (2) is evidently satisfied for $n+1$, and if not, then (2) is also satisfied since then $p \in B_{n+1} \cap C_{n+1}$. We conclude that $C_{n+1}$ is as required.
If $\bigcup_{n=1}^\infty C_n = X$ then the interiors of the $C_n$'s cover $X$ by (1). The compactness of $X$ then implies that one of the $C_n$'s is equal to $X$. But this contradicts statement (3). Pick an arbitrary point $y \in X \setminus \bigcup_{n=1}^\infty C_n$. By assumption, $y$ is a cut point. There consequently exist nonempty, disjoint and open sets $E$ and $F$ such that $X \setminus \{y\}$ is equal to $E \cup F$. We may assume without loss of generality that $x \in E$. Since $y \notin C_n$ we have that $C_n \subseteq E$ for every $n$. Hence $\bigcup_{n=1}^\infty C_n$ is contained in $E$ and so misses the nonempty open set $F$. But it is dense by (2). This gives us the contradiction we are after.

▶**Exercise A.11.3:**
Let $\{U_n : n \in \mathbb{N}\}$ be a countable collection of neighborhoods of $\{\mathbb{N}\}$ in $\mathbb{R}/\mathbb{N}$ and let $\pi \colon \mathbb{R} \to \mathbb{R}/\mathbb{N}$ be the natural quotient map. For every $n$ we have that the set $\pi^{-1}[U_n]$ is an open neighborhood of $\mathbb{N}$ in $\mathbb{R}$. For every $n \in \mathbb{N}$ let $a_n < n < b_n$ be such that $(a_n, b_n)$ is a proper subset of $(n - 1/4, n + 1/4) \cap \bigcap_{i=1}^n \pi^{-1}[U_i]$. Observe that the collection of intervals $\{(a_n, b_n) : n \in \mathbb{N}\}$ is pairwise disjoint. The set
$$V = \bigcup_{n=1}^\infty (a_n, b_n)$$
is a neighborhood of $\mathbb{N}$ such that $\pi^{-1}[\pi[V]] = V$, hence $\pi[V]$ is an open neighborhood of the point $\{\mathbb{N}\}$ in $\mathbb{R}/\mathbb{N}$. Assume that there exists $N \in \mathbb{N}$ such that $U_N$ is

contained in $\pi[V]$. Then $\pi^{-1}[U_N] \subseteq V$. But this is impossible since $(a_N, b_N)$ is a proper subset of $\pi^{-1}[U_N] \cap (N - 1/4, N + 1/4)$ and

$$(N - 1/4, N + 1/4) \cap (n - 1/4, n + 1/4) = \emptyset$$

for every $n \neq N$.

▶**Exercise A.11.5:**
Let $\mathcal{C}$ be the collection of all components of $S$. By Corollary A.11.3, $S/\mathcal{C}$ has a base consisting of clopen sets. Let $q \colon S \to S/\mathcal{C}$ be the natural quotient map. Observe that $S/\mathcal{C}$ is compact since $S$ is, and that by our assumptions,

$$q[A \cap S] \cap q[B \cap S] = \emptyset.$$

Since $q$ is closed (Exercise A.5.5), the sets $q[A \cap S]$ and $q[B \cap S]$ are closed in $S/\mathcal{C}$. Since they are also disjoint, for every $x \in S/\mathcal{C}$ we may pick a clopen neighborhood $U_x$ such that

$$U_x \cap q[A \cap S] = \emptyset \quad \text{or} \quad U_x \cap q[B \cap S] = \emptyset.$$

The cover

$$\mathcal{U} = \{U_x : x \in S/\mathcal{C}\}$$

has by compactness of $S/\mathcal{C}$ a finite subcover, say

$$\{U_{x_1}, \ldots, U_{x_n}\}.$$

For every $i \leq n$ now put

$$V_i = U_{x_i} \setminus \bigcup_{j < i} U_{x_j}.$$

It is easy to verify that the collection

$$\mathcal{V} = \{V_1, \ldots, V_n\}$$

consists of pairwise disjoint clopen sets, and covers $S/\mathcal{C}$ (cf. Exercise A.1.3). In addition, $V_i \cap q[A \cap S] = \emptyset$ or $V_i \cap q[B \cap S] = \emptyset$ for every $i$. So the collection

$$\mathcal{N} = \{q^{-1}[V_i] : i \leq n\}$$

is as required.

▶**Exercise A.11.8:**
By Lemma A.5.6 there is a (countable) collection $\{(A'_n, B'_n) : n \in \mathbb{N}\}$ of pairs of disjoint closed subsets of $aF$ such that for any pair $(A, B)$ of disjoint closed subsets in $aF$ there is an $m \in \mathbb{N}$ such that $A \subseteq A'_m$ and $B \subseteq B'_m$. For every $n$, let

$$A_n = A'_n \cap F, \quad B_n = B'_n \cap F,$$

respectively. All the $A_n$'s and $B_n$'s are closed subsets of $X$.
By the same reasoning, there is a (countable) collection $\{(H'_n, G'_n) : n \in \mathbb{N}\}$ of pairs of disjoint closed sets in $\gamma X$ such that for any pair $(H, G)$ of disjoint closed subsets in $\gamma X$ there is an $m \in \mathbb{N}$ such that $H \subseteq H'_m$ and $G \subseteq G'_m$. For every $n$ put

$$H_n = H'_n \cap X, \quad G_n = G'_n \cap X,$$

respectively.
By Exercise A.9.3 there is a compactification $\beta X$ of $X$ such that for every $n$ the closures of $A_n$ and $B_n$ as well as the closures of $H_n$ and $G_n$ are disjoint. This implies by Lemma A.8.3 that the identity function $X \to X$ can be extended to a

continuous function $f\colon \beta X \to \gamma X$. In addition, the identity function $F \to F$ can be extended to a continuous function $\xi\colon F^* \to aF$. Here $F^*$ means the closure of $F$ in $\beta X$.

Since $aF \geq \overline{F}$ there is a continuous function $\eta\colon aF \to \overline{F}$ which extends the identity on $F$. Observe that the continuous functions $\eta \circ \xi$ and $f \upharpoonright F^*$,

$$F^* \xrightarrow{\xi} aF \xrightarrow{\eta} \overline{F}, \quad F^* \xrightarrow{f \upharpoonright F^*} \overline{F},$$

both restrict to the identity on $F$. Since $F$ is dense in $F^*$, both functions are equal by Exercise A.1.9.

It is now clear that the adjunction space $bX = \beta X \cup_\xi F^*$ is the required compactification of $X$.

▶**Exercise A.12.8:**
To the contrary, assume that $\psi_n$ is inessential. By Theorem A.12.8 there exists a continuous function $\varphi\colon \mathbb{S}^1 \to \mathbb{R}$ such that

$$\psi_n(z) = z^n = e^{2\pi i \varphi(z)}$$

for all $z \in \mathbb{S}^1$ (the constant '$2\pi$' is added for technical reasons). Fix $\theta \in \mathbb{R}$. Then for $z = e^{2\pi i \theta}$ we have

$$(e^{2\pi i \theta})^n = e^{2\pi i \varphi(z)}$$

i.e.,

$$e^{2\pi i (\varphi(z) - n\theta)} = 1.$$

From this we conclude that the continuous function $f\colon \mathbb{R} \to \mathbb{R}$ defined by

$$f(\theta) = \varphi(e^{2\pi i \theta}) - n\theta$$

takes its values in $\mathbb{Z}$ only. By connectivity of $\mathbb{R}$ it therefore follows that $f$ is constant. So there exists $N \in \mathbb{Z}$ such that

$$\varphi(e^{2\pi i \theta}) = n\theta + N \quad (\forall \theta \in \mathbb{R}).$$

From this we conclude that

$$n \cdot 1 + N = \varphi(e^{2\pi i \cdot 1}) = \varphi(e^{2\pi i \cdot 0}) = n \cdot 0 + N,$$

so that $n = 0$, which is a contradiction.

▶**Exercise A.12.10:**
For (1), let $f\colon \mathbb{I} \to \mathbb{I}$ be continuous. If $f$ has no fixed-point, then $\mathbb{I}$ is the disjoint union of the open sets $E = \{x \in \mathbb{I} : x < f(x)\}$ and $F = \{x \in \mathbb{I} : f(x) < x\}$. By connectivity of $\mathbb{I}$, at most one can be non-empty, say $E$. But then $1 < f(1)$ contradicts the fact that the range of $f$ is contained in $\mathbb{I}$.

The proof of (2) is trivial.

For (3), let $f\colon X \to X$ be continuous. Put $E = \{(0,y) : -1 \leq y \leq 1\}$. If $f[E] \subseteq E$ then we get a fixed-point in $E$ by (1). So assume that $f[E]$ is not contained in $E$. Then $f[E] \cap E = \emptyset$ since by Exercise A.2.11 it follows that $f[E]$ is locally connected and no locally connected subcontinuum of $X$ intersects both $E$ and $X \setminus E$. Observe that by compactness the function $f$ is closed (Exercise A.5.5(1)). There is a closed neighborhood $V$ of $E$ such that $f[V] \cap E = \emptyset$. We may assume that

$$V = X \setminus \{(x, \sin \tfrac{1}{x}) : \varepsilon < x \leq 1/\pi\}$$

for certain $\varepsilon > 0$. Let $p = (\varepsilon, \sin 1/\varepsilon)$. Then $f(p) \notin E$ and so by the same argument as above, the image of the arc $\{(x, \sin 1/x) : \varepsilon \leq x \leq 1/\pi\}$ misses $E$. So $f[X]$ misses $E$ and hence is contained in a subarc of $X$. Then $f$ maps this arc into itself, and hence we find a fixed-point there by again applying part (1).

▶**Exercise A.12.11:**
To the contrary, assume that $f(x) \neq x$ for every $x \in D$. Define $r \colon \mathbb{C} \setminus \{0\} \to \mathbb{S}^1$ and $g \colon \mathbb{S}^1 \to \mathbb{S}^1$ by $r(z) = z/|z|$ and $g(u) = r(u - f(u))$, respectively. For every $t \in \mathbb{I}$ and $u \in \mathbb{S}^1$ we have
$$H(u, t) = u - t f(u) \neq 0.$$
For $t = 1$ this is clear and for $t < 1$ observe that
$$|tf(u)| = t|f(u)| \leq t < 1 = |u|.$$
Consequently, $r \circ H$ is a homotopy connecting the identity function on $\mathbb{S}^1$ and the function $g$.
Observe that for every $t \in \mathbb{I}$ and $u \in \mathbb{S}^1$ we have
$$G(u, t) = tu - f(tu) \neq 0. \tag{$*$}$$
We conclude that $r \circ G$ is a homotopy from $\mathbb{S}^1$ to $\mathbb{S}^1$ connecting the constant function with value $r(-f(0))$ and the function $g$.
By transitivity of the homotopy relation (Lemma A.12.1(3)), it therefore follows that the identity on $\mathbb{S}^1$ is homotopic to a constant function, which violates Exercise A.12.9.

▶**Exercise A.13.4:**
Let us first assume that $X$ is absolutely Borel and Baire. Let $aX$ be a compactification of $X$ (Corollary A.4.8). So $X \in \mathcal{B}(aX) \subseteq \mathbf{C}(aX)$ by Corollary A.13.9. There consequently exist an open subset $U \subseteq aX$ and nowhere dense subsets $S_i, T_i \subseteq aX$ such that
$$X = \left(U \setminus \bigcup_{i=1}^{\infty} S_i\right) \cup \bigcup_{i=1}^{\infty} T_i.$$
Then $G = U \setminus \bigcup_{i=1}^{\infty} \overline{S_i}$ is dense in $U$ by the Baire Category Theorem A.6.6. Since every $T_i$ is nowhere dense in $aX$ and $X$ is dense in $aX$ it follows that every $T_i \cap X$ is nowhere dense in $X$. So $G$ is a dense subset of $X$. Since it is a $G_\delta$-subset of the compact space $aX$ it is topologically complete by Theorem A.6.3.
If $X$ contains a dense topologically complete subspace then it is a Baire space by Exercise A.6.7.

APPENDIX C

# Notes and comments

We will be brief regarding historical notes. The books of ENGELKING [153, 155] are excellent sources for the reader interested in the history of topology. See also AULL and LOWEN [30], and JAMES [212].

▶ **§1.1:**
Exercise 1.1.10 is due to Kuratowski [239] and Wojdysławski [411]. Its generalization Corollary 1.1.8 was proved by Arens and Eels [23] and Blumenthal and Klee [63]. Their proofs yield isometric imbeddings of metric spaces in normed linear spaces of the form $\ell_\infty(A)$, for some set $A$. The elegant proof of Lemma 1.1.6 is basically due to Michael [286]. It was brought to my attention by Dobrowolski that Michael's arguments can be improved to get imbeddings in function spaces of the form $C(A)$, where $A$ is a compact space. With his permission, his observation is published here. The Anderson Theorem from 1966 about the homeomorphy of $\ell^2$ and $s$ that was mentioned on Page 9 is due to Anderson [15] and gave the answer to a question posed by Fréchet [167] in 1928 and also by Banach [38] in 1932. Theorem 1.1.13 is due to Banach. The proof presented here was taken from Brown and Page [79, pp. 316–317]. As in the proof of Theorem 1.1.13, separability is not used in the proof of Theorem 1.1.15. So the result is true for arbitrary Banach spaces. This will be used in Chapter 6. For many more classical results on Banach spaces, see Lindenstrauss and Tzafriri [261]. Theorem 1.1.22 is due to Michael [287]. Continuous selections are important tools in topology and functional analysis. For the first book containing a systematic and comprehensive study of the field of continuous selections of multivalued mappings, see Repovš and Semenov [351]. A more general result than Corollary 1.1.23 is due to Bartle and Graves [44]. The homeomorphism in that result can in general not be chosen to be linear. For let $E$ be a separable Banach space not isomorphic to $\ell^1$. There is a continuous linear surjection $T\colon \ell^1 \to E$. Then $\operatorname{Ker} T \times E$ is not isomorphic to $\ell^1$ since all complemented linear subspaces of $\ell^1$ are isomorphic to $\ell^1$ by Pełczyński [335]. In view of our remarks at the end of this section, the question naturally arises whether the linear space $C(\mathbb{I})$ admits an equivalent norm $\|\cdot\|$ such that $(C(\mathbb{I}), \|\cdot\|)$ is an inner product space. The answer to this question is in the negative. The proof of this fact requires techniques that are outside the scope of this book. The interested reader may consult for example Taylor [386, p. 195]. Exercise 1.1.13 shows that the formula $\|x\| = \left(\sum_{n=1}^\infty |x_n|^p\right)^{1/p}$ does not define an admissible norm on $\ell^p$ for $0 < p < 1$; it is known that if for $p \geq 1$ the set $\ell^p$ is defined similarly then this formula does define an admissible norm on it.

## ▶§1.2:

Lemma 1.2.1 and Theorem 1.2.2 are due to Dugundji [140]. AR's and ANR's were introduced by Borsuk [64, 65]. The material on them presented in this section is standard. For more information, see Borsuk [70], Hu [197] and Chapter 4 of this monograph. In Exercise 1.2.4 we presented a very simple example of a contractible space which is not an ANR. There is also an example of a contractible LC space which is not an ANR, but this is much more complicated. See Borsuk [70, p. 124] for details. This is of particular interest, since in §4.2 we will show that contractibility and local contractibility implies the AR-property in *finite dimensional* spaces. Exercises 1.2.10 and 1.2.11 are due to van de Vel. See van Mill [298, Theorem 1.4.13] for more details. It is tempting to believe that the property in Exercise 1.2.12 characterizes the AR's among the linear spaces. But this is not true. Cauty [87] proved that there is a linear space $L$ which is an AR having a closed linear subspace which is not an AR. It is not clear whether the AR-property in linear spaces can be characterized in terms of selections.

## ▶§1.3:

All results in this section are standard.

## ▶§1.4:

Theorem 1.4.2 and its applications are due to Borsuk [68]. That the cone of an ANR is an AR is well-known. The nice proof of this fact, which was presented here in Theorem 1.4.6, was communicated to me by Dobrowolski who learned it from Toruńczyk.

## ▶§1.5:

The Cantor middle-third set C was introduced by Cantor [85]. The characterization Theorem 1.5.5 was proved by Brouwer [74]. Theorem 1.5.10 is due to Alexandroff and Urysohn [11]. The first space filling curve was constructed by Peano in 1890. Theorem 1.5.12 and Corollary 1.5.13 are due to Souslin [375]. Sets $A$ and $B$ such as in Corollary 1.5.14 are called *Bernstein sets*. Corollary 1.5.15 is due to van Douwen [133]. The proof presented here is taken from Becker, van Engelen and van Mill [45]. It is also due independently to Pol. Example 1.5.18 was established in Erdős [157]. Theorem 1.5.19 was proved in Sierpiński [367]. Theorem 1.5.22 is due to Mazurkiewicz [275]. Finally, Exercise 1.5.11 is due to Hausdorff [192] and de Groot [177].

## ▶§1.6:

The Inductive Convergence Criterion 1.6.2 is due to Fort [165] and was later rediscovered by Anderson [16]. Theorem 1.6.6 is due to Keller [219]. The proof presented here is due to Fort [165]. Theorems 1.6.7 and 1.6.9 are due to Bennett [46]. Exercises 1.6.1 and 1.6.2 can be found in Anderson, Curtis and van Mill [20]. Observe that Exercise 1.6.2 generalizes Theorem 1.6.9. The question naturally arises whether every strongly locally homogeneous Baire space is countable dense homogeneous. This question was answered in the negative by van Mill [294].

## ▶§1.7:

Bing's Shrinking Criterion 1.7.2 is of course due to Bing [58]. It is one of the most powerful tools in geometric topology. It is easy to realize the cone of $Q$ as a convex subset of $\ell^2$. So Theorem 1.7.5 is a direct consequence of Keller's Theorem from [219] that all compact convex infinite-dimensional subspaces of $\ell^2$ are homeomorphic (to $Q$). Exercise 1.7.1 is due to Schori [362]. Exercise 1.7.2 is folklore. It was observed in the late sixties while thinking about the following well-known problem of de Groot: *does there exist a homogeneous continuum homeomorphic to its own cone but which is not the Hilbert cube?* This problem is still unsolved. A related open problem is the one due to Bing and Borsuk [61]: *is every homogeneous finite dimensional compact* ANR *a manifold?*.

▶ **§1.8:**
The 'graph method' of extending homeomorphisms used in the proof of Theorem 1.8.2 is due to Klee [221].

▶ **§1.9:**
Lemma 1.9.1 is due to van Mill [293]. Sierpiński [369] proved the characterization of $\mathbb{Q}$ in Theorem 1.9.6. Theorem 1.9.8 is due to Alexandroff and Urysohn [10] who in the same paper also characterized the product

$$\mathbb{Q} \times \mathsf{C}$$

as the topologically unique $\sigma$-compact zero-dimensional space which is nowhere countable. The product

$$\mathbb{Q} \times \mathbb{P}$$

was topologically characterized in the same spirit by van Mill [292]: it is the topologically unique zero-dimensional space which is a countable union of closed topologically complete subspaces and which in addition is nowhere topologically complete and nowhere $\sigma$-compact. It was shown subsequently by van Engelen [148] that there are 'only' $\omega_1$ many topologically distinct homogeneous zero-dimensional absolute Borel sets, and that they can all be topologically characterized. His proofs make use of complicated results from descriptive set theory. Corollary 1.9.13 is due to Hurewicz [202] and its proof is to van Douwen [129]. For results in the same spirit, see e.g., van Engelen and van Mill [150]. Exercises 1.9.1 and 1.9.2 are due to van Mill [293]. Call a space *rigid* if the identity is its only homeomorphism. The real line can also be decomposed into two homeomorphic rigid sets, as was shown independently by van Engelen [149] and Shelah [365]. This is much more complicated than the construction in Exercises 1.9.1 and 1.9.2. It is a natural question (due to van Douwen) whether there exist rigid zero-dimensional absolute Borel sets. The zero-dimensionality in this question is essential for there are simple examples of rigid continua (de Groot and Wille [180]). It was shown in van Engelen, Miller and Steel [151] that rigid zero-dimensional absolute Borel sets do not exist.

▶ **§1.10:**
Inverse sequences and their limits are used extensively in topology. Especially for the construction of various counterexamples. They are in fact nothing but generalizations of products. Most of the general results on inverse limits that are mentioned here are standard. The dyadic solenoid is one of a family of peculiar spaces. Everything we said here about $\Sigma_2$ also works if instead of the function $f(z) = z^2$ in the inverse sequence one considers the function $f^p(z) = z^p$. Here $p = 2, 3, \ldots$. One then gets the so-called $p$-adic solenoid $\Sigma_p$ as inverse limit. The solenoids are

due to van Dantzig [108]. They can be characterized topologically as the unique homogeneous continua having only arcs as proper subcontinua. This interesting result is due to Hagopian [185] (see also Bing [59]). The results we presented on indecomposable continua are all standard. See Kuratowski [243, pp. 208–214] for more details. The hereditarily indecomposable continua are in may respects generic objects, playing an essential role in dimension theory, cf. Mazurkiewicz [278], Krasinkiewicz [230], Sternfeld [382], Levin [253], Levin and Sternfeld [258, 257, 259], Lewis [260] and Nadler [320]. Theorem 1.10.18 is due to van Douwen [126]. The space in Lemma 1.10.20 is due to van Mill [291]. Theorem 1.10.23 is due to Brown [80] and was widely used in the early days of infinite-dimensional topology. For information on dyanmical systems, see e.g., Devaney [110] and de Vries [401]. Exercise 1.10.8 is due to Farah [158] and Sierpiński [372]. Exercise 1.10.13 is also due to Brown [80]. Exercise 1.10.23 can be stated more precisely: the number of composants is in fact $\mathfrak{c}$.

▶ §1.11:
Hyperspaces were first considered in the early 1900's in the work of Hausdorff and Vietoris. For general information on hyperspaces see Nadler [320] and Illanes and Nadler [209]. Corollary 1.11.4 is due to Ważewski [404]. For a different proof, see van Mill [298, Proposition 4.7.2] and Exercise 1.11.15. Engelking [154] informed the author that the idea in the proof of [298, Proposition 4.7.2] is close to Kuratowski's proof of the Generalized Bolzano-Weierstrass Theorem (see [242, p. 340]). The Vietoris topology on $2^X$ was introduced in Vietoris [400]. Whitney maps were in a context different from hyperspaces first constructed by Whitney [409]. The construction in the proof of Proposition 1.11.15 is due to Krasinkiewicz [227]. For other constructions and results on Whitney maps, see Nadler [320] and Illanes and Nadler [209]. The function $e$ in Exercise 1.11.19 is called an *expansion homotopy* on $2^X$.

▶ §2.1:
The results in this section are standard. Simplicial complexes are very important in ANR-theory and algebraic topology. See e.g., Spanier [376] and Munkres [319].

▶ §2.2:
The results in this section are standard as well.

▶ §2.3:
Nerves of covers were introduced by Alexandroff [8], and their $\kappa$-mappings by Kuratowski [238]. Freudenthal's Approximation Theorem is due to Freudenthal [168]. The proof presented here was taken from Nagata [324]. In fact, a much more general statement is true (and proved by Freudenthal) than the one presented here. See Engelking [155, Theorem 1.3.12] for details.

▶ §2.4:
Theorem 2.4.3 is due to Sperner [377]. Theorem 2.4.5 is due to Brouwer [75]. The approach in this section for proving the Brouwer Fixed-Point Theorem is due to Knaster, Kuratowski and Mazurkiewicz [223]. A stronger result than Exercise 2.4.5 is due to Schauder [361]. In fact, Schauder claimed to have proved that every compact convex set in a linear space has the fixed-point property. But an inspection of

his proof showed that it is valid for locally convex linear spaces only. For decades it was a formidable open problem whether the local convexity condition in Schauder's Theorem could be dropped. It was shown only recently by Cauty [89] that this is indeed the case. For more details, see also Dobrowolski's very informative notes in [119]. Exercises 2.4.8 and 2.4.9 are due to Borsuk [66]. The converse is also true. That is, if a compact subspace $A \subseteq \mathbb{R}^n$ does not separate $\mathbb{R}^n$ then every continuous function $f \colon A \to \mathbb{S}^{n-1}$ is nullhomotopic. Similarly for $\mathbb{S}^n$.

▶ §2.5:

The Borsuk-Lusternik-Schnirelman Theorem is due to Borsuk [67] and Lusternik and Schnirelman [263]. For more information, see the very informative notes in Dugundji and Granas [143, p. 171].

▶ §3.1:

Motivated by work of Poincaré [337, 338], Brouwer [77] presented the first definition of a topological dimension function. See §3.10 for details. Corollary 3.1.2 is due to Brouwer [77]. The covering dimension dim in terms of finite open covers was formally introduced in Čech [94]. Our definition of dim is usually called 'The Theorem on Partitions' and was proved in Eilenberg and Otto [145]. The results in §3.1 are well-known.

▶ §3.2:

The proof of Theorem 3.2.5 consists mostly of small modifications of arguments that can be found in Engelking [155, Chapter 1]. Theorem 3.2.8 for the case of the small inductive dimension function ind (which takes the same value as dim, cf. Theorem 3.4.10) was proved for compact spaces by Menger [282] and by Urysohn [398]; finally, it was established in full generality by Tumarkin [393] and Hurewicz [199]. The Subspace Theorem 3.2.9 is a triviality for the small inductive dimension function ind. Theorem 3.2.12 and Corollary 3.2.13 are due to Brouwer [77]. Corollary 3.2.13 is also due to Brouwer [77]. Theorem 3.2.14 is due to Dowker [137] and Kuratowski [238]. Exercise 3.2.1 shows that within the class of separable metrizable spaces, the dimension cannot be raised by the adjunction of a single point. If one drops the separability condition, then this is no longer true, as was shown by van Douwen [125], and independently by Przymusiński [348]. Almost zero-dimensional spaces were introduced by Oversteegen and Tymchatyn [330]. They proved that almost zero-dimensional spaces are at most one-dimensional, and used this result to conclude that the homeomorphism groups of various interesting spaces such as Sierpiński's Carpet and Menger's Universal Curve, are one-dimensional. For a simpler proof that almost zero-dimensional spaces are at most one-dimensional, see Levin and Pol [254] and Theorem 3.5.10.

▶ §3.3:

That every $n$-dimensional space can be imbedded in $\mathbb{R}^{2n+1}$ is due independently to Nöbeling [327], Pontrjagin and Tolstowa [347] and Lefschetz [247]; that $\mathfrak{N}_n$ is universal for $n$-dimensional spaces is due to Nöbeling [327]. Corollary 3.3.8 is due to Hurewicz [200]. Corollary 3.3.12 is due to Tumarkin [393].

▶ §3.4:

That dim, ind and Ind take the same values, is a consequence of the fact that we deal here with separable metrizable spaces exclusively. Even for the well-behaved class of metrizable spaces this does not hold in general. The famous result of Roy [355, 356] shows that for completely metrizable spaces, ind and dim need not agree (for recent developments, see Mrówka [317, 318] and Kulesza [235]). On the other hand, dim and Ind take the same values on metrizable spaces, as was shown independently by Katětov [216] and Morita [314]. The small inductive dimension function ind was first defined by Urysohn [396] and Menger [281]. The Addition Theorem 3.4.4 is due, for compact spaces, to Urysohn [398] and for general spaces to Tumarkin [393] and Hurewicz [199]. The large inductive dimension function Ind is related to Brouwer [77] but was formally first defined by Čech [93]. The Coincidence Theorem 3.4.10 is partly due to Hurewicz [200, 198], Brouwer [78], Menger [282], Urysohn [398] and Tumarkin [393]. Theorem 3.4.12 is due to Menger [283]. An example with similar properties as Erdős' space in Example 3.4.13 cannot be compact. Hurewicz [204] proved that if $X$ is compact and $Y$ is an arbitrary space and $\dim X \leq 1$ then $\dim(X \times Y) = \dim X + \dim Y$ (see also [155, Problem 1.9(E)]). This result is 'best possible'. There exist compact spaces $X$ and $Y$ such that $\dim X = \dim Y = 2$, while $\dim(X \times Y) = 3$, Pontrjagin [346]; see also Kodama [224]. Examples as the one in Example 3.4.13 exist in every dimension, as was shown by Anderson and Keisler [21]. Their spaces are not complete however. Topologically complete spaces with the same properties were constructed later by Kulesza [234]. For more information on dimension modulo a class of spaces, see Aarts and Nishiura [4]. The compactness degree and compactness deficiency are notions due to de Groot [176]. Exercises 3.4.2 and 3.4.3 are due to de Groot [176]. (See also de Groot and Nishiura [179].) He raised the problem whether $\operatorname{cmp} X = \operatorname{def} X$ is true for all spaces $X$. This problem, posed in 1942, was finally solved in the negative by Pol [341] in 1982 (a simpler example was recently constructed by Levin and Segal [256]). See Aarts and Nishiura [4] for a comprehensive study of de Groot's compactification problem. It can be shown that the space $Y$ in Exercise 3.4.4 also has the property that $\operatorname{cmp} Y = n$. See [4, Example 5.10(d)] for details.

▶§3.5:

The results on compactifications in this section are well-known. They follow for example from the work of Kuratowski [241]. The part of Corollary 3.5.7 not dealing with colorability is due to Engelking [152]. Theorem 3.5.8 and Corollary 3.5.9 are due to Pol and Levin [254]. Theorem 3.5.10 is due to Oversteegen and Tymchatyn [330]. Their proof, based on the notion of $R$-trees, is rather complicated. The proof presented here is due to Levin and Pol [254]. It is not the most economical proof of Theorem 3.5.10. The proof in [254, §2] is shorter and more elegant and avoids the use of Corollary 3.5.9. However, Corollary 3.5.9 is of independent interest; its full strength will be used in §3.11 as well. Exercise 3.5.2 is due to Nagami [321]. Exercise 3.5.6 is due to van Mill and Pol [308]

▶§3.6:

Theorem 3.6.5 is basically due to Alexandroff [9]; contributions were also made by Hurewicz [205] and Hurewicz and Wallman [208]. Theorem 3.6.8 is due to Brouwer [76]. The proof presented here was taken from Hurewicz and Wallman [208]. Theorem 3.6.9 is due to Curtis and was published in [296]. Theorem 3.6.10 is for compact spaces due to Hurewicz [201] and for general spaces to Hurewicz and Wallman [208]. The fact that open maps on compact spaces can raise dimension is due to Kolmogoroff [225]. Anderson [13] proved that every Peano continuum is an open continuous image of the universal Menger curve. Theorem 3.6.14 is a corollary of this result. The proof presented here was brought to my attention by Uspenskiĭ. It is due to Dranišnikov and is close to a construction of Kozlovskiĭ [226], see also Fedorchuk [159, p. 186]. Exercise 3.6.1 is due to Dijkstra [113]. Exercise 3.6.4 is due to Borsuk [69].

▶§3.7:
Theorem 3.7.1 and Corollaries 3.7.2 and 3.7.3 are due to Menger [282] and Urysohn [397]. Proposition 3.7.4 can be found in Rubin, Schori and Walsh [357]. Theorem 3.7.6 is due to Mazurkiewicz [277]. The proof presented here is due to Rubin, Schori and Walsh [357]. The notion of a Cantor-manifold was introduced by Urysohn [397]. Theorem 3.7.8 was proved independently by Hurewicz and Menger [207] and by Tumarkin [394]. The proof presented here is due to Hurewicz [206]. A similar proof was given by Freudenthal [168]. Exercise 3.7.4 is due to Hart, van Mill and Pol [188].

▶§3.8:
The interesting Theorem 3.8.1 is due to Bing [57]. It is difficult to follow Bing's arguments. The proof presented here can be found in Hart, van Mill and Pol [188]. It is in fact close to Bing's original reasoning but is much easier to follow; variations of it circulated among various specialists in continua theory. The first Henderson compactum was constructed by Henderson [194]. Simpler examples can be found in Henderson [195], Bing [60], Zarelua [412], Rubin, Schori and Walsh [357] and Levin [252]. See §3.13 for information on stronger results. The proof of Corollary 3.8.8 is due to Levin [252]. Proposition 3.8.9 is due to Bing [57]. A nonmetrizable infinite-dimensional homogeneous indecomposable continuum was constructed by van Mill [299].

▶§3.9:
Theorem 3.9.2 is due to Bourbaki [72, p. 144, Exercise 9a]. The proof presented here is independently due to Kulesza [234] and van Mill [298, Theorem 4.7.3] and is inspired by Parthasarathy [332, Theorem 4.1]. The first topologically complete one-dimensional totally disconnected space was constructed by Sierpiński [370]. Theorem 3.9.3 is due to Mazurkiewicz [276]. The proof presented here is due to Rubin, Schori and Walsh [357] and Pol [340]. The technique in the proof goes back to Mazurkiewicz [276] and Knaster [222] and has been used by several authors: see Lelek [250], Zarelua [412], Rubin, Schori and Walsh [357], Kulesza [234], Ivanov [210], Pol [343], van Mill and Pol [303] and Krasinkiewicz [228, 229]. These papers contain further references. The interesting Theorem 3.9.5 is due to Kulesza [234]. Without the completeness condition, it was proved earlier by Anderson and Keisler [21].

▶§3.10:
Theorem 3.10.3 is due to Fedorchuk, Levin and Shchepin [161]. Theorem 3.10.4 is due to Fedorchuk and van Mill [162]. Every $n$-dimensional space $X$ can be imbedded in $\mathbb{R}^{2n+1}$ by Theorem 3.3.5 and hence in a product of $2n+1$ one-dimensional spaces.

The number $2n + 1$ is not optimal. It was proved by Lipscomb [262] that every $n$-dimensional space imbeds in a product of a family of at most $n+1$ one-dimensional spaces. See also Sternfeld [381], Bowers [73] and Olszewski [329]. Nagata [323, p. 163] asked whether every $n$-dimensional space can be imbedded in a product of at most $n$ one-dimensional spaces. This question was answered in the negative by Borsuk [71]. He proved (essentially) the result stated in Exercise 3.10.5. The solution to this exercise presented here which is simpler than Borsuk's original arguments, is due to van Mill and Pol [306].

▶ **§3.11:**
The proof of Lemma 3.11.5 presented here seems to be new. It follows immediately from Hurewicz's Theorem (Kuratowski [243, §45, statement IX]) that the zero-dimensional maps $g \colon K \to \mathbb{I}^{n+1}$ are dense in the function space $C(K, \mathbb{I}^{n+1})$ (if $p \colon \mathbb{I}^{n+1} \to \mathbb{I}$ is the projection and $g \colon K \to \mathbb{I}^{n+1}$ is zero-dimensional then

$$\dim[(p \circ g)^{-1}(t)] \leq n$$

for all $t \in \mathbb{I}$). Alternatively, one can also use Nagata's metric on $K$ (Nagata [325, Theorem V.4]). The first examples of weakly $n$-dimensional spaces were given by Sierpiński [370] and Mazurkiewicz [277]. A simpler construction can be found in Tomaszewski [389]. Theorem 3.11.8 is due to van Mill and Pol [303]. Theorem 3.11.10 is due to van Mill and Pol [307]. That the product of two weakly one-dimensional spaces is one-dimensional, is due to Tomaszewski [389]. Corollary 3.11.12 is due to van Mill and Pol [308]. The proof presented in the exercises of the fact that the product of two weakly one-dimensional spaces is one-dimensional, is basically due to to Tomaszewski [389]. (See van Mill and Pol [307].)

▶ **§3.12:**
The notion of a coloring of a map is motivated on the one hand by the Lusternik-Schnirelman-Borsuk Theorem (see §2.5) and on the other hand by results in combinatorial set theory due to de Bruijn and Erdős [81], Katětov [217], and others. See also Frolík [170]. Theorem 3.12.1 is due to van Douwen [131]. The Ellentuck topology on $[\mathbb{N}]^\omega$ was defined by Ellentuck in [147]. Corollary 3.12.6 of which the proof was taken from Kechris [218, §19], is a special case of the so-called Galvin-Prikry Theorem. The same conclusion can be derived if the $P_i$'s are merely Borel subsets of $[\mathbb{N}]^\omega$. See Galvin and Prikry [171] and [218, §19] for details. Theorem 3.12.7 was first stated in [231, Theorem 3.4] and is due to Mazur. Theorem 3.12.8 is a special case of a result of van Douwen [131]. He showed that if $X$ is finite dimensional and if $f \colon X \to X$ is closed, continuous and fixed-point free such that

$$\sup\{|f^{-1}(x)| : x \in X\} < \infty$$

then $f$ can be colored. (Theorem 3.12.7 shows that this result is sharp.) Theorem 3.12.10 is due to van Mill [301]. The proof of Theorem 3.12.12 makes use of a technique independently due to Krawczyk and Steprāns [231, Lemma 2.1] and Błaszczyk and Kim [62]. Corollary 3.12.15 is due to Aarts, Fokkink and Vermeer [3]. The proof presented here is due to van Mill [301]. As was already remarked in Remark 3.12.13, it is sharp for all $n$ by results of Steinlein [380, 379]. It is of interest to note that it can be shown that if $f$ is a fixed-point free involution on an at most $n$-dimensional space then $f$ can be colored with at most $n + 2$ colors. For this the central thing one needs to verify for involutions is that in Theorem 3.12.10

the $n + 3$ can be reduced to $n + 2$. For a generalization of Theorem 3.12.15 to nonmetrizable spaces, see van Hartskamp and Vermeer [190]. See also van Hartskamp [189]. The inverse limit trick used in the proof of Theorem 3.12.17 is due to Pol. Theorem 3.12.18 is due to Aarts and Fokkink [2].

▶ §3.13:
Theorem 3.13.7 is due to Lelek [249]. The proof presented here is due to Engelking and Pol [156]. The solution of Alexandroff's Problem presented in Theorem 3.13.8 is due to Pol [340]. Theorem 3.13.10 is due to Walsh [403]. The proof presented here was taken from Pol [343]. The first example of a Henderson compactum is due to Henderson [194], cf. §3.8. It is an interesting question whether there is a Henderson compactum containing a subspace with positive but finite dimension. The existence of such an example was established by Chatyrko and Pol [97]. There are even hereditarily indecomposable such spaces, as was recently proved by Pol [339]. The subspace with positive but finite dimension is one-dimensional in both examples. The problem whether similar spaces can be constructed containing subspaces with arbitrarily high finite dimension seems to be open. A stronger result than Exercise 3.13.2 is due to Rubin [358]. In Theorem 3.13.7 we saw that every topologically complete space can be compactified through the addition of a countable dimensional remainder. Exercise 3.13.4 shows that the completeness assumption is essential. In view of Corollary 3.5.7 the question naturally arises whether every homeomorphism of a topologically complete space can be 'compactified' through the addition of a countable dimensional remainder. The answer to this question is in the negative. Let us represent $s$ by $\mathbb{R}^{\mathbb{Z}}$ and let $\alpha \colon s \to s$ be the 'left shift' on $s$, i.e., $\alpha(x)_i = x_{i+1}$ for $i \in \mathbb{Z}$. It was shown by Dijkstra and van Mill [115] that if $\gamma s$ is a compactification of $s$ such that $\alpha$ can be extended to a continuous function $\bar{\alpha} \colon \gamma s \to \gamma s$ then $\gamma s \setminus s$ contains strongly infinite-dimensional continua. Adjan and Novikov [5] constructed (answering a question of Lusin) the first upper semicontinuous function on $\mathbb{I}$ that is not countably continuous. A similar construction was performed also by Sierpiński [373]. Jackson and Mauldin [211] proved, using notions from recursion theory, that Lebesgue measure $\lambda$ on the hyperspace $2^{\mathbb{I}}$ is not countably continuous (it is easy to see that it is upper semi-continuous). For a direct proof of this interesting fact avoiding recursion theory, see van Mill and Pol [305]. Exercise 3.13.5 is also due to them.

▶ §3.14:
The result in §3.14 that the existence of a two-dimensional hereditarily indecomposable continuum implies the existence of an infinite-dimensional hereditarily indecomposable continuum is due to Kelley [220]. (See Levin and Rogers [255] for an improvement.) His method was used by Levin and Sternfeld [257, 258] to prove the interesting result that the hyperspace of nonempty subcontinua of an at least two-dimensional compactum is infinite-dimensional. The observation in §3.14 that Brouwer in $\mathbb{I}^3$ implies Brouwer in all dimensions is due to Hart, van Mill and Pol [188]. Exercise 3.14.1 is due to Rogers [353].

▶ §4.1:
Most of the results in this chapter are well-known. Absolute (Neighborhood) Retracts were first defined by Borsuk [64, 65]. For a much more comprehensive study

of the subject, see Borsuk [70], Hu [197] and van Mill [298]. Theorems 4.1.1 and 4.1.13 are due to Hanner [187] and Theorem 4.1.9 is due to Dugundji [141] and Lefschetz [248]. Our notion of ANR-pair seems to be new. There are similar notions in the literature but they are different from ours, see e.g., Mardešić and Segal [274], Moszyńska [316] and van der Bijl [54]. Our notion was motivated by Toruńczyk's concept of a *locally homotopy negligible set* from [390]. It was a famous open problem due to Borsuk whether every compact ANR has the homotopy type of a (compact) polyhedron. The answer to this problem is in the affirmative, as was shown by West [406].

▶§4.2:

Theorem 4.2.1 is essentially due to Dugundji [141] and Lefschetz [247]. Our proof of Theorem 4.2.1 follows the exposition of Hu [197, Chapter IV §4]. Theorem 4.2.14 is essentially due to Toruńczyk [390]. Lemma 4.2.22 is due to Curtis and Nhu [106]. Wojdysławski [411] proved that the hyperspace of every Peano continuum is an AR and conjectured that in fact $2^X \approx Q$ if and only if $X$ is a Peano continuum. In [364], Schori and West verified this conjecture in the case $X = \mathbb{I}$. Their result was used in [363] to prove that $2^G \approx Q$ for every connected graph. Based on these results of Schori and West, in [107] Curtis and Schori verified Wojdysławski's conjecture. The proof that $2^X$ has the disjoint-cells property presented in this section is implicit in Curtis and Nhu [106]. Using the existence of convex metrics on Peano continua it is possible to give different proofs of this fact (Toruńczyk [391] and Curtis (private communication)). Corollary 4.2.24 is due to Curtis [105]. The proof presented here, based on our notion of an ANR-pair, seems to be new. Theorem 4.2.25 is due to Toruńczyk [391]. He not only topologically characterized $Q$, but also manifolds modeled on it. In addition, he proved a corresponding result for $\ell^2$-manifolds in [392] (see also Bestvina, Bowers, Mogilski and Walsh [52]). There are similar results for finite dimensional manifolds, but they are not as elegant as their infinite-dimensional counterparts since delicate algebraic obstructions enter the picture. See Daverman [109] for details. Finally, there are also similar results for manifolds defined on Menger compacta by Bestvina [51]. Theorem 4.2.30 is due to Kuratowski [240].

▶§4.3:

Theorems 4.3.1 and 4.3.5 are due to Hanner [187]. Our use of cones simplifies the proof of Theorem 4.3.5 slightly. Theorem 4.3.7 and its corollary 4.3.8 are due to Toruńczyk [390]. The proof presented here is due to Dobrowolski and Marciszewski [122].

▶§5.1:

Infinite-dimensional topology is the creation of Anderson. Several books were written on the subject, or deal with aspects of infinite-dimensional topology. See e.g., Chapman [96], Bessaga and Pełczyński [50], van Mill [298], Chigogidze and Fedorchuk [160], Dijkstra [112], Gladdines [175], Chigogidze [98] and Banakh, Radul and Zarichnyi [41]. The highlights of infinite-dimensional topology are the theorems of Anderson [15] on the homeomorphy of $\ell^2$ and $s$, of Chapman [95] on the invariance of Whitehead torsion, of West [406] on the finiteness of homototopy types of compact ANR's and of Toruńczyk [391, 392] on the topological characterization of manifolds modeled on various infinite-dimensional spaces. A large collection of open problems is West's paper [407]. The subjects that are being touched upon there range from absorbing sets and function spaces to ANR-theory.

The central concept of a $Z$-set in infinite-dimensional topology is due to Anderson [16]. Intuitively, $Z$-subsets of $X$ are 'small' in $X$. Here smallness has a different meaning than being 'small' with respect to category or measure. It is 'small' in the sense of homotopy (this will be made precise in Exercise 5.1.7). The notion of an $A$-set (the terminology is due to the author) is due to Anderson [16]. In the early days of infinite-dimensional topology, this was the central concept. But it soon became clear that it is equivalent to the notion of a $Z$-set defined in this section which turned out to be simpler to deal with in 'practical' situations.

▶ §5.2:

Theorem 5.2.4 is due to Anderson [16] and Barit [43]. The technique used in the proof is motivated by work of Klee [221]. The solution to Exercise 5.2.1 by elementary means is not simple. It is a consequence of a known and stronger homeomorphism extension result than the one we derived in this section. However, the proof of this result requires the development of some nontrivial facts that play no further role in this book. So we decided to skip that. See Anderson and Chapman [19], Chapman [96] and van Mill [298, §7.4] for details.

▶ §5.3:

Theorem 5.3.7 is due to Anderson [16] and Barit [43]. The other results in this section are standard applications of their result. Exercise 5.3.3 is due to Anderson [16]. It was asked in Anderson and Bing [18, Question 17] whether the condition $Q_0 \cap Q_1$ is a $Z$-set both in $Q_0$ and $Q_1$ in this result can be dropped. For some classes of Hilbert cubes, the answer was subsequently shown to be affirmative. For example, if both $Q_0$ and $Q_1$ are Keller cubes in $\ell^2$ (Quinn and Wong [350] and Mogilski [311]). It was shown by Handel [186] that the condition can be relaxed to $Q_0 \cap Q_1$ is a $Z$-set in $Q_0$, or in $Q_1$. This is of interest since it immediately implies the following old result of Anderson [14]: the product of the letter $T$ and $Q$ is a Hilbert cube. That is, $T$ is a so-called *Hilbert cube factor*, i.e., a space whose product with the Hilbert cube is the Hilbert cube. It was finally shown by Sher [366] that the answer to the Anderson-Bing question is in the negative.

▶ §5.4:

The notion of a capset is independently due to Anderson [12] and Bessaga and Pełczyński [49]. Many results in this section are standard. For references, see e.g., Bessaga and Pełczyński [50]. A subset $B \in \mathcal{Z}_\sigma(Q)$ is called a *boundary set* provided that $Q \setminus B \approx s$. This notion is due to Curtis [104] who characterized the way boundary sets are topologically placed in $Q$. The standard example of a boundary set is $B(Q)$. For 'pathological' examples of boundary sets, see e.g. Henderson and Walsh [196] and van Mill [295]. The statement in Corollary 5.4.2 comes close to saying that a capset is a so-called *absorber* or *absorbing set* in $Q$ for the class of all compact spaces. Absorbers are related to capsets and were first defined by West [405]. (See §5.5 for more information.) The influential paper by Bestvina and Mogilski [53] revived the interest in absorbers. Since then they are used quite effectively to prove the homeomorphy of various incomplete linear spaces. They are now a major tool in infinite-dimensional topology. See the monographs by Dijkstra [112] and Banakh, Radul and Zarichnyi [41] for more information. Theorem 5.4.9 is due to Anderson [12]. The interesting Theorem 5.4.10 is due to Curtis [104]. The Capset Characterization Theorem proved in Corollary 5.4.11 is also due to Curtis [104]. It was claimed earlier in Kroonenberg [233] that it is enough to replace condition (iv) by the weaker statement that for every $\varepsilon > 0$ there exists for some $n$ a map $\eta\colon Q \to B_n$ with $\hat\varrho(\eta, 1) < \varepsilon$. But this is not true, as was shown by Anderson, Curtis and van Mill [20]. The argument given in Kroonenberg [233] breaks down at the attempted application of the Homeomorphism Extension Theorem 5.3.7 for the copies $B_n$ of $Q$, using the restriction of the metric $\varrho$ on $Q$, and these restrictions may be highly nonconvex. Theorem 5.4.20 and Corollary 5.4.21 are due to Kroonenberg [233]. They were the first results of this type. Subsequently, many similar results were obtained by a variety of techniques, see e.g., Curtis [105] and Curtis and Nhu [106] for details. Exercise 5.4.6 is due to Anderson [17].

▶§5.5:
The results in this section are due to Dijkstra, van Mill and Mogilski [117]. For related results, see Bestvina and Mogilski [53]. A generalization of Theorem 5.5.7 can be found in Dobrowolski and Rubin [124]. Lemma 5.5.14 is an adaptation to our needs of Proposition 3.2 in Dobrowolski, Gul'ko and Mogilski [120]. A stronger result than Corollary 5.5.16 was first proved independently by Dobrowolski, Gul'ko and Mogilski [120] and Cauty [86] using the Bestvina-Mogilski approach from [53]. We will come back to this in Chapter 6. The first result in the spirit of Corollary 5.5.16 seems to be van Mill [297].

▶§6.1:
The largest compactification of a space $X$ is its so-called Čech-Stone compactification $\beta X$. For more information on $\beta X$, see e.g., Gillman and Jerison [174], Comfort and Negrepontis [102].

▶§6.2:
Theorem 6.2.5 is due to Arhangel'skiĭ [24]. Banakh and Cauty [40] proved that if $I$ is countable and $C_p(X_i)$ is analytic for every $i \in I$ then for $X = \bigoplus_{i \in I} X_i$ we have

$$C_p^*(X) \approx \prod_{i \in I} C_p^*(X_i). \qquad (*)$$

Moreover, examples in Marciszewski and van Mill [272] show that if $I$ is countably infinite then $(*)$ in general is not true.

▶§6.3:
It was shown in Lutzer and McCoy [264] that if $C_p(X)$ is a $G_\delta$-subset of $\mathbb{R}^X$ then $X$ is discrete. The stronger result mentioned in Remark 6.3.5 and its simple proof are due to Dijkstra, Grilliot, Lutzer and van Mill [114]. Theorems 6.3.6 and 6.3.8 are also due to these authors. The simple proof of Theorem 6.3.8 presented here is due to van Mill [300]. Theorem 6.3.8 was used in Cauty, Dobrowolski and Marciszewski [91] to prove the interesting result that if $X$ is countable and $C_p(X)$ is a Borel subset of $\mathbb{R}^X$ then it is of multiplicative Borel class. For more results on the Borel complexity of function spaces, see e.g., Calbrix [83, 84], Lutzer, van Mill and Pol [265].

## ▶§6.4:

A partial characterization of $C_p(X)$ being a Baire space is due to Lutzer and McCoy [264]. Theorem 6.4.3 is due independently to van Douwen [135], Pytkeev [349] and Tkachuk [388]. Our exposition of the proof of Theorem 6.4.3 closely followed McCoy and Ntantu [280, Theorem 5.3.8]. That there is a countable dense-in-itself space $X$ such that $C_p(X)$ is Baire is due to Michalewski [289, Page 22]. The interesting Theorem 6.5.3 is due to Talagrand [384]. With the aid of topological game theory, results in the same spirit were obtained by Lutzer and McCoy [264]. Their proofs were simplified by Dobrowolski, Marciszewski and Mogilski [123]. Theorem 6.5.3 in the context of function spaces was used for the first time in Dobrowolski and Marciszewski [121]. Theorem 6.12.1 and Corollary 6.5.4 and their consequences are due to Dobrowolski, Marciszewski and Mogilski [123]. That $C_p(\mathbb{N}_{\mathcal{F}})$ is a Baire space for an ultrafilter $p$ is due to Lutzer and McCoy [264].

## ▶§6.6:

Theorem 6.6.2 is essentially due to Dugundji [140]. Proposition 6.6.6 and Corollaries 6.6.8 and 6.6.10 are folklore. They were probably first proved by Arhangel'skiĭ.

## ▶§6.7:

The results in this section are folklore. They were used quite frequently in the Russian literature without explicit references and proofs, see e.g., Pavlovskiĭ [333] and Pestov [336]. For another approach, see Baars and de Groot [35, §1.3]. See also Arhangel'skiĭ [27].

## ▶§6.8:

The concept of support of a point is due to Arhangel'skiĭ [24]. The alternative description of supports for continuous linear functions is due independently to Arhangel'skiĭ and Pelant. It was of crucial importance to Baars and de Groot [35] in their analysis of function spaces of countable spaces. Most of Lemma 6.8.2 is due to Baars and de Groot [35, Chapter 1]. Theorem 6.8.3 and its Corollary 6.8.4 were proved by Arhangel'skiĭ [24]. Corollary 6.8.5 is due to Baars and de Groot [35, Lemma 1.2.11]. Proposition 6.8.6 for normal spaces is Lemma 1.4.5 in Baars and de Groot [35]. The suggestion that it is also true for arbitrary spaces was made to the author by Marciszewski.

## ▶§6.9:

Theorem 6.9.1 are Corollary 6.9.2 are due to Arhangel'skiĭ [24]. The interesting Theorem 6.9.4, which is due to Pestov [336], is in fact true for the class of all spaces, and not only for separable metrizable spaces. It has motivated the natural question whether dimension is preserved by $t$-equivalence. This question is unsolved. See Remark 6.9.5 for more details. Theorem 6.9.7 is due to Baars, de Groot and Pelant [37]. For generalizations, see Marciszewski and Pelant [273] and Marciszewski [271]. It is also true, see Baars, de Groot and Pelant [37], that if $X$ and $Y$ are metrizable and $\varphi \colon C_p^*(X) \to C_p^*(Y)$ is a continuous linear surjection then $X$ completely metrizable implies $Y$ completely metrizable. The proof of this is much more complicated than the proof of its corresponding result Theorem 6.9.7.

## ▶§6.10:

The results in this section are taken from Baars, de Groot, van Mill and Pelant [36].

The proof of Theorem 6.10.8 is a generalization of a method due to Pelant [334]. For other results in the same spirit, see e.g., Baars and de Groot [33, 34, 35].

▶ **§6.11:**
It is an intriguing open problem whether $t$-equivalent spaces have the same dimension. Marciszewski [271] proved the interesting Theorem 6.11.3, modifying Cauty's technique from [90] and using an idea from Gul'ko [182].

▶ **§6.12:**
It is an interesting question which topological properties of the filter $\mathcal{F}$ (as subspace of $\mathcal{P}(\mathbb{N})$) translate into 'nice' topological properties of the function space $c_\mathcal{F}$. There are many of such results. We will mention only the result of Calbrix [83, 84] that $c_\mathcal{F}$ is an absolute Borel set if and only if $\mathcal{F}$ is. The proof of Theorem 6.12.15 presented here contains elements that can be found in Dobrowolski, Marciszewski and Mogilski [123], Baars, Gladdines and van Mill [32] and Dijkstra and Mogilski [118]. However, the proof in the latter two papers has a gap. This gap was bridged in the paper Dijkstra and van Mill [116]. As we remarked at the beginning §6.12, the consequence of Theorem 6.12.15 that all $F_{\sigma\delta}$ function spaces are homeomorphic was first shown by Dobrowolski, Marciszewski and Mogilski [123].

▶ **§6.13:**
Example **E11** is due to Dijkstra, Grilliot, Lutzer and van Mill [114]. That there is an almost disjoint family of $\mathfrak{c}$ subsets of $\omega$ is due to Sierpiński [371]. The Disjoint refinement Lemma 6.13.5 is a well-known result from set theory. It is in fact true in a much more general setting. The example $\Delta$ is due to van Mill and Pol [304]. It is based on a space constructed earlier in van Douwen and Pol [136]. For an example with similar properties, see Marciszewski [267].

▶ **§A.1:**
Our notation is standard and all the results in this chapter are classical. For information on ultrafilters, see Comfort and Negrepontis [102]. Exercise A.1.2 is due to van Douwen [130]. A space without isolated points and satisfying one of the equivalent statements in Exercise A.1.2 is called *perfectly disconnected*. It is not clear at all that such spaces exist. They were first defined and constructed by van Douwen [134, 130]. It will be shown in §6.4 that a countable dense-in-itself perfectly disconnected space $X$ has the property that its function space $C_p(X)$ is Baire. So $X$ is not Baire but its function space *is* Baire, which is rather curious. Exercise A.1.4 is usually called the $\Delta$-System Lemma. See Juhász [214] for this and many other related results. Exercise A.1.10 and its solution are well-known.

▶ **§A.2:**
Exercise A.2.13 is the so-called Cantor-Bendixson Theorem which was proved by Cantor and Bendixson independently in 1883 for subsets of $\mathbb{R}$.

▶ **§A.3:**
It is a natural question to ask when the limit of a sequence of continuous functions exists and is continuous. This basic question is studied in this section.

▶ **§A.4:**
Lemma A.4.1 can be viewed to be some sort of an extension result. For let $Y$ be

the subspace $A \cup B$ of $X$. It is clear that both $A$ and $B$ are clopen in $Y$. So the function $f: Y \to \mathbb{I}$ defined by $f \upharpoonright A \equiv 0$ and $f \upharpoonright B \equiv 1$ is continuous. Hence Lemma A.4.1 tells us among other things that $f$ can be extended to a continuous function $\alpha: X \to \mathbb{I}$. The possibility of extending continuous functions is very important in topology. We will come back to this in §1.2. Theorem A.4.6 can be found in Tietze [387] and Urysohn [399]. See also Hausdorff [191] for a related approach. In Corollary A.4.4 we proved in fact that every regular $T_1$-space with a countable base can be imbedded in $Q$ and hence is metrizable, being homeomorphic to a subspace of a metrizable space. This is usually called *Urysohn's Metrization Theorem*. The most famous and important metrization theorem is the one due to Nagata, Smirnov and Bing. See e.g., Engelking [153, Chapter 4] for more details. The Alexandroff one-point compactification was first considered in Alexandroff [7]. Exercise A.4.4 is due to Berney [47].

▶§A.5:

The notion of compactness is one of the most important in topology. As proved in this section, in the realm of metrizable spaces compactness can be characterized in terms of sequences. This simplifies life considerably (see for example the proof of Corollary 1.11.4). Lemma A.5.3 is essentially due to Lebesgue [246]. It was shown independently by Bing [56] and Moise [312] that every Peano continuum admits an admissible convex metric. This is a highly nontrivial result. So Exercise A.5.20 follows from their result.

▶§A.6:

Theorem A.6.1 is due to Fréchet [166]. Theorem A.6.3 is due to Alexandroff [6] and Kuratowski [242]. The important class of Baire spaces was introduced by Bourbaki bourbaki. Theorem A.6.6 is due to Hausdorff [193]. Theorem A.6.10 and its generalization Exercise A.6.11 are due to Oxtoby [331].

▶§A.7:

The results in this section are standard. Corollary A.7.6 is a classical result due to Dowker [138], Katětov [215] and Dieudonné [111].

▶§A.8:

Lemma A.8.1 is due to Kuratowski [243, §15, XIII, p. 122]. Lemma A.8.3 was proved independently by Taĭmanov [385] and by Eilenberg and Steenrod [146]. Theorem A.8.5 is due to Lavrentieff [245]. See van Douwen [132] for a partial converse to his theorem. In Engelking [153, Theorem 4.3.21] the reader can find more information on Lavrentieff's Theorem. The solution to Exercise A.8.1 is due to van Douwen [132]. Exercise A.8.3 is implicit in Marciszewski [271].

▶§A.9:

Wallman compactifications were first defined by Frink [169], who was inspired by the work of Wallman [402]. In this section we discussed separable metrizable Wallman compactifications only, but from the construction it is clear that there also is a theory for general (Tychonoff) spaces. Frink [169] asked whether every compactification is a Wallman compactification. This was a central problem for some time, until it was solved in the negative by Ul′janov [395]. That every metric compactification is a Wallman compactification (Corollary A.9.8), is due independently to

Aarts [1] and Steiner and Steiner [378]. Different proofs of the same result were given by Berney [47], van Douwen [127], Bandt [42], and others. We closely followed Berney's proof.

▶ **§A.10:**
Theorem A.10.6 is due to Sierpiński [368]. It has a nontrivial generalization due to Dijkstra [113]. See Exercise 3.6.1 for more details. The compactness of the the $C_n$'s in Proposition A.10.7 is essential, see Example 1.10.11. Exercise A.10.6 is due to Moore [313]. Exercise A.10.7 is due to Urysohn. For a description of the troubles this example caused in the early days of dimension theory, see e.g., Johnson [213]. It can even be shown that the set $X \setminus \bigcup \mathcal{F}$ in Exercise A.10.8 has cardinality $\mathfrak{c}$. For the set $X \setminus \bigcup \mathcal{F}$ is topologically complete, and an uncountable topologically complete space has size $\mathfrak{c}$ (Corollary 1.5.13).

▶ **§A.11:**
The notion of adjunction space discussed in this section is very important in topology. We did not discuss it in its most general form. For more details, see Dugundji [142]. Null-sequences are used extensively in topology for creating decomposition spaces with various interesting properties. See e.g., Daverman [109] for more information.

▶ **§A.12:**
The material in this section is folklore. The approach to some of the results here is due to Burckel [82]. Theorem A.12.8 remains true for any space $X$. This is due to Eilenberg [144]. It can be shown that every cube $\mathbb{I}^n$ has the fixed-point property. This is the Brouwer Fixed-Point Theorem, see §2.4.

▶ **§A.13:**
Borel subsets of $\mathbb{R}$ were introduced by Borel. The results in this section are all classical. The proof of Theorem A.13.5 is due to van Engelen [148, Lemma A.2.4]. In view of Corollary A.13.12, it is a natural question whether every analytic subset of a topologically complete space is Borel. This is nontrivial and was answered in the negative by Souslin. See Hurewicz [203] and Mazurkiewicz [279] for 'explicit' examples of such spaces. The classical proof of Theorem A.13.13 goes as follows. First one proves that the Baire property is invariant under the so-called $\mathcal{A}$-*operation* (this is due to Nikodym [326]), and then one proceeds to prove that the analytic subsets of a topologically complete space $X$ coincide with the sets that one obtains by performing the $\mathcal{A}$-operation on all closed subsets of $X$. For details, see e.g., Kuratowski and Mostowski [244]. The direct proof of Theorem A.13.13 presented here is motivated by these classical constructions. Call a subset of a space $X$ *coanalytic* if its complement is analytic. Since $\mathbf{C}(X)$ is a $\sigma$-algebra, it contains every coanalytic set by Theorem A.13.13. Since there are coanalytic spaces which are not analytic (see e.g., Kuratowski and Mostowski [244] or Kechris [218]), this shows that the converse to Theorem A.13.13 does not hold, i.e., that in general $\mathbf{C}(X)$ does not coincide with the collection of all analytic subsets of $X$. Exercise A.13.4 is a special case of a result due to Levi [251]. In Exercise A.13.6 we proved that there are subsets of $\mathbb{R}$ which are not Borel. If $X = \mathbb{N}$ with the discrete topology, then clearly all subsets of $X$ are Borel. A more interesting example is $X = \mathbb{Q}$: simply observe that every subset of $\mathbb{Q}$ is countable, hence $F_\sigma$, and hence Borel. In view of

this, it is tempting to believe that every uncountable space has a non-Borel subset. A so-called $Q$-set shows that this need not be true, see e.g., Miller [309].

# Bibliography

[1] J. M. Aarts, *Every metic compactification is a Wallman-type compactification*, Proc. Int. Symp. on Top. and its Appl., Herceg-Novi (Yugoslavia), 1968, Beograd, 1969, pp. 29–34.

[2] J. M. Aarts and R. J. Fokkink, *An addition theorem for the color number*, 1999, to appear in Proc. Amer. Math. Soc.

[3] J. M. Aarts, R. J. Fokkink, and J. Vermeer, *Variations on a theorem of Lusternik and Schnirelmann*, Topology **35** (1996), 1051–1056.

[4] J. M. Aarts and T. Nishiura, *Dimension and Extensions*, North-Holland Mathematical Library, vol. 48, North-Holland Publishing Co., Amsterdam, 1993.

[5] S. I. Adjan and P. S. Novikov, *On a semicontinuous function*, Moskov. Gos. Ped. Inst. Uchen. Zap. **138** (1958), 3–10, In Russian.

[6] P. Alexandroff, *Sur les ensembles de la première classe et les ensembles abstraits*, C. R. Acad. Paris **178** (1924), 185–187.

[7] P. Alexandroff, *Über die Metrisation der im Kleinen kompakten topologischen Räume*, Math. Ann. **92** (1924), 294–301.

[8] P. Alexandroff, *Une définition des nombres de Betti pour un ensemble fermé quelconque*, C.R. Acad. Paris **184** (1927), 317–319.

[9] P. Alexandroff, *Dimensionstheorie. Eein Beitrag zur Geometrie der abgeschlossenen Mengen*, Math. Ann. **106** (1932), 161–238.

[10] P. Alexandroff and P. Urysohn, *Über nuldimensionale Punktmengen*, Math. Ann. **98** (1928), 89–106.

[11] P. Alexandroff and P. Urysohn, *Mémoire sur les espaces topologiques compacts*, Verh. Akad. Wetensch. Amsterdam **14** (1929).

[12] R. D. Anderson, *On sigma-compact subsets of infinite-dimensional manifolds*, unpublished manuscript.

[13] R. D. Anderson, *A characterization of the universal curve and a proof of its homogeneity*, Annals of Math. **67** (1958), 313–324.

[14] R. D. Anderson, *The Hilbert cube as a product of dendrons*, Notices Amer. Math. Soc. **11** (1964), 572.

[15] R. D. Anderson, *Hilbert space is homeomorphic to the countable infinite product of lines*, Bull. Amer. Math. Soc. **72** (1966), 515–519.

[16] R. D. Anderson, *On topological infinite deficiency*, Mich. Math. J. **14** (1967), 365–383.

[17] R. D. Anderson, *Strongly negligible sets in Fréchet manifolds*, Bull. Amer. Math. Soc. **75** (1969), 64–67.

[18] R. D. Anderson and R. H. Bing, *A complete elementary proof that Hilbert space is homeomorphic to the countable infinite product of lines*, Bull. Amer. Math. Soc. **74** (1968), 771–792.

[19] R. D. Anderson and T. A. Chapman, *Extending homeomorphisms to Hilbert cube manifolds*, Pac. J. Math. **38** (1971), 281–293.

[20] R. D. Anderson, D. W. Curtis, and J. van Mill, *A fake topological Hilbert space*, Trans. Amer. Math. Soc. **272** (1982), 311–321.
[21] R. D. Anderson and J. E. Keisler, *An example in dimension theory*, Proc. Amer. Math. Soc. **18** (1967), 709–713.
[22] R. Arens, *Extension of functions on fully normal spaces*, Pac. J. Math. **1** (1951), 353–367.
[23] R. Arens and J. Eels, *On embedding uniform and topological spaces*, Pac. J. Math. **6** (1956), 397–403.
[24] A. V. Arhangel′skiĭ, *On linear homeomorphisms of function spaces*, Soviet Math. Doklady **25** (1982), 852–855.
[25] A. V. Arhangel′skiĭ, *A survey of $C_p$-theory*, Questions and Answers in Gen. Top. **5** (1987), 1–109.
[26] A. V. Arhangel′skiĭ, *$C_p$-theory*, Recent Progress in General Topology (M. Hušek and J. van Mill, eds.), North-Holland Publishing Co., Amsterdam, 1992, pp. 1–56.
[27] A. V. Arhangel′skiĭ, *Topological function spaces*, Math. Appl., vol. 78, Kluwer Academic Publishers, Dordrecht, 1992.
[28] A. V. Arhangel′skiĭ, *Some observations on $C_p$-theory and bibliography*, Top. Appl. **89** (1998), 203–221.
[29] W. Arveson, *An invitation to $C^*$-algebras*, Springer-Verlag, Berlin, 1976.
[30] C. E. Aull and R. Lowen, eds., *Handbook of the History of General Topology*, Kluwer Academic Publishers, Dordrecht, 1997.
[31] J. Baars, *On the $\ell_p^*$-equivalence of certain locally compact spaces*, Top. Appl. **52** (1993), 43–57.
[32] J. Baars, H. Gladdines, and J. van Mill, *Absorbing systems in infinite-dimensional manifolds*, Top. Appl. **50** (1993), 147–182.
[33] J. Baars and J. de Groot, *An isomorphical classification of function spaces of zero-dimensional locally compact separable metric spaces*, Comm. Math. Univ. Carolinae **29** (1988), 577–595.
[34] J. Baars and J. de Groot, *On the $\ell$-equivalence of metric spaces*, Fund. Math. **137** (1991), 25–43.
[35] J. Baars and J. de Groot, *On topological and linear equivalence of certain function spaces*, CWI Tract, vol. 86, Centre for Mathematics and Computer Science, Amsterdam, 1992.
[36] J. Baars, J. de Groot, J. van Mill, and J. Pelant, *An example of $\ell_p$-equivalent spaces which are not $\ell_p^*$-equivalent*, Proc. Amer. Math. Soc. **119** (1993), 963–969.
[37] J. Baars, J. de Groot, and J. Pelant, *Function spaces of completely metrizable spaces*, Trans. Amer. Math. Soc. **340** (1993), 871–879.
[38] S. Banach, *Théorie des opérations linéaires*, PWN, Warszawa, 1932.
[39] T. Banakh, *Some properties of the linear hull of the Erdős set in $\ell^2$*, Bull. Polon. Acad. Sci. Sér. Math. Astronom. Phys. **47** (1999), 385–392.
[40] T. Banakh and R. Cauty, *Universalité forte pour les sous-ensembles totalement bornés. Applications aux espaces $C_p(X)$*, Colloq. Math. **73** (1997), 25–33.
[41] T. Banakh, T. Radul, and M. Zarichnyi, *Absorbing sets in infinite-dimensional manifolds*, Mathematical Studies, vol. 1, VNTL Publishers, Lviv, 1996.
[42] C. Bandt, *On Wallman-Shanin-compactifications*, Math. Nachr. **77** (1977), 333–351.
[43] B. Barit, *Small extensions of small homeomorphisms*, Notices Amer. Math. Soc. **16** (1969), 295.
[44] R. G. Bartle and L. M. Graves, *Mappings between function spaces*, Trans. Amer. Math. Soc. **72** (1952), 400–413.
[45] H. Becker, F. van Engelen, and J. van Mill, *Disjoint embeddings of compacta*, Mathematika **41** (1994), 221–232.
[46] R. Bennett, *Countable dense homogeneous spaces*, Fund. Math. **74** (1972), 189–194.

[47] E. S. Berney, *On Wallman compactifications*, Notices Amer. Math. Soc. **17** (1970), 215.
[48] C. Bessaga and A. Pełczyński, *Spaces of continuous functions IV (on isomorphical classification of spaces of continuous functions)*, Studia Math. **19** (1960), 53–62.
[49] C. Bessaga and A. Pełczyński, *The estimated extension theorem homogeneous collections and skeletons, and their application to the topological classification of linear metric spaces and convex sets*, Fund. Math. **69** (1970), 153–190.
[50] C. Bessaga and A. Pełczyński, *Selected topics in infinite-dimensional topology*, PWN—Polish Scientific Publishers, Warsaw, 1975, Monografie Matematyczne, Tom 58.
[51] M. Bestvina, *Characterizing k-dimensional universal Menger compacta*, Mem. Amer. Math. Soc. **71** (1988), no. 380, vi+110.
[52] M. Bestvina, P. Bowers, J. Mogilski, and J. J. Walsh, *Characterizing Hilbert space manifolds revisited*, Top. Appl. **24** (1986), 53–69.
[53] M. Bestvina and J. Mogilski, *Characterizing certain incomplete infinite-dimensional absolute retracts*, Michigan Math. J. **33** (1986), 291–313.
[54] J. van der Bijl, *Extension of continuous functions with compact domain*, Ph.D. thesis, Vrije Universiteit, Amsterdam, 1991.
[55] R. H. Bing, *A homogeneous indecomposable plane continuum*, Duke Math. J. **15** (1948), 729–742.
[56] R. H. Bing, *Partitioning a set*, Bull. Amer. Math. Soc. **55** (1949), 1101–1110.
[57] R. H. Bing, *Higher-dimensional hereditarily indecomposable continua*, Trans. Amer. Math. Soc. **71** (1951), 267–273.
[58] R. H. Bing, *A homeomorphism between the 3-sphere and the sum of two solid horned spheres*, Annals of Math. **56** (1952), 354–362.
[59] R. H. Bing, *A simple closed curve is the only homogeneous bounded plane continuum that contains an arc*, Pac. J. Math. **12** (1960), 209–230.
[60] R. H. Bing, *A hereditarily infinite dimensional space*, Proceedings of the Second Prague Topological Symposium, Academia, Praha, pp. 56–62.
[61] R. H. Bing and K. Borsuk, *Some remarks concerning topologically homogeneous spaces*, Annals of Math. **81** (1965), 100–111.
[62] A. Błaszczyk and D. Y. Kim, *A topological version of a combinatorial theorem of Katětov*, Comm. Math. Univ. Carolinae **29** (1988), 657–663.
[63] L. M. Blumenthal and V. L. Klee, *On metric independence and linear independence*, Proc. Amer. Math. Soc. **6** (1955), 732–734.
[64] K. Borsuk, *Sur les rétractes*, Fund. Math. **17** (1931), 152–170.
[65] K. Borsuk, *Über eine Klasse von lokal zusammenhängende Räumen*, Fund. Math. **19** (1932), 220–242.
[66] K. Borsuk, *Über Schnitte der n-dimensionalen Euklidischen Räume*, Math. Ann. **106** (1932), 230–248.
[67] K. Borsuk, *Drei Sätze über die n-dimensionale Euclidische Sphäre*, Fund. Math. **20** (1933), 177–190.
[68] K. Borsuk, *Sur les prolongements des transformations continus*, Fund. Math. **28** (1936), 99–110.
[69] K. Borsuk, *Un théorème sur les prolongements des transformations*, Fund. Math. **29** (1937), 161–166.
[70] K. Borsuk, *Theory of retracts*, Państwowe Wydawnictwo Naukowe, Warsaw, 1967, Monografie Matematyczne, Tom 44.
[71] K. Borsuk, *Remark on the Cartesian product of two 1-dimensional spaces*, Bull. Polon. Acad. Sci. Sér. Math. Astronom. Phys. **23** (1975), 971–973.
[72] N. Bourbaki, *Topologie Générale (2$^e$ ed.)*, Hermann, Paris, 1958.

[73] P. Bowers, *General position properties satisfied by finite products of dendrites*, Trans. Amer. Math. Soc. **228** (1985), 739–753.
[74] L. E. J. Brouwer, *On the structure of perfect sets of points*, Proc. Akad. Amsterdam **12** (1910), 785–794.
[75] L. E. J. Brouwer, *Über Abbildung von Mannigfaltigkeiten*, Math. Ann. **71** (1912), 97–115.
[76] L. E. J. Brouwer, *Invariantz des n-dimensionalen Gebiets*, Math. Ann. **71–72** (1912–1913), 305–313; 55–56.
[77] L. E. J. Brouwer, *Über den natürlichen Dimensionsbegriff*, J. Reine Angew. Math. **142** (1913), 146–152.
[78] L. E. J. Brouwer, *Bemerkungen zum natürlichen Dimensionsbegriff*, Proc. Akad. Amsterdam **27** (1924), 635–638.
[79] A. L. Brown and A. Page, *Elements of functional analysis*, Van Nostrand Reinhold, London, 1970.
[80] M. G. Brown, *Some applications of an approximation theorem for inverse limits*, Proc. Amer. Math. Soc. **11** (1960), 478–483.
[81] N. G. de Bruijn and P. Erdős, *A colour problem for infinite graphs and a problem in the theory of relations*, Indag. Math. **13** (1951), 371–373.
[82] R. B. Burckel, *Inessential maps and classical Euclidean topology*, Yearbook: Surveys of mathematics 1981, Bibliographisches Inst., Mannheim, 1981, pp. 119–137.
[83] J. Calbrix, *Classes de Baire er espaces d'applications continues*, C. R. Acad. Sci. Paris **301** (1985), 759–762.
[84] J. Calbrix, *Filtres boreliens sur l'ensemble des entiers et espaces des applications continues*, Rev. Roumaine Math. Pure Appl. **33** (1988), 655–661.
[85] G. Cantor, *Über unendliche, lineare Punktmannichfaltigkeiten*, Math. Ann. **21** (1883), 545–591.
[86] R. Cauty, *L'espace des fonctions continues d'un espace métrique dénombrable*, Proc. Amer. Math. Soc. **113** (1991), 493–501.
[87] R. Cauty, *Un espace métrique linéaire qui n'est pas un rétracte absolu*, Fund. Math. **146** (1994), 85–99.
[88] R. Cauty, *La classe Borélienne ne détermine pas le type topologique de $C_p(X)$*, Serdica Math. J. **24** (1998), 307–318.
[89] R. Cauty, *Solution du problème de point fixe de Schauder*, 1999, Preprint.
[90] R. Cauty, *Sur l'invariance de la dimension infinie forte par t-équivalence*, Fund. Math. **160** (1999), 95–100.
[91] R. Cauty, T. Dobrowolski, and W. Marciszewski, *A contribution to the topological classification of the spaces $C_p(X)$*, Fund. Math. **142** (1993), 269–301.
[92] M. Valdivia, *Products of Baire topological vector spaces*, Fund. Math. **125** (1985), 71–80.
[93] E. Čech, *Sur la théorie de la dimension*, C.R. Acad. Paris **193** (1931), 976–977.
[94] E. Čech, *Prispevek k theorii dimense*, Casopis Pest. Mat. Fys. **62** (1933), 277–291.
[95] T. A. Chapman, *Topological invariance of Whitehead torsion*, Amer. J. Math. **96** (1974), 488–497.
[96] T. A. Chapman, *Lectures on Hilbert cube manifolds*, American Mathematical Society, Providence, R. I., 1976, Expository lectures from the CBMS Regional Conference held at Guilford College, October 11-15, 1975, Regional Conference Series in Mathematics, No. 28.
[97] V. A. Chatyrko and E. Pol, *Continuum many Fréchet types of hereditarily strongly infinite-dimensional Cantor manifolds*, Proc. Amer. Math. Soc. **128** (2000), 1207–1213.
[98] A. Chigogidze, *Inverse spectra*, North-Holland Mathematical Library, vol. 53, North-Holland Publishing Co., Amsterdam, 1996.

[99] G. Choquet, *Lectures on analysis. Vol. I: Integration and topological vector spaces*, W. A. Benjamin, Inc., New York-Amsterdam, 1969, Edited by J. Marsden, T. Lance and S. Gelbart.
[100] J. P. R. Christensen, *Topology and Borel structure*, North-Holland Publishing Co., Amsterdam, 1974.
[101] P. E. Cohen, *Products of Baire spaces*, Proc. Amer. Math. Soc. **55** (1976), 119–124.
[102] W. W. Comfort and S. Negrepontis, *The theory of ultrafilters*, Grundlehren der mathematischen Wissenschaften, vol. 211, Springer-Verlag, Berlin, 1974.
[103] H. Cook, *Continua which admit only the identity mapping onto nondegenerate subcontinua*, Fund. Math. **60** (1967), 241–249.
[104] D. W. Curtis, *Boundary sets in the Hilbert cube*, Top. Appl. **20** (1985), 201–221.
[105] D. W. Curtis, *Hyperspaces of finite subsets as boundary sets*, Top. Appl. **22** (1986), 97–107.
[106] D. W. Curtis and N. T. Nhu, *Hyperspaces of finite subsets which are homeomorphic to $\aleph_0$-dimensional linear metric space*, Top. Appl. **19** (1985), 251–260.
[107] D. W. Curtis and R. M. Schori, *Hyperspaces of Peano continua are Hilbert cubes*, Fund. Math. **101** (1978), 19–38.
[108] D. van Dantzig, *Über topologisch homogene Kontinua*, Fund. Math. **15** (1930), 102–125.
[109] R. J. Daverman, *Decompositions of manifolds*, Academic Press, New York, 1986.
[110] R. L. Devaney, *An introduction to chaotic dynamical systems*, Addison-Wesley Publishing Co., Reading, 1989.
[111] J. Dieudonné, *Une généralisation des espaces compacts*, J. de Math. Pures et Appl. **23** (1944), 65–76.
[112] J. J. Dijkstra, *Fake topological Hilbert spaces and characterizations of dimension in terms of negligibility*, CWI Tract, vol. 2, Centre for Mathematics and Computer Science, Amsterdam, 1984.
[113] J. J. Dijkstra, *A generalization of the Sierpiński theorem*, Proc. Amer. Math. Soc. **91** (1984), 143–146.
[114] J. J. Dijkstra, T. Grilliot, J. van Mill, and D. J. Lutzer, *Function spaces of low Borel complexity*, Proc. Amer. Math. Soc. **94** (1985), 703–710.
[115] J. J. Dijkstra and J. van Mill, *On the dimension of Hilbert space remainders*, Proc. Amer. Math. Soc. **124** (1996), 3261–3263.
[116] J. J. Dijkstra and J. van Mill, *The ambient homeomorphy of certain function and sequence spaces – addendum*, 2001.
[117] J. J. Dijkstra, J. van Mill, and J. Mogilski, *The space of infinite-dimensional compact spaces and other topological copies of $(\ell_f^2)^\omega$*, Pac. J. Math. **152** (1992), 255–273.
[118] J. J. Dijkstra and J. Mogilski, *The ambient homeomorphy of certain function and sequence spaces*, Comm. Math. Univ. Carolinae **37** (1996), 595–611.
[119] T. Dobrowolski, *Cauty's proof of Schauder's fixed point theorem*, 1999, Unpublished notes.
[120] T. Dobrowolski, S. P. Gul'ko, and J. Mogilski, *Function spaces homeomorphic to the countable product of $\ell_2^f$*, Top. Appl. **34** (1990), 153–160.
[121] T. Dobrowolski and W. Marciszewski, *Classification of function spaces with the pointwise topology determined by a countable dense set*, Fund. Math. **148** (1995), 35–62.
[122] T. Dobrowolski and W. Marciszewski, *Rays and the fixed point property in noncompact spaces*, Tsukuba J. Math. **21** (1997), 97–112.
[123] T. Dobrowolski, W. Marciszewski, and J. Mogilski, *On topological classification of function spaces $C_p(X)$ of low Borel complexity*, Trans. Amer. Math. Soc. **678** (1991), 307–324.

[124] T. Dobrowolski and L. R. Rubin, *The hyperspaces if infinite-dimensional compacta for covering dimension and cohomological dimension are homeomorphic*, Pac. J. Math. **164** (1994), 15–39.

[125] E. K. van Douwen, *The small inductive dimension can be raised by the adjunction of a single point*, Indag. Math. **35** (1973), 434–442.

[126] E. K. van Douwen, *Prime numbers, number of factors and binary operations*, Dissertationes Math. (Rozprawy Mat.) **199** (1981), 1–35.

[127] E. K. van Douwen, *Special bases for compact metrizable spaces*, Fund. Math. **111** (1981), 201–209.

[128] E. K. van Douwen, *A compact space with a measure that knows which sets are homeomorphic*, Adv. Math. **52** (1984), 1–33.

[129] E. K. van Douwen, *Closed copies of the rationals*, Comm. Math. Univ. Carolinae **28** (1987), 137–139.

[130] E. K. van Douwen, *Applications of maximal topologies*, Top. Appl. **51** (1993), 125–139.

[131] E. K. van Douwen, *$\beta X$ and fixed-point free maps*, Top. Appl. **51** (1993), 191–195.

[132] E. K. van Douwen, *Nonextendible functions characterize dense $G_\delta$'s in metrizable spaces*, Top. Appl. **51** (1993), 187–190.

[133] E. K. van Douwen, *Uncountably many pairwise disjoint copies of one metrizable compactum in another*, Top. Appl. **51** (1993), 87–91.

[134] E. K. van Douwen, *Simultaneous extension of continuous functions*, Ph.D. thesis, Collected papers I, II (J. van Mill, ed.), North-Holland Publishing Co., Amsterdam, 1994, p. 140.

[135] E. K. van Douwen, *Unpublished results*, Collected papers I, II (J. van Mill, ed.), North-Holland Publishing Co., Amsterdam, 1994, p. 34.

[136] E. K. van Douwen and R. Pol, *Countable spaces without extension properties*, Bull. Polon. Acad. Sci. Sér. Math. Astronom. Phys. **25** (1977), 987–991.

[137] C. H. Dowker, *Mapping theorems for non-compact spaces*, Amer. Journ. of Math. **69** (1947), 200–242.

[138] C. H. Dowker, *On countably paracompact spaces*, Can. J. Math. **3** (1951), 219–224.

[139] A. N. Dranišnikov, *On a problem of P.S. Alexandrov*, Matem. Sbornik **135** (1988), 551–557.

[140] J. Dugundji, *An extension of Tietze's theorem*, Pac. J. Math. **1** (1951), 353–367.

[141] J. Dugundji, *Absolute neighborhoods retracts and local connectedness in arbitrary metric spaces*, Compositio Math. **13** (1958), 229–246.

[142] J. Dugundji, *Topology*, Allyn and Bacon, Boston, 1966.

[143] J. Dugundji and A. Granas, *Fixed point theory. I*, Państwowe Wydawnictwo Naukowe (PWN), Warsaw, 1982.

[144] S. Eilenberg, *Transformations continues en circonférences et la topologie du plan*, Fund. Math. **26** (1936), 61–112.

[145] S. Eilenberg and E. Otto, *Quelques propriétés caractéristiques de la dimension*, Fund. Math. **31** (1938), 149–153.

[146] S. Eilenberg and N. Steenrod, *Foundations of algebraic topology*, Princeton University Press, Princeton, New Jersey, 1952.

[147] E. Ellentuck, *A new proof that analytic sets are Ramsey*, J. Symb. Logic **39** (1974), 163–165.

[148] A. J. M. van Engelen, *Homogeneous zero-dimensional absolute Borel sets*, CWI Tract, vol. 27, Centre for Mathematics and Computer Science, Amsterdam, 1986.

[149] F. van Engelen, *A decomposition of $\mathbb{R}$ into two homeomorphic rigid parts*, Top. Appl. **17** (1984), 275–285.

[150] F. van Engelen and J. van Mill, *Borel sets in compact spaces: some Hurewicz-type theorems*, Fund. Math. **124** (1984), 271–286.

[151] F. van Engelen, A. W. Miller, and J. Steel, *Rigid Borel sets and better quasi-order theory*, Logic and combinatorics (Arcata, Calif., 1985), Amer. Math. Soc., Providence, R.I., 1987, pp. 199–222.
[152] R. Engelking, *Sur la compactification des espaces métriques*, Fund. Math. **48** (1960), 321–324.
[153] R. Engelking, *General topology*, Heldermann Verlag, Berlin, second ed., 1989.
[154] R. Engelking, *Letter to the author (2-10-92)*, 1992.
[155] R. Engelking, *Theory of dimensions finite and infinite*, Heldermann Verlag, Lemgo, 1995.
[156] R. Engelking and R. Pol, *Compactifications of countable-dimensional and strongly countable-dimensional spaces*, Proc. Amer. Math. Soc. **104** (1988), 985–987.
[157] P. Erdős, *The dimension of the rational points in Hilbert space*, Annals of Math. **41** (1940), 734–736.
[158] I. Farah, *Powers of* $\mathbb{N}^*$, 2000, to appear in Proc. Amer. Math. Soc.
[159] V. V. Fedorchuk, *The Fundamentals of Dimension Theory*, Encyclopedia of Mathematical Sciences (A. V. Arhangel'skiĭ and L. S. Pontrjagin, eds.), vol. 17, Springer-Verlag, Berlin, 1990, pp. 91–202.
[160] V. V. Fedorchuk and A. Chigogidze, *Absolyutnye retrakty i beskonechnomernye mnogoobraziya*, "Nauka", Moscow, 1992.
[161] V. V. Fedorchuk, M. Levin, and E. V. Shchepin, *On the Brouwer definition of dimension*, Uspekhi Mat. Nauk **54** (1999), 193–194.
[162] V. V. Fedorchuk and J. van Mill, *Dimensionsgrad for locally connected Polish spaces*, Fund. Math. **163** (2000), 77–82.
[163] W. G. Fleissner and K. Kunen, *Barely Baire spaces*, Fund. Math. **101** (1978), 229–240.
[164] A. Flores, *Über n-dimensionale Komplexe die im* $\mathbb{R}_{2n+1}$ *absolut selbstverschlungen sind*, Ergebnisse eines math. Koll. **6** (1935), 4–6.
[165] M. Fort, *Homogeneity of infinite products of manifolds with boundary*, Pac. J. Math. **12** (1962), 879–884.
[166] M. Fréchet, *Les ensembles abstraits et le calcul fonctionnel*, Rend. del Circ. Mat. di Palermo **30** (1910), 1–26.
[167] M. Fréchet, *Les espaces abstraits*, Hermann, Paris, 1928.
[168] H. Freudenthal, *Entwicklungen von Räumen und ihren Gruppen*, Compositio Math. **4** (1937), 145–234.
[169] O. Frink, *Compactifications and semi-normal spaces*, Amer. J. Math. **86** (1964), 602–607.
[170] Z. Frolík, *Fixed points of maps of extremally disconnected spaces and complete Boolean Algebras*, Bull. Polon. Acad. Sci. Sér. Math. Astronom. Phys. **16** (1968), 269–275.
[171] F. Galvin and K. Prikry, *Borel sets and Ramsey's Theorem*, J. Symb. Logic **38** (1973), 193–198.
[172] K. Geba and Z. Semadeni, *Spaces of continuous functions V*, Studia Mathematica **19** (1960), 303–320.
[173] I. M. Gel'fand and A. N. Kolmogoroff, *On rings of continuous functions on topological spaces*, Comptes Rendus (Doklady) de l'Académie des Sciences de l'URSS **22** (1939), 11–15.
[174] L. Gillman and M. Jerison, *Rings of continuous functions*, Van Nostrand, Princeton, 1960.
[175] H. Gladdines, *Absorbing systems in infinite-dimensional manifolds and applications*, Ph.D. thesis, Vrije Universiteit, Amsterdam, the Netherlands, 1994.
[176] J. de Groot, *Topologische Studiën*, Ph.D. thesis, Groningen, the Netherlands, 1942.

[177] J. de Groot, *Non-archimedean metrics in topology*, Proc. Amer. Math. Soc. **7** (1956), 948–953.//
[178] J. de Groot, *Subcompactness and the Baire category theorem*, Indag. Math. **25** (1963), 761–767.//
[179] J. de Groot and T. Nishiura, *Inductive compactness as a generalization of semicompactness*, Fund. Math. **58** (1966), 201–218.//
[180] J. de Groot and R. J. Wille, *Rigid continua and topological group-pictures*, Arch. Math. (Basel) **9** (1958), 441–446.//
[181] S. P. Gul'ko, *Spaces of continuous functions on ordinals and ultrafilters*, Mat. Zametki **47** (1990), 26–34.//
[182] S. P. Gul'ko, *On uniform homeomorphisms of spaces of continuous functions*, Proc. Steklov Inst. Math. **3** (1992), 87–93.//
[183] S. P. Gul'ko and T. E. Khmyleva, *Compactness is not preserved by t-equivalence*, Mat. Zametki **39** (1986), 895–903, in Russian.//
[184] S. P. Gul'ko and G. A. Sokolov, *P-points in $\mathbb{N}^*$ and the spaces of continuous functions*, Top. Appl. **85** (1998), 137–142.//
[185] C. F. Hagopian, *A characterization of solenoids*, Pac. J. Math. **68** (1977), 425–435.//
[186] M. Handel, *On certain sums of Hilbert cubes*, Gen. Top. Appl. **9** (1978), 19–28.//
[187] O. Hanner, *Some theorems on absolute neighborhood retracts*, Arkiv Mat., Svenska Vetens. Akad. **1** (1951), 389–408.//
[188] K. P. Hart, J. van Mill, and R. Pol, *Remarks on hereditarily infinite-dimensional continua*, 2000, manuscript submitted for publication.//
[189] M. A. van Hartskamp, *Colorings of Fixed-Point Free Maps*, Ph.D. thesis, Vrije Universiteit, Amsterdam, the Netherlands, 1999.//
[190] M. A. van Hartskamp and J. Vermeer, *On colorings of maps*, Top. Appl. **73** (1996), 181–190.//
[191] F. Hausdorff, *Über halbstetige Fuktionen und deren Verallgemeinerung*, Math. Zeitschr. **5** (1919), 292–309.//
[192] F. Hausdorff, *Über innere Abbildungen*, Fund. Math. **23** (1934), 279–291.//
[193] F. Hausdorff, *Grundzüge der Mengenlehre*, Chelsea, New York, 1949.//
[194] D. W. Henderson, *Finite dimensional subsets of infinite dimensional spaces*, Topology Seminar Wisconsin 1965, Ann. of Math. Studies (**60**), 1965, pp. 141–146.//
[195] D. W. Henderson, *An infinite-dimensional compactum with no positive-dimensional compact subsets-a simpler construction*, Amer. J. Math. **89** (1967), 105–121.//
[196] J. P. Henderson and J. J. Walsh, *Examples of cell-like decompositions of the infinite-dimensional manifolds $\sigma$ and $\Sigma$*, Top. Appl. **16** (1983), 143–154.//
[197] S. T. Hu, *Theory of retracts*, Wayne State University Press, Detroit, 1965.//
[198] W. Hurewicz, *Über stetige Bilder von Punktmengen*, Proc. Akad. Amsterdam **29** (1926), 1014–1017.//
[199] W. Hurewicz, *Normalbereiche und Dimensionstheorie*, Math. Ann. **96** (1927), 736–764.//
[200] W. Hurewicz, *Über das Verhältnis separabler Räume zu kompakten Räumen*, Proc. Akad. Amsterdam **30** (1927), 425–430.//
[201] W. Hurewicz, *Über stetige Bilder von Punktmengen (Zweite Mitteilung)*, Proc. Akad. Amsterdam **30** (1927), 159–165.//
[202] W. Hurewicz, *Relativ perfekte Teile von Punktmengen und Mengen (A)*, Fund. Math. **12** (1928), 78–109.//
[203] W. Hurewicz, *Zur Theorie der analytischen Mengen*, Fund. Math. **15** (1930), 4–17.//
[204] W. Hurewicz, *Sur la dimension des produits Cartésiens*, Annals of Math. **36** (1935), 194–197.//
[205] W. Hurewicz, *Über Abbildungen topologischer Räume auf die n–dimensionale Sphäre*, Fund. Math. **24** (1935), 144–150.

[206] W. Hurewicz, *Ein einfacher Beweis des Hauptsatzes über Cantorsche Mannigfaltigkeiten*, Prace Mat. Fiz. **34** (1937), 289–292.
[207] W. Hurewicz and K. Menger, *Dimension und Zusammenhangsstuffe*, Math. Ann. **100** (1928), 618–633.
[208] W. Hurewicz and H. Wallman, *Dimension theory*, Van Nostrand, Princeton, N.J., 1948.
[209] A. Illanes and S. B Nadler, *Hyperspaces: Fundamentals and Recent Advances*, Marcel Dekker, New York, 1999.
[210] A. V. Ivanov, *An example concerning a theorem of Mazurkiewicz*, Seminar in General Topology, Moskov. Gos. Univ., Moskow, 1981, pp. 49–51.
[211] S. Jackson and R. D. Mauldin, *Some complexity results in topology and analysis*, Fund. Math. **141** (1992), 75–83.
[212] I. M. James, ed., *History of Topology*, North-Holland Publishing Co., Amsterdam, 1999.
[213] D. M. Johnson, *The problem of the invariance of dimension in the growth of modern topology, Part II*, Arch. Hist. Exact Sci. **25** (1981), 85–267.
[214] I. Juhász, *Cardinal functions in topology*, Mathematical Centre Tract, vol. 34, Mathematical Centre, Amsterdam, 1975.
[215] M. Katětov, *On real-valued functions in topological spaces*, Fund. Math. **38** (1951), 85–91.
[216] M. Katětov, *On the dimension of non-separable spaces*, Czech. Math. J. **2** (1952), 333–368.
[217] M. Katětov, *A theorem on mappings*, Comm. Math. Univ. Carolinae **8** (1967), 431–433.
[218] A. S. Kechris, *Classical descriptive set theory*, Grad. Texts Math., vol. 156, Springer-Verlag, Berlin, 1994.
[219] O. H. Keller, *Die Homoiomorphie der kompakten konvexen Mengen in Hilbertschen Raum*, Math. Ann. **105** (1931), 748–758.
[220] J. L. Kelley, *Hyperspaces of a continuum*, Trans. Amer. Math. Soc. **52** (1942), 22–36.
[221] V. L. Klee, *Some topological properties of convex sets*, Trans. Amer. Math. Soc. **78** (1955), 30–45.
[222] B. Knaster, *Sur les coupures biconnexes des espace euklidens de dimension $n > 1$ arbitrare*, Mat. Sbornik **19** (1946), 9–18.
[223] B. Knaster, K. Kuratowski, and S. Mazurkiewicz, *Ein Beweis des fixpunktsatzes für n-dimensionale Simplexe*, Fund. Math. **14** (1929), 132–137.
[224] Y. Kodama, *Cohomological dimension theory*, Appendix to: K. Nagami, Dimension theory (New York) (1970).
[225] A. Kolmogoroff, *Über offene Abbildungen*, Annals of Math. **38** (1937), 36–38.
[226] I. M. Kozlovskiĭ, *Dimension-raising open mappings of compacta onto polyhedra as spectral limits of inessential mappings of polyhedra*, Soviet Math. Doklady **33** (1986), 118–121.
[227] J. Krasinkiewicz, *Shape properties of hyperspaces*, Fund. Math. **101** (1978), 79–91.
[228] J. Krasinkiewicz, *Essential mappings onto products of manifolds*, Geometric and Algebraic Topology (H. Toruńczyk, S. Jackowski, and S. Spież, eds.), (Banach Center Publications volume 18) PWN, Warszawa, 1986, pp. 377–406.
[229] J. Krasinkiewicz, *Homotopy separators and mapping into cubes*, Fund. Math. **131** (1988), 149–154.
[230] J. Krasinkiewicz, *On mappings with hereditarily indecomposable fibers*, Bull. Polon. Acad. Sci. Sér. Math. Astronom. Phys. **44** (1996), 147–156.
[231] A. Krawczyk and J. Steprāns, *Continuous colourings of closed graphs*, Top. Appl. **51** (1993), 13–26.

[232] M. R. Krom, *Cartesian products of metric Baire spaces*, Proc. Amer. Math. Soc. **42** (1974), 588–594.

[233] N. Kroonenberg, *Pseudo-interiors of hyperspaces*, Compositio Math. **32** (1976), 113–131.

[234] J. Kulesza, *The dimension of products of complete separable metric spaces*, Fund. Math. **135** (1990), 49–54.

[235] J. Kulesza, *New properties of Mrówka's space $\nu\mu_0$*, 1999, Preprint.

[236] K. Kunen, *Weak P-points in $\mathbb{N}^*$*, Topology, Vol. II (Proc. Fourth Colloq., Budapest, 1978), North-Holland Publishing Co., Amsterdam, 1980, pp. 741–749.

[237] K. Kuratowski, *Sur la puissance de l'ensemble des "nombres de dimension" de M. Fréchet*, Fund. Math. **8** (1925), 201–208.

[238] K. Kuratowski, *Sur un théorème fondamental concernant le nerf d'un système d'ensembles*, Fund. Math. **20** (1933), 191–196.

[239] K. Kuratowski, *Quelques problèmes concernant les espaces métriques non-séparables*, Fund. Math. **25** (1935), 534–545.

[240] K. Kuratowski, *Sur les espaces localement connexes et péaniens en dimension $n$*, Fund. Math. **24** (1935), 269–287.

[241] K. Kuratowski, *Quelques théoremes sur le plongement topologique des espaces*, Fund. Math. **30** (1938), 8–13.

[242] K. Kuratowski, *Topology I*, Academic Press, New York, 1966.

[243] K. Kuratowski, *Topology II*, Academic Press, New York, 1968.

[244] K. Kuratowski and A. Mostowski, *Set theory*, North-Holland Publishing Co., Amsterdam, revised ed., 1976, With an introduction to descriptive set theory, Translated from the 1966 Polish original, Studies in Logic and the Foundations of Mathematics, Vol. 86.

[245] M. Lavrentieff, *Contribution à la théorie des ensembles homéomorphes*, Fund. Math. **6** (1924), 149–160.

[246] H. Lebesgue, *Sur les correspondances entre les points de deux espaces*, Fund. Math. **2** (1921), 256–285.

[247] S. Lefschetz, *On compact spaces*, Annals of Math. **32** (1931), 521–538.

[248] S. Lefschetz, *Topics in Topology*, Annals of Mathematical Studies 10, Princeton University Press, Princeton, 1942.

[249] A. Lelek, *On the dimension of remainders in compact extensions*, Dokl. Akad. Nauk SSSR **160** (1965), 534–537.

[250] A. Lelek, *Dimension inequalities for unions and mappings of separable metric spaces*, Coll. Math. **23** (1971), 69–91.

[251] S. Levi, *On Baire cosmic spaces*, Proceedings of the Fifth Prague Topological Symposium, Heldermann Verlag, Berlin, pp. 450–451.

[252] M. Levin, *A short construction of hereditarily infinite dimensional compacta*, Top. Appl. **65** (1995), 97–99.

[253] M. Levin, *Bing maps and finite-dimensional maps*, Fund. Math. **151** (1996), 47–52.

[254] M. Levin and R. Pol, *A metric condition which implies dimension $\leq 1$*, Proc. Amer. Math. Soc. **125** (1997), 269–273.

[255] M. Levin and J. T. Rogers, Jr., *A generalization of Kelley's Theorem for $C$-spaces*, Proc. Amer. Math. Soc. **128** (1999), 1537–1541.

[256] M. Levin and J. Segal, *A subspace of $\mathbb{R}^3$ with $\mathrm{Cmp} \neq \mathrm{def}$*, Top. Appl. **95** (1999), 165–168.

[257] M. Levin and Y. Sternfeld, *Hyperspaces of two dimensional continua*, Fund. Math. **150** (1996), 17–24.

[258] M. Levin and Y. Sternfeld, *The space of subcontinua of a 2-dimensional continuum is infinite dimensional*, Proc. Amer. Math. Soc. **125** (1997), 2771–2775.

[259] M. Levin and Y. Sternfeld, *Atomic maps and the Chogoshvili-Pontrjagin claim*, Trans. Amer. Math. Soc. **350** (1998), 4623–4632.
[260] W. Lewis, *The pseudo-arc*, Boll. Soc. Math. Mexicana **5** (1999), 25–77.
[261] J. Lindenstrauss and L. Tzafriri, *Classical Banach spaces, I and II*, Springer-Verlag, Berlin, 1996.
[262] S. L. Lipscomb, *On imbedding finite-dimensional metric spaces*, Trans. Amer. Math. Soc. **211** (1975), 143–160.
[263] L. Lusternik and L. Schnirelman, *Topological methods in variational calculus (Russian)*, Moscow, 1930.
[264] D. J. Lutzer and R. McCoy, *Category in function spaces. I*, Pac. J. Math. **90** (1980), 145–168.
[265] D. J. Lutzer, J. van Mill, and R. Pol, *Descriptive complexity of function spaces*, Trans. Amer. Math. Soc. **291** (1985), 121–128.
[266] W. Marciszewski, *A function space $C(K)$ not weakly homeomorphic to $C(K) \times C(K)$*, Studia Mathematica **88** (1988), 129–137.
[267] W. Marciszewski, *On analytic and coanalytic function spaces $C_p(X)$*, Top. Appl. **50** (1993), 241–248.
[268] W. Marciszewski, *A function space $C_p(X)$ not linearly homeomorphic to $C_p(X) \times \mathbb{R}$*, Fund. Math. **153** (1997), 125–140.
[269] W. Marciszewski, *On topological embeddings of linear metric spaces*, Math. Ann. **308** (1997), 21–30.
[270] W. Marciszewski, *P-filters and hereditary Baire function spaces*, Top. Appl. **89** (1998), 241–247.
[271] W. Marciszewski, *On properties of metrizable spaces $X$ preserved by t-equivalence*, 1999, to appear in Mathematika.
[272] W. Marciszewski and J. van Mill, *An example of $t_p^*$-equivalent spaces which are not $t_p$-equivalent*, Top. Appl. **85** (1998), 281–285.
[273] W. Marciszewski and J. Pelant, *Absolute Borel sets and function spaces*, Trans. Amer. Math. Soc. **349** (1997), 3585–3596.
[274] S. Mardešić and J. Segal, *Shape theory*, North-Holland Publishing Co., Amsterdam, 1982.
[275] S. Mazurkiewicz, *O arytmetyzacji continuów*, C.R. Varsovie **6** (1913), 305–311.
[276] S. Mazurkiewicz, *Sur les problème $\kappa$ et $\lambda$ de Urysohn*, Fund. Math. **10** (1927), 311–319.
[277] S. Mazurkiewicz, *Sur les ensembles de dimension faible*, Fund. Math. **13** (1929), 210–217.
[278] S. Mazurkiewicz, *Sur l'hyperespace d'un continu*, Fund. Math. **18** (1932), 171–177.
[279] S. Mazurkiewicz, *Über die Menge der differenzierbaren Funktionen*, Fund. Math. **27** (1936), 244–249.
[280] R. A. McCoy and I. Ntantu, *Topological properties of spaces of continuous functions*, Lectute Notes in Mathematics, vol. 1315, Springer-Verlag, 1988.
[281] K. Menger, *Über die Dimension von Punktmengen I*, Monatsh. für Math. und Phys. **33** (1923), 148–160.
[282] K. Menger, *Über die Dimension von Punktmengen II*, Monatsh. für Math. und Phys. **34** (1026), 137–161.
[283] K. Menger, *Dimensionstheorie*, Leipzig und Berlin, 1928.
[284] K. Menger, *Bemerkungen über dimensionelle Feinstruktur und Produktsatz*, Prace Mat.-Fiz. **38** (1930), 77–90.
[285] E. A. Michael, *Some extension theorems for continuous functions*, Pac. J. Math. **3** (1953), 789–806.
[286] E. A. Michael, *Local properties of topological spaces*, Pac. J. Math. **21** (1954), 163–171.

[287] E. A. Michael, *Continuous selections I*, Annals of Math. **63** (1956), 361–382.
[288] E. A. Michael, *Continuous selections and countable sets*, Fund. Math. **111** (1981), 1–10.
[289] H. Michalewski, *Przestrzenie funkcij ciągłych i przestrzenie dziedzicznie Baire'a (Master's Thesis)*, 1998.
[290] A. A. Miljutin, *Isomorphisms of the spaces of continuous functions over compact sets of the cardinality of the continuum (Russian)*, Teor. Funkcii Funkcional Anal i Prilozen. (Kharkov) **2** (1966), 150–156.
[291] J. van Mill, *A Peano continuum homeomorphic to its own square but not to its countable infinite product*, Proc. Amer. Math. Soc. **80** (1980), 703–705.
[292] J. van Mill, *Characterization of some zero-dimensional separable metric spaces*, Trans. Amer. Math. Soc. **264** (1981), 205–215.
[293] J. van Mill, *Homogeneous subsets of the real line*, Compositio Math. **45** (1982), 3–13.
[294] J. van Mill, *Strong local homogeneity does not imply countable dense homogeneity*, Proc. Amer. Math. Soc. **84** (1982), 143–148.
[295] J. van Mill, *A boundary set for the Hilbert cube containing no arcs*, Fund. Math. **118** (1983), 93–102.
[296] J. van Mill, *Domain invariance in infinite-dimensional linear spaces*, Proc. Amer. Math. Soc. **101** (1987), 173–180.
[297] J. van Mill, *Topological equivalence of certain function spaces*, Compositio Math. **63** (1987), 159–188.
[298] J. van Mill, *Infinite-dimensional topology: prerequisites and introduction*, North-Holland Publishing Co., Amsterdam, 1989.
[299] J. van Mill, *An infinite-dimensional homogeneous indecomposable continuum*, Houston J. Math. **10** (1990), 195–201.
[300] J. van Mill, $C_p(X)$ *is not* $G_{\delta\sigma}$*: a simple proof*, Bull. Polon. Acad. Sci. Sér. Math. Astronom. Phys. **47** (1999), 319–323.
[301] J. van Mill, *Easier proofs of coloring theorems*, Top. Appl. **97** (1999), 155–163.
[302] J. van Mill and R. Pol, *The Baire Category Theorem in products of linear spaces and topological groups*, Top. Appl. **22** (1986), 267–282.
[303] J. van Mill and R. Pol, *On the existence of weakly n-dimensional spaces*, Proc. Amer. Math. Soc. **113** (1991), 581–585.
[304] J. van Mill and R. Pol, *A countable space with a closed subspace without measurable extender*, Bull. Polon. Acad. Sci. Sér. Math. Astronom. Phys. **41** (1993), 279–283.
[305] J. van Mill and R. Pol, *Baire 1 functions which are not countable unions of continuous functions*, Acta Math. Hungar. **66** (1995), 289–300.
[306] J. van Mill and R. Pol, *Remark on products of 1-dimensional compacta*, Q&A in General Topology **13** (1995), 97–98.
[307] J. van Mill and R. Pol, *Note on weakly n-dimensional spaces*, 2000, (to appear in Monatsh. Math.).
[308] J. van Mill and R. Pol, *In preparation*, 2001.
[309] A. W. Miller, *Special Subsets of the Real Line*, Handbook of Set-Theoretic Topology (K. Kunen and J.E. Vaughan, eds.), North-Holland Publishing Co., Amsterdam, 1984, pp. 201–233.
[310] A. W. Miller, *Descriptive Set Theory and Forcing*, Lecture Notes in Logic, vol. 4, Springer-Verlag, Berlin, 1995.
[311] J. Mogilski, *Unions of infinite-dimensional compact convex subsets of the Fréchet space*, Bull. Polon. Acad. Sci. Sér. Math. Astronom. Phys. **24** (1976), 1097–1102.
[312] E. E. Moise, *Grille decompositions and convexification theorems for compact locally connected continua*, Bull. Amer. Math. Soc. **55** (1949), 1111–1121.
[313] R. L. Moore, *Concerning simple continuous curves*, Trans. Amer. Math. Soc. **21** (1920), 333–347.

[314] K. Morita, *Normal families and dimension theory for metric spaces*, Math. Ann. **128** (1954), 350–362.
[315] Y. N. Moschovakis, *Descriptive Set Theory*, North-Holland Publishing Co., Amsterdam, 1980.
[316] M. Moszyńska, *Generalization of the theory of retracts on the pairs of spaces*, Bull. Polon. Acad. Sci. Sér. Math. Astronom. Phys. **13** (1965), 13–19.
[317] S. Mrówka, *Small inductive dimension of completions of metric spaces*, Proc. Amer. Math. Soc. **125** (1997), 1545–1554.
[318] S. Mrówka, *Small inductive dimension of completions of metric spaces. II*, Proc. Amer. Math. Soc. **128** (2000), 1247–1256.
[319] J. R. Munkres, *Elements of Algebraic Topology*, Addison-Wesley Publishing Co., Menlo Park, 1984.
[320] S. B. Nadler, *Hyperspaces of sets*, Marcel Dekker, New York and Basel, 1978.
[321] K. Nagami, *Finite-to-one closed mappings and dimension, II*, Proc. Japan Acad. **35** (1959), 437–439.
[322] J. Nagata, *On lattices of functions on topological spaces and functions on uniform spaces*, Osaka Math. J. **1** (1949), 166–181.
[323] J. Nagata, *Modern dimension theory*, Interscience Publishers John Wiley & Sons, Inc., New York, 1965, Bibliotheca Mathematica, Vol. VI. Edited with the cooperation of the "Mathematisch Centrum" and the "Wiskundig Genootschap" at Amsterdam.
[324] J. Nagata, *Modern general topology*, North-Holland Mathematical Library, vol. 33, North-Holland Publishing Co., Amsterdam, 1985.
[325] J. I. Nagata, *Modern dimension theory*, Heldermann Verlag, Berlin, 1983, (revised and extended edition).
[326] O. Nikodym, *Sur une propriété de l'opération A*, Fund. Math. **7** (1925), 149–154.
[327] G. Nöbeling, *Über eine n-dimensionale Universalmenge in $\mathbb{R}^{2n+1}$*, Math. Ann. **104** (1931), 71–80.
[328] O. G. Okunev, *Weak topology of a dual space and a t-equivalence relation*, Math. Notes **46** (1989), 534–538.
[329] W. Olszewski, *Embeddings of finite-dimensional spaces into finite products of 1-dimensional spaces*, Top. Appl. **40** (1991), 93–99.
[330] L. G. Oversteegen and E. D. Tymchatyn, *On the dimension of certain totally disconnected spaces*, Proc. Amer. Math. Soc. **122** (1994), 885–891.
[331] J. Oxtoby, *Cartesian products of Baire spaces*, Fund. Math. **49** (1961), 157–166.
[332] K. R. Parthasarathy, *Probability measures on metric spaces*, Academic Press, New York, 1967.
[333] D. Pavlovskiĭ, *On spaces of continuous functions*, Soviet Math. Doklady **22** (1980), 34–37.
[334] J. Pelant, *A remark on spaces of bounded continuous functions*, Indag. Math. **91** (1988), 335–338.
[335] A. Pełczyński, *Projections in certain Banach spaces*, Studia Math. **19** (1960), 209–228.
[336] V. G. Pestov, *The coincidence of the dimension* dim *of ℓ-equivalent topological spaces*, Soviet Math. Doklady **26** (1982), 380–383.
[337] H. Poincaré, *L'espace et ses trois dimensions*, Revue de Metaph. et de Morale **11** (1903), 281–301 and 407–429.
[338] H. Poincaré, *Pourquoi l'espace a trois dimensions*, Revue de Metaph. et de Morale **20** (1912), 483–504.
[339] E. Pol, *On hereditarily indecomposable continua, Henderson compacta and a question of Yohe*, 2000, (to appear).
[340] R. Pol, *A weakly infinite-dimensional compactum which is not countable-dimensional*, Proc. Amer. Math. Soc. **82** (1981), 634–636.

[341] R. Pol, *A counterexample to J. de Groot's problem* cmp = def, Bull. Polon. Acad. Sci. Sér. Math. Astronom. Phys. **30** (1982), 461–464.
[342] R. Pol, *An infinite-dimensional pre-Hilbert space not homeomorphic to its own square*, Proc. Amer. Math. Soc. **90** (1984), 450–454.
[343] R. Pol, *Countable dimensional universal sets*, Trans. Amer. Math. Soc. **297** (1986), 255–268.
[344] R. Pol, *On metrizable E with $C_p(E) \not\cong C_p(E) \times C_p(E)$*, Mathematika **42** (1995), 49–55.
[345] R. Pol, *Private communication (20-09-99)*, 1999.
[346] L. S. Pontrjagin, *Sur une hypothèse fondamentale de la théorie de la dimension*, C.R. Acad. Paris **190** (1930), 1105–1107.
[347] L. S. Pontrjagin and G. Tolstowa, *Beweis des Mengerschen Einbettungssatzes*, Math. Ann. **105** (1931), 734–747.
[348] T. C. Przymusiński, *A note on dimension theory of metric spaces*, Fund. Math. **85** (1974), 277–284.
[349] E. G. Pytkeev, *The Baire property of spaces of continuous functions*, Math. Zametki **38** (1985), 726–740.
[350] J. Quinn and R. Y. T. Wong, *Unions of convex Hilbert cubes*, Proc. Amer. Math. Soc. **65** (1977), 171–176.
[351] D. Repovš and P. V. Semenov, *Continuous selections of multivalued mappings*, Mathematics and its Applications, vol. 455, Kluwer Academic Publishers, Dordrecht, 1998.
[352] C. A. Rogers, J. E. Jayne, C. Dellacherie, F. Topsøe, J. Hoffman–Jørgensen, D. A. Martin, A. S. Kechris, and A. H. Stone, *Analytic sets*, Academic Press, New York, 1980.
[353] J. T. Rogers, Jr., *Dimension of hyperspaces*, Bull. Polon. Acad. Sci. Sér. Math. Astronom. Phys. **20** (1972), 177–179.
[354] J. T. Rogers, Jr., *Orbits of higher-dimensional hereditarily indecomposable continua*, Proc. Amer. Math. Soc. **95** (1985), 483–486.
[355] P. Roy, *Failure of equivalence of dimension concepts for metric spaces*, Bull. Amer. Math. Soc. **68** (1962), 609–613.
[356] P. Roy, *Nonequality of dimensions for metric spaces*, Trans. Amer. Math. Soc. **134** (1968), 117–132.
[357] L. Rubin, R. M. Schori, and J. J. Walsh, *New dimension-theory techniques for constructing infinite-dimensional examples*, Gen. Top. Appl. **10** (1979), 93–102.
[358] L. R. Rubin, *Noncompact hereditarily strongly infinite dimensional spaces*, Proc. Amer. Math. Soc. **79** (1980), 153–154.
[359] M. E. Rudin, *A new proof that metric spaces are paracompact*, Proc. Amer. Math. Soc. **20** (1969), 603.
[360] W. Rudin, *Homogeneity problems in the theory of Čech compactifications*, Duke Math. J. **23** (1956), 409–419.
[361] J. Schauder, *Der Fixpunktsatz in Funktionalräumen*, Studia Math. **2** (1930), 171–180.
[362] R. M. Schori, *Hyperspaces and symmetric products of topological spaces*, Fund. Math. **63** (1968), 77–88.
[363] R. M. Schori and J. E. West, *Hyperspaces of graphs are Hilbert cubes*, Pac. J. Math. **53** (1974), 239–251.
[364] R. M. Schori and J. E. West, *The hyperspace of the closed interval is a Hilbert cube*, Trans. Amer. Math. Soc. **213** (1975), 217–235.
[365] S. Shelah, *Decomposing topological spaces into two rigid homeomorphic subspaces*, Israel J. Math. **63** (1988), 183–211.
[366] R. B. Sher, *The union of two Hilbert cubes meeting in a Hilbert cube need not be a Hilbert cube*, Proc. Amer. Math. Soc. **63** (1977), 150–152.

[367] W. Sierpiński, *L'arc simple comme un ensemble de points dans l'espace à m dimensions*, Ann. Mat. Pur. Appl. **26** (1917), 131–150.
[368] W. Sierpiński, *Un théorème sur les continus*, Tôhoku Math. **13** (1918), 300–303.
[369] W. Sierpiński, *Sur une propriété topologique des ensembles dénombrables denses en soi*, Fund. Math. **1** (1920), 11–16.
[370] W. Sierpiński, *Sur les ensembles connexes et non connexes*, Fund. Math. **2** (1921), 81–95.
[371] W. Sierpiński, *Sur une décomposition d'ensembles*, Monatsh. für Math. und Phys. **35** (1928), 239–242.
[372] W. Sierpiński, *Remarque sur la courbe péanienne*, Wiadom. Mat. **42** (1937), 1–3.
[373] W. Sierpiński, *Sur un problème concernant les fonctions semi-continues*, Fund. Math. **28** (1937), 1–6.
[374] R. Sikorski, *On the Cartesian product of metric spaces*, Fund. Math. **34** (1977), 288–292.
[375] M. Souslin, *Sur une défenition des ensembles mesurables B sans nombres transfinis*, Compt. Rend. Acad. Sci. **164** (1917), 89.
[376] E. Spanier, *Algebraic topology*, Springer-Verlag, Berlin, 1982.
[377] E. Sperner, *Neuer Beweis für die Invarianz des Dimensionszahl und des Gebietes*, Abh. Math. Semin. Hamburg. Univ. **6** (1928), 265–272.
[378] A. K. Steiner and E. F. Steiner, *Proucts of compact metric spaces are regular Wallman*, Indag. Math. **30** (1968), 428–430.
[379] H. Steinlein, *Borsuk-Ulam Sätze und Abbildungen mit kompakten Iterierten*, Dissertationes Math. (Rozprawy Mat.) **177** (1980), 113.
[380] H. Steinlein, *On the theorems of Borsuk-Ulam and Ljusternik-Schnirelmann-Borsuk*, Canad. Math. Bull. **27** (1984), 192–204.
[381] Y. Sternfeld, *Mappings in dendrites and dimension*, Houston J. Math. **19** (1993), 483–497.
[382] Y. Sternfeld, *Stability and dimension—a counterexample to a conjecture of Chogoshvili*, Trans. Amer. Math. Soc. **340** (1993), 243–251.
[383] A. H. Stone, *Paracompactness and product spaces*, Bull. Amer. Math. Soc. **54** (1948), 977–982.
[384] M. Talagrand, *Compacts de fonctions mesurables et filtres non mesurables*, Studia Math. **67** (1980), 13–43.
[385] A. D. Taĭmanov, *On extension of continuous mappings of topological spaces*, Mat. Sb. (N.S.) **31** (1952), 459–463.
[386] A. E. Taylor, *Introduction to functional analysis*, John Wiley and Sons, London, 1964.
[387] H. Tietze, *Über Funktionen die auf einer abgeschlossenen Menge stetig sind*, J. Reine Angew. Math. **145** (1915), 9–14.
[388] V. V. Tkachuk, *Characterization of Baire property in $C_p(X)$ by the properties of a space $X$*, The mappings and the extensions of topological spaces (Ustinov, ed.), 1985, in Russian, pp. 21–27.
[389] B. Tomaszewski, *On weakly n-dimensional spaces*, Fund. Math. **103** (1979), 1–8.
[390] H. Toruńczyk, *Concerning locally homotopy negligible sets and characterizations of $\ell_2$ manifolds*, Fund. Math. **101** (1978), 93–110.
[391] H. Toruńczyk, *On CE-images of the Hilbert cube and characterizations of Q-manifolds*, Fund. Math. **106** (1980), 31–40.
[392] H. Toruńczyk, *Characterizing Hilbert space topology*, Fund. Math. **111** (1981), 247–262.
[393] L. A. Tumarkin, *Beitrag zur allgemeinen Dimensionstheorie*, Mam. Cb. **33** (1926), 57–86.

[394] L. A. Tumarkin, *Sur la structure dimensionelle des ensembles fermés*, C.R. Acad. Paris **186** (1928), 420–422.

[395] V. M. Ul′janov, *Solution of a basic problem on compactifications of Wallman type*, Soviet Math. Doklady **18** (1977), 567–571.

[396] P. Urysohn, *Les multiplicités Cantoriennes*, C.R. Acad. Paris **175** (1922), 440–442.

[397] P. Urysohn, *Mémoire sur les multiplicités Cantoriennes*, Fund. Math. **7** (1925), 30–137.

[398] P. Urysohn, *Mémoire sur les multiplicités Cantoriennes (suite)*, Fund. Math. **8** (1926), 225–359.

[399] P. Urysohn, *Über die Mächtigkeit der zusammenhängenden Mengen*, Math. Annalen **94** (25), 262–295.

[400] L. Vietoris, *Bereiche zweiter Ordnung*, Monatsh. für Math. und Phys. **32** (1922), 258–280.

[401] J. de Vries, *Elements of topological dynamics*, Kluwer Academic Publishers, Dordrecht, 1992.

[402] H. Wallman, *Lattices and topological spaces*, Annals of Math. **39** (1938), 112–126.

[403] J. J. Walsh, *Infinite-dimensional compacta containing no n-dimensional ($n \geq 1$) subsets*, Topology **18** (1979), 91–95.

[404] T. Ważewski, *Sur un continu singulier*, Fund. Math. **4** (1923), 214–245.

[405] J. E. West, *The ambient homeomorphy of an incomplete subspace of an infinite-dimensional Hilbert space*, Pac. J. Math. **34** (1970), 257–267.

[406] J. E. West, *Mapping Hilbert cube manifolds to ANR's: a solution to a conjecture of Borsuk*, Annals of Math. **106** (1977), 1–18.

[407] J. E. West, *Problems in Infinite-dimensional Topology*, Open Problems in Topology (J. van Mill and G. M. Reed, eds.), North-Holland Publishing Co., Amsterdam, 1990, pp. 523–597.

[408] J. H. C. Whitehead, *Simplicial spaces nuclei, and m-groups*, Proc. London Mathematical Soc. **45** (1939), 243–327.

[409] H. Whitney, *Regular families of curves, I*, Proc. Nat. Acad. Sci. **18** (1932), 275–278.

[410] E. Wimmers, *The Shelah P-point independence theorem*, Israel J. Math. **43** (1982), 28–48.

[411] M. Wojdysławski, *Rétractes absolus et hyperespaces des continus*, Fund. Math. **32** (1939), 184–192.

[412] A. V. Zarelua, *Hereditarily infinite-dimensional spaces*, Theory of sets and topology (in honour of Felix Hausdorff, 1868–1942), VEB Deutsch. Verlag Wissensch., Berlin, 1972, pp. 509–525.

# Special Symbols

$A^d$, 417
$[a, A]$, 239
$\overline{A}$, 460
AE, 25
aff$(S)$, 111
$\alpha_v(x)$, 114
$\alpha X$, 472
$A/n$, 240
ANE, 25
ANR, 24
AR, 24

$B(A, \varepsilon)$, 459
$B(x, \varepsilon)$, 459
$B^n$, 461
$B_\varepsilon(x)$, 459
$B_\varrho(A, \varepsilon)$, 459
$\beta X$, 368
$B(Q)$, 60
$\mathcal{B}(X)$, 517

C, 42
car $x$, 119
$C_\varepsilon(X, Y)$, 34
$C(f)$, 250
$c_{\mathcal{F}}$, 435
$\chi_A$, 458
cmp $X$, 182
$\mathfrak{c}$, 459
conv$(A)$, 1
conv$_\infty(A)$, 2
conv$_n(A)$, 1
$C(X)$, 4
$C(X, Y)$, 460
$C^*(X)$, 7
$C_p(X)$, 4, 368
$C_{p,A}(X)$, 396
$C^*_{p,A}(X)$, 396
$C_\varrho(X, Y)$, 29
$C^*_u(X)$, 418

$C^*_{u,A}(X)$, 418
$c_0$, 8

$D(A, \varepsilon)$, 459
$D(x, \varepsilon)$, 459
$D_\varepsilon(x)$, 459
$D_\varrho(A, \varepsilon)$, 459
def $X$, 183
$\partial|\sigma|$, 114
diam$(A)$, 459
dim $X$, 152
dim$_k(X)$, 340
dim$_{\geq k}(X)$, 340
dim$_{\leq k}(X)$, 340

$\varepsilon$-map, 34

$f \equiv t$, 458
$F\colon X \Rightarrow Y$, 12
$f \simeq g$, 510
$f_n^\infty$, 81
$f^{-1}(y)$, 458
$f_m^m$, 81
$f_n^m$, 81
$\mathcal{F}_n(X)$, 100
Fr$(A)$, 460
$F_\sigma$, 466
$\mathcal{F}_\sigma$, 347
$F_{\sigma\delta}$, 518
$f \cup g$, 458
$\mathcal{F}(X)_\infty$, 100

$G_\delta$, 466
$G_{\delta\sigma}$, 518
$\mathcal{G}_\varepsilon(X, Y)$, 34
$\mathcal{G}_{n,k}(X)$, 340

$H^\infty$, 239
$\mathcal{H}(X)$, 35
$\mathcal{H}(X, Y)$, 35

$\mathbb{I}$, 457

# SPECIAL SYMBOLS

$1_X$, 458
Ind $X$, 180
ind $X$, 177
Int($A$), 460
$I(x,y)$, 19
$\mathfrak{I}(X,Y)$, 35

$\mathbb{J}$, 457

$\kappa(X)$, 418
$[K, U]$, 19

$\Lambda(X)$, 227
LC, 516
$\varprojlim(X_n, f_n)_n$, 80
$\lin(S)$, 111
$\ell^2$, 8
lsc, 488
$L(X)$, 399

$\mathfrak{M}_\delta$, 348
mesh($\mathcal{U}$), 459
$M^Q$, 326

$\mathbb{N}$, 457
$\hat{\mathbb{N}}$, 387
$[\mathbb{N}]^\omega$, 238
$\mathfrak{N}_n$, 168

$\omega$, 459
$\omega(X, \mathcal{F})$, 497

$\mathbb{P}$, 42, 457
$\pi_E$, 457
$P$-point, 393
$\prod_{i \in I} X_i$, 457
$\mathcal{P}(X)$, 458

$Q$, 461
$\mathbb{Q}$, 42, 457

$\mathbb{R}$, 457
$\varrho$, 460
$\hat{\varrho}(f, g)$, 468
$\varrho_H$, 96

$s$, 3, 60
$(SC)_{abstact}$, 116
$(SC)_{geometric}$, 117
$(SC)$, 116
$sd^{(0)} S$, 130
$sd^{(1)} S$, 130
$sd^{(n)} S$, 130
$\mathcal{S}_\varepsilon(X,Y)$, 34
$\Sigma$, 46

$\dot{\sigma}$, 125
$\sigma$, 3
$|\sigma|^\circ$, 114
$\langle \sigma, \tau \rangle$, 238
$\Sigma_2$, 85
$\mathbb{S}^n$, 26
$\mathbb{S}^n$, 461
St $x$, 119
$\bigoplus$, 371
supp($A$), 405
$supp_\varphi(y)$, 404

$|\mathfrak{T}|$, 118
$\tau X$, 278
$\Theta_n$, 354
$\triangle(X)$, 515
$T_U$, 409
$T_\mathcal{U}$, 409

$\overline{\mathcal{U}}$, 460
$\mathcal{U} < \mathcal{V}$, 458
usc, 489

$V(A, S)$, 462

$W_n^\theta$, 462

$|X|$, 459
$X^{(\alpha)}$, 417
$X \approx Y$, 460
$X \sim_\ell Y$, 370
$X \sim_{\ell^*} Y$, 370
$X_\infty$, 348
$[X]^k$, 427
$X/\mathcal{P}$, 505
$X_{(n)}$, 183
$X^*$, 369
$[X, Y]$, 511

$\mathbb{Z}$, 457
$\underline{0}$, 2
$\mathbf{0}$, 461
$\mathcal{Z}_\sigma(X)$, 307
$\mathcal{Z}(X)$, 307

# Author Index

Aarts, 584, 586, 587, 594, 597
Adjan, 587, 597
Alexandroff, 253, 254, 580–582, 584, 587, 593, 597
Anderson, 9, 325, 579, 580, 584, 585, 588–590, 597, 598
Arens, 396, 579, 598
Arhangel'skiĭ, 367, 372, 411, 426, 451, 590, 591, 598, 603
Arveson, 455, 598
Aull, 579, 598

Baars, x, 367, 411, 425, 434, 591, 592, 598
Banach, 579, 598
Banakh, 426, 437, 445, 588, 590, 598
Bandt, 594, 598
Barit, 589, 598
Barov, x
Bartle, 579, 598
Becker, 49, 580, 598
Bendixson, 592
Bennett, 580, 598
Berney, 593, 594, 599
Bessaga, 371, 588, 589, 599
Bestvina, 434, 588–590, 599
van der Bijl, 588, 599
Bing, 215, 580–582, 585, 589, 593, 597, 599
Błaszczyk, 586, 599
Blumenthal, 579, 599
Borel, 594
Borsuk, 301, 580, 581, 583, 585–588, 599
Bourbaki, 585, 593, 599
Bowers, 586, 588, 599, 600
Brouwer, 221–223, 580, 582–584, 600
Brown, 579, 582, 600
de Bruijn, 586, 600
Burckel, 594, 600

Calbrix, 446, 590, 592, 600

Cantor, 580, 592, 600
Cauty, 24, 26, 264, 367, 414, 426, 445, 446, 456, 580, 583, 590, 592, 598, 600
Čech, 583, 584, 600
Chapman, 588, 589, 597, 600
Chatyrko, 587, 600
Chigogidze, 588, 600, 603
Choquet, 384, 601
Christensen, 426, 601
Cohen, 384, 601
Comfort, 590, 592, 601
Cook, 447, 601
Curtis, 291, 325, 580, 584, 588–590, 598, 601

van Dantzig, 582, 601
Daverman, 588, 594, 601
Dellacherie, 610
Devaney, 582, 601
Dieudonné, 593, 601
Dijkstra, x, 434, 585, 587, 588, 590, 592, 594, 601
Dobrowolski, ix, x, 367, 434, 446, 579, 580, 583, 588, 590–592, 600–602
van Douwen, 75, 384, 580–583, 586, 591–594, 602
Dowker, 583, 593, 602
Dranišnikov, 24, 585, 602
Dugundji, 23, 24, 29, 149, 372, 457, 580, 583, 588, 591, 594, 602

Eels, 579, 598
Eilenberg, 583, 593, 594, 602
Ellentuck, 239, 586, 602
van Engelen, 49, 580, 581, 594, 598, 602, 603
Engelking, 174, 368, 369, 394, 412, 415, 428, 457, 465, 579, 582–584, 587, 593, 603
Erdős, 580, 586, 600, 603

Farah, 582, 603
Fedorchuk, 585, 588, 603
Fleissner, 384, 603
Flores, 174, 603
Fokkink, 586, 587, 597
Fort, 580, 603
Fréchet, 221, 579, 593, 603
Freudenthal, 582, 585, 603
Frink, 593, 603
Frolík, 586, 603

Galvin, 586, 603
Geba, 394, 451, 603
Gel′fand, 371, 603
Gillman, 590, 603
Gladdines, 434, 588, 592, 598, 603
Granas, 149, 583, 602
Graves, 579, 598
Grilliot, 590, 592, 601
de Groot, J., 373, 580, 581, 584, 603, 604
de Groot, J.A.M., 367, 411, 591, 592, 598
Gul′ko, 393, 411, 412, 414, 447, 456, 590, 592, 601, 604

Hagopian, 582, 604
Handel, 589, 604
Hanner, 588, 604
Hart, x, 585, 587, 604
van Hartskamp, x, 587, 604
Hausdorff, 580, 582, 593, 604
Henderson, D.W., 585, 587, 604
Henderson, J.P., 589, 604
Hoffman–Jørgensen, 610
Hu, 580, 588, 604
Hurewicz, 222, 224, 581, 583–586, 594, 604, 605
Hušek, 598

Illanes, 582, 605
Ivanov, 585, 605

Jackowski, 605
Jackson, 587, 605
James, 579, 605
Jayne, 610
Jerison, 590, 603
Johnson, 222, 594, 605
Juhász, 592, 605

Katětov, 584, 586, 593, 605
Kechris, 517, 586, 594, 605, 610
Keisler, 584, 585, 598
Keller, 580, 581, 605
Kelley, 587, 605

Khmyleva, 411, 412, 456, 604
Kim, 586, 599
Klee, 579, 581, 589, 599, 605
Knaster, 582, 585, 605
Kodama, 584, 605
Kolmogoroff, 371, 585, 603, 605
Kozlovskiĭ, 585, 605
Krasinkiewicz, 582, 585, 605
Krawczyk, 586, 605
Krom, 384, 606
Kroonenberg, 590, 606
Kulesza, 584, 585, 606
Kunen, 384, 386, 603, 606, 608
Kuratowski, 80, 447, 459, 517, 579, 582–584, 586, 588, 593, 594, 605, 606

Lavrentieff, 593, 606
Lebesgue, 593, 606
Lefschetz, 583, 588, 606
Lelek, 585, 587, 606
Levi, 594, 606
Levin, 582–585, 587, 603, 606, 607
Lewis, 582, 607
Lindenstrauss, 579, 607
Lipscomb, 586, 607
Lowen, 579, 598
Lusin, 587
Lusternik, 583, 607
Lutzer, 367, 446, 456, 590–592, 601, 607

Marciszewski, ix, x, 18, 367, 393, 398, 414, 426, 434, 445–447, 456, 588, 590–593, 600, 601, 607
Mardešić, 588, 607
Martin, 610
Mauldin, 587, 605
Mazur, 586
Mazurkiewicz, 580, 582, 585, 586, 594, 605, 607
McCoy, 367, 590, 591, 607
Menger, 234, 583–585, 605, 607
Michael, 396, 452, 579, 607, 608
Michalewski, x, 591, 608
Miljutin, 370, 608
van Mill, 9, 49, 64, 294, 325, 367, 384, 426, 434, 445–447, 456, 580–582, 584–592, 598, 601–604, 607, 608, 612
Miller, 517, 581, 595, 603, 608
Mogilski, ix, 367, 434, 588–592, 599, 601, 608
Moise, 593, 608
Moore, 594, 608
Morita, 584, 609

# AUTHOR INDEX

Moschovakis, 517, 609
Mostowski, 517, 594, 606
Moszyńska, 588, 609
Mrówka, 584, 609
Munkres, 582, 609

Nadler, 582, 605, 609
Nagami, 584, 609
Nagata, 372, 457, 582, 586, 593, 609
Negrepontis, 590, 592, 601
von Neumann, 455
Nhu, 588, 590, 601
Nikodym, 594, 609
Nishiura, 584, 597, 604
Nöbeling, 583, 609
Novikov, 587, 597
Ntantu, 367, 591, 607

Okunev, 426, 609
Olszewski, 586, 609
Otto, 583, 602
Oversteegen, 583, 584, 609
Oxtoby, 384, 593, 609

Page, 579, 600
Parthasarathy, 585, 609
Pavlovskiĭ, 591, 609
Peano, 580
Pelant, 591, 592, 598, 607, 609
Pełczyński, 371, 579, 588, 589, 599, 609
Pestov, 591, 609
Poincaré, 221, 583, 609
Pol, E., 587, 600, 609
Pol, R., x, 220, 384, 446, 447, 456, 580, 583–587, 590, 592, 602–604, 606–610
Pontrjagin, 583, 584, 603, 610
Prikry, 586, 603
Przymusiński, 583, 610
Pytkeev, 591, 610

Quinn, 589, 610

Radul, 437, 588, 590, 598
Reed, 612
Repovš, 579, 610
Rogers, 455, 610
Rogers, Jr., 216, 587, 606, 610
Roy, 584, 610
Rubin, 585, 587, 590, 602, 610
Rudin, M.E., 369
Rudin, W., 393, 610

Salomon, x
Schauder, 582, 610

Schnirelman, 583, 607
Schori, 291, 581, 585, 588, 601, 610
Segal, 584, 588, 606, 607
Semadeni, 394, 451, 603
Semenov, 579, 610
Shchepin, 585, 603
Shelah, 393, 581, 610
Sher, 589, 610
Sierpiński, 580–582, 585–587, 592, 594, 611
Sikorski, 384, 611
Smirnov, 593
Sokolov, 393, 604
Souslin, 580, 594, 611
Spanier, 582, 611
Sperner, 582, 611
Spież, 605
Steel, 581, 603
Steenrod, 593, 602
Steiner, A.K., 594, 611
Steiner, E.F., 594, 611
Steinlein, 245, 586, 611
Sternfeld, 582, 586, 587, 606, 607, 611
Stone, 369, 610, 611

Taĭmanov, 593, 611
Talagrand, 591, 611
Taylor, 579, 611
Tietze, 593, 611
Tkachuk, 591, 611
Tolstowa, 583, 610
Tomaszewski, 586, 611
Topsøe, 610
Toruńczyk, 263, 291, 294, 580, 588, 605, 611
Tumarkin, 583–585, 611, 612
Tymchatyn, 583, 584, 609
Tzafriri, 579, 607

Ul′janov, 593, 612
Urysohn, 222, 580, 581, 583–585, 593, 594, 597, 612
Uspenskiĭ, 585
Ustinov, 611

Valdivia, 384, 600
Vaughan, 608
van de Vel, 580
Vermeer, 586, 587, 597, 604
Vietoris, 582, 612
de Vries, x, 582, 612

Wallman, 222, 224, 584, 585, 593, 605, 612
Walsh, 585, 587–589, 599, 604, 610, 612
Ważewski, 582, 612
West, 291, 588, 589, 610, 612
Whitehead, 132, 612
Whitney, 582, 612
Wille, 581, 604
Wimmers, 393, 612
Wojdysławski, 579, 588, 612
Wong, 589, 610

Yankov, 455

Zarelua, 585, 612
Zarichnyi, 437, 588, 590, 598
Zorn, 80, 459

# Subject Index

▶A:

Absolute (Neighborhood) Extensor, 25
Absolute Neighborhood Retract, 24–29,
    38, 39, 124, 145, 148, 263–265, 270,
    274, 276, 284, 288, 289, 301–305,
    580, 581, 588
  pair, 266, 267, 273, 276, 277, 285, 290,
    301, 304, 305, 588
Absolute Retract, 24–27, 29, 39, 145, 276,
    288, 289, 294, 301, 305, 580
  hyperspace, 292, 588
  pair, 266, 288, 290
    hyperspace, 292
absolute value, $512$
absorber, 347, 348, 589
  Bestvina-Mogilski type, 434, 589, 590
  Dijkstra-van Mill-Mogilski type, 347
absorbing
  sequence, 347, 348
  system, 346, 347
accumulation point, $462$
Addition Theorem, 178
adjunction space, $507$, $508$
affine
  combination, 111, 114
  coordinates, 114, 125, 126, 134
  function, 112–114
    continuous, 113
  hull, 111
  subspace, 111–113, 124
Alexandroff
  compactification, $472$
  problem, 253, 254, 587
almost zero-dimensional, 167, 168, 189,
    192, 583
analytic set, see set, analytic
Anderson Theorem, 9, 579, 588
antipodal
  map, 148, 238, $461$
  points, 149, $461$

preserving, 149
$\mathcal{A}$-operation, 594
Approximation Theorem
  Brown, 90
  Freudenthal, 137

▶B:

Baire Category Theorem, $482$
Baire space, 3, 36, 66, 78, 79, 373, 378,
    379, 383–385, 387, 391, 392, 446,
    $482$–$485$, $526$, 580, 591–593
  hereditary, 393, 446
  not topologically complete, $484$
ball
  closed, $459$
  open, $459$
Banach space, 3, 4, 9, 10, 12, 17, 21, 370,
    394, 579
barycenter, 125, 126, 128, 129, 138, 139
barycentric
  subdivision, 129–132
  triangulation, 129, 130
base, 41, 64, 122, 153, 167, 177, 238, 239,
    285, 288, 369, $465$, $467$, $497$, $507$
  closed sets, $494$, $497$
  collection of sets, 441
  countable, 593
  Ellentuck topology, 250, 251
  hyperspace, 101
  inverse limit, 81
  local, 19, 183, 368, $467$, $468$
  Wallman, 93, 185, $494$–$500$
basic core set, 342, 343
Bernstein set, 580
Bing
  compactum, 210, 212–215
  Shrinking Criterion, 67, 580
bonding map, 80, 84, 93
Borel
  complexity, 445, 590

set, 80, 373, 455, 456, *517–519*, *523*, *526*, 586, 590, 595
  absolute, 18, 437, 456, *526*, 592
  homogeneous, 581
  rigid, 581
Borsuk
  Antipodal Theorem, 149
  Example, 301, 580
  Homotopy Extension Theorem, 38
    controlled, 265
  Problem, 588
Borsuk-Ulam Theorem, 149, 174
boundary, 20, 125, 196, *460*, *464*
  geometric, 114
  preserving homeomorphism, 311
  set, 589
bounded
  component, *505*
  function, 29, 174, 368, 445
  metric, *459*, *463*
  set in a normed linear space, 3, 7, 20, 21
  subspace, 407, 410, *477*, *478*
Brouwer
  Dimensionsgrad, 221
  Fixed-Point Theorem, 143, 145, 148, 149, 260, 587, 594
  Invariance of Domain Theorem, 197
  $\mathbb{R}^n \not\approx \mathbb{R}^m$, 166

▶C:
Cantor set, 42–48, 75, 176, 229, 238, 254, 257, 387, 388, 434, 436, 581
Cantor-Bendixson Theorem, *467*
Cantor-manifold, 207, 208
capset, 329–331, 333, 334, 337, 338, 341–343, 347, 354, 589
carrier, 119, 134, 284
Cauchy sequence, 4, 33, 58, 96, *464*, *468*, *469*
Cauty Examples, *see* Example(s), Cauty
Čech-Stone compactification, 368, 590
centered, *495*
  maximal, *495*
characterization
  absolute $F_{\sigma\delta}$, *519*
  analytic set, 77
  ANR, 284
  A(N)R-pair, 267, 277
  AR, 39, 289
  boundary point, 196
  Cantor set, 43

capset, 334, 337, 590
compact, *460*
$C_p(X)$ Baire, 379
$C_p(X) \approx B(Q)^\infty$, 444
$C_p(X)$ metrizable, 369
dimension, 160, 166, 174, 182, 195
finite dimensional manifold, 588
homogeneous Borel set, 77
indecomposable continuum, 86
  hereditarily, 94
inessential map, *514*
$\mathbb{I}$, 51
$LC^n$, 296, 298
$LC^n$ and $C^n$, 300
$\ell^2$-manifold, 588
meager filter, 389, 434
Menger manifold, 588
$\mathbb{P}$, 76, 77
$Q$, 294
$Q$-manifold, 588, 589
$\mathbb{Q}$, 76
topologically complete, *480*
zero-dimensional, 58, 581
$C$-imbedded, 410
class of spaces
  closed hereditary, 344
  topological, 344
clopen, 41, 50, 57, 58, 63, 443, *460*, *500*
closed
  ball, *459*
  base, *494*
  map, *see* map, closed
  shrinking, *485*
closure, *460*
coanalytic set, 594
color, 187
  number, 250
  open, 251
  open or closed, 251
colorable, 188
coloring, 187, 188, 238, 241, 242, 244, 245, 248, 249, 251, 586
combinatorially equivalent, 123
compact-open topology, 19, 20, 31
compactification, 57, 64, 87, 93, 174, 182–184, 188, 193, 207, 234, 253, 257, *472*, *473*, *478*, *500*, *510*, *516*
  Alexandroff, *472*
  Čech-Stone, 368, 590
  dimension preserving, 183
  equivalent, *472*
  one-point, *472*

## SUBJECT INDEX

remainder, 183
Wallman, 93, 185, *494*, *498*, *499*, 593
compactness
  deficiency, 183, 204
  degree, 182, 183
compactum
  Bing, see Bing compactum
  Henderson, see Henderson compactum
complete
  metric, see metric, complete
  topologically, see topologically complete
  with respect to a norm, 16
completely Ramsey set, 239
complex conjugate, *512*
component, 37, 156, 167, 204, 209, 285, *467*, *500*, *501*, *505*, *507*
  path, *503*
composant, 87, 94
cone, 21, 39, 70, 89, 302, *515*, *516*, 581
connected, 54, 55, 57, 58, 64, 85, 102, 156, 206, 207, 209, *473*, *503*–*505*
  in dimension $n$, 295
  locally, see locally, connected
  path, see path-connected
continuous image
  C, 46
  complete space, *522*
  Menger curve, 585
  one-dimensional space, 203
  $\mathbb{P}$, 77
  zero-dimensional space, 57, 58
continuous logarithm, *513*
continuum, 37, 51, 58, 85–89, 94, 95, 105, 176, 209, 210, 216, 261, *501*, *502*, *505*, *515*, 581, 593
  $\sin(1/x)$, 29, 57, *503*, *505*, *517*
  Cook, 447
  decomposable, 86
  from $A$ to $B$, 206, *504*
  hereditarily indecomposable, 87, 94, 106, 108, 212, 213, 258, 260, 587
    homogeneous, 215, 216
  indecomposable, 86, 87, 94
    homogeneous, 215
  Peano, 58, 226, 292, 294, 295, 300, 340, 585, 588
  unicoherent, 226, *514*
Continuum Hypothesis, 47, 384, 393
continuum-connected, 206, 207, 209
contractible, 29, 39, 70, 146–148, 274, 285, 289, 301, *511*, *512*, *515*, *516*, 580

locally, see locally, contractible
contraction, *511*
convex, 1, 2, 15, 16, 19, 20, 23, 25, 29, 124, 125, 148, 288, 581, 582
  combination, 1, 2
  hull, 1
  metric, *477*, *479*, 588
coordinate
  functions of a simplex, 114
  space of inverse sequence, 80
countable
  closed sum theorem, 163
  dense homogeneous, 63–66, 580
    not homogeneous, 63
  dimensional, 155, 156, 221, 252, 253, 428
countably
  compact, *473*
  continuous, 257
cover, *458*
  locally finite, *486*
  refinement, *458*
  star-finite, 135, 136
  star-refinement, *460*
covering dimension, 152, 160, 583
$C_p(X)$, 4, 368–375, 377–379, 383–387, 393–399, 402–404, 406, 407, 409–415, 425–428, 441, 442, 444–447, 452, 455, 456, 590–592
$C_p^*(X)$, 368, 371, 394, 396–399, 418, 421, 425, 426, 445, 590, 591
Criterion
  Bing Shrinking, 67
  Inductive Convergence, 59, 65, 325
Curtis-Schori-West
  Hyperspace Theorem, 291, 295
cut, 222, *504*
  need not be a partition, 221, *505*
  point, 51, *504*, *505*
$C(X)$, 4, 5, 19, 20, 31, 368, 370–372, 393, 394, 418
$c_0$, 8, 354, 360, 435, 436

▶**D:**

decomposable continuum, 86
deformation, *510*
  property, 334, 337, 343
  through a subset, 266–268, 276, 294, 327, 355, 360, 361
degree
  compactness, 182, 183
  partition, 153

δ-multiplicative, *517*, *522*
derivative, 417
derived set, 417
descriptive
  complexity, 373–375, 377, 446, 590
  set theory, *483*, *517*, 581
diameter, *459*
dimension
  at a point, 227, 230
  coloring, 242
  compactification, 183, 184, 186, 187
  component, 167, 204
  countable dimensional, 155, 214, 252, 254
  covering dimension, 152, 160, 583
  $C_p(X)$, 413, 414, 428, 456
  Dimensionsgrad, 221–224, 226
  hyperspace, 261
  $\mathbb{I}^n$, 165
  infinite-dimensional, 156, 157, 168, 221, 252, 260, 456
    hereditarily, 254, 256
  inverse limit, 167, 193
  large inductive dimension, 180
  modulo a class, 183
  $\mathfrak{N}_n$, 168
  product, 181–183, 220, 226
  raising
    closed map, 199
    open map, 203
  $\mathbb{R}^n$, 165
  simplex, 114
  small inductive dimension, 177
  $\mathbb{S}^n$, 165, 194, 195, 226
  strongly infinite-dimensional, 155, 156, 213, 214, 218, 251, 253, 257, 414, 426
    hereditarily, 257
  theorem
    coincidence, 180
    countable closed sum, 163
    fundamental, 165
    $G_\delta$-enlargement, 175
  totally disconnected, *see* totally disconnected
  weakly $n$-dimensional, *see* weakly $n$-dimensional
  weakly infinite-dimensional, 251, 252, 254, 257
  $X_{(n)}$, 183
dimensional kernel, 227, 228
Dimensionsgrad, 221–224, 226

counterexample, 224
discrete, *466*
disjoint-cells property, 294, 588
dominating space, 274, 276
  controlled, 274
Dranišnikov Example, 24
dual, 399, 400, 402–404
Dugundji
  system, 22, 23, 29, 278
  theorems, 23, 29, 394, 451, 588
dyadic solenoid, 85, 86, 94, 215
dynamical system, 91, 92, 249

▶E:
Ellentuck topology, 239, 240, 250, 251
endface, 310, 320, 322, 324, 326, 334, *462*, *473*
enlargement
  A(N)R, 304, 305
  theorem, 175
ε-map, 34, 66, 67, 90, 135
equiconnecting function, 276
equicontinuous, 418
Erdős space, *see* Example(s), Erdős
essential
  continuous function, *512*, *516*
  family, *see* family, essential
euclidean
  space, 3, 64, 65, 149, 151, *461*
  topology, 19, 20
evaluation, 399
Example(s)
  Bing, 210, 212
  Borsuk, 580
  Cantor, 42
  Cauty, 24, 26, 264, 445, 456, 580, 583
  Cook, 447
  Dranišnikov, 24
  Erdős, 542, 547
  Erdős, 50, 58, 160, 167, 182, 183, 437, 584
  Henderson, 214
  Kulesza, 220
  Kuratowski, 447
  Marciszewski, 447
  Mazur, 241
  Pol, 254, 447, 450, 587
  Roy, 584
  Sher, 589
  Urysohn, 221, 594
expansion
  homotopy, 582

## SUBJECT INDEX

hyperspace, 291, 292, 294
exponential function, *512*
extender, 393
  continuous, 454
    linear, 394–396
  linear, 393
  measurable, 455
extension, 21, 23, 24, 38, 40, 46, 72, 98, 148, 188, 193–196, 204, 265, 270, 273, 276, 277, 284, 288, 296–298, 300, 319, 325, 394, 451, *491–494*

▶**F:**
face
  of a simplex, 114, 125, 126, 128, 132, 138, 139, 174
  opposite, 146, 147
  proper, 114
factorization of $C_p^{(*)}(X)$, 396–399
faithful indexing, *458*
family
  Borel sets, *517*
  Cantor sets, 45
  closed under supersets, 389
  composants, 94
  $\delta$-multiplicative, *517*
  equicontinuous, 418
  essential, 151–153, 155–157, 206, 209, 251
  inessential, 151, 152, 156, 201, 202
  open sets, 278
  pairwise disjoint homeomorphs, 48
  $\sigma$-additive, *517*
  solenoids, 581
  strongly discrete, 377–379
fiber, *458*
filter, 387, 389, 390, 434–436, 439, 441, *458*
  fixed, *459*
  Fréchet, 435
  free, *459*
  generated by, *459*
finite dimensional, 153
fixed-point, 145, 148
  free, 60, 92, 148, 188, 241, 249–251
    homeomorphism, 242, 245, 248
    involution, 238, 586
  property, 92, 144–146, 148, *516, 517*
  theorem
    Brouwer, *see* Brouwer FPT
    Schauder, *see* Schauder FPT

Freudenthal's Approximation Theorem, 137
$F_\sigma$, 87, 227, 230, 374, *466, 467, 517, 518, 522*, 595
  absolute, *518*
$\mathcal{F}_\sigma$-absorber, 347, 350, 361
$F_{\sigma\delta}$, 377, 442, 445, 446, *518*, 592
  absolute, 344, *518–520*
$\mathcal{F}_{\sigma\delta}$-absorber, 350, 353, 360, 361, 442, 444
full
  realization, 270, 273, 277, 284, 296
  simplex, 141
$\mathcal{F}$-ultrafilter, *495, 496*
function vanishing on a set, 396, 418
functional, 399, 400, 402
Fundamental Theorem of Dimension Theory, 165

▶**G:**
Galvin-Prikry Theorem, 586
$G_\delta$, 35, 172, 175, 217, 220, 221, 229, 231, 238, 257, 372, 373, 375, 427, 428, 434, 454, *464, 466, 467, 480, 483, 485, 493, 494, 517, 526*, 590
  absolute, *518*
  enlargement, 175, 304
  selection, 217
$G_{\delta\sigma}$, *518*
  absolute, 344, *518–520*
general position, 168, 169
geometric
  boundary, 114
  dependence, 111, 112
  independence, 112, 125, 126
  interior, 114, 125
  realization of a simplicial complex, 116
  simplex, 114, 131, 138
graph
  method of Klee, 581
  of a function, 12, *458, 463*
group
  homeomorphisms, 35
  topological, *see* topological group

▶**H:**
Hamel basis, 18, 19
  for $L(X)$, 400
Hausdorff
  distance, 95
  metric, 95, 96, 106, 107
  space, *463*
height, 418

Henderson compactum, 213, 214, 254, 587
hereditarily indecomposable continuum, *see* continuum, hereditarily indecomposable
Hilbert cube, 20, 26, 57, 60, 63, 66, 69, 73, 144, 147, 151, 176, 251–253, 257, 291, 294, 295, 309, 310, 325–331, 333, 334, 337, 338, 343, 345, 346, 348, 354, 355, 360, 361, 414, 426, 428, 436, 439, 441, 444, *461, 462, 470, 473, 479, 526*, 581, 589, 590
  manifold, 588
Hilbert space, 8, 9, 20, 21, 197, 437, 579, 588, 589
  manifold, 588
Homeomorphism Extension Theorem, 320, 325
homogeneous, 63, 64, 66, 73, 75, 80, 94, 147, 168, 215, 216, *462, 464*, 581, 582
  countable dense, 63–66, 580
  isotopically, 92, 94, 328
  strong, 75
  strong local, 64, 66, 73, 75, 580
homotopic, 38, 195, 263, 265, *510*
homotopically trivial, 285, 288–290, 301, 305, 310
homotopy, 70, 71, 148, *510*
  expansion, 582
  level, 70, *510*
  limited by a cover, 263
  type, 276, 588
hyperspace, 95, 96, 101, 102, 107, 108, 156, 292, 587
  Absolute Retract, 292
  continua, 103, 261, 587
  diam is continuous, 107
  expansion, 291, 292, 294
  finite sets, 100, 291, 294
  Hilbert cube, 295
  map, 98, 100
  topologically complete, 98
  ⟨𝒰⟩, 101
  union operator, 99

▶ I:
imbedding, 5, 19, 107, 172, 174, 176, 197, 323, 384, 402, *460*
  AE, 25
  C, 45
  closed, *460*
  dense, *460*
  $\mathfrak{N}_n$, 174
  open, *460*
  $\mathbb{P}$, 45
  product, 586
  $Q$, *470*
  $\mathbb{R}^\infty$, *480*
  $\mathbb{R}^{2n+1}$, 583
  $Z$, *see* $Z$-imbedding
indecomposable continuum, 86, 87, 94
  homogeneous, 215, 216
Inductive Convergence Criterion, 59, 65
inessential
  continuous function, *512*
  family, 151, 152, 156, 201, 202
infinite left product, 59
infinite-dimensional, *see* dimension, infinite-dimensional
  strongly, *see* dimension, strongly infinite-dimensional
Infinite-dimensional topology, 588
initial segment, 46
inner product space, 18
integers, *457*
interior, 124, *460*
  geometric, 114
interval
  $\mathbb{I}$, *457*
  $\mathbb{J}$, *457*
invariance
  domain, 197
  Whitehead torsion, 588
invariant, 151, 242, 250, 251
inverse
  limit, 80, 82–85, 90–94, 137, 167, 193, 220, 249
  sequence, 80–85, 90–94, 137, 167, 220, 249
  subsequence, 83
involution, 238, 586
irrational numbers, 42, 45, 58, 76, 77, 168, *457, 482, 483, 494*, 581
irreducible
  function, *477, 479, 526*
  partition, 155, 157, 193, 226
isometry, 19, 96, *460*
isotopically homogeneous, 92, 94, 328
isotopy, 70, 71, 73, 320, 328

▶ J:
joined by a path, *503, 504*
Jordan Curve Theorem, 174

## SUBJECT INDEX

▶K:

κ-function, 134, 135, 275, *488*, 582
Kuratowski-Wojdysławski Isometric Imbedding Theorem, 19
Kuratowski-Zorn Lemma, 18, *459*

▶L:

large inductive dimension, 180
Lavrentieff Theorem, *493*
Lebesgue
  measure, 587
  number, *475*
Lemma
  △-system, 592
  Disjoint Refinement, 453
  Kuratowski-Zorn, 18, *459*
length of a sequence, 46
$L_\varepsilon$-enlargement, 189
$\ell$-equivalent, 370, 371, 399, 413, 425, 445, 456
level
  homotopy, 70, *510*
  preserving, 71
  Whitney, 105, 106, 108, 109, 258
lexicographic order, 216
$L$-imbedded, 189, 193, 234
limited by a cover, 65, 263, 265, 267
linear
  function, 12, 17, 20, 21
  homeomorphism, 12, 20, 21, 369, 370, 398, 403, 411–413, 447, 450
  hull, 111
  order, 126, 216
  space, 1–4, 8, 10, 12, 15–21, 26, 29, 111, 113, 114, 116, 123, 130, 131, 148, 197, 252, 288, 361, 363, 364, 369, 402, 435, 452
    incomplete, 589
    not ANR, 264, 580
    not locally convex, 20
    unit sphere, 20
    subspace, 20
linear space, 29
linearly
  homeomorphic, 369, 370, 398, 403, 404, 447, 450
  independent, 5, 7, 18, 20, 112
Lipschitz function, 99, *490*
local base, 19, 183, 368, *467*, *468*
locally
  compact, 58, 64, 87, 121, 425, 456, *473*, *478*, *482*

  nowhere, 57, 58, 76
connected, 55–58, 86, 102, 222, 224, 226, 285, 288, *467*, *505*, *516*
  in dimension $n$, 296
contractible, 28, 296, 301, *516*
  at a point, *516*
convex, 2, 18, 19, 23–25, 29, 288, 402, 452, 583
$\ell^p$ is not, 20
finite, 22, 108, 136, 138, 157, 158, 160, 409, *486*, *487*, *490*
  simplicial complex, 118, 121, 125, 273, 277, 284
homotopy negligible, 263, 588
path-connected, 29, *503*, *504*
uncountable, 568
lower semi-continuous, 404, 406, *488*
  set-valued function, 13
$\ell^*$-equivalent, 370, 418, 419, 421, 425, 456
Lusternik-Schnirelman-Borsuk Theorem, 148, 149, 174, 586
$L(X)$, 399, 400, 402–404

▶M:

map
  closed, 109, 241, *460*, *464*, *467*
    dimension raising, 199
  $\varepsilon$, 34, 66, 67, 90, 135
  essential, *512*
  inessential, *512*
  irreducible, *477*
  linear, *see* linear, function
  monotone, *504*
  open, 107, 108, *464*, *477*
  quotient, 108, *505*, *506*, *508*, *509*
  Sperner, 141
  $\mathcal{U}$, 169
  Whitney, 103–106, 108, 109, 258
Mazurkiewicz Theorem, 55
meager, 66, 87, 447, 449, *483*, *520*, *522*, *526*
  at a point, *526*
  collection of sets, 389
  filter, 389–391, 434, 436
Menger
  compacta, 588
  curve, 583, 585
mesh
  collection, *459*
  simplicial complex, 130
  subdivision, 130, 131

metric, 30, 90, 361, 363, *463*, *464*, *478*, *484*
  admissible, *460*
  bounded, *459*, *463*
  complete, 3, 21, 33, 35, 58, 96, 98, *463*, *474*, *477*, *479*, *480*, *485*
  convex, 108, *477*, *479*, 588, 593
  derived from a norm, 2, 3
  Euclidean, 3
  Hausdorff, 95, 96, 106, 107
  Nagata, 586
  non-Archimedean, 57, 58
  nonconvex, 325
  *s*, 19
  totally bounded, 96, 98, *479*, *484*
Metrization Theorem
  Nagata-Smirnov-Bing, 593
  Urysohn, *465*, 593
Miljutin Theorem, 370
monotone map, *504*
multi-valued function, 404

▶ N:
Nagata-Smirnov-Bing Theorem, 593
natural numbers, *457*
near homeomorphism, 66, 67, 70, 90, 94
neighborhood retract, *512*
nerve, 132, 133
net, *484*
Nöbeling's universal space, 168, 172, 174
nodec space, 385
norm, 2, 3, 579
  Euclidean, 3
normal set in the sense of Fréchet, 221
nowhere
  countable, 581
  dense, 86, 197, 387, 388, 436
  locally compact, 57, 58, 76
  $\sigma$-compact, 581
  topologically complete, 581
null-sequence, *508*
nullhomotopic, 145, 148, 149, 204, 300, *510*, *511*, *514*

▶ O:
open
  ball, *459*
  Ellentuck topology, 240
  map, 10, 57, 107, 108, *460*, *464*, *477*
  problem, see problem, open
  shrinking, *485*
  swelling, 157, 158
order

cover, 160
  lexicographic, 216
  linear, 126, 216
  preserving indexed collection, 344
origin of $Q$, *461*

▶ P:
paracompact, 125, 369, 396, *487*, *490*
partial realization, 270, 273, 277, 281, 282, 284, 296
partition, 146, 153–155, 157, 175, 179, 182, 193, 206, 209, 221, 252, *504*, *505*
  Bing, 210, 213
  continuum, *502*
  cut is, 222
  degree, 153
  irreducible, 155, 157, 193, 226
  $\mathbb{R}$, 80
  space, 125, 257, 443, *509*
  unity, *487*, *488*
path, *503*
path-component, *503*
path-connected, 29, 55, 57, 58, 94, 285, *503*–*505*, *511*
  locally, 29, 57, *503*, *504*
Peano
  continuum, see continuum, Peano
  map, 580
perfectly disconnected, 385, 592
point
  accumulation, *462*
  cut, 51
  $P$, 393, 446
  weak $P$, 386
pointwise convergence topology, see topology, pointwise convergence
Pol Examples, 447, 450, 587
polyhedron, 124, 136, 274, 310, 588
polytope, 123, 132, 136, 166, 270, 274, 276, 304
power set, *458*
$P$-point, 393, 446
problem
  Alexandroff, 253, 254, 587
  Anderson-Bing, 589
  Banach and Fréchet, 579
  de Groot, 584
  open, 145, 161, 216, 236, 260, 404, 414, 425, 446, 447, 580, 581, 587, 592
  Schauder, 582

# SUBJECT INDEX

product, 26, 41, 58, 82, 235, 339, 368, 371, 450, *461*, *467*, *480*, *516*, *519*
  Cartesian, *457*
  weak, 355
projection, 70, 81, 84, 398, *457*, *477*, *478*
property
  closed hereditary, 84
  countably productive, 84
  fixed-point, 92, 145, 146, 148, *516*
property of Baire, 434, *520–523*, 594
pseudo
  arc, 215
  boundary, 60
  interior, 60
punctiform, 221

▶ **Q:**
$Q$-set, 595
quotient map, 108, *505*, *506*, *508*, *509*
quotient topology, 398, *505*, *506*, *509*
  not metrizable, *509*

▶ **R:**
Ramsey
  set, 239
    completely, 239, 240
  theory, 238, 240
rational numbers, 42, 58, 76, 78–80, 182, 384, 416, *457*, *483*, *494*, *517*, *519*, *520*, 581, 594
  not topologically complete, *483*
real numbers, *457*
realization
  full, 270, 273, 277, 284, 296
  geometric, 116
  partial, *see* partial realization
refinement, 160, 166, 176, 263, 273, 277, 284, 296, *458*
  disjoint, 41, 168, 453, *463*
  star, 272, *460*
  star-finite, 136, 166, 272
reflexively universal, 345
retract, 146, 148, 276, *512*, *516*, *517*
  neighborhood, 26, *512*
retraction, *512*
rigid space, 447, 581

▶ **S:**
$s$, 3, 9, 19, 20, 26, 60, 66, 257, 310, 311, 314, 319, 320, 322–324, 339, 342, 343, 364, 579, 587, 588
scattered, 417, 419, 456
  height, 418

Schauder Fixed-Point Theorem, 148, 583
segment
  initial, 46
  straight-line, 19
selection, 13, 29
  $G_\delta$, 217
  continuous, 13, 16
semi-continuous
  lower, 13, 406, *488*
  upper, 257, *489*
separated, *491*
separating set, 148
separator, *504*
sequence
  absorbing, 350
  compacta, *478*
  $\mathcal{F}_\sigma$-absorbing, 352
  functions, *469*
  homeomorphisms, 59, 65
  inverse, *see* inverse, sequence
  null, 58, *508*, *509*
sequentially compact, *473*, *474*
set
  analytic, 48, 49, 77, 435, 436, 439, 441, 455, *522*, *523*, *526*, 590, 594
  Borel, *see* Borel set
  Cantor, *see* Cantor set
  coanalytic, 594
  derived, 417
  meager, *483*
  power, *458*
  property of Baire, 434, *520–523*, 594
  Ramsey, *see* Ramsey, set
  scattered, 417
  $\sigma Z$, *see* $\sigma Z$-set
  $Z$, *see* $Z$-set
set-valued function, 12
shift
  inverse limit, 91, 92, 249
  $\mathbb{R}^\mathbb{Z}$, 587
shrinkable, 66–68, 70
shrinking, 159, 160, 167, 244, *485*, *486*
  closed, *485*
  open, *485*
Sierpiński
  Carpet, 583
  Theorem, *502*
$\sigma$, 257
$\sigma$-additive, *517*, *522*
$\sigma$-algebra, 455, *458*, *521*

$\sigma$-compact, 58, 66, 226, 228, 339, 347, 361, 411, 412, 426, *479, 518, 519*, 581
  nowhere, 581
$\sigma$-discrete, 369
$\sigma Z$-set, 307, 310, 334, 340, 346, 361, 436, 441
simple chain, 54
simplex, 114, 124
  $n$-dimensional, 114
  full, 141
  geometric, 114
simplicial complex, 116, 117
  locally finite, 118
simplicially homeomorphic, 123
$\sin(1/x)$-continuum, 29, 57, *503, 505, 517*
skeleton, 117, 122, 271
solenoid
  dyadic, 85
  $p$-adic, 581
space, *465*
  adjunction, *507, 508*
  Baire, see Baire space
  Banach, see Banach space
  euclidean, 3, 64, 65, 149, 151, *461*
  infinite-dimensional compacta, 350
  linear, see linear, space
  normal, 121
  perfectly disconnected, 385, 592
  rigid, 447, 581
  scattered, 417
  topological, *465*
  Tychonoff, 367
space filling curve, 46
Sperner
  Lemma, 141
  map, 141
sphere, 3, 9, 20, 26, 29, 63, 85, 92, 146–149, 194–196, 201, 204, 207, 226, 288, 291, 297, 310, *461, 473, 511, 516*
square, 87, 89, 93, 384, 450
standard
  simplex, 539
  triangulation, 117, 139
star, 119, *460*
  finite, 135, 138, 166
  refinement, 272, *460*
straight-line segment, 19
  parametrization, 361
strong local homogeneity, 64, 66, 73, 75, 580

strongly
  discrete family, 377–379
  homogeneous, 75
  infinite-dimensional, see dimension, strongly infinite-dimensional
subbase, 19, 31
subcomplex, 117
subpolyhedron, 124
subpolytope, 124
Subspace Theorem, 164
sum, 371
sup-norm, 4
support, 404–407
  homeomorphism, 64
swelling, 157, 158
symmetric difference, *457*

▶T:
$t$-equivalent, 371, 414, 426, 428, 445, 456, 591, 592
$t^*$-equivalent, 426, 456
Theorem
  Addition, 178
  Anderson, 9, 579, 588
  Borsuk, 149
  Borsuk-Ulam, 149, 174
  Brouwer, 143, 145, 148, 149, 157, 260, 587, 594
  Brown, 90
  Cantor-Bendixson, *467*
  Cauty, 24, 583
  Chapman, 588
  Closed Graph, 12
  Coincidence of Dimension Functions, 180
  Countable Closed Sum, 163
  Curtis-Schori-West, 291
  Dobrowolski-Marciszewski-Mogilski, 444
  van Douwen, 48
  Dugundji, 23, 29, 394, 451, 588
  van Engelen, 581
  Freudenthal, 137
  Galvin-Prikry, 586
  Jordan, 174
  Keller, 581
  Kuratowski-Wojdysławski, 19
  Lavrentieff, *493*
  Lusternik-Schnirelman-Borsuk, 149, 174
  Marciszewski, 398, 447
  Mazurkiewicz, 55
  Michael, 16
  Miljutin, 370

Nagata-Smirnov-Bing, 593
Pol, 254
Schauder, 148, 583
Sierpiński, *502*
Souslin, 47
Sperner, 141
Subspace, 164
Tietze, 24
Toruńczyk, 291, 294, 588, 589
Urysohn, *465*
West, 588
topological group, 1, 37, 60, 182, 215, *462*
  homeomorphisms, 35–37, 583
  homogeneous, *464*
  $Q$ is not, 60
  $\mathbb{S}^1$, 85
  $\Sigma_2$, 85
topological property
  closed hereditary, 84
  countably productive, 84
topological space, *465*
topologically complete, 47–49, 55, 57, 65, 66, 76, 77, 183, 222, 226, *480*, *482*, *484*, *493*, *518*, *519*, *522*, *523*, *526*, 581, 594
  A(N)R enlargement, 305
  Baire, *482*
  compactification, 253
  Continuum Hypothesis, 47
  $C_p^*(X)$, 591
  $C_p(X)$, 415
  $C(X, Y)$, 33
  dense subspace, *484*, *526*
  dimension, 220, 224, 584
  $\ell^2$, 9
  $\mathcal{H}(X, Y)$, 35
  hyperspace, 98
  inverse limit, 84
  nowhere, 581
  $\mathbb{P}$, *482*
  $\mathbb{Q}$ is not, *483*, *517*
  topological sum, *484*
  totally disconnected, 183, 218, 221
  weakly $n$-dimensional, 231
topology
  compact-open, 19, 20, 31
  coordinatewise convergence, 9, *467*
  euclidean, 19, 20
  pointwise convergence, 4, 368, 372
  quotient, *see* quotient topology
  Tychonoff product, 368
  uniform convergence, 7, 31
  Vietoris, 101, 108, 582
  weak, 354
  Whitehead, 118, 119, 132
Toruńczyk Theorems, 291, 294, 588
totally disconnected, 50, 58, 167, 183, 218, 221
tower, 337
  deformation property, 334, 343
  expansive, 334
translation, *462*
triangulation, 117
  standard, 117
trivial homotopy class, *511*
Tychonoff
  product topology, 368
  space, 367
type
  homotopy, 276
  of a point, 63

▶**U:**
$\mathcal{U}$-close, 263
ultrafilter, 391–393, *459*, *462*, *463*, *495*, *496*
$\mathcal{U}$-map, 135, 136, 166, 169, 176
unbounded component, *505*
unicoherent, 204, 226, *514*, *515*
uniformly continuous, *477*, *478*
union operator, 99
unit
  ball, 3
    $\mathbb{R}^n$, 3, *461*
  interval, *457*
  sphere, 3
    $\mathbb{R}^n$, 3, *461*
universal
  curve, 583, 585
  $\mathcal{F}_\sigma$, 354
  $\mathcal{F}_{\sigma\delta}$, 439
  $\mathcal{M}_\Gamma$, 344, 345
  reflexively, 345, 350, 355, 364, 439
  space
    compact, 176
    Nöbeling, 168, 172, 174
    weakly $n$-dimensional, 232
  strongly
    $\mathcal{F}_{\sigma\delta}$, 361
    $\mathcal{M}$, 348, 361
    $\mathcal{M}_\delta$, 348
    $\mathcal{M}_\Gamma$, 344, 346
upper semi-continuous, 257, *489*

decomposition, 106, *501*, *505–509*
Urysohn
  function, *469*
  Metrization Theorem, *465*, 593
usual topology on $[\mathbb{N}]^\omega$, 239

▶ **V:**
vertex, 116–118, 280
Vietoris topology, 101, 108, 582

▶ **W:**
Wallman
  base, 93, 185, *494–500*
  compactification, 93, 185, *494*, *498*, *499*, 593
Warsaw circle, 301
  homotopically trivial, 285
weak
  Cartesian product, 355
  topology, 354, *465*
weakly
  infinite-dimensional, 251, 252, 254, 257
    not countable dimensional, 254
  $n$-dimensional, 228, 231, 586
    universal, 232
weakly one-dimensional, 235
weak $P$-point, 386
weight, 369
Whitehead
  topology, 118, 119, 132
  torsion, 588
Whitney
  level, 105, 106, 108, 109, 258
  map, 103–106, 108, 109, 258

▶ **Z:**
zero-dimensional, 41–45, 57, 58, 73, 75, 76, 153, 154, 157, 165, 174, 177, 178, 182, 210, 230, 241, 253, 425, 581
  almost, 167, 168, 189, 192, 583
$Z$-imbedding, 327, 328, 338, 343, 360, 361
$Z$-map, 327
$Z$-set, 307–311, 323–331, 333, 334, 337, 338, 343, 589